HISTORICAL DICTIONARY

The historical dictionaries present essential information on a broad range of subjects, including American and world history, art, business, cities, countries, cultures, customs, film, global conflicts, international relations, literature, music, philosophy, religion, sports, and theater. Written by experts, all contain highly informative introductory essays on the topic and detailed chronologies that, in some cases, cover vast historical time periods but still manage to heavily feature more recent events.

Brief A–Z entries describe the main people, events, politics, social issues, institutions, and policies that make the topic unique, and entries are cross-referenced for ease of browsing. Extensive bibliographies are divided into several general subject areas, providing excellent access points for students, researchers, and anyone wanting to know more. Additionally, maps, photographs, and appendixes of supplemental information aid high school and college students doing term papers or introductory research projects. In short, the historical dictionaries are the perfect starting point for anyone looking to research in these fields.

HISTORICAL DICTIONARIES OF PROFESSIONS AND INDUSTRIES

Jon Woronoff, Series Editor

Journalism, by Ross Eaman, 2009.
U.S. Maritime Industry, by Kenneth J. Blume, 2012.
Librarianship, by Mary Ellen Quinn, 2014.
Japanese Business, Second Edition, by Stuart D. B. Picken, 2016.
Fashion Industry, Second Edition, by Francesca Sterlacci and Joanne Arbuckle, 2017.
Petroleum Industry, Second Edition, by Marius S. Vassiliou, 2018.

Historical Dictionary of the Petroleum Industry

Second Edition

Marius S. Vassiliou

ROWMAN & LITTLEFIELD
Lanham • Boulder • New York • London

Published by Rowman & Littlefield
An imprint of The Rowman & Littlefield Publishing Group, Inc.
4501 Forbes Boulevard, Suite 200, Lanham, Maryland 20706
www.rowman.com

Unit A, Whitacre Mews, 26-34 Stannary Street, London SE11 4AB

British Library Cataloguing in Publication Information Available

Library of Congress Cataloging-in-Publication Data

Name: Vassiliou, Marius S., author.
Title: Historical dictionary of the petroleum industry / Marius S. Vassiliou.
Description: Second edition. | Lanham : Rowman & Littlefield, 2018. | Series: Historical dictionaries of professions and industries | Includes bibliographical references.
Identifiers: LCCN 2017061840 (print) | LCCN 2018001876 (ebook) | ISBN 9781538111604 (electronic) | ISBN 9781538111598 (hardcover : alk. paper)
Subjects: LCSH: Petroleum engineering—History—Dictionaries. | Gas engineering—History—Dictionaries. | Petroleum industry and trade—History—Dictionaries.
Classification: LCC TN865 (ebook) | LCC TN865 .V38 2018 (print) | DDC 338.2/728203—dc23
LC record available at https://lccn.loc.gov/2017061840

∞ ™ The paper used in this publication meets the minimum requirements of American National Standard for Information Sciences Permanence of Paper for Printed Library Materials, ANSI/NISO Z39.48-1992.

Printed in the United States of America.

To my wife, Cynthia K. Vassiliou,
to my mother, Avra S. Vassiliou,
and
to the memory of my father, Simos G. Vassiliou.

Contents

Editor's Foreword

We are living in the oil age. Petroleum is used for a vast and ever-growing range of purposes, most of which the developed world could barely survive without and, alas, for which there are not many readily accessible—let alone reasonably cheap—alternatives. These many uses include fuel for automobiles and airplanes, grease for sophisticated machinery, natural gas for heating and fuel, and plastics used in the manufacture of thousands of different articles and product packaging. Since we have lived in the oil age all our lives, it's difficult to imagine a post-oil age. But it must be done because resources are dwindling.

This second edition of the *Historical Dictionary of the Petroleum Industry*, like the first, both examines the past and attempts to predict the future. It includes essential new entries on fracking, shale oil, and tight gas, as well as important people in the oil industry and oil producers. In addition, many entries from the first edition have been updated. All other sections have been revised and updated as well, including the list of acronyms and abbreviations, introduction, appendixes, and bibliography.

The author of both editions is Marius Vassiliou, whose distinguished career has mixed practice and research. He was trained in geophysics, computer science, and business, and he worked in the industry as senior research geophysicist at Arco Oil and Gas. After that, he researched oil industry problems and other concerns at the Rockwell International Science Center, where he eventually rose to the position of executive director. He then became an analyst at the Institute for Defense Analyses, where one of his concerns was oil depletion. Here he presents the important and complicated issues associated with the long history of oil in accessible language, with the insight of decades of serious thought and study.

Jon Woronoff
Series Editor

Preface

This historical dictionary covers the modern petroleum industry from the mid-19th century to the first decade of the 21st century, but it also contains a number of entries relevant to earlier periods. Although it is about petroleum in general, it emphasizes oil more than gas. And although it covers the entire spectrum of the industry, it emphasizes upstream more than downstream operations. Upstream operations consist of the exploration and production of petroleum, while downstream operations consist of refining, transportation, and distribution.

Each entry—on companies, people, events, technologies, phenomena, countries, cities, and regions related to the petroleum industry—is intended to be largely self-contained. However, related entries are indicated with cross-references, and I encourage their consultation for a fuller account. The importance of the United States in the global petroleum industry is reflected in this book, but I have attempted to include as much information as possible from the rest of the world. Russia in particular has been crucial in the history of the industry, and many works have not given it the credit and attention it deserves. Hopefully, this book will contribute to remedying the situation.

In the eight years since the first edition of this book was published, there have been monumental changes in the petroleum industry. Hydraulic fracturing (fracking) and horizontal drilling have enabled the extraction of oil and gas from impermeable formations, especially shales, to an extent that can only be viewed as revolutionary. The United States has been at the center of these developments, but they may spread to the many other countries endowed with such unconventional hydrocarbon resources. Meanwhile, major discoveries continue to be made offshore in very deep water and under very difficult conditions, especially in Brazil.

The industry has thus far been able to offset the inevitable decline in "easy oil" by going after "difficult oil." The same is true for natural gas. Meanwhile, there have been serious setbacks, such as the *Deepwater Horizon* spill. This second edition incorporates all these and other developments and updates historical information as new data and perspectives have emerged.

Acknowledgments

I am grateful to Rasoul Sorkhabi and Mir-Yusif Mir-Babayev for valuable correspondence. I thank Alexander Matveichuk for providing invaluable information on the early Russian petroleum industry and for very kindly sending me copies of his authoritative books and articles on the subject. I also thank my colleague Jenya Macheret for translating some of Dr. Matveichuk's material from Russian into English. Anna Rubino reviewed the entry on the journalist Wanda Jablonski and made valuable suggestions. David McKain did the same for the entry on West Virginia. I thank Bob Tippee of the *Oil and Gas Journal*, Tom Wallin of *Petroleum Intelligence Weekly*, and Mary Kay Grosvald of the American Association of Petroleum Geologists for providing me with historical material on their respective institutions. I also acknowledge a debt of gratitude to the U.S. Library of Congress and the libraries of George Mason University, Georgetown University, and the American University. Finally, I thank my wife, Cynthia, and my son, Simos, for their patience and their support.

Acronyms and Abbreviations

NAMES AND TERMS

AAPG	American Association of Petroleum Geologists
AAR	Alfa Access Renova
ACG	Azeri–Chirag–Guneshli (Azerbaijan)
AD	Acción Democrática (Venezuela)
ADCO	Abu Dhabi Company for Onshore Oil Operation
ADMA	Abu Dhabi Marine Areas
ADMA-OPCO	Abu Dhabi Marine Areas Operating Company
ADNOC	Abu Dhabi National Oil Company
ADPC	Abu Dhabi Petroleum Company
AGIP	Azienda Generale Italiana Petroli
AIME	American Institute of Mining and Metallurgical Engineers
AIOC	Anglo–Iranian Oil Company
AIOC	Azerbaijan International Oil Company
ALNAFT	Agence Nationale pour la Valorisation des Ressources en Hydrocarbures (Algeria)
Aminoil	American Independent Oil Company
Amoco	American Oil Company (later did not stand specifically for this)
ANP	Agência Nacional do Petróleo, Gás Natural e Biocombustíveis (Brazil)
ANWR	Arctic National Wildlife Refuge
API	American Petroleum Institute
APOC	Anglo–Persian Oil Company
APQ	aggregate programmed quantity
Aramco	Arabian American Oil Company
Arco	Atlantic Richfield Company

ASCOOP	Association Coopérative (Algeria)
ASEAN	Association of Southeast Asian Nations
Banoco	Bahrain National Oil Company
Bapco	Bahrain Petroleum Company
BHP	Broken Hill Proprietary
BNITO	Batum Neftepromushlenoe i Torgovoe Obchshestvo (Batumi Oil Production and Trading Company)
BNOC	British National Oil Company
BOD	British Oil Developments
BP	British Petroleum originally; the initials now do not stand for anything specific
BPC	Basra Petroleum Company
BP Migas	Badan Pelaksana Kegiatan Usaha Hulu Minyak Dan Gas Bumi (Executive Agency for Upstream Oil and Gas Activity) (Indonesia)
Branobel	Tovarishchestvo Nephtanavo Proisvodtsva Bratiev Nobel (Nobel Brothers Petroleum Production Company)
BRP	Bureau de Recherches des Pétroles
BTC	Baku–Tbilisi–Ceyhan (Pipeline)
Caltex	California–Texas Company
Canol	Canadian Oil (Project)
Casoc	California Arabian Standard Oil Company
CCPDP	Chad Cameroon Pipeline Development Project
CEO	chief executive officer
CEPE	Corporación Estatal Petrolera Ecuatoriana
CFP	Compagnie Française des Pétroles
CIA	Central Intelligence Agency (United States)
CIS	Commonwealth of Independent States
CNOOC	China National Offshore Oil Company
CNP	Conselho Nacional do Petróleo (Brazil)
CNPC	China National Petroleum Company
CO&T	Continental Oil and Transportation Company
COMEX	New York Commodities Exchange

Conoco	Continental Oil Company
CPC	Caspian Pipeline Consortium
CPPC	Calgary Petroleum Products Company
CRP	Centro de Refinación de Paraguaná
CSA	Canadian Security Administrators
CVP	Corporación Venezolana de Petróleo
DPC	Dubai Petroleum Company
EBN	Energie Beheer Nederland
Ecopetrol	Empresa Colombiana de Petróleos
ECX	European Climate Exchange
EIA	Energy Information Administration (U.S. Department of Energy)
Enarsa	Energía Argentina Sociedad Anónima
ENI	Ente Nazionale Idrocarburi; later did not stand specifically for this
EOR	enhanced oil recovery
EPSA	exploration and production-sharing agreement (Libya)
ERAP	Entreprise de Recherches et d'Activités Pétrolières
EU	European Union
EUR	estimated ultimate recovery
FARC	Fuerzas Armadas Revolucionarias de Colombia (Revolutionary Armed Forces of Colombia)
FERC	Federal Energy Regulatory Commission
FLEC	Front for the Liberation of the Enclave of Cabinda
FLN	Front de Libération Nationale (Algeria)
GAIL	Gas Authority of India, Limited
GCC	Gulf Cooperation Council
GDP	gross domestic product
GECF	Gas Exporting Countries Forum
GRC	Geophysical Research Corporation
GSI	Geophysical Service, Incorporated
HAIS	Hartley/Anglo–Iranian/Siemens
HAMEL	Hammick and Ellis

ICE	Intercontinental Exchange
ICSID	International Center for the Settlement of Investment Disputes
IEA	International Energy Agency
IEOC	International Egypt Oil Company
IEP	International Energy Programme
ILSA	Iran–Libya Sanctions Act
IMINOCO	Iranian Marine International Oil Company
INH	Instituto Nacional de Hidrocarburos (Spain)
IOC	Indian Oil Corporation
IOC	Interstate Oil Compact
IOOC	Iranian Oil Offshore Company
IOP	Iranian Oil Participants
IPAC	Iran Pan American Oil Company
IPC	Iraq Petroleum Company
IPE	International Petroleum Exchange
IPSA	Iraqi Pipeline across Saudi Arabia
ISA	Iran Sanctions Act
ITIO	Indian Territory Illuminating Oil Company
ITT	Ishpingo–Tapococha–Tipufuni (Ecuador)
KBR	Kellogg Brown and Root
KChNTO	Kaspiysko Chernjmjrskoe Neftepromushlenoe I Torgovoe Obchshestvo (Caspian and Black Sea Oil Production and Trading Company)
KMG	KazMunaiGaz
KNPC	Kuwait National Petroleum Company
KOC	Kuwait Oil Company
KPC	Kuwait Petroleum Corporation
KRG	Kurdistan Regional Government
L&M	Laboratory and Manufacturing
LAPCO	Lavan Petroleum Company
LNG	liquefied natural gas
LPG	liquefied petroleum gas

LWD	logging while drilling
MAP	Marathon Ashland Petroleum
MEND	Movement for the Emancipation of the Niger Delta
MPC	Mosul Petroleum Company
MPI	Ministry of the Petroleum Industry (China)
NAM	Nederlandse Aardolie Maatschappij
NATO	North Atlantic Treaty Organization
NEDC	Near East Development Corporation
NELP	New Exploration License Policy (India)
NGL	natural gas liquids
NGPL	natural gas plant liquids
NIOC	National Iranian Oil Company
NIRA	National Industrial Recovery Act
NNOC	Nigerian National Oil Company
NNPC	Nigerian National Petroleum Company
NOSR	Naval Oil Shale Reserve (U.S.)
NPD	Norwegian Petroleum Directorate
NPR	Naval Petroleum Reserve (U.S.)
NYMEX	New York Mercantile Exchange
OAPEC	Organization of Arab Petroleum Exporting Countries
OCP	Oleoducto de Crudos Pesados (Ecuador)
OGJ	*Oil and Gas Journal*
OGJ100	*Oil and Gas Journal* list of 100 largest petroleum companies headquartered outside the United States
OGJ200	*Oil and Gas Journal* list of 200 largest petroleum companies headquartered in the United States (list does not always contain 200 companies)
OMV	Österreichischen Mineralölverwaltung Aktiengesellschaft (Austrian Oil Corporation; no longer stands directly for this)
ONGC	Oil and Natural Gas Commission (India)
OPEC	Organization of the Petroleum Exporting Countries
OSA	Oil Sand Areas (Canada)
OSA	operating service agreement

OSS	Office of Strategic Services (U.S.)
OSCE	Organization for Security and Co-operation in Europe
PAD	Petroleum Administration for Defense (U.S.)
PADD	Petroleum Administration for Defense District (U.S.)
Partex	Participations and Explorations Corporation
PAW	Petroleum Administration for War (U.S.)
PDO	Petroleum Development (Oman)
PDQ	Petroleum Development (Qatar)
PDTC	Petroleum Development (Trucial States)
PdVSA	Petróleos de Venezuela, Sociedad Anónima
PEI	Petroleum Exploration Incorporated
Pemex	Petróleos Mexicanos
Petrobrás	Petróleo Brasileiro
Petroecuador	Empresa Estatal Petróleos del Ecuador
Petronas	Petroliam Nasional Berhad (National Petroleum Limited) (Malaysia)
PIW	*Petroleum Intelligence Weekly*
PLUTO	Pipeline Under the Ocean
PNR	Partido Nacional Revolucionario (National Revolutionary Party) (Mexico)
PNZ	Partitioned Neutral Zone (Saudi Arabia and Kuwait)
PRC	Petroleum Reserves Corporation
PRI	Partido Revolucionario Institucional (Institutional Revolutionary Party) (Mexico)
Pró-Álcool	Programa Nacional do Álcool (National Alcohol Program) (Brazil)
PRT	Petroleum Revenue Tax (United Kingdom)
PSA	production-sharing agreement
PT ETMSU	Exploitasi Tambang Minyak Sumatra Utara (Northern Sumatra Oil Exploitation)
PTMN-RI	Perusahaan Tambang Minyak Negara Republik Indonesia (National Oil Company of the Republic of Indonesia)
QERL	Queensland Energy Resources Limited
QGPC	Qatar General Petroleum Company

QPC	Qatar Petroleum Company
RMOTC	Rocky Mountain Oilfield Testing Center
RPSA	risk/profit-sharing agreement
RRC	Railroad Commission of Texas
SA	strategic association
SEC	Securities and Exchange Commission (U.S.)
SEG	Society of Exploration Geophysicists
SHT	Société des Hydrocarbures du Tchad
SIRIP	Société Irano–Italienne des Pétroles
SNAM	Società Nazionale Metanodotti
SNEA	Société Nationale Elf Aquitaine
SNPA	Société Nationale des Pétroles d'Aquitaine
SNPC	Société Nationale des Pétroles du Congo
SNPG	Société Nationale Petrolière de Gabon
Socal	Standard Oil Company of California
Socar	State Oil Company of the Azerbaijan Republic
Socony	Standard Oil Company of New York
SOFIRAN	Société Française de Pétroles d'Iran
Sohio	Standard Oil Company of Ohio
Sonangol	Sociedade Nacional de Combustíveis de Angola
Sonatrach	Société Nationale de Transport et de Commercialisation des Hydrocarbures; no longer stands specifically for this
SOTE	Sistema Oleoducto Trans-Ecuatoriano
SPDC	Shell Petroleum Development Company of Nigeria
SPE	Society of Petroleum Engineers
Stanvac	Standard–Vacuum Oil Company
SUMED	Suez–Mediterranean (Pipeline)
Tapline	Trans-Arabian Pipeline
TEPPCO	Texas Eastern Products Pipeline Corporation
TETCO	Texas Eastern Transmission Corporation
TI	Texas Instruments
TNK	Tyumenskaya Neftyanaya Kompaniya (Tyumen Oil Company)

TNT	Tri-nitro Toluene
TPC	Turkish Petroleum Company
UAE	United Arab Emirates
UK	United Kingdom
UK-SORP	United Kingdom Statement of Recommended Practices
UN	United Nations
UNCLOS	United Nations Convention on the Law of the Sea
UNFC	United Nations Framework Committee
UNITA	União Nacional para a Independência Total de Angola
URR	ultimately recoverable resource
U.S.	United States
USGS	United States Geological Survey
USS	United States Steel
USSR	Union of Soviet Socialist Republics (Soviet Union)
WPC	World Petroleum Council
WTI	West Texas Intermediate
WTO	World Trade Organization
YPF	Yacimientos Petrolíferos Fiscales
YPFB	Yacimientos Petrolíferos Fiscales Bolivianos
ZIC	Zone d'Interêt Commun (Congo and Angola)

MEASUREMENTS

b	barrel(s) (1 b = 42 U.S. gallons)
Bb	billions of barrels
bbl	barrels
Bbo	billions of barrels of oil
Bboe	billions of barrels of oil equivalent
Bcf	billions of cubic feet (of natural gas)
Bcfpd	billions of cubic feet (of natural gas) per day
bo	barrel(s) of oil
boe	barrel(s) of oil equivalent

boepd	barrels of oil equivalent per day
bopd	barrels of oil per day
bpd	barrels per day
Btu	British thermal unit
cf	cubic feet (of natural gas)
cp	centipoise
DWT	deadweight tons
EUR	estimated ultimate recovery
Mmb	millions of barrels
Mmbo	millions of barrels of oil
Mmboepd	millions of barrels of oil equivalent per day
Mmbopd	millions of barrels of oil per day
Mmbpd	millions of barrels per day
Mmcf	millions of cubic feet (of natural gas)
nm	nautical miles
Tcf	trillions of cubic feet (of natural gas)

Chronology

c. 1875 BC References to an oil trade in ancient Sumeria appear in official records of King Hammurabi's government.

c. 450 BC Greek historian Herodotus describes oil and salt production from springs and wells in Persia.

c. 400 BC Bamboo pipes wrapped with waxed cloth are used to transport natural gas for lighting in China.

211 BC Natural gas is discovered in Chi-lui-ching, Szechuan, China, while drilling for salt and is used for heat and light.

c. 50 BC Greek historian Diodorus records roaring gas seeps at Kirkuk and presence of large quantities of asphalt in Babylonia.

347 Wells are percussion-drilled in China using bamboo poles with attached bits. Depths of almost 800 feet are achieved. Some of these are salt wells with incidental oil discoveries.

642 Byzantine emperor Heraclitus destroys a number of temples near Baku where locals worship before burning gas wells.

885–886 Premodern oil industry in Azerbaijan and the North Caucasus. Caliph of Baghdad grants revenues from oil springs to the citizens of Darband (Derbent).

1132 Wells in China reach 3,000 feet in depth.

1184 Ibn Jubayr records large oil seeps near Mosul.

1188 Giraldus Cambrensis describes rocks in Wales that appear to be oil bearing and speculates on the possibility of extracting oil from them somehow.

1298 Marco Polo observes the exploitation of oil from surface seeps on the Abseron Peninsula. He notes that the oil is used for fuel and medicinal purposes and is transported on camels.

1460 Francisco Ariosto writes of surface oil springs near Modena, Italy. The oil is used for medicinal purposes.

1526 G. Fernández de Oviedo y Valdéz uses pitch to caulk his ships near Havana, Cuba.

1539 Spanish ship *Santa Cruz* transports a barrel of Venezuelan oil to Spain, reportedly to help treat King Charles I's gout.

1543 Hernando de Soto describes oil seeps in Mexico.

1545 Georgius Agricola mentions oil seeps near Hanover, Germany, near the site of the modern oil fields of Reitling and Hänigsen. He describes the use of oil for illumination, lubrication, and waterproofing.

1594 An inscription inside a 115-foot oil well in Balakhany on the Abseron Peninsula identifies the proprietor as Allah Yar Mammad Nuroghlu.

1595 Walter Raleigh observes the pitch lake in what is now Trinidad and Tobago.

1596 Jan Huygen van Linschoten reports oil seeps and a primitive oil industry in Sumatra.

1625 Johan Volck discusses possible uses of oil in the Pechelbronn district of Alsace, but no commercial development occurs.

1636 German diplomat Adam Oleari observes 30 oil wells in the Baku area and remarks that some of them are gushers.

1683 Engelbert Kaempfer, secretary of the Swedish embassy in Persia, visits Baku and observes the extraction, use, transport, and export of Abseron Peninsula oil.

1702 Russian Tsar Peter I (the Great) orders oil samples to be sent to him from Siberia and orders further exploration of the vast region.

1717 Tsar Peter I's physician Gottlieb Schober describes oil sources in Russia's Volga–Urals area.

1732 A small rudimentary refinery is built at Baku. Its products serve the local area.

1734 Jean-Théophile Hoeffel completes a doctoral thesis in which he describes the results of distilling light fractions from Pechelbronn bitumen and oil sands in Alsace.

1737 Fifty-two hand-dug wells are producing oil at the site of the present-day giant Balakhany field on the Abseron Peninsula.

1746 Louis Pierre Ancillon de la Sablonnière and Jean d'Amascéne Eyrénis begin mining oil sands in the Pechelbronn area of Alsace. Oil sands are brought to the surface, and the oil is washed out of them. An oil company is formed, with 40 shares issued. Fyodor Pryadunov builds a rudimentary installation to collect oil from the surface of the Ukhta River in Russia's Timan–Pechora area.

1748 Peter Kalm of Sweden publishes a map showing oil seeps in Oil Creek, in what is now Pennsylvania.

1788 Peter Pond reports that natives in the area of the Athabasca oil sands (present-day Alberta) use bitumen to caulk their canoes.

1795 M. A. Symes visits the Yenangyaung oil field in Burma, a country with a large oil industry that has existed perhaps for centuries by this time. He describes the export of oil via boat.

1806 David and Joseph Ruffner drill a salt well using a spring pole, drive pipe, casing, and tubing near the Kanawha River in western Virginia (today West Virginia) and strike oil instead.

1813 Manual research drilling is conducted at the Pechelbronn field in Alsace to depths of 640 feet to guide development of the oil sands mining operation.

1815 Oil is remarked as an undesirable by-product from Pennsylvania brine wells.

1821 A natural gas well is drilled in Fredonia, New York, and some local residences are served with gas for lighting.

1825 An order from Russian finance minister Yegor Kankrin places all oil wells under the control of the state.

1827 British geologist John Crawford investigates the Yenangyaung oil field in Burma. He describes the hand-dug wells as shafts about four feet square, cribbed with wood. Crawford estimates annual output at the equivalent of 250,000–300,000 barrels. He notes seeing about 200 boats waiting to load oil. Oil is used in the area for many purposes, including illumination, preserving wood, and boat building.

1830 Johann Lenz investigates the Baku oil industry for the Imperial Academy of St. Petersburg. He reports annual production equivalent to 30,145 barrels from 80 wells or springs.

1833 Benjamin Silliman (Senior) performs an experimental distillation of crude oil.

1835 Nikolai Ivanovich Voskoboinikov of the Corps of Mining Engineers in Baku designs and implements a ventilation system to improve the efficiency and safety of hand-dug oil pits.

1836 Gian Bianconi studies the geology of oil and gas in Italy. Natural gas is used to light the baths at Poretta.

1837 A pilot refinery designed by Nikolai Ivanovich Voskoboinikov is built near Baku.

1838 Oil is produced on Cheleken Island in the Caspian Sea from 3,500 hand-dug pits or surface seeps.

1844 A report detailing ideas developed by Nikolai Ivanovich Voskoboinikov on drilling for oil rather than digging pits by hand is presented to the Russian government in the Caucasus by Vasili Nikolaevich Semyonov.

1845 Prince Mikhail Vorontsov, first viceroy of the Caucasus, authorizes funds for oil drilling based on Voskoboinikov's ideas.

1846 A 69-foot well is drilled to explore for oil in the Bibi–Eibat area near Baku and strikes a small quantity of crude. The well is drilled under the command of Major Alexeev, based on the ideas of Voskoboinikov.

1849 Samuel Kier establishes a small refinery in Pittsburgh to process and bottle oil that is a by-product from his family's brine wells at Tarentum, Pennsylvania. He markets "Kier's Rock Oil" as a patent medicine.

1850–1856 Andrés Pico distills oil from hand-dug pits in southern California to produce lamp oil and illuminate the San Fernando Mission.

1851 Abraham Gesner develops a distillation process yielding an illuminating oil from Trinidad asphalt.

1853 Ignacy Łukasiewicz distills kerosene from crude oil and develops a lamp to burn it. On 31 July, one of his lamps provides light for an operation in Lvov's hospital. Over the next several years, Łukasiewicz produces and refines oil from hand-dug wells in the Bóbrka Forest (present-day Poland).

1854 Pennsylvania Rock Oil Company is established by George Bissell and Jonathan Eveleth. This is one of the first modern oil companies. Natural gas from a water well in Stockton, California, is used to light the Stockton courthouse.

1855 Benjamin Silliman Jr. of Yale University reports to the Pennsylvania Rock Oil Company on his distillation experiments on Pennsylvania crude oil. He concludes that crude oil can be distilled to obtain a range of useful products, including kerosene for illumination.

1857 James Miller Williams strikes oil with the Williams No. 1 well, Enniskillen Township, Ontario, and begins refining the oil to produce kerosene in 1857 or 1858. Some call Williams No. 1 the first modern well in North America, but it is not clear from Williams's records if his early wells were drilled or were dug by hand and cribbed. Michael Dietz invents a kerosene lamp.

1858 Seneca Oil Company of New Haven, Connecticut, is formed as a result of disputes within the Pennsylvania Rock Oil Company and takes over the petroleum exploration effort in Pennsylvania. Seneca hires Edwin Drake. The world's first modern natural gas company, the Fredonia Gas and Light Company, is established.

1859 Successful oil wells are drilled in West Virginia by Robert Hazlett's Virginia Petroleum Company and by Charles Shattuck. Edwin L. Drake strikes oil at a depth of about 69 feet with his well near Titusville, Pennsylvania, for the Seneca Oil Company.

1860 The steamboat *Venango* carries the first load of petroleum to Pittsburgh.

1861 Thomas Sterry Hunt and E. B. Andrews independently elucidate the anticlinal theory of petroleum trapping in geological formations. McClintock No. 1, the oldest continuously producing oil well in the world, begins producing near Oil Creek in Pennsylvania, not far from the Drake Well.

1865 Latin America's first successful oil well is drilled in Peru's northwestern coastal region, discovering the Zorritos oil field.

1866 Atlantic Petroleum Storage Company, the ancestor of Atlantic Refining Company (eventually part of Atlantic Richfield), is established. Vacuum Oil Company, an ancestor of Mobil, is established. Oil is collected from tunnels dug at Sulphur Mountain in Ventura County, California, by the brothers of railroad baron Leland Stanford.

1868–1873 The world's first giant oil field, La Brea (also called Negritos), is discovered in the northwestern coastal region of Peru.

1870 Standard Oil Company is founded by John D. Rockefeller in Cleveland, Ohio. Oil City Oil Exchange is established in Pennsylvania.

1871 Bradford field is discovered in Pennsylvania near the New York border. Rangoon Oil Company (antecedent to Burmah Oil) is established.

1872 Tsar Alexander II establishes new rules for the oil industry, auctioning oil leases to private investors, including foreigners. A boom begins in the Russian oil industry.

1874 Atlantic Refining becomes affiliated with Standard Oil.

1875 Continental Oil and Transportation Company is founded in Ogden, Utah. This company eventually becomes Continental Oil, Conoco, and ConocoPhillips. The Nobel brothers (Ludvig and Robert, with Alfred of Nobel Prize fame as a partner) begin activities in Baku oil industry.

1876 First commercial oil field in California is discovered at Pico Canyon in Los Angeles County.

1877 *Zoroaster*, the first modern oil tanker, designed by Ludvig Nobel, is constructed. It and more advanced models will form a fleet plying the Caspian Sea for the Nobel brothers.

1879 Nobel Brothers Petroleum Production Company (Branobel) is established in Baku. The world's first long-distance crude oil pipeline is completed in Pennsylvania, a six-inch-diameter line running 109 miles from the Bradford oil field to Allentown. Standard Oil acquires Vacuum Oil. U.S. Geological Survey is established. Pacific Coast Oil Company is established in San Francisco. The company forms a core for what will eventually become Standard Oil Company of California (later Chevron).

1882 The Standard Oil trust is fashioned from the alliance of companies hitherto comprising Standard Oil. An Ohio court orders the Standard Oil trust dissolved, but it is reincorporated in New Jersey. The Nobel brothers' factories begin continuous distillation of oil, based on ideas proposed by Dmitri Mendeleev. Herman Frasch sells his process for reducing the sulfur content of crude oil to Imperial Oil.

1883 Transcaucasian railroad is completed; tanker cars transport Baku oil to the Black Sea port of Batumi. Batumi Oil Production and Trading Company (Russian acronym BNITO) is established, the forerunner to the Rothschild's company. Manuel Antonio Pulido Pulido strikes small amounts of oil in Táchira state, Venezuela, south of Lake Maracaibo. His company continues small-scale production until the 1930s.

1884 George Westinghouse strikes natural gas in the backyard of his home in Pittsburgh and decides to sell gas to local customers. By 1887, his company will supply about 5,000 residential and 470 industrial customers.

1885 Continental Oil and Transportation Company is absorbed into Standard Oil. The Rothschilds establish the Caspian and Black Sea Oil Production and Trading Company.

1886 Gasoline-powered automobiles are introduced in Europe by Karl Benz and Wilhelm Daimler. Burmah Oil Company is established as a public joint-stock corporation. The tanker *Gluckauf* makes a transatlantic voyage to deliver U.S. oil at Geestemunde, Germany. Zeinalabdin Tagiyev strikes a gusher in the Bibi–Eibat oil field near Baku.

1887 Canadian senate committee issues report on Alberta oil sands. Frasch process for reducing the sulfur content of crude oil is patented. Ohio Oil Company is established in Lima, Ohio. The Rothschilds gain control of the Mazut company.

1889 Ohio Oil Company is acquired by Standard Oil, which also forms the Standard Oil Company of Indiana and the South Penn Oil Company.

1890 U.S. Congress passes the Sherman Antitrust Act to limit monopolistic and anticompetitive practices by large American corporations. Standard Oil is reincorporated in New Jersey as a holding company. Royal Dutch Petroleum Company is founded. Union Oil Company of California is established.

1891 Shukhov patent for thermal cracking of crude oil is granted. The process can increase gasoline yields, although at this time production of gasoline is not a priority. A similar process patented later by Burton (1913) is the one that will achieve commercial success. Railroad Commission of Texas is established.

1892 Oil is discovered in Los Angeles, California, by Edward Doheny and his partner Charles Canfield. Norman No. 1 well in Neodesha, Kansas, opens up the vast Mid-Continent Oil Region.

1895 The Pure Oil Company is founded. First important commercial oil is discovered in Texas, at Corsicana, while drilling for water.

1897 Nellie Johnstone No. 2, the discovery well of the giant Bartlesville-Dewey field, is drilled in Oklahoma.

1900 Standard Oil controls more than 90 percent of U.S. oil refining. Standard Oil acquires Pacific Coast Oil Company.

1901 Anthony Lucas drills a gusher at Spindletop, Texas, initiating the Texas oil boom. William Knox D'Arcy obtains a concession from Shah Muzzafar al-Din of Persia to explore for petroleum in a 400,000-square-mile area of the country. Alberta's first commercial gas field is developed at Medicine Hat. J. M. Guffey Petroleum Company and Gulf Refining Company (antecedents to Gulf Oil Corporation) are established. Edward Doheny discovers Panuco–Ebano field in Mexico. Indian Territory Illuminating Oil Company is founded. The Texas Fuel Company is founded; it soon reorganizes to the Texas Company and eventually will become Texaco. The *Oil and Gas Journal* (initially called *Oil Investors Journal*) begins publication in Beaumont, Texas.

1902 Ida Tarbell begins publishing a muckraking series of articles on Standard Oil in *McClure's Magazine*.

1903 Anchor Oil (later Phillips Petroleum) established in Bartlesville, Oklahoma, by Frank Phillips.

1905 British Royal Navy, under First Sea Lord Jackie Fisher, begins converting from coal to oil. British government persuades Burmah Oil Company to invest in D'Arcy's exploration venture in Iran. Burmah and D'Arcy sign an agreement on 5 May establishing Concessions Syndicate Ltd. to finance the effort. Glenn Pool is discovered in Oklahoma.

1906 Kerosene pipeline is completed between Baku and the Black Sea port of Batumi (522 miles).

1907 Royal Dutch combines operations with Shell Transport and Trading Company to form the Royal Dutch Shell Group. Gulf Oil Corporation is formed from the J. M. Guffey Petroleum Company and the Gulf Refining Company. J. M. Guffey is bought out for $3 million. First oil discovery in Argentina, by a government team, opens up the giant Comodoro Rivadavia field.

1908 First major oil discovery in Iran (Persia), at Masjid-e-Suleiman, by a team led by geologist George Reynolds, working for the D'Arcy subsidiary of the Burmah Oil Company. Weetman Pearson establishes Mexican Eagle Petroleum Company (Compañía Mexicana de Petróleo el Aguila), a vertically integrated oil company in Mexico.

1909 Burmah Oil establishes the Anglo–Persian Oil Company (antecedent to BP). Patents are granted for the Hughes drill bit, capable of drilling through hard rock. Standard Oil is found guilty of violations of the Sherman Antitrust Act. It appeals the verdict to the U.S. Supreme Court.

1910 Mexican Eagle strikes a gusher at Potrero del Llano No. 4 well near Veracruz, in Mexico's "Golden Lane" (Faja de Oro).

1911 The U.S. Supreme Court orders the dissolution of Standard Oil into 34 smaller, independent competing companies, among which are Standard Oil of New Jersey (later Exxon and ExxonMobil), Standard Oil Company of New York (later Mobil and ExxonMobil), Vacuum Oil Company (later Mobil and ExxonMobil), Standard Oil of Indiana (later Amoco, eventually merged into BP), Standard Oil of California (later Chevron), Atlantic Refining Company (later Atlantic Richfield, Arco, eventually acquired by BP), Continental Oil Company (later Conoco and ConocoPhillips), South Penn Oil Company (later Pennzoil, eventually acquired in part by Royal Dutch Shell), Standard Oil Company of Ohio (Sohio, eventually acquired by BP), and Ohio Oil Company (later Marathon). Magnolia Petroleum Company is established. Humble Oil Company, antecedent to the Humble Oil and Refining Company (later part of Standard Oil of New Jersey, ultimately Exxon), is established.

1912 Anglo–Persian Oil Company begins operating a refinery in Abadan, Iran. Turkish Petroleum Company consortium is established to seek a concession in Mesopotamia (Iraq). The venture is half owned by the Anglo–Persian Oil Company. Royal Dutch Shell and Deutsche Bank each own 22.5 percent, and entrepreneur Calouste Gulbenkian owns 5 percent. Royal Dutch Shell acquires Rothschild petroleum interests in Russia.

1913 Burton patent for thermal cracking of crude oil is granted. The Burton process sharply increases gasoline yields. Note earlier (1891) Shukhov patent, which was not placed in as immediate and widespread practice. The American Institute of Mining Engineers (AIME) Standing Committee on Oil and Gas is established. This will evolve into the Society of Petroleum Engineers. Global oil production exceeds 1 million barrels per day. Walter Snelling obtains a patent for a method to capture natural gas liquids.

1914 British government acquires 51 percent of Anglo–Persian Oil Company for £2.2 million. Oil is discovered at Turner Valley, Alberta. Venezuela's first significant oil discovery is made near Lake Maracaibo. Royal Dutch Shell builds an oil refinery on Curaçao, prompted by the discovery of oil in Venezuela.

1915 Empire Gas Company (Cities Service Company) discovers the El Dorado oil field in Kansas with its Stapleton No. 1 well. El Dorado production will provide a significant edge for the Allies during World War I.

1916 Germany invades Romania to gain control of oil resources and processing facilities, mostly located around Ploieşti. Harry Sinclair establishes the Sinclair Oil Corporation. Secret Sykes–Picot Agreement between Britain and France divides Arab portions of the Ottoman Empire into zones of British and French influence, to take effect when World War I is over. The future Iraq is to go largely to the British.

1917 Cabinas, Venezuela's first giant field and part of the Bolivar Coastal Complex, is discovered. Anchor Oil becomes Phillips Petroleum. American Association of Petroleum Geologists, operating informally since 1915, is formally established. Humble Oil and Refining Company (later part of Standard Oil of New Jersey) is established, based on the Humble Oil Company. Article 27 of the Mexican constitution reserves ownership of subsoil resources to the nation.

1918 Royal Dutch Shell acquires control of Mexican Eagle Company.

1919 American Petroleum Institute is established. Standard Oil of New Jersey (later Exxon) acquires half interest in Humble Oil and Refining Company.

1920 Occidental Petroleum is established in California. The San Remo Conference, held to determine the future of the Ottoman Empire's territories, restructures the ownership of the Turkish Petroleum Company. German shares are given to France. Compagnie Financière Belge des Pétroles (later Petrofina) is established.

1922 American Oil Company (later part of Amoco and the source of the Amoco name) is incorporated. Yacimientos Petroliferos Fiscales (YPF), the national oil company of Argentina and the first vertically integrated national oil company, is formed. Uqayr Conference establishes Neutral Zone between Saudi Arabia and Kuwait.

1923 Edward Doheny's Pan American Petroleum and Transport Company acquires a half interest in American Oil. U.S. establishes a 22-million-acre National Petroleum Reserve near Prudhoe Bay, Alaska. German researchers Franz Fischer and Hans Tropsch, working at the Kaiser Wilhelm Institute, develop a process for converting coal to gas, which is then used to make synthetic fuels. Ibn Saud issues Frank Holmes and his Eastern and General Syndicate the first petroleum concession in Saudi Arabia. (The concession does not bear fruit and is canceled by Ibn Saud in 1928.)

1924 Compagnie Française des Pétroles (later Total) is established. First oil discovery using a geophysical method is made: Nash Dome in Texas, using gravity techniques. First oil discovery using seismic methods is made: Orchard Dome in Texas, using refraction seismology. U.S. Senate investigates Teapot Dome affair.

1925 Standard Oil of Indiana (later Amoco) acquires Edward Doheny's Pan American Petroleum and Transport Company (except California operations). Magnolia Petroleum Company becomes a wholly owned subsidiary of Standard Oil Company of New York (later Mobil).

1926 Azienda Generale Italiana Petroli (AGIP) is established by the Fascist government in Italy.

1927 First oil discovery using reflection seismology is made: Maud Field in Oklahoma. The giant Kirkuk field is discovered in Iraq by Turkish Petroleum Company (later Iraq Petroleum Company) at Baba Gurgur, near the city of Kirkuk. This is the first major discovery in an Arab country. U.S. chemist Graham Edgar develops the octane number for characterizing gasoline. Conrad and Marcel Schlumberger record the first electrical resistivity well log at Pechelbronn in Alsace.

1928 Tia Juana field (Bolivar Coastal Complex) is discovered in Venezuela. Indian Territory Illuminating Oil Company discovers Oklahoma City field. Red Line Agreement is signed by members of the Turkish Petroleum Company (Iraq Petroleum Company) consortium. The agreement prohibits signatories from seeking independent concessions in the area of the former Ottoman Empire, including Saudi Arabia. As-Is Agreement (Achnacarry Agreement) is signed by major Western petroleum companies to limit competition and control the oil market. Signatories include Royal Dutch Shell, Standard Oil of New Jersey (later Exxon), and Anglo–Persian Oil Company (later BP). American major oil companies join the Turkish Petroleum Company consortium, via the Near East Development Corporation.

1929 Continental Oil merges with Marland Oil (combined company later becomes part of Conoco and ConocoPhillips).

1930 East Texas oil field is discovered, leading to a collapse in oil prices and attempts by the Railroad Commission of Texas to regulate production. Society of Exploration Geophysicists is established in Tulsa, Oklahoma (original name: Society of Economic Geophysicists).

1931 Railroad Commission of Texas issues its first proration order in an attempt to stabilize the market after the East Texas field. After resistance from some producers and several legal battles over the commission's charter, the commission will ultimately prove successful. Standard Oil Company of New York (Socony) and Vacuum Oil Company merge to form Socony–Vacuum (later Mobil).

1932 Standard Oil of California (Socal, later Chevron) discovers oil in Bahrain. Ivan Strizhov discovers the Yarega (Yaregskoye) field in the Timan–Pechora area of Russia (Soviet Union).

1933 Saudi Arabia grants Standard Oil of California (Socal) an exclusive petroleum concession. Socal forms the California Arabian Standard Oil Company (Casoc, predecessor to Aramco) to drill for oil in Saudi Arabia. Shah Reza Pahlavi terminates the 1901 concession granted to William Knox D'Arcy that has been exploited by the Anglo–Persian Oil Company. A new concession is negotiated.

1934 Gulf Oil Corporation and Anglo–Persian Oil Company form the Kuwait Oil Company and obtain a concession in Kuwait.

1935 Anglo–Persian Oil Company becomes Anglo–Iranian Oil Company. Iraq Petroleum Company partners form Petroleum Development (Trucial States) to operate in what is now the United Arab Emirates. United States establishes Connally Hot Oil Act to restrict flow of oil produced in excess of legal quotas. Cities Service Company acquires Indian Territory Illuminating Oil Company.

1936 Texaco partners with Socal, acquiring half of Casoc. The two companies form Caltex. The collaboration between them is sometimes called the "Blue Line Agreement." DeGolyer and MacNaughton is established in Dallas, Texas. Eugene Houdry opens a plant to produce high-octane gasoline via his catalytic cracking process. The plant has financial backing from Sun Oil and Socony Vacuum (later Mobil).

1937 Bolivia expropriates Standard Oil of New Jersey's petroleum interests in the country without compensation. Anglo–Iranian Oil Company (BP) and Royal Dutch Shell form a joint venture named Shell/D'Arcy to explore in Nigeria. Global oil production exceeds 5 million barrels per day.

1938 Casoc makes the first significant oil discovery in Saudi Arabia with its Dammam No. 7 well near the Persian Gulf. Kuwait Oil Company discovers Burgan, which will prove to be the world's second-largest oil field. Mexico nationalizes its petroleum industry.

1939 The first Saudi Arabian crude oil is shipped from Ras Tanura.

1941 Shah Reza Pahlavi of Iran is forced by the British and Russian armies to abdicate in favor of his son Mohammed Reza Pahlavi.

1943 Big Inch Pipeline (1,811 miles) between oil fields of East Texas and U.S. East Coast is completed. Major allied air raid on the Romanian oil fields and processing facilities around Ploieşti, held by the Nazis. (Other raids occurred in 1941 and 1942.)

1944 California Arabian Standard Oil Company (Casoc) changes name to Arabian American Oil Company (Aramco). Everette DeGolyer reports back from Saudi Arabia, where he has been investigating the country's oil potential under the auspices of the U.S. Petroleum Reserves Corporation; he predicts that the center of gravity of the world's petroleum industry will eventually move to this region.

1946 First test of Stanolind's "Hydrafrac" process performed in the Klepper No. 1 well in the Hugoton natural gas field in southwestern Kansas. This is a precursor to modern fracking methods.

1947 Leduc No. 1 gusher initiates Alberta and Canada as a major producer. Strike of refinery workers in Abadan, Iran. Kerr McGee completes an offshore well in the Gulf of Mexico out of sight of land. Soviet Union discovers Romashkino, Russia's second-largest field, in the Volga–Urals area.

1948 Aramco discovers Ghawar, which will prove to be the world's largest oil field. Standard Oil of New Jersey (later Exxon) and Socony Vacuum (later Mobil) become partners in Aramco, joining Standard Oil of California (later Chevron) and Texaco. Venezuela adopts 50–50 profit sharing with foreign oil companies.

1949 Getty Oil Company acquires a 60-year concession in Saudi Arabia's part of the Neutral Zone for $9.5 million. The People's Republic of China issues its "Nine-Dash Line" laying claim to most of the South China Sea.

1950 Saudi Arabia adopts 50–50 profit sharing with foreign oil companies. Aramco completes the Trans-Arabian Pipeline (Tapline) linking the oil fields of Saudi Arabia's Eastern Province to the Mediterranean Sea via Lebanon. Global oil production exceeds 10 million barrels per day.

1951 Aramco discovers Safaniya, which will prove to be the world's largest offshore oil field and the third-largest oil field overall by some estimates. Iran under Prime Minister Mohammed Mossadegh nationalizes its oil industry and forms the National Iranian Oil Company (NIOC). Amerada Petroleum (later Hess) discovers first oil in North Dakota with Iverson No. 1 well.

1953 Italian parliament creates ENI; Enrico Mattei is the first president. Iran's Prime Minister Mossadegh is removed by a coup engineered by the U.S. Central Intelligence Agency (CIA). Shah Mohammed Reza Pahlavi, who briefly fled the country, returns with his power solidified. Iranian oil remains nationalized. Petroleum Development (Trucial States) discovers Bab oil field in Abu Dhabi. Brazilian government forms Petróleo Brasileiro (Petrobrás).

1954 Consortium of Western petroleum companies (Iranian Oil Participants Ltd.) is formed to operate in Iran's nationalized oil industry. Anglo–Iranian Oil Company becomes British Petroleum. Abu Dhabi Marine Areas, two-thirds owned by BP and one-third by Compagnie Française des Pétroles, is formed to explore offshore Abu Dhabi.

1955 Socony Vacuum becomes Socony Mobil. Libya issues its first petroleum law, Royal Decree No. 80, declaring subsurface minerals the property of the state and establishing rules for concessions. Petrofina makes the first discovery in Angola, in the onshore Kwanza basin.

1956 Suez crisis results in closure of the Suez Canal to navigation; it will reopen in 1957. Shell geologist M. King Hubbert publicly and accurately predicts that U.S. oil production for the lower 48 states will peak around 1970. Compagnie Française des Pétroles (later Total) discovers Hassi Messaoud oil field in Algeria. Shell/D'Arcy discovers Oloibiri oil field in Nigeria. The field is relatively small but is Nigeria's first commercially viable discovery. Shell/D'Arcy changes its name to Shell-BP.

1957 Armand Hammer acquires Occidental Petroleum. Iran passes a law allowing NIOC to enter into joint ventures to explore and develop areas not leased to Iranian Oil Participants Ltd. ENI and NIOC sign a contract creating Société Irano–Italienne des Pétroles (SIRIP), a joint venture to explore for and develop petroleum in Iran. ENI offers NIOC true joint management and favorable terms, enraging the major oil companies. Compagnie Française des Pétroles (later Total) discovers Hassi R'Mel gas and oil field in Algeria. Society of Petroleum Engineers is established.

1958 Standard Oil Company of Indiana (later Amoco) creates a joint venture with NIOC named Iran Pan-American Oil Company (IPAC).

1959 China discovers Daqing oil field. Socony Mobil (later Mobil) and Gelsenberg of Germany discover Amal, Libya's largest oil field. Anadarko Petroleum is founded. *Universe Apollo* becomes the first oil tanker to surpass 100,000 DWT. Representatives from Venezuela, Saudi Arabia, Iran, Kuwait, and Iraq sign the secret and nonbinding Mehdi Pact at the First Arab Petroleum Congress in Cairo. The pact presages the formation of the Organization of the Petroleum Exporting Countries (OPEC). Standard Oil of New Jersey consolidates all its domestic operations with those of Humble Oil and Refining Company, whose acquisition it completed a year earlier after decades of increasing its stake. The Groningen gas field is discovered in the Netherlands.

1960 Standard Oil of New Jersey (later Exxon) cuts posted price of oil unilaterally and without warning by 7 percent and is followed by other companies. Oil-producing countries are infuriated. This action is one immediate catalyst for the formation of OPEC. Iraq, Iran, Kuwait, Saudi Arabia, and Venezuela form the Organization of the Petroleum Exporting Countries (OPEC).

1961 Iraq passes Law No. 80, revoking 99.5 percent of Iraq Petroleum Company's concession area but allowing it to keep operating its producing interests. BP and Bunker Hunt discover Sarir field in Libya. Qatar joins OPEC.

1962 Ohio Oil Company acquires Plymouth Oil and changes its name to Marathon Oil Company. Libya and Indonesia join OPEC.

1963 Pennzoil is formed via the merger of South Penn Oil Company and Zapata Petroleum. Petroleum Development (Trucial States) changes name to Abu Dhabi Petroleum Company. Algeria establishes Sonatrach. Global oil production exceeds 25 million barrels per day.

1964 Modern Canadian oil sands industry is launched. Esso Canada (Imperial Oil) begins an extraction project at Cold Lake, Alberta, using cyclic steam injection. The world's longest pipeline, the 2,500-mile Druzhba (Friendship) Pipeline between the Urals and Eastern Europe, begins operation. It will be extended in 1970.

1965 Union Oil Company of California acquires Pure Oil Company. Algeria and France sign Algiers Accords, allowing French companies to keep concessions, but only in cooperation with the national company Sonatrach. Abu Dhabi Marine Areas discovers Zakum field, Abu Dhabi/UAE's largest. Abu Dhabi negotiates a 50–50 deal with Abu Dhabi Petroleum Company. Arco, Murphy Oil, Sun Oil Company, and the Union Oil Company of California partner with NIOC in the joint venture Lavan Petroleum Company (LAPCO). ENI, Phillips Petroleum, and the Indian Oil and Natural Gas Commission partner with NIOC in the joint venture Iranian Marine International Oil Company (IMINOCO). Soviet Union discovers Samotlor, Russia's largest oil field, in western Siberia.

1966 Atlantic Refining Company acquires Richfield Oil Company and becomes Atlantic Richfield Company (Arco). Abu Dhabi negotiates a 50–50 deal with Abu Dhabi Marine Areas. Gulf makes the first offshore discovery in Angola. Soviet Union discovers Urengoy, Russia's largest gas field, in western Siberia. The first tanker surpassing 200,000 DWT, a midrange VLCC, is delivered. Socony Mobil becomes Mobil Oil Corporation.

1967 Arco and Humble Oil and Refining Company (part of Standard Oil of New Jersey, later Exxon) discover Prudhoe Bay oil field in Alaska. Sun Oil (later Suncor Energy) completes Great Canadian Oil Sands plant, an integrated mining and upgrading operation, north of Fort McMurray, Alberta. *Torrey Canyon* oil spill occurs off the southwest coast of England. Abu Dhabi (United Arab Emirates) joins OPEC. Algeria nationalizes the interests of five U.S. companies after the Six-Day War. U.S. begins testing of well stimulation via peaceful nuclear explosions, with Project Gasbuggy, detonating a 29-kiloton device in New Mexico's San Juan basin.

1967–1975 Suez Canal closed to navigation because of hostilities between Israel and Arab neighbors, beginning with the Six-Day War.

1968 Organization of Arab Petroleum Exporting Countries (OAPEC) is formed by Saudi Arabia, Kuwait, and Libya.

1969 Peru nationalizes U.S. petroleum interests. A consortium led by Phillips Petroleum discovers the Ekofisk oil field in the Norwegian sector of the North Sea. Atlantic Richfield Company (Arco) acquires Sinclair Oil Corporation. Santa Barbara oil spill occurs. The 332,178 DWT *Universe Iran*, the first ULCC, is introduced. Algeria joins OPEC.

1970 Algeria, Bahrain, Qatar, and Abu Dhabi (United Arab Emirates) join OAPEC.

1971 Tehran Agreement is concluded between major oil companies and OPEC countries of the Persian Gulf region, raising prices and increasing producer countries' profit shares. Tripoli Agreement is concluded between major oil companies and OPEC countries delivering oil to the Mediterranean region, raising prices and increasing producer countries' profit shares. Algeria nationalizes 51 percent of French petroleum interests in the country. Nigeria joins OPEC and forms Nigerian National Oil Company (NNOC). Royal Dutch Shell discovers the megagiant North Field offshore Qatar. This is the world's largest gas field and the largest known single hydrocarbon deposit in energy-equivalent terms. Royal Dutch Shell and Standard Oil Company of New Jersey (Exxon) discover the Brent oil field in the British sector of the North Sea. Abu Dhabi establishes the Abu Dhabi National Oil Company (ADNOC).

1972 First Geneva Agreement is concluded, adjusting OPEC oil price for the devaluation of the U.S. dollar. Norwegian Petroleum Directorate and the Norwegian national oil company Statoil are established. Participation agreements are concluded between major Western oil companies and some OPEC countries of the Persian Gulf, allowing for gradual acquisition of petroleum operations by national governments. Iraq and Syria join OAPEC. Standard Oil Company of New Jersey becomes Exxon. Global oil production exceeds 50 million barrels per day.

1973 Second Geneva Agreement is concluded between major oil companies and eight OPEC nations, adjusting oil prices upward to compensate for devaluation of the U.S. dollar. Saudi Arabia acquires 25 percent of Aramco under the participation framework. Abu Dhabi acquires 25 percent of Abu Dhabi Petroleum Company and Abu Dhabi Marine Areas under the participation framework. NNOC acquires 35 percent interest in foreign petroleum operations in Nigeria. Libya nationalizes 51 percent of the interests of ENI, Oasis, Exxon, Mobil, and Occidental. It completely nationalizes the interests of Amoseas/Texaco, Royal Dutch Shell, Arco, and the partnership of BP and Bunker Hunt. In the wake of the Yom Kippur War, Persian Gulf OPEC nations raise the price of Arab Light from $2.90 to $5.11 a barrel. OAPEC nations begin a 5 percent per month cut in collective oil production until Israel agrees to withdraw to pre-1967 borders, and impose a selective embargo on countries supporting Israel (including the United States). These OPEC and OAPEC actions precipitate the First Oil Shock. Ecuador joins OPEC. Egypt joins OAPEC.

1974 Most of the Arab oil-producing nations end their embargo against the United States. Sixteen countries of the Organisation for Economic Co-operation and Development (OECD) conclude the International Energy Programme Agreement and establish the International Energy Agency to guard against future supply disruptions by stockpiling and sharing oil. Saudi Arabia increases its stake in Aramco from 25 percent to 60 percent. Kuwait acquires a 60 percent stake in Kuwait Oil Company. Abu Dhabi raises its stake in Abu Dhabi Petroleum Company and Abu Dhabi Marine Areas to 60 percent. The companies are then restructured as Abu Dhabi Company for Onshore Oil Operation (ADCO) and Abu Dhabi Marine Operating Company (ADMA-OPCO). Petrobrás discovers Garoupa gas field, the first find in the offshore Campos basin. Military government in Peru expropriates assets of the International Petroleum Company (a subsidiary of Standard Oil Company of New Jersey [later Exxon]).

1975 Venezuela nationalizes its oil industry and establishes Petróleos de Venezuela, S.A. (PdVSA) as the national oil company. Iraq completes nationalization of its petroleum industry. Kuwait completes its acquisition of Kuwait Oil Company. Gabon joins OPEC.

1976 UK government establishes British National Oil Company (BNOC) as an instrument of public participation in the North Sea. (BNOC's oil production interests will be privatized as Britoil beginning in 1982 and will come under the full control of BP in 1988. BNOC's remaining functions will be abolished in 1986.) Pemex discovers the Cantarell oil field in the Bay of Campeche. Elf Aquitaine is established. Angolan government establishes national oil company Sonangol.

1977 The first oil from Prudhoe Bay reaches the port of Valdez through the Trans-Alaska Pipeline. Suez–Mediterranean (Sumed) Pipeline linking the Red Sea to the Mediterranean begins operation. Nigerian National Oil Company becomes Nigerian National Petroleum Corporation (NNPC).

1978 *Amoco Cadiz* runs aground off the French coast, spilling more than 1.6 Mmb of crude. BP acquires majority interest in Sohio.

1978–1979 Second Oil Shock: the Iranian Revolution causes oil prices to double. Ayatollah Ruhollah Khomeini formally assumes power in Iran in December 1979.

1979 Iran cancels consortium agreement (Iranian Oil Participants). Egypt is suspended from OAPEC for signing Camp David Accords. Nigeria nationalizes BP's interest in Shell–BP. Shell–BP becomes the Shell Petroleum Development Company of Nigeria (SPDC), the dominant company in Nigeria's sector.

1980 Average oil price nears $37/b, a level not again reached in nominal terms until 2004 (in 2016 dollars, 1980 prices averaged over $107/b). U.S. Congress establishes Arctic National Wildlife Refuge (ANWR). Most of the refuge is closed to petroleum exploration and development, with the exception of the "1002 Area" near the Beaufort Sea. Saudi Arabia completes acquisition of Aramco. Iran–Iraq War begins when Iraq invades Iran. Iran acquires the assets of all four successful joint ventures (SIRIP, IPAC, LAPCO, and IMINOCO) that were outside the framework of the 1954 consortium agreement. The assets are taken over by the new Iranian Oil Offshore Company (IOOC). Elf Aquitaine begins tests of horizontal drilling in European oil fields.

1981 Aramco completes construction of the Petroline pipeline from Abqaiq to the Red Sea port of Yanbu.

1982 Exxon and Tosco's Colony Oil Shale Project in Colorado is canceled because of technical difficulties and falling oil prices. U.S. Steel acquires Marathon Oil Company. Occidental Petroleum acquires Cities Service Company. Tunisia joins OAPEC.

1983 Ali al-Naimi becomes the first Saudi president of Aramco. *Oil and Gas Journal* begins the OGJ400 (later OGJ300 and OGJ200) tabulation of the largest U.S.-based petroleum companies. The journal will also publish the OGJ100 list of the largest petroleum companies headquartered outside the United States. Union Oil Company of California becomes Unocal.

1984 Standard Oil of California acquires Gulf Oil Corporation and changes its name to Chevron. Pennzoil agrees to acquire three-sevenths of Getty Oil. Texaco steps in and acquires all of Getty. Pennzoil sues Texaco for tortuous interference. This will result in a record jury award of $10.5 billion in damages to Pennzoil.

1985 Compagnie Française des Pétroles becomes Total Compagnie Française des Pétroles. Standard Oil Company of Indiana becomes Amoco. Petrobrás discovers Marlim and Marlim Sul in the offshore Campos basin.

1985–1986 Oil Countershock: Saudi Arabia gives up singlehandedly trying to defend the price against market pressures. It increases production and offers attractive netback pricing. Oil prices drop to a low of about $8/b during 1986.

1986 Iraq bombs Iran's Kharg Island oil installations and damages them severely. Tunisia requests suspension of OAPEC membership. First successful multifracture horizontal well is drilled in Wayne County, West Virginia.

1987 BP completes acquisition of Sohio. Pennzoil settles with Texaco for about $3 billion.

1988 Aramco changes its name to Saudi Arabian Oil Company, or Saudi Aramco. The "Aramco" name continues to be used informally. Iran–Iraq War ends.

1989 *Exxon Valdez* oil spill occurs in Prince William Sound, Alaska. Soviet Ministry of the Gas Industry is reorganized into a company named Gazprom. Egypt is reinstated in OAPEC. Aubrey McClendon and Tom Ward establish Chesapeake Energy with an initial investment of $50,000.

1990 Iraq invades Kuwait, precipitating the 1991 Persian Gulf War. A $120 million judgment is issued against Amoco in the case of the 1978 *Amoco Cadiz* oil spill. PdVSA acquires Citgo.

1991 U.S.-led coalition drives Iraq out of Kuwait. Total Compagnie Française des Pétroles becomes Total. USX (formerly U.S. Steel) creates a separate stock for its Marathon Oil subsidiary, USX-Marathon. Mitchell Energy under George P. Mitchell performs the first horizontal fractured well in the Barnett Shale in Texas.

1992 Ecuador suspends its OPEC membership. Gazprom becomes a joint stock company in Russia. Argentinean government begins privatizing YPF. BP discovers Cusiana field in Colombia.

1993 Russia establishes Lukoil, formerly the Soviet Union's first integrated oil company, as a joint stock company. Russia establishes Yukos, combining a number of Soviet-era assets. Russia forms Rosneft from assets previously held by Rosneftegaz, the successor to the Soviet Ministry of Oil and Gas.

1994 Gabon suspends its OPEC membership. Russia signs its first-ever production-sharing agreement with a consortium led by Royal Dutch Shell to develop oil and gas off Sakhalin Island. The license will be issued in 1996. In 2007 Gazprom will take majority control of the project. Exxon is ordered to pay $287 million in compensatory damages and $5 billion in punitive damages for the *Exxon Valdez* oil spill. A scandal breaks out in France involving corruption and embezzlement at Elf Aquitaine (later part of Total), unfolding over several years and eventually reaching to the highest levels of the French government.

1995 Venezuela begins the Apertura Petrolera by launching a suite of new policies meant to attract foreign investment in the country's upstream petroleum sector. Russia forms Sibneft by combining the Soviet-era oil company Noyabrskneftegaz, the Omsk refinery, a distribution network, and other entities. Loans for Shares scheme begins in Russia under the government of Boris Yeltsin. It will allow a small group of oligarchs to gain control of large state enterprises, including petroleum companies, for relatively small sums. Sibneft and Yukos, among others, pass into private control in this fashion.

President Bill Clinton issues Executive Order 12959, banning U.S. trade with and investment in Iran. Clinton blocks a $1 billion contract between Conoco and Iran to develop a large offshore oil tract in the Persian Gulf.

1996 First Iraqi oil is shipped under the United Nations "Oil for Food" program. The program will operate until the beginning of the Iraq War in 2003. U.S. passes Iran–Libya Sanctions Act (ILSA), attempting to punish energy investments by non-U.S. companies in Iran and Libya (U.S. companies have already been forbidden from such investments). Petrobrás discovers Roncador, Brazil's largest field yet, in the offshore Campos basin. SPDC discovers Bonga field offshore Nigeria (deepwater); this is Nigeria's largest oil field.

1998 Amoco merges with BP to form BP Amoco. Clinton administration waives ILSA sanctions on a Total-led consortium to develop Iran's mega-giant South Pars gas field. Hugo Chávez is elected president of Venezuela.

1999 Exxon and Mobil merge to form ExxonMobil in an $83 billion stock transaction. Total acquires Petrofina for $11 billion and becomes TotalFina. Phillips Petroleum acquires Arco's Alaska operations for $7 billion, clearing the way for BP to acquire Arco. Repsol acquires 97 percent of YPF for $13.4 billion. NIOC discovers Azadegan, one of the largest oil fields found in two decades.

2000 BP Amoco shortens its name to BP. Repsol completes acquisition of YPF and becomes Repsol YPF. TotalFina acquires Elf Aquitaine, forming TotalFinaElf. BP acquires Atlantic Richfield Company (Arco) in a $27 billion deal. BP acquires Burmah Castrol in a $4.7 billion deal. Kashagan field in the offshore Caspian of Kazakhstan is one of the world's most important oil discoveries in two decades. Hugo Chávez wins a second term as president of Venezuela in an election required by a new constitution.

2001 Chevron merges with Texaco in a deal valued at $38.3 billion in stock and $6.7 billion in debt. The merged company is initially called ChevronTexaco. Venezuela under Hugo Chávez passes a new hydrocarbons law; the country begins to reverse the Apertura Petrolera. Caspian Pipeline Consortium begins pumping crude oil through a 990-mile pipeline from Kazakhstan's Tengiz field to the Russian Black Sea port of Novorossiysk. Gas Exporting Countries Forum (GECF) is established. USX spins off U.S. Steel and renames USX-Marathon as Marathon Oil Company. Statoil is partially privatized. Consortium led by Total discovers Girassol, Angola's largest deepwater discovery. Punitive damages for the *Exxon Valdez* spill are halved on appeal to $2.5 billion.

2002 Conoco and Phillips Petroleum merge to form ConocoPhillips. Production begins at Muskeg River oil sands mine, Fort McMurray area, Alberta. There is an ultimately unsuccessful coup against President Hugo Chávez of Venezuela. There are strikes, disarray, management shake-ups, and mass firings at the national oil company PdVSA.

2003 Iraq War begins. TotalFinaElf becomes Total, SA. Yukos head Mikhail Khodorkovsky is arrested and imprisoned, accused of tax evasion and other charges. Global oil production exceeds 70 million barrels per day. Tyumen Oil Company (TNK) and BP form joint venture in Russia. *Oil and Gas Journal* begins counting the Alberta oil sands in its reserves estimate for Canada, increasing the estimate by a factor of 37. Kazakhstan, Azerbaijan, and Russia sign a three-way agreement delimiting adjacent sections of the Caspian.

2004 Russia auctions Yukos's main subsidiary, Yuganskneftegaz, for $9.4 billion as a way to collect alleged back taxes. The winning bidder, Baikal Finance, is acquired by Rosneft. Range Resources opens up the modern Marcellus Shale play with its Renz No. 1 well in Pennsylvania. ConocoPhillips and Lukoil begin a strategic alliance; ConocoPhillips acquires a 7.59 percent stake in Lukoil.

2005 Baku–Tbilisi–Ceyhan Pipeline (1,099 miles) connecting Caspian oil fields to the Mediterranean is completed. ChevronTexaco shortens its name to Chevron. China National Offshore Oil Corporation (CNOOC) attempts unsuccessfully to buy Unocal. Chevron acquires Unocal. Former Yukos head Mikhail Khodorkovsky is found guilty of tax evasion and other charges and sentenced to nine years in prison. Gazprom acquires control of Sibneft from Roman Abramovich and Boris Berezovsky.

2006 Attempted attack by Al Qaeda on Abqaiq oil processing facility in Saudi Arabia. Sibneft is renamed Gazprom Neft. Bolivia under Evo Morales nationalizes its gas industry; government forces occupy the largest gas fields. Chevron's Jack-2 well is completed in the Gulf of Mexico in 7,000 feet of water and more than 20,000 feet below the sea floor, discovering a significant oil field. Hugo Chávez is reelected to a third term as president of Venezuela. ConocoPhillips increases its stake in Lukoil to 20 percent. Ecuador expels Occidental Petroleum and transfers its operations to Petroecuador. A new Iraqi constitution enters into force, with recognition of northern Kurdistan as a special region.

2007 Statoil merges with Norsk Hydro's oil and gas business to form StatoilHydro. Ecuador rejoins OPEC. Angola joins OPEC. Rosneft completes acquisition of Yukos's remaining assets. Gazprom takes majority control over the formerly Shell-led Sakhalin II project for $7.35 billion; Petrobrás and partners discover the giant Tupi oil field (later renamed Lula) in the offshore Santos basin. ConocoPhillips turns over its operations in Dubai to the emirate government. Chávez government in Venezuela begins taking majority control of Orinoco Belt heavy oil projects. Bolivia bars foreign companies from exporting petroleum products from the country. Kurdistan Regional Government begins signing its own contracts with petroleum companies, setting the stage for continuing tensions with the Iraqi central government in Baghdad.

2008 Third Oil Shock: oil prices reach record highs, exceeding $135/b in July. Prices fall back below $50 later in the year as the world economy slides into recession. Additional major offshore discoveries are announced in Brazil, although there is some controversy about their true extent. A struggle for control of the TNK–BP joint venture occurs between BP and Russian partners; the Russian government is accused of strong-arm tactics. Iraq invites Western oil companies back into the country as service providers. U.S. Supreme Court reduces *Exxon Valdez* punitive damages of $2.5 billion to match actual damages estimated at closer to $500 million.

2009 Indonesia suspends OPEC membership. Evo Morales is elected to a second term as president of Bolivia. Nigerian government announces amnesty for the anti-oil militants of the Movement for the Emancipation of the Niger Delta (MEND). Suncor Energy acquires Petro-Canada.

2010 *Deepwater Horizon* blowout in the Gulf of Mexico begins an oil spill of 5 million barrels into the environment that takes five months to stop. EU begins sanctions against Iran, targeting its petroleum industry, with the motive of persuading Iran to stop its nuclear program. ConocoPhillips sells back its stake in Lukoil. A team led by Noble Energy discovers the Leviathan gas field in Israel's portion of the Levant basin in the eastern Mediterranean.

2011 Texas overtakes the federal offshore areas in the Gulf of Mexico and once again becomes the top oil-producing state in the U.S. *Oil and Gas Journal* begins including very heavy oil in the Orinoco Belt in Venezuela's reserves estimates, eventually making Venezuela first in the world in deemed oil reserves. South Sudan becomes an independent country; oil production collapses. A treaty between Norway and Russia delimiting the two countries' respective zones in the Barents Sea enters into force. A team led by Noble energy discovers the Aphrodite Gas Field in Cyprus's portion of the Levant basin in the eastern Mediterranean. Chevron acquires Atlas Energy and thereby enters the Marcellus Shale. Statoil acquires Brigham Resources and thereby enters the Bakken Shale. Marathon spins off downstream operations into Marathon Petroleum. Rosneft and BP agree to joint exploration in the South Kara Sea. An Ecuadorean court issues a judgment of $18 billion against Chevron, later reduced to $9.5 billion, for alleged environmental damage arising from Texaco's Ecuadorean operations between 1964 and

1990. Government of Gabon forms a new national oil company, Gabon Oil Company, to participate in new ventures. U.S. withdraws most forces from Iraq. Revolt breaks out in Libya; Muammar Qaddafi is killed.

2012 New U.S. and EU economic sanctions target Iran's petroleum industry. The cumulative effect of sanctions reduces Iranian petroleum production around 20 percent between 2011 and 2013. Argentina expropriates 51 percent of Repsol's shares in YPF. Maersk Oil discovers first oil in the Angola deepwater pre-salt layer. Hugo Chávez wins reelection to a fourth term as president of Venezuela. Boris Berezovsky loses a case he brought against Roman Abramovich over the ownership of Sibneft. ICSID rules that Ecuador should pay Occidental $1.77 billion in connection with the company's 2006 expulsion from the country. EU launches antitrust investigation on Gazprom.

2013 Mexico reforms its petroleum industry, ending the monopoly of state-owned Pemex. Rosneft acquires TNK-BP. Venezuelan president Hugo Chávez dies and is succeeded by Nicolás Maduro, who continues his policies. Mikhail Khodorkovsky is pardoned by President Vladimir Putin of Russia and released from prison. Lukoil acquires Samara-Nafta, Hess Corporation's Russian subsidiary. Fracking pioneer George P. Mitchell dies.

2014 Wide-ranging corruption scandal breaks out in Brazil, with Petrobrás at its center. Repsol accepts a settlement offer of $5 billion in Argentine government bonds for its expropriated shares of YPF. Evo Morales is elected to a third term as president of Bolivia. Det Norske Oljeselskap acquires Marathon Oil Norge. ICSID awards ExxonMobil $1.6 billion in connection with Venezuela's 2007 assumption of majority control of Orinoco Belt heavy oil projects. Baker Hughes and Halliburton announce an intent to merge, which is never fulfilled. Gazprom signs a contract with the China National Petroleum Corporation to deliver 1.342 Tcf of gas annually to China for 30 years beginning in 2018. Kurdistan Regional Government takes control of Kirkuk. China deploys its *Hai Yang Shi You* mobile offshore oil platform in disputed waters near the Paracel Islands in the South China Sea, sparking riots in Vietnam.

2015 Iran reaches an agreement with the U.S. and other powers to rein in its nuclear program. It obtains relief from many economic sanctions, including ones targeting its petroleum industry. Shale gas production exceeds 50 percent of all U.S. natural gas production. China opens a 1,500-mile pipeline from Maday Island in Burma (Myanmar) to Kunming in China to reduce its dependence on the Strait of Malacca. Representatives from 196 countries adopt Paris Climate Accord. ENI discovers Egypt's supergiant Zohr gas field in the eastern Mediterranean. Schlumberger agrees to plead guilty to violating U.S. sanctions against Iran and Sudan and pay $232.7 million in fines and forfeited profits.

2016 Saudi Arabia announces it is considering listing shares in Saudi Aramco. Gabon rejoins OPEC. Indonesia rejoins OPEC, then leaves again before the year ends. Brazilian president Dilma Rousseff is removed from office because of the Petrobrás Scandal, even though she is not accused of having any personal role in it. An international tribunal in The Hague rules against China's claims to most of the South China Sea, upholding the Exclusive Economic Zones of surrounding countries as defined by the UN Convention on the Law of the Sea. Bolivian referendum denies Evo Morales the chance to run for a fourth term as president in 2020. Ecuador agrees to pay $980 million to Occidental in connection with its 2006 expulsion from the country, a lower figure than the $1.77 billion imposed by ICSID four years earlier. Aubrey McClendon, shale gas pioneer and former head of Chesapeake Energy, dies in a single car crash one day after being indicted by a federal grand jury for bid rigging and other charges. Halliburton and Baker Hughes abandon merger plans. A court in Chad fines a consortium led by ExxonMobil an astonishing $75 billion (nearly seven times the country's GDP) for an alleged $819 million in unpaid royalties.

2017 Equatorial Guinea joins OPEC. U.S. president Donald Trump announces intention to withdraw from Paris Climate Accord. Kurdish area of Iraq votes for independence in a referendum. Iraqi army takes back control of Kirkuk from the Kurdistan Regional Government. ExxonMobil settles with the government of Chad. General Electric acquires Baker Hughes. Repsol announces Horseshoe, a 1.2 Bb oil discovery in Alaska's North Slope, possibly the largest onshore find in the United States in three decades. The UK's Premier Oil becomes the first private sector company to drill an offshore well in Mexico and announces a discovery of 1 Bb. Former ExxonMobil CEO Rex Tillerson is sworn in as U.S. secretary of state. Gulf Cooperation Council enters a period of crisis as Saudi Arabia, the United Arab Emirates, and Bahrain, along with nonmember Egypt, cut diplomatic relations with Qatar over its alleged support for militant groups.

Introduction

The world as we have known it for the past century would have been very different without petroleum. Petroleum, particularly in the form of crude oil, has been central to all aspects of modern industrial society and has been a major strategic geopolitical objective for nations. The 20th century was the age of oil, and at least part of the 21st century will be as well. Petroleum is used as an energy source and as a raw material for the production of an immense variety of chemicals and synthetic materials. Petroleum in the form of natural gas is important in the production of fertilizer. Almost all the world's food relies on petroleum for fertilizer, pesticides, cultivation, or transport. Petroleum has been particularly dominant as a source of transportation fuels, an application for which cost-effective and equally performing substitutes will be especially difficult to find.

Although the energy intensity of advanced economies (the energy required to produce a unit of gross domestic product, GDP) has dropped since 1950, it would be erroneous to deduce from this that energy has become less necessary. Energy, particularly oil, is still a major foundation on which the entire edifice of modern society is built. In this, petroleum may be somewhat comparable to food: agriculture is an ever-decreasing percentage of advanced economies, but without it, all other activity would collapse.

TYPES OF PETROLEUM

"Petroleum" has no universally recognized standard definition. Many people use the term as a synonym for crude oil. Here, the word means any naturally occurring hydrocarbon mixture, whether in a solid, liquid, or gaseous form. By this definition, petroleum encompasses a number of varieties of both crude oil and natural gas.

Usually the unqualified term "crude oil" refers to what is more precisely called conventional crude oil. This crude oil exists as a liquid mixture in natural underground reservoirs and remains liquid at atmospheric pressures. Conventional crude usually ranges in density from 20–40°API gravity and flows relatively easily. The overwhelming majority of crude oil produced to date has been conventional crude.

Unconventional crudes include extra-heavy oil; oil sands; tight oil in impermeable formations, especially shale oil; and kerogen in oil shale (not to be confused with oil trapped in impermeable shales). Some of these varieties are generally processed to produce an upgraded or synthetic crude more similar to conventional oil. In the case of tight oil, the oil may be "conventional" in its properties but require special techniques, such as hydraulic fracturing and horizontal drilling, to extract it. These techniques have enabled a revolution in oil and gas production in the 21st century, especially in the United States. Extra-heavy oil is exploited in Venezuela's Orinoco Oil Belt, and oil sands are mined and processed in the Canadian province of Alberta. Synthetic oil production from oil shale has occurred on a relatively limited basis, including in the United States, Estonia, and Scotland.

Other things being equal, unconventional crudes are more expensive than conventional crude—in both money and energy—to extract into a useful form. However, it should be noted that even conventional crude can be difficult and expensive to produce. Deposits of conventional crude can be found in remote locations, such as the Arctic, under deep water offshore, or in relatively deep reservoirs. As sources of shallow and favorably located conventional crude become scarcer and prices rise, interest increases in more remote conventional crudes and in unconventional crude varieties.

Natural gas, which is mostly methane, is sometimes found associated with crude oil and sometimes alone. "Conventional" natural gas is found in pressurized reservoirs and recoverable with relative ease. Unconventional natural gas is more difficult to recover and comes in a number of forms. Coal-bed methane is found in coal deposits and must typically be recovered by fracturing. Tight gas reservoirs are ones in which the gas is held in low-permeability formations and cannot move easily. As with tight oil, hydraulic fracturing and horizontal drilling have enabled major increases in production in the years since the first edition of this book was published. Another type of unconventional gas is geopressurized or ultra-deep gas, dissolved in deep aquifers. Methane hydrates are largely submarine deposits of ice-like solids with trapped gas. Methane hydrate resources are believed to be immense but thus far have not been systematically exploited. Methane hydrate is not generally classified as petroleum.

Natural gas reservoirs often have recoverable natural gas liquids (NGL) associated with them. These should not be confused with liquefied natural gas (LNG). LNG is natural gas liquefied at a processing plant for shipment. Natural gas liquids are recoverable hydrocarbon liquids obtained from a gas reservoir or in a gas processing plant. Liquids recovered at the well or in small field separators are called "lease condensate." Lease condensate consists primarily of pentanes and heavier molecules. Liquids recovered at natural gas processing plants are referred to as natural gas plant liquids or liquefied petroleum gases (LPG). These consist of such hydrocarbons as propane, butane, and isobutane. Gas reservoirs without natural gas liquids are referred to as dry. Wet gas can have NGL contents of 200 barrels per thousand cubic feet of gas.

EVOLUTION OF PETROLEUM USE

In premodern times, crude oil was used to caulk boats, lubricate cartwheels, impregnate wood to make torches, and for other applications, including some medicinal ones of varying degrees of efficacy. Natural gas was used as an energy source and an illuminant. Exploitation was generally opportunistic, relying on naturally occurring surface seeps. In the 19th century, engineers and entrepreneurs began drilling deliberately for petroleum, and the modern industry was born. The main use for crude oil at that time was as a source of kerosene to replace increasingly scarce whale oil in lamps. With the invention and diffusion of the automobile, the major thrust of the industry shifted to producing gasoline and other transportation fuels. Gasoline production in the United States surpassed that of kerosene around 1915.

During the 20th century, the United States led the world in adopting the automobile. In 1900, the United States had about 8,000 passenger cars in circulation. By 1930, it had over 23 million, out of a world total of about 29 million. The petroleum industry was taken somewhat by surprise and struggled to produce enough gasoline, which in the 19th century had been treated as a waste product. Advances in refining technology helped the industry meet demand. A major leap occurred with the development of thermal cracking to increase the yield of gasoline from a given quantity of crude oil. William Burton of the Standard Oil Company of Indiana (later Amoco) patented a thermal cracking process in 1913 that saw widespread use. Another important milestone in refining came in 1936, when Eugene Houdry's catalytic cracking process, which increased both the yield and quality of gasoline, began commercial production.

By 1940, there were nearly 36 million passenger cars in the world, of which over 27 million, or 75 percent, were in the United States. In 2013, the world had almost 1.2 billion passenger cars, with about 253 million, or 21 percent, in the United States. The rest of the world has thus been catching up to the United States in automobile use, but there is still considerable room for growth. The United States had 790 passenger cars per 1,000 people in 2013, while the world average was 174. India still had only about 20 cars per 1,000 people, while China had 91 (up from 16 only eight years earlier). The implications for future petroleum demand are significant. In 2004, nearly 58 percent of all oil consumed in the world was used for transport, up from about 24 percent in 1973. Transport has been a growing slice of a growing pie. In developed countries, about 85 percent of oil consumed for transport is used for road transportation, mostly in passenger cars. In the United States, road transportation accounted for over half of all oil consumed in the mid-2000 decade. By comparison, air transport accounted for about 5 percent. If China had a car ownership rate similar to that of the United States, it would have well over a billion vehicles for its current population—almost as many as the whole world in 2013.

As a primary energy source, oil was negligible in 1900. Even in 1950, it only accounted for 30.3 percent of the world's consumption, with natural gas accounting for another 9.9 percent, while 55.5 percent of the world's primary energy still came from coal. Oil surpassed coal in 1964. In that year, 40.8 percent of the world's energy came from oil, 39.7 percent from coal, and 15.8 percent from natural gas. As a proportion of total world primary energy consumption, oil peaked at 49.4 percent in 1974. By then, coal was 25.6 percent, and natural gas had risen to 19.5 percent. Thereafter natural gas continued to rise, as did nuclear and hydroelectric energy. In addition to residential heating and cooking applications, natural gas found increasing use in electricity generation. In 2016, oil accounted for 34.1 percent of global primary energy consumption, natural gas 22.7 percent, and coal 26.5 percent. It is important to note that although oil accounted for a lower percentage of primary energy consumption in 2016 than it did in 1974, oil consumption increased greatly in absolute terms—from around 55 million barrels a day of oil and liquids to over 96.

Apart from its usefulness as an energy source, petroleum has also transformed the world through petrochemicals and a multitude of new synthetic materials, including plastics, synthetic fibers, synthetic rubber, explosives, pesticides, and many others. In 2015, about 12 percent of the world's crude oil production was used as a petrochemical feedstock. Some of the refining techniques developed in the early 20th century to increase gasoline production also enabled the development of petrochemistry. By the 1930s, petrochemical products were already being introduced, and they played a significant role in World War II—for example, in the use of synthetic rubber for tires. In the 1950s, the petrochemical revolution accelerated as widely abundant and inexpensive plastics spread throughout society. Shah Mohammed Reza Pahlavi of Iran stated in the 1970s that oil should not be burned, because it is too valuable as a source of petrochemicals. Although he made his comments in the political context of pushing for higher oil prices, they are perhaps still worth noting.

THE WORLD'S ENDOWMENT OF PETROLEUM

"Reserves" of petroleum are those known quantities that are recoverable using present technology, under current economic and regulatory conditions. "Resources" include quantities that may not be currently recoverable but might be so in the future under different conditions. Resources also include as-yet undiscovered quantities of petroleum that might be technically recoverable if known. Estimates of reserves are necessarily more accurate than those of resources, but even they can be subject to considerable uncertainty, variation, and political manipulation. Estimates of conventional oil reserves or resources sometimes include natural gas liquids and sometimes do not. Very roughly, natural gas liquids can add 20 percent or even more.

In 2016, the world was estimated to have reserves of conventional crude oil (including natural gas liquids) in the amount of about 1,319 billion barrels (Bb). This excludes about 165 Bb of oil recoverable from Canadian oil sands and 222 Bb of extra-heavy oil in Venezuela's Orinoco Belt, although the *Oil and Gas Journal* has included the former in its reserves estimates since 2003 and the latter since 2011. Moving beyond known reserves and considering the likely amounts of remaining technically recoverable resources, estimates are that the world has about 2,400 Bb of conventional oil, with another 480 Bb recoverable using enhanced recovery technology. The world may have another 1,500 Bb in extra-heavy oil and oil-sand resources, over 1,000 Bb in oil derivable from kerogen in oil shale, 250 Bb in tight oil, including shale oil, and over 250 Bb in oil in highly inaccessible places, including the Arctic and the ultra-deepwater offshore. All told, the world may have nearly 6,000 Bb of technically recoverable oil resources.

Estimated global reserves of conventional natural gas in 2016 were 6,589 trillion cubic feet (Tcf), or about 1,167 billion barrels of oil equivalent (Bboe). This is of the same order of magnitude as estimated global oil reserves, in energy terms. Beyond known reserves, global conventional gas resources could be an additional 2,800 Bboe, and unconventional shale gas resources are estimated at about 7,500 Tcf, or 1,330 Bboe, excluding methane hydrate. Methane hydrate resources are difficult to estimate but might be immense, perhaps a hundred times larger than proven conventional gas reserves.

HISTORY OF GLOBAL OIL PRODUCTION

By the end of 2016, the world had produced about 1,315 Bb of crude oil, almost all of it conventional crude (with lease condensate included). This is very close to the amount of remaining global reserves, suggesting that the world has already produced about half of its oil. However, this is misleading, since total remaining resources, even of conventional crude oil, likely significantly exceed remaining reserves. Global production in 2016 was about 78.3 million barrels per day (Mmbpd), or 28.6 billion barrels (Bb) for the year.

World oil production in 1860, at the dawn of the modern industry, was less than 400 barrels per day (bpd). It surpassed 10,000 bpd in 1866 and 100,000 bpd in 1885. By 1900, it was nearly 410,000 bpd. The dominant producers in the late 19th century were the United States (particularly Pennsylvania) and Russia (particularly the Baku region of Azerbaijan). Although Russian production exceeded that of the United States in some years, the United States was the top producer in the 19th century as a whole, with Russia second. These two countries were also in first and second place in total production from 1900 to 1940.

World production surpassed 1 Mmbpd in 1913 and 2 Mmbpd in 1921. By 1941, it exceeded 6 Mmbpd. During the first third of the 20th century, several new countries became important producers, including Mexico, Venezuela, Indonesia, Romania, and Iran. From 1901 to 1920, Mexico was third in total production over the interval, followed by Indonesia. From 1921 to 1940, Venezuela was third, followed by Mexico. The United States and the Soviet Union were first and second. Within the United States, California and Oklahoma traded places as the top annual producers between 1903 and 1928, when they were overtaken by Texas. The vast oil reserves of the Persian Gulf region, particularly Saudi Arabia, began to be discovered in the 1930s. Progress was interrupted by World War II, but after 1950 the Persian Gulf region became a more important producer. Between 1941 and 1960, Saudi Arabia was the fourth-largest total producer, after the United States, Venezuela, and the Soviet Union. Iran was fifth. In the next 20 years, Saudi Arabia moved into third place after the United States and the Soviet Union. Iran and Venezuela were fourth and fifth. From 1981 to 2000, the United States was first and Saudi Arabia was second, followed by the Soviet Union (Russia after 1991), Mexico, and Iran. By this time, Norway and the United Kingdom had also become important producers, from fields in the North and Norwegian Seas. From 2000 to 2016, Saudi Arabia was first in total production over the interval, followed by Russia, the United States, Iran, and China.

From the foregoing, it should be obvious that the largest individual producers in a given interval of time are not necessarily the countries that now have the largest total reserves. The United States ranked 10th in the world in reserves of oil and liquids in 2016, with less than 3 percent of the world's total. However, it dominated global oil production for over a century. In the early years of the modern industry from 1860 to 1874, it accounted for over 90 percent of the world's annual production. It also accounted for over half the world's annual production in every year from 1903 to 1952 and was even the world's largest producer during the year 2016. The United States has accounted for just under a fifth of all the oil produced in the world so far. Clearly, even with the resurgence in U.S. production enabled by fracking and horizontal drilling, the United States will have difficulty maintaining such a position in the long run.

When the Organization of the Petroleum Exporting Countries (OPEC) was formed in 1960, it accounted for 41.4 percent of world oil production. Over the course of the next decade, OPEC increased its membership, and its members increased their production faster than the rest of the world. As a proportion of world oil production, OPEC reached a peak of 55 percent in 1973. This coincided with its increased pricing and market power. However, during the 1970s, new sources of oil outside OPEC were developed, particularly in the North Sea and in Mexico. Alaska production also ramped up, slowing the decline in U.S. output. OPEC's share of world oil production fell to about 45 percent in 1980 and reached a low of under 28 percent in 1985 as non-OPEC production soared and Saudi Arabia restrained output to defend prices. By this point, OPEC's pricing power was almost gone, and when Saudi Arabia increased production in late 1985 and 1986, oil prices collapsed. OPEC gained back share in the late 1980s as non-OPEC production stumbled somewhat. After 1991, OPEC accounted for over 40 percent of the world's annual oil production. In 2016, it still had nearly 42.7 percent of global production.

World oil production surpassed 10 Mmbpd in 1950, 30 Mmbpd in 1965, and 50 Mmbpd in 1972. World production peaked at over 62 Mmbpd in 1979, then sank back to a low of about 53 Mmbpd in 1983, as demand in the industrialized world decreased somewhat in the aftermath of the price shocks of the 1970s, and OPEC, particularly Saudi Arabia, also sought to impose discipline on supply, as noted above. By 1990, world production again exceeded 60 Mmbpd, and it surpassed 70 Mmbpd in 2004. In 2016, world production was more than 78 Mmbpd.

THE MOVE TO MORE DIFFICULT ENVIRONMENTS

As the "easier" areas of the world became intensively explored, attention increasingly turned to harsher and more difficult environments. Both the United States and the Soviet Union intensified activity in their arctic and subarctic regions, for example. The Prudhoe Bay oil field on Alaska's North Slope, discovered in 1967, is still the largest field found in the United States.

Global efforts also moved offshore, in increasing water depths. In the beginning of the 1980s, "deepwater" meant about 800 feet of water. The early major discoveries in the North Sea had been in water depths less than 500 feet. By 2007, anything less than 1,500 feet was commonly considered shallow. Definitions vary, but 1,500 feet and more is now considered deepwater, with depths over 7,000 feet often called "ultra-deepwater." Offshore oil has been an increasingly important component of global oil production in the last two decades. As late as 1972, it accounted for only roughly 5 percent of crude oil produced in the world; by 1984 the proportion was about 27 percent, and in 2000 it was nearly 35 percent. The years between 2000 and 2015 saw similar percentages. Global production of onshore crude oil between 1982 and 2006 was essentially flat, hovering somewhat above or somewhat below 40 Mmbpd. The growth in global crude production, which stood over 78 Mmbpd in 2016, was from offshore oil and unconventional oil.

HISTORY OF GLOBAL NATURAL GAS PRODUCTION

Global production of natural gas has generally trailed oil production in energy-equivalent terms. Gas is more difficult to capture, store, and transport. Gas that was associated with oil was often wasted in the past, disposed of by burning, which is referred to as "flaring." This was because oil was the objective and gas was of no interest, or because markets for gas were too distant and no infrastructure existed for storage or distribution. Flaring is less common today but is still sometimes practiced on offshore platforms and in some parts of the Middle East. In 1950, the world produced about 3.2 million barrels of oil equivalent per day (Mmboepd) worth of natural gas, compared to 10.4 Mmbpd of oil. In energy-equivalent terms, natural gas production was about 30 percent of oil. The proportion rose over the next 30 years. In 1980, gas production was 24.6 Mmboepd, or about 41 percent of oil production, which stood at around 59.4 Mmbpd. Oil production stumbled somewhat in the 1980s, not regaining its 1979 high until 1995–1996. During this period, gas production continued to grow fairly steadily; by 1996, global natural gas production (37.5 Mmboepd) was about 59 percent of oil production. In 2016, the world produced 60.7 Mmboepd of gas, amounting to 77 percent of oil production in energy-equivalent terms.

The United States was the world's dominant producer of natural gas for a large part of the 20th century. In 1950, the United States produced 90 percent of the world's natural gas. The United States continued to produce more than half the world's gas until 1972. During the interval, the Soviet Union became a major producer. In 1972, the United States and the Soviet Union accounted for nearly 73 percent of global gas production. Production in the rest of the world grew rapidly, but the United States and the Soviet Union (or components of the former Soviet Union) taken together still accounted for more than half the world's annual gas production until 2002. In 2016, the world's top gas-producing countries were (in decreasing order) the United States, Russia, Iran, Qatar, Canada, China, Norway, Saudi Arabia, Algeria, and

Australia. Together, they accounted for two-thirds of world production. The United States alone produced 21 percent of the world's gas in 2016 and Russia over 16 percent.

Because of its dependence on a fixed distribution network and for other reasons, gas does not have as well developed a global market as oil. Most gas contracts are long term, with prices set in advance. The relatively recent use of liquefied natural gas (LNG) transported by tanker has begun to change patterns, introducing more of a global market for gas and making it more like oil. LNG still represents a small proportion of total gas sold, but its share is increasing. In 2016, LNG accounted for about 10.1 percent of total global gas consumption; this was almost triple the proportion in 1990. Over half of all LNG exports in 2016 went to just three countries: Japan, South Korea, and China, with Japan alone accounting for 32.3 percent. Top LNG exporters in 2016 were Qatar, Australia, Malaysia, Nigeria, Indonesia, Algeria, and Russia, in that order. Qatar accounted for about 30 percent of total LNG exports.

AN UNEVEN ENDOWMENT AND ITS IMPLICATIONS

To the chagrin of many nations, the world's endowment of petroleum is rather unevenly distributed. The countries of the Persian Gulf had over 60 percent of the world's conventional oil reserves in 2016. Saudi Arabia alone had a fifth of the world's total. Countries of the former Soviet Union together had about 11 percent. North and South America together had close to 13 percent. The picture changes somewhat if one follows the *Oil and Gas Journal* and includes the Canadian oil sands and the Venezuelan Orinoco Belt in world reserves. North and South America's percentage then jumps to 32.5, but the Persian Gulf's is still nearly half at 47.5 percent.

Natural gas is also unevenly distributed, although somewhat differently from crude oil. Russia had 17.3 percent of the world's conventional natural gas reserves in 2016; countries of the former Soviet Union taken together had nearly 28.3 percent. The Persian Gulf region had 42.3 percent, with 18 percent in Iran and 13 percent in Qatar. Qatar's North Field represents the largest single hydrocarbon deposit yet discovered. North and South America together had 10 percent of the world's conventional gas reserves.

The consumption of petroleum is also unevenly distributed. The rich countries of the West, including the United States, the European Union, and Japan, are among the highest consumers of petroleum. Among the largest consuming nations, the United States had a truly major petroleum endowment of its own. However, it has already produced a large portion of its reserves and has been a net importer since 1948, still importing 40 percent of its consumption in 2016. U.S. crude oil production peaked in 1970 and began a long-term decline, mitigated first by Alaska and then by federal offshore Gulf of Mexico production. It is true that the revolution in fracking and horizontal drilling turned U.S. production around beginning in 2007 and that U.S. production in 2016 was at its highest point since 1972. Yet the fact still remains that the United States, with less than 5 percent of the world's population, consumes about 20 percent of the world's oil output. The United States' 2016 import percentage of 40 percent was down from a peak of 50 percent in 2008, but still higher than its level of 20 percent in 1985.

In the 21st century, India and China emerged as major consumers and accounted for much of the world's new demand growth. China became a net importer of oil after 1993, and the gap between its production and consumption has increased steadily. In 2016, it consumed 8.4 Mmbpd more than it produced. India consumed 3.6 Mmbpd more than it produced.

The uneven distribution of petroleum reserves and resources around the globe has had profound implications. The petroleum endowments of the Middle East, Caucasus, and Romania were major strategic and tactical objectives in both world wars. The Middle East's oil caused the British and later the Americans to meddle significantly in the affairs of the region, with often disastrous consequences. The U.S.-engineered and British-supported overthrow of the Mossadegh government in Iran in the wake of that country's 1951 nationalization of its oil industry echoed into the 21st century. One can draw a line from that event, which reinstalled and strengthened an often-brutal shah, to the Islamic Revolution of 1979 and the taking of American hostages in Tehran during the Carter administration. Those events placed Iran on a collision course with the United States and created tensions that seem unabated in 2016. And oil reserves in Iraq have contributed to a veritable U.S. obsession with that country. Both the Persian Gulf War of 1991 and the Iraq War that began in 2003 were at least partly about attempting to protect access to, and a measure of control over, the petroleum resources of the region.

The uneven distribution of petroleum has also led to large transfers of wealth from consuming to producing countries. During times of high oil prices, rich consuming countries have often suffered recessions, inflation, and other dislocations. Poorer consuming countries have suffered even more profoundly and been plunged into debt. Meanwhile, oil-producing countries have often found themselves deluged with more wealth than their economies could productively absorb. The Venezuelan oil minister Juan Pablo Pérez Alfonso, a founding father of OPEC, observed that a country can become ill not only from hunger, but also from indigestion. He worried greatly about the effects of indigestible sudden wealth on his country, and he decried the unwise and wasteful spending that resulted. He even published a book titled *Hundiéndos en el Excremento del Diablo*—"Sinking in the Devil's Excrement," where the devil's excrement is petroleum.

In the particular case of the United States, the effect of the wealth transfer to producing countries has been partially mitigated by the fact that the wealth has tended to come back as recycled "petrodollars." During the period of high price increases in the 1970s, oil exporting countries found themselves awash in dollars—oil was (and still is, in 2016) priced in dollars. The oil exporting countries received sums far exceeding their domestic investment needs. The oil exporting nations invested the surplus petrodollars with Western banks, typically American or British ones. Petrodollar recycling was still taking place into the 21st century, although in a somewhat different form. Increasingly, the oil exporting nations of the Middle East used petrodollars to import goods from the rest of Asia, particularly Japan and China, as well as the European Union. Petrodollars earned from oil sales to Japan and China were thus largely recycled back to those countries, which then typically invested them in U.S. treasury securities. It is estimated that in 2006, 45 percent of the huge U.S. current account deficit was financed by recycled petrodollars.

EVOLUTION OF THE MODERN INDUSTRY

The modern petroleum industry began in the early 19th century in a variety of locations, including the United States (Pennsylvania and West Virginia), Russia (the Baku area), Romania, Austro-Hungarian Galicia, and Canada. Edwin Drake's 1859 well near Titusville, Pennsylvania, is popularly considered the "first modern oil well." This is an oversimplification, as there were contemporaneous and even earlier efforts, particularly in West Virginia and the Russian Empire. However, Drake's well sparked the first true modern oil boom. Pennsylvania completely dominated U.S. and world oil production for a considerable part of the late 19th century, until Russian production began on a large scale and other sources were discovered in the United States.

In the late 19th century, the Standard Oil empire nearly monopolized oil refining in the United States and was a major influence throughout the world. Standard Oil achieved dominance through shrewd management, efficient operations, and the application of a number of anticompetitive practices. As the company grew, it was able to negotiate secret rebates from freight companies, thus lowering its costs and further increasing its advantage against competitors. It selectively lowered prices in markets where competitors operated, while keeping prices high elsewhere. It forced competitors to choose between going bankrupt and being acquired by Standard Oil. Standard was thus able to achieve significant horizontal integration of its operations. It also became vertically integrated, controlling refining, transportation, and marketing, and eventually moving upstream into oil production. Standard's business practices earned it considerable public loathing and mistrust. Much of the petroleum industry's unfavorable popular reputation through the decades originated with Standard Oil.

Standard Oil's virtual monopoly helped inspire the passage of the Sherman Antitrust Act in 1890, and attempts to break the company up began soon after. Eventually, after lengthy litigation, Standard Oil was dissolved in 1911 by order of the U.S. Supreme Court. Among its successor companies were Exxon (Standard Oil of New Jersey), Mobil (Standard Oil of New York and Vacuum Oil), and Chevron (Standard Oil of California). These three, along with Royal Dutch Shell, Gulf, Texaco, and BP (Anglo–Persian/Anglo–Iranian Oil Company), constituted the so-called Seven Sisters.

The Seven Sisters—along with Compagnie Française des Pétroles (CFP, later Total)—dominated much of the world industry through a series of interlocking consortia controlling production in the Middle East. BP, Shell, Exxon, Mobil, and CFP were all members of the Iraq Petroleum Company (IPC) consortium, which virtually monopolized petroleum production in Iraq for much of the 20th century. IPC also obtained dominant positions in Abu Dhabi (by far the most important oil producer in what became the United Arab Emirates) and Qatar. BP and Gulf were equal partners in the Kuwait Oil Company, which controlled the industry in Kuwait. Chevron, Texaco, Exxon, and Mobil were partners in Aramco, which dominated petroleum production in Saudi Arabia. All of the Seven Sisters and CFP were members of the Iran Consortium, which effectively controlled Iran's industry between 1954 and 1979, despite the fact that the industry had been nationalized in 1951.

OPEC, formed in 1960, broke much of the power of these major multinational companies in the 1970s. By 1980, all the various consortia were dismantled by nationalizations, both within and outside of OPEC's participation framework. Control of the industry passed from multinational companies to producer countries, and pricing power in the 1970s moved decisively to OPEC. In the 1980s, OPEC in turn lost much of its own power as the market was flooded with non-OPEC production. In the first decade of the 21st century, OPEC temporarily saw something of a resurgence, although neither OPEC nor the multinational companies would dominate the market so completely again.

OIL PRICES

Perhaps no statistic is watched as keenly as the price of a barrel of oil. Although oil prices fluctuated tremendously in the short term throughout the industry's history, annual average prices for much of the 20th century were remarkably stable. Between 1921 and 1970, they largely moved in a band between one and two dollars a barrel in money of the day. In estimated constant (2016) dollars, they moved largely between $13 and $20. There was a major collapse in prices in the U.S. Southwest after the discovery of the East Texas oil field in 1930, leading to successful stabilizing actions by the Railroad Commission of Texas and the U.S. government. Between the end of World War II and the early 1970s, the Seven Sisters largely succeeded in maintaining reasonable price stability.

The 1970s were a decade of "oil shocks," large increases in the price of oil. During the First Oil Shock from 1973 to 1974, prices rose about fourfold, with a large part of the increase taking place between October and December of 1973. The immediate catalyst was the Arab reaction to the Yom Kippur War. OPEC increased prices immediately by 76 percent, and the Organization of Arab Petroleum Exporting Countries (OAPEC) began a selective embargo on countries, including the United States, that supported Israel. Other factors also contributed to the shock, including the peaking of lower-48 U.S. conventional oil production in 1970. In 1979, in the Second Oil Shock, average prices doubled from what was already a relatively high level. The catalyst of the Second Oil Shock was the Iranian Revolution and the withdrawal of large amounts of Iranian crude from the global market. The shock was exacerbated by panic buying (also a factor in the First Oil Shock). All in all, average annual prices of Arab light crude rose about 20-fold over the decade, from $1.80 per barrel in 1970 to $36.83 in 1980, in money of the day.

The high prices collapsed in the 1980s, during the oil "countershock." Large quantities of non-OPEC oil entered the market, for example from the North Sea and Mexico. Oil had also begun to flow from Alaska, thus alleviating some of the U.S. import requirement. OPEC attempted to support the price but was unable to maintain production discipline. Saudi Arabia tried to do so alone but eventually gave up. Prices decreased nearly 70 percent in the six months after September 1985. Prices, and OPEC's power, made a partial recovery later in the decade. This was due to a number of factors, including high taxes in the United Kingdom, underinvestment in Mexico, and the implosion of the Soviet Union. Still, many of the underlying factors that contributed to the countershock were felt through the 1990s. In that decade, there was also increased production from a variety of countries, such as Venezuela, that liberalized their petroleum sectors. Prices dipped below $10 a barrel at some points during 1998 and 1999 before finally beginning what seemed like an inexorable increase.

The Third Oil Shock began around 2003. Prices more than doubled by 2006. There was then a particularly rapid increase in 2007 and 2008. Average world prices for all crudes (FOB, weighted by export volume) were $54.63 a barrel in the first week of 2007 and by July 2008 exceeded $135. The causes of the rapid increase are still debated. The inelasticity of both the global demand for, and the global supply of, crude oil certainly played a role, in the face of increased demand from India, China, and other countries. The weakness of the U.S. dollar was also a factor. Some observers have also claimed that prices were driven up by speculators. With interest rates low and the stock market not offering attractive returns, oil and other commodities became attractive to investors. A number of people believed that the Third Oil Shock was an indication that the world was approaching "peak oil" in 2008. This conclusion proved unjustified. Oil prices began to decrease after their July 2008 highs. In the fall of 2008, the world began to experience a global financial crisis, with lower economic activity and lower demand for oil. By December 2008, oil prices dropped below $50 a barrel. There was another period of high prices (well over $100 a barrel) between 2011 and 2014, but by 2016, prices averaged about $43 a barrel and people talked of a continuing oil "glut."

PEAK OIL?

If oil is a finite and nonrenewable resource, it stands to reason that continuing consumption will eventually exhaust it, and some sort of peak in production will be encountered along the way. The main points of contention are how much oil is really left and when the peak will be reached. The same is true for natural gas.

M. King Hubbert predicted in 1956 that oil production in the lower 48 states of the United States would peak around 1970. Hubbert applied a mathematical model of resource exhaustion and fit bell-shaped logistic curves to the data available to him. Despite the rigor of his analysis, many found his prediction astonishing, as oil production was still rising. However, U.S. conventional oil production in the lower 48 states did peak in 1970 at 9.408 Mmbpd and declined thereafter, roughly symmetrically with its rise. The Hubbert decline was partially offset by an infusion of oil from two new provinces, Alaska and the federal offshore areas in the Gulf of Mexico, but overall U.S. production was still declining until 2007. In the years since then, the application of fracking technology, along with horizontal drilling, enabled huge increases in the output of shale oil (and shale gas) in the United States, and dramatically reversed, at least temporarily, the long decline in U.S. hydrocarbon production. By 2015, U.S. production of shale oil represented nearly 49 percent of U.S. oil production and brought production levels to their highest point since 1972. (U.S. production of tight gas and shale gas in 2015 accounted for nearly 69 percent of total gas production, bringing levels to their highest point in history.)

Hubbert's prediction was, as we observed above, for conventional crude oil as he understood it at the time, and was limited to the lower 48 states of the United States. His prediction was correct within those restrictions. Since then, many investigators have attempted to apply Hubbert's method to world oil production and have issued predictions of the peak year of world oil production, but the years have come and gone without any real peak. In 1977, Hubbert himself predicted inaccurately that world conventional oil production would peak in the 1990s. The pattern of world oil production does not follow the rising portion of a bell-shaped curve exactly. Fitting such a curve to the data is not a reliable method for inferring a peak year and the level of the world's ultimately recoverable resource (URR) of oil.

If one makes an assumption about the level of the URR based on geological data, one can further constrain the problem. However, the inferred peak year is then quite sensitive to errors in the URR—which must, of necessity, be rather large. One must also bear in mind that the URR is a function of price, technology, and other variables. Improved technology can increase the URR. Higher prices can also do so, by making currently marginal resources economical to extract. Also, the lower 48 states of the United States, on which Hubbert based his prediction, constituted a province subject to extensive and relatively unconstrained exploration. This is not true of the world as a whole.

A number of indicators do not bode well for future oil production, however. The most serious are the depletion of the world's giant oil fields, the decreasing rate of discovery of giant fields, and the failure of discoveries as a whole to replace consumed reserves. Giant fields—ones with over 500 Mmb of recoverable oil—are crucial for the world's oil supply. Only about 1 percent of the world's fields are giants, but they account for about two-thirds of the world's estimated recoverable oil. The hundred largest oil fields accounted for about 45 percent of the world's oil production in the mid-2000s. Geologically, the largest fields have tended to involve anticlinal traps, with sedimentary reservoirs (mostly sandstones). Since large anticlinal structures are relatively easier to spot during exploration, this has led some people to assume that the largest and most obvious giants in the world have already been found.

Giant field discoveries have been declining since the 1940s, when they peaked. In the 1940s, 30 giant fields were discovered, with an average estimated ultimate recovery (EUR) of 8.6 Bb. In the 1990s, 50 giant fields were discovered—but their average size was 1.2 Bb. The very largest fields, where a disproportionate share of the oil resides, are very old and have been significantly depleted. Saudi Arabia's Ghawar, the largest oil field ever known, was discovered in 1948. The second-largest field, Kuwait's Burgan, was discovered in 1938. The fields of Venezuela's Bolivar Coastal Complex were discovered beginning in 1917. Not one of the world's top 20 fields has been discovered since 1980. Of course, this can change at any time, but the trends are worrying.

Overall, oil discoveries have not been replacing consumed reserves since about 1981. Although some years since then have seen new discoveries exceeding production by a small margin, most years have seen the opposite. Global oil production was about 28.6 Bb in 2016. Replacing that would entail discovering the equivalent of a new Kirkuk field. Replacing five years of production at that level would entail discovering the equivalent of a new Ghawar (if one accepts high estimates of Ghawar's oil). It will not be easy to achieve such feats. One should note, however, that oil exploration is a costly undertaking. It is difficult to predict what might be accomplished if prices rise high enough, and the need (and potential profit) is great enough. The fracking revolution is a case in point.

ALTERNATIVES TO PETROLEUM

Whether global oil production peaks in 2035 or 2050 or later, alternatives to oil must be developed, both because of the need of modern economies for energy and for environmental reasons. Alternatives to crude oil—certainly ones that can perform all its functions—will not be easy to develop. Conventional crude oil is (so far) abundant, storable, transportable, and has a high energy density. Its energy density is considerably higher than coal's, and much higher than that of such traditional energy sources as wood or peat.

Crude oil has also been cheap to extract in energy terms; that is, it has yielded far more energy than has been required to produce it. This is a crucial point. As conventional crude oil is depleted and humanity turns to unconventional crudes, the average energy return on investment (energy obtained for energy expended) will decrease. This is even true of conventional crude itself. The "easy" oil of the early Middle Eastern industry has increasingly given way to more difficult oil that requires more energy for extraction. A lower energy return on investment also applies to such alternative transportation fuels as alcohol, or synthetic gasoline and diesel equivalents obtained from natural gas or coal.

Solar and wind energy thus far have been aimed primarily at electricity generation. They thus offset coal and natural gas more than crude oil. Until the day when something like a completely solar-electric car can be effectively produced, transportation is likely to depend on liquid fuels derived from oil. Hydrogen is expensive in both money and energy. Alcohol can be expensive in energy, and some of the processes for producing it have the further disadvantage of diverting agricultural land from food production.

The future well-being of the human race will depend on disciplined and concerted research on a variety of energy alternatives as well as effective conservation measures. It will also depend crucially on a continued search for, and exploitation of, remaining petroleum resources.

A

AAPG. *See* AMERICAN ASSOCIATION OF PETROLEUM GEOLOGISTS (AAPG).

ABADAN. City (pop. 212,700, 2011) in **Iran**; a major center of the Iranian oil industry. Abadan is located in oil-rich **Khuzestan** Province, near the border with **Iraq**. It lies on Abadan Island, which is near the eastern bank of the Shatt al Arab River. Abadan was disputed between the Ottoman Empire and Iran. Iran acquired it in 1847. Abadan remained a relatively obscure village until the discovery of oil at Masjid-e-Suleiman in 1909 by the antecedent of the Anglo–Persian Oil Company (APOC).

APOC was later renamed the **Anglo–Iranian Oil Company**. In 1951, its properties were nationalized and turned over to the National Iranian Oil Company; the remaining Anglo–Iranian Oil Company became British Petroleum (**BP**). The squalid conditions of the oil workers in Abadan, culminating in major strikes in 1947 that provoked a brutal response from the British, contributed greatly to the discontent that ultimately led to nationalization. After nationalization, the British were expelled from Abadan, leading Iran and Britain to the brink of war. The period of tension between 1951 and 1953, when Prime Minister Mohammed **Mossadegh** was overthrown in a coup engineered by the British and the Americans, is sometimes referred to as the Abadan crisis. (*See* AJAX, OPERATION (1953); PLAN BUCCANEER.)

APOC began operating a refinery in Abadan with a capacity of about 2,400 bpd in 1912, after the British government reached an agreement with **Khaz'al Khan**. In 1913, a **pipeline** was completed from the Masjid-e-Suleiman field to Abadan. Abadan's economy became dominated by oil refining and shipping. The refinery was expanded. For a time in the 1970s, it may have been the largest refinery in the world. By 1980, it had a capacity in the range of 480,000–635,000 bpd (sources vary).

During the **Iran–Iraq War**, Abadan came under heavy bombardment. Abadan's oil infrastructure was severely damaged and operations were interrupted. The refinery was essentially destroyed. **Refining** and **petrochemical** production began again after the war ended in 1988, after major reconstruction efforts. The reconstruction was accomplished by Iran without foreign contractors. A capacity of about 135,000 bpd was restored by 1989. Capacity reached about 260,000 bpd in 1994 and about 330,000 bpd in 1995. By 2001, the Abadan refinery complex had a capacity of about 400,000 bpd. Abadan is linked to Tehran by the Trans-Iranian Pipeline, completed in January 1957. Another pipeline runs to Shiraz. In 2007, plans were announced for a pipeline linking Abadan to **Basra** in Iraq. Iran would purchase Iraqi crude and send refined products back to Iraq. The plans would require an understanding between the **United States** and Iran on Iraq.

ABERDEEN. City (pop. 197,000, 2016) in Scotland, **United Kingdom**, astride the rivers Don and Dee on the **North Sea** coast. Aberdeen is a major center for the UK petroleum industry and has been called the "Oil Capital of Europe." The North Sea boom that began in the 1970s attracted more than 200 new companies to the city.

ABIOGENIC THEORY. Theory that petroleum is formed from nonbiological sources of hydrocarbons located deep in the earth's crust and mantle, and not from the decay of biological organisms as is commonly believed. The theory was largely developed in **Russia**. One early abiogenic theory was proposed by **Dmitri Mendeleev**. The major developer of the abiogenic theory was **Nikolai Kudryavtsev**, who published his findings in the 1950s. The theory has been largely discounted in the West, with the notable exception of astrophysicist **Thomas Gold**. If the theory is even partially correct, petroleum may be much more abundant than commonly believed (although not necessarily more accessible).

ABQAIQ. Site of a major oil processing facility in **Saudi Arabia**, near a **giant** oil field of the same name. The compound in which the facility is located was built by **Aramco** in the 1940s and lies about 40 miles southwest of **Dhahran**. In 2005, it had a population of about 30,000. The oil processing facility accepts sour oil (refer to appendix 5) by **pipeline**. It reduces the hydrogen sulfide and the vapor pressure of the oil, making it safer to ship in **tankers**. The Abqaiq facility is the largest of its kind in the world. After processing, the oil is sent by pipeline to **Ras Tanura** or **Yanbu**, where it is either refined further or exported directly. Security analysts worry about the possibility of a terrorist attack on the Abqaiq processing facility crippling the export of Saudi crude. Such an attack was attempted by Al Qaeda on 24 May 2006. Abqaiq also has a **natural gas liquids** plant extracting the liquids from crude and stabilizing them. There is a 290,000 bpd natural gas liquids pipeline from Abqaiq to Yanbu on the Red Sea.

ABRAMOVICH, ROMAN (1966–). Russian billionaire who once had a controlling interest, along with **Boris Berezovsky**, of the oil company Sibneft (later **Gazprom Neft**). Roman Arkadyevich Abramovich was born 24 October 1966 in Saratov, **Russia**. He was orphaned at an early age and attended the Ukhta Industrial Institute. He served in the Soviet army and later earned a degree by correspondence from Moscow State Law Academy. During perestroika, he began his business career. After the dissolution of the Soviet Union in 1991, he founded several companies, concentrating particularly on petroleum and petroleum products trading.

Between 1995 and 1997, Abramovich and Berezovsky acquired a controlling interest in Sibneft for about $100 million under the corrupt **Loans for Shares** scheme established by the government of Boris Yeltsin. In 2006, **Gazprom** bought the Sibneft stake for $13 billion. Sibneft was renamed Gazprom Neft. Abramovich has many other business interests and in 2007 was ranked by *Forbes* magazine as the 16th-richest person in the world, with a net worth of $18.7 billion. He is also the governor of the Chukotka autonomous region of Russia, despite spending a large proportion of his time living in the **United Kingdom**. He was elected to the governorship in 2000 and reappointed in 2005 after President Vladimir Putin eliminated elections for regional government. Chukotka is believed to have oil potential, and Sibneft announced plans to explore there in 2001. In 2007, Abramovich invested $1 billion to buy shares of **Rosneft** on the London Stock Exchange.

ABSERON PENINSULA (ABSHERON, APSHERON). Peninsula in **Azerbaijan** extending about 37 miles eastward into the **Caspian Sea**; site of one of the world's earliest oil industries, with activity continuing today. The peninsula reaches a maximum width of about 19 miles. Azerbaijan's capital, **Baku**, lies on the peninsula's southern coast. Today, the 11 districts of Greater Baku comprise nearly the whole peninsula. The exploitation of oil from surface seeps in the Abseron Peninsula extends back to ancient times, and hand-dug wells for extracting oil date back at least to the Middle Ages. An important early oil well was drilled here in 1846 in the **Bibi–Eibat** area by a group under the command of Major Alexeev of the Russian tsar's Corps of Mining Engineers, 13 years before **Edwin Drake**'s well in **Titusville, Pennsylvania**. *See also* FIRST OIL WELL; VOSKOBOINIKOV, NIKOLAI (1801–C. 1861).

ABU DHABI. One of the constituent emirates of the **United Arab Emirates** (UAE) and by far the most important oil producer. Abu Dhabi has about 92.2 Bb (more than 94 percent) of the UAE's 97.8 Bb of oil **reserves** (2015). Abu Dhabi joined the **Organization of the Petroleum Exporting Countries** (OPEC) in 1967 and the **Organization of Arab Petroleum Exporting**

Countries (OAPEC) in 1970. In 1935, the **Iraq Petroleum Company** (IPC) consortium formed a company with the same ownership structure as IPC called Petroleum Development (Trucial States) Ltd., commonly referred to as PDTC. The Trucial States (roughly today's UAE) were within the area covered by the **Red Line Agreement**. PDTC obtained petroleum **concessions** in the Trucial States, including Abu Dhabi, which granted its concession in January 1939 for a period of 75 years. The IPC group did not begin significant activity in Abu Dhabi right away. At first, it concentrated on **Qatar** through a parallel consortium there, but **World War II** interrupted most of its efforts.

PDTC began serious exploration in Abu Dhabi in 1950. Conditions were difficult and success was not immediate. In July 1953, PDTC drilled the discovery well of the **supergiant** Bab field in Murban (**EUR** 10.28 Bb and 29.3 Tcf gas). Although this would ultimately prove to be the second-largest field in the UAE, PDTC initially abandoned it because of operational difficulties. It was 1960 before PDTC fully understood the potential of the field. In 1962, PDTC discovered Bu Hasa (EUR 6.52 Bb and 8.3 Tcf gas) and, in contrast to the Bab case, began exploitation immediately. Exports from the field began in 1963. PDTC changed its name to Abu Dhabi Petroleum Company (ADPC) that year and relinquished its concessions in the other Trucial States.

In 1951, Abu Dhabi determined that PDTC's (later ADPC's) concession did not cover **offshore** areas. It granted a different concession for the offshore to the International Marine Oil Company, which failed. In 1954, the concession passed to Abu Dhabi Marine Areas Ltd. (ADMA), a new company two-thirds owned by **BP** and one-third by **Compagnie Française des Pétroles** (CFP, later **Total**). In 1959, ADMA discovered the supergiant Umm Shaif field (EUR 3.9 Bb and 30.7 Tcf gas) about 80 miles offshore in the **Persian Gulf**, near Das Island. Production began in 1962. ADMA's success continued with the discovery of the UAE's largest field, the supergiant Zakum field (EUR 17.2 Bb and 12.4 Tcf gas), in 1965.

Abu Dhabi negotiated a **50–50 agreement** for profit sharing with ADPC and ADMA in 1965 and 1966, respectively. In 1971, the emirate established the Abu Dhabi National Oil Company (ADNOC). ADNOC acquired 25 percent of ADPC and ADMA under OPEC's **participation agreement** framework. It raised its stake to 60 percent of both companies via a separate agreement in December 1974. ADPC and ADMA were then restructured as the Abu Dhabi Company for Onshore Oil Operation (ADCO) and Abu Dhabi Marine Operating Company (ADMA-OPCO). Unlike other OPEC members, Abu Dhabi did not undertake to remove foreign ownership completely from its petroleum sector. ADNOC maintained its strategic alliances with Western firms and is considered one of the better managed national oil companies. Of Abu Dhabi's 19 **giant** fields, 17 are predominantly oil. All were discovered before 1981, and all but seven were discovered before 1970.

ACHNACARRY AGREEMENT. *See* AS-IS AGREEMENT (1928).

ADAMS, KENNETH S. (1899–1975). President of **Phillips Petroleum** 1938–1951. Kenneth Stanley Adams was born in Kansas on 31 August 1899. In 1920, his final year at the University of Kansas, he began his employment at Phillips Petroleum as a clerk in the warehouse division. He advanced rapidly. In 1922, he became district operation manager in the production department. He was treasurer by 1925 and in 1932 became assistant to the president. Five years later, he was president of the company. In his 13 years as president, Adams raised Phillips's gross income from just under $78 million to more than $520 million. At his retirement, Phillips had the world's largest **reserves** of **natural gas**. Adams turned Phillips into a vertically integrated company, from oil production to marketing. He stepped up exploration, built new refineries in the U.S. Midwest, and laid more than 3,000 miles of **pipeline**. He also expanded the company's operations to **Mexico**, **Colombia**, **Venezuela**, and **Canada**. He died on 30 March 1975.

AGGREGATE PROGRAMMED QUANTITY. *See* IRANIAN OIL PARTICIPANTS LTD.

AGIP. Azienda Generale Italiana Petroli (Italian General Oil Company), a national petroleum company established in 1926 by Italy's Fascist government; later a subsidiary of **ENI**. AGIP explored for hydrocarbons in Italy and participated in foreign ventures in **Romania**, Albania, and **Iraq**. In 1929, it discovered the Fontevivo oil field near Parma, Italy, and in 1931 signed an agreement to buy crude oil from the Soviet Union. It began downstream operations in 1936. In 1939, it discovered **natural gas** in the Po Valley, at Podenzano. In 1941, AGIP joined with Ente Nazionale Metano, Salsomaggiore Regie Terme, and Surgi to form the Società Nazionale Metanodotti (SNAM), with the purpose of building natural gas **pipelines**. The first pipeline was completed in 1943, supplying gas to Lodi and Milan from the Salsomaggiore field.

In 1944, AGIP made a large gas strike in the Po Valley, the Caviaga field near Milan. In 1945, **Enrico Mattei** was named provisional administrator of AGIP with a mission to liquidate it. Instead Mattei moved to expand the Italian petroleum industry. In this, he was aided by additional gas discoveries in the Po Valley. In 1953, the Italian government created ENI with Mattei as head. The AGIP brand with its distinctive "six-legged dog" logo was absorbed into the new entity.

AJAX, OPERATION (1953). A coup engineered by the governments of the **United States** and the **United Kingdom** to depose **Iran**'s prime minister **Mohammed Mossadegh** in the wake of his nationalization of Iran's oil industry in May 1951. The British, angered by the expropriation of the assets of the **Anglo–Iranian Oil Company** (later **BP**), wanted Mossadegh removed. They embargoed Iran's oil and considered military action (*see* PLAN BUCCANEER). U.S. president Harry S. Truman's administration restrained them. The administration of President Dwight D. Eisenhower was more receptive to Mossadegh's ouster. Secretary of State John Foster Dulles and his brother Allen Dulles, head of the Central Intelligence Agency (CIA), both gave the green light for a coup. The ostensible motivation was fear of instability in Iran and the possibility of a communist takeover there. The execution of the plan, dubbed Operation Ajax, began in August 1953. The operational leader was **Kermit Roosevelt**, grandson of Theodore Roosevelt. The operation was supposed to begin with Shah **Mohammed Reza Pahlavi** dismissing Mossadegh and naming Fuzlollah Zahedi prime minister. The order was delayed, and Mossadegh somehow became informed of it. He then launched a counteroperation. The shah left the country, going first to Baghdad and then to Rome.

Roosevelt made another attempt despite instructions from his superiors to leave the country. He had a network of agents and provocateurs, and he had in his possession the *firmans* (orders) dismissing Mossadegh signed by the shah. He sought out General Zahedi, who had gone into hiding, and urged him to circulate copies of the *firmans*. Roosevelt had his agents promulgate the falsehood that Mossadegh had conspired to seize the shah's throne, thus making the victim of the intended coup seem like the aggressor. The lie was printed in newspapers, with only a few media outlets publishing the truth. Roosevelt then used his agents to hire crowds of provocateurs and stage riots, which were then put down by other staged "patriotic" counterriots aided by police. The engineered riots took on a life of their own and expanded. Eventually pro-shah elements and the paid thugs on their side carried the day.

The shah was able to return to Tehran before the end of August 1953, and Mossadegh was placed under arrest. Zahedi became prime minister and received an immediate $5 million from the CIA to help set up his new government, as well as $1 million for his own pocket. Iran's oil remained nationalized, but Western companies were allowed back in (*see* IRANIAN OIL PARTICIPANTS LTD.) and essentially ran the country's petroleum sector for the next two decades, despite nationalization.

The shah would remain in power for 26 years, until the Iranian Revolution of 1979 created the Islamic republic and brought Ayatollah **Ruhollah Khomeini** to power. The shah's regime was a Western-oriented one but also brutally repressive. Support for the shah would later haunt the United States and the West as Iran was driven into an increasingly uncompromising anti-Western stance after 1979.

ALASKA. Oil production 2015: 0.483 Mmbpd (5.13 percent of U.S.); **reserves** 2.034 Bb (6.29 percent of U.S.); gas production: 0.326 Tcf (1.2 percent of U.S.); reserves 4.566 Tcf (1.48 percent of U.S.). Early commercial interest in Alaska centered around the Kenai Peninsula in the southern part of the state. Richfield Oil (later **Arco**) discovered a significant deposit on 19 July 1957 with its Swanson River No. 1 well. However, the most important accumulation of oil was found on the North Slope, with the discovery of the **Prudhoe Bay oil field** in 1967 by Arco and **Humble Oil and Refining Company** (later **Exxon**). From 1988 to 1996, Alaska was very nearly tied with **Texas** as the largest oil-producing state in the **United States**. By 1999, the federal **offshore** areas (mostly in the **Gulf of Mexico**) had overtaken both. Alaska's oil production, driven mostly by Prudhoe Bay, rose rapidly from 0.174 Mmbpd in 1976 to a peak of 2.02 Mmbpd in 1988. It declined thereafter. Alaska's oil production in 2015 was barely above 1977 levels, before the boom.

Since 1981, the North Slope has accounted for more than 97 percent of Alaska's oil production. The North Slope's oil production has been mostly from Prudhoe Bay, but the Kuparuk River field has also contributed. The Kuparuk River field, discovered in 1969, has an **EUR** of about 2.6 Bb, approximately a fifth of Prudhoe Bay's. Since 1977, North Slope oil has been delivered 800 miles southward to the ice-free port of **Valdez** by the **Trans-Alaska Pipeline**, for transshipment mostly to the lower 48 states. In 2017 **Repsol** announced Horseshoe, a 1.2 Bb oil discovery in the North Slope, possibly the largest onshore find in the United States in three decades.

Alaska may have considerable quantities of undiscovered oil and **natural gas** in the **Arctic National Wildlife Refuge** (ANWR). Estimates vary but range as high as 16 Bboe. The idea of drilling in ANWR is a continuing source of political controversy in the United States. *See also* ANDERSON, ROBERT O. (1917–2007); BP; SOHIO.

ALASKA PIPELINE. *See* TRANS-ALASKA PIPELINE SYSTEM (TAPS).

ALBERTA. Canada's foremost oil-producing province. In 2003, the ***Oil and Gas Journal*** began including the **Alberta oil sands** in Canada's **reserves**. If the oil sands are included, then Alberta's total oil reserves in 2015 were 167.4 Bb, or 98.6 percent of Canada's total. Alberta's **conventional crude oil** reserves stood at 1.8 Bb, or about 43.9 percent of Canada's conventional total. Alberta accounted for about three-fourths of Canada's cumulative output of conventional crude oil through 2001. During the 2000s, the fraction of Canadian conventional crude produced in Alberta was lower, accounting for only 37.4 percent in 2016. This was caused both by the depletion of Alberta's major fields and by increasing production elsewhere in Canada. However, increasing production from the oil sands reaffirmed Alberta's prominence. In 2016, Alberta produced 2.85 Mmbpd from the oil sands, including **syncrude** and marketable **bitumen**, for a total provincial production of 3.296 Mmbpd, including conventional crude. This amounted to nearly 81.3 percent of total Canadian production.

Alberta's modern petroleum history dates back to at least 1883, when the Canadian Pacific Railway first discovered **natural gas** near Medicine Hat (then in Canada's Northwest Territories but later in Alberta). Other discoveries followed in 1890, and the first commercial gas field was developed in 1901 near Medicine Hat. In the same year, the first commercial oil well was drilled near Cameron Creek (also then in the Northwest Territories but later part of Alberta).

The first important discoveries of oil were in the Turner Valley, in the foothills of the Rockies, where gas seeps had long been noticed. Archibald Dingman, cofounder and head driller of the Calgary Petroleum Products Company (CPPC), made the first find in what would become the Turner Valley oil field in 1914, about 35 miles southwest of **Calgary**. CPPC, needing capital, formed a partnership with **Imperial Oil**, in which the Standard Oil Company of New Jersey (later **Exxon**) had a majority interest, to develop the discovery. By 1930, output exceeded 2,700 bpd, and it peaked at about 29,000 bpd in 1942. Turner Valley production spurred the creation of the Alberta Petroleum and Natural Gas Conservation Board to manage the industry.

In 1946, the Conrad and Taber oil fields near Lethbridge began producing, as did the Loydminster field straddling the Alberta-Saskatchewan border. However, the truly important discovery of the time was the **giant** Leduc field, about 60 miles south of **Edmonton**. Imperial Oil drilled the first Leduc well on 20 November 1946, after applying scientific **geophysical prospecting** techniques. By 13 February 1947, Leduc was already producing an initial 1,000 bpd. In January 1948, the Woodbend field was discovered across the Saskatchewan River from Leduc and was shown to be linked to it. Within five years, there were more than 800 wells in a giant field (**EUR** 0.5 Mmb) known as Leduc-Woodbend. Imperial oil discovered an even larger field,

Redwater (EUR 0.832 Mmb), in 1948. In 1953, the Seaboard Oil Company and the Socony–Vacuum Company (later **Mobil**) discovered Alberta's largest field, Pembina (EUR 1.8 Mmb). By 1956, Pembina was Canada's largest producer, at 100,000 bpd.

Alberta's oil production began its rapid rise in 1948, after Leduc. In that year, it was about 30,000 bpd. In 1955, it surpassed 0.5 Mmbpd; in 1971, it exceeded 1 Mmbpd. From 1971 to 1984, the province's conventional oil production mostly stood above 1 Mmbpd. After 1984, it entered a period of long-term decline. Alberta oil sands production, however, rose from under 250,000 bpd in 1985 to more than 2.85 Mmbpd in 2016, as noted (including syncrude and marketable bitumen).

Excluding the oil sands, Alberta has 11 giant fields, nine predominantly oil and two predominantly gas. All were discovered before 1980, and all but three were discovered before 1960. One of the more recent was the giant Elmworth gas field (EUR 20 Tcf), discovered in 1976 by John Masters and Jim Gray of Canadian Hunter Exploration, when the company was just about to lose all its backing. The Elmworth reservoir is a **tight**, or low-permeability, gas sand.

ALBERTA OIL SANDS (TAR SANDS). Deposits of crude **bitumen** in the **Canadian** province of **Alberta**, constituting one of the largest known hydrocarbon accumulations in the world. Bitumen is a viscous mixture of heavy hydrocarbons. It does not normally flow through a well like **conventional crude oil**; it must be mined and processed. The production of synthetic crude oil (**syncrude**) from bitumen requires energy-intensive processing that can use significant amounts of **natural gas**, consumes large quantities of freshwater, and can require lighter hydrocarbons as diluents. Crude oil prices generally need to be more than $35–$50 per barrel (in 2016 money) for the processes to be competitive.

The bitumen generally has an **API gravity** of 8–10°, with a high sulfur content (3–5 percent), and occurs in sands and carbonates. The total resource could range as high as 1,600–2,500 Bb. The estimated initial established **reserves** are 165.5 Bb, with perhaps 315 Bb in **ultimately recoverable resources**. The ***Oil and Gas Journal*** began to include oil sands in its reserves estimate for Canada beginning in 2003. This had the effect of increasing the reserve estimate by a factor of 37, from 4.858 Bb in 2002 to 180.021 Bb in 2003, of which 174.8 Bb were oil sands at the time. The decision by *Oil and Gas Journal* reflected the increasing importance of oil sands to Canada's oil production. By 2006, syncrude and marketable bitumen from the oil sands accounted for roughly 45 percent of Canadian production, versus about 22 percent a decade earlier. In 2016 the proportion was 66 percent.

The government of Alberta designated three specific "Oil Sand Areas" (OSAs) totaling about 55,000 square miles: Athabasca, Cold Lake, and Peace River. Athabasca contains about 80 percent of the deposits of the three areas. Athabasca contains both shallow deposits (less than 250 feet) that can be mined by open-pit techniques and deeper ones that require in situ production techniques. Within Athabasca, the best deposits lie in the area north of **Fort McMurray**. They occur there in high cliff banks along the Athabasca River and its tributaries.

The first known European contact with the oil sands was in 1719, when a Cree native brought a sample of the Athabasca sands to a Hudson's Bay trading post. In 1778, a fur trader named Peter Pond reported that the natives in the area used the bitumen to caulk their canoes. No systematic attempt was made to exploit the oil sands until the early 20th century, and most failed. In 1913, Sidney Ells, an engineer with the Canadian Department of Mines, tried using hot water flotation to separate bitumen from the Athabasca oil sands. He experimented with obtaining a suitable material for road paving but could not produce one for less than the cost of imported asphalt.

The modern oil sands industry was launched in 1964, when Esso Canada (**Imperial Oil**, controlled by **ExxonMobil**) began a project at Cold Lake using cyclic steam injection. In 1967, **Sun Oil** completed an integrated mining and upgrade plant in the Athabasca area north of Fort McMurray. The plant was then known as the Great Canadian Oil Sands plant, later **Suncor**. By 1999, the Suncor plant was producing 105,600 bpd of syncrude. Expansion projects increased production to 260,000 bpd of syncrude by 2005. In 1978, the Lake Mildred plant about 25 miles north of Fort McMurray began production. Lake Mildred is operated by Syncrude, a joint venture with 10 member companies. Imperial Oil, with 25 percent of the venture, is the largest shareholder. Production reached 223,000 bpd of syncrude from open-pit mined sands. In 2000, Syncrude opened the Aurora mine, an extension to Lake Mildred. The Aurora mine's output is partially processed on-site and then pipelined to the upgrader at Lake Mildred, which has also been expanded.

Additional development in the Athabasca area has recently taken place at the Muskeg River mine about 47 miles north of Fort McMurray. The mine began producing in 2002. It is operated by the Albian Sands Energy joint venture between Shell Canada (**Royal Dutch Shell**), **Chevron** Canada, and Western Oil Sands, with 60 percent, 20 percent, and 20 percent of the shares respectively.

ALCOHOL. *See* ALTERNATIVE ENERGY.

ALEKPEROV, VAGIT (1950–). President of **Lukoil** (1993–) who played an important role in the company's creation. Vagit Alekperov was born in **Baku** on 1 September 1950, the son of an oil worker. He graduated in 1974 from the **Azerbaijan** Oil and Chemistry Institute. As a student, he worked at Kaspmorneft, a petroleum company operating in the **Caspian** region; he continued to work there after graduation. In 1979, he became deputy head of one of the production units. Between 1979 and 1985, he worked at Surgutneftegaz in western **Siberia**. In 1987, he was named general director of a new production company, Kogalymneftegaz.

By 1990, Alekperov was the **Soviet Union**'s deputy minister of the oil and gas industry. He played an important role in the formation of Lukoil, the Soviet Union's first vertically integrated petroleum company. Initially Lukoil reported to the Soviet Ministry of Fuel and Energy. However in 1993, after the breakup of the Soviet Union, Lukoil was established as a joint stock company with Alekperov as its president.

Forbes magazine estimated Alekperov's net worth at $13.8 billion in 2017. Vagit Alekperov also published a very useful history of Russian oil, *Oil of Russia*, in 2011.

ALEXANDER II (1818–1881). Alexander Nikolaevich Romanov, tsar of **Russia** 1855–1881. Alexander, born on 29 April 1818, is best known for emancipating the serfs in 1861. He undertook other domestic reforms, although he also initiated a period of repression after 1866 that bred revolutionary fervor and ultimately led to his own assassination on 13 March 1881 in St. Petersburg.

On 17 February 1872, Alexander promulgated new rules for the oil industry. Beginning 1 January 1873, the old feudal system governing oil lands in the **Abseron Peninsula** was abolished. Instead, 24-year oil leases would be auctioned by the government to private parties, including foreign investors. Winning bidders would pay annual fees and be obliged to begin oil production within two years. The reform, along with a further one on 6 June 1877 abolishing excise taxes on oil and petroleum products, would lead to the creation of a modern petroleum industry in Russia.

ALEXEEV, MAJOR. *See* VOSKOBOINIKOV, NIKOLAI (1801–C. 1861).

ALFA ACCESS RENOVA (AAR). *See* TNK–BP.

ALFONSO, JUAN PABLO PÉREZ. *See* PÉREZ ALFONSO, JUAN PABLO (1903–1979).

ALGERIA. Oil production 2015: 1.112 Mmbpd (1.42 percent of world); **reserves** 12.2 Bb (0.74 percent of world); gas production: 2.99 Tcf (2.40 percent of world); reserves 159.1 Tcf (2.62 percent of world). Algeria has been a member of the **Organization of the Petroleum Exporting Countries** (OPEC) since 1969. Although it also joined the **Organization of Arab Petroleum Exporting Countries** (OAPEC) in 1970, it continued to export oil to the **United States** during the Arab oil embargo of 1973 (*see* FIRST OIL SHOCK). Algeria is an increasingly important net exporter of oil and in 2016 was the world's sixth-largest exporter of **liquefied natural gas**. Algeria is a member of the **Gas Exporting Countries Forum**. Algeria is also estimated to have the world's third-largest endowment, after **China** and **Argentina**, of **unconventional** technically recoverable **shale gas** resources. These are as yet unexploited.

Oil production in Algeria was minuscule before the 1950s. In 1944, it was 11 bpd, and it did not surpass 100 bpd until 1951. Production rose considerably after **Compagnie Française des Pétroles** (CFP, later **Total**) discovered the country's most important fields, Hassi Messaoud and Hassi R'Mel, in 1956 and 1957 respectively. Hassi Messaoud has an **EUR** of 3.84 Bb and 7.0 Tcf of natural gas.

Hassi R'Mel was far less rich in oil (about 57 Mmb) but was a major gas **giant** (100 Tcf) very rich in condensate (about 4 Bb). Algerian production reached 9,000 bpd in 1957. By 1960, it had surpassed 185,000 bpd and in 1963 exceeded 0.5 Mmbpd. It surpassed 1 Mmbpd in 1970. Between 1972 and 1999, it stayed between 1 and 1.2 Mmbpd, reflecting both OPEC quotas and the depletion of the 1950s fields. However, it grew rapidly after 1999, reaching as high as 1.7 Mmbpd in some years. Much of Algerian oil is of extremely high quality, very light and sweet (refer to appendix 5), and is highly sought after. The high-quality oil has been marketed over the years as "Saharan Blend," which forms part of the **OPEC Basket**.

The petroleum discoveries of the 1950s gave France an increased impetus to try to retain Algeria as a colony. Although France was unable to accomplish this, it was nevertheless able to secure its hydrocarbon interests for a time. The Evian Accords signed on 18 March 1962 between the French government and the Front de Libération Nationale (FLN) set a framework for decolonization that included the preservation of the colonial petroleum code that allowed the French relatively unfettered operation in the Sahara. It did not take long for the Algerians to challenge the petroleum code and create a new petroleum policy with nationalization as its ultimate aim. In December 1963, the government formed its national oil company, the Société Nationale de Transport et de Commercialisation des Hydrocarbures, or Sonatrach. Sonatrach's initial role was to build a third export **pipeline** from Hassi Messaoud to the Arzew oil terminal on the Mediterranean. However, the government wanted the company to do much more, and in 1965 it conducted its first independent exploration for petroleum. The government lengthened the company's name in 1966 to Société Nationale pour la Recherche, la Production, le Transport, la Transformation, et la Commercialisation des Hydrocarbures to reflect its desire for the company to control all phases of hydrocarbon exploitation in the country. The company would always be universally known by its shortened form Sonatrach. In July 1981, it officially became Entreprise Nationale Sonatrach.

On 29 July 1965, Algeria and France signed the Algiers Accords, which allowed French petroleum companies to keep their petroleum **concessions**, but only in cooperation with Sonatrach. An Association Coopérative (ASCOOP) was formed between Sonatrach and a French national company that became ERAP (later **Elf Aquitaine**, ultimately Total). The French were to pay 60 percent of costs and a $2.08/b royalty. In 1967, Algeria also purchased **BP**'s interests in the country. After the Six-Day War that same year, it nationalized the interests of five U.S. companies. In August 1967, it took over the assets of Standard Oil Company of New Jersey (later **Exxon** and **ExxonMobil**) and Socony Mobil (later **Mobil** and ExxonMobil). **Getty Oil Company** negotiated an agreement in 1968 giving Algeria 51 percent of its interests. On 24 February 1971, Algeria under President Houari Boumedienne finally nationalized 51 percent of French petroleum interests and took over all pipelines not already owned by Sonatrach. The French banned Algerian imports into their country, but this was ineffective, as the Algerians found other customers with relative ease. By the end of 1971, the two countries settled their differences.

After nationalization, Algeria still found itself needing foreign expertise. A new law of 12 April 1971 abolished the concession system but allowed foreign companies to enter into service contracts and joint ventures with Sonatrach. Sonatrach would have a 51 percent interest in all joint ventures. All gas discovered would belong to Sonatrach; 49 percent of any crude oil discovered (net of royalties and taxes) would belong to the foreign operator. If oil was found, Sonatrach would also refund 15 percent of exploration costs and assume 51 percent of future development costs. CFP, Elf Aquitaine, **Amoco**, Hispanoil, Deminex, and Petrobrás all entered into exploration joint ventures with Sonatrach.

In 1986 and 1991, the government passed legislation liberalizing the petroleum sector and making joint ventures more attractive to foreign companies. **Anadarko** invested considerable sums in Algeria and made many discoveries. It began operating in the country in 1990 and made its first discoveries in 1993. It discovered the giants Ourhoud (EUR 2 Bb) and Hassi Berkine (EUR 0.8 Bb) in 1994 and 1995 respectively. The two fields provided a sustainable stream of more than 500,000 bpd. Other operators in Algeria in the 1990s included **Arco**, Mobil, and **Occidental**.

While not neglecting oil, Sonatrach paid close attention to developing the country's natural gas industry. It entered into long-term supply contracts, the most important of which was signed in October 1977 with Italy's **ENI**. The contract provided for about 0.45 Tcf of gas per year for 25 years through a trans-Mediterranean pipeline system reaching ultimately from Hassi R'Mel to Bologna. The pipeline is now called the **Enrico Mattei** Pipeline. BP contracted with Sonatrach to develop known gas fields south of In Salah in 1995. Sonatrach also made gas deals with Total and **Repsol**. The Gazoduc Maghreb–Europe (GME) between Algeria and Spain was completed in 1996. Gas production in Algeria has been on a long-term growth trend since 1980. In that year, it was 0.411 Tcf. By 2006, it was 3.072 Tcf, or 1.4 Mmboepd—almost as important as oil in energy-equivalent terms. Algeria was the world's first producer of liquefied natural gas, with facilities beginning operation at Arzew in 1964.

In 2001, the Algerian government proposed further liberalization of the petroleum sector, ending Sonatrach's monopoly in the domestic market. The legislation was delayed by opposition from Sonatrach and from labor unions but was reintroduced in 2004 and passed in 2005. Sonatrach had heretofore functioned both as a petroleum company and as a government agency regulating the petroleum sector and administering exploration leases. The law took these regulatory functions away and placed them in two new agencies. The Agence Nationale pour la Valorisation des Ressources en Hydrocarbures (ALNAFT) would henceforth administer exploration leases and collect taxes and royalties. The Agence Nationale de Contrôle et de Régulation des Activités dans le Domaine des Hydrocarbures would administer and regulate the petroleum sector overall. Initially, Sonatrach was also envisioned as having to compete on equal terms for licenses with all other companies, which would no longer be required to form joint ventures. In July 2006, however, the government restored Sonatrach's right to a 51 percent stake.

In 2016 a legal battle loomed between Algeria and Total. Total accused Algeria of changing the profit-sharing terms on contracts in the mid-2000s. Sonatrach initially vowed significant effort to defend the case, but in 2017 the two sides reached a settlement, allowing plans for a gas project in the Sahara to move forward.

Of Algeria's 12 giant fields, seven are primarily gas, while five are primarily oil. One of the gas giants, Hassi Messaoud, is also the country's largest oil discovery. Three of the 12 giants were discovered in the 1990s, the most important being Ourhoud; the rest were all discovered before 1965.

AL SAUD, ABDUL AZIZ IBN ABDUL RAHMAN IBN FAISAL IBN TURKI. *See* IBN SAUD (1876?–1953).

ALTERNATIVE ENERGY. Energy sources other than fossil fuels, nuclear energy, or hydroelectric power; some use the term to refer to anything except fossil fuels. Alternative energy includes solar, wind, wave, tidal, geothermal, hydrogen, and other sources. It also generally includes ethanol and methanol. Alternative energy sources are intended to be environmentally friendly and are often renewable. World consumption of renewable energy as a whole, including hydroelectric, has grown at about 2.2 percent annually since 1990, while total primary energy consumption has grown at 1.9 percent per year. By 2014 the renewable share of global primary energy consumption was 13.9 percent; excluding hydroelectric it was 11.5 percent. Excluding biofuels and waste, and including only geothermal, wind, solar, and tidal energy, the proportion was quite low, at 1.3 percent.

Petroleum companies have invested significant sums in alternative energy research, and some have run successful alternative energy businesses. Although the investments have been relatively small compared to the companies' core areas, they have sometimes been significant in absolute terms. **Arco** had a successful company manufacturing photovoltaic modules and offering solar power solutions during the 1980s. Arco sold the business to Siemens in February 1990. When **BP** acquired **Amoco** in 1998, it also acquired Amoco's share of Solarex, a solar energy joint venture with Enron. BP acquired Solarex in toto a year later. Known later as BP Solar, the division became profitable in 2004. BP's alternative energy subsidiary also has activities in wind energy and in hydrogen for fuel-cell vehicles. In the mid-2000s, BP planned $8 billion in alternative energy investments through the year 2015. It even capitalized on its activities in an advertising campaign suggesting that "BP" stands for "Beyond Petroleum." **Royal Dutch Shell** also has alternative energy businesses, including in wind power, hydrogen, biofuels, and more recently electric vehicle technology. Other petroleum companies have also made investments.

Many alternative energy sources, for example wind and solar power, are aimed at electricity generation. They are potential replacements largely for coal and **natural gas**, which are the important fossil fuels in that arena. In the future, they may also find increased application in the transportation sector, as electric vehicles are improved. However, in that sector, it is also important to develop alternative liquid fuels to replace those derived from crude oil. Alcohol, particularly ethanol, has received considerable attention, although it is not a new idea. Some early internal combustion engines were designed to burn ethanol, and ethanol was visible enough for **Standard Oil** to oppose it. In the 1930s, a blend of gasoline and alcohol called Agrol was marketed in the midwestern states of the **United States**. Petroleum companies opposed it. In Germany, by contrast, the Nazi government encouraged fuel ethanol use, and Germany produced 1.5 Mmb in 1935.

The world's leading fuel ethanol producers in recent times have been the United States and **Brazil**, with Brazil the more successful thus far. In 1975, in the aftermath of the **First Oil Shock**, Brazil launched its Programa Nacional do Álcool (National Alcohol Program), or Pró-Álcool, to produce ethanol from sugarcane and use it as a motor fuel. The ethanol manufacturing process in Brazil is relatively efficient, producing 8–10 times as much energy as it consumes. Brazilian service stations sell both pure ethanol and a blend of **gasoline** and ethanol (up to 25 percent ethanol). In 2006, Brazil produced 102 Mmb of ethanol. In 2015 the figure was 190.4. In 1986, 76 percent of cars manufactured in Brazil were ethanol capable. By 2000, in the wake of lower oil prices and significant Brazilian offshore oil discoveries, only 0.7 percent were. By 2007, the percentage was back up to 71.9, with the market dominated by flex-fuel vehicles capable of using ethanol and gasoline in any blend.

Ethanol production in the United States has been from corn. Thus far, the process has been inefficient, with some calculations indicating that it is a net energy consumer. It also has the disadvantage of diverting food resources to fuel use and driving up some food prices. Future cellulosic processes using other types of vegetation may be more promising. U.S. ethanol in the 2000s was competitive only because of tax subsidies. U.S. ethanol is generally blended with gasoline in proportions low enough to be usable by unmodified cars. In 2016, the United States produced a record 1.043 Mmbpd of ethanol fuel; gasoline consumption was 9.342 Mmbpd.

Methanol as a fuel also has proponents, including Nobel laureate George Olah. Methanol can be made from processes using natural gas or coal as a feedstock. Its manufacturing process can also consume undesirable carbon dioxide, removing it from the atmosphere. It is nonagricultural and can be used in fuel cells as well as consumed directly. *See also* GAS TO LIQUIDS.

ALYESKA PIPELINE SERVICE COMPANY. *See* TRANS-ALASKA PIPELINE SYSTEM (TAPS).

AMERADA HESS. *See* AMERADA PETROLEUM CORPORATION; HESS, LEON (1914–1999).

AMERADA PETROLEUM CORPORATION. The Amerada Corporation (later Amerada Hess, eventually Hess Corporation) was founded in 1919 by the British entrepreneur **Weetman Pearson**, with the help of **Everette DeGolyer**, to explore for oil in the **United States**, **Canada**, and Central America. DeGolyer had previously worked for Pearson in Mexico, making major discoveries there in 1910–1911. DeGolyer was Amerada's general manager. The company made significant discoveries in Kansas and **Oklahoma** in the early 1920s. These included the Urschel field in Kansas and the Osage and Cromwell fields in Oklahoma. The company was also one of a number that made early discoveries in the greater Seminole field in Oklahoma. Much of this success is attributable to DeGolyer's pioneering use of **geophysical** exploration methods. He organized the Geophysical Research Corporation (GRC) as an Amerada subsidiary in 1925 under John Karcher.

Amerada used geophysical methods and systematic geologic reasoning to make discoveries. It pioneered the use of reflection seismology, which would come to be the dominant geophysical exploration method in the petroleum industry. In 1927, it discovered the Maud field in Oklahoma. This was not the first discovery made by geophysical or even seismic techniques, but it was the first made by reflection seismology. GRC undertook a survey for its parent Amerada in 1927 and 1928, resulting in the discovery of oil deposits in the Wilcox Sands that had been missed by a number of other companies. DeGolyer became president in 1929 and chairman in 1930; he left the company in 1932 to become an independent consultant, later cofounding the firm of **DeGolyer and MacNaughton**.

In 1933, Amerada and Standard Oil Company of Indiana (later **Amoco**) conducted a seismic survey and discovered the Katy fields in **Texas**. Amerada also discovered fields in the West Texas **Permian basin** and in North Dakota's Williston basin, again applying geophysical methods. In 1941, Amerada reorganized, merging with its principal operating subsidiary and becoming Amerada Petroleum Corporation. During **World War II**, the Bank of England acquired 10 percent of Amerada. In 1948, Amerada Petroleum Corporation, Continental Oil Company (later **Conoco**, eventually **ConocoPhillips**), and the Ohio Oil Company (later **Marathon**) formed the Conorada Petroleum Corporation to explore outside the United States and Canada. Conorada obtained concessions in **Egypt** near the Libyan border. In January 1963, Amerada bought out Conoco and Marathon and became sole owner of Conorada. In 1964, Amerada formed the Oasis Petroleum Consortium in **Libya** with Marathon, Conoco, and **Royal Dutch Shell**. Oasis controlled about half of Libyan production.

Amerada's **reserves** made it attractive to independent oil executive **Leon Hess**, whose Hess Oil and Chemical focused on downstream operations and needed crude supplies. In 1966, Hess acquired the Bank of England's 10 percent stake in Amerada with the intention of increasing his position. Amerada attempted to stop the takeover, first by trying to merge with Ashland Oil and then by making an agreement to be acquired by **Phillips Petroleum**. When Hess offered a substantial premium for Amerada's stock, Phillips withdrew from the bidding. Amerada shareholders approved the merger, and in 1969 the Amerada Hess Company was formed. (For history after 1969, *see* HESS, LEON (1914–1999).)

AMERICAN ASSOCIATION OF PETROLEUM GEOLOGISTS (AAPG). Professional society headquartered in **Tulsa, Oklahoma**, dedicated to advancing the science of geology, particularly as it relates to the exploration for and development of petroleum deposits. The AAPG is an international association, with about 40,000 members in 2017. It publishes the monthly professional geological journal *AAPG Bulletin* as well as the monthly magazine *AAPG Explorer*. It also publishes an extensive range of other occasional technical publications and books. The AAPG's Division of Professional Affairs conducts certification of petroleum geologists, petroleum geophysicists, and coal geologists.

Formation of the AAPG began in October 1915, when **Everette DeGolyer** visited the University of Oklahoma and discussed the possibility of forming a technical association of geologists in the U.S. Southwest with Professor Charles Taylor. About 30 geologists attended a dinner soon after, and there were other preliminary meetings. On 7 and 8 January 1916, a group of geologists met in Norman, Oklahoma, and presented technical papers. This is generally accepted as the first meeting of the association, even though it would not be formally organized as AAPG until a year later at a meeting in Tulsa on 9–10 February 1917. The association had 122 founding members, and its first president was J. Elmer Thomas. The *AAPG Bulletin* began as an annual publication, with print runs of only 400–500 in the early years.

The AAPG grew rapidly, reaching about 2,600 members in 1931. Membership growth reversed during the Great Depression, with levels dipping back under 2,000 in 1936. This was followed by a period of rapid and uninterrupted growth until the early 1960s, when membership reached about 15,500. Growth stagnated until the 1970s, when high oil prices caused it again to accelerate. In 1947, the *Bulletin* became a monthly publication. On 20 March 1953, the AAPG dedicated a new permanent headquarters in Tulsa.

The AAPG's technical excellence and contributions to the field of geology are undoubted, but the association came under some criticism for its 1999 position statement that discounted the likelihood of human influence on climate change. Critics charged that this position reflected oil company politics more than considered scientific reasoning. In 2006, the association was further criticized for presenting its journalism award to Michael Crichton for his book *State of Fear*, a novel espousing a skeptical view of global warming. The AAPG issued a statement in 2007 that respected the majority scientific opinions on the reality of global warming but added that current trends could fall within documented natural variations in past climate. *See also* ENVIRONMENTAL IMPACTS.

AMERICAN INDEPENDENT OIL COMPANY. *See* AMINOIL (AMERICAN INDEPENDENT OIL COMPANY).

AMERICAN INSTITUTE OF MINING AND METALLURGICAL ENGINEERS (AIME). *See* SOCIETY OF PETROLEUM ENGINEERS (SPE).

AMERICAN OIL COMPANY. Originator of the **Amoco** brand; later absorbed into Standard Oil Company of Indiana (later Amoco, eventually part of **BP**). The American Oil Company's roots go back to 1910, when **Louis Blaustein** and his son Jacob sold kerosene from a tank wagon. It was formally incorporated in 1922. The company marketed an innovative antiknock **gasoline** under the Amoco brand. Amoco gasoline was sold at Lord Baltimore filling stations in the Mid-Atlantic region of the **United States**.

In 1923, Blaustein sold a half interest in American Oil to the Pan American Petroleum and Transport Company of oil magnate **Edward Doheny**. Part of the deal was that he would obtain a guaranteed supply of crude oil and petroleum products. However, Pan American was acquired in 1925 by Standard Oil of Indiana (later Amoco). Under the terms of the acquisition, Blaustein's guaranteed supply would last only until 1933. This led to litigation. The Blaustein family eventually received a sizable holding in the Standard Oil Company of Indiana, which adopted the Amoco brand. In 1954, American Oil became a wholly owned subsidiary of the Standard Oil Company of Indiana.

AMERICAN PETROLEUM INSTITUTE (API). Petroleum industry trade organization founded 20 March 1919 in New York City. In 1969, API moved its headquarters to Washington, D.C. API's stated mission is to influence public policy in support of a strong, viable **United States** oil and **natural gas** industry. The API engages in lobbying at U.S. federal and state levels and works with the industry and its various associations to formulate policies.

The API grew out of the effort to ensure adequate petroleum supplies for the U.S. military in **World War I**. This effort was coordinated by the National Petroleum War Service Committee, a quasi-governmental organization. The wartime activities highlighted the general need for cooperation across the industry, and API was established to fill the need in 1919. The API embarked on several areas of activity. In 1920, it began to collect and issue weekly industry statistics, beginning with crude oil production but expanding to encompass other variables.

The API also played a role in standards development. During World War I, the industry attempted to improve efficiency by pooling equipment. The effort was stymied by a lack of uniformity in field equipment, for example in pipe size, threading, and coupling. API developed the first industry standards, publishing them in 1924. It has developed more than 500 standards and recommended practices for the oil and gas industry. The API has also worked with the U.S. Treasury Department and with state governments to develop oil and gas taxation policies. *See also* FUEL ADMINISTRATION.

AMERIVEN. *See* ORINOCO OIL BELT.

AMINOIL (AMERICAN INDEPENDENT OIL COMPANY). A consortium of American **independent** oil companies formed to bid on a **concession** from **Kuwait** for the **Neutral Zone**. The companies were **Phillips Petroleum**, Ashland Oil, Signal Oil and Gas, J. S. Abercrombie, Sunray Mid-Continent Oil, Globe Oil and Refining, and Pauley Petroleum Inc. Aminoil won the concession in 1948. Working with **Getty Oil Company** (then called Pacific Western Company), which had the **Saudi** rights for the Neutral Zone, it made major discoveries, beginning with the **giant** Wafra field in 1953. Aminoil and Getty continued to work together in the Neutral Zone under the terms of a 1960 joint operating agreement.

In 1970, tobacco giant R. J. Reynolds acquired Aminoil for $600 million, as part of a major diversification program. In 1975, Kuwait sought to increase the sums payable to it by Aminoil. Aminoil essentially refused and continued to make payments on the basis of an agreement reached two years earlier. In 1977, Kuwait issued Decree Law No. 124, nationalizing all of Aminoil's assets in Kuwait (including the Neutral Zone), for compensation to be assessed later. Aminoil sought arbitration, allowed under the terms of the original 1948 concession agreement. In the end, Kuwait owed Aminoil $179 million, including adjustments for interest and inflation. In 1984, Reynolds sold Aminoil for $1.7 billion to Phillips Petroleum.

AMOCO. Organized in 1889 as the Standard Oil Company of Indiana; eventually part of **BP**. After the 1911 breakup of **Standard Oil**, Standard of Indiana became an independent company with the exclusive rights to the Standard name in the U.S. Midwest and with strong refining and marketing operations. In 1925, it acquired control of Pan American Petroleum and Transport's operations outside **California** for almost $38.6 million, the largest oil company consolidation up to that time. Pan American had a 50 percent interest in **American Oil**, founded in 1910 by **Louis Blaustein**. Blaustein had marketed antiknock **gasoline** under the Amoco brand name, which now became a signature brand for Standard Oil of Indiana. Pan American was **Edward Doheny**'s company and had interests in **Mexico** and **Venezuela**. Its California operations were segregated into Pan American Western Corporation and remained with Doheny. In 1932, Standard Oil Company of Indiana, fearful of making large foreign investments and also worried (correctly) that it might lose Pan American's Mexican properties to nationalization, sold Pan American's foreign interests to Standard Oil of New Jersey (later **Exxon**).

Standard Oil of Indiana formed Amoco Chemicals in 1945. In the 1950s and 1960s, Standard Oil of Indiana purchased several refineries. In 1958, it formed a joint venture with the National Iranian Oil Company named Iran Pan American Oil Company (IPAC). The venture successfully developed some important oil fields in **Iran**, including Fereidoon. The company had an environmental disaster in 1978 when its supertanker ***Amoco Cadiz*** ran aground off the French coast, spilling 120,000 tons of crude. This ultimately resulted in a $128 million judgment against the company in 1990. Standard Oil of Indiana diversified into minerals via its purchase of Cyprus Mines in 1979 but divested those interests by 1985. In that year, it changed its name to Amoco, the brand by which it was already known. In 1988, it bought the Canadian company Dome Petroleum. In 1992, Amoco began exploration in **China**, the first foreign company to do so on the mainland. In 1996, it began production in the South China Sea. In the mid-1990s, it also made significant oil discoveries in the **Gulf of Mexico**, **Colombia**, and the **North Sea**, as well as large **natural gas** discoveries near Trinidad.

In 1998, Amoco merged with BP. The combined company was known briefly as BP Amoco but by 2000 was called by the shortened name of BP PLC. In 1997, the last full year before the merger, Amoco controlled 2.417 Bb of proved oil **reserves** and 21.46 Tcf of natural gas. It produced 0.216 Mmbpd of oil and 4.1 Bcfpd of gas.

AMOCO CADIZ. Oil **tanker** owned by **Amoco** that ran aground on 16 March 1978 on Portsall Rocks (three miles off the coast of Brittany, France), resulting in the largest recorded oil spill from a tanker up to that time; as of 2017, it was the fourth-largest such spill. The ship suffered a steering failure as it entered the English Channel off Brittany.

The ship could not be moved from the site where it ran aground. It drifted into rocky shoals and split in half eight days later. Thousands of sea birds died, and 30 percent of France's seafood production was threatened. The total amount of crude spilled was about 1.64 Mmb. About 14 percent of the oil released into the environment was incorporated within the water column, 28 percent was washed into the intertidal zone, and up to 40 percent evaporated. Other than some inlets, most of the impact did not last longer than three years. Refer to appendixes 8 and 9. *See also EXXON VALDEZ*; SANTA BARBARA OIL SPILL; *TORREY CANYON.*

ANADARKO PETROLEUM CORPORATION. Large **independent** petroleum company founded in Kansas in 1959, initially to explore for and produce **natural gas**. Anadarko has had oil and gas interests in the **United States**, **Algeria**, **Venezuela**, **Canada**, and the Middle East. When the company was established in 1959, it was owned by the Panhandle Eastern Pipe Line Company and concentrated on gas in the Anadarko basin. It was successful and expanded to other areas and to oil exploration and production. Anadarko bought Ambassador Oil of Fort Worth, **Texas**, in 1965 and moved to that city; in 1974, it moved to the **Houston** area. In 1970, it began to explore offshore in the **Gulf of Mexico**. It made major discoveries with **Amoco** on Matagorda Island, Texas, about a decade later.

Anadarko was spun off as an independent company in 1986. It began operating in Algeria in 1989 in partnership with Sonatrach. In 1991, Anadarko expanded to the South **China** Sea and the North Slope of **Alaska**. In 1996, it discovered the Mahogany oil field in the **offshore** Louisiana area. The next year, it expanded exploration to Tunisia and the North Atlantic, and in 1998 it made additional significant discoveries in the Gulf of Mexico. The company decided to focus further on the Gulf of Mexico and partnered with **Texaco** to do so in 1999.

Beginning in 2000, Anadarko acquired a number of companies. It bought Union Pacific Resources for $5.7 billion in stock. In 2001, it bought Canada's Berkley Petroleum for $1 billion and Gulfstream Resources Canada. Berkley operated in western Canada, where Anadarko already had a presence. Gulfstream operated in **Qatar** and **Oman**, a new area for Anadarko. In 2006, Anadarko acquired Kerr McGee for $16.4 billion and Western Gas Resources for $4.7 billion. It also sold several properties, including Anadarko Canada to Canadian Natural Resources for $4.2 billion, and a number of interests in the Gulf of Mexico. It continued selling properties in 2007 to cut debt. It sold its gas fields in Louisiana to EXCO Resources for $1.6 billion and agreed to sell petroleum interests in Wyoming, West Texas, the Gulf of Mexico, and **Oklahoma** to other companies. Anadarko owned a 25 percent share in the Macondo Prospect in the Gulf of Mexico and thus had some exposure to the consequences of the ***Deepwater Horizon* oil spill**. In 2016, Anadarko was the fourth-largest U.S.-based petroleum company in total assets. It controlled 0.985 Bb of liquid **reserves** and 4.424 Tcf of gas. Its liquid and gas production was 0.581 Mmbpd and 2.112 Bcfpd respectively.

ANCHOR OIL. *See* PHILLIPS PETROLEUM.

ANDERSON, ROBERT O. (1917–2007). Creator of **Arco** from the merger of Atlantic Refining and Richfield Oil. Robert Orville Anderson was the motive force behind the 1967 discovery of oil at **Prudhoe Bay** in **Alaska**. He also led the effort to develop the **Trans-Alaska Pipeline**. Anderson was born in Chicago on 13 April 1917. He graduated from a two-year program at the University of Chicago in 1939 and began his career at a subsidiary of the **Pure Oil Company**. In 1941, he joined with his brothers to buy a refinery in New Mexico, with backing from their father. With no experience and a very tight deadline, he managed to produce 91-octane fuel (*see* GASOLINE) for the war effort. He went on to buy several other refineries and widened his focus upstream, discovering the Empire Abo field in New Mexico in 1957. In 1958, he decided to concentrate only on upstream operations. He sold all his refining interests and reorganized his holdings as the Hondo Oil and Gas Company. In 1963, he merged Hondo into the Atlantic Refining Company. Because of Hondo's strength and Atlantic's relative weakness, the merger was favorable to him, and he became Atlantic's largest shareholder. In 1965, he became president. Even with Hondo's oil, Atlantic still needed more supplies of crude to feed its downstream operations. In May 1966, Anderson acquired more crude **reserves** by merging Atlantic with Richfield Oil to form Arco (Atlantic Richfield Company). In 1969, he further enlarged Arco by acquiring **Sinclair Oil**.

Perhaps Anderson's most notable achievement was the discovery of the Prudhoe Bay oil field. The merger with Richfield brought rights to explore on Alaska's North Slope, and Anderson partnered Arco with **Humble Oil and Refining Company** (later **Exxon**) to do so. He was persistent in continuing to explore in Alaska when other companies were giving up—even **BP**, which had spent eight years looking for oil on the North Slope. Despite several failures, Anderson kept Arco in Alaska and even bought more leases. By April 1967, Arco's drilling rig was the only one remaining on the North Slope. Anderson had it moved to Prudhoe Bay, striking oil at 9,000 feet on 26 December 1967. A second well drilled seven miles away in 1968 struck another part of the same field, confirming a huge find—the largest ever in North America, with an **EUR** of 13 Bb. To bring this remote oil to market, Arco proposed constructing a **pipeline**. It joined with seven other companies to form the consortium that completed the Trans-Alaska Pipeline in 1977 to transport oil from Prudhoe Bay to the ice-free southern Alaska port of **Valdez**. In 1982, Anderson negotiated the first American drilling effort in **China**, in the **South China Sea**.

Anderson retired from Arco in 1986 and started another Hondo Oil and Gas Company, which was not successful. In addition to being an oil executive, Anderson was the largest landowner in the **United States**, with more than a million acres under his control. He became involved in publishing, rescuing two famous but financially troubled publications: the *Observer* newspaper in the **United Kingdom** (1977) and *Harper's* magazine (1980). Anderson was also an environmentalist, speaking out on global warming and donating large sums to environmental causes. The Anderson School of Management at the University of New Mexico is named for him. He died on 2 December 2007 in Roswell, New Mexico.

ANGLO–IRANIAN OIL COMPANY (AIOC). Called Anglo–Persian Oil Company 1909–1935; British Petroleum after 1954; later **BP**. On 28 May 1901, British engineer **William Knox D'Arcy** obtained a **concession** to explore for oil in a 400,000-square-mile area of Persia (**Iran**) from Shah Muzzafar al-Din. D'Arcy had exclusive rights for exploration, development, production, and export. In return, he agreed to pay Iran £40,000, half in stock and half in cash, plus 16 percent of his annual profits for 60 years. D'Arcy's operation discovered evidence of oil in 1903. By 1904, he had a producing well at Chia Surkh. But the operations were costly and difficult. In financial straits, he appealed to the British government for help. The government, fearing that D'Arcy might sell his concession to a foreign government, pressured the **Burmah Oil Company**, a major British firm with Admiralty contracts, to invest in D'Arcy in 1905. Burmah and D'Arcy signed an agreement on 5 May 1905 forming Concessions Syndicate Ltd. to finance the work in Iran. D'Arcy's company became a subsidiary of Burmah, and D'Arcy became a director of the subsidiary.

In 1908, D'Arcy's geologist **George Reynolds** made a major oil discovery at Masjid-e-Suleiman. On 14 April 1909, Burmah formed the Anglo–Persian Oil Company (APOC) with an initial capital of £2 million to develop what would turn out to be an enormous find (**EUR** about 1.13 Bb). A million ordinary, fully paid shares were apportioned as follows: 570,000 to Burmah Oil, 400,000 to Concessions Syndicate, and 30,000 to **Donald Smith** (Lord Strathcona), the company chairman. There were also a million preference shares. D'Arcy received 170,000 shares of Burmah Oil. He was appointed honorary director of APOC but no longer had a say in running the company. That same year, the British government asked Percy Cox, British resident at Bushehr, to negotiate an agreement with **Khaz'al Khan** for APOC to obtain a site on **Abadan** Island for a refinery, depot, storage tanks, and other operations. Khaz'al was the sheikh of Mohammerah (now Khorramshahr) and held sway over the Arab tribes of **Khuzestan** Province, which enjoyed considerable autonomy from the central government in Tehran.

The refinery was built and began operating in 1912. However, APOC needed more capital to expand. At the time, First Lord of the Admiralty **Winston Churchill** was continuing the conversion of the Royal Navy's fleet from coal to oil, and he wanted to secure a steady, long-term supply of the fuel at a reasonable price. A fairly involved series of events culminated in the British government's acquisition of 51 percent of APOC for £2.2 million on 20 May 1914. APOC continued to operate in Iran and supplied the British government with fuel during **World War I**.

APOC/AIOC also developed interests outside Iran. As a partner in the **Iraq Petroleum Company** consortium, it had an interest in the first oil discovery in **Iraq** in 1927, as well as later interests in Iraq, **Abu Dhabi**, **Qatar**, and **Oman**. As a partner in the Kuwait Oil Company, it had an interest in **Kuwait**. It was a party to the **As-Is Agreement** of 1928 to limit competition and control the market for oil.

In Iran, there was considerable discontent over the arrangements with APOC. The parliament of the Constitutional Revolution of 1905–1911 failed to ratify the 1901 D'Arcy concession. APOC also apparently did not keep its promises to the Iranian government. It is estimated that between 1914 and 1920, APOC paid Iran only about a third of the money owed to it under the terms of the concession. There was also considerable resentment at the squalid living and working conditions of Iranian oil workers. In 1933, Shah **Reza Pahlavi** terminated the 1901 concession. There was an outcry from the British, who appealed to the League of Nations. Iranian representatives countered that APOC had falsified accounts to cheat Iran out of rightful payments. Shah Reza negotiated a new agreement directly with **John Cadman**, head of APOC. The area of the concession was reduced from 400,000 to 100,000 square miles. The new agreement also provided for a minimum annual payment of £225,000–£300,000 in the form of a tax on production, and an additional royalty of 4 shillings per ton. Iran would also receive 20 percent of the net profit over and above a guaranteed dividend of £671,250. Under the agreement, the company's name was changed to the Anglo–Iranian Oil Company (AIOC). While the new agreement's terms appeared harsher to the company in some ways than those of the original 1901 concession, there were also some advantages. The 1901 concession was due to expire in 1961, while the new agreement was supposed to last until 1993. The new agreement was also ratified by the Iranian parliament and was more formal than the 1901 concession, which was essentially a "handshake" between Shah Muzzafar al-Din and William Knox D'Arcy.

The revised concession did not prove satisfactory to the Iranians, and there were growing calls for further modifications. Between 1945 and 1950, AIOC earned £250 million in profit. The Iranian government got only £90 million in royalties, less than the British government collected from AIOC in taxes. To compound the injury, AIOC also paid dividends to the British government, its majority shareholder. Iranians suspected that AIOC was hiding profits, a suspicion compounded by the fact that AIOC would not let them audit its books. Discontent continued to be fueled by the poor living conditions of Iranian workers, compared to the relatively lavish living standards enjoyed by British expatriates. In addition, the Iranians felt aggrieved at the lack of educational and advancement opportunities for their citizens in what they believed was their own oil industry. The British became hated in Iran, to the extent that there was some sympathy for the Nazis in **World War II**.

Shah Reza Pahlavi was overthrown by the British and Russians for his allegedly pro-Nazi stance in 1941. He was replaced by his son **Mohammed Reza Pahlavi**, who was regarded by many as a British stooge.

The overall discontent in the oil industry led to a major strike in 1947 by Abadan refinery workers. The British countered by organizing mobs of paid provocateurs, giving AIOC a pretext to respond with force. In that same year, the Majlis (the Iranian parliament) passed a law calling for a renegotiation of the AIOC concession. The main promoter was **Mohammed Mossadegh**, who would later become prime minister. In 1948, **Venezuela** adopted a **50–50 agreement**, under which the government would obtain half the profits of oil companies operating in the country. Pressure was mounting to adopt 50–50 agreements in other oil-producing countries. AIOC stood resolutely against this. The **United States**, ever fearful of driving Iran into the hands of communists, pressured the British to increase royalties to Iran, but AIOC under its stubborn leader **William Fraser** resisted. Fraser softened his stance when he heard that a 50–50 deal was in the works for **Saudi Arabia**. AIOC reluctantly negotiated a "supplemental agreement" more favorable to Iran, but it was too late. The Majlis rejected the supplemental agreement in November 1950 even though the shah had supported it. One month later, Saudi Arabia's 50–50 agreement was concluded. Prime Minister Ali Razmara abrogated his previous support for the supplemental agreement. Mossadegh's calls for outright nationalization of Iran's oil industry, which had begun in earnest more than a year earlier, grew more strident. Razmara declared his opposition to nationalization and was murdered four days later.

The Majlis passed a nationalization resolution, and on 28 April 1951 Mossadegh took over as prime minister. The nationalization law went into effect in May 1951; the National Iranian Oil Company (NIOC) would henceforth own all of AIOC's assets in Iran. The Harry S. Truman administration in the United States initially sympathized with Mossadegh, on grounds of opposition to imperialism and because of fears that Iran might be driven into the arms of the Soviet Union if it were not treated carefully. The British worried less about the Soviets and looked at Mossadegh and Iran with a mixture of fear and contempt. Winston Churchill, who would become prime minister once again in October 1951, was shocked by the U.S. attitude. The British began a successful embargo of Iranian oil, with AIOC threatening legal action against anyone who tried to buy or transport Iranian crude, on the grounds that the oil belonged to AIOC. The British considered activating **Plan Buccaneer**, a contingency plan for military intervention in Iran. The United States was opposed to any such action, and it attempted to mediate. But mediation efforts led by Averell Harriman failed, and on 25 September 1951 Mossadegh gave the remaining British personnel of AIOC one week to leave the country. The Americans continued to try to mediate, but the embargo and the crisis dragged on, and the Iranian economy neared collapse.

Mossadegh nevertheless remained immensely popular. The British suggested a coup to remove him. The Truman administration had lost patience with Mossadegh by now but still demurred. The new administration of Dwight D. Eisenhower that assumed power in January 1953 was more receptive. Secretary of State John Foster Dulles and his brother Allen Dulles, the head of the Central Intelligence Agency (CIA), both gave approval for a coup. Secretary Dulles feared that Mossadegh was becoming increasingly dictatorial as the Iranian economy suffered the effects of two years without oil revenue; he predicted Iran would become a dictatorship and then suffer a communist takeover. The execution of the plan, dubbed **Operation Ajax**, began in August 1953. After stumbling initially, the coup was successful. By the end of August, the shah, who had left the country after attempting to dismiss Mossadegh, was back and Mossadegh was under arrest.

However, AIOC would not be returning to conduct business as usual. Iran's oil remained nationalized. A consortium of foreign companies was allowed back into the country in October 1954. The consortium, **Iranian Oil Participants**, had rights to explore for and produce oil in the "Agreement Area" in southwestern Iran but would then have to buy that oil from NIOC for resale or refinement. AIOC, renamed British Petroleum (BP), was allowed to be a member of the consortium, with a 40 percent interest. AIOC eventually received compensation from Iran and the other members of the consortium. In practical terms, the consortium would effectively run Iran's petroleum industry for the next two decades, despite nationalization. *See also* FISHER, JOHN ARBUTHNOT (1841–1920); GREENWAY, CHARLES (1857–1934); ROYAL COMMISSION ON FUEL AND ENGINES (1912–1913); SLADE COMMISSION (1913–1914).

ANGLO–PERSIAN OIL COMPANY. *See* ANGLO–IRANIAN OIL COMPANY (AIOC).

ANGLO–SAXON PETROLEUM COMPANY. Holding company of the amalgamated **Royal Dutch Shell** Group, created in 1907. Anglo–Saxon Petroleum was incorporated under English law. **Shell Transport and Trading Company** owned 40 percent of the shares, and the **Royal Dutch Company** owned 60 percent.

ANGOLA. Oil production 2015: 1.766 Mmbpd (2.25 percent of world); **reserves** 8.273 Bb (0.5 percent of world); gas reserves 10.88 Tcf (0.16 percent of world). The country of Angola is an increasingly significant petroleum producer, not least because it still has considerable undiscovered potential, and reserves and production are still rising. Angola produced 150,000 bpd of oil in 1980, 475,000 bpd in 1990, and 746,400 bpd in 2000. Between

2000 and 2006, production nearly doubled. In 2016, Angola was the second-largest producer of crude oil and liquids in Africa, after **Nigeria**, and is expected to surpass Nigeria in the near future. In fact, counting only crude oil and condensate, Angola already exceeded Nigeria in 2016. It exports 90 percent of its crude, mostly to the **United States** and **China**. In January 2007, Angola joined the **Organization of the Petroleum Exporting Countries** (OPEC) and was granted a production quota of 1.9 Mmbpd, effective January 2008. In 2016 the quota was 1.751 Mmbpd, and Angola produced at approximately that level.

It is unknown when or if the indigenous inhabitants of the area that is now Angola made use of oil seeps. Portuguese colonists did notice such seeps in the late 1700s and shipped some oil to Portugal and Brazil for use in caulking ships. In 1915 the Portuguese company Companhia de Pesquisias Mineras de Angola began drilling for oil in the Dande River area and struck a small amount of oil (about 6 barrels per day) with the Dande No. 4 well in 1916. The well was not commercially viable. The first commercially successful discovery in Angola was in 1955, in the onshore Kwanza basin, by **Petrofina**. Onshore potential was limited, partly for political reasons. Beginning in 1966, the focus of exploration moved **offshore**. The first discovery was the Malongo field off **Cabinda**, by the Cabinda **Gulf** Oil Company (CABGOC, now part of **Chevron**). Over the next four decades, most of Angola's exploration and development was offshore, in progressively deeper waters.

In 1976, the government created a national oil company, Sociedade Nacional de Combustíveis de Angola (Sonangol). Sonangol acquired 51 percent of all onshore concessions and the Cabinda offshore concession. Foreign operating companies retained management rights. Sonangol evolved to play a number of roles: it acts as a sector regulator, it administers and participates in **production-sharing agreements**, it markets government crude, and it pays taxes to the government. Some of these roles conflict with each other at times. Other parts of the government also play a role in the industry. The Ministry of Finance assesses and collects royalties and taxes, while the Ministry of Petroleum, among other things, determines prices for fiscal purposes. The legal framework in Angola includes the older concession agreements and newer production-sharing agreements of various vintage.

In 1978–1979, the government began dividing its exploration and development area into blocks. Initially there were 14 shallow-water blocks, 0–13. These were followed by **deepwater** blocks 14–30 and ultra-deepwater blocks 31–40. New ultra-deepwater blocks are also being offered. Blocks 0 and 15 have been fairly successful. Block 0 is offshore Cabinda, led by Chevron (CABGOC), and includes the Takula (**EUR** about 2.5 Bb, discovered 1971), Numbi, and Kokongo oil fields. The Block 15 complex was discovered in 2003 by a consortium led by **ExxonMobil**, with EUR of about 3.5 Bb. In 1996 **Elf Aquitaine** (later **Total**) discovered the important Girassol field (about 880 Mmbo, in 4,593 feet of water) in the deepwater Block 17. Other important discoveries in Block 17, led by TotalFinaElf (later Total) include Dalia in August 1997; Rosa and Lirio in 1998; Tulipa, Orquidea, and Cravo in 1999; and Jasmim and Perpetua in 2000.

In 1999, the government, anxious to obtain an advantage in its war against the União Nacional para a Independência Total de Angola (UNITA), demanded up-front payments from oil companies wishing to engage in exploration. These "signature bonuses" raised more than $900 million that the government used to purchase arms to fight UNITA. Large companies operating in Angola include Total, **BP**, Chevron, ExxonMobil, **Statoil**, Sinopec, **ConocoPhillips**, **ENI**, and others.

An important development for the future of Angola's oil petroleum industry was the discovery of oil in the **pre-salt** layers of deepwater offshore **Brazil**. The margins of what are now Angola and Brazil were joined when what is today the South Atlantic began to open in the Cretaceous period, about 140 million years ago. Brazil's Campos and Santos basins were once conjoined with Angola's Benguela and Kwanza basins. Thus the Angolan pre-salt oil represents an exciting prospect, although some analysts do not expect it to be as prolific as Brazil's. In 2011 Sonangol awarded 11 deepwater and ultra-deepwater blocks to a number of Western oil companies, especially Statoil. In January 2012 the Danish company Maersk Oil drilled the first well (Azul-1 in Block 23) to penetrate into the deepwater pre-salt, at a depth of around 14,500 feet into the earth, under more than 3,000 feet of water. Maersk assessed the capacity at 3,000 bpd. In 2012 Cobalt International Energy also penetrated into the pre-salt with the Cameia-1 well in Block 21, assessing its potential at 20,000 bpd. Exploring the pre-salt is very costly and risky, and some subsequent attempts by other companies including Statoil were not successful. However deepwater offshore Angola is still considered to have significant potential.

APERTURA PETROLERA. Suite of policies adopted by **Venezuela** beginning around 1995 to attract foreign investment to the country's upstream oil sector. The Apertura Petrolera (Petroleum Opening) was proposed and actively promoted by **PdVSA** under the leadership of **Luis Giusti**. The agreements concluded under the Apertura fell mostly into three broad categories: (1) *operating service agreements* (OSAs), under which foreign companies operated oil fields for PdVSA, which paid them a fee and purchased the oil; (2) *risk/profit-sharing agreements* (RPSAs), under which foreign companies would explore for oil in designated areas, would bear the entire cost of exploration, and would spend at least $40–60 million per block (if a company discovered oil, PdVSA would have the option to purchase a 35 percent stake); and (3) *strategic associations* to produce extra-heavy crude in the **Orinoco Oil Belt**, with royalties set as low as 1 percent.

Thirty-three operating service agreements, eight risk/profit-sharing agreements, and four strategic associations were initiated. The Apertura reversed a significant decline in Venezuelan oil production. Production had peaked at about 3.7 Mmbpd in the early 1970s and had been declining ever since, to about 1.8 Mmbpd in 1986–1989. Production increased steadily through the 1990s, reaching about 3.5 Mmbpd in 1999. The Orinoco strategic associations led to major projects valued at $30 billion for upgrading heavy crude, producing 600,000 bpd of heavy oil.

During his 1998 presidential campaign, **Hugo Chávez** had harsh words for the Apertura, which he deemed an undesirable step away from full national sovereignty over petroleum, and he pledged to reverse it. After his election as president, Chávez continued the rhetoric but at first left the Apertura contracts substantially untouched. In 2001, using his recently granted temporary power to rule by decree, he enacted a new hydrocarbons law, beginning the reversal of the Apertura. Under the Apertura, royalties had ranged from 1 percent to 17 percent, depending on the project. They would now be raised to the 20–30 percent range. The law required that all future foreign participation in Venezuela's petroleum industry be in the form of joint ventures with PdVSA, which would have a majority share. Previous Apertura OSAs and RPSAs were transferred to the PdVSA subsidiary Corporación Venezolana de Petróleo (CVP). Although preexisting Apertura agreements were supposed to continue as before, they were transformed retroactively to the new joint venture structure beginning in 2005. PdVSA, through CVP, would have a 60 percent share in most ventures. In 2006, royalties for the four strategic associations in the Orinoco Belt were increased to 33.3 percent, and income tax rates to 50 percent. Venezuelan oil production decreased from its recent high in 1999 to about 2.5 Mmbpd during 2007.

In May 2007, Chávez announced intentions to take majority control of the Orinoco heavy oil projects. **ExxonMobil**, **ConocoPhillips**, and **Petro-Canada** withdrew from the heavy crude projects rather than submit to the new requirements. ExxonMobil sought arbitration with the International Center for Settlement of Investment Disputes (ICSID). Years later, on 9 October 2014, ICSID awarded ExxonMobil $1.6 billion. Other major companies, including **Chevron**, **Statoil**, **BP**, and **Total**, ultimately accepted Venezuela's new terms and remained in the Orinoco projects, transferring 60 percent to 83 percent of their interests to PdVSA.

While the Apertura was reversed for oil, a new Apertura was initiated for **natural gas**. In 1999, Venezuela instituted a gas hydrocarbons law to open all aspects of the Venezuelan gas industry to private investment, even allowing 100 percent private ownership. In 2005, the Ministry of Energy and Mines awarded three licenses for offshore gas exploration on the northwest coast (Falcón state). **Gazprom** received two of the licenses and Chevron the other.

API GRAVITY. A measure of the density of crude oil and petroleum products devised by the American Petroleum Institute. API gravity is expressed in °API (degrees API), often denoted simply by the degree symbol (°). Higher API gravities indicate a lighter oil. Water has a gravity of 10°. Some extra-**heavy** oils have gravities less than 10°. **Conventional crude oil** tends to range between 20° and 40°, with 20°API indicating a heavy conventional crude and 40°API a light one. Refer to appendix 1.

ARABIAN AMERICAN OIL COMPANY. *See* ARAMCO.

ARABIAN GULF. *See* PERSIAN GULF.

ARABIAN OIL COMPANY. *See* NEUTRAL ZONE.

ARAB LIGHT. A variety of **Saudi Arabian** crude, from such fields as **Abqaiq**, **Ghawar**, Abu Hadriyah, Qatif, and others, long used as a pricing reference by the **Organization of the Petroleum Exporting Countries** (OPEC) in the era of **posted prices**. Arab Light has an **API gravity** of 33.4° and a sulfur content of 1.8 percent on average, making it medium light and sour. In 1987, OPEC adopted a more diversified group of crudes, the **OPEC Basket** (refer to appendix 6), as a pricing reference. At first the basket, of which Arab Light is a component, was used to help establish fixed prices. After 1988, OPEC abandoned the fixed price system in response to the development of a true international free market for oil and used the basket instead for monitoring that market. Arab Light has never been a true **benchmark crude** in the same sense as **Brent Blend**, **West Texas Intermediate**, or **Dubai crude**, because it is not freely and fully traded in a free market as those oils are.

ARAMCO. Arabian American Oil Company; later Saudi Aramco or Saudi Arabian Oil Company. Saudi Aramco, often still referred to simply by its older name Aramco, is the national oil company of **Saudi Arabia** and the world's top oil producer, controlling by far the largest **conventional** oil **reserves**. The company's origins date back to 1933. The **Red Line Agreement** of 1928 prevented many of the major oil companies from seeking independent concessions in Saudi Arabia. Standard Oil Company of California (Socal, later **Chevron**) was not a signatory to that agreement and was already operating in **Bahrain**, where it had set up the Bahrain Petroleum Company (Bapco). Desiring to expand into Saudi Arabia, it obtained a petroleum concession there in May 1933. Socal created a subsidiary, the California Arabian Standard Oil Company (Casoc), to operate the concession. Casoc began exploration in September 1933.

Conditions were extremely difficult, and success eluded the company for years. Socal considered abandoning Saudi Arabia and could not continue without a partner. Since it also needed marketing outlets for the oil it was producing successfully in Bahrain, Socal decided to pool its resources with the Texas Corporation (known before 1926 and after 1941 as the Texas Company, and after 1959 as **Texaco**), which had marketing networks in Africa and Asia but no crude there.

In 1936, Socal sold 50 percent of Casoc and 50 percent of Bapco to the Texas Corporation for $21 million and 50 percent of the Texas Corporation's marketing operations in the Eastern Hemisphere. The marketing operations were made into a subsidiary of Bapco named **Caltex**. Caltex would not become part of Aramco but would stay with Socal and Texaco, eventually becoming part of the merged company ChevronTexaco, later simply Chevron.

On 3 March 1938, Casoc made the first significant oil find in Saudi Arabia with its Dammam No. 7 well, under the leadership of then chief geologist **Max Steineke**. The first Saudi crude was exported from the shipping terminal at **Ras Tanura** on 1 May 1939. Four months later, **World War II** broke out, and Casoc's operations were significantly slowed down. However, Saudi Arabia's immense potential was becoming recognized. U.S. secretary of the interior **Harold Ickes**, head of the **Petroleum Administration for War**, was pessimistic about future domestic petroleum production in the **United States**. He wanted the government to acquire interests in reserves abroad. He sent a team to Saudi Arabia that included the prominent geologist **Everette DeGolyer** to assess its potential. DeGolyer concluded it was huge. There were plans for the **Petroleum Reserves Corporation** (PRC), which Ickes also headed, to purchase a majority interest in Casoc outright. The plans were shelved, but the U.S. government remained very interested in Saudi Arabia and its petroleum potential. Herbert Feis, a government representative who had played a role in the PRC's deliberations and negotiations concerning the purchase of an interest in Casoc, suggested that Socal and the Texas Company (as it was by then again known) change Casoc's name to something more flattering to the Saudis. Casoc became Arabian American Oil Company (Aramco) in January 1944.

In 1948, Aramco discovered **Ghawar**, which would eventually prove to be the world's largest oil field complex. The full scale of Ghawar would not be understood until years later, in 1956. In December 1948, two new partners joined Aramco as co-owners: Standard Oil Company of New Jersey (Jersey Standard, later **Exxon**) and Socony–Vacuum (later **Mobil**), with respectively 30 percent and 10 percent interest. They were able to do so because they were able to withdraw from the relevant provisions of the Red Line Agreement that year, and they were only allowed in once the Saudi king, **Ibn Saud**, had been completely convinced that neither company was under any surreptitious British control. Originally, Jersey Standard and Socony–Vacuum had each been offered 20 percent, but Socony–Vacuum had balked at the magnitude of the investment and reduced its share to 10 percent. Jersey Standard picked up Socony–Vacuum's remaining 10 percent for a total of 30 percent, making it an equal partner with Socal and the Texas Company. The new partners brought significant capital to the company. In 1950, Aramco completed the 1,063-mile **Trans-Arabian Pipeline** (Tapline), linking the oil fields of the Eastern Province to the Mediterranean Sea via Lebanon. Ras Tanura was also significantly developed, and exploration continued. In 1951, Aramco discovered **Safaniya**, which would prove to be the world's largest **offshore** oil field and possibly the third largest overall. In 1961, Aramco began processing and shipping liquefied petroleum gas from Ras Tanura. In 1966, Sea Island, a new offshore crude loading platform, began operating.

Aramco was closely linked with the Saudi Arabian government and generally sensitive to its interests, although it was accused of deliberately obfuscating profits in the 1950s so it could cheat on payments to the government, and it was also not always enlightened in its treatment of its Arab workforce. It followed the rules of the **Organization of Arab Petroleum Exporting Countries** (OAPEC) oil embargo of 1973, even though the United States was one of the targets. The Aramco partners also regularly lobbied on behalf of Saudi Arabia in the United States. Despite the good relationship, though, American companies could not maintain such a controlling role in Saudi Arabia's birthright forever. Taking advantage of the **participation agreement** framework championed by Sheikh **Ahmad Zaki Yamani**, Saudi Arabia acquired 25 percent of Aramco in 1973, increasing its share to 60 percent in 1974. By the end of that year, the Saudis told Aramco that they would not be satisfied with 60 percent, even though this was already greater than the 51 percent target originally conceived in the participation agreement, to be reached supposedly in 1983. The Saudis wanted 100 percent, and in 1976 reached another agreement with Aramco to achieve it. By 1980, Saudi Arabia owned 100 percent of Aramco. This was not a confiscation but an acquisition with compensation to the partner companies at net book value. The partner companies were reluctant sellers, but they were sellers. Links continued between the now Saudi-owned company and the Aramco partners. They continued to provide services to Saudi Arabia and were slated to receive 21 cents a barrel. They also received the right to market 80 percent of Saudi production.

In 1981, the company constructed the east–west pipelines—*see* PETROLINE (EAST–WEST PIPELINE)—for crude oil and **natural gas liquids** from the Eastern Province fields to **Yanbu** on the Red Sea. In 1983, **Ali al-Naimi** became the first Saudi president of Aramco. He held the position until 1995, when he became the Saudi minister of petroleum and mineral resources, and was succeeded by Abdullah S. Jumah. Beginning in 1984, Aramco acquired its own supertankers. By 1995, it completed construction of 15 very large crude carriers (VLCCs; refer to appendix 8).

In 1988, Aramco officially became the Saudi Arabian Oil Company, or Saudi Aramco. In the next year, it discovered oil and gas south of Riyadh. This was the first discovery outside Aramco's original operating area. Saudi Aramco then began to expand its downstream operations overseas. In 1989, it launched a joint refining and marketing venture with Texaco named Star Enterprises. In 1998, it partnered with **Royal Dutch Shell** and Texaco to create Motiva Enterprises, a refining and marketing operation in the southern and eastern United States. In 1993, it acquired 35 percent of South Korean refiner Ssang Yong, and in 1994 acquired 40 percent of Filipino refiner Petron. In 2000, Aramco created a new unit, Aramco Gulf Operations Ltd., to manage the Saudi government's petroleum interests in the **Neutral Zone**. The company continued on major expansion and construction projects. In 2001, it completed the Hawiyah gas plant, with a daily processing capacity of 1.6 Bcf of nonassociated gas, and in 2003 it completed the Haradh gas plant. Since the 1980s, Aramco has executed a program of deliberate Arabization, reducing its dependence on foreign employees. It has trained and employed Saudi citizens and let contracts to Saudi businesses. By 2000, 85 percent of its employees were Saudi nationals.

In 2016, Crown Prince Muhammad Bin Salman announced that Saudi Arabia was considering listing shares in Saudi Aramco. Some informal valuations suggest that Saudi Aramco could be the world's most valuable company. *See also* HOLMES, FRANK (1874–1947); PHILBY, HARRY ST. JOHN (JACK) (1885–1960).

ARBUSTO ENERGY. *See* BUSH, GEORGE W. (1946–).

ARCHBOLD, JOHN D. (1848–1916). Vice president of **Standard Oil**, second only to **John D. Rockefeller** in his influence in the company; first president of Standard Oil Company of New Jersey (later **Exxon**). John Dustin Archbold was born 26 July 1848 in Leesburg, Ohio. His father died when he was 11 years old, and he was forced to start work at an early age. He began working in a store that was close to the new oil regions of **Pennsylvania**, and by 1864 he was working in the oil fields. In 1867, he became a partner in W. H. Abbott's oil company, which soon faced the problem of Standard Oil's predatory pricing. Archbold helped organize the independent oil companies of **Titusville** to resist Standard Oil; eventually, however, he allowed his company to be subsumed by the giant. He operated his company as Acme Oil, but it was owned by Standard. When the Standard Oil trust was formed in 1882, Archbold became a trustee.

Archbold became the second most important person in Standard Oil, after Rockefeller himself, concentrating on improving the efficiency of operations and the distribution of products and acting as the company's major spokesman. After Rockefeller's de facto retirement in 1896, Archbold exercised effective control of the company, although he continued to hold the title of vice president. After the Supreme Court ordered the dissolution of the company in 1911, Archbold became the first president of the most important of the daughter companies, Standard Oil of New Jersey. Archbold remained president of Jersey Standard until his death on 5 December 1916.

ARCO. Atlantic Richfield Company; American integrated oil company headquartered in Los Angeles, **California**; acquired in 2000 by BP Amoco (later **BP**). At the time of its acquisition, Arco had 2.945 Bb in proved oil **reserves**, mostly in the **United States**, and those mostly in **Alaska**. It also had 9.9 Tcf in **natural gas** reserves. In 1999, the year before its acquisition, Arco had a net income of $1.422 billion on total revenues of $13.055 billion and produced 0.625 Mmbpd of oil and 2.378 Bcfpd of gas.

Arco was formed from two different legacies: the **Standard Oil**–affiliated Atlantic Refining Company and the **independent** Richfield Oil Company. Atlantic Refining was founded as the Atlantic Petroleum Storage Company in 1866 by Charles Lockhart and his partners, all veterans of the early **Pennsylvania** oil industry. The name was changed to Atlantic Refining Company in 1870. By 1874, it had come under the expanding umbrella of Standard Oil. It became independent once again in 1911, when Standard Oil was dismantled by order of the U.S. Supreme Court. Atlantic explored for oil in **Iraq** in the 1920s. It is credited with the design of the first all-welded ship. It made several acquisitions in the 1950s and 1960s. The most notable of these was Hondo Oil and Gas in 1963, led by **Robert O. Anderson**, who then became the chief executive officer of Atlantic.

Atlantic expanded on the U.S. West Coast and became an important retailer there. In 1966, Atlantic acquired Richfield Oil Company and changed its name to Atlantic Richfield, or Arco. In 1969, Arco also acquired **Sinclair Oil Corporation**. The 1966 acquisition of Richfield was in many ways to prove the making of the company. Richfield had rights to explore at Prudhoe Bay on Alaska's North Slope. It was there that Arco, partnering with **Humble Oil and Refining Company** (later **Exxon**), discovered the largest accumulation of oil in North America in 1967. The **Prudhoe Bay oil field** began producing oil for shipment in 1977. The oil was far removed from markets, however. Arco had joined with seven other companies to build the **Trans-Alaska Pipeline** to transport oil from Prudhoe Bay to the ice-free southern Alaska port of **Valdez**.

In 1977, Arco bought Anaconda, a uranium and copper mining company, but sold it in 1985. Anderson retired in 1986. In the 1990s, Arco expanded international operations, buying 8 percent of **Lukoil** in 1997 and forming the Lukarco joint venture. It purchased oil assets in Tunisia from **Elf Aquitaine** (later **Total**). In 1998, it bought Union Texas Petroleum for $3.3 billion. By this time, Arco had arguably too many interests worldwide and not enough capital to handle them. Although Arco had a fiercely independent legacy and had previously displayed an unwillingness to be acquired by another company, it allowed itself to be taken over by BP Amoco (later BP). The $27 billion deal was completed in 2000. In order for the deal to be approved by regulators, Arco sold its Alaska interests to **Phillips Petroleum** for $7 billion. *See also* COOK, LODWRICK M. (1928–).

ARCTIC NATIONAL WILDLIFE REFUGE (ANWR). A 19.8-million-acre area of northeastern **Alaska** established as a wildlife refuge by the **United States** in 1980. The Arctic National Wildlife Refuge has been a source of conflict between energy interests and environmentalists. Part of the refuge is thought to have considerable oil and gas potential. A special exemption reserves the right to drill for petroleum in about 1.5 million acres of the refuge near the Beaufort Sea. This area is often called the "1002 area" after the relevant legislative provision. The 1002 area is a critical habitat for a rich biological community. The amount of oil and gas is unknown, but some industry estimates range as high as 16 Bb of oil. Oil found in the 1002 area could be transported via the **Trans-Alaska Pipeline**. Proponents of drilling argue that exploration will take place in winter, when most wildlife is hibernating or absent. Opponents argue that construction, noise, and pollution will permanently damage the environment there. The area's Native Americans have differing viewpoints on ANWR drilling. The Gwich'in, who live in the southern part of the refuge, are generally opposed, fearing detrimental effects on migrating caribou. The coastal Inupiat are generally in favor, as they stand to benefit from oil revenues. Congress passed a budget in 1995 that contained a provision allowing ANWR drilling, but it was vetoed by President Bill Clinton. In 2002, the House of Representatives passed an energy bill allowing ANWR drilling, but the Senate refused to allow it in its own energy bill. The provision was abandoned, despite the support of President **George W. Bush**. As of 2017, there was still no drilling activity in ANWR.

ARCTIC REGIONS. Although there are already significant producing areas north of the Arctic Circle, the Arctic remains largely unexplored. Interest in the Arctic is increasing as the ice pack melts and more of the region becomes accessible. Areas where petroleum production has occurred include **Alaska** (especially the **Prudhoe Bay oil field**, North America's largest), the

Beaufort Sea, the **Barents Sea**, and portions of the **Timan–Pechora** and western **Siberia** areas of **Russia**. Estimates of the potential undiscovered resources of the Arctic range widely and are based on sketchy information. An oft-quoted figure is that the Arctic contains 25 percent of the world's undiscovered **conventional** petroleum resources; however, this figure appears to be based on a misunderstanding of the **United States Geological Survey**'s (USGS) 2000 world petroleum resources assessment. The USGS assessment for the Arctic included the East Siberian basin of Russia, which lies south of the Arctic Circle. Removing that from the assessment would reduce the estimate to 14 percent. On the other hand, the USGS assessment did not include all potential Arctic basins. In 2008 the USGS issued an estimate that the Arctic has 90 Bb of undiscovered technically recoverable oil resources, perhaps 13 percent of the world's undiscovered oil. The USGS also estimated 1,670 Tcf of technically recoverable undiscovered gas.

In 2007, Russia caused a stir by sending a submarine under the Arctic ice cap and planting a flag on the seabed. On 28 May 2008, representatives from Russia, the **United States**, **Norway**, Denmark (of which Greenland is a self-governing territory), and **Canada** signed the Ilulissat Declaration, affirming that the 1982 **United Nations Convention on the Law of the Sea** should govern allocation of Arctic resources. A similar approach has been used in other areas such as the **North Sea**. It is unlikely that the approach will resolve all potential disputes without significant negotiation. Denmark and Russia dispute control of the Lomonosov Ridge. There is also a dispute between Denmark and Canada concerning the exact frontier between Greenland and Canada's Northwest Territories. Both countries claim Hans Island in the Kennedy Channel. It should also be noted that the United States, despite signing the Ilulissat Declaration, has never ratified the Law of the Sea Treaty. *See also* ARCTIC NATIONAL WILDLIFE REFUGE (ANWR).

ARGENTINA. Oil production 2015: 0.532 Mmbpd (0.68 percent of world); **reserves** 2.19 Bb (0.13 percent of world); gas production: 1.278 Tcf (1.02 percent of world); reserves 11.174 Tcf (0.16 percent of world). The first modern commercial discovery of oil in Argentina was in December 1907 by a government team reportedly drilling for water near the southern port city of Comodoro Rivadavia. This first discovery opened up Argentina's only **giant** oil field, **Comodoro Rivadavia**, a mainstay of Argentinean production for decades. President José Figueroa Alcorta declared that ownership of subsoil resources was vested in the state. His government created a state oil commission and exploited the field itself. The Argentine navy managed **tanker** transportation of the field's oil to Buenos Aires, more than 1,000 miles to the north.

On 7 June 1922, President Hipólito Yrigoyen signed a decree creating the world's first vertically integrated national oil company, Yacimientos Petrolíferos Fiscales (YPF). The example of YPF would have an impact reaching far beyond Argentina's borders, particularly in Latin America. In 1935, Argentina promulgated its first oil law, giving provinces the right to administer **concessions** and collect a 12 percent royalty, whether the producer was YPF or a private company. In a move aimed at preventing Standard Oil Company of New Jersey (later **Exxon**) from exerting undue influence, the law restricted provincial ability to grant **pipeline** concessions. In practice, most Argentinean oil production remained under the control of YPF. In 1958, Law No. 14,773 gave the nation exclusive ownership of its oil fields.

In the late 1950s, the government of President Arturo Frondizi, who had once been an outspoken oil nationalist, signed contracts with some foreign companies in order to expedite the development of the national petroleum industry. Argentinean oil production had reached 76,000 bpd in 1943 but had then languished. By 1958 it was only 98,000 bpd. The contracts were of three types. The first, and most controversial, was a set of contracts for foreign companies to drill in areas where oil had already been discovered and partly developed by YPF. The government would buy oil produced by the foreign companies at low prices. Companies in this first category of contracts included **Amoco**, **Cities Service**, and Tennessee Gas. The second category of contracts was that of drilling service contracts. The third category was composed of pure exploration contracts. Companies signing such contracts included Standard Oil Company of New Jersey, **Royal Dutch Shell**, and the **Union Oil Company of California**.

The contracts with foreign oil companies had their intended effect. Output almost tripled from 1958 to 1963 to 0.265 Mmbpd, making Argentina self-sufficient in oil. The output of the contractors rose from zero in 1958 to about 0.081 Mmbpd in 1963, or 47 percent of YPF's output. Nevertheless, the contracts offended nationalist sensibilities. President Frondizi was removed by the military in March 1962. Although his removal was not for petroleum-related reasons, his successor, Arturo Illia, elected in 1963 after a military interregnum, annulled the contracts with foreign companies. The government settled with the various companies over a period of years, eventually paying a total of about $872 million in compensation.

Oil production flattened in the wake of the contract cancellations but began growing again in 1966. It dipped between 1972 and 1976 but reached 0.491 Mmbpd in 1980. It was almost flat for the next decade. In 1991, it was 0.485 Mmbpd. That year, YPF was Argentina's largest enterprise, and many considered it bloated and inefficient. However the government of Carlos Saúl Menem had already begun to deregulate and restructure the petroleum sector. Production rose rapidly, peaking at 0.847 Mmbpd in 1998. After that, it declined unevenly, with another rise after 2000. By 2006, however, it was down significantly to 0.696 Mmbpd, and in 2015 it was even lower at 0.532 Mmbpd.

In 1992, the government began privatizing YPF, selling off assets or allowing private companies to operate them under concession, and raised $1.4 billion. In 1993, the government made a global offering for YPF, raising an additional $3 billion. This was the largest privatization in Latin America to that time. Within three months, YPF shares were trading at a level 50 percent higher than the initial offer price. In January 1999, **Repsol** acquired the Argentine government's remaining 15 percent share in YPF for $2 billion. Repsol paid $38 a share, $10 above current market value. By June 1999, Repsol had acquired a total of 97 percent of the stock for $13.4 billion, paying $44.78 a share (representing a 25 percent premium). Repsol paid about $4.65 per barrel of reserves for YPF, considerably less than the $7.40/b **BP** paid for **Arco** and the $8/b Exxon paid for **Mobil**. In 1998, the last year before Repsol acquired it, YPF controlled 1.517 Bb of oil reserves and 10.387 Tcf of gas, and it produced oil and gas at the rate of 0.518 Mmbpd and 1.421 Bcfpd respectively. By acquiring YPF, Repsol more than tripled its oil reserves and increased its gas reserves more than sevenfold. Repsol renamed itself Repsol YPF, which dominated the oil industry in Argentina, although the sector was still open to other private companies, including foreign ones. Other producers in Argentina included Pan American Energy, **Chevron**, and Petrobrás.

In 2001, the pendulum began swinging back somewhat toward nationalism. Domestic prices were reregulated, and exports were subjected to greater control and taxation. There was a decline in exploration investment, contributing to the decline in production in the 2000 decade mentioned above. In October 2004, the government also formed a new national oil company, Energía Argentina Sociedad Anónima (Enarsa), to promote oil exploration in Argentina. Enarsa signed some joint exploration agreements with foreign companies. In 2012, Argentina expropriated 51 percent of Repsol's shares of YPF. Repsol sued, reaching a settlement in 2014 (*see* REPSOL).

Gas production in Argentina is arguably more significant than oil production. It has risen almost continuously since 1980, when it stood at 0.28 Tcf for the year. The 2015 level of 1.278 Tcf represents roughly 0.620 Mmboepd, greater than oil production that year. Argentina's gas giants include Ramos (**EUR** 3 Tcf), discovered in 1928, and four others. The largest of these is Loma de Lata (8.7 Tcf and 0.208 Bb of condensate), discovered in 1977. Another important field is San Pedrito (5 Tcf and 0.235 Bb condensate), discovered in 1996.

Argentina is thought to have 802 Tcf of technically recoverable resources of **shale gas**, second only to **China** in the world. As of 2017 these resources have not been exploited, but Total announced its intention to invest half a billion dollars over four years to develop shale gas in the Vaca Muerta Formation.

ASIAN PREMIUM. Observed difference between the price of crude oil sold in East Asian markets and that sold in Europe and the **United States**. The United States and Europe have, in practice, a broader choice of suppliers, whereas East Asia is more dependent on the **Persian Gulf**, particularly **Saudi Arabia**. Also, Asian customers have exhibited a greater tendency to worry about supply security and are more willing to pay a premium to ensure it. They typically purchase via term contracts rather than relying on the spot market. The Asian premium has averaged about $1.00–$1.50 a barrel for Arab Light since 1957.

ASIATIC PETROLEUM COMPANY. A joint distribution company for the Far East formed in 1902 by the **Royal Dutch Company**, **Shell Transport and Trading Company**, and the **Rothschilds**. **Henri Deterding** was its managing director.

AS-IS AGREEMENT (1928). Agreement between major Western petroleum companies to limit competition and control the market for oil. The As-Is Agreement is also called the Achnacarry Agreement. **Henri Deterding**, **Walter Teagle**, and **John Cadman**, the leaders respectively of **Royal Dutch Shell**, Standard Oil of New Jersey (later **Exxon**), and the Anglo–Persian Oil Company (APOC, later the **Anglo–Iranian Oil Company** and eventually **BP**), met secretly at Achnacarry Castle in Scotland and agreed to limit production, fix prices, and divide markets among themselves. Even though the agreement, signed 17 September 1928, was concluded two years before the discovery of the **East Texas oil field**, oil was already flooding the market, particularly from the Soviet Union. From the companies' point of view, the overproduction and destructive competition needed to be addressed. Cadman had already approached Teagle earlier in the year and had had a draft agreement produced internally at APOC.

The core of the agreement was that each company would be allocated a quota in various markets. The percentage share of total sales would be based on the share in 1928—thus, things would be left "as is." Markets would be supplied from the nearest geographical source. Prices would be based on the **United States** Gulf Coast price, plus freight to the endpoint. If the actual freight were lower, since markets were supplied from the nearest geographical source, the freight difference would go straight to profit. A year later, the cartel managed to secure the cooperation of the Soviet Union as well. However, the U.S. market remained off limits because of the fear of antitrust prosecution.

Overall, the As-Is Agreement did not work well, at least not for very long. Too many **independent** oil companies were operating outside the agreement, and too many new sources of oil were soon developed, including the East Texas field.

ATHABASCA OIL SANDS (TAR SANDS). *See* ALBERTA OIL SANDS (TAR SANDS).

ATLANTIC REFINING COMPANY. *See* ARCO.

ATLANTIC RICHFIELD COMPANY. *See* ARCO.

AUSTRALIA. Oil production 2015: 0.318 Mmbpd (0.41 percent of world); **reserves** 1.821 Bb (0.11 percent of world); gas production: 2.555 Tcf (2.05 percent of world); reserves 70.23 Tcf (1.02 percent of world). Australia is a net importer of oil. Since 1989, it has been a net exporter of **natural gas**. In 2016 it was the world's second-largest exporter of **liquefied natural gas**, after **Qatar**.

Australia's first **giant** oil discoveries were in the Bass Strait, off the coast of Victoria. This was an area of rough water that had long been deemed unpromising. It was adjacent to an onshore area where 140 dry holes had been drilled in 40 years. However the Standard Oil Company of New Jersey (later **Exxon** and **ExxonMobil**) and BHP (later **BHP Billiton**) concluded a partnership in 1964 and found several major **offshore** fields in the eastern part of the Bass Strait, within about 50 miles of the Victoria shore. In 1965, they found the Barracouta field (**EUR** 0.5 Bb, 1.8 Tcf gas, and 0.14 Bb of condensate). In 1967, they found the Halibut (0.7 Bb) and Kingfish (1.25 Bb) oil fields, and in 1968 they discovered the Snapper gas field (2.4 Tcf). Kingfish, lying under about 250 feet of water, is Australia's largest oil field.

Other than two onshore gas giants near the border of Queensland and South Australia, the rest of the country's major fields—mostly gas—lie along the country's long northwest coast. The most important recent discovery there was Jansz in 2000. This is a gas field with an EUR of about 20 Tcf, discovered by ExxonMobil under up to 4,400 feet of water. There have been no giant oil discoveries since the 1960s, although gas discoveries have been more evenly spread over time: of 26 gas giants, six were discovered in the 2000s, four in the 1990s, and three in the 1980s. In 2004, the Australian government introduced tax incentives to encourage further offshore exploration.

As can be expected from the pattern of discoveries, Australian oil production rose rapidly after 1969, from 43,000 bpd that year to 390,000 bpd in 1973. It then rose unevenly, through several peaks and valleys, to reach 0.772 Mmbpd in 2000. Oil production has been declining since. The 2015 level of 0.318 Mmbpd was lower than the level in 1980. Gas production, on the other hand, has risen fairly steadily since 1980, from 0.313 Tcf that year to 2.555 Tcf, or 1.239 Mmboepd in 2015. In 2015, gas production was nearly quadruple oil production in energy terms.

Australia has significant **oil shale** (not **shale oil**) resources in Queensland, estimated as high as 30 Bb. Southern Pacific Petroleum tried unsuccessfully to develop the resources. Queensland Energy Resources Ltd. (QERL) took over the project and has been assessing its economic viability.

AZERBAIJAN. Oil production 2015: 0.839 Mmbpd (1.07 percent of world); **reserves** 7 Bb (0.43 percent of world); gas production: 0.621 Tcf (0.5 percent of world); reserves 35 Tcf (0.51 percent of world). Azerbaijan has been an important **Caspian** petroleum producer as part of the **Russian** Empire and the Soviet Union, and since 1991 as an independent nation. Petroleum has been known and used in Azerbaijan since ancient times. Azerbaijan's **Baku** region was one of the cradles of the modern oil industry. Led by **Branobel**, the **Rothschilds**, and others, the industry propelled Russia to the status of the second-largest and sometimes the top oil producer in the world in pre-Bolshevik times. The **Baku Oil Society**, formed in 1874, was possibly the world's first vertically integrated oil company. The early modern Azerbaijan industry was also a highly innovative one, spearheading the use of **tankers** and making great progress in storage and **refining**, as well as offshore **drilling** and production in the Caspian Sea.

A number of Azerbaijan's **giant** oil fields were discovered and exploited in tsarist times. **Bibi–Eibat**, with an **EUR** of 2 Bb, was discovered in the modern sense in 1871. However the general Bibi–Eibat area was exploited earlier and was the site of one of the world's first deliberately drilled modern oil wells—*see* FIRST OIL WELL; VOSKOBOINIKOV, NIKOLAI (1801–C. 1861)—in 1846. The Surakhany field, with an EUR of 0.9 Bb, was discovered in 1870. Balakhany, with an EUR of 2.4 Bb, was discovered in 1896. Major discoveries continued in Soviet times, particularly **offshore**: Neft Dashlary, Neftyanyye Kamni (EUR 1.2 Bb), was found in 1949; Azeri (1.25 Bb) was discovered in 1987. A major discovery of post-Soviet times was **BP**'s offshore Shah Deniz field in 1999. This field has an EUR of 22.1 Tcf of gas and 0.8 Bb of condensate, for a total of nearly 4.5 Bboe. Another was **Total**'s offshore discovery of the Abseron gas field in 2011. The field has an estimated 10.6 Tcf.

The Shah Deniz field mentioned above is part of a larger Azeri–Chirag–Guneshli (ACG) complex that may have reserves of 5–7 Bb. The ACG complex is operated by the Azerbaijan International Oil Company (AIOC, not to be confused with the Anglo–Iranian Oil Company). AIOC is a consortium composed of a number of international oil companies. These include BP, **Chevron**, **ExxonMobil**, **Statoil**, **Hess**, and others. The consortium also includes Azerbaijan's postindependence national oil company Socar (State Oil Company of the Azerbaijan Republic). Socar was established in September 1992 via the merger of two government oil companies, Azerineft and Azneftkimiya. Socar is responsible for all phases of the Azerbaijani petroleum industry except for management of the **Baku–Tbilisi–Ceyhan Pipeline** (operated by BP). Socar is a member of all new international consortia developing petroleum resources in Azerbaijan.

Despite the age of Azerbaijan's oil industry and the exhaustion of its older fields, the country continues to be an important oil producer; production has grown since independence from the Soviet Union, particularly since 2004. In 1992, oil production was about 0.213 Mmbpd, about a quarter of 2015 levels. Production declined somewhat during the mid-1990s, reaching a low point of about 0.173 Mmbpd in 1997, but then began to increase with foreign investment. By 2000, production was 0.280 Mmbpd, and by 2004 it reached 0.311 Mmbpd. From 2004 to 2006, it more than doubled, and it exceeded 1 Mmbpd for the first time in 2009 and 2010. That growth was driven largely by growth in the output of the Azeri–Chirag–Guneshli complex. Gas production was 0.275 Tcf in 1992 and decreased over the next decade or so, standing at 0.177 Tcf in 2004. Since 2004, it has increased, exceeding 0.99 Tcf in 2011. *See also* ABSERON PENINSULA (ABSHERON, APSHERON); BAKU–BATUMI PIPELINES; BATUMI (BATUM); MANTASHEV (MANTASHIAN), ALEXANDER (1842–1911); MANTASHEV COMPANY; MAZUT; MENDELEEV, DMITRI (1834–1907); NOBEL, EMANUEL (1859–1932); NOBEL, LUDVIG (1831–1888); NOVOROSSIYSK; PETER I (PETER THE GREAT) (1672–1725); SEMYONOV, VASILI (1801–1863); TAGIYEV, ZEINALABDIN (1823–1924).

AZIENDA GENERALE ITALIANA PETROLI. *See* AGIP.

B

BAHRAIN. Oil production 2015: 0.051 Mmbpd (0.065 percent of world); **reserves** 0.125 Bb (0.01 percent of world); gas production: 0.548 Tcf (0.44 percent of world); reserves 3.25 Tcf (0.05 percent of world). Bahrain was never a member of the **Organization of the Petroleum Exporting Countries** (OPEC) but joined the **Organization of Arab Petroleum Exporting Countries** (OAPEC) in 1970. The Standard Oil Company of California (Socal, later **Chevron**), which was not bound by the **Red Line Agreement**, formed the Bahrain Petroleum Company (Bapco) in 1929 and began operating in Bahrain, taking over the interests of Red Line signatory **Gulf Oil Corporation**. Bapco was registered in **Canada** so as to comply with the **British Nationality Clause**. It discovered the **giant** Awali (Bahrain) field (**EUR** 0.9 Bbo and 6.8 Tcf gas) in 1932 and began commercial production. In 1936, Socal entered into a joint agreement with the Texas Corporation (later **Texaco**), by which the latter company acquired half of Bapco—*see* CALTEX (CALIFORNIA–TEXAS COMPANY).

Oil production in Bahrain rose from minimal levels in the early 1930s to a **Hubbert**-like peak of 0.0766 Mmbpd in 1970, after which it declined. In 2006, production was 0.035 Mmbpd, about the level in 1957. It rose somewhat to 0.051 Mmbpd in 2015. Gas production has been on a long-term growth trend since 1980, rising from 0.1 Tcf that year to 0.548 Tcf in 2015. Bahrain's gas production in 2015 was equivalent in energy terms to 0.266 Mmboepd—significantly greater than its oil production.

In 1975, Bahrain acquired more than 60 percent of Bapco, procuring the rest in 1980. It founded the Bahrain National Oil Company (Banoco) in 1976 to control the petroleum sector. It bought out **Caltex**'s remaining share in the country's industry in 1997 and in 1998 merged Bapco and Banoco into a single entity, the Bahrain Petroleum Company—also commonly known as Bapco.

Bahrain has been a minor petroleum producer. Its greatest historical significance lies in having been the entry point for Socal into the **Persian Gulf** region. From its springboard in Bahrain, Socal moved into **Saudi Arabia**, eventually forming the company that became **Aramco**. *See also* HOLMES, FRANK (1874–1947).

BAKER HUGHES INC. (Later part of General Electric.) Oil field services company formed from the 1987 merger of Baker International (formerly **Baker Oil Tools**) and **Hughes Tool Company**. The merged company was organized in three major groups: **drilling** equipment, production tools, and mining equipment. The Reed Tool Company, a drill-bit manufacturer acquired by Baker in 1975, had to be divested because of antitrust concerns in the drilling equipment segment. The company consolidated significantly after the merger, cutting $90 million in annual costs by 1988. It sold its mining equipment business in 1989 for $155 million and acquired Eastman Christensen, a maker of horizontal and directional drilling equipment, for $550 million in 1990. In 1998, Baker Hughes acquired Western Atlas, which had been spun off by Litton Industries, for $3.3 billion in stock and the assumption of $1.3 billion in debt. Further consolidation followed. In 2000, Baker Hughes and **Schlumberger** combined their seismic services in a joint venture, **WesternGeco** (acquired in toto by Schlumberger in 2006). Baker Hughes's reputation suffered in 2000–2001 when the SEC accused it of bribing an Indonesian government official. The company settled with the U.S. Securities and Exchange Commission (SEC) in 2001 without admitting guilt. In March 2002, a former employee sued Baker Hughes, claiming that he had been dismissed in 2001 for refusing to bribe a **Nigerian** official. In 2014 Baker Hughes and Halliburton announced an intention to merge, but the plans were abandoned in 2016. In 2017 Baker Hughes was acquired by General Electric.

BAKER INTERNATIONAL. *See* BAKER OIL TOOLS.

BAKER OIL TOOLS. Oil field services company later known as Baker International and **Baker Hughes**. Baker was founded in Coalinga, **California**, by Reuben Carlton "Carl" Baker in 1913 as the Baker Casing Shoe Company. The Baker casing shoe was a device Baker had patented on 16 July 1907. It improved the driving of cable tool casing down a borehole. Baker had licensed the invention to other manufacturers and founded his company initially to protect his patents and collect royalties. In 1912, Baker also patented another important invention, the Baker cement retainer. In the 1920s, the company began manufacturing its own tools in Huntington Park, California. In 1928, it changed its name to Baker Oil Tools. It suffered during the Great Depression but prospered thereafter. In 1942, motivated partly by a wartime steel shortage, it introduced the Model D Packer, a device enabling multiple completions in the same well. The Model D Packer was still sold in 2017. Carl Baker retired in 1956, dying the next year.

After Baker's death, the company expanded overseas, and it went public in 1961. In 1965, E. H. Clark became the chief executive of Baker. He continued the trend of international diversification and also set about acquiring a number of oil field service companies. In 1975, Baker merged with Reed Tool Company, a drill-bit manufacturer, and in 1976 changed its name to Baker International. Baker suffered during the oil glut of the 1980s (*see* OIL COUNTERSHOCK). In 1986, its revenues declined 18 percent. On 3 April 1987, it merged with **Hughes Tool Company** to form **Baker Hughes Inc.**

BAKKEN FORMATION. Shale and other rock formation in the U.S. states of **North Dakota** and Montana, extending into the **Canadian** provinces of **Saskatchewan** and Manitoba. The first discovery in the Bakken was in 1951, but large-scale development took place after the application of **fracking** and **horizontal drilling** technology.

The Bakken produced about 20,000 barrels of **tight oil** per day in 2000, rising explosively to over a million barrels per day in 2014. Production was still over a million barrels per day in 2016 but was lower than 2015 production, leading to some concerns about the future. Production from the Bakken Formation catapulted North Dakota into the front rank of oil-producing states in the **United States**, from producing under 2 percent of the U.S. total in 2005 to nearly 12 percent in 2016, behind only **Texas** and the U.S. federal **offshore** area in the **Gulf of Mexico**.

BAKU. Capital city (pop. 3.2 million, 2012) of **Azerbaijan** and site of one of the world's earliest oil industries. Baku lies on the south shore of the **Abseron Peninsula**, which extends for 37 miles into the **Caspian Sea**. Today the 11 districts of Greater Baku comprise nearly the entire peninsula, as well as some islands off the tip of the peninsula. The exploitation of oil from surface seeps in the Baku region extends back to ancient times, and hand-dug wells for extracting oil date back to at least the Middle Ages. An important early oil well was drilled here in 1846 in the **Bibi–Eibat** area by a group under the command of Major Alexeev of the **Russian** tsar's Corps of Mining Engineers, 13 years before **Drake**'s well in **Titusville**, **Pennsylvania**—*see* FIRST OIL WELL; VOSKOBOINIKOV, NIKOLAI (1801–C. 1861). Baku was the major center of the early Russian–Azerbaijani oil industry in the 19th and early 20th centuries. In the beginning of the 20th century, the Baku oil region was the world's largest, and it remained the largest in the Soviet Union until the 1940s. Baku was a site of communist agitation and significant unrest in the early 1900s, with a general strike of the oil workers in December 1904 and riots in February 1905 that led to large-scale arson in the oil fields.

Baku oil was historically shipped by **tanker** across the Caspian and up the Volga River. It was also exported from the Black Sea port of **Batumi**, which it reached by rail or via a **pipeline** (for kerosene) completed in 1906 (*see* BAKU–BATUMI PIPELINES). A pipeline for crude oil was completed in 1930. The Western Early Oil Pipeline, completed in 1999, takes oil from Baku to the port of **Supsa**, about 30 miles north of Batumi. The Northern Early Oil Pipeline runs from Baku to the Russian Black Sea port of **Novorossiysk**. However, the most important new pipeline out of Baku is the **Baku–Tbilisi–Ceyhan** (BTC), which takes Caspian crude to the southern Mediterranean coast of Turkey and bypasses the Black Sea and the **Bosporus**. *See also* ALEXANDER II (1818–1881); BAKU OIL SOCIETY; BRANOBEL; MANTASHEV (MANTASHIAN), ALEXANDER (1842–1911); MANTASHEV COMPANY; MAZUT; MENDELEEV, DMITRI (1834–1907); NOBEL, EMANUEL (1859–1932); NOBEL, LUDVIG (1831–1888); ROTHSCHILD COMPANY; SEMYONOV, VASILI (1801–1863); TAGIYEV, ZEINALABDIN (1823–1924).

BAKU–BATUMI PIPELINES. Pipelines from **Baku** to the Black Sea port of **Batumi**. One was an early kerosene pipeline built by an alliance of business executives in the **Russian**/Azeri oil industry, including at various times the **Nobel** brothers, the **Rothschilds**, **Alexander Mantashev**, and **Zeinalabdin Tagiyev**. Another was a later Soviet pipeline between the two cities. Batumi gave Baku oil more direct access to European markets, and the existing mode of rail transportation was inefficient.

The idea of a pipeline to transport refined kerosene between Baku and Batumi was discussed and debated in the 1880s and championed by **Dmitri Mendeleev**. There was stiff opposition to the pipeline from individuals and organizations that were earning considerable fees—legal and otherwise—from the transport of oil via the Baku–Batumi railroad. Eventually a compromise was reached whereby the pipeline would coexist with the railroad. Construction began in earnest in 1896 and finished in 1906. The completed pipeline had an 8-inch diameter and was 522 miles long, with 16 pumping stations driven by steam and diesel engines.

On 13 February 1930, the Soviet Union opened a crude oil pipeline between Baku and Batumi. At about 514 miles, it was the world's longest crude pipeline at the time. The pipeline incorporated a number of ideas from the 1870s developed by Russian engineer Sigismund Wojslaw. For example, it eliminated intermediate tanks at pumping stations, a source of loss of light hydrocarbons. The arrangement of pumping stations was also optimized.

BAKU OIL SOCIETY. Possibly the world's first vertically integrated oil company. The Baku Oil Society was founded in 1874 by Vasily Kokarev and Pyotr Gubonin. At this time, **Standard Oil** in the **United States** was integrating both horizontally and vertically, but not yet completely into upstream operations; it was a refiner and marketer. The Baku Oil Society's Antecedent was the Trans-Caspian Trade Partnership, in which Kokarev and Gubonin were also principal shareholders. The Trans-Caspian Trade Partnership, founded in 1865, established and ran a refinery at Surakhany near **Baku**. After Tsar **Alexander II** promulgated new rules for the oil industry in 1872, auctioning leases to private investors, Kokarev and Gubonin deliberately set out to form a joint stock company that would cover the full spectrum of oil operations from exploration and production to refining and marketing. The initial capitalization of the new company was 2.5 million rubles.

The Baku Oil Society had oil leases in the Balakhany area and elsewhere acquired by the Trans-Caspian Trade Partnership, as well as the refinery and a fleet of schooners and barges. It had marketing outlets in Baku, Moscow, Saratov, Samara, Tsaritsyn, Nizhni Novogorod, Astrakhan, and a number of other cities. By 1876, the Baku Oil Society accounted for more than 32 percent of **Russia**'s oil production and was the country's leading oil company. In 1879, the Baku Oil Society completed a kerosene pipeline from the Surakhany refinery to the Zykh wharf on the **Caspian Sea**. It also commissioned an oil **tanker**, *Surakhany*, from the Swedish Abo shipyard.

Beginning in 1879, the Baku Oil Society began to face formidable competition, first from **Branobel** and later from the **Rothschilds**, as well as other companies. Gradually, it lost its leadership position, despite continuing to increase its oil production. By 1899, its production of roughly 7,500 bpd was only about a quarter of Branobel's figure. In the early 1900s, the Baku Oil Society tried to recapture its leadership of the Russian oil industry, but without success. On 1 June 1918, all its assets were nationalized by the Soviets.

BAKU–TBILISI–CEYHAN (BTC) PIPELINE. Major **pipeline** carrying **Caspian** oil to the Mediterranean, with a maximum capacity of 1 Mmbpd. The pipeline runs northwest from **Baku** in **Azerbaijan** to Tbilisi in Georgia, then turning eventually southwest to reach the port of **Ceyhan** on Turkey's southern Mediterranean coast. At 1,099 miles, the BTC Pipeline is the second-longest oil pipeline in the world, after the **Druzhba Pipeline**. The motivation behind the pipeline was to find a way to transport Caspian oil to Western markets that avoided the Black Sea and the **Bosporus**, and did not necessitate crossing **Iranian** or **Russian** territory.

The decision to build the pipeline was reached at the Istanbul meeting of the Organization for Security and Co-operation in Europe (OSCE) on 18 November 1999. Construction began in September 2002 and was completed in May 2005. The pipeline represents a considerable engineering achievement. It crosses mountain ranges nearly 9,300 feet high as well as rivers, railways, and utility lines. The pipeline is completely underground. Its diameter is 42.1 inches for most of its length; it narrows to 34 inches as it approaches Ceyhan. The first oil was pumped from Baku on 10 May 2005 and reached Ceyhan on 28 May 2006, where it was loaded onto the ship *British Hawthorn*. The pipeline is operated by **BP** and is owned by a consortium of oil companies led by BP. BP has a 30.1 percent ownership share, and Socar (the national oil company of Azerbaijan) has 25 percent. **Chevron** has 8.9 percent and **Statoil** has 8.71 percent. The remaining partners all have smaller shares.

BARD, THOMAS ROBERT. *See* UNION OIL COMPANY OF CALIFORNIA.

BARENTS SEA. In 2017 the **Norwegian Petroleum Directorate** estimated that the Barents Sea holds a possible 17.6 Bb of oil equivalent. The sea has been explored to a degree by both **Norway** and **Russia**. The two countries concluded a treaty, which entered into force 7 July 2011, delimiting their respective zones in the sea. Both countries have discovered large fields, mostly **natural gas**. The sea is an outlying portion of the Arctic Ocean covering a relatively shallow continental shelf. The average depth is about 750 feet. Norway's Snøhvit complex, discovered in 1984–1988 in up to 1,100 feet of water and operated by **Statoil**, has an **EUR** of 4 Tcf of gas and about 80 Mmb of oil and condensate. In 2000, a consortium led by **ENI** discovered the Norwegian Goliat field, about 30 miles southeast of Snøhvit. Unlike other Barents fields, Goliat is primarily an oil field with about 179 Mmb. The Soviet Union's first discovery was the Murmansk gas field in 1984. In 1988, the Soviets (Sevmorneftegaz) discovered the huge Shtokman field, with an EUR of 60 Tcf of gas and about 100 Mmb of condensate (some estimates range to 100 Tcf), under more than 1,100 feet of water. Other gas finds include Ludlov (1990) and Ledovoye-Barents (1991). *See also* ARCTIC REGIONS; TIMAN–PECHORA.

BARNETT SHALE. Formation in **Texas** underlying, among others, the city of Fort Worth, with estimated technically recoverable resources of **natural gas** of about 13 Tcf. The Barnett is significant because it was a crucible for the development of modern **fracking** and **horizontal drilling** techniques to extract **shale gas**—*see* MITCHELL, GEORGE P. (1919–2013). In 2015 the province produced 3.23 Bcfpd of gas.

BARREL. Standard unit of volume in the petroleum industry. A barrel of petroleum is equivalent to 42 U.S. gallons. (This is different from the accepted definitions outside the petroleum industry. A standard U.S. barrel for liquid goods is 31.5 U.S. gallons, and a British one is 36 imperial gallons.) The standard petroleum barrel originated in the **Pennsylvania** oil fields in the 19th century. Barrels were originally 40 gallons and then grew to 42. The common abbreviation for barrel as well as barrels is *bbl*, or *b* when used in combination with other measures, as in *bpd* (barrels per day). Some investigators have reported that *bbl* stands for "blue barrel," describing the particular barrels used by **Standard Oil** in the 19th century. However, it appears that this explanation may be apocryphal. The origin of *bbl* is not clear, but it was used much earlier than the beginnings of the Pennsylvania oil industry. Refer to appendix 1.

BASRA. City (pop. 2.15 million, 2017) in southern **Iraq**; Iraq's principal port and a major center of the country's petroleum industry. Basra has a refinery with a capacity of about 140,000 bpd, down from about 200,000 bpd before the **Iraq War**. The city lies on the west bank of the Shatt al Arab, about 70 miles from the **Persian Gulf**. The area around Basra is a prolific oil-producing region. There are 11 **giant** fields within about 60 miles of the city. The first giant discovery was Nahr Umr (6.5 Bb and 9.9 Tcf gas) in 1948, followed by Zubair (8.2 Bb and 5.9 Tcf gas) in 1949. The largest discovery was the **supergiant** Rumaila North and South, with 30 Bb and 20 Tcf of gas, for a total of 33.3 Bboe, in 1953. Discoveries continued through the succeeding decades. The most recent giant discovery near Basra was the Subba field in 1989.

The British occupied Basra in **World War I**, and during **World War II** the Allied powers used Basra to help supply the Soviet Union. Both Basra and its oil facilities have been ravaged by wars since 1980. During the **Iran–Iraq War**, the refinery sustained heavy damage. The Iranians advanced to within six miles of the city in 1987 and destroyed many buildings in Basra with their artillery. Basra also suffered considerable damage in 1991 during the **Persian Gulf War**. There was further destruction after the war as a Shia revolt against **Saddam Hussein** was brutally subdued. The area near Basra also saw heavy fighting between March and May 2003, during the beginning of the Iraq War. British forces took the city on 6 April 2003. Basra was plagued by internecine violence during and after the Iraq War.

BASRA PETROLEUM COMPANY (BPC). *See* IRAQ PETROLEUM COMPANY.

BATAAFSCHEPETROLEUM MAATSCHAPPIJ. Batavian (Dutch) Petroleum Company; holding company of the amalgamated **Royal Dutch Shell** Group, created in 1907. It was incorporated under Dutch law. **Shell Transport and Trading Company** owned 40 percent of the shares, and **Royal Dutch Company** owned 60 percent.

BATUMI (BATUM). Port (pop. 153,000, 2014) on the Black Sea coast of Georgia, about 12 miles from the Turkish border. Batumi is the capital of the Adjara autonomous republic in Georgia. The city has an oil refinery and oil export terminal. In 2003, about 72 Mmbo were transshipped through Batumi. Batumi was a major port in the early **Russian**/Azeri oil industry, offering a window to the Mediterranean via the Black Sea and **Bosporus**. It was the terminus for an early kerosene **pipeline** from **Baku** completed in 1906, built close to the Transcaucasian railroad line (*see* BAKU–BATUMI PIPELINES). During the 2000s, oil was transported to Batumi by rail, and only from Baku. The **Baku–Tbilisi–Ceyhan Pipeline**, completed in 2005, has decreased Batumi's importance. A modern pipeline from Baku (the Western Early Oil or Baku–Supsa Pipeline, completed in 1999) does not go to Batumi but rather to the port of **Supsa**, about 30 miles to the north. In 2003, about 45 Mmbo were shipped through Supsa.

BATUMI OIL PRODUCTION AND TRADING COMPANY. Oil company founded on 13 July 1883 by Sergei Palashkovsky and Andrei Bunge; by 1885 it came under the control of the **Rothschilds**, giving them their toehold in the **Russian** petroleum industry. The company's Russian name, Batum Neftepromushlenoe i Torgovoe Obchshestvo, gave rise to the acronym BNITO, often used to refer to the Rothschilds' subsequent company. Palashkovsky and Bunge were railroad engineers who had supervised the construction of the Baku–Tbilisi–Batumi railway. They acquired tank cars, built storage tanks, and concentrated on transporting crude oil and petroleum products from the oil fields of the **Baku** area to the Black Sea port of **Batumi**. In 1884, they handled 44.5 percent of all oil and petroleum products exported from Batumi. That year, they also began to move upstream, drilling in the Balakhany area near Baku. Their ambitious expansion necessitated more capital, and they were authorized in November 1884 by Tsar Alexander III to issue more bonds. The new bonds ended up largely in the hands of the Paris Rothschilds, who soon after acquired all the company's indebtedness and effectively took over the business.

BEARSTED, VISCOUNT. *See* SAMUEL, MARCUS (1853–1927).

BEATON, RALPH. *See* CORSICANA OIL FIELD.

BEATY, AMOS (1870–1939). President of **Texaco** 1920–1926; president of the **American Petroleum Institute** in the early 1930s. Amos Leonidas Beaty was born in Red River County, **Texas**, on 1 September 1870. He was admitted to the Texas bar in 1891 at the age of 21. He had not attended a formal law school but had read law in a law office in Clarksville. He became a junior partner in a law office in Sherman and handled a number of high-profile oil industry cases. He became president of the Texas Bar Association in 1906 and a year later joined the Texas Company (later **Texaco**) as its attorney. He helped create the legal environment for the Texas Company to operate in **Oklahoma**.

In 1920, Beaty became president of the Texas Company. In that role, he expanded the company's downstream operations, focusing on **gasoline** production and marketing. He also successfully defended the Texas Company in a federal lawsuit charging restraint of trade. In 1926, he became chairman of the board but resigned in a dispute a year later. In 1929, he became chairman of Transcontinental Oil Company, and in 1931 he joined the executive committee of **Phillips Petroleum**.

Beaty was active in the American Petroleum Institute (API) from the time of its founding in 1919. In 1925, he led an API committee to reduce waste, in cooperation with the Oil Conservation Board. In 1926, he became treasurer of the API and in 1931 was elected president. Although he was generally antagonistic to government regulation, he advocated it as necessary to control the severe overproduction of the early 1930s, in the wake of the discovery of the **supergiant East Texas oil field**. This did not endear him to his API constituency, and he was not reelected as president. In 1933, he was appointed to the U.S. government's petroleum planning and coordination committee, again focusing on controlling overproduction. In the last years of his life, he was president of his own Amos Beaty Oil Company. He died 27 April 1939.

BENCHMARK CRUDE. A benchmark crude, or marker crude, is a crude oil that is widely accepted as a pricing reference. Other crudes trade at a discount or premium with respect to the price of the benchmark. Benchmarks serve to simplify the negotiation process, which could otherwise be cumbersome, given the wide range in the quality of crude oils of different provenance available on the market. Arguably, the first benchmark crude was **Pennsylvania** crude oil, traded at the **Oil City Oil Exchange** beginning in 1874.

Examples of modern benchmark crudes are **Brent Blend**, **Dubai crude** (also called Fateh crude), and **West Texas Intermediate** (WTI). Brent Blend is used to price **North Sea**, African, **Russian**, and **Caspian**/Central Asian oil, as well as **Persian Gulf** oil headed for Europe. It has served as a de facto international standard. Dubai crude has been used as a benchmark for oil exports from the Persian Gulf to Asia. WTI is used to price domestic crudes in the **United States** as well as imports into the United States.

The use of benchmarks began in earnest after the oil price crash of 1985–1986. In the postwar era up until the mid-1980s, oil was sold at fixed, **posted prices** determined first by **major oil companies** and then by the **Organization of the Petroleum Exporting Countries** (OPEC). OPEC used **Arab Light** crude as its reference for pricing crude. OPEC members negotiated with each other to determine the premium or discount for their own crudes with respect to Arab Light. However, Arab Light was not a benchmark in the same sense that Brent Blend is today, because it was not fully traded in a free market.

In the 1980s, non-OPEC producers competed increasingly with OPEC and traded their crudes on the spot market, without fixed prices. In 1985, **Saudi Arabia** began to implement a **netback pricing** system in partial response to market pressures. In this system, it priced crude oil based on its value relative to refined products. In effect, this guaranteed a margin to refiners and encouraged purchases. This put further downward pressure on prices. Although Saudi Arabia gained back market share, prices went into freefall. In 1986, prices averaged $14.32 a barrel, down from $27.53 a year earlier. Saudi Arabia was forced to give up netback pricing and moved back to fixed official prices (*see* OIL COUNTERSHOCK).

However the market had changed forever and fixed prices were unsustainable. By the fall of 1987, freely traded benchmarks began to be widely used in pricing. Brent Blend became gradually established in Europe, and Asia began using Dubai crude as noted above. In the United States, pricing was done initially using **Alaska** North Slope oil. However, the benchmark was changed to West Texas Intermediate in 1991. Falling output in Alaska meant that most North Slope oil was going to the West Coast of the United States and hence that the crude was not widely traded.

This points out a serious problem with benchmarks. The suitability of a benchmark is determined largely by its tradability and price transparency, not necessarily by its production volume. However, if production falls below a minimum threshold, the market in the benchmark becomes subject to distortion, and this undermines the benchmark's credibility and suitability. Such a distortion was observed in 1999 when the price of Dubai crude rose higher than that of **Oman** crude. Normally this should not occur, because Oman crude is lighter and sweeter than Dubai crude. Distortions were bound to occur, however, because of Dubai crude's falling production. From 1990 to 1995, production of Dubai crude averaged about 400,000 bpd. By 2004, it was about 120,000 bpd and fell even further thereafter. Oman crude was added to the Dubai benchmark in 2001. Production of West Texas Intermediate and Brent Blend is also falling, and eventually new benchmarks may need to be adopted. WTI showed itself vulnerable to distortions in 2007, when a surplus of WTI at the delivery hub of **Cushing, Oklahoma**, caused its price to fall to artificially low levels. There were other pricing distortions between 2011 and 2013, when WTI traded at an unusually cheap price compared to Brent Blend.

Russian Urals crude has been suggested as a possible future benchmark (*see* VOLGA–URALS). However, there are problems with this choice. For example, a significant quantity of Urals crude is exported through the **Bosporus**, a major choke point whose deliberate or accidental closure could have a significant extraordinary effect on the price of Urals crude; this would be highly troublesome for world markets if Urals were a benchmark.

BEREZOVSKY, BORIS (1946–2013). A Russian oligarch who once controlled Sibneft (later **Gazprom Neft**). Boris Abramovich Berezovsky was born in Moscow on 23 January 1946. He studied electronics and computer science and obtained a doctorate in decision-making theory in 1983. He established his business empire in the waning years of the Soviet Union. In 1989, he set up Logovaz, the country's first capitalist car dealership. He bought Soviet-made cars at low prices set for export and sold them internally at the higher prices they could fetch.

After the breakup of the Soviet Union, Berezovsky was well connected in **Russia**'s government led by Boris Yeltsin and is reported to have helped frame the **Loans for Shares** scheme that allowed a small group of oligarchs to acquire former Soviet state enterprises at rock-bottom prices. Under the scheme, he gained control of **Sibneft**, formed by combining Noyabrskneftegaz with the Omsk refinery. The paperwork to create the new company was allegedly pushed through in only three days in November 1995. Berezovsky acquired 51 percent of Sibneft for $100.9 million, which worked out to about 6 cents per barrel in the ground at a time when in-ground crude was valued at between $1 and $6 per barrel, and **Brent Blend** was selling for $17 a barrel. Sibneft floundered under Berezovsky's control. The company did better after 2000, when Berezovsky sold his interest to **Roman Abramovich**, who had been his partner in the original acquisition.

Berezovsky had many other business interests, including a media empire that he put at the service of Boris Yeltsin's reelection in 1996. But by the year 2000, Berezovsky had fallen afoul of the new government of Vladimir Putin. While not necessarily opposed to the oligarchs as a class, the government appears to have taken a dim view of individual oligarchs exercising too much political power. Berezovsky went into exile in the **United Kingdom**. In 2011, he still had an estimated net worth of nearly $1 billion, according to *Forbes* magazine. In 2012 Berezovsky lost a case he brought against Roman

Abramovich over the ownership of Sibneft. Berezovsky was found dead in his home in England on 23 March 2013. A coroner's inquest ultimately issued an open verdict.

BETANCOURT, RÓMULO (1908–1981). President of **Venezuela** 1945–1948 and 1958–1964; a reformer whose administration oversaw a major improvement of Venezuela's terms of business with the **major oil companies**. Rómulo Betancourt was born 22 February 1908 in eastern Venezuela, in modest circumstances. In 1928, he was one of the leaders of the failed student revolution against the dictator **Juan Vicente Gómez**. After a short jail term, Betancourt went into exile in **Colombia**, traveling around Latin America and flirting briefly with communism, which he later repudiated. He returned to Venezuela after Gómez's death in December 1935.

Betancourt led the effort to organize a democratic party. Venezuela had a military government at the time, and his political activities soon resulted again in exile in 1939. He returned in 1941 after the democratic party, now known as Acción Democrática (AD), was legalized. When the erstwhile president Isaías Medina Angarita reneged on a promise to allow free elections, the AD and some military leaders mounted a successful coup, and Rómulo Betancourt became provisional president of Venezuela in 1945. He instituted a number of reforms, including rent controls.

Minister of Production **Juan Pablo Pérez Alfonso** obtained major improvements in terms from the international oil companies in Venezuela. Following the law of 12 November 1948, the companies would pay 50 percent of their profits to the Venezuelan government. This **50–50 agreement** set a new standard, and some important oil producers of the Middle East, including **Saudi Arabia**, soon demanded and received equally favorable terms from the major oil companies. Betancourt established the Venezuelan Development Corporation and through it used the increased oil revenues to build schools and hospitals and to undertake other projects for the public benefit.

Betancourt's first presidency may have begun with a coup, but it ended after free elections in 1947; he was succeeded by Rómulo Gallegos in 1948. Unfortunately, there was a military coup that same year. Betancourt found himself exiled again, working to oppose the military dictatorship. It was overthrown in 1958, and Betancourt was elected that year to a second term as president. He undertook further reforms, including the distribution of land to families and nurturing organized labor. He also built roads, power stations, and other infrastructure, effectively modernizing the country. It was during Betancourt's second administration that Venezuela, particularly Minister of Mines and Hydrocarbons Juan Pablo Pérez Alfonso, took a leading role in the formation of the **Organization of the Petroleum Exporting Countries** (OPEC). Betancourt's writings were considered influential in the creation of that organization. Long established as an author, Betancourt wrote several books after his second presidency. Some were on oil and were translated into English, including *Venezuela's Oil* (1978) and *Venezuela: Oil and Politics* (1979).

BHP BILLITON. A diversified resources company formed from the merger of **Australia**'s BHP (Broken Hill Proprietary) and the UK-based Billiton mining group. Broken Hill Proprietary was established in 1885 and concentrated on mining, steel, and related businesses. It did not enter the petroleum sector until the 1960s. In 1964, it entered a 50–50 partnership with Esso Standard, the Australian subsidiary of the Standard Oil Company of New Jersey (later **Exxon** and **ExxonMobil**). The partnership discovered major oil and gas fields in the Bass Strait off the coast of Victoria. From 1986 to 1988, BHP acquired Monsanto Oil, Hamilton Oil, and Gulf Energy Development, extending its Pacific interests and moving into the **North Sea**. By 1992, it ranked 10th among the world's oil companies. Its position slipped during the 1990s, and in 2004 it made a renewed push. It acquired Atlantis, based in the **Gulf of Mexico**, for $1 billion and also moved to acquire smaller producers in Australia. During the U.S. boom in **shale gas** and **shale oil**, BHP Billiton acquired assets in the various shale plays including the Haynesville, Eagle Ford, and Fayetteville. In 2017 it announced its intention to sell those assets, while retaining interests in conventional oil and gas in the Gulf of Mexico, Australia, and elsewhere. It also decided to change its name back to BHP.

BIBI–EIBAT OIL FIELD. Early **giant** oil field in an area of the same name in greater **Baku**, **Azerbaijan**, covering 2,125 acres of land and 1,137 acres of water surface. The year of discovery is usually given as 1871, and the first gusher was in 1878. A huge gusher on the property of **Zeinalabdin Tagiyev** was drilled in 1886. It blew uncontrolled for more than two weeks, drenching the area and wasting about 1.4 Mmbo. Exploration activity in the Bibi–Eibat area actually began much earlier. An important early oil well was drilled in 1846 by a group under the command of Major Alexeev of the Russian tsar's Corps of Mining Engineers, 13 years before **Edwin Drake**'s well in **Titusville, Pennsylvania**—*see* FIRST OIL WELL; VOSKOBOINIKOV, NIKOLAI (1801–C. 1861). However, this is not considered the discovery well of the Bibi–Eibat oil field. Exploitation of **offshore** oil seeps in Bibi–Eibat may have occurred as early as 1798. Overwater oil drilling began in 1877. By 1924, the Bibi–Eibat offshore operation was relatively sophisticated, with drilling and production in waters as deep as about 250 feet. The wells were erected on timber piled in Bibi–Eibat Bay. The Bibi–Eibat field has an **EUR** of about 2 Bb. Estimated remaining recoverable resources range from near 0 to about 300 Mmbo.

BIG INCH AND LITTLE BIG INCH. Two **pipelines** carrying oil from **Texas** to the Northeast, built by the **United States** during **World War II** to help avoid German submarine attacks on oil **tankers**. W. Alton Jones, president of **Cities Service Company** and a strong advocate of the project, observed that no one ever sank a pipeline; his company, like many others, lost tankers during the war. Jones served as president of the War Emergency Pipelines organization.

Big Inch, a 1,811-mile crude oil pipeline, was 24 inches in diameter and was designed for a capacity of 300,000 bopd. The companion Little Big Inch product pipeline was 20 inches in diameter and was designed to transport 190,000 bpd of refined petroleum products. Pipeline construction was urged strongly in 1940 and 1941 by Secretary of the Interior **Harold Ickes**, petroleum coordinator for national defense and later head of the **Petroleum Administration for War**. Ickes asked the Federal Allocation Board for steel to build the pipelines but was turned down in September 1941 and then again in November. After the attack on Pearl Harbor, Ickes requested steel from the War Production Board but was again rejected.

Finally the War Production Board approved the first leg of Big Inch from East Texas to Norris City in southern Illinois on 10 June 1941. Construction began on 3 August 1942, and the second leg to Phoenixville, **Pennsylvania**, was approved 26 October. First oil flowed in Norris City on 13 February 1943. The third and final leg, 20 inches in diameter, was completed to Philadelphia and New York City on 14 August 1943. Little Big Inch was approved in early 1943 and completed on 2 March 1944. Little Big Inch ran from the refinery complex between **Houston** and Port Arthur, Texas, to Linden, New Jersey. The total cost of the two pipelines was $146 million.

The U.S. government sold the pipelines on 14 November 1947 to the Texas Eastern Transmission Corporation (TETCO) for a bit over $143 million. TETCO converted the pipelines to transport natural gas. Later TETCO expanded both pipelines and converted Little Big Inch back to petroleum products. Little Big Inch was then operated by a different company, the Texas Eastern Products Pipeline Corporation (TEPPCO). TEPPCO and TETCO later came under the umbrella of Duke Energy. *See also* EAST TEXAS OIL FIELD.

BISSELL, GEORGE HENRY (1821–1884). Founder of the **Pennsylvania Rock Oil Company**, one of the world's first modern oil companies; Bissell's vision eventually led to the drilling of the **Edwin Drake** well. George Henry Bissell was born on 8 November 1821 in Hanover, New Hampshire. He graduated from Dartmouth College in 1845, after which he held a number of jobs, including journalist, high school principal, and professor of languages.

While working, he also studied law, obtaining a degree and gaining admission to the New York bar in 1853. He then began practicing law in New York City.

Bissell passed through the oil regions of western **Pennsylvania** on a trip to the Northeast from New Orleans, where he worked as an educator in the 1840s, and became interested in the surface seeps of oil. His interest grew when he happened to see a sample of Pennsylvania oil on a visit to Dartmouth. The sample had been brought to the medical school there in 1853 for analysis by Francis Beattie Brewer, a Massachusetts physician who was also a Dartmouth alumnus. Two years earlier, Brewer had moved to **Titusville** to join his father's lumber firm, Brewer, Watson, and Company, and had there also become intrigued by the seeps of oil on the company's land.

Bissell's New York law partner Jonathan C. Eveleth went to Pennsylvania to investigate the oil-bearing lands near Titusville from which the sample had come. The land was mostly in **Venango County**, between high mountains in the valley of Oil Creek. It included the Hibbard farm and was owned by Brewer, Watson, and Company. Bissell and Eveleth bought or leased about 1,200 acres of land from Brewer, Watson, for $5,000. On 30 December 1854, they established the Pennsylvania Rock Oil Company of New York, which intended initially to produce oil from surface seeps or via hand-dug wells.

Bissell later had the idea of drilling for oil. Allegedly, the idea came to him in 1856 after he saw an advertisement for rock oil medicine that had a picture of drilling derricks that were used to drill for salt. Some petroleum used for medicinal purposes was a by-product of salt drilling. Pennsylvania oil was marketed as a patent medicine, notably by **Samuel Kier**. Similar activities were occurring in neighboring **West Virginia**. Kier and others believed oil could be useful as an illuminant, and Bissell had apparently reached a similar conclusion. Neither idea—drilling for oil deliberately or using petroleum as an illuminant—was actually original, although Bissell may not have known it. **Russia** had already drilled a well for oil in 1846—*see* FIRST OIL WELL; VOSKOBOINIKOV, NIKOLAI (1801–C. 1861), and **Ignacy Łukasiewicz** in Austro-Hungarian **Galicia** was already making kerosene and lamps in 1853.

Needing more capital for the company, Bissell attempted to interest a group of investors from New Haven, Connecticut. They were interested but cautious. An analysis of an oil sample by **Benjamin Silliman** of nearby Yale University helped convince them. In 1855, Silliman issued his *Report on the Rock Oil, or Petroleum from Venango Co., Pennsylvania, with Special Reference to Its Use for Illumination and Other Purposes*. The New Haven investors, among them James Townsend, the president of the City Savings Bank, agreed to buy stock if the company was restructured under Connecticut law, which was more liberal. The Pennsylvania Rock Oil Company of Connecticut was formed on 18 September 1855, with a capitalization of $300,000.

The company acquired all the lands owned by Bissell and Eveleth that showed oil, and it leased the rest for 99 years. The amount paid is not certain but is believed to have been $24,000.

Conflicts soon arose between Bissell and Eveleth on the one hand and the New Haven group on the other. Bissell wanted to move fast, and the New Haven investors were conservative and parsimonious. The company did little in 1855 and 1856. Bissell and Eveleth convinced the New Haven investors to buy some of their stock, promising to find another party to work the prospect. On 6 November 1856, the property was leased to David H. Lymans and Rensellaer Havens of New York. However, the following year was a depression year, with collapsing commodity prices. Lyman and Havens were able to back out of their lease. Some of the New Haven group unjustly accused Bissell of having devised nothing more than an elaborate scheme to unload some of his shares. Bissell and Eveleth attempted a reconciliation, suggesting the formation of a new company. The New Haven investors were not interested. Townsend and others from New Haven acquired more stock and gradually marginalized Bissell and Eveleth. They organized a new company, the Seneca Oil Company, to take up the lease that Lyman and Havens had abandoned.

In 1858, Townsend employed Edwin Drake to drill for oil on behalf of Seneca. Drake struck oil on 27 August 1859, launching a modern oil boom. Seneca would later fall victim to competition and bad management. Drake, who was president for a time, was forced to resign in March 1860. After several more management changes and continuing poor performance, the company sold its Pennsylvania interests to George W. Steele of Brooklyn, New York, on 7 March 1864 for $10,000. The company that helped begin the modern oil industry in America then ceased operating. None of the New Haven investors made any money from the venture. James Townsend had paid in about $50,000 and gotten a similar amount back.

Jonathan Eveleth, Bissell's partner, died in 1862. The New Haven investors ceased their involvement in the oil regions, but Bissell did not. He left his law practice in New York and moved to Franklin, Pennsylvania, where he remained until 1864. Along with a number of financial partners from New York, he invested heavily in oil lands around Franklin, on the McClintock Farm, and elsewhere. He was very successful and also participated in many other commercial developments, including railroads. He died in New York City, a wealthy man, on 18 November 1884. Despite his success, he reportedly remained bitter toward Townsend and the rest of the New Haven group.

BITOR. *See* ORINOCO OIL BELT.

BITUMEN. A form of degraded crude oil that has lost most of its lighter fractions and is both dense and very viscous. Bitumen generally ranges in **API gravity** from 5° to 10° and has a viscosity greater than 10,000 centipoise (cp), ranging up to hundreds of thousands of cp. Bitumen is generally immobile in the reservoir and must be extracted by mining. Bitumen can be processed to yield synthetic crude oil (**syncrude**). Bitumen is sometimes called "extra-heavy oil," but that term should probably be reserved for dense crude oils with much lower viscosity (*see* HEAVY OIL). Large deposits of bitumen occur in **oil sands**, particularly the **Alberta oil sands** of **Canada**.

BLACK GIANT. *See* EAST TEXAS OIL FIELD.

BLAUSTEIN, LOUIS (1868–1937). Founder of the **American Oil Company** and developer of the **Amoco** brand; innovator in **gasoline** technology and marketing. Louis Blaustein was born on 16 January 1868 in Lithuania, immigrating to the **United States** in 1884. He settled in Baltimore in 1891 and began selling kerosene as part of a grocery business. He was the first to begin selling kerosene from a tank wagon. In 1910, he founded the American Oil Company and was joined by his son Jacob. He incorporated formally in 1922. This company originated the Amoco brand and was later absorbed by Standard Oil of Indiana (which changed its name formally to Amoco in 1985).

Blaustein is credited with a number of innovations in the oil industry, including possibly the first drive-in gasoline station and the first visible gasoline pump. He was also a pioneer in the use of **tankers** to transport refined products from refinery to port. One of his most important inventions was an antiknock gasoline, which he developed with Jacob and marketed under the Amoco brand. He referred to his gasoline blend, which contained benzol and other ingredients, as "high test" gasoline. Blaustein was an important adviser to the U.S. government on petroleum issues during **World War II**. He was named by **Harold Ickes**, secretary of the interior and head of the **Petroleum Administration for War**, to serve on two of the administration's petroleum industry committees, the Marketing Committee and the Supply and Distribution Committee.

BLUE LINE AGREEMENT (1936). Agreement in 1936 between Standard Oil of California (Socal, later **Chevron**) and the Texas Corporation (later **Texaco**) to pool their resources in the Eastern Hemisphere (east of Suez). The name of the agreement references the **Red Line Agreement**, which involved many of the **major oil companies** of the time but not Socal or the Texas Corporation. The Texas Corporation acquired 50 percent of Casoc, Socal's subsidiary in **Saudi Arabia**, and 50 percent of the Bahrain Petroleum

Company (Bapco), its Bahrain subsidiary. Since Casoc was the forerunner to **Aramco**, Texaco became an Aramco partner. Socal acquired cash and a half interest in the Texas Corporation's Eastern Hemisphere downstream operations. The joint downstream operations were placed in a Bapco subsidiary called **Caltex**. Aramco would ultimately be acquired by the Saudi government. Caltex remained with Socal and Texaco, ultimately becoming part of the merged company ChevronTexaco (later simply Chevron) in 2001.

BOLIVAR COASTAL COMPLEX. A set of oil fields in **Venezuela** located in the area of Lake **Maracaibo**. By some estimates, the Bolivar complex is the third- to fifth-largest aggregation of oil in the world, after **Ghawar**, **Burgan**, and possibly Rumaila and **Safaniya** in the **Persian Gulf** region. The first **giant** discovery was Cabimas in 1917. The largest, Tia Juana, was found in 1928. Lagunillas and Bachaquero, both **supergiants** but not as large as Tia Juana, were discovered respectively in 1926 and 1930. The most recent giant discovery was Lama in 1957. The **EUR** of the whole complex is approximately 30 Bboe. Estimates of remaining recoverable resources range from about 200 Mmboe to 11 Bboe.

BOLIVIA. Oil production 2015: 0.051 Mmbpd (0.064 percent of world); **reserves** 0.211 Bb (0.01 percent of world); gas production: 0.730 Tcf (0.59 percent of world); reserves 10.45 Tcf (0.15 percent of world). Bolivia is a minor oil producer but a significant **natural gas** producer. It is a member of the **Gas Exporting Countries Forum**. The history of its hydrocarbon industry is one of repeated nationalizations, denationalizations of varying degrees, and renationalizations.

Oil was known to exist in Bolivia in colonial times, but serious exploration did not begin until 1916, by which year the government had granted a large **concession** to Braden interests. In 1920, it granted further large concessions to Richmond Levering and Company and to Jacobo Backus. The Backus concession was sought by **Royal Dutch Shell** but abandoned in 1924 for lack of development. The Braden and Richmond concessions were taken over by Standard Oil of New Jersey (later **Exxon**), which thereby managed to bypass a 1921 law limiting new concessions to 100,000 hectares. Jersey Standard formed Standard Oil of Bolivia to explore and exploit in the concession areas. Standard intended to produce oil from fields discovered in the southeastern part of the country and export it via **pipeline** to the Parana River in **Argentina**. Bureaucratic barriers in Argentina led Standard to scale back production, and this annoyed the Bolivian government.

From 1932 to 1935, Bolivia fought the bitter **Chaco War** with Paraguay. Oil formed an important part of the background. During the war, Paraguay briefly occupied some of Bolivia's oil regions but was repelled. Ultimately,

Bolivia lost the war. The conflict increased oil nationalism in the country and intensified resentment of Standard Oil, which had adopted a position of neutrality in the war.

After the Chaco War, a military government came to power through a coup. In March 1937, President Toro expropriated Standard Oil's interests without compensation and placed them under the control of the new national oil company, Yacimientos Petrolíferos Fiscales Bolivianos (YPFB). YPFB had been established in December 1936, following the model of Argentina's Yacimientos Petrolíferos Fiscales (YPF; *see* REPSOL). Bolivia's nationalization presaged **Mexico**'s more significant one the following year. The Franklin D. Roosevelt administration in the **United States**, not wanting to create a pretext for Axis influence in Bolivia, took no retaliatory action.

Oil production increased in the 1940s but remained at very low levels, never exceeding 1,300 bpd. A 1938 joint exploration agreement with **Brazil** never came to fruition. In 1955, a left-wing government that was nevertheless supported by the Dwight D. Eisenhower administration reopened the country's hydrocarbon industry to foreign participation. The 1955 law was called the Davenport Code, after the American law firm that helped write it. A number of foreign companies took advantage of the new law, but only **Gulf** was successful. It struck oil in 1961 and by 1968 was producing about 33,000 bpd.

On 17 October 1969, there was a renationalization. The government of General Alfredo Ovando Candia expropriated all of Gulf's interests in Bolivia. The United States retaliated by cutting aid. In 1972, a new hydrocarbons law passed by the government of General Hugo Banzer Suarez partially rolled back nationalization. Foreign companies were allowed to enter into contracts with YPFB, splitting profits evenly. The **Union Oil Company of California** and **Occidental Petroleum** both entered Bolivia. Oil production reached 0.047 Mmbpd in 1973, a level not exceeded until 2004. From 1980 to 2006, production of crude oil and natural gas liquids averaged about 0.035 Mmbpd. Bolivia was variously a small net oil importer and a small net oil exporter in the 20th century, most recently the latter.

Bolivia's oil reserves stood at a modest 0.211 Bb in 2015. Gas reserves, on the other hand, are more substantial. At 10.45 Tcf, or 1.85 Bb, they are almost nine times higher than oil reserves in energy-equivalent terms. Bolivia liberalized its hydrocarbon industry again in 1996, encouraging foreign investment, and this had an effect on gas reserves and production. A number of **giant** gas fields were discovered in 1998 and 1999, including Margarita, Ipati, San Alberto, Sabalo (San Antonio), and Itau. The largest foreign interests were held by Petrobrás, Repsol YPF, and **Total**. **BP** also had significant interests.

In May 2006, the government of **Evo Morales** renationalized the petroleum industry. In a theatrical show of force, government troops occupied the largest gas fields, and President Morales gave a speech from one of them (San Alberto). Foreign companies would henceforth be operators of fields for YPFB. Royalties increased immediately from 50 percent to 82 percent. By November 2006, the key foreign players had accepted the new terms. As part of the renationalization, YPFB also reacquired refineries it had previously sold to Petrobrás. Shares of YPFB business units privatized in 1996 were renationalized. Although compensation was offered, its cost to the government would be offset by the increased royalties on large gas fields. In effect, private companies would be paying their own compensation over time.

After the renationalization, exports of gas increased, but exploration suffered. The industry lived largely on consuming its reserves without making much effort to replace them. The high government take disincentivized the kind of foreign investment that would likely be necessary for renewed exploration.

BONNY (IBANI, UBANI). Oil port (pop. 9,200, 2005) in **Nigeria**. Nigerian oil refined at **Port Harcourt** is exported from Bonny. The port was a major commercial center in the 18th and 19th centuries, exporting palm oil, and was the administrative center of the British Oil River Protectorate between 1885 and 1894. Since 1961, Bonny has been the main export port for oil from the Niger Delta. A harbor enlargement completed in 1964 permitted vessels of up to 35-foot draft.

BOSPORUS AND DARDANELLES. Narrow straits separating European and Asian Turkey. The Bosporus links the Black Sea to the Sea of Marmara, while the Dardanelles leads from the Sea of Marmara to the Aegean Sea and Mediterranean. At its narrowest point, the Bosporus is only about half a mile wide. Approximately 3.2 Mmbpd passed through the straits in 2016. The Bosporus is an important choke point in crude oil transportation and has been considered a potential terrorist target since 11 September 2001. Commercial shipping has the right of free peacetime passage through the straits. However, Turkey reserves the right to regulate shipping for environmental or safety purposes. With 5,500 oil **tankers** transiting the straits annually, restrictions seem increasingly likely.

Russian and **Caspian** oil has been passing through the straits since the early days of the modern petroleum industry. **Baku** oil and petroleum products transported by rail or **pipeline** to the Black Sea port of **Batumi** entered the Mediterranean through the Bosporus and Dardanelles. Caspian oil is also piped to the Russian Black Sea port of **Novorossiysk**, from which it may transit the Bosporus. The **Baku–Tbilisi–Ceyhan Pipeline** (BTC) offers direct access to the Mediterranean, bypassing the Bosporus, for Caspian Oil. Proposed pipelines between the Bulgarian Black Sea port of Burgas and Mediterranean ports (e.g., Vlore, Albania, or Alexandroupolis, Greece) may also offer alternatives to the straits, as would the proposed Samsun–Ceyhan Pipeline across Turkey. The Burgas project was suspended in 2011.

BP. A **major oil company**; one of the **Seven Sisters**. *BP* once stood for "British Petroleum," but the initials are now intended to stand on their own. The official name of the company has been BP PLC since 2000. BP began life as the Anglo–Persian Oil Company in 1909. The British government became its majority owner in 1914. After 1935, Anglo–Persian was known as the **Anglo–Iranian Oil Company**; as its name implied, it relied on **Iran** for most of its oil. Anglo–Iranian changed its name to British Petroleum in 1954 after the nationalization of the Iranian oil industry. It continued to do business in Iran, with a 40 percent interest in the consortium **Iranian Oil Participants Ltd.** However, the company needed to diversify and proceeded to do so.

In 1965, BP discovered **natural gas** in the **North Sea**, and in 1970 it discovered oil there. It discovered and developed oil fields in many different parts of the world, including **Prudhoe Bay** in **Alaska**.

The Alaskan strike in 1969 followed a decade of dry holes in the area and happened just as the company was ready to give up. After discovering oil in Alaska, BP traded some Alaskan **reserves** for a 25 percent interest in the Standard Oil Company of Ohio (**Sohio**), to ensure refinery capacity for Alaskan oil. It later increased its stake to a controlling interest, completing the acquisition in 1987. In 1977, the British government began selling shares of BP to the public, and by the late 1980s the company had been completely privatized. During a later stage of this privatization, the **Kuwait** Investment Office acquired a substantial stake in BP. Despite some assurances to the contrary, the British feared the Kuwaitis would seek board representation. The British government—whose commitment to Thatcherite free market ideology apparently had its limits—blocked Kuwait from gaining control, using its authority to prevent foreign ownership that was "against the public interest."

In 1987, BP acquired Britoil PLC, a privatized British oil company formed from the previously nationalized **British National Oil Company** (BNOC) in 1982. Britoil was an important North Sea producer and accounts for a significant part of BP's North Sea activity. In 1998, BP merged with **Amoco** in a $52 billion deal, becoming BP Amoco, at the time the largest oil company in the world. In 2000, BP Amoco acquired Atlantic Richfield Company (**Arco**) in a $27 billion deal. In order to gain regulatory approval for the acquisition, it sold Arco's Alaskan properties to **Phillips Petroleum** for $7 billion. In the same year, it bought **Burmah** Castrol for $4.7 billion and adopted the shortened name of BP PLC. By purchasing Burmah Castrol, BP was acquiring the descendant of the original entity that had founded the Anglo–Persian Oil Company—which in turn eventually became BP.

BP also developed considerable interests in **Russia**. In 2003, it merged those interests with those of the Tyumen Oil Company (Tyumenskaya Neftyanaya Kompaniya, or TNK), creating the **TNK–BP** joint venture. Among TNK–BP's assets was **Samotlor**, Russia's largest oil field. In 2008, BP alleged that its Russian partners were trying to use strong-arm tactics to take control, with the complicity of the Russian government. TNK–BP continued to operate as a major producer in Russia but was eventually acquired by **Rosneft** in 2013.

In 2010 BP was involved in the disastrous ***Deepwater Horizon* oil spill**.

In 2016, BP controlled 9.879 Bb of oil reserves and 43.368 Tcf of gas. It produced 2.041 Mmbpd of oil and 7.575 Bcfpd of gas. In total assets, it was the sixth-largest private sector petroleum company in the world.

BP AMOCO. *See* BP.

BRADFORD OIL FIELD. Early **giant** oil field in **Pennsylvania**, discovered in 1871. The Bradford field was the most prolific and also the last of the major fields discovered in Pennsylvania. The northernmost field in the Pennsylvania oil region, it lies mainly in the north central part of McKean County and extends slightly into Cattaraugus County in the state of New York. It has already produced as much oil as its **EUR** of 658 Mmb and may be nearly exhausted. It is not an exceptionally large field by the standards set in the 20th century but was a major find by the standards of its time. It reached a peak production of 22.945 Mmb in 1881 (62,863 bopd). In 1880, 1881, and 1882, it produced 63 Mmb (average of 57,534 bopd). This was 74 percent of all the oil produced in the **United States** and 64 percent of all the oil produced in the world (some sources give even higher figures). No other oil field since then has dominated global production to such an extent. As Bradford production increased, prices dropped, from about $4 a barrel in 1876 to $0.70 in 1878. The Bradford field was discovered only months after Henry E. Wrigley, head of the Pennsylvania Geological Survey, predicted that Pennsylvania—and global—oil production had peaked and would soon decline rapidly.

Production declined steeply from 1881 to 1887, when it was about 4.5 Mmb (about 12,300 bpd). After that, production declined more gradually. Many wells were abandoned, and a number were not properly plugged, so that freshwater accidentally entered. In some cases, this flooding led to improved production in nearby wells, and the industry began using intentional flooding to increase production. The practice was illegal in Pennsylvania but was legalized in 1921. This led to the development of scientific flooding methods to enhance production. The Bradford field became a testing ground for the development of an important secondary **oil recovery** mechanism. Production rose until it reached 16.739 Mmb in 1937 (45,859 bopd), or 85 percent of the total production of Pennsylvania. Since then, production has declined.

BRANOBEL. Leading and highly innovative **Russian** integrated oil company in the late 19th and early 20th centuries, with production primarily in the **Abseron Peninsula** (**Baku** area); founded and run by the Nobel family. "Branobel" is the shortened name for Tovarishchestvo Nephtanavo Proisvodtsva Bratiev Nobel (Nobel Brothers Petroleum Production Company), used for cable communications. **Ludvig Nobel** headed the company from 1879 to 1888; his son **Emanuel Nobel** headed it from 1888 to 1920.

The roots of Branobel extend back to 1873, when Ludvig's brother Robert Nobel went to the Caucasus to make a large purchase of walnut for the family business in St. Petersburg to use in manufacturing rifle stocks for the Russian army. Robert already had some experience trading kerosene in Finland, and he became interested in the petroleum industry of the Abseron Peninsula. He used the entire 25,000 rubles intended for the walnut purchase to buy some oil-bearing land and a refinery from the Dutch De Boer brothers. The 1872 oil industry reforms of Tsar **Alexander II**, allowing private investors to bid on oil leases, improved business prospects. Robert got the business started, and in 1876 Ludvig and his son Emanuel visited the Abseron Peninsula. Ludvig and later Emanuel would assume leadership of the enterprise. Branobel was formally established as a joint stock company in 1879. Its initial capitalization was 3 million rubles. Ludvig Nobel invested 1.61 million rubles and had a controlling interest.

The company adopted mechanical **drilling** methods and grew rapidly. In its first year, it accounted for about 5 percent of Russia's production. Two years later in 1881, it accounted for 13.5 percent, and by 1887 it was producing 18.5 percent of Russia's oil and 7.5 percent of the world's. By 1916, it would control about a third of Russian crude oil production, and 40 percent of Russian production of refined products. That year, Branobel supplied two-thirds of the domestic market. Branobel faced stiff competition from the **Rothschilds** as well as from **independents** such as **Mantashev**. The Rothschilds gained significant market share, but Branobel was more profitable. It invested more consistently and was more effectively integrated. Branobel encompassed exploration, production, **refining**, storage, transportation, and marketing.

In 1878, the Nobels completed Russia's first **pipeline**, 5.6 miles long and 2 inches in diameter, with a capacity of nearly 10,000 bpd, and the company deployed the world's first modern oil **tanker**, the ***Zoroaster***, which was designed by Ludvig Nobel. By the mid-1880s, Branobel had a fleet of tankers transporting Baku oil across the **Caspian**, to be shipped on company ships and barges up the Volga River. Thanks again to Ludvig's inventiveness, some of the company's tankers heated their boilers with a type of fuel oil when the rest of the world was using coal. Ludvig Nobel also developed a continuous distillation unit (Russian Patent No. 9206, 17 December 1882), incorporating some ideas from **Dmitri Mendeleev**, and the company became the first in the world to perform continuous distillation on a large scale. Emanuel Nobel continued in his father's innovative tradition, introducing rotary drilling techniques and building the world's first diesel engine plant. He installed diesel engines in power plants, tankers, tugboats, and pumping stations on the **Baku–Batumi Pipeline**.

Branobel was also an exceptionally progressive employer by the standards of its time. Ludvig Nobel abolished child labor and shortened daily working hours from 14 to 10.5. He built decent housing for workers and provided them with free medical care, technical training, and elementary education for their children. He also created a cooperative bank for workers. Unlike the Rothschilds, who sold out their interests to **Royal Dutch Shell** in 1912, the Nobels saw half of theirs nationalized by the Soviets in 1920 (they had sold half to Standard Oil Company of New Jersey, later **Exxon**). Emanuel and his brothers were able to escape eventually to Sweden. They fought unsuccessfully for more than a decade to recover their assets from the Soviet government.

BRAZIL. Oil production 2015: 2.43 Mmbpd (3.10 percent of world); **reserves** 13 Bb (0.79 percent of world); gas production: 0.803 Tcf (0.64 percent of world); reserves 15.184 Tcf (0.22 percent of world). Brazil became self-sufficient in oil in 2006 and is now a major producer with considerable future potential. Brazil has also been a successful producer and user of alcohol as an alternative transportation fuel (*see* ALTERNATIVE ENERGY).

The first modern attempts to find oil in Brazil took place on a small scale in the late 19th century and did not result in notable successes. In 1917, the Brazilian government's Geographical and Mineralogical Service established an oil exploration department, but again there were no notable successes. Nevertheless, the idea persisted that Brazil was bound to have major reserves of petroleum. On 29 April 1938, President Getúlio Vargas established the Conselho Nacional do Petróleo (CNP; National Petroleum Council), via Law No. 395, to set national policies and control the development of Brazil's oil industry.

The CNP made its first discovery at Lobato, just outside Salvador in the northern state of Bahia, in 1941. The find was modest and was not commercially exploited. The CNP discovered Brazil's first commercial oil field in Cadeias, Bahia, also in 1941. Although Bahia would later be eclipsed in importance by the southern **offshore**, discoveries continued to be made there and in the neighboring state of Sergipe. Brazil's first **giant** field, Miranga (**EUR** about 600 Mmb), was discovered near Salvador in 1965. The first giant offshore field, the gas field Guaricema, was discovered in 1968 off Sergipe. The onshore gas giant Riacheulo was discovered in Sergipe in 1972.

Petroleum has long been linked in Brazil with notions of national sovereignty and national security. Getúlio Vargas once declared that "whoever hands over petroleum to foreigners threatens our independence." A nationalistic campaign, with the tagline "O petróleo é nosso" (the petroleum is ours), began with students but spread to the wider society. During his second presidential administration (1951–1954), Vargas acted on these sentiments and proposed the creation of a national oil company. Law No. 2,004 of October 1953 created Petróleo Brasileiro (Petrobrás), to execute all activities in the petroleum sector in Brazil on behalf of the federal government. Petrobrás assumed the assets of the CNP, which continued in a supervisory role. In 1960, the CNP was incorporated into the Ministério de Minas e Energia.

Petrobrás succeeded in raising Brazilian production. In 1950, Brazil produced about 950 bpd of oil. By 1956, the figure was 6,500, and 10 years later it was 150,000. In 1980, it was about 182,000 bpd. Brazilian production increased dramatically after 1980. Production in 2006 was nearly 10 times 1980 production (and 2016 production was 44 percent higher than in 2006). Gas production also increased greatly, about eightfold between 1980 and 2006, but remained all the while at a much lower level (energy equivalent) than oil.

The dramatic increases in Brazilian production came from offshore discoveries in the Campos basin and later the Espirito Santo and Santos basins. In 2006, offshore production accounted for more than 80 percent of Brazilian oil production. The oil reserves in these basins lie about 60 to 90 miles from the coasts of the southern states of São Paulo, Rio de Janeiro, and Espirito Santo, often in very deep water. Petrobrás became a leader in deepwater drilling and production. It made its first discovery in the Campos basin in 1974, when it found the Garoupa field. First production was in 1977, and by the mid-1980s the Campos basin was producing about 370,000 bpd, or about 60 percent of the national total. Petrobrás's first offshore giant in the Campos basin was Albacora (EUR about 0.67 Bb), discovered in 1984 in water depths ranging from about 500 feet to nearly 3,500 feet. Albacora was followed in 1985 by Marlim and Marlim Sul (EUR 2.43 and 1.29 Bb, excluding gas), in water depths from 3,500 feet to 8,500 feet. In 1996, Petrobrás discovered Roncador, the largest field to that date, with an EUR of about 3.2 Bb. Roncador lies under water depths from 4,900 to 6,600 feet. In 2001, three explosions sank Roncador's production platform, killing 11 workers and severely interrupting operations.

In 1997, Law No. 9,378 partially privatized Petrobrás and opened the Brazilian petroleum industry to private investment. The law also created the Agência Nacional do Petróleo, Gás Natural, e Biocombustíveis (ANP) to regulate the sector. The Brazilian government now owns just over 50 percent of Petrobrás. Since 1997, Petrobrás has continued to dominate the Brazilian industry, but other players have entered, including Brazilian **independents** and international **major oil companies**. **ExxonMobil**, **Chevron**, **Royal Dutch Shell**, and others have since developed interests in Brazil. After 1997, Petrobrás also made certain strategic changes. It centralized decision making and increased investment and training in exploration. It also deliberately prospected beyond the Campos basin, into the Espirito Santo basin to the north and the Santos basin to the south. The **United States Geological Survey** (USGS) estimated huge possible resources in the Santos basin, perhaps 23 Bb.

In 2001, Petrobrás discovered the giant Jubarte field in the northern part of the Campos basin, and in 2003 it discovered another giant, Golfinho, in the Espirito Santo basin. In the Santos basin, it discovered a giant oil field designated 1-RJS-539 and Mexilhao, a gas giant.

However the most spectacular finds came from exploration in the **pre-salt** layer. In October 2006, Petrobrás discovered the Tupi oil field in the Santos basin, in partnership with the BG Group of the **United Kingdom** and Galp Energia of Portugal, under 6,600 to 9,800 feet of water. The field was renamed Lula in 2010. *Lula* is Portuguese for "squid," but we note that the 35th president of Brazil (from 2003 to 2011), Luiz Inácio Lula da Silva, is also known simply as "Lula." The Lula field is thought to contain up to 8 Bb of recoverable oil and began producing in 2009. On 21 January 2008, Petrobrás announced another major pre-salt discovery in the Santos basin, this one in partnership with Galp: the Júpiter field, estimated at 1.6 Bboe of gas and condensate. It lies under about 7,200 feet of water and is owned by Petrobrás (80 percent) and Galp Energia (20 percent). Petrobrás also discovered other huge fields in the pre-salt in partnership with various companies, including BG Group, Repsol-Sinopec (a joint venture between **Repsol** and Addax Petroleum UK Limited, a subsidiary of Sinopec, the China Petrochemical Corporation), and others. The largest was the Libra field in 2010, thought to contain 8 Bb of recoverable oil, possibly larger than Lula. Other giants are Iara, discovered in 2008 and containing over 3 Bb, and Búzios, discovered in 2010 and also containing over 3 Bb. Not all the giant discoveries had begun production as of 2017, but overall pre-salt production has indeed taken off in Brazil. During the year 2017, pre-salt production began to exceed production from other sources. In 2016 Petrobrás was the fourth-largest petroleum company in the world by total assets, excluding wholly state-owned national oil companies. It controlled 8.167 Bb of oil reserves and 8.604 Tcf of gas, producing oil at the rate of 2.102 Mmbpd and gas at 2.449 Bcfpd.

One obstacle to Brazil's continuing growth as a major oil producer is the perception and reality of corruption in its oil industry, radiating out to other sectors of the economy and political parties. The **Petrobrás scandal** shocked and consumed the country beginning in 2014 and resulted in the removal from office in 2016 of President Dilma Rousseff (despite no allegations that she was personally involved).

BRENT BLEND. A **benchmark crude** used since the 1980s as a pricing reference for **North Sea**, African, **Russian**, and Central Asian oil, as well as **Persian Gulf** oil headed for Europe. It emerged gradually as a de facto international standard over the last two decades. As its name implies, it is a blend of different crude oils. These come from various oil fields in the North Sea's **Brent** and Ninian systems. Brent Blend is light (**API gravity** of 38.3°) and sweet (0.37 percent sulfur). It is a bit heavier and sourer than **West Texas Intermediate** but lighter and sweeter than **Dubai crude**. As is the case with other benchmarks, falling production of Brent Blend is a problem that may lead to its replacement as a benchmark in the long term. In 2002, Platt's pricing agency began supplementing Brent with some similar North Sea grades from the Forties and Oseberg systems.

BRENT OIL FIELD. Early **giant** oil field in the **United Kingdom** sector of the **North Sea**; operated by **Royal Dutch Shell** and in the process of ending production and entering its decommission phase. Crudes from oil fields related to Brent still constitute a major pricing **benchmark** known as **Brent Blend**. Brent was discovered in 1971 by Royal Dutch Shell and the Standard Oil Company of New Jersey (renamed **Exxon** a year later). The first **tanker** was loaded with Brent oil on 13 December 1975. The field is located under about 460 feet of water, roughly 110 miles northeast of Scotland's Shetland Islands and a like distance west of the **Norwegian** coast. The name originates from the Brent goose; Shell started out naming its UK North Sea oil fields after waterbirds. Oil is transported to **Sullom Voe** in the Shetland Islands via undersea **pipeline**. Brent produced about 4 Bboe in oil and gas over its lifetime. In 2000, it produced about 60,000 bpd, versus about 440,000 bpd in 1985. Some observers predicted exhaustion by the late 1990s, but a large redevelopment project completed in 1997 extended the life of the field.

BREWER, FRANCIS BEATTIE. *See* BISSELL, GEORGE HENRY (1821–1884).

BRITISH NATIONALITY CLAUSE. A required clause in petroleum **concession** agreements in some **Persian Gulf** sheikhdoms, including **Kuwait** and **Bahrain**, that development be carried out by companies from the **United**

Kingdom. The British government had secured this requirement in its agreements with local sheikhs before **World War I**. The original intent was to keep Germany from developing petroleum resources in the Persian Gulf, but by the 1930s the main effect was to limit penetration by the **United States**. Although neither the British government nor the Anglo–Persian Oil Company (APOC, later the **Anglo–Iranian Oil Company**, still later **BP**) believed that the Arabian Peninsula had good oil prospects, they still wanted to keep competitors out of what they regarded as APOC's "sphere of influence." In the late 1920s and early 1930s, Standard Oil of California (Socal, later **Chevron**) had a concession in Bahrain, and **Gulf** wanted to explore in Kuwait. Both were hamstrung by the nationality clause.

In the Bahrain case, the British government relented in 1929, under diplomatic pressure from the United States. Socal could proceed with exploration and development, provided the British controlled all communication with the emir. In the case of Kuwait, the British government was more stubborn, fearing not only the incursion of American companies but also the possible establishment of an American military presence in the Persian Gulf to protect U.S. interests. However, in Kuwait too it relented, in the early part of 1932, especially since APOC was not showing any interest in exploring in Kuwait. It retained the right to review bids and make recommendations to the emir.

When Socal's subsidiary Bahrain Petroleum Company (Bapco) discovered oil in Bahrain on 31 May 1932, APOC became much more interested in Kuwait. A bidding competition ensued between APOC and Gulf, amid American suspicions that the British government would try to steer the bid to the British company. The emir of Kuwait, Sheikh Ahmad, was shrewd and encouraged the competition. Eventually Gulf and APOC formed the Kuwait Oil Company (KOC) as a joint venture and obtained a concession in 1934. However, the British government forced an agreement to keep actual operations in British hands, despite the joint ownership of the venture. KOC went on to discover **Burgan** in 1938, which in 2017 was still the second-largest oil field ever found in the world. *See also* HOLMES, FRANK (1874–1947).

BRITISH NATIONAL OIL COMPANY (BNOC). Petroleum company established in 1976 by the government of the **United Kingdom**. The company was privatized as Britoil beginning in 1982. **BP** acquired full control in 1988. The Labour government of Harold Wilson established BNOC on 1 January 1976 as an instrument of public participation and control in the development of **North Sea** oil and gas. BNOC would obtain a seat and a vote, but not a veto, on the operating committees of consortia operating in the North Sea. It also had an option to purchase 51 percent of the oil produced in a venture at market prices. In 1976, BNOC also purchased the North Sea interests of the **Burmah Oil Company**, which was experiencing serious financial trouble. The government wanted to ensure that the Thistle field, which Burmah operated, and the significant Ninian field, in which Burmah had a major share, would continue to be developed. BNOC also acquired the staff of Burmah's North Sea subsidiary.

The election of the Tory government of Margaret Thatcher changed British petroleum policy. Thatcher was ideologically opposed to government involvement in areas such as petroleum exploration and production, which she felt properly belonged in the private sector. She embarked on a program of privatization of state-owned enterprises. The government established Britoil in August 1982 and transferred BNOC's oil production interests to it. BNOC was left as a government agency that still traded in North Sea crude. In November 1982, the government sold off 51 percent of Britoil's shares. In August 1985, the government made a second offering to dispose of the remaining shares. BP took control of Britoil in February 1988. The government abolished BNOC on 27 March 1986. As of 2017, Britoil was still a subsidiary of BP based in **Aberdeen**.

BRITISH OIL DEVELOPMENTS LTD. *See* IRAQ PETROLEUM COMPANY.

BRITISH PETROLEUM. The British Petroleum Company was established in 1906 as a subsidiary of the German oil company Europäische Petroleum Union. It developed considerable marketing operations in Britain and was acquired in 1917 by the Anglo–Persian Oil Company, later named the **Anglo–Iranian Oil Company**. After 1954, the Anglo–Iranian Oil Company was called British Petroleum, but the company was known widely by its initials, **BP**, and eventually adopted the initials as its name.

BRITOIL. *See* BRITISH NATIONAL OIL COMPANY (BNOC).

BUFFALO BAYOU. *See* HOUSTON SHIP CHANNEL.

BUREAU DE RECHERCHES DES PÉTROLES (BRP). *See* ELF AQUITAINE.

BURGAN OIL FIELD. World's second-largest oil field (after **Ghawar**), located in **Kuwait**. Burgan was discovered on the night of 23–24 February 1938 by the Kuwait Oil Company (KOC), a joint venture between the **Anglo–Iranian Oil Company** (later **BP**) and the **Gulf Oil Corporation** (later part of **Chevron**). KOC drilled eight more wells at Burgan before July 1942, when it suspended all operations and plugged all completed wells because of **World War II**. After the war, operations resumed, and production began in 1946.

The Greater Burgan complex consists of the Burgan field and the Magwa and Ahmadi fields discovered in the 1950s. The complex has an **EUR** of about 61 Bb, although some estimates range to more than 70 Bb. The complex also has an EUR of 42.8 Tcf of gas. Production peaked in 1972 at 2.4 Mmbpd, greater than 70 percent of Kuwait's production at the time. In the 2000 decade, Greater Burgan produced in the range of 1.35–1.6 Mmbpd. The complex is estimated to have produced about 28 Bb through 2002.

BURMA (MYANMAR). Oil production 2015: 0.014 Mmbpd (0.017 percent of world); **reserves** 0.139 Bb (0.01 percent of world); gas production: 0.694 Tcf (0.56 percent of world); reserves 22.5 Tcf (0.33 percent of world). *See also* BURMAH OIL COMPANY.

BURMAH CASTROL. *See* BURMAH OIL COMPANY.

BURMAH OIL COMPANY. Early British oil company that began in the country of **Burma** (formerly spelled "Burmah" and now known as Myanmar); later Burmah Castrol, eventually part of **BP**. Burmah Oil was instrumental in establishing the Anglo–Persian Oil Company (APOC, later **Anglo–Iranian Oil Company**), which was BP's antecedent. Burmah Oil eventually sold out its interest in Anglo–Persian but was merged into Anglo Persian's descendant, BP, decades later.

The country of Burma was the site of an old oil operation with hand-dug wells that had been worked perhaps for centuries. The wells were observed in 1795 by Michael Symes, who was sent on a mission to Burma by the governor general of India. In 1871, the Rangoon Oil Company was founded to refine and market the oil collected there. The company was registered in Edinburgh, Scotland. Within a few years, it had built up shipments to India to an annual average of about 260 barrels.

The company nevertheless collapsed and was taken over in 1876 by Scottish entrepreneur **David Cargill**. In 1881, he was granted a prospecting license covering about four square miles adjacent to the hand-dug wells. He took the company public in 1886 as Burmah Oil Company Ltd. In the same year, Burma became a province in the British Empire. Burmah mechanized **drilling** and introduced other innovations, including a 275-mile **pipeline** between the oil fields and the Rangoon refinery in 1909.

The British government developed an interest in Burmah Oil, which was then the only large oil company in the empire. In 1905, the British Royal Navy under the direction of Admiral **John Arbuthnot Fisher** began converting its ships from coal to oil and entered into a long-term contract with Burmah. At the request of the British government, Burmah Oil became involved in **Iran**. It purchased a large **concession** initially granted by Persia

(Iran) to **William Knox D'Arcy**. The government feared that the cash-short D'Arcy might otherwise sell the concession to non-British entities. D'Arcy's company became a Burmah subsidiary. It struck oil in Iran in 1908 at Masjid e Suleiman, and Burmah Oil established the Anglo–Persian Oil Company (APOC) in 1909. Refining and transporting the Iranian oil proved too costly for Burmah. In 1914, APOC sold a 51 percent interest to the British government—*see* CHURCHILL, WINSTON (1874–1965). Burmah retained a minority interest in APOC.

Burmah concentrated on supplying petroleum products to the Indian subcontinent. It concluded a joint distribution agreement for **India** with **Royal Dutch Shell**, forming Burmah Shell. When the Japanese army invaded Burma in 1942, Burmah Oil's managing director, Robert I. Watson, ordered the destruction of the company's oil field installations and its Rangoon refinery to prevent Burmese oil supplies from falling into enemy hands. After the war, the company restored its equipment and operations. It sought but failed to obtain compensation from the British government for the wartime damage. In 1963, it sold all its Burmese oil interests to the government of Burma.

Burmah then diversified its exploration and production interests into Peru and **Ecuador** and acquired additional refining capacity. In 1966, it acquired Castrol, a major supplier of lubricating oil. It reorganized in 1968 as Burmah Castrol and two other groups. In the 1970s, the company went through financial difficulties and almost collapsed. It received a large loan from the Bank of England in 1974 and shortly afterward sold the bank its remaining shares in BP. In 1976, it sold its **North Sea** interests to the **British National Oil Company**. It sold off its exploration and production units in the **United Kingdom** and many other divisions. In 1987, Burmah Castrol changed its name to Castrol, but in 1990 it was Burmah Castrol again. It moved more decisively into lubricants and chemicals and in 2000 was acquired by BP Amoco, later BP PLC.

BURMAH SHELL. *See* INDIA.

BURTON, WILLIAM M. (1865–1954). American chemist and oil company executive who developed a process for thermal cracking, increasing the yield and quality of **gasoline** from a given quantity of crude oil. William Merriam Burton was born in Cleveland, Ohio, on 17 November 1865. In 1889, he earned one of the first chemistry PhDs in the **United States**, from Johns Hopkins University. He then went to work as **Herman Frasch**'s assistant at **Standard Oil**. In 1890, he moved to a new Standard Oil refinery at Whiting's Crossing, Indiana, near Chicago. He redirected his career toward management and rose rapidly. He became a vice president of Standard Oil of Indiana in 1903. At the time, Standard of Indiana was a subsidiary of the still-whole Standard Oil.

The early years of the 20th century saw an increasing demand for gasoline, which was to become a far more important petroleum product than the kerosene for illumination that had dominated the 19th-century industry. Typical crude oils yielded only about 20 percent gasoline, and increasing this yield was a major research problem. Burton was interested in the problem and once again became active in science. Working with Robert E. Humphreys and F. M. Rogers at the Whiting laboratory, Burton developed a thermal cracking process. By subjecting petroleum to both high temperature and high pressure, the researchers were able to double the yield of gasoline from a given quantity of petroleum.

Burton asked Standard Oil's management for $1 million to scale up the process, but they refused. After **John D. Rockefeller**'s retirement in 1897, the company had become more risk averse, and the high-pressure, high-temperature work was physically dangerous as well as expensive. The break-up of Standard Oil in 1911 left Standard Oil of Indiana (later **Amoco**, and still later **BP**) an independent company, which embraced Burton's work. The effort continued, and Burton filed for a patent on 3 July 1912. U.S. Patent No. 1,049,667 was granted on 7 January 1913. One of the crucial innovations was subjecting both the boiler and the condenser to high pressure, not only the boiler.

It should be noted that a thermal cracking process for petroleum similar to Burton's was developed and patented in **Russia** in 1891 by **Vladimir Shukhov**. Unlike Burton's work, Shukhov's was not immediately applied commercially to the important problem of increasing gasoline yields. Burton died in Miami, Florida, on 29 December 1954.

BUSH, GEORGE H. W. (1924–). Oil executive and 41st president of the **United States** (1989–1993). George Herbert Walker Bush was born 12 June 1924 in Milton, Massachusetts. He served in **World War II**, winning the Distinguished Flying Cross, and later attended Yale University, graduating in 1948.

Bush moved to **Texas** and began a career in the petroleum industry selling oil field supplies. In 1951, he and John Overbey formed the Bush-Overbey Oil Development Company. Two years later, Bush and Overbey joined with the brothers J. Hugh Liedtke and William Liedtke to establish **Zapata Petroleum**. In 1954, Zapata reorganized into Zapata Drilling Company and Zapata Offshore Company. Bush led Zapata Offshore until he was elected to the U.S. Congress in 1966. In 1963, the Zapata companies were merged with **South Penn Oil Company** to form **Pennzoil**.

Bush served as U.S. ambassador to the United Nations in the Richard M. Nixon administration and then as chief of the U.S. Liaison Office in **China** under President Gerald R. Ford. In 1976, he was director of the Central Intelligence Agency. In 1979, he ran for the Republican presidential nomination but lost to Ronald Reagan, whom he then served as vice president of the United States 1981–1989. In 1986, global oil prices collapsed as **Saudi Arabia** attempted to gain market share from the rest of the **Organization of the Petroleum Exporting Countries** (OPEC), and OPEC as a whole tried to gain market share back from **North Sea** and other producers (*see* NETBACK PRICING; OIL COUNTERSHOCK). Vice President Bush visited Saudi Arabia and warned that its current policies were producing "ramifications among its allies." He was speaking largely of the effects of low prices on the higher-cost U.S. petroleum industry. His visit did not appear to produce any immediate effects in the market.

In 1989, Bush became president of the United States. During his presidency, communism collapsed in the Soviet Union, and Germany was reunified. When **Iraq** invaded **Kuwait** in 1990, Bush skillfully assembled a coalition including European and Arab countries as well as the United States to restore Kuwaiti independence—*see* PERSIAN GULF WAR (1991). Bush ran for reelection in 1992 but was defeated by Democratic candidate Bill Clinton. Bush's son **George W. Bush** became the 43rd president of the United States in 2001.

BUSH, GEORGE W. (1946–). The 43rd president of the **United States** (2001–2009); son of **George H. W. Bush**, the 41st president. George Walker Bush was born 6 July 1946 in New Haven, Connecticut. In 1968, he graduated from Yale University. In 1975, he received an MBA from Harvard Business School.

Like his father, Bush began his business career in the oil industry around Midland, **Texas**. He started out as a "landman," organizing drilling ventures by bringing together geologists, property owners, and investors. In 1977, he formed Arbusto Energy in Midland (*arbusto* is a Spanish word meaning "bush" or "shrub"). He changed the company's name to Bush Exploration in 1982. The company was valued at less than $400,000 that year; however, a Bush family friend bought a 10 percent stake for $1 million.

Unfortunately, oil prices were dropping (*see* OIL COUNTERSHOCK), and by 1984 Bush Exploration was in financial trouble. It was rescued by a merger with Spectrum 7 Energy Corporation, which then made Bush its chief executive officer. Despite reporting a $1.5 million loss in 1985, Spectrum 7 was purchased in 1986 for $2.2 million by Harken Energy, which also assumed $3.1 million in debt. Bush joined the Harken board. Harken got a contract to drill **offshore Bahrain** despite having no international experience. George W. Bush sold most of his Harken shares on 22 June 1990, just before the end of a second quarter that produced major losses. The U.S. Securities and Exchange Commission investigated the sale but decided to take no action against Bush. Nevertheless Ann Richards, Bush's opponent in the 1994 Texas gubernatorial election, made the affair an issue during the campaign. Bush won the election nevertheless and became governor of Texas in 1995. He was reelected in 1998 but did not complete his second term, having been elected president of the United States in 2000. He was reelected president in 2004.

Bush's presidency was soon dominated by the terrorist attacks on the World Trade Center and the Pentagon on 11 September 2001. The United States invaded Afghanistan in 2001 in retaliation for the 11 September attacks. In September 2002, the Bush administration issued a new National Security Strategy incorporating the doctrine of preventive war: the United States would henceforth apply military force at will to forestall attacks from terrorists or rogue states possessing weapons of mass destruction. Although neither **Iraq** nor its leader had had anything to do with the attacks of 11 September, Bush ordered the invasion of that country in 2003, initiating the **Iraq War** and the U.S. occupation of Iraq. President Bush was succeeded in 2009 by Barack Obama.

C

CABINDA. Major port (pop. 598,000, 2014) of **Angola**. Cabinda is a northern exclave of Angola, bordered by the Republic of the **Congo** (Congo–Brazzaville) to the north and separated from the rest of Angola by a portion of the Democratic Republic of the Congo to the south. The main city, also called Cabinda, is an oil port on the right bank of the Bele River. The Malongo oil fields discovered from 1966 to 1969 are in the near **offshore**, and several **giant** oil fields discovered from the 1970s to the 1990s are about 60 miles west. The giant Takula oil field, with **EUR** of about 2.5 Bb, was discovered in 1971. Offshore discoveries further to the south continued into the 2000s. From the mid-1970s on, the exclave was plagued by guerrilla activities by groups such as the Front for the Liberation of the Enclave of Cabinda (Frente para a Libertação do Enclave de Cabinda—FLEC) demanding independence for the area.

CADMAN, JOHN (1877–1941). British mining engineer and executive; chairman of the Anglo–Persian Oil Company (later **Anglo–Iranian Oil Company**, eventually **BP**), 1927–1941. Born in Silverdale, England, John Cadman was trained in mining and geology at Durham University, graduating with first-class honors in 1899. In 1900, he became a fellow of the Geological Society and in 1904 became chief inspector of mines in Trinidad. Cadman was a professor of mining in Birmingham in 1908, and in 1912 he developed the **United Kingdom**'s first course in petroleum geology. He advised the British government on the importance of oil supplies before **World War I**, as the Royal Navy converted from coal to oil—*see* CHURCHILL, WINSTON (1874–1965); FISHER, JOHN ARBUTHNOT (1841–1920). He was a member of the **Slade Commission** that went to **Iran** to report on the oil fields there for the Admiralty. He was awarded a knighthood for these and related services in 1918. In 1921, he became a technical adviser of the Anglo–Persian Oil Company. Cadman was a champion of **geophysical prospecting** techniques and helped initiate their use in Iran.

Cadman rose rapidly through the ranks of APOC, becoming director in 1923, deputy chairman in 1925, and chairman in 1927. He believed in combinations and alliances between companies to limit destructive competition. In February 1928, he approached **Walter Teagle**, the head of Standard Oil of New Jersey (later **Exxon**), with an idea for establishing a "small 'clearing-house' for matters of the very highest policy" involving the West's three **major oil companies**: APOC, Standard Oil of New Jersey, and **Royal Dutch Shell**. Later in the year, he had an internal draft of a collaboration agreement produced. The draft formed the basis of the **As-Is Agreement** secretly concluded by the three companies to limit competition and control the global oil market.

When Shah **Reza Pahlavi** terminated Anglo–Persian's concession in Iran in 1933, Cadman directly negotiated a new agreement with him. In 1937, Cadman became Baron Cadman of Silverdale. He also became a fellow of the Royal Society. Cadman was one of the founders of the Institute of Petroleum and was its president in 1915 and 1936. He died in Bletchley, England, on 31 May 1941.

CALDERA RODRÍGUEZ, RAFAEL (1916–2009). President of **Venezuela** (1969–1974; 1994–1999) and reformer of petroleum laws. Rafael Antonio Caldera Rodríguez was born in Yaracuy, Venezuela, on 24 January 1916. He entered politics in 1936 when he was elected to the Chamber of Deputies. In 1946, he participated in the establishment of the Christian Democratic Party. In 1957, he was jailed and exiled by the military government of Marcos Pérez Jiménez. Caldera ran for president a number of times, finally succeeding in 1968.

Caldera was president during the **First Oil Shock**. In keeping with the times, he moved aggressively to enhance Venezuela's position vis-à-vis the **major oil companies**. He nationalized the **natural gas** industry, raised taxes on oil production, and enacted strict laws regulating foreign oil companies in Venezuela. The government adopted a reversion law providing for a return of all **concessions** to the state after expiration, beginning in 1983. This measure would be overtaken by events, with a full nationalization of Venezuela's industry occurring in 1975–1976 under Caldera's successor, **Carlos Andrés Pérez**. After leaving office, Caldera pursued a variety of scholarly and public projects with distinction. He served another term as president from 1994 to 1998, succeeding Carlos Andrés Pérez. Caldera was succeeded in turn by **Hugo Chávez**.

CALGARY. City (pop. 1.2 million, 2016) in the province of **Alberta, Canada**, and a major center of Canada's oil industry. The city is in the foothills of the Canadian Rockies, about 160 miles north of the border with the **United**

States. Approximately 400 oil companies have operations in Calgary. Calgary was founded in 1875 by the Royal Northwest Mounted Police as a fort at the confluence of the Bow and Elbow Rivers. It was originally named Fort Brisebois but was renamed Fort Calgary a year later. The oil and gas fields of the Turner Valley lie to the city's southwest. Like **Edmonton**, Canada's other major oil center, Calgary boomed in the 1970s and went into recession in the 1980s, closely following the trend in oil prices. The city later recovered economically and was booming again in the first decade of the 21st century. Its economy is still dominated by the petroleum industry, although there has been some diversification into high-tech manufacturing and tourism.

CALIFORNIA. Oil production 2015: 0.551 Mmbpd (5.9 percent of U.S.); **reserves** 2.33 Bb (7.2 percent of U.S.); gas production: 0.219 Tcf (0.81 percent of U.S.); reserves 1.832 Tcf (0.6 percent of U.S.). California became the **United States**' largest oil-producing state in 1903. It was overtaken by **Oklahoma** in 1907, then took the lead again from 1909 to 1914. Between 1915 and 1928, it traded places with Oklahoma several times for the leading position, holding it in 1919 and again in 1923–1926. In 1928, both California and Oklahoma were overtaken by **Texas**, which led in U.S. oil production for almost all of the rest of the 20th century. California oil production followed a complex pattern over time, with many peaks and valleys, but was on a long-term growth trend for the first half of the 20th century, reaching 1 Mmbpd in 1953. Between 1953 and 1981, it stayed mostly in a band between 0.8 Mmbpd and just over 1 Mmbpd. It then increased to its all-time high of 1.16 Mmbpd in 1985, after which it began a long-term decline. California's 2016 oil production was lower than it was in 1930. California has 17 **giant** fields, 15 predominantly oil and two predominantly **natural gas**. Four of these fields were discovered in the 19th century. All but one were discovered before 1950, and all but four were discovered before 1930.

California's Native Americans were aware of oil seeps and used material from tar pools to caulk their boats. Such pools were observed by Spaniards in the 1500s. Tar pools still exist in the city of Los Angeles, where the La Brea Tar Pits are a famous attraction. In the 1850s, Andrés Pico—who had served as the Mexican governor of Alta California and was later a state senator after California became a U.S. state—collected crude oil from seeps found in the Pico Canyon, named for him, in present-day Los Angeles County. He distilled it to create illuminating oil for the San Fernando Mission. There were other early refining operations in the 1850s, although records are spotty. C. S. Gilbert had a commercial refinery in southern California that may have been operating as early as 1857, producing roughly 7–10 barrels a week.

California's first oil boom was in the 1860s. The first well intentionally drilled for oil was probably in Humboldt County, in northern California, in the early part of the decade. It was unsuccessful. George Gilbert drilled wells in the Ojai Valley, which **Benjamin Silliman Jr.** of Yale University proclaimed would yield large quantities of oil. Silliman's reports contributed to a speculative oil bubble in southern California in 1865. A number of companies failed. The most successful well was drilled by Thomas R. Bard, a future president of the **Union Oil Company of California**, in 1866. It produced about 15–20 bpd. Although southern California did have enormous quantities of oil and Silliman would ultimately be proved correct, the first boom of the 1860s was a failure.

In the 1870s, a more stable industry began to take hold. In 1876, C. A. Mentry of the California Star Oil Works Company drilled Pico Canyon Well No. 4, which turned out to be the discovery well of the Newhall field. It flowed at 25 bpd and is usually cited as California's first true commercial well. This early activity in Pico Canyon formed the seed for what was to become the Standard Oil Company of California, later **Chevron**. Another important California company, the Union Oil Company of California, also had some of its roots in Pico Canyon, beginning in 1884.

In 1885, the Hardison and Stewart company, a Union Oil predecessor, began commercial development in Adams Canyon in Ventura County, to the west of Los Angeles. In 1892, Union Oil brought in California's first gusher, in Adams Canyon. The well, No. 28, flowed 40,000 barrels before it was brought under control. Union did not have enough infrastructure to capture the oil. The great majority flowed down the Santa Clara River into the Pacific Ocean. The next year, **Edward Doheny** and his partner Charles Canfield discovered the Los Angeles City oil field and initiated the Los Angeles oil and land boom. In 1896, the first **offshore** wells in the United States were drilled from piers built over the water close to the shore, to exploit an offshore extension of the Summerland oil field in Santa Barbara County.

Major discoveries in the late 19th century included Brea (1884), Coalinga (1887), Kern River (1899), and Midway–Sunset—California's second-largest field, and the fourth largest in the United States. Midway–Sunset is a complex of 22 identifiable pools, located northwest of Los Angeles in Kern County. Discoveries in the Sunset part of the complex date back to 1890. The first wells were small. Development accelerated after 1900 with the discovery of the Midway part of the complex. In 1910, Union Oil, in partnership with the Lakeview Company, drilled the Lakeview gusher in the Midway–Sunset field. This was the longest-lasting (18 months) and the most productive (9 Mmb) gusher in U.S. history. Discoveries in the field continued throughout the 20th century. Union Oil and Chevron were involved in the field from the earliest days. **Royal Dutch Shell** and numerous other companies have also had interests. The Midway–Sunset complex had produced a total of 2.4 Bb by 1999. With an **EUR** of about 2.7 Bb and 0.5 Tcf of gas, it is nearing the end of its useful life.

Between 1909 and 1928, eight giant fields were discovered in California. The largest were South Belridge in 1911 (EUR about 1.38 Bb), Elk Hills in 1919 (EUR about 1.41 Bb), and Huntington Beach in 1920 (EUR about 1.14 Bb). Elk Hills was discovered in an area set aside as a U.S. **Naval Petroleum Reserve**. The discovery of the improper leasing of Elk Hills to Edward Doheny in exchange for a bribe would later form part of the **Teapot Dome scandal**. Elk Hills lay largely idle until 1976, when President Gerald Ford authorized production from the Naval Reserves. Elk Hills produced at a peak rate of 181,000 bpd in 1981 and for a time was the top-producing field in the United States outside **Alaska**. On 5 February 1998, the U.S. Department of Energy sold Elk Hills to **Occidental Petroleum Corporation** for $3.65 billion.

California's largest field, and the third-largest field discovered in the United States (after the **East Texas** and **Prudhoe Bay** fields), is Wilmington. It is located within the city of Long Beach, near Los Angeles. Wilmington was discovered in 1932. The first large well in the field was General Petroleum Corporation's Terminal No. 1 in December 1936, which flowed initially at 1,500 bpd. The Wilmington field was developed by a multitude of companies and produced 2.7 Bb by 2013. An assessment by the **United States Geological Survey** that year estimated that the field may contain between 0.2 to nearly 2 Bb of recoverable oil. Previous estimates of the EUR were lower than the amount of oil already actually produced.

California may have **tight oil** resources in its Monterey Shale Formation, although estimates of the technically recoverable resources have varied widely. The U.S. Energy Information Administration initially estimated 13.7 Bb in 2011, but revised that number drastically downward in 2014 to 0.6 Bb after a more careful consideration of the available technology. In any case, California is not a friendly environment for the **fracking** techniques typically required to extract tight oil. Several counties have banned the practice. The sixth county to do so, Monterey County, sparked a legal challenge at the end of 2016. Chevron and Aera Energy, with interests in the San Ardo fields, sued the county to block implementation of the ban, on the grounds that it places undue restrictions on how they can use their property. *See also* SANTA BARBARA OIL SPILL; STEWART, LYMAN (1840–1923).

CALIFORNIA ARABIAN STANDARD OIL COMPANY (CASOC). *See* ARAMCO.

CALTEX (CALIFORNIA–TEXAS COMPANY). Joint venture formed 30 June 1936 between Standard Oil of California (Socal, later **Chevron**) and the Texas Corporation (known before 1926 and after 1941 as the Texas Company, and after 1959 as **Texaco**; ultimately part of Chevron). Socal had not yet found oil in **Saudi Arabia**, but it had in **Bahrain**. It had oil, but no marketing outlets in the region. The Texas Corporation had marketing outlets in Africa and Asia but no crude oil in the Eastern Hemisphere. The two companies decided to work together, pooling their resources "east of Suez." Socal sold the Texas Corporation 50 percent of its Casoc subsidiary (California Arabian Standard Oil Company, the predecessor to **Aramco**) operating in Saudi Arabia, and 50 percent of Bahrain Petroleum Company (Bapco), its Bahrain subsidiary. The Texas Corporation paid $21 million and 50 percent of its Eastern Hemisphere marketing operations. The marketing operations were made into a subsidiary of Bapco named **Caltex**. The area of operation for the joint venture was delineated by the **Blue Line Agreement**, a reference to the **Red Line Agreement** in which neither company participated. Caltex also operated both upstream and downstream in **Indonesia**.

Casoc went on to become Aramco. In 1948, the original Aramco partners Socal and the Texas Company (as it was now again known) were joined by new ones, Standard Oil of New Jersey (later **Exxon**) and Socony–Vacuum (later **Mobil**). In 1973, the Saudi government began acquiring Aramco under the **participation** framework of the **Organization of the Petroleum Exporting Countries** (OPEC), and by 1980 the process was complete. Caltex was not part of it. It continued to be a downstream venture owned by Socal and Texaco, expanding its operations globally over the decades. It moved its headquarters from Dallas to Singapore in 1999. In 2001, its two parents merged into ChevronTexaco, and Caltex became part of the merged company. In 2005, ChevronTexaco formally changed its name simply to Chevron.

CAMPOS BASIN. *See* BRAZIL.

CANADA. Oil production 2015: 3.67 Mmbpd (4.72 percent of world); **reserves** 169.7 Bb (10.31 percent of world); gas production: 5.256 Tcf (4.22 percent of world); reserves 77.07 Tcf (1.12 percent of world). The ***Oil and Gas Journal*** began to include the **Alberta oil sands** in its reserves estimate for Canada beginning in 2003. This had the effect of increasing the reserves estimate by a factor of 37, from 4.858 Bb in 2002 to 180.021 Bb in 2003, of which 174.8 Bb were oil sands. The decision by *Oil and Gas Journal* reflected the increasing importance of oil sands to Canada's oil production. Canada's 2015 reserves of **conventional** crude were 4.132 Bb.

A recognizably modern oil industry began to take shape in the province of Ontario in the 1850s. By 1857 or 1858, one to two years before **Edwin Drake**'s **Pennsylvania** well, **James Miller Williams** was **refining** petroleum in Ontario. The petroleum was possibly obtained from hand-dug wells. Oil fields were developed in the province's southwestern peninsula, between Lakes Huron and Erie, in the second half of the 19th century. Production peaked at just under 2,200 bpd in 1890. Canada did not become a major oil producer until the 1940s, when large discoveries were made in **Alberta**. Alberta accounted for about three-fourths of cumulative Canadian conventional crude production through 2001 and has assumed renewed importance with the development of the oil sands.

Outside Alberta, one of the most important discoveries was the Hibernia field in the Atlantic off Newfoundland. The Hibernia field, Canada's largest conventional oil field (**EUR** 1.85 Bb), was discovered by Standard Oil Company of California (later **Chevron**) in 1979 and began producing in 1997. Partners in the development of the field include Chevron, **ExxonMobil**, **Petro-Canada**, Murphy Oil, and **StatoilHydro**. Hibernia lies under about 260 feet of water and has been developed using a massive concrete structure designed to withstand icebergs. Another important discovery in Atlantic Canada was the Terra Nova oil field in 1984, made by Petro-Canada.

Canada's oil production has been systematically recorded since 1863. It remained at a low level for more than 80 years. As late as 1930, it was just over 4,000 bpd. After the Leduc discovery in Alberta in 1946, however, Canadian production began to rise rapidly. In 1948, it was 0.033 Mmbpd. It surpassed 1 Mmbpd in 1968 and 1.5 Mmbpd in 1972. It reached a peak of nearly 1.8 Mmbpd in 1973. Between 1975 and 1981, it stayed mostly between 1.3 and 1.6 Mmbpd and then entered a period of long-term growth. By 2015, production was 3.67 Mmbpd. Of this, 25.3 percent was **syncrude** and 34 percent was marketable **bitumen** from the oil sands. Broken down another way, 81.2 percent of Canada's oil production came from Alberta, 11.1 percent from Saskatchewan, and the remainder from the rest of Canada. Canada's **natural gas** production grew from 2.76 Tcf in 1980 to 6.47 Tcf in 2000. In 2015 it was 5.256 Tcf.

Excluding the oil sands, Canada has 22 **giant** fields (11 of which are in Alberta); 13 are predominantly oil (nine of which are in Alberta), and nine are predominantly gas. All of Canada's giants were discovered before 1982. *See also* CALGARY; CANOL PROJECT; EDMONTON; IMPERIAL OIL.

CANADIAN OIL SANDS. *See* ALBERTA OIL SANDS (TAR SANDS).

CANADIAN TAR SANDS. *See* ALBERTA OIL SANDS (TAR SANDS).

CANFIELD, CHARLES. *See* DOHENY, EDWARD (1856–1935).

CANOL PROJECT. A project to develop oil in **Canada**'s Northwest Territories as a fuel supply for U.S. military forces in **Alaska** during **World War II**. The Japanese capture of the Aleutian Islands of Attu and Kiska in June 1942 highlighted the strategic importance of Alaska. The **United States** and Canada quickly began construction of the Alaska Highway. The U.S. War Department also moved rapidly to secure potential oil supplies in the region. The decision to go ahead with the Canol (Canadian Oil) project was taken rather precipitously on 19 April 1942, without consulting the secretary of the navy or **Harold Ickes**, the secretary of the interior and head of the **Petroleum Administration for War**.

The project sought to develop existing oil **reserves** at Norman Wells on the Mackenzie River. Oil had been reported by Europeans as early as 1789, but the first modern discovery was in 1920. By 1939, a small refinery was producing 840 bpd to supply the local area. Norman Wells was the northernmost-producing field in North America at the time. The Noble Drilling Corporation of **Tulsa**, **Oklahoma**, drilled new wells to expand the field. Bechtel Price-Callahan, the project's main contractor, constructed a 557-mile, 4-inch **pipeline** under very difficult conditions from Norman Wells to a new refinery at Whitehorse in the Yukon Territory. The pipeline was completed on 16 February 1944, along with an additional leg from Whitehorse to the U.S. Army Air Corps base in Fairbanks, Alaska. First oil flowed to Whitehorse in April 1944.

The project was continually criticized for cost overruns and accused of poor planning. Still, the achievement was impressive, although the project did not enjoy a long useful life. The feared Japanese invasion of Alaska never happened, and the pressing need for Norman Wells oil disappeared by the time the project was completed. The Whitehorse refinery was shut down on 1 April 1945. The total cost of Canol was $134 million.

After the war, when **Imperial Oil** discovered the giant Leduc field near **Edmonton** in February 1947, a refinery was urgently needed. Designing and building one would take three years. Imperial Oil decided instead to move the Whitehorse refinery to Edmonton.

CANTARELL. The largest oil field complex in **Mexico**, located about 50 miles **offshore** in the Bay of Campeche (in the **Gulf of Mexico**) under about 130 feet of water, now nearly exhausted. Cantarell is composed of four major fields: Akal (the largest, with about 90 percent of the **reserves**), Nohoch, Chac, and Kutz. The Sihil field is often included in the complex as well. Cantarell was discovered in 1976 and began production in 1979. The discovery traces its roots back to March 1971, when a fisherman named Rudesindo

Cantarell first noticed oil floating to the surface of the Bay of Campeche. By 1981, production was nearly 1.2 Mmbpd. It dropped to 1 Mmbpd by 1995. Beginning in 2000, the field was rejuvenated by nitrogen injection (42 Mmcf per day), and by 2003 production was an astounding 2.1 Mmbpd, about 62 percent of Mexico's output that year. However, the spike was short lived. After a similar figure in 2004, production began to decline rapidly. Cantarell lost 0.101 Mmbopd of production in 2005, 0.234 Mmbpd in 2006, and 0.304 Mmbpd in 2007, when production was under 1.5 Mmbpd. By 2009, production was down to 0.5 Mmbpd, and by 2014 it was 0.340 Mmbpd. Cantarell has an **EUR** of 17–19 Bb, making it a **supergiant**, although a nearly exhausted one.

CÁRDENAS, LÁZARO (1895–1970). President of **Mexico** 1934–1940; nationalized the Mexican oil industry, 1938. Lázaro Cárdenas del Rio was born 21 May 1895 in Jiquilpan, Mexico. He fought in the Mexican Revolution and became a general in the Mexican army. In 1928, he was elected governor of Michoacan, his native state. In 1929, he became president of the Partido Nacional Revolucionario (PNR, later to become PRI) and in 1934 was elected president of Mexico. As president, Cárdenas embarked on a wide-ranging reform program. He distributed land, organized workers and peasants, and made loans available to poor common people. He also expropriated and nationalized industries dominated by foreign ownership, including the petroleum industry.

The Mexican constitution of 1917 declared in Article 27 that underground resources belonged to the Mexican people and not to the owner of the corresponding surface property. This was not vigorously enforced in the 1920s, but it was never forgotten. Cárdenas was particularly bothered by the heavy foreign presence in Mexico's oil industry. He resented the attitude of the foreign oil companies, many of which treated the Mexican government as subordinate. In the wake of an oil workers' strike in 1937, Cárdenas appointed a presidential commission to examine the foreign oil companies. The commission recommended higher wages for oil workers and a large number of new benefits. It also called for foreign technical personnel to be replaced by Mexicans within two years. The companies appealed, but the government sided with the commission. The administration of Franklin D. Roosevelt in the **United States** did not give the American companies in Mexico any support. With war on the horizon, the U.S. government wanted to ensure good relations with Mexico.

On 18 March 1938, Cárdenas signed the order to nationalize the oil industry and expropriate foreign assets. Affected companies included **Mexican Eagle**, owned by **Royal Dutch Shell**; Standard Oil Company of New Jersey (later **Exxon**); **Sinclair**; **Cities Service**; and **Gulf**. Partial compensation for expropriated assets came later, after Cárdenas's presidency. Lázaro Cárdenas withdrew from public life in 1940 when his presidential term was over, but the outbreak of **World War II** drew him back in. He was minister of national defense from 1943 to 1945 and commander in chief of the Mexican army in 1945. He died on 19 October 1970 in Mexico City.

CARGILL, DAVID (1826–1904). Founder of the **Burmah Oil Company** (later Burmah-Castrol and eventually part of **BP**). David Sime Cargill was born in Scotland on 9 April 1826. From 1844 to 1861, he worked in Ceylon (Sri Lanka) for a trading firm. In 1876, he visited Burma (the early spelling was "Burmah," and the country is now known as Myanmar) and then bought the operations of the ailing Rangoon Oil Company. Rangoon Oil had been established in 1871 to refine and market oil gathered from existing hand-dug wells in Burma. These wells had been observed in 1795 by Westerner Michael Symes and may have been operating for centuries. The Rangoon Oil Company had run into financial difficulties because the independent king of Upper Burma acted as a monopolist and overcharged for any crude oil found on his territory. In 1881, Cargill was granted a prospecting license on four square miles adjacent to the hand-dug wells. His company continued to incur losses, which he covered personally. The British annexed Upper Burma in 1886, and in that year Cargill took his company public as the Burmah Oil Company Ltd. The British authorities granted the company prospecting licenses. Cargill modernized operations, introducing mechanized **drilling** and improving the Rangoon refinery. The company found success marketing kerosene for illumination in India and Burma. In 1904, it sold more than 2 Mmb of refined product and made a profit of £264,000. David Cargill died on 25 May 1904 near Glasgow, Scotland.

CARLL, JOHN F. (1828–1904). Pioneering petroleum geologist and engineer; relatively unrecognized during his lifetime, but his work and ideas had a lasting impact. John Franklin Carll was born 7 May 1828 on a farm in what is now Brooklyn, New York. He attended Union Hall Academy in Flushing and had no further formal education. He worked in a variety of occupations, including newspaper publishing, surveying, civil engineering, and wire manufacture. After the tragic death of his wife and two young children in 1859 and the 1864 destruction of the wire works that employed him, he moved to **Venango County**, **Pennsylvania**, to work in the oil business. He diligently studied the industry and geology of the region and was soon in demand as a geologist. In 1874, the Pennsylvania legislature ordered a second geological survey, the first having been conducted in 1836–1841, well before the oil boom. Carll joined the survey and produced several reports that were models of modern scientific petroleum geology, at a time when such a systematic approach was not yet common. The emphasis of the survey eventually turned away from the petroleum regions, and Carll resigned in 1882, returning to the petroleum industry as a consultant. He remained a petroleum industry consultant except for 1885–1888, when he returned to the survey.

Carll made many innovations in petroleum geology. He was the first to use drill cuttings to perform the important task of correlation between wells. He also devised the strip log, a graphic representation of rock penetrated in a well bore, annotated with lithological, mineralogical, and paleontological information. He realized that the Venango County oil sands primarily originated as beach sands or other shallow marine accumulations. Importantly, Carll recognized that oil occurs primarily in the pores of the reservoir rocks themselves, rather than in underground caverns as was commonly believed. Further, he recognized the importance of gas pressure as a driver of oil movement in a reservoir. He also saw that subsurface oil occurrence was not always correlated with surface landscape features. This may have led him to miss the importance of anticlinal traps, which are not dominant in Pennsylvania but have constituted a mainstay of petroleum geology in general. However, some observers have pointed out that Carll understood anticlinal traps but that his reports were censored by the Pennsylvania Geological Survey director, J. P. Lesley, who was a passionate opponent of the anticlinal theory. In addition to his geological work, Carll also invented a variety of oil field equipment.

In 1886, Carll issued an inaccurate prediction that Pennsylvania's oil fields, having produced an estimated 330 Mmbo, were nearing exhaustion. Pennsylvania went on to produce a cumulative 1.4 Bb by 1999. Carll was hardly alone in making this particular mistake, which would be repeated time and again in the oil industry all over the world. Carll died in Waldron, Arkansas, on 13 March 1904, after 15 years of poor health. His contributions remained largely unrecognized. Today, he is seen as one of the pioneers of both petroleum geology and petroleum engineering. The **Society of Petroleum Engineers** named a medal after him in 1957, awarding it annually to a member whose outstanding achievements have advanced petroleum engineering or the application of engineering principles to petroleum development and discovery.

CARTER, JAMES EARL, JR. ("JIMMY") (1924–). The 39th president of the **United States** (1977–1981). James Earl Carter Jr. was born 1 October 1924 in Plains, Georgia. He served in the Georgia State Senate (1962–1966) and as governor of Georgia (1971–1975) and was elected U.S. president in 1976. As president, he espoused a pessimistic view of the future of global petroleum supplies, advocated energy independence for the United States, and promoted conservation. He established a national energy policy and created the U.S. Department of Energy in 1977. He also enunciated the **Carter Doctrine**, declaring the **Persian Gulf** to be an area of vital U.S.

interest. Carter was president during the **Second Oil Shock** and the taking of American hostages by revolutionaries in **Iran**. In the immediate aftermath of the hostage crisis, he stopped imports of Iranian oil into the United States and froze Iranian assets in the country.

Carter ran for reelection but lost to Ronald Reagan in 1980. After leaving office, he maintained a high profile as an advocate for peace and for democratic elections around the world. He was awarded the Nobel Peace Prize in 2002.

CARTER DOCTRINE. Doctrine enunciated by U.S. President **Jimmy Carter** declaring the **Persian Gulf** an area of vital U.S. interest. Carter announced the doctrine during his State of the Union address in January 1980. He declared that the **United States** would regard any attempt to gain control of the Persian Gulf area as an assault on its vital interests and would respond by military force if necessary. This had been U.S. policy for decades, but the declaration was important for its explicitness. It echoed a similar statement, the **Lansdowne Declaration**, made by the **United Kingdom** in 1903.

CASPIAN AND BLACK SEA OIL PRODUCTION AND TRADING COMPANY. *See* ROTHSCHILD COMPANY.

CASPIAN PIPELINE CONSORTIUM (CPC). *See* CASPIAN SEA; KAZAKHSTAN; NOVOROSSIYSK.

CASPIAN SEA. The Caspian is the world's largest enclosed body of water, variously classed as an inland sea or the world's largest lake. The 1982 Convention on the Law of the Sea defined it as a unique "special inner sea." The Caspian has a surface area of 143,244 square miles. Its depth ranges from very shallow (about 16 feet) in the north to greater than 3,000 feet in the south. In the central area, it is roughly 600 feet deep.

The Caspian was one of the cradles of the modern petroleum industry, with deliberate exploration and development stretching back to the early 19th century, and premodern exploitation going back much further. The Caspian has also been a focus of renewed attention from the petroleum industry and from politicians since the breakup of the Soviet Union. The 2000 discovery of **Kazakhstan**'s **offshore** 10 Bb Kashagan oil field, perhaps the most important find anywhere in the world from the 1970s up to that time, caused considerable excitement. Some have seen the greater Caspian region as the new indispensable alternative to the **Persian Gulf**. Others believe that its potential resources have been wildly overestimated. Potential recoverable resources of more than 200 Bb have regularly been quoted, but such estimates are not well grounded. In 2013, the U.S. Department of Energy's Energy Information Administration published an estimate of 48 Bb and 292 Tcf of gas. Other estimates from informed sources have ranged as low as 8 Bb.

The Caspian is bordered by **Azerbaijan**, **Russia**, Kazakhstan, Turkmenistan, and **Iran**, all of which are interested in the basin's petroleum resources. The division of the Caspian between these powers is unresolved and can stymie further development. If the Caspian is regarded as a sea, it makes sense to divide it according to the principle of median lines, as was done in the **North Sea**. If it is regarded as a lake, on the other hand, one might make the argument that it belongs equally to all the littoral states. Iran has tended to favor the latter approach, while the other states have tended to favor the former. Iran refers to previous treaties with Russia and the Soviet Union (1921 and 1940), asserting that the Caspian belongs to both nations. After 1991, however, Iran and Russia no longer shared a border, and three new nations came into the mix. Also, the actual behavior of Iran and the Soviet Union during communist times belied the notion of condominium on the Caspian. The USSR and Iran carried out their activities in the Caspian independently. They explored for and exploited oil resources without consulting each other or claiming joint ownership. The USSR, for its part, divided the Caspian internally among its littoral republics according to the median line principle, and the Soviet Oil Ministry ordered petroleum deposits distributed accordingly.

The nations of the former Soviet Union proceeded to conclude agreements among themselves, without Iran. In 1997, Kazakhstan and Azerbaijan agreed to adhere to borders drawn according to a median line principle until a comprehensive convention on the Caspian's legal status could be concluded. That same year, Kazakhstan and Turkmenistan signed a communiqué pledging to divide their interests in the Caspian using median lines. In 1998, Russia and Kazakhstan concluded a bilateral agreement dividing the northern Caspian seabed along median lines and pledging to develop resources along the boundary jointly. At the 2001 summit of the Commonwealth of Independent States (CIS), Kazakhstan and Azerbaijan followed up their 1997 understanding with a bilateral agreement based on a modified median line division. They signed an agreement on 29 November 2001 and an implementation protocol on 27 February 2003, In January 2001, Russia and Azerbaijan agreed on a median line division of the seabed, which they finalized on 24 September 2002. On 14 May 2003, Kazakhstan, Azerbaijan, and Russia signed a three-way agreement delimiting adjacent sections of the Caspian. Problems remain between Azerbaijan and Turkmenistan because Azerbaijan's **Abseron Peninsula** juts sharply out into the Caspian.

As of 2017, there was still no comprehensive agreement including Iran. Bilateral treaties had fully defined the Russian sector and mostly defined the Kazakh and Azeri sectors. The undefined parts of the Kazakh sector were not disputed, although those of the Azeri sector still were. Turkmenistan's interests were only partially defined, and Iran's were not settled at all. Iran did not recognize the bilateral agreements and favored a single multilateral agreement that would give each nation a 20 percent share (a larger share than Iran could obtain by any reasonable application of median lines). *See also* BAKU; BAKU–TBILISI–CEYHAN (BTC) PIPELINE.

CASTROL. *See* BURMAH OIL COMPANY.

CERRO NEGRO PROJECT. *See* ORINOCO OIL BELT.

CEYHAN. City (pop. 108,000, 2012) in Adana Province on Turkey's southeast Mediterranean coast; terminus of the **Baku–Tbilisi–Ceyhan Pipeline** and the **Kirkuk–Ceyhan Pipeline** (although the terminus of the Kirkuk–Ceyhan Pipeline is technically the nearby port of Yumurtalik). The city is located on the Ceyhan River, about 27 miles east of Adana. Ceyhan's oil terminal contains seven storage tanks and a jetty that can simultaneously load two **tankers** up to 300,000 deadweight tons (DWT) each. Since first oil from the Baku–Tbilisi–Ceyhan Pipeline reached Ceyhan in May 2006, the city has become a crucial port for distributing oil from the landlocked **Caspian Sea**. The Kirkuk–Ceyhan Pipeline has operated irregularly since the beginning of the **Iraq War** in 2003.

CFP. *See* COMPAGNIE FRANÇAISE DES PÉTROLES (CFP).

CHACO WAR. War fought between **Bolivia** and Paraguay from 9 September 1932 to 12 June 1935, in which oil formed an important part of the background. This bitter conflict was the most intense war of the 20th century, measured by battlefield deaths as a proportion of the civilian populations of the belligerent countries. During the war, Paraguay briefly occupied some of Bolivia's oil regions but was repelled. Ultimately, Bolivia lost the war. Rumors spread that **Standard Oil** of Bolivia (a subsidiary of Standard Oil of New Jersey, later **Exxon**) and **Royal Dutch Shell** had started the war deliberately in order to gain control of hypothetical oil resources in the disputed Chaco region, and these suspicions persisted. Some commentators even argued that it had been a proxy war between Standard Oil and Shell. Scholar Stephen Cote largely dismisses such claims but argues that the Chaco War was an oil war nevertheless. The unfolding of the war intensified Bolivia's awareness of the precious nature and the vulnerability of its natural resources and increased its sense of oil as a strategic resource that should not be in foreign hands. This impression was fortified by Standard Oil's position of neutrality in the conflict and its refusal to increase production to support Bolivia's war effort.

CHAD. Oil production 2015: 0.153 Mmbpd (0.153 percent of world); **reserves** 1.5 Bb (0.09 percent of world). The country of Chad is a relatively new oil producer. In 1979 and 1980, it briefly produced a modest 1,500 bpd from the Sédigui region north of Lake Chad. Exploration resumed in the early 1990s in the Doba basin in the southern part of the country, and significant oil reserves were identified. Chad signed an agreement in 1995 with Cameroon for a **pipeline** from Doba to the Cameroonian port of Kribi. Construction of the 670-mile pipeline began in 2000 and finished in 2003. Its capacity is about 0.225 Mmbpd. The pipeline is operated by the Chad Cameroon Pipeline Development Project (CCPDP) consortium involving **ExxonMobil** (40 percent), **Chevron** (25 percent), and **Malaysia**'s Petronas (35 percent). The consortium also operates the Doba oil fields, mainly Bolobo, Komé, and Miandoun. The pipeline made Chad a net oil exporter. Chad's oil production began in 2003 at 36,000 bpd and increased nearly fivefold to 0.177 Mmbpd by 2005. In July 2006, Chad created a national oil company, the Société des Hydrocarbures du Tchad (SHT), with the goal of controlling 60 percent of the country's oil sector and entering the CCPDP as a fourth partner. In 2016 a court in Chad fined the consortium led by ExxonMobil an astonishing $75 billion for an alleged $819 million in unpaid royalties. This sum is nearly seven times the country's gross domestic product. ExxonMobil disagreed with the decision and negotiated with the government, reaching a settlement in 2017.

CHARTERHOUSE PETROLEUM. *See* PETROFINA.

CHÁVEZ, HUGO (1954–2013). Four-term president of **Venezuela** (1999–2001, 2001–2007, 2007–2013, and January–March 2013) with a highly nationalistic outlook on the petroleum industry. Hugo Rafael Chávez Frías was born 28 July 1954 in Sabaneta, Venezuela. He became a military officer and in 1992 attempted a coup against President **Carlos Andrés Pérez**. The coup failed and Chávez was imprisoned. He was pardoned by President **Rafael Caldera** in 1994 and released. He became politically active again, founding the Fifth Republic Movement.

Chávez was elected president of Venezuela in 1998 and initiated a "Bolivarian Revolution," a socialistic program of industry nationalization and redistribution of wealth. It was coupled with strong criticism of globalization, neoliberal economics, and the **United States**. (Ironically, Simón Bolívar was a proponent of the free market and an admirer of the American Revolution.) Under a new constitution, elections were required again in 2000. These were held, and Chávez won, beginning a second term in 2001. In 2000, Venezuela's National Assembly passed the Enabling Act, which allowed Chávez to rule by decree for one year. He used this power to enact 49 decrees, including a new hydrocarbons law that would reverse many aspects of the **Apertura Petrolera**. The Apertura reforms had allowed foreign participation in the Venezuelan petroleum industry and had resulted in a significant increase in oil production after decades of decline. Chávez was strongly opposed to the Apertura, regarding it as an unacceptable infringement of national sovereignty over petroleum.

Chávez took an activist attitude toward the national oil company **PdVSA**. He accused it of acting as a "state within a state" and wanted to bring it under control. His actions led to discontent and work stoppages at the end of 2001. The PdVSA stoppage led to a general strike. In the chaos, there was a short-lived coup against Chávez on 11 April 2002. He was held at a military base, and Pedro Carmona was appointed interim president. Carmona issued a decree reversing Chávez's reforms. This was met with massive pro-Chávez uprisings, and by 13 April Chávez was president again. He alleged U.S. involvement in the coup, which the U.S. government denied. Chávez reportedly had advance knowledge of a possible coup from his former energy minister, Ali Rodriguez, then secretary general of the **Organization of the Petroleum Exporting Countries** (OPEC), and had prepared some troops to help him in the eventuality.

In December 2002, there was a strike at PdVSA instigated by anti-Chávez elements in the management. They wanted to block Chávez's access to oil revenue and thus force him out of office. PdVSA's operations were brought to a virtual standstill and Venezuelan oil exports nosedived. There were also significant domestic shortages. Chávez fired PdVSA's upper management and 18,000 PdVSA employees, largely skilled workers. Meanwhile, Chávez continued with ambitious socialist reforms, including adult literacy programs, programs to protect indigenous peoples, promises of free higher education, and many other projects. In 2004, he survived a recall vote. The disarray in the Venezuelan oil industry wrought by his interference in PdVSA and his reversals of the Apertura was masked somewhat by sharp increases in the price of oil (*see* THIRD OIL SHOCK). Oil revenues helped consolidate Chávez's position. He focused on foreign relations, stepping up his anti-U.S. rhetoric and engaging with such leaders as Mahmoud Ahmadinejad of **Iran** and Alyaksandr Lukashenka of Belarus, among others. He dubbed himself, Fidel Castro, and **Evo Morales** the "Axis of Good," in mocking reference to U.S. president **George W. Bush**'s 2002 designation of Iran, North Korea, and **Iraq** as part of an "Axis of Evil." His government increasingly began to buy weapons from **Brazil**, **Russia**, **China**, and Spain rather than the United States. In 2005, he began a program to provide discounted heating oil for low-income families in the northeastern United States.

Chávez was reelected president of Venezuela on 3 December 2006 for a third term. He obtained new powers to rule by decree on 31 January 2007 for a period of 18 months. The government purchased stakes in foreign electricity and telecommunications firms. In May 2007, he announced intentions to take majority control of the Apertura-initiated heavy crude projects in the **Orinoco Oil Belt**. Venezuela followed through. **Total**, **Chevron**, **Statoil**, and **BP** gave up majority control of their projects in the country. **ExxonMobil** is an example of a company that refused to comply with government demands, and its assets were expropriated. The company sued and was awarded $1.6 billion in 2014 by the International Center for Settlement of Investment Disputes.

In December 2007, Chávez lost a referendum on a package of constitutional changes that included a provision allowing him to run for reelection indefinitely. However he won a different referendum in 2009 that abolished the two-term limit for all public offices. On 7 October 2012 he won election to a fourth term as president. However, he died on 5 March 2013, serving only three months. He was succeeded by Vice President Nicolás Maduro. The actions of Hugo Chávez and those of his successor, whatever their motivations, are widely regarded as causing significant damage to the Venezuelan oil industry.

CHESAPEAKE ENERGY. *See* McCLENDON, AUBREY (1959–2016).

CHEVRON. A **major oil company**; one of the **Seven Sisters**. Chevron's origins date back to the earliest days of the **California** petroleum industry. In 1875, driller C. A. Mentry, employed by the Star Oil Works, struck small amounts of oil in Pico Canyon north of Los Angeles. In 1876, Star Oil Works was reorganized as the California Star Oil Works Company, and Mentry drilled Pico Canyon Well No. 4. This well flowed 25 bpd and is considered the discovery well of the Newhall field, marking the beginning of the modern California oil industry. On 10 September 1879, Charles N. Felton, Lloyd Tevis, George Loomis, and others organized the Pacific Coast Oil Company, which acquired the assets of California Star Oil Works.

By 1900, Pacific Coast had several producing wells in the Newhall field, a refinery near San Francisco, its own railroad tank cars, and a **tanker** named for George Loomis, Pacific Coast's first president. In that year, it was acquired by **Standard Oil** for $761,000. In 1906, Standard Oil merged Pacific Coast with its Standard Oil Company of Iowa, a marketing subsidiary active in California and a major customer of Pacific Coast. The merged entity was named Standard Oil Company (California), or California Standard. Standard increased the company's capitalization from about $1 million to $25 million, adding refinery capacity and using California Standard to supply Asian markets, saving on freight costs.

After the 1911 dissolution of Standard Oil, California Standard became an independent company, with strength in **refining** and marketing, a network of **pipelines**, a small tanker fleet, and $14 million in retained earnings. Its first president was Demetrius Scofield, who had been involved with California Star Oil Works and Pacific Coast. In 1912, the company moved into newly built headquarters on Bush Street in San Francisco, where it would remain for several decades; current headquarters are in nearby San Ramon. Like other former Standard companies, it was stronger downstream than upstream, and it needed new sources of crude. Under the guidance of Fred Hillman, the director of the new company's Producing Department, California Standard moved from sixth place among California's oil producers in 1911 to first place in 1916, with several discoveries in Kern County and the Los Angeles basin. The company also bought some Murphy Oil Company Southern California holdings, in West Coyote and East Whittier, in 1913. In 1919, the company also made several discoveries in Elk Hills, in the San Joaquin Valley. It also expanded its retail operations, with 218 service stations in 1919 and 735 in 1926, making it the leading **gasoline** marketer in the western **United States**.

In 1926, Standard Oil Company (California) acquired Pacific Oil Company, which managed the oil properties of the Southern Pacific Railroad. This increased California Standard's production capacity by nearly 50 percent. The company created a new corporate structure and renamed itself Standard Oil Company of California, or Socal. Socal intensified a search for overseas crude that it had begun in the early 1920s. It had tried and failed to find major oil fields in the Philippines and **Colombia**. It had also tried and failed in **Alaska**. In the late 1920s, it turned its attention to the Middle East. There, it interacted with two companies—**Gulf** and the Texas Corporation (known before 1926 and after 1941 as the Texas Company, and after 1959 as **Texaco**)—that it would eventually absorb decades later.

In 1925, New Zealander **Frank Holmes** had obtained a concession in **Bahrain**. After some difficulty peddling the concession to oil companies, he convinced Gulf Oil to take it over in 1927. In 1928, Gulf found itself enjoined by the **Red Line Agreement** from operating independently in Bahrain. Gulf transferred its interest to Socal, which was not a party to the Red Line Agreement, for reimbursement of expenses incurred in Bahrain. Holmes became Socal's agent in Bahrain. After some difficulties, Socal struck oil on 31 May 1932 in Jabal ad Dukhan (Hill of Smoke) via its new subsidiary, the Bahrain Petroleum Company (Bapco). Socal wanted to move on to neighboring **Saudi Arabia**, which most oil companies found unpromising and in any case too forbidding an environment. It obtained a **concession** in May 1933 for what is now the Eastern Province.

Socal created a new subsidiary, California Arabian Standard Oil Company (Casoc), to operate the concession. Casoc would eventually become **Aramco**, and Saudi Arabia would eventually become known as the country with the world's largest endowment of **conventional** oil. However, in September 1933 when Casoc began exploration, the future was anything but clear. Conditions were extremely difficult, and success eluded the company for years. Socal considered abandoning Saudi Arabia but instead found a partner. Since it needed marketing outlets for the oil it was producing successfully in Bahrain, Socal decided to pool its resources with the Texas Corporation, which had marketing networks in Africa and Asia but no crude there. In 1936, Socal sold 50 percent of Casoc and 50 percent of Bapco to the Texas Corporation for $21 million and 50 percent of the Texas Corporation's marketing operations in the Eastern Hemisphere. The marketing operations were made into a subsidiary of Bapco named **Caltex**. On 3 March 1938, Casoc's team under chief geologist **Max Steineke** made the first significant oil find in Saudi Arabia with its Dammam No. 7 well. The first Saudi crude was exported from the shipping terminal at **Ras Tanura** on 1 May 1939.

During the 1930s, Socal also made discoveries in Louisiana and expanded operations in **Mexico** and Central America. During **World War II**, it became a major supplier to the Allied war effort in the Pacific. It spent $57 million to expand its production of 100-octane aviation fuel (*see* GASOLINE) at its Richmond, Bakersfield, and El Segundo refineries (all in California). It also spent $9 million to boost production of synthetic toluene, necessary for the manufacture of TNT explosive.

In January 1944, Casoc became Aramco. In 1948, Aramco discovered **Ghawar**, which would eventually prove to be the world's largest accumulation of oil. The full scale of Ghawar would not be understood until years later, in 1956. In December 1948, the two Aramco partners, Socal and the Texas Company (as it was once again called), needing capital, sold a 30 percent interest to Standard Oil of New Jersey (later **Exxon**) and a 10 percent interest to Socony–Vacuum (later **Mobil**). Socal made a number of other important discoveries in the decade after the war. In the United States, it discovered the Kelly–Snyder field (West **Texas**), the Rangely field (Colorado), and a number of fields in the **Gulf of Mexico offshore**. Abroad, it discovered the Acheson field near **Edmonton** in **Canada** and the Boscan field in **Venezuela**. In 1949, Socal also found out that an oil field discovered by Japanese occupying forces in **Indonesia**, using equipment left behind by Caltex after wartime evacuation, was a **giant**. The Minas field of central Sumatra held more than 1 Bb and nearly three times as much oil equivalent of **natural gas**. In 1954, Socal along with the other Aramco partners obtained an 8 percent interest (later reduced to 7 percent) in **Iranian Oil Participants** (IOP), the consortium established to operate in **Iran**.

Socal revenues in 1951 surpassed $1 billion for the first time. By 1961, they exceeded $2 billion and in 1969 neared $6 billion. The company expanded its retail operations beyond its original area of California and four other western states. Among other actions, Socal acquired the Standard Oil Company of Kentucky in 1961, which had an extensive marketing network in the southeastern states.

In the 1970s, Socal lost its ownership interest in Aramco. Saudi Arabia began acquiring the company under the terms of the 1972 **participation** framework and by 1980 had exceeded the terms of the agreement and acquired it all. Socal also suffered from the nationalization of a number of Caltex interests in the Middle East. In 1978, Caltex Oil Refining of **India** was nationalized by the Indian government. Socal was also losing market share in its retail operations. Socal reorganized, merging all its domestic oil and gas divisions into a single entity, Chevron USA. Socal had been using Chevron as a trade name since the 1930s. Socal had missed out on the discoveries in Alaska and the early **North Sea** discoveries in the 1960s. However, during the 1970s it was involved in some major finds, including the West Pembina field in **Alberta**, the Ninian field in the North Sea, the Middleton and North Apoi fields in **Nigeria**, and the Hibernia field offshore Newfoundland, Canada.

Discoveries continued into the glut years of the 1980s (*see* OIL COUNTERSHOCK), including the Point Arguello field offshore southern California and finds in **Louisiana**, **Wyoming**, **Australia**, and Sumatra. However, the company still worried about its dependence on the Middle East. It alleviated that problem by acquiring Gulf Oil in 1984. Although struggling, Gulf was still the sixth-largest oil company in the United States and had exploration and production operations in the Gulf of Mexico, Canada, the North Sea, and West Africa. In August 1983, Mesa Petroleum, headed by corporate raider **T. Boone Pickens**, made a hostile attempt to take over Gulf. Mesa built up an 11 percent share. To deter Pickens, Gulf opened itself to takeover by other parties. On 5 March 1984, the Gulf board voted to be acquired by Socal for $13.2 billion. This was the largest corporate merger up to that time. It nearly doubled Socal's **reserves** and briefly made it the largest U.S. oil company by assets. After acquiring Gulf, Socal changed its name to Chevron Corporation.

By acquiring Gulf, Chevron took on about $12 billion in debt at a time when oil prices were collapsing. The 1980s and early 1990s were a time of cost cutting and consolidation. In 1993, Chevron also made a notable agreement with the government of **Kazakhstan** to develop the **supergiant** Tengiz field. Later, in the 2000s, Chevron invested $7 billion in expanding production capacity at Tengiz using sour gas injection.

In the late 1990s, low oil prices prompted a wave of mergers and consolidations in the industry. Chevron attempted but failed to acquire **Arco** and **Amoco**, both of which were acquired instead by **BP**. In 1999, Chevron entered into merger talks with Texaco, but they initially broke down, partly because of disagreements over who should run the merged company. In 2001, the merger finally took place. The merger was structured as a Chevron takeover of Texaco for $38.3 billion in stock and $6.7 billion in Texaco debt. Both Texaco and Caltex became wholly owned subsidiaries of the new ChevronTexaco Corporation. In 2005, ChevronTexaco became simply Chevron, although the Texaco name would continue as a subsidiary and a brand.

In 2000, the last full year before the merger, Chevron had a net income of $5.185 billion on revenues of $52.129 billion, and it controlled 5 Bb of proven oil reserves and 9.55 Tcf of gas. It produced 1.162 Mmbpd of oil and 2.477 Bcfpd of gas. In 2002, the first full year after the merger, the enlarged company had a net income of $1.132 billion on revenues of $99.049 billion, and it controlled 8.668 Bb of proven oil reserves and 19.335 Tcf of gas. It produced 0.692 Mmbpd of oil and 4.375 Bcfpd of gas.

In 2005, Chevron acquired Unocal for $18.4 billion in cash and stock. Unocal, formerly the **Union Oil Company of California**, had originated in the early California petroleum industry around the same time as Chevron's ancestor company, Pacific Coast. The Unocal acquisition strengthened Chevron's position in the **Caspian**, the **Gulf of Mexico**, and East Asia. Chevron advanced the frontiers of deepwater and ultra-deepwater exploration and development with its **Jack** well in the Gulf. In May 2009, it also began producing oil from the Tahiti field, with the Gulf's deepest producing well. In 2015 Chevron began producing in the Lianzi project, a joint development between **Angola** and the **Congo** (Brazzaville), and was involved in other African products.

Chevron has been active in **shale gas** and **tight oil** production. It acquired Atlas energy in 2011, thereby entering the **Marcellus Shale** play. It has also been involved in other shale plays in the United States and has begun development in **Argentina**'s Vaca Muerta Shale.

In 2016, Chevron controlled 4.131 Bb of proven liquid reserves and 25.432 Tcf of gas. It produced 1.381 Mmbpd of liquids and 4.814 Bcfpd of gas. In 2016, Chevron was the third-largest private sector petroleum company in the world by total assets, behind **Royal Dutch Shell** and **ExxonMobil**.

CHINA. Oil production 2015: 4.2924 Mmbpd (5.48 percent of world); **reserves** 25.62 Bb (1.56 percent of world); gas production: 4.818 Tcf (3.86 percent of world); reserves 183.42 Tcf (2.66 percent of world). Petroleum use in China dates to premodern times. The word for petroleum, *shiyou*, appeared in an 11th-century Song dynasty text on natural philosophy. Crude oil from seeps near the Yanshui River in Shanxi Province was used both for fuel and to produce pitch for sealing ship hulls. Bamboo pipes wrapped with waxed cloth were used as early as 400 BCE to transport **natural gas** for lighting and for use as fuel in making salt from brine. In the late 19th and early 20th centuries, China had a large market for kerosene, dominated by foreign companies. **Royal Dutch Shell** and **Standard Oil** both had extensive operations in the country. American exports of kerosene to China rose from about 12,000 barrels in 1870 to nearly 4 Mmb in 1920.

Modern petroleum exploration began in the early 20th century. The Qing government invited Japanese teams to explore near Yancheng in Shanxi Province (northern central China), where surface seeps had been observed. The Japanese, as well as Standard Oil, failed to produce oil in commercial quantities, even though significant reserves were later proven in the area. In 1935, the Red Army took over the petroleum activities there and established the Yancheng Petroleum Factory, producing about 22,000 barrels between 1939 and 1946. The nationalist government also began development in far western Xinjiang Province, at Dushaniz, where oil seeps were also known. There was limited success in the 1930s; more significant development did not occur until after 1950.

In 1906, the government surveyed the area of what would become the Yumen fields of Gansu Province, toward the northwestern part of the country. A renewed effort by the Nationalist government in the mid-1930s led to a 1939 oil discovery at Laojunmiao, leading ultimately to the development of the productive Yumen fields. The effort yielded about 3.3 Mmb between 1939 and 1944. The oil was refined at Lanzhou, which became a center of the Chinese petroleum industry. The Academy of Petroleum was established there.

After the founding of the People's Republic of China in 1949, exploration and development proceeded aggressively, with significant technical assistance from the Soviet Union. The Yancheng fields were confirmed, and the Yumen fields were developed. Between 1955 and 1963, the giant Karamay complex was discovered in Xinjiang, with an **EUR** of 5.027 Bb of oil. In 1959, one of the world's largest oil accumulations, the **Daqing** complex (EUR about 18 Bb), was discovered. China's rift with the Soviet Union meant that the field was developed without outside assistance, using vast armies of laborers. Daqing accounted for a large proportion of China's crude production over the next five decades. Even as late as 2005, it accounted for nearly a quarter. A long-distance **pipeline** was completed between Daqing and Fushian in 1973 and was later extended to Beijing.

The influence of the petroleum industry within the communist government varied significantly over time. Before 1955, the industry was run by the Petroleum Administration Department of the Ministry of the Fuel Industry. In 1955, the department was promoted to form the Ministry of the Petroleum Industry (MPI). In 1988, the MPI was dissolved and the new China National Petroleum Company (CNPC) was established. Some considered this a demotion, as electricity and coal achieved ministerial status. On 5 November 1999, CNPC established PetroChina as a joint stock international petroleum company. It was listed on the New York and Hong Kong Stock Exchanges in 2000 and the Shanghai Stock Exchange in 2007. At the end of 2007, CNPC owned 86.29 percent of PetroChina shares. In 2006, PetroChina had a net income of $18.734 billion on revenues of $86.398 billion, and it controlled 11.618 Bb of oil reserves and 53.469 Tcf of gas. It produced 2.276 Mmbpd of oil and 3.759 Bcfpd of gas.

Before 1980, China's petroleum activity was mostly onshore, although the country had shown serious interest in the **offshore** beginning in the 1960s. To explore its offshore potential fully, China needed access to foreign technology and equipment. This was not practical in the political environment of the 1960s and early 1970s, but after Mao Zedong's death in 1976, the country began to open its economy. In March 1978, the Chinese government invited representatives of foreign oil companies to Beijing for discussions on offshore exploration and development. The Chinese proceeded slowly, wanting to learn more about **production-sharing agreements** throughout the world before making commitments. In May 1980, they signed contracts with the French (**Elf Aquitaine**, later **Total**) and the Japanese to explore the Bohai and Heibu gulfs respectively. They also signed an agreement with **Arco** for the exploration of the Yellow Sea.

In 1982, China established the China National Offshore Oil Company (CNOOC) and tendered production-sharing agreements for the **South China Sea**. Between 1982 and 2001, about 6 Bb of proven recoverable reserves were discovered offshore China, with several Western companies participating, including **BP**, **Texaco** (later **Chevron**), and **Phillips Petroleum** (later **ConocoPhillips**). Chinese offshore oil production was roughly 25,000 bpd in 1990. By 2005, it was about 550,000 bpd, about a seventh of the country's total. More significantly, nearly 80 percent of China's production increase between 1993 and 2002 came from offshore oil. The South China Sea and its resources are the focus of a serious political dispute between China and its neighbors in the 21st century (*see* SOUTH CHINA SEA). In 2005, CNOOC attempted to buy Unocal (formerly **Union Oil Company of California**) but gave up its bid in the face of a negative political and popular reaction in the **United States**. Unocal was acquired by Chevron instead.

Chinese oil and gas production have increased fairly steadily since 1971. Oil production in 2015 was over double the level in 1980. China does not appear to be reaching an obvious peak of production. While China is a significant producer, it is as a consumer that it has its largest impact on the world market. China became a net importer of oil after 1993, and the gap between production and consumption has increased steadily. In 2015, it consumed about 7.7 Mmbpd more oil than it produced. Its oil consumption and imports that year were second only to those of the United States. Moreover, its potential for consumption growth is significant. Its per capita consumption is less than a sixth that of the United States. China's increasing demand for petroleum is thought to have been a factor in the **Third Oil Shock**.

China is endowed with the world's largest estimated technically recoverable resources (1,115 Tcf—80 percent more than the United States) of **shale gas**, although significant exploitation has not yet occurred. China may also have 32 Bb of **tight oil**.

CHINA NATIONAL OFFSHORE OIL COMPANY (CNOOC). *See* CHINA.

CHINA NATIONAL PETROLEUM CORPORATION (CNPC). *See* CHINA.

CHURCHILL, WINSTON (1874–1965). British prime minister 1940–1945 and 1951–1955. Winston Churchill was a major historical figure who led the **United Kingdom** to victory in **World War II**. Perhaps less well known was the important role he played in the oil industry and oil geopolitics.

When Churchill became first lord of the Admiralty on 24 October 1911, the Royal Navy was already partially reliant on oil. Admiral **John Arbuthnot Fisher** had initiated a policy of conversion from coal, as well as other modernizations, during his term as first sea lord from 1904 to 1910. **Marcus Samuel** of **Shell Transport and Trading Company** and later **Royal Dutch Shell** was also an early and vocal advocate. Churchill recognized the advantages of oil: it was a more powerful fuel, cleaner burning, and required fewer workers for handling and engine operation. Churchill wanted to complete the Royal Navy's conversion to oil and ensure long-term supplies. His charter was to prepare the fleet for possible war with Germany. The German naval law of 1900 provided for large numbers of new battleships and cruisers and was an explicit challenge to British maritime supremacy. Churchill relied on Fisher, who was now retired, as an adviser and made him head of the **Royal Commission on Fuel and Engines** to study and report on all the ramifications of oil as a fuel.

The commission and the Admiralty were lobbied by the Anglo–Persian Oil Company (APOC, later the **Anglo–Iranian Oil Company**, eventually **BP**), notably by the managing director, **Charles Greenway**. The forerunner to APOC had discovered oil in Persia (**Iran**) in 1908. APOC was still a young company. It had spent its capital on initial exploration and development and needed a financial infusion to expand and thrive. APOC argued that without financial assistance, it would be taken over by foreign interests. In fact, the "foreign" company in question was Shell, which was domiciled in London and had a majority of British directors on its board. However, Shell was part of the Royal Dutch Shell group, which was only 40 percent British owned overall, the majority 60 percent stake being in Dutch hands. The group's Dutch ownership was exploited by APOC. The implication was that the Dutch owners would be subject to German influence or coercion. Marcus Samuel and **Henri Deterding** counterlobbied unsuccessfully.

Although the Admiralty would continue to purchase oil from Shell, it wanted a long-term contract with a purely British company. Churchill sent a group of experts to Persia (Iran) in late 1913 to examine the oil fields and operations on-site. This group, the **Slade Commission**, reported back favorably and advised the Admiralty to enter into a long-term contract with APOC and to provide the necessary financial assistance, while maintaining some control over the company's strategic direction. Various possible financial arrangements were considered. Ultimately, the British government purchased a 51 percent stake in APOC for £2.2 million. Churchill had skillfully managed Parliament and the various government departments to make this happen. The agreement was signed 20 May 1914. There was considerable criticism in the press from a variety of angles. Some observers thought the government should not risk public money in such a fashion. Others worried about the effect on the British coal industry. Some worried that the investment would inevitably lead to military intervention in Persia. These and other objections were all swept aside when **World War I** broke out in August 1914.

It is not clear that Churchill deliberately set out to put the British government in the oil business. His goal was to obtain a favorable long-term contract for the Royal Navy to get oil supplies, and in this one can argue that he succeeded. Churchill continued on as head of the Admiralty until 1915, when he was forced out in the wake of the disastrous campaign in the Dardanelles. Throughout his career, Churchill remained convinced of the centrality of oil and its link to national security. He was an imperialist to the end. During his second premiership, Churchill was strongly in favor of removing **Mohammed Mossadegh** from power in Iran after he nationalized the country's petroleum industry. Beginning in the early 20th century, Churchill helped establish a pattern of intervention by Western governments in the Middle East for access to resources. *See also* AJAX, OPERATION (1953).

CITGO. *See* CITIES SERVICE COMPANY; PDVSA.

CITIES SERVICE COMPANY. Integrated petroleum company founded in 1910; acquired by **Occidental Petroleum** in 1982. Cities Service Company was founded in 1910 by **Henry L. Doherty** to supply **natural gas** and electricity to small public utilities. Doherty soon expanded Cities Service from utilities into petroleum production. In 1913, Cities Service acquired the Empire Gas Company, which discovered the El Dorado field in Kansas in 1915. The field played a major role in supplying oil during **World War I**, yielding about 100 Mmb of crude by 1919. Cities Service increased its production from about 8,200 bpd in 1915 to almost 0.274 Mmbpd in 1917.

In 1935, Cities Service acquired the Indian Territory Illuminating Oil Company (ITIO), founded in 1901 by H. V. Foster. ITIO had made major discoveries in **Oklahoma** in the 1920s. In 1926, it was one of the companies that opened up the Greater Seminole field, and in 1928 it discovered the giant Oklahoma City field, with an **EUR** of 829 Mmbo. In its retail operations, Cities Service introduced high-octane “Koolmotor” **gasoline** in 1927, along with “Koolmore” motor oil. In 1936, Cities Service had 24 oil companies, 15 natural gas companies, and 45 electric utility companies. Doherty died in December 1939 and was succeeded by W. Alton Jones. Jones lobbied the U.S. government to build the **Big Inch** Pipeline from **Texas** to the U.S. East Coast during **World War II**. Cities Service had lost **tankers** to attacks by German submarines; Jones argued that “nobody ever sank a pipeline.” He served as the president of War Emergency Pipelines. Big Inch was completed in 1943.

During the war, Cities Service began disposing of its utility assets, a process it completed by 1958. During the 1950s, it discovered oil in the **Gulf of Mexico** and explored in Texas and in the Middle East (**Oman**). By 1960, Cities Service’s annual revenues exceeded $1 billion and it produced 47 Mmb of crude. In 1965, it adopted Citgo as a brand name. In December 1982, after repelling a takeover attempt by **T. Boone Pickens**, Cities Service was acquired by Occidental Petroleum for $4.3 billion. Occidental placed Cities Service’s downstream operations into a separate subsidiary named Citgo and sold them to Southland Corporation in 1983. Southland wanted an assured supply of gasoline for its chain of 7-Eleven convenience stores but was unable to operate Citgo profitably and sold 50 percent of it to **PdVSA** in 1985. PdVSA completed the acquisition in January 1990. PdVSA gained a secure outlet for its crude and access to U.S. consumers, although it also gained, to its chagrin, reporting responsibilities to the U.S. Securities and Exchange Commission. Citgo gained a guaranteed supply of crude oil to feed its refineries and marketing outlets.

CLEVELAND MASSACRE. *See* SOUTH IMPROVEMENT COMPANY.

CLIMATE CHANGE. *See* ENVIRONMENTAL IMPACTS.

COHEN, ROBERT WALEY. *See* WALEY COHEN, ROBERT (1877–1952).

COLD LAKE OIL SANDS (TAR SANDS). *See* ALBERTA OIL SANDS (TAR SANDS).

COLOMBIA. Oil production 2015: 1.0 Mmbpd (1.28 percent of world); **reserves** 2.0 Bb (0.12 percent of world); gas production: 0.402 Tcf (0.32 percent of world); reserves 4.36 Tcf (0.06 percent of world). The early modern Colombian petroleum industry was dominated by two **concessions** granted early in the 20th century: the Barco concession and the De Mares concession. The Barco concession was granted on 16 October 1905 to Colonel Virgilio Barco, the grandfather of the Virgilio Barco who was president of Colombia from 1986 to 1990. The concession was for the Catatumbo region in the state of Norte de Santander, bordering **Venezuela**. In 1918, Barco sold the concession to the American-owned Carib Syndicate. A year later, the Colombian Petroleum Company, owned by **Cities Service Company**, acquired a 75 percent interest. Cities Service was unsuccessful and in 1926 sold the concession to **Gulf Oil Corporation**. That year, in a climate of increasing nationalism, the Colombian government withdrew the concession for noncompliance. In 1931, the government reauthorized the concession, provided that Gulf construct a **pipeline** to the sea and pay a 6 percent royalty to the government and a 3.5 percent royalty to Barco's successors on all oil delivered. Gulf discovered oil but had difficulty upholding its promise to build a pipeline and move the oil for export. In 1936, the Texas Corporation (later **Texaco**) and Socony–Vacuum (later **Mobil**) bought the concession and proceeded to construct a 263-mile pipeline to the Caribbean port of Coveñas, in Sucre state. First oil flowed in 1939. The Barco concession reverted to the Colombian government in 1955.

The De Mares concession, which would become the more important of the two, was granted on 7 March 1906 to Roberto de Mares. It covered the Infantas region near Barrancabermeja, Santander, for 30 years. It was withdrawn on 22 October 1909 for noncompliance. The De Mares group began work nevertheless, and the concession was reauthorized in 1916. That year, De Mares sold the concession to Tropical Oil Company. In 1918, Tropical discovered the **giant** Infantas–La Cira field (**EUR** about 0.5 Bb; La Cira portion discovered in 1926). In 1920, Standard Oil Company of New Jersey (Jersey Standard, later **Exxon**) used its International Petroleum Company subsidiary, headquartered in **Canada**, to acquire Tropical Oil. IPC created the Andian National Company to build a pipeline to Coveñas; the pipeline was completed in 1925. The Infantas–La Cira field became a mainstay of Colombian oil in the first part of the 20th century, reaching a peak production of almost 60,000 bpd in the mid-1940s. It would continue producing into the 21st century, although 2004 production was down to about 5,400 bpd. The De Mares concession reverted to the Colombian government on 9 January 1951.

The entity that assumed control of the two producing concessions was the new Empresa Colombiana de Petróleos, or Ecopetrol, authorized by Law 165 of 1948 and established on 9 January 1951. Although Ecopetrol began by taking over foreign assets, it remained relatively friendly with foreign oil companies, unlike the case in many other countries. Foreign oil companies continued to be allowed to operate in Colombia. In 1948, the Colombian government offered Jersey Standard a continuing minority stake in the De Mares/Tropical Oil concession, which Jersey Standard declined. However, it continued to support Ecopetrol through a service contract and even loaned Ecopetrol $10 million to expand the capacity of Tropical Oil's (i.e., Jersey Standard's) refinery.

Colombia's old fields were being depleted by the 1960s. Ecopetrol undertook secondary recovery efforts for Infantas–La Cira using water flooding. More importantly, it began an intensive exploration program, concentrating on the Llanos basin. It partnered with **Occidental Petroleum** and **Royal Dutch Shell**, and in 1983 the consortium discovered the giant Caño Limón field (EUR 1.07 Bb) in Arauca state. The consortium also built the 460-mile Caño Limón Pipeline across the Andes to Coveñas. In 1992, an Ecopetrol consortium led by **BP** discovered Colombia's most important oil field, the giant Cusiana (EUR 1.45 Bb). Cusiana is also in the Llanos basin, in Casanare state, about 150 miles southwest of Caño Limón. A year later, BP discovered the nearby Cupiagua field (EUR 0.728 Bb), sometimes considered part of the same complex. The Cusiana/Cupiagua system peaked in 1999 at 440,000 bpd and declined rapidly thereafter. In 2006, its production was about 170,000 bpd, an indication that the ultimate recovery may have been overestimated. Cusiana and Cupiagua are served by the 500-mile Ocensa Pipeline to Coveñas. Cusiana and Cupiaga crudes, as well as crude from the Orito field, constitute Colombia's main export crudes and are lighter and sweeter than other important Latin American crudes, ranging between 28° and 36° **API gravity**.

Colombia's oil production reached a peak in 1999 at 0.816 Mmbpd after nearly two decades of growth. It declined rapidly over the next three years, reaching 0.541 Mmbpd in 2002. Thereafter it stayed more or less level for a while and then increased, exceeding the 1999 peak in 2011 and two years later breaking through 1 Mmbpd. Colombia was a net exporter of oil from the late 1920s to 1977, when it began to import. In 1984, it became a net exporter again. Preventing Colombia from once again becoming a net importer was considered an important national goal. In order to help realize it, the government undertook a number of reforms since 1999 that created a favorable investment climate for petroleum and sparked new exploration and drilling. After the reforms, foreign companies were allowed to own 100 percent stakes in ventures. The government also lowered royalty rates and allowed longer exploration licenses. It also forced Ecopetrol to compete with private operators. In the 1990s, it began to privatize some of Ecopetrol's assets in **natural gas** distribution and transportation. In 2003 the government created the Agencia National de Hidrocarburos as public administrator of the country's petroleum sector, relieving Ecopetrol of those duties. In 2007, the government partially privatized Ecopetrol as an entity. Ecopetrol sold 10.1 percent of itself to Colombian investors.

Colombia's 2015 gas production of 0.402 Tcf corresponds roughly to 0.195 Mmboepd, or about a fifth of its oil production. Gas has been on an upward trend since 1980. Colombia's largest gas field, Volcanera (EUR 5 Tcf, and 0.25 Bb condensate), was discovered in the Llanos basin in 1993, not far from the Cusiana and Cupiagua fields. The Llanos basin has most of Colombia's gas reserves, although most current production is from the Guajira basin, where the giant Chuchupa gas field was discovered in 1973. In 2003, Colombia and Venezuela agreed to build a 0.484 Bcfpd natural gas pipeline linking Colombia's Guajira basin to Venezuela's Lake **Maracaibo** area. It began operating in 2007.

One serious problem facing Colombia's petroleum industry has been the decades-long civil war between government forces and the Fuerzas Armadas Revolucionarias de Colombia (FARC, the Revolutionary Armed Forces of Colombia). Pipelines have been frequently attacked, disrupting operations and causing companies to lose profits and the government to lose export revenue. The Caño Limón Pipeline, for example, was breeched so many times that some local residents began calling it "the flute." At least 3 Mmb of crude oil leaked out over the years. This is almost twice as much oil as was spilled by the ***Amoco Cadiz***, and more than 10 times as much as that spilled by the ***Exxon Valdez***. In November 2016 the Colombian Congress ratified a peace treaty between FARC and the government, officially ending the conflict. A difficult process of implementation followed.

COLONIAL PIPELINE. Major petroleum product **pipeline** system in the **United States**, with 5,500 miles of pipeline running in a network from **Houston, Texas**, to New York City. The pipeline crosses most of what were the original 13 colonies of the United States, hence the name. The main line was constructed between 1962 and 1964 by the Colonial Pipeline Company, motivated originally by a strike of the maritime union. It is the world's largest petroleum products pipeline, transporting a daily average of more than 2 Mmbpd of refined products, including **gasoline**, aviation fuel, and home heating oil.

COLONY OIL SHALE PROJECT. *See* EXXON; GREEN RIVER FORMATION; OIL SHALE.

COMECON PIPELINE. *See* DRUZHBA PIPELINE.

COMODORO RIVADAVIA OIL FIELD. Argentina's only **giant** oil field (there are a number of gas giants), discovered in December 1907 by a government team reportedly drilling for water. Some observers have disputed the "accidental" nature of the find. The field has an **EUR** of 3.2 Bb. It is located about 15 miles north of the port city of the same name, in the state of Chubút. The Argentine government decided to exploit most of the field itself and created a State Oil Commission. The navy provided its **tanker** *Ministro Ezcurra*, and the chartered tanker *Waneeta*, to transport Comodoro oil to Buenos Aires, more than 1,000 miles to the north. The first voyage was on 29 July 1914. In 1917, the navy purchased two more tankers, *Aristóbulo del Valle* and *Ingeniero Luis Huergo*. By 1920, Comodoro Rivadavia was producing 4,564 bpd from 39 wells. More tankers were added in 1921–1922. On 17 October 1921, the tanker fleet passed to the civilian oil commission, which became Yacimientos Petrolíferos Fiscales (YPF) in 1922. YPF (later **Repsol YPF**) was the world's first vertically integrated national oil company. Oil from Comodoro Rivadavia is still shipped to Buenos Aires by tanker, and **natural gas** is transported there via a 1,100-mile **pipeline**.

COMPAGNIE FINANCIÈRE BELGE DES PÉTROLES. *See* PETROFINA.

COMPAGNIE FRANÇAISE DES PÉTROLES (CFP). A **major oil company** considered an "eighth sister" of the **Seven Sisters**; later **Total**. CFP was established in 1924. At its formation, it had a 22.5 percent share in the Turkish Petroleum Co. (TPC, later **Iraq Petroleum Company**), acquired by France from Germany after **World War I** at the **San Remo Conference**. The French government acquired a 25 percent share of CFP in 1929, two years after the discovery of oil near **Kirkuk** in **Iraq** made the company a potential force in the global oil industry. In 1931, the government increased its share to 35 percent but did not attempt to manage the company. CFP became a vertically integrated oil company with its own downstream operations before **World War II**. During the war, France was occupied by Germany, and

CFP's operations were curtailed. It was still a partner in TPC, which by then was called the Iraq Petroleum Company (IPC). CFP's share was held for it by its partners.

In 1948, the IPC partners abrogated provisions of the **Red Line Agreement** limiting competition among themselves in the Middle Eastern lands of the former Ottoman Empire; CFP was against ending the agreement. CFP made a deal in 1947 to obtain crude oil from **Venezuela**. In 1954, CFP introduced the Total brand name. It made a number of significant discoveries in **Algeria** in 1956 and 1957. Algeria was still a French colony at the time. New discoveries in **Indonesia** and the **North Sea** softened the blow to CFP when a now independent Algeria nationalized its oil and gas industry in 1971. In 1985, the company changed its name to Total Compagnie Française des Pétroles, truncating it to Total in 1991. In 1991, the company was also listed on the New York Stock Exchange, and a year later the French government began a drastic reduction in its stake, ultimately privatizing the company. Total eventually acquired **Petrofina** and **Elf Aquitaine**.

COMPAÑÍA MEXICANA DE PETRÓLEO EL ÁGUILA (MEXICAN EAGLE PETROLEUM COMPANY). Pioneering petroleum company in **Mexico**. *See also* PEARSON, WEETMAN (1856–1927).

CONCESSION. An agreement between a host government and a petroleum company allowing the company to explore for and develop hydrocarbon deposits within a predetermined territory for a set period of time. The company pays all costs and generally manages its affairs as it sees fit, paying royalties and taxes to the government if successful. Concession-type agreements over limited areas are used in advanced countries such as the **United States**, **Canada**, the **United Kingdom**, and **Norway**.

In the early to mid-20th century, concessions in the developing world could cover vast territories and last for several decades. In exchange for the right to explore and exploit, companies would pay the host government some combination of up-front fees, rent, royalty per unit of oil produced, and taxes on profits. Under the concession system, which was imperialist in spirit if not always in name, companies generally operated with little or no interference from the host government and essentially ran the country's petroleum sector.

The first major concession in the Middle East was that of **William Knox D'Arcy** in **Iran** in 1901, granted for 60 years. That concession and modifications thereof were to dominate the country's industry for 50 years and would nurture the growth of **BP**'s antecedent companies, the Anglo–Persian Oil Company and **Anglo–Iranian Oil Company**. The Turkish Petroleum Company (**Iraq Petroleum Company**) obtained its concession in **Iraq** in 1925. Early concessions, beginning before **World War I**, were also granted in

Mexico and **Venezuela**. The sheikhdoms of **Bahrain**, **Kuwait**, and **Qatar** granted concessions in 1930, 1934, and 1935 under British rule. **Saudi Arabia** granted its first successful concession, which would ultimately lead to the formation of **Aramco** and the discovery of the world's largest oil **reserves**, in 1933.

The old international concession system dominated global oil for the first half of the 20th century, but it eventually produced discontent in the host countries and was dismantled. When market prices exceeded **posted prices**, host governments' relative shares of the profits declined, leading to considerable resentment. Local populations also resented the poor working conditions and lack of opportunity offered by oil companies under the old concession system. The regime of huge, long-lasting concessions was brought to an end by nationalizations of the oil industries in a number of countries, by the **50–50 agreements** for profit sharing that began in the 1940s, and by the creation of the **Organization of the Petroleum Exporting Countries** (OPEC).

Concession agreements still exist, but since the 1960s they have involved much smaller blocks of territory and much shorter time periods—often with considerably higher royalty rates. Other than collecting fees and enforcing laws, the host country may not be very actively involved. However, these modern concession agreements also share the competitive space with other types of agreements involving higher degrees of participation on the part of the host government. One such type is the participation-sharing agreement or **production-sharing agreement**, which often creates an explicit joint venture between a host country's national oil company and foreign petroleum companies. Another is a service agreement, or contractual agreement, whereby a foreign company is paid a fee for performing specified services for a national oil company. Such services may include exploration, development, or technical services, often requiring skilled labor not readily available in the host country.

CONGO REPUBLIC (CONGO–BRAZZAVILLE). Oil production 2015: 0.269 Mmbpd (0.34 percent of world); **reserves** 1.6 Bb (0.10 percent of world); gas production: insignificant; reserves 3.2 Tcf (0.05 percent of world). Oil in Congo-Brazzaville accounts for about 65 percent of GDP and 90 percent of exports. Congo's first discovery was the Point Indienne field in 1951, with production beginning in 1957. Congo has four **giant** oil fields, each with an **EUR** of roughly 500 Mmb: Emeraude Marin, discovered 1969; Loango, discovered 1971; N'Kossa, discovered 1983; and Moho Marine, discovered 1995. Congo's highest production levels (about 60,000 bpd in 2003) come from a smaller field, Kitina, discovered 1991 with an EUR of about 145 Mmb.

Congo's national oil company, Société Nationale des Pétroles du Congo (SNPC), regulates exploration and production and develops **production-sharing agreements**, under which foreign companies typically bear all investment costs. A third of all crude produced goes to the government and is sold by SNPC. Important operators in Congo include **Total**, **ENI**, and Perenco of the **United Kingdom**. In March 2003, Congo and **Angola** created a zone of common interest, the Zone d'Interêt Commun (ZIC). This joint development area includes part of Angola's Block 14 and Congo's Haute Mer (led by Total). Congo and Angola will share revenues from ZIC production equally. In December 2004, **Chevron**, operating from Angola, announced a significant discovery (Lianzi-1) in the ZIC. Congo is often cited as a country where petroleum has undermined democracy. The current president, Denis Sassou-Nguesso, came to power by force in 1997, reportedly supported by France and **Elf Aquitaine** (later acquired by Total).

CONNALLY HOT OIL ACT (1935). A federal law enacted in the **United States** to restrict the flow of **hot oil** (oil produced in excess of legal quotas). The discovery of the **East Texas** oil field in 1930 flooded the market with oil and caused a collapse in prices. The **Railroad Commission of Texas** attempted to impose production quotas, which were widely ignored by producers who continued to overproduce and smuggle hot oil. The National Industrial Recovery Act of 16 June 1933 contained a section prohibiting the interstate transport of oil produced in excess of state quotas, but this provision was struck down by the U.S. Supreme Court in January 1935.

In response, Senator Thomas Connally of Texas sponsored a bill that would control hot oil using a slightly different method. The Connally Hot Oil Act, which became law on 22 February 1935, gave the president the power to regulate the interstate shipment of petroleum and petroleum products, and to require certificates of clearance before such shipments could take place. The law provided for boards to be established that would issue the certificates. The boards could conduct investigations and hold hearings. People or companies that shipped oil across state lines without a permit could be fined up to $2,000 and imprisoned for up to six months. The contraband oil could be confiscated by the federal government. The law was set to expire on 16 June 1937, but it was extended. Four federal courts upheld the act in 1937.

CONOCO. Petroleum company founded in 1875 in Ogden, Utah, by Isaac Elder Blake as the Continental Oil and Transportation Company (CO&T); eventually part of **ConocoPhillips**. In 1885, CO&T merged with **Standard Oil** and was reincorporated as Continental Oil in Colorado. Standard Oil was broken up in 1911, and Continental Oil became an independent oil company two years later. Continental entered the **gasoline** marketing business relatively early, establishing a service station in 1914. In the decade that followed, it acquired and merged with production, **refining**, and distribution companies. In 1929, it merged with Marland Oil. In the 1930s, it built a pipeline from **Oklahoma** to Chicago and Minnesota. From 1950 to 1970, it acquired oil fields all over the world, including Africa, South America, and the Middle East.

In the 1960s, the company began diversifying away from petroleum into coal, chemicals, and minerals, though it retained its petroleum business. In 1979, it changed its name to Conoco. Two years later, it was acquired by DuPont. In the decade that followed, Conoco made significant oil discoveries in the **North Sea**, **Ecuador**, **Indonesia**, and the **Gulf of Mexico**. In 1991, it entered a joint venture with **Lukoil** to explore in **Russia**'s arctic region. In 1998, DuPont spun Conoco off, creating an independent company once again. This was the largest initial public offering in the **United States** up to that time. Conoco made significant discoveries in the Gulf of Mexico in 1999 and in the North Sea in 2000, and it bought North Sea properties from **Norsk Hydro**. In 2001, it bought Gulf Canada Resources.

In 2002, Conoco merged with **Phillips Petroleum**, forming the combined company ConocoPhillips in a deal valued at $15.12 billion. In 2001, the last full year before the merger, Conoco controlled 2.142 Bb in proved oil **reserves** and 8.619 Tcf of **natural gas**. It produced 0.433 Mmbpd of oil and 2.03 Bcfpd of gas. In total assets, it was the fourth-largest petroleum company in the United States, behind **ExxonMobil**, ChevronTexaco (later **Chevron**), and Phillips.

CONOCOPHILLIPS. One of the largest petroleum companies in the world, formed by the 2002 merger of **Conoco** and **Phillips Petroleum**. The merger reportedly generated $1.75 billion in cost savings. In 2003, the first full year after the merger, ConocoPhillips controlled 5.171 Bb in proved liquid **reserves** and 16.06 Tcf of gas. That year it produced 1.036 Mmbpd of liquids and 3.855 Bcfpd of gas. In 2016, it controlled 2.504 Bb in proved liquid reserves and 10.844 Tcf of gas; it produced 0.721 Mmbpd of liquids and 3.353 Bcfpd of gas.

ConocoPhillips was the third-largest petroleum company in the United States in 2016 by total assets, after **ExxonMobil** and **Chevron**.

CONORADA PETROLEUM. *See* AMERADA PETROLEUM CORPORATION; MARATHON OIL CORPORATION.

CONTINENTAL OIL AND TRANSPORTATION COMPANY. *See* CONOCO.

CONVENTIONAL OIL AND GAS. Conventional crude oil exists as a liquid mixture in natural underground reservoirs and remains liquid at atmospheric pressures. Conventional crude usually ranges in **API gravity** from 20° to 40°, flows relatively easily, and accounts for the overwhelming majority of crude oil produced to date. Conventional gas is gas that flows easily through its reservoir and can be extracted through naturally occurring pressure or ordinary pumping mechanisms. As sources of conventional oil and gas become scarcer and prices rise, the industry becomes increasingly interested in **unconventional** oil and gas, which may require new and advanced technology and be more costly and difficult to extract.

COOK, LODWRICK M. (1928–). President of **Arco** 1985–1995, succeeding **Robert O. Anderson**. Lodwrick M. Cook further developed Arco's **reserves** in **Alaska**'s North Slope. He also spearheaded the development of cleaner-burning **gasoline**. During his tenure, Arco doubled its market share in the western **United States**.

COOPERATION COUNCIL FOR THE ARAB STATES OF THE GULF. *See* GULF COOPERATION COUNCIL (GCC).

CORPORACIÓN VENEZOLANA DEL PETRÓLEO (CVP). National oil company founded in 1961 in **Venezuela** to develop petroleum **reserves** not already being worked by foreign companies. The Corporación Venezolana del Petróleo (Venezuelan Petroleum Corporation) produced about 2 percent of Venezuela's oil by the early 1970s. When Venezuela nationalized its industry in 1975, **PdVSA** became the national petroleum company.

CORPOVEN. Autonomous subsidiary of **PdVSA** formed in 1975 to take over the petroleum operations in **Venezuela** of various companies (other than **Royal Dutch Shell** and **Exxon**) after nationalization of the Venezuelan petroleum industry. The subsidiary was merged back into PdVSA in 1998. The subsidiary participated in joint ventures in the **Orinoco Oil Belt** during the **Apertura Petrolera**. *See also* CORPORACIÓN VENEZOLANA DEL PETRÓLEO (CVP).

CORSICANA OIL FIELD. The first commercial oil field in **Texas**, discovered 9 June 1894. The modern Texas petroleum industry is often said to have begun with the 1901 **Spindletop** discovery. However, Corsicana, while not quite as seminal, preceded Spindletop by seven years. Even Corsicana did not represent the first modern attempt at petroleum exploitation in the state. Drilling occurred at Oil Springs near Nacogdoches in East Texas as early as

1866, and there was a brief period of production there in the 1880s. However, the activity at Oil Springs had no permanent effects on the industry. Corsicana, on the other hand, led to a sustained local boom.

Corsicana in 1890 was a cotton processing and trade center. Its civic leaders wanted to diversify the industrial base and determined that better water supplies were necessary. While drilling for water, the contractor (American Well and Prospecting Company) discovered oil on 9 June 1894. On 6 September, local citizens Ralph Beaton and H. G. Damon organized the Corsicana Oil Development Company with John Davidson, a **Pennsylvania** oil entrepreneur, and leased oil rights near the well. In 1895, Corsicana Oil Development entered into an agreement with the Guffey and Galey Company of Pittsburgh—*see* GUFFEY, JAMES McCLURG (1839–1930). Guffey and Galey agreed to drill five test wells and would obtain a half interest in the leases of the Corsicana company. By August 1896, four of the five wells were producing in small quantities, averaging 20–40 bpd. This touched off an oil boom, which was unfortunately chaotic and wasteful, with wells too closely spaced. Guffey and Galey sold their interest to local investors in September 1897.

Drilling and production in the Corsicana field continued. Production increased sharply from 1,450 barrels in 1896 to 65,974 barrels in 1897. In 1898, the field expanded to encompass a broader area of the countryside surrounding Corsicana, and production shot up to 546,070 barrels. By 1993, the Corsicana field had produced 44 Mmbo. **Joseph S. Cullinan**, the future founder of the Texas Company (later **Texaco**), visited Corsicana in 1897 to inspect the oil field at the invitation of the mayor. He stayed and formed the J. S. Cullinan Company, which would later form part of the **Magnolia Petroleum Company** (eventually part of **Mobil**).

COWDRAY (LORD). *See* PEARSON, WEETMAN (1856–1927).

CREOLE PETROLEUM. A company formed in 1920 as the Creole Syndicate to deal in petroleum **concessions** in **Venezuela**. Standard Oil Company of New Jersey (later **Exxon**) acquired it in 1928. By the 1950s, Creole accounted for nearly half of Venezuela's oil production. Creole's holdings in Venezuela were taken over by **PdVSA** subsidiary **Lagoven** after the nationalization of the industry in 1975.

CRUDE OIL. *See* CONVENTIONAL OIL AND GAS; HEAVY OIL; OIL SANDS; OIL SHALE; SHALE OIL; TIGHT OIL.

CULLINAN, JOSEPH S. (1860–1937). Founder of the Texas Company (later **Texaco**) and its president 1902–1913; founder of the J. S. Cullinan Company, eventually part of **Mobil**. Joseph Steven Cullinan was born 31 December 1860 near Sharon, **Pennsylvania**. He began working in the Pennsylvania oil fields at the age of 14 and joined **Standard Oil** in 1882. He rose through the ranks at Standard and then left in 1895. He moved to **Corsicana, Texas**, where oil had been discovered in 1894, and was a consultant to producers there. He formed the J. S. Cullinan Company and completed a refinery on 25 December 1898. The J. S. Cullinan Company was antecedent to one of the firms that were merged together to form the **Magnolia Petroleum Company** in 1911. Magnolia eventually became part of Mobil.

Cullinan moved to Beaumont, Texas, in 1901 after oil was discovered at **Spindletop**. That year, he formed the Texas Fuel Company with an initial capitalization of $50,000 to purchase crude and resell it to refineries. With the help of Arnold Schlaet, a New York investment manager, he sought additional investments. In 1902, he and Schlaet reorganized the Texas Fuel Company as the Texas Company, with a capitalization of $3 million. The Texas Company would later be known to the world as Texaco.

In 1905, Cullinan moved the company to **Houston**. He increased the Texas Company's capitalization by a factor of 10, to $30 million, during his tenure as president from 1902 to 1913. He was known for his willingness to take large risks. In 1903, with opposition from Schlaet, he risked the company on an exploration venture in Sour Lake, about 20 miles from Beaumont. With resources to drill three wells, the company drilled two dry holes but struck oil with the third in January 1903. Cullinan also reportedly once risked $6 million on a **pipeline** in **Oklahoma** when the Texas Company had assets of less than $8 million. In 1913, he lost control of the company in a proxy battle. After he left the Texas Company, he founded 10 more petroleum companies in Texas. Cullinan was president of the **Houston** Chamber of Commerce from 1913 to 1919. He supported the development of the **Houston Ship Channel** and made other contributions to the city's infrastructure and cultural life. He died on 11 March 1937 in Palo Alto, **California**.

CUSHING. City (pop. 7,800, 2010) in **Oklahoma**; price settlement point for **West Texas Intermediate** (WTI) crude oil on the **New York Mercantile Exchange**. Cushing is a former center of oil exploration and production, but it is now more famous as a transportation hub, with both north–south and east–west **pipelines** running through there. As the pricing settlement point for WTI crude, Cushing has provided an important gauge of world oil prices. However, the WTI **benchmark** showed itself vulnerable to distortions in 2007, when a surplus of WTI at Cushing caused its price to fall to artificially low levels. There were other pricing distortions between 2011 and 2013, when WTI traded at an unusually cheap price compared to **Brent Blend**.

CYPRUS. *See* LEVANT BASIN.

D

DAQING COMPLEX. Supergiant agglomeration of oil fields in the Songliao basin, northeastern **China**. The first discovery in the Daqing complex was the Songji No. 3 gusher at Datongzhen on 26 September 1959. This was just before the 10th anniversary of the founding of the People's Republic of China, 1 October. The well and the field were renamed Daqing (Great Joy). The Daqing agglomeration includes a number of oil fields, including three **giants**. These include Taching (1959, **EUR** 4.823 Bb), Saertu (1960, EUR 9.418 Bb), and Xingshugang (1960, EUR 4.258 Bb). The Daqing complex overall is among the 10 or 20 largest oil accumulations in the world, depending on which estimates one uses. In 1960, Daqing accounted for 18.6 percent of all China's crude production. By 1965, the figure was 73.7 percent. This was a high point, but the proportion of China's crude coming from Daqing stayed above 40 percent until 1995. Even in 2005, it was 24.8 percent. China developed the Daqing field with a severe shortage of materials, equipment, and technology. It was no longer receiving technical and other assistance from the Soviet Union. It employed large numbers of laborers, up to 40,000 at any one time, working under military-style discipline (the "Daqing Method"). They performed tasks by brute human force that might otherwise have been done by heavy equipment.

D'ARCY, WILLIAM KNOX (1849–1917). British entrepreneur whose company struck oil in **Iran** and effectively initiated the Iranian and Middle Eastern petroleum industries. William Knox D'Arcy was born 11 October 1849 in Devonshire, England. He immigrated with his father to **Australia** and became involved in mining there. He was highly successful financially in the 1880s and returned to England. In 1901, with assistance from the government of the **United Kingdom**, D'Arcy obtained a **concession** to explore for oil in a 400,000 square mile area of Persia (Iran) from Shah Muzzafar al-Din. His representatives began exploring for oil, and D'Arcy had a producing well by 1904. Lacking the financial wherewithal to pursue further exploration, he appealed to the British government for help. The government persuaded the

Burmah Oil Company, a major British firm with Admiralty contracts, to invest in D'Arcy in 1905. D'Arcy's company became a subsidiary of Burmah, and D'Arcy became a director of the subsidiary.

In 1908, D'Arcy's geologist **George Reynolds** made a major oil discovery at Masjid-e-Suleiman. In 1909, Burmah formed the Anglo–Persian Oil Company (APOC) to develop what would turn out to be an enormous find, with D'Arcy receiving some shares but not a controlling interest. He lost his voice in company affairs and retired soon after. APOC was later renamed the **Anglo–Iranian Oil Company**. In 1951, its properties in Iran were nationalized and turned over to the National Iranian Oil Company; the remaining Anglo–Iranian Oil Company would become **BP**. William Knox D'Arcy died 1 May 1917 in England.

DARDANELLES. *See* BOSPORUS AND DARDANELLES.

DEEPWATER. There is no universally accepted definition of "deepwater" offshore oil and gas, and the definition changes with time and the progress of technology. A common definition of a deepwater play is one that lies beneath more than 1,500 feet of water. *See also* ULTRA-DEEPWATER.

***DEEPWATER HORIZON* OIL SPILL.** One of the worst environmental disasters in history, involving a spill of 4.9 million barrels of oil into the **Gulf of Mexico** in 2010. The *Deepwater Horizon* was a semisubmersible mobile offshore platform owned by Transocean and being operated by **BP** about 40 miles off the **Louisiana** coast in the Macondo Prospect. On 20 April 2010 there was a blowout caused by a natural gas surge, leading to an explosion and a fire that killed 11 workers and injured 17. The platform sank on 22 April, causing an oil spill. BP tried for months to seal the well, finally succeeding in September. Over a thousand miles of shoreline in Louisiana, Mississippi, Alabama, and Florida were fouled. More than a third of U.S. federal waters were closed to fishing, and the U.S. government imposed a drilling moratorium until November 2010 (lifted a month early). A variety of investigations placed most of the blame on BP. A U.S. district court in September found that BP's actions were "reckless," while those of **Halliburton** (the cementer) and Transocean were negligent. The court allocated 67 percent of the blame to BP, 30 percent to Transocean, and 3 percent to Halliburton. BP's costs for the cleanup, environmental damages, economic damages, fines, and penalties were over $50 billion. There were both civil and criminal penalties.

DEGOLYER, EVERETTE (1886–1956). Petroleum geologist and entrepreneur who led in the promotion and application of **geophysical prospecting** techniques. Everette Lee DeGolyer was born outside Greenburg, Kansas, on 9 October 1886. His parents were homesteaders. He entered the University of **Oklahoma** in 1906 to study geology, working part of the time for the **United States Geological Survey**. He left in 1909 before graduating to work for **Weetman Pearson**'s Mexican Eagle Oil Company as a geologist. He made a major impact very quickly. He selected the location of the Potrero del Llano No. 4 well in Veracruz, **Mexico**, in 1910. A gusher was struck there on 27 December that would take two months to bring under control. The well would produce 140 Mmbo over its lifetime. DeGolyer also sited the first successful well in the **giant** Los Naranjos field, which has produced more than 1.2 Bb. In 1911, he returned to the University of Oklahoma to complete a bachelor's degree in geology.

DeGolyer left Mexican Eagle in 1914 and began a career as an independent consultant. In 1919, Weetman Pearson—who had left Mexico and sold Mexican Eagle to **Royal Dutch Shell**—hired him again, this time to establish and lead the **Amerada Petroleum Corporation** (later Amerada Hess, eventually Hess Corporation). DeGolyer made Amerada into a prominent **independent** oil company.

After **World War I**, DeGolyer contacted members of the Physics Department at Cambridge University about the possibility of using geophysical techniques to locate potential petroleum-bearing structures. He was interested in adapting artillery location methods developed during the war and was also interested in gravity methods. As a test case, he ordered a torsion balance survey of the **Spindletop** field, which confirmed the detectability of the salt dome structure. This was the first geophysical survey of an oil field in the **United States**. In 1924, Amerada discovered the Nash salt dome field in Fort Bend County, **Texas**, using the torsion balance.

Through Cambridge, DeGolyer met **Reginald Fessenden**, who had patented a scheme for using seismic waves to locate ore bodies. DeGolyer organized the Geophysical Research Corporation (GRC) as an Amerada subsidiary in 1925 under John Clarence Karcher. GRC licensed Fessenden's patent and constructed its own seismic instruments. It was able to locate several potential oil-bearing structures. DeGolyer wanted to sell GRC's data and results to other **major oil companies**, but Amerada's president overruled him. (Although DeGolyer was Amerada's general manager and operational leader, he was not yet the chief executive.) This led DeGolyer secretly to finance the formation of **Geophysical Service Inc.** (GSI) in 1930. GSI would become one of the largest and most important geophysical contractors in the world; it would also spawn Texas Instruments. Soon DeGolyer became president and chairman of the board of Amerada, and then he retired in 1932 at the young age of 46.

He returned to consulting and in 1936 founded the firm of **DeGolyer and MacNaughton** with former Amerada colleague Lewis W. MacNaughton. DeGolyer and MacNaughton became perhaps the world's leading oil and gas appraisal and verification firm. During **World War II**, DeGolyer served as director of conservation and as assistant deputy administrator of the **Petroleum Administration for War**. He traveled to **Saudi Arabia** under the auspices of the **Petroleum Reserves Corporation** to assess that country's oil potential. He reported back in early 1944 with characteristic prescience that there was likely the greatest accumulation the world had ever known and that the center of gravity of the entire global oil industry would eventually move to that region. DeGolyer was an organizer of the **American Association of Petroleum Geologists** and served as its president in 1925. He was also president of the American Institute of Mining and Metallurgical Engineers in 1927. He died on 14 December 1956 in Dallas, Texas.

DEGOLYER AND MACNAUGHTON. Leading independent oil and gas appraisal and verification firm, located in Dallas, **Texas**. DeGolyer and MacNaughton was established in 1936 by **Everette Lee DeGolyer** and Lewis W. MacNaughton. It was incorporated in 1949. It is owned by employees and has no external shareholders. It is most famous for **reserves** assessments and company appraisals. It also offers services in reservoir simulation, engineering analysis, geophysics, geology, and other areas. DeGolyer and MacNaughton also publishes *20th Century Petroleum Statistics*, an important annual statistical review of the oil industry. This compendium was first published in 1945 by the U.S. government, specifically the U.S. Navy's Office of **Naval Petroleum and Oil Shale Reserves**. The director of that office, Commodore W. G. Greenman, requested that the compendium be maintained and updated. DeGolyer and MacNaughton took over its publication in 1946 and has produced it annually ever since. The compendium emphasizes the **United States** but also contains some global data.

DETERDING, HENRI (1866–1939). Creator of the **Royal Dutch Shell** Group and prime mover in making it one of the world's **major oil companies**. Henri Wilhelm August Deterding was born 19 April 1866 in Amsterdam. When he was four years old, his father died, leaving his large family in a precarious financial position. Deterding's formal education ended when he was 16, and he began working at Twentsche Bank. At 22, he went to the Dutch East Indies (present-day **Indonesia**) to work for Nederlandsche Handel-Maatschappij (Netherlands Trading Company), first at Medan and then as subagent for the more important branch at Penang. He distinguished himself quickly, making profits by exploiting a variety of business and arbitrage opportunities. When he was at the Penang branch, he was approached by

Jean-Baptist Kessler, the local head of the **Royal Dutch Company**. Royal Dutch was severely short of capital in the early 1890s and was unable to get a loan anywhere else. Kessler and Deterding already knew each other from the Netherlands. Deterding took a chance and loaned Royal Dutch the money it needed, using kerosene in inventory as collateral.

In 1895, Deterding went to work for Royal Dutch to run its new trading organization. By the end of 1900, he was its managing director, succeeding Kessler, who had died that December. Deterding had a vision to make Royal Dutch, then a minor player, into a major force in the petroleum industry. He believed in the value of global alliances among companies and in the orderly exploitation of the market. He abhorred destructive competition and price wars. **Standard Oil** had initiated such a price war in Asia beginning in 1894, sometimes giving oil lamps away for free.

Deterding approached **Marcus Samuel**, the founder and head of the **Shell Transport and Trading Company**, a number of times to propose an alliance. Shell had oil fields in East Asia and an advanced transportation organization. Both Shell and Royal Dutch had been takeover targets for Standard Oil but had remained independent. Kessler had already had discussions with Shell in the 1890s to combine marketing operations, but the two companies had not come to a formal agreement. This was largely because Shell had wanted to do more than combine marketing operations; it had wanted to acquire Royal Dutch. In 1903, the two companies finally agreed to cooperate. Deterding and Samuel joined the **Rothschilds** to form the Asiatic Petroleum Company, a joint distribution concern for their oil products. Deterding moved to England to become Asiatic's managing director.

In 1907, Deterding formed the Royal Dutch Shell Group by combining Royal Dutch with Samuel's Shell. Deterding became the Royal Dutch Shell Group's managing director while Samuel became a nonexecutive and largely titular chairman. Samuel was reluctant but had little choice. Shell's financial position had grown increasingly fragile by then. Unlike Deterding, who focused purely on commerce, Samuel had many other social and political interests with a claim on his energies. Deterding took the upper hand: Royal Dutch would control 60 percent of the group's shares, while Shell would control the minority 40 percent. The capital structure was complex. Two holding companies were created: the Anglo–Saxon Petroleum Company and the Bataafsche Petroleum Maatschappij (Batavian [Dutch] Petroleum Company). The former was incorporated under English law and the latter under Dutch. Shell owned 40 percent of each, and Royal Dutch owned 60 percent of each. This basic capital structure would continue until 2005, when the group was fused into a single Royal Dutch Shell.

Forever trying to create a stable and orderly global market, Deterding tried to reach an agreement with Standard Oil, and its descendant Standard Oil of New Jersey (later **Exxon**), to split the global market outside the **United**

States. Standard rebuffed him with price wars. As a result, Deterding moved decisively into Standard's home turf in the Western Hemisphere. In 1913, he acquired operations in **California** and **Venezuela**. He made several other moves, including having Royal Dutch Shell join the Turkish Petroleum Company (later **Iraq Petroleum Company**) consortium in 1912 with a 22.5 percent interest. Not all Deterding's actions were ultimately successful. Two of his moves proved disastrous: in 1912 he acquired the Rothschild petroleum interests in **Russia**, and in 1918 he acquired control of **Weetman Pearson**'s Mexican Eagle Petroleum Company in **Mexico**. Both of these interests would be lost when nationalized by the Soviets in 1918 and Mexico in 1938.

Deterding wanted to do business with the British government, supplying the Royal Navy with fuel. He was unhappy when the British government bought a 51 percent stake in the Anglo–Persian Oil Company (APOC; later **Anglo–Iranian Oil Company**, still later **BP**) in 1914. He clashed with Anglo–Persian's **Charles Greenway**, who attempted to paint both Deterding and Royal Dutch Shell as foreign and unreliable, despite the fact that Shell, the 40 percent partner, was British. Nevertheless, Deterding cooperated with the British government during **World War I** and was given an honorary knighthood for his service.

After the war, the petroleum industry entered another phase of oversupply (due partly to Soviet oil) and competition of the type that Deterding abhorred. In an attempt to control the market, he hosted a meeting at Achnacarry Castle in Scotland attended by **John Cadman** and **Walter Teagle**, heads respectively of APOC and Standard Oil of New Jersey. It should be noted that Cadman had already approached Teagle and had a draft agreement produced internally. The three magnates concluded the secret **As-Is Agreement** to limit competition and control the global oil market. The agreement would not prove successful in the long run. Too many **independent** oil companies were operating outside the agreement, and too many new sources of oil were soon developed, including the giant **East Texas** field discovered two years later.

In 1936, Deterding retired from Royal Dutch Shell. Many observers have criticized Deterding for directly supporting the Nazi regime in Germany during the 1930s, possibly under the influence of his German second wife. He did appear unfortunately smitten by Nazi rhetoric, although it is not clear that he was motivated by all of the same ideologies or by fervent anti-Semitism. He was decidedly anticommunist and may perhaps have seen Germany as part of a bulwark against the Soviets. In 1935, he proposed that Royal Dutch Shell supply Germany with a year's worth of oil on credit, and this was viewed negatively in both Britain and the Netherlands. The fact that he had a German residence in Mecklenburg where he spent considerable periods of time also did not help his image. However, he was not always pleased by the actions of the Nazi regime. He was, for example, incensed when they began buying oil from Mexico after Royal Dutch Shell's interests had been nationalized there. Henri Deterding died in St. Moritz, Switzerland, on 4 February 1939.

DHAHRAN (ZAHRAN). Town (pop. 121,000, 2010) on the **Persian Gulf** coast of **Saudi Arabia**, in the Eastern Province. Dhahran hosts the administrative headquarters of Saudi **Aramco** and is a major center of the Saudi oil industry. The town is part of a greater metropolitan area including Dammam and Khobar, and it is connected to **Bahrain**, offshore to the east, by a 15-mile causeway. Dhahran was built around the Dammam oil field, near the site of the first modern commercial oil discovery in Saudi Arabia, which occurred in 1938. The famous well is known as Dammam No. 7. Dhahran is the site of the King Fahd University of Petroleum and Minerals and has an important Saudi air force base, the King Abd'al Aziz Base, which was originally established by the **United States** for its own air force in 1945. Dhahran was a major deployment point for U.S. forces during the 1991 **Persian Gulf War**.

DOHENY, EDWARD (1856–1935). Early oil entrepreneur in **California** and **Mexico**. Edward Laurence Doheny was born 10 August 1856 in Fond du Lac, Wisconsin. After several years in Mexico and the southwestern **United States**, he arrived in Los Angeles in 1892. Seeing a man hauling tar in a wagon one day, Doheny asked what it was, and the man said it was "breer" (from Spanish *brea*, for "pitch"). This supposedly convinced Doheny to prospect for oil. He had almost no financial resources but persuaded his friend Charles Canfield to put up $400 for a three-acre parcel of land near downtown Los Angeles. The two men began sinking a shaft by hand at a site close to where present-day Glendale Boulevard approaches Second Street. At a depth of seven feet, they encountered a seep of light oil. A bit deeper, the oil-soaked shale crackled from bubbling gas. At 155 feet, the two men had to stop digging because they could no longer bear the fumes. They then used a sharpened eucalyptus tree as a makeshift drill bit and struck oil and gas in November 1892. Doheny and Canfield had discovered the Los Angeles City oil field and triggered a land boom.

By mid-1894, the field was producing about 120 bpd. Between 1892 and 1900, almost a thousand wells were drilled in the area. Doheny went on to discover oil throughout the Los Angeles basin and southern California.

Doheny began prospecting in Mexico around the turn of the century. He was first invited by railway builders who had just built a line between San Luis Potosí and Tampico to exploit petroleum seeps along their right-of-way, hoping to use petroleum to fire their locomotive boilers. Doheny's first discovery was in 1901 at El Ebano near Tampico, but the oil was initially too heavy for the railroad's use. Doheny drilled more wells and was eventually able to refine fuel oil. He was also able to export the heavy oil to the United States for use in paving. He continued to explore without much success, sinking $6 million of his personal fortune into his new Mexican Petroleum Company. In 1905, he acquired new properties in the Tuxpan district farther south. Doheny created the Huasteca Petroleum Company to hold these new properties and explored them intensively. His persistence paid off in 1910 when he drilled several producing wells at Casiano that yielded a lighter and better-quality oil than his properties at El Ebano. His gusher Casiano No. 7 released 402,000 barrels before it was brought under control. Casiano was in the rich oil area that would be known as the Faja de Oro, or Golden Lane. Huasteca began to export crude. In 1911, Doheny concluded an export contract with **Standard Oil**.

During the Mexican Revolution (1910–1920), Doheny created a private army of roughly 6,000 to protect his interests. Doheny formed the Pan American Petroleum and Transport Company in 1916. Incorporated in Delaware, Pan American became a holding company for the Mexican Petroleum Company, the Huasteca Petroleum Company, and Doheny's petroleum interests in California. Doheny's old partner Charles Canfield was a vice president; he died in 1919. By 1918, Pan American had a net profit of $5 million on sales of about $17 million. In 1920, the assets of the company totaled $103 million. By 1921, it had a fleet of 31 **tankers**. Pan American also expanded into **Venezuela**. In 1925, the Standard Oil Company of Indiana (later **Amoco**) acquired a controlling interest in Pan American's operations outside California. The company's California properties were placed in a new company, Pan American Western Corporation, and remained with Doheny. The deal with Standard Oil of Indiana, valued at nearly $38.6 million, was the largest transaction of its type up to that time, and Doheny was reputed to be the richest person in the United States, with a fortune thought to exceed $100 million. In 1924, he ran unsuccessfully for the Democratic Party's vice-presidential nomination.

The Doheny family became major philanthropic donors in Los Angeles, helping to construct the St. Vincent de Paul Church and providing the funds for the Edward L. Doheny Memorial Library at the University of Southern California, built in 1932. Doheny Drive, running through Beverly Hills and West Hollywood, is named for him. However, Doheny was also ruthless and was alleged to have engaged in a number of corrupt practices in his career. His reputation suffered considerably because of his involvement in the **Teapot Dome scandal** (Elk Hills Scandal), although he was formally acquitted of charges. The 2007 film *There Will Be Blood* appears to have been at least partially inspired by aspects of Doheny's life. Doheny died in Los Angeles on 8 September 1935.

DOHERTY, HENRY L. (1870–1939). Founder of **Cities Service Company**. Henry Latham Doherty was born in Columbus, Ohio, on 15 May 1870. Forced by his father's early death to leave school at the age of 12, he began working for the Columbia gas company and rose rapidly through the ranks. He developed his own strategic vision, realizing early on that illumination could no longer be the dominant use for gas, given the advent of electric lighting systems, and new applications would have to be found. By the end of the 1890s, Doherty was chief engineer for Emerson McMillan, Columbia's parent. In 1905, Doherty struck out on his own and formed the Henry L. Doherty Company, providing engineering and business services to utility companies. In 1910, he bought three companies and placed all his operations under a new entity called Cities Service. By 1914, Cities Service had 170 operating companies, typically once-ailing firms that Doherty restored to health. Doherty gradually moved upstream in the gas business and eventually moved into oil as well.

In 1915, the Empire Company, a Cities Service subsidiary, discovered significant quantities of oil near El Dorado, Kansas. Doherty established Empire Gas and Fuel as an oil exploration subsidiary of Cities Service. He moved to increase production to help meet the needs of the **United States** and its allies during **World War I**. By 1918, El Dorado was producing more oil than the **Romanian** and **Galician** fields seized by the Germans. The company developed the Empire, then the world's largest dehydrator, to remove water from oil. Despite its new oil orientation, Cities Service also maintained its focus on gas, utilities, and service. In 1931, it completed the first long-distance, high-pressure natural gas **pipeline** system in the United States, between Amarillo, **Texas**, and Chicago. Apart from his role as an energy executive, Doherty was also involved in the **American Petroleum Institute** (API). He helped establish the organization in 1919 and was its vice president 1919–1924. He remained as a director until 1931. Doherty died 26 December 1939.

DOLL, HENRI. *See* SCHLUMBERGER LTD; WELL LOGGING.

DOLLAR, U.S. The **United States** has had a unique role in the international oil market because oil is priced in U.S. dollars. If a country wants to buy oil, it must—for the most part—acquire dollars first, generally through export earnings. Only the United States can always buy oil in its own currency. As of 2017, this was still true, although cracks were beginning to form in the system and there was mounting pressure to accommodate more currencies. Until the 21st century, there had been no viable challenger to the U.S. dollar, but the 1999 introduction of the euro by the European Union (EU) changed the situation, and there are other potential contenders as well.

The United States enjoyed unprecedented global economic leadership from the end of **World War II** to the 1970s. The Bretton Woods system established the dollar as an international reserve currency backed by gold. The United States had the largest gold reserves in the world. The growing economic strength of Europe and Japan, along with U.S. deficit financing of the Vietnam War, put an end to the Bretton Woods order. Lack of confidence in the U.S. dollar led to increased redemptions of dollars for gold reserves by foreign countries, including France and the **United Kingdom**. On 11 August 1971, the Richard M. Nixon administration abandoned the link of the dollar to gold. This was followed by devaluations of the dollar in 1971 and 1973. The **Organization of the Petroleum Exporting Countries** (OPEC) reacted by demanding that oil prices be corrected in accordance with the value of the dollar, resulting in the **Geneva Agreements** of 1971 and 1973.

OPEC also began to consider pricing oil using a basket of major currencies. The United States was anxious to avoid this eventuality. The Nixon administration convinced **Saudi Arabia** to price international oil sales in dollars only. The result was that OPEC continued to price oil in dollars. The U.S. dollar was no longer backed by gold, but it was effectively backed by oil. In 1974, the Saudi government also secretly purchased \$2.5 billion in U.S. government bonds, setting the pattern for **petrodollar recycling**, whereby U.S. dollars earned by oil-producing countries would find their way back into the U.S. financial system. In more recent years, petrodollar recycling has taken the form of Middle Eastern oil-producing countries using their wealth to buy goods from Japan, China, and the European Union, which then effectively send the petrodollars back to the United States by purchasing U.S. treasury securities. It is estimated that in 2006, 45 percent of the massive U.S. current account deficit was financed by recycled petrodollars.

The introduction of the euro in 1999 created a serious rival to the dollar. Another potential rival may be the Asian currency unit (ACU), a notional unit of exchange based on the currencies of Japan, China, South Korea, and the 10 member countries of the Association of Southeast Asian Nations (ASEAN). The United States is still strongly committed to maintaining the dollar as the international reserve currency and the currency of the oil trade. But it is becoming increasingly difficult to explain the necessity of a dollar-based pricing structure when other strong currencies are as viable or more so.

In 2003, Iran began demanding euros from EU and Asian customers. This did not constitute actual euro pricing for Iran's oil. The pricing structure was and still is based on the three major **benchmark crudes** (**West Texas Intermediate**, **Brent Blend**, and **Dubai crude**), denominated in dollars. True euro pricing would necessitate that those benchmark crudes be denominated in euros. This is unlikely to happen in the short term. However, it is possible that other benchmark crudes will be introduced, and those may be denominated in euros or other currencies. **Russia** has signaled interest in euro payments and in a stronger role for the ruble. Russia's **Volga–Urals** crude is a potentially viable benchmark.

In 2004, Iran announced plans to open an **Iranian Oil Bourse** (sometimes called the International Oil Bourse). The Bourse opened in 2008, trading at first in petrochemicals, in currencies other than the U.S. dollar. There has been a continuing intent, still not realized, eventually to introduce a new benchmark crude with nondollar pricing.

Some observers believe that U.S. saber rattling over Iran's nuclear program has been motivated partly by a desire to prevent Iran from executing a successful campaign to undermine the dollar's role in the oil markets. Some of the same observers point out that **Iraq**'s **Saddam Hussein** began demanding euros in payment for oil in September 2000, and that this was one possible reason for launching the **Iraq War** in 2003. Others dismiss such contentions as unfounded conspiracy theories.

DONNELL, JAMES C. (1854–1927). President of the Ohio Oil Company (later **Marathon Oil**) 1911–1927. James C. Donnell was born in Ireland on 20 April 1854. When he was two years old, his family moved to Waterford, **Pennsylvania**. At 18, Donnell began working in the new Pennsylvania oil industry, transporting oil. After the **Bradford oil field** was discovered, he started his own oil company in that area. He then moved his operations to the **Lima–Indiana** field in Ohio and began working with **Standard Oil**. In 1887, he drilled Standard's first significant gas well in Indiana. In 1889, Standard gained control of the Ohio Oil Company, in which Donnell would spend the rest of his career, beginning as a manager of field operations. In 1901, he became vice president and general manager of Ohio Oil. Ten years later, Ohio Oil became an independent company upon the dissolution of Standard Oil. Donnell became president, succeeding **John D. Archbold**.

As president, Donnell expanded Ohio Oil's operation to **Mexico** and 16 states of the **United States**. In 1924, he acquired Lincoln Oil Refining for the company. At the end of his tenure, Ohio Oil controlled about 5 million acres of oil-producing properties and had assets of $108 million and a net income of $17 million. The company had significant oil production operations in Ohio, Indiana, **Oklahoma**, Kansas, **Texas**, and Louisiana as well as in Mexico. James C. Donnell died 10 January 1927. He was succeeded as president of Ohio Oil by his son **Otto D. Donnell**.

DONNELL, OTTO D. (1883–1961). President of the Ohio Oil Company (later **Marathon Oil**) 1927–1948. Otto Dewey Donnell was born in Ohio on 26 September 1883. He graduated as a mechanical engineer from the Case School of Applied Science in 1906. He joined the Ohio Oil Company, where his father, **James C. Donnell**, was then vice president and general manager. He rose rapidly through the ranks. When his father became president in 1911, Donnell became a director of the company. When his father died in 1927, he succeeded him as president. He remained president until 1948, when he was succeeded in turn by his own son, James C. Donnell II.

Donnell began a major expansion program in 1930. He bought a majority interest in Transcontinental Oil Company as well as several interests from other oil companies. He also purchased service stations. By 1948, he had nearly doubled Ohio Oil's assets. Donnell increased annual profits by nearly a factor of 10 during his tenure, to $68 million. In 1936, he absorbed a former subsidiary, Marathon Oil Company, into the parent company. Ohio Oil would adopt Marathon as its company name in 1962. Otto Donnell died 9 April 1961.

DRAKE, EDWIN L. (1819–1880). Supervised drilling an early oil well near **Titusville, Pennsylvania**, helping to launch the modern American oil industry. Drake's well is popularly considered the "first modern oil well," although this is an oversimplification. There were contemporaneous and even earlier efforts (*see* FIRST OIL WELL), but Drake's well and its consequences are regarded as more prototypical of the modern industry than most of the other efforts, important as they also were.

Edwin Laurentine Drake was born 29 March 1819 in Greenville, New York, and grew up in Castleton Corners, Vermont. He held several different jobs before becoming a conductor for the New York and New Haven Railroad. In 1857, he fell ill and resigned from that job, settling in New Haven to recuperate. During this period, he bought some shares of stock in the Seneca Oil Company. Seneca was an offshoot of **George Henry Bissell**'s **Pennsylvania Rock Oil Company**, one of the world's first modern oil companies. The Pennsylvania Rock Oil Company had obtained samples of crude oil from surface seeps near Titusville and sent them to Yale University for analysis. Professor **Benjamin Silliman Jr.** had reported in 1855 that the crude oil could be distilled to obtain useful products, including kerosene for illumination. The depression of 1857 stalled the company's plans. Several of the shareholders organized a new company, Seneca, and leased land near Titusville from the old company.

Drake accepted a job with Seneca to travel to Titusville and report on the site to the stockholders. He made a brief trip in December 1857 and reported positively, having witnessed the skimming of oil from Oil Creek. The stockholders sent him back in May 1858 as the general agent of the Seneca Oil Company, with the fictitious title of "Colonel" to lend him more prestige. His mission was to find a way to produce oil in quantity. Drake spent several weeks digging in the main oil spring on the Hibbard farm. Unfortunately he only struck water. Deciding to drill for oil, he traveled to Tarentum, near

Pittsburgh, to observe the drilling of salt wells. He acquired a six-horsepower steam engine to power the drill, which was a cable tool that operated via percussion and not the modern rotary type (*see* DRILLING TECHNOLOGY). Drake spent the winter of 1858 constructing a derrick to support the drill. He hired William A. "Uncle Billy" Smith, a blacksmith and experienced salt driller. Their first efforts were stymied by groundwater flooding, until they cased the well by driving cast-iron pipe down the hole with an oak battering ram. They resumed drilling in August 1859, by which time the company was nearly out of money and wanted to shut down the operation.

On 27 August 1859, work stopped after the drill slipped through a crevice at a depth of 69 feet. The next day, Uncle Billy noticed oil floating at the top of the well, confirming that they had made a discovery. Drake drilled a second well nearby, striking oil at 480 feet. The first well produced about 20 bpd, the second one about 24, which are extremely low rates of production by the standards that would soon be established. A frantic oil boom followed Drake's discoveries, launching the modern U.S. oil industry.

Drake left the oil industry in 1863. He had served for a time as president of the Seneca Oil Company but was not a good manager and had been forced to resign in 1860. Although he had paved the way for others to become wealthy, he made only a modest sum, and he managed to lose even that through poor investments. In 1873, his friends in Titusville convinced the Pennsylvania legislature to provide him with a pension in recognition of his contributions to the state's economy. Drake lived with debilitating muscular illness in Bethlehem, Pennsylvania, until his death on 8 November 1880.

DRESSER, SOLOMON ROBERT. *See* DRESSER INDUSTRIES.

DRESSER INDUSTRIES. Oil field services company; later part of **Halliburton**. Dresser was founded by Solomon Dresser (1842–1911), who invented the Dresser cap packer to keep oil and water separated in a well. He did not invent the first packer, but he used rubber for a tight fit and received a patent in 1880. He sold his product in the industry surrounding the **Bradford oil field** in **Pennsylvania**. In 1885, he invented the Dresser coupling for joining pipes so that gas would not leak. Again he used rubber. The Dresser coupling helped enable long-distance **pipelines** for **natural gas**.

As the gas industry expanded, so did the company's fortunes. Dresser's descendants ran the firm after his death in 1911, and by 1927 annual revenues were $3.7 million. In 1928, the company went public. In the 1930s, it acquired a number of companies making valves, pumps, and other mechanical equipment. After **World War II**, it diversified into other oil field products, including drill bits, drilling mud, and derricks. In 1950, the company moved its headquarters to Dallas, **Texas**.

Acquisitions and diversification continued until the 1980s, when the downturn in the petroleum industry (*see* OIL COUNTERSHOCK) forced a refocusing of the organization. Dresser Industries divested divisions in areas such as insurance, construction equipment, and mining. But after 1987–1988, it resumed acquisition and diversification. In 1987, it entered into a joint venture with Litton Industries to form **Western Atlas**, which offered clients the full spectrum of upstream services from exploration to production, including **geophysical prospecting** and **well logging**. In the early 1990s, Litton bought back Dresser's share of Western Atlas and in 1993 spun it off as an independent company (later to be acquired by **Baker Hughes** and still later to be partially acquired by **Schlumberger**).

In 1988, Dresser acquired M. W. Kellogg and entered into a joint agreement with Komatsu of Japan to manufacture construction equipment. In 1994, it acquired oil field services firm Baroid Corporation and pump manufacturer Wheatey TXT. In 1998, Dresser Industries merged with Halliburton.

DRILLING TECHNOLOGY. Drilling technology originated in ancient times, motivated by the search not for petroleum but for water. Percussion drilling methods were developed in **China** about 2,000 years ago. Bamboo framework structures, similar to modern derricks, were employed to lift and drop a heavy chiseling tool. The modern cable tool operates on the same basic principle. The Chinese drilled wells as deep as 3,000 feet, although completion may have taken several generations.

At **Pechelbronn** (Alsace, France), as many as 140 manual wells were drilled in 1813 at an early oil sands mining operation that had been operating since the 18th century. The wells were not drilled to obtain oil directly but rather to guide the development of oil sands mining. In the 19th-century **United States**, salt drove the development of drilling technology. Between 1806 and 1808, David and Joseph Ruffner completed a 58-foot salt well near the Kanawha River, not far from Charleston, in what is today **West Virginia**. They used a spring pole. This consisted of a long timber pole, anchored at one end and propped on a fulcrum. A rope holding heavy chisel-type drilling tools was suspended from the free end of the pole. An operator would use stirrups to mount the pole and push it down with his legs, so that the drill bit would hit the target. The spring pole would bring the tools back up, so a new stroke could be started.

The Ruffners ushered in a vigorous salt well industry that inspired the petroleum industry. For one thing, oil was sometimes an undesirable waste product in salt drilling. **Samuel Kier**, who began marketing petroleum as a patent medicine in 1848, obtained it from his family's brine wells in Tarentum, **Pennsylvania**, near Pittsburgh.

The salt drilling industry was where the early petroleum industry obtained its drilling technology. In the 1830s and 1840s, in what is now West Virginia, salt and oil drilling operations were sometimes water powered. **Edwin Drake**, the driller of an early oil well near **Titusville** that sparked the Pennsylvania oil boom, traveled to Tarentum to observe salt drilling techniques. He also tapped the salt industry for an experienced driller, William A. "Uncle Billy" Smith, to act as a rig mechanic and blacksmith.

Drake's cable tool was descended from the spring-pole technique and the methods of ancient China. A wooden derrick, "Drake's yoke," about 30 feet tall, supported the tools. The steel bit, chisel pointed and about four feet long, was suspended from rope ("cable") and was raised and dropped repeatedly to pulverize the rock. The raising of the bit was accomplished by rocking a "walking beam," driven by a steam engine. Every three to eight feet drilled, the bit was raised and a bailer lowered into the hole to remove debris. Drake also employed casing in the hole to guard against flooding by groundwater and to keep the sides of the well from collapsing. Cable tools continued to dominate throughout the 19th century and are still used in some applications today. The "Pennsylvania" type rig had a wooden derrick, eventually standardized at 72 feet or 82 feet in height. By the early 20th century, cable tool rigs were more often a kind used in **California**. The "California" type had iron and steel derricks supporting heavier tools for deeper drilling. Pennsylvania-type rigs could drill to about 2,000 feet, whereas California-type rigs could drill down to 10,000 feet. Cable tool rigs were effective and relatively simple to operate, but they were slow. They averaged about 25 feet per day, depending on the type of rock.

In the 20th century, rotary drilling largely supplanted cable tool drilling. Rotary rigs eliminated the laborious bailing process and could drill faster and deeper. Rotary rigs can drill hundreds or even thousands of feet per day. Elementary rotary drilling techniques were used as far back as Egyptian times. In 1844, Robert Beard obtained a U.S. patent for a rotary drilling system with hollow drilling rods and circulating fluid to remove the cuttings. The idea underwent many improvements during the 19th century, and a rotary tool was used in **Corsicana, Texas**. One was also employed by **Anthony Lucas** at **Spindletop**. Initially, rotary rigs could only be used in soft rock formations. This problem was solved by the invention of the roller cone drill bit by **Howard Hughes Sr.** in 1908. The bit used 166 cutting edges distributed over the surfaces of two opposingly mounted metal cones. The bit ground the rock rather than chipping or scraping it and could be used in hard formations.

The 20th century saw many innovations in all areas of rotary drilling: collars to prevent well deviation from vertical, advances in drilling fluids, blowout prevention and control, and techniques of well cementing. The ability to control blowouts enabled deeper wells. In 1938, a depth of 15,000 feet was reached. In 1949, a depth of 20,000 feet was reached. In 1958, **Phillips Petroleum** drilled down to 25,340 feet in West Texas. The 30,000-foot barrier was breached in 1974. In 2007, **Exxon**'s Russian subsidiary Exxon Neftegaz drilled a 37,016-foot well on **Sakhalin** Island.

Another important innovation is extended-reach drilling: **horizontal drilling** from a borehole begun vertically. Extended-reach techniques can be used to drill multilateral wells, where several directional boreholes fan out from a single well. Multilateral wells are useful in **offshore** environments, where a large area must be covered from a limited number of well positions on a platform. Multilateral wells are also useful in producing from discontinuous and compartmentalized reservoirs. These are often considered a recent innovation, but a type of multilateral well was drilled in the Bashkiria field in the Soviet Union in 1955. Horizontal drilling can also parallel a relatively thin bed and thus expose much more of it to the borehole than if the borehole were slicing vertically through it. This, combined with **fracking**, has facilitated the extraction of previously inaccessible **tight oil** (**shale oil**), **shale gas**, and **tight gas** trapped in low-permeability formations.

DRUZHBA PIPELINE. One of the world's longest oil **pipelines** (2,500 miles), transporting oil from the **Volga–Urals** and western **Siberia** fields of **Russia** and from **Kazakhstan** to Eastern Europe. The Druzhba Pipeline (Friendship Pipeline, Comecon Pipeline) begins in Samara, Russia, and runs west to Mozyr, Belarus, where it splits into two main branches. The northern branch goes to Poland and Germany. The southern branch goes to Ukraine, Slovakia, the Czech Republic, and Hungary. The pipeline was built by the Soviet Union to supply its own needs and those of its communist-bloc partners, thereby also tightening the integration of the bloc and helping to buttress Soviet leadership. The Soviet-bloc Council for Mutual Economic Assistance recommended building the pipeline at its meeting in Prague in December 1958. The decision to build was announced on 23 August 1959 by the Soviet news agency Tass. By 1962, first oil reached what was then Czechoslovakia. In September 1963, it reached Hungary, and two months later Poland. In October 1964, the pipeline was fully operational.

The pipeline was hailed in the Soviet bloc as a socialist victory, but it relied on significant imports of critical materials from the West. In particular, 40 percent of the needed 40-inch-diameter pipe was imported from Sweden, Italy, and (then) West Germany, despite recommendations from the North Atlantic Treaty Organization (NATO) not to export 40-inch steel pipe to the Soviet Union. The pipeline system has diameters ranging from 20.9 inches to 40.2 inches, and 31 pumping stations. It crosses mountain ranges as well as the Volga, Dnieper, Danube, and Tisza Rivers. The pipeline is entirely underground. Current capacity is 1.2 to 1.4 Mmbpd. The pipeline is now operated by the Russian state pipeline monopoly Transneft.

DUBAI. One of the constituent emirates of the **United Arab Emirates** (UAE). Dubai has about 4 Bb, or roughly 4.1 percent, of the UAE's 97.8 Bb of oil **reserves** (2015), making it a distant second to **Abu Dhabi**. Dubai joined the **Organization of Arab Petroleum Exporting Countries** (OAPEC) in 1970. **Dubai crude**, or Fateh crude, constitutes a major **benchmark crude** in the international market. The first **concession** in Dubai was granted in 1937 to Petroleum Development (Trucial States), or PDTC, a partnership with the same ownership structure as the **Iraq Petroleum Company**. (The present-day UAE was within the territory covered by the **Red Line Agreement**). PDTC relinquished its concession in Dubai by 1963 in order to concentrate on Abu Dhabi.

The concession was taken over by the Dubai Petroleum Company (DPC), owned by **Conoco** (later **ConocoPhillips**). DPC discovered the **giant** Fateh field (**EUR** about 1.1 Bb) in 1966. Dubai began exporting crude in 1969. In 1970, DPC discovered Fateh Southwest (EUR about 1.14 Bb). Other discoveries included Rashid and Falah. Dubai's oil production peaked in 1991 at about 0.41 Mmbpd and has declined steadily since. On 2 April 2007, Conoco-Phillips turned over all operations to the government of Dubai, although the concession was not due to expire until 2012. Negotiations on the renewal of the concession had failed. A new government entity, the Dubai Petroleum Establishment, took over operations.

DUBAI CRUDE. A principal **benchmark crude** used as a reference especially for pricing **Persian Gulf** oil exported to Asia. Dubai crude, also called Fateh crude, is medium heavy, with an **API gravity** of 31.4°, and it is medium sour, at 2 percent sulfur. It is heavier and sourer than **Brent Blend** and **West Texas Intermediate**, two other important benchmarks, and more typical of the grades of oil produced in the Persian Gulf region. For many years, Dubai was the only oil in the Middle East that was truly freely traded. This was a major factor in its adoption as a benchmark. As is the case with other benchmarks, falling production of Dubai has been a problem. From 1990 to 1995, production of Dubai crude averaged about 400,000 bpd. By 2004, it was about 120,000 bpd and fell even further thereafter.

As less Dubai crude becomes available in the market, pricing distortions may occur, making it less reflective of overall supply and demand. Such a distortion was observed in 1999 when the price of Dubai crude rose higher than that of **Oman** crude. Normally this should not occur, because Oman crude is lighter and sweeter than that of Dubai. Oman crude was added to the Dubai benchmark in 2001. With falling production, it also became problematic that Dubai crude traditionally traded only in large lots—typically in increments of a full cargo of 500,000 barrels. Both Brent Blend and West Texas Intermediate have traded in lots as small as 1,000 barrels. In 2004, mechanisms were devised to allow Dubai to trade in lots as small as 25,000 barrels. Dubai crude also formed part of the **OPEC Basket** from 1 January 1987 to 15 June 2005.

DUTCH DISEASE. A situation arising when a country discovers and exploits a new natural resource, such as petroleum, and then suffers a decline in the competitiveness of its traditional economic sectors. This can happen, for example, if there is an export boom in the resource, causing an appreciation of the currency; this in turn can lead to declines in exports other than the resource and possibly in import-competing industries. The phenomenon derives its name from the experience of the Netherlands. The discoveries of the **supergiant Groningen gas field** in 1959 and the **giant** Bergen gas field in 1964 are believed to have resulted in slow growth and high unemployment into the 1970s. Dutch disease may have occurred in **Nigeria**, **Indonesia**, and other oil exporting countries after the oil price boom of the 1970s (*see* FIRST OIL SHOCK; SECOND OIL SHOCK). The term "Dutch disease" can be applied more generally to harmful economic conditions and deindustrialization that occur from any sudden increase in a country's wealth accompanied by large inflows of foreign exchange.

E

EASTERN AND GENERAL SYNDICATE. *See* HOLMES, FRANK (1874–1947).

EASTERN GAS SHALES PROJECT. Partnership (1976–1992) between private gas companies in the **United States** and the U.S. Department of Energy to develop new **drilling** and well stimulation technologies and to evaluate the **natural gas** potential of shale formations in the eastern part of the country. The project successfully applied the two crucial technologies of **fracking** and **horizontal drilling** in combination to increase production of gas in shale formations, particularly the Big Sandy Gas Field in Kentucky and **West Virginia**. The project helped set the stage for the commercial developments of **shale gas** and **shale oil**.

EAST TEXAS OIL FIELD. Supergiant oil field in eastern central **Texas**, discovered 5 October 1930. The East Texas oil field, which some call the "Black Giant," has produced about 5.2 Bb thus far. Most observers believe the field to be nearing exhaustion, although some estimate significant remaining **reserves**. However, production in 2002 was about 0.016 Mmbpd versus about 1 Mmbpd in 1931. The East Texas field, first discovered by **Dad Joiner**, was the largest known field in history at that time, and it is still the largest field ever discovered in the **United States** outside **Alaska**. Its areal extent is about 140,000 acres, and more than 30,000 wells have been drilled in it. Uncontrolled overproduction in the East Texas field led to a collapse of oil prices in the 1930s, prompting action by the **Railroad Commission of Texas** and the U.S. government to stabilize prices. During **World War II**, the field became the point of origin for the **Big Inch and Little Big Inch** Pipelines that transported its crude and products to the East Coast.

EAST–WEST PIPELINE. *See* PETROLINE (EAST–WEST PIPELINE).

ECUADOR. Oil production 2015: 0.541 Mmbpd (0.69 percent of world); **reserves** 8.27 Bb (0.5 percent of world); gas reserves 0.385 Tcf (0.01 percent of world). Ecuador is a net exporter of crude oil, with a large proportion of its exports going to the **United States**. Ecuador produces two main types of crude: 29°**API gravity** Oriente and 19°API Napo. Oriente is medium sour, while the heavy Napo is sour. Ecuador does not have significant reserves of **natural gas**. The country joined the **Organization of the Petroleum Exporting Countries** (OPEC) in 1973, only a year after its first exports. Ecuador left the organization in 1992 because of membership costs and because it wanted a higher production quota than OPEC would grant. In 2007, it rejoined the organization.

In the 2000s the national oil company Empresa Estatal Petróleos del Ecuador (Petroecuador) accounted for nearly half of the country's oil production. Foreign and private oil companies operating in the country in 2015 included **Repsol**, Andes Petroleum, **ENI (AGIP)**, Tecpetrol of **Argentina**, and Andes Petroleum, a consortium led by the Chinese National Petroleum Company (CNPC). CNPC acquired assets from EnCana and Perenco in 2005.

Oil exploration in Ecuador dates back to 1917, with the **drilling** of some low-output wells on the Santa Elena Peninsula, near Guyaquil. However, early Native Americans in that area had long known of oil seeps. They dug pits and allowed oil to flow into them, then waited for evaporation to turn it into tar they could use to caulk their canoes. The Spaniards also adopted the practice. Standard Oil of New Jersey (later **Exxon**) obtained a **concession** in 1920 to explore for and develop oil in the Amazon region in the eastern part of the country (the Oriente). They were largely unsuccessful. In 1949, President Galo Plaza Lasso declared, "the Oriente is a myth."

President Lasso was proven wrong in 1967, when **Texaco** and **Gulf** began to make significant discoveries in northern Sucumbios Province, in the Oriente/Amazon region. The **giant** Sacha and Sushufindi fields were discovered in 1969. Sacha, now operated by Petroecuador, has an **EUR** of 650 Mmbo, of which a mean estimate of 238 Mmbo may be remaining. Sushufindi, also now operated by Petroecuador, has an EUR of 586 Mmbo, of which 236 Mmbo may be remaining. Other significant fields include Dorinde, operated by Andes Petroleum, and Eden Yuturi, formerly operated by **Occidental Petroleum** but now operated by Petroecuador.

The 310-mile Sistema Oleoducto Trans-Ecuatoriano (SOTE) Pipeline was completed in 1972 to carry oil to the Balao terminal near Esmeraldas on the Pacific coast. First oil flowed on 26 June 1972, and Ecuador began exporting oil that year. SOTE has a capacity of 0.4 Mmbpd. It has suffered frequent interruptions because of landslides and other natural disasters. It was shut down by an earthquake in 1987, and Ecuador's oil production declined that year to 0.174 Mmbpd from 0.293 in 1986. SOTE also suffered a number of terrorist attacks in 2000.

In 1983, the government enacted reforms to encourage investment by foreign petroleum companies, and a number of them signed service contracts with Petroecuador's antecedent, the Corporación Estatal Petrolera Ecuatoriana (CEPE). Production rose from 0.237 Mmbpd in 1983 to 0.302 Mmbpd in 1988. The increase was steady except for the significant dip in 1987 mentioned above. Petroecuador was established in 1989. Its charter was to increase national control over downstream operations. It took over SOTE. However, declining revenues soon led to further market liberalizations, and new **production-sharing agreements** giving private companies the opportunity to share in finds were signed in the 1990s. By 2001, production was up to 0.415 Mmbpd. After a dip, it stood at 0.5 Mmbpd in 2007.

Despite the market reforms, Ecuador has a noticeable nationalist streak in petroleum matters, and disagreements arise between private companies and the government. A serious dispute occurred between the government and Occidental Petroleum in 2002, culminating in Occidental's ejection from the country in 2006 and the transfer of all its operations to Petroecuador. Occidental accused the government in 2002 of refusing to grant value-added tax rebates to which the company claimed it was legally entitled. The company sought international arbitration, further angering the government. Later, Occidental sold its share of a concession to EnCana (which later sold it to CNPC as noted above). The government accused Occidental of making the sale without asking permission. The **United States** reacted negatively to Occidental's expulsion from Ecuador and claimed that Ecuador was obliged by treaty to compensate the company. The United States suspended negotiations with Ecuador for a free trade agreement, resulting in charges from Ecuador of interference in its internal affairs. The case went to arbitration, and in 2012 the International Center for Settlement of Investment Disputes ruled that Ecuador should pay Occidental $1.77 billion. In 2016 Ecuador settled by agreeing to pay $980 million. There were other disputes. In 2011 an Ecuadorean court issued a judgment of $18 billion against **Chevron**, later reduced to $9.5 billion, for alleged environmental damage arising from Texaco's Ecuadorean operations between 1964 and 1990. Chevron acquired Texaco in 2001.

In addition to the tension between the government and private companies, there has been significant opposition to the petroleum industry by Amerindian groups. Indigenous groups are against further exploration and development in the Amazon region. Development of Ecuador's most promising source of new production, the Ishpingo–Tapococha–Tipufuni (ITT) bloc, was stalled between 2007 and 2013. ITT is estimated to hold proven reserves of 900 Mmbo. Protesters have frequently shut down oil production and disrupted **pipeline** flow. Since 2013, activity has been sporadic.

A new pipeline, the Oleoducto de Crudos Pesados (OCP), was completed in 2003. It largely parallels the route of SOTE. OCP roughly doubles Ecuador's transport capacity and removes a significant constraint on production and exports. OCP is used mostly by private companies, while Petroecuador relies primarily on SOTE. Ecuador also exports some oil through the international Transandino Pipeline to the Colombian Pacific port of Tumaco.

EDMONTON. Capital city (pop. 933,000, 2016) of the Canadian province of **Alberta**, and arguably **Canada**'s oil capital (another major center being **Calgary**, Alberta). Edmonton is Canada's northernmost major city, about 350 miles from the border with the **United States**, and is known as Canada's "Gateway to the North." It was founded as a trading post in 1795 by the Hudson's Bay Company. The University of Alberta, founded 1906, is located here. The city is within 50 miles of three **giant** oil fields: Leduc and Bonnie Glen to the south, and Redwater to the north. A cluster of four giants lies a bit farther to the northwest. The Leduc field, discovered on 13 February 1947, was the first major oil discovery in Alberta and a milestone for Canada's oil industry.

Major discoveries in northern and central Alberta continued into the 1950s, and Edmonton boomed. From 1950 to 1960, the population increased from 149,000 to 269,000. By 1981, it was greater than half a million, following the high oil prices of the 1970s. The city suffered a major recession in the 1980s during the era of low oil prices, not recovering until the late 1990s. Edmonton is also a center of business activity for northern Alberta's **oil sands** projects.

EGYPT. Oil production 2015: 0.687 Mmbpd (0.88 percent of world); **reserves** 4.4 Bb (0.27 percent of world); gas production: 1.57 Tcf (1.26 percent of world); reserves 77.2 Tcf (1.12 percent of world). Egypt has been important in the history of petroleum in a number of ways. It was the first country in the Middle East and North Africa to be systematically explored for oil. Although it did not become a major oil producer by Middle Eastern standards, its production has been significant. Egypt became a net oil exporter in 1976 and has recently been a fast-growing gas producer. In 2005, it began exporting **liquefied natural gas** (LNG). Egypt is a member of the **Gas Exporting Countries Forum**. Finally, Egypt has seen significant growth as a transit corridor. The **Suez Canal** and the **Sumed Pipeline** are both strategic routes for **Persian Gulf** oil. Egypt also has had considerable political influence on other oil-producing countries as a wellspring of Arab nationalism. Egypt never joined the **Organization of the Petroleum Exporting Countries** (OPEC) but in 1973 became a member of the **Organization of Arab Petroleum Exporting Countries** (OAPEC). OAPEC suspended Egypt's membership in 1979 for signing the Camp David Accords and making peace with Israel. The organization readmitted Egypt in 1989. Egypt lost its Sinai oil fields to Israel after the Six-Day War in 1967 but regained them in 1975.

About half of Egypt's oil production comes from the Western Desert, and the rest from other areas including the Gulf of Suez, the Eastern Desert, the Sinai Peninsula, and **offshore** in the Mediterranean. The earliest discoveries were around the Gulf of Suez. Khedive Ismail, ruler of Egypt from 1863 to 1879, embarked on a program of modernization that included mineral exploration. In 1869, the French company Société Soufrière des Mines de Jemsah et de Ranga discovered oil seeps at Ghubbat al Jamsah (Gemsa), along the coast of the Gulf of Suez, while exploring for sulfur. The company asked permission to explore specifically for petroleum, but the government refused, wishing to preserve the prerogative for itself. The company sued but lost. Between 1883 and 1888, the government employed a number of Western experts, including the Belgian M. de Bay and the American Herbert Tweddle, to explore in the Jamsah area. Despite success in finding oil, particularly in Tweddle's case, the government deemed further exploration and development too expensive and terminated the efforts in 1888.

In June 1907, the Egyptian Oil Trust Company was formed in London. It obtained a **concession** of 100 square miles west of the Red Sea, including the Jamsah area. In 1909, the company struck a gusher at Jamsah that flowed about 3,000 bpd. Exports began in 1912. The country's production in 1911 was only 58 bpd, rising past 1,000 bpd by 1916. The Jamsah field was relatively limited, producing oil commercially until 1927. In 1911, Anglo–Egyptian Oilfields (AEO), a joint venture between the Anglo–Persian Oil Company (later **Anglo–Iranian Oil Company**, eventually **BP**) and **Royal Dutch Shell** made Egypt's first major discovery, the Hurghada field, also in the Gulf of Suez region. Hurghada peaked in 1931 at nearly 5,000 bpd. The 1920s saw no significant discoveries, even though AEO began an early application of **geophysical** gravimetric techniques. AEO's next major discovery, Ras Gharib (about halfway between Hurghada and Suez), was in 1938. Egyptian production, which had stayed mostly under 5,000 bpd before Ras Gharib, reached 13,000 bpd in 1939 and continued to rise during **World War II**. In 1948, the government banned exports of crude oil, allowing only refined products to be exported. This had the effect of curbing exploration until 1953, when the law was liberalized. **ENI** entered Egypt in 1955. In that year, the Southern California Petroleum Corporation discovered the **giant** Belayim field (**EUR** about 0.5 Bb). Belayim Marine, with an additional EUR of about 0.8 Bb in the Gulf of Suez, was discovered in 1961.

In 1956, the government created the General Petroleum Authority (GPA) to oversee the sector. In 1962, the GPA became the Egyptian General Petroleum Corporation (EGPC). Between 1961 and 1964, both AEO and Shell Egypt (which conducted marketing and distribution) were nationalized.

EGPC entered into a number of joint ventures with Western companies and developed a reputation as a well-managed national oil company. One of its earliest partners was Standard Oil Company of Indiana (later **Amoco**, eventually BP). EGPC and Standard of Indiana formed the Gulf of Suez Petroleum Company (GUPCO) and discovered the giant El Morgan field (EUR 1.5 Bb) in 1965. A year later **Phillips Petroleum** and EGPC discovered the El Alamein field in the Western Desert. Egyptian production, which had surpassed 0.1 Mmbpd in 1963, accelerated after the El Morgan discovery, reaching 0.326 Mmbpd in 1970. It declined to 0.147 Mmbpd in 1974. The giant October field (EUR about 0.5 Bb) was discovered in the Gulf of Suez in 1977. Egyptian production grew rapidly to 0.887 Mmbpd in 1985. It stayed mostly between 0.8 and 0.9 Mmbpd between 1985 and 2000, peaking at 0.922 Mmbpd in 1996. Production in 2006 was lower than in 1983.

Egyptian gas production was about 0.03 Tcf in 1980. It grew steadily to 0.518 in 1999, after which it grew more rapidly, reaching 1.87 Tcf in 2006. In 2015 it was 1.57 Tcf. Egypt's largest gas fields before 2015 were in the Nile Delta. The largest, El Temsah (EUR about 4.5 Tcf), was discovered in 1977. Ha'py (about 3 Tcf) and Scarab-Saffron (about 4 Tcf) were discovered in 1996 and 1998 respectively. In August 2015, ENI discovered the **supergiant** Zohr gas field in the offshore Mediterranean. With an estimated 30 Tcf of gas, Zohr is the Mediterranean's largest discovery. The Western Desert also has significant gas reserves.

EKOFISK OIL FIELD. The **North Sea**'s first **giant** oil field, discovered in 1969 by a group led by **Phillips Petroleum**, with first production in 1971. The Ekofisk field belongs to **Norway** and has an **EUR** of 3.8 Bb and 3.9 Tcf of gas, for a total of 4.452 Bboe. Estimates of the remaining recoverable oil and gas range from about 0.6 to almost 3 Bboe. The name also refers to a larger complex of Norwegian fields including Ekofisk itself, Ekofisk West (1969), Tor (1970), and Albuskjell (1972). The complex also includes Cod, the original gas discovery of 1968. The Ekofisk complex is located halfway between Norway and the **United Kingdom**, under about 240 feet of water. Crude oil is piped to Teesside in the United Kingdom, while gas is piped to Emden, Germany. The reservoir rocks are the Ekofisk and Tor chalk formations (upper Paleocene and lower Cretaceous).

EL DORADO OIL FIELD. *See* CITIES SERVICE COMPANY; DOHERTY, HENRY L. (1870–1939); KANSAS; WORLD WAR I (1914–1918).

ELF AFFAIR. Complex and multifaceted corruption scandal involving the French petroleum company **Elf Aquitaine** (later part of **Total**). The scandal began in 1994 with allegations that top Elf executives were enriching themselves with kickbacks, or by embezzling money under the cover of commissions on large oil or property deals, with funds routed circuitously through offshore accounts. Government investigations soon expanded into many other lines of inquiry. Elf was also found to pay large bribes to African leaders. The bribes were intended to serve two purposes: to ensure Elf got access to African oil over companies from other countries, and to cement the African leaders' allegiance to France, keeping former colonies within the French sphere of influence. Elf essentially acted almost as an activist intelligence agency for the government. This was especially true in **Gabon**, which France, through Elf, used as a base for military activities and espionage in Africa. Elf was also accused of paying bribes in Europe.

Another side of the scandal involved the sale of naval frigates to Taiwan by the French defense contractor Thomson, also state owned at the time. Thomson allegedly turned to Elf for its experience in executing illegal bribes and commissions to help the sale go through, and Elf obliged, with its participants taking a cut. Possibly over $100 million went to Taiwanese officials. Some of the money may have gone to officials in **China** as well, since China had to be persuaded to drop objections to the deal. A few million also allegedly went to people associated with French foreign minister Roland Dumas to secure French government assent. Still another facet of the scandal was the alleged creation of well-paid no-show jobs for French politicians by Elf Aquitaine International, a subsidiary based in Switzerland. Former minister Dumas was convicted and sentenced to prison, but he remained free and was found not guilty on appeal. Thirty-seven other criminal defendants, including former Elf chairman Loik Le Floch-Prigent and former director Alfred Sirven, were found guilty in 2003 of siphoning €350 million from Elf between 1989 and 1993. Le Floch and Sirven were sentenced to five years in prison.

ELF AQUITAINE. French integrated petroleum and mining company; now part of **Total**. Elf's origins lie in three French companies. Régie Autonome des Pétroles (RAP) was established in 1939 with a charter to explore for and develop petroleum in France. The Société Nationale des Pétroles d'Aquitaine (SNPA) was established in 1941 with government majority ownership to exploit a gas discovery in the Aquitaine region. The Bureau de Recherches de Pétroles (BRP) was established four years later.

In December 1965, the Entreprise de Recherches et d'Activités Pétrolières (ERAP) was formed by merging RAP and BRP and placing SNPA and some other companies into it as subsidiaries. SNPA was the largest subsidiary, and the French government was the majority owner. ERAP was a vertically integrated company, and in 1967 it launched the Elf brand for all its products. The company led the consortium discovering the Frigg field in the **North Sea** in 1971. In 1976, it became the Société Nationale Elf Aquitaine (SNEA), later Elf Aquitaine. From 1980 to 1983 Elf conducted successful tests of

horizontal drilling technology in European oil fields. This type of technology, improved by many others and combined with fracking, would ultimately become very important in the production of **shale oil** and **shale gas** in the 21st century. Elf Aquitaine was listed on the New York Stock Exchange in 1991. In 1993, it was awarded the exclusive contract to operate **Iraq**'s oil fields by president **Saddam Hussein**. In 1996 Elf discovered the significant Girassol field in **offshore Angola**.

Beginning in 1994 Elf was tainted by scandal when it was revealed that top managers were using company money for lavish personal expenditures as well as political favors. The scandal spawned wide-ranging investigations and kept growing with new revelations, becoming "l'affaire Elf" (the **Elf affair**).

In 1994, the government began privatizing Elf, selling its entire interest by 1996. In 1999, TotalFina (as Total was known briefly after its acquisition of **Petrofina**) launched a $43 billion hostile takeover bid for Elf. Despite Elf's counterbid, TotalFina acquired 95 percent of Elf in 2000 and renamed itself TotalFinaElf. Later in the year, TotalFinaElf completed the acquisition. The combined company changed its name to Total SA in 2003. In 1998, Elf had a net income of $591.4 million on revenues of $35.919 billion. It controlled 2.756 Bb of oil **reserves** and 6.170 Tcf of gas. It produced oil and gas at the rate of 0.799 Mmbpd and 1.208 Bcfpd respectively.

ELK HILLS. *See* NAVAL PETROLEUM AND OIL SHALE RESERVES.

ELK HILLS SCANDAL. *See* TEAPOT DOME SCANDAL (1922–1930).

EMPRESA COLOMBIANA DE PETRÓLEOS (ECOPETROL). *See* COLOMBIA.

ENI. Originally Ente Nazionale Idrocarburi, an Italian national petroleum company, now mostly privatized. ENI was created by the Italian government in 1953 with **Enrico Mattei** as its first president. The bill creating it was passed on 10 February, two years after its introduction. ENI brought together the government's petroleum interests, including **AGIP**. ENI focused at first on **natural gas** production and distribution in Italy, and it succeeded in putting Italy ahead of the rest of Europe in natural gas usage. By the end of the 1950s, Italy had the world's most advanced natural gas distribution network, after the **United States** and the Soviet Union. Among other things, this generated capital for international expansion.

Mattei and ENI had a significant impact on the evolution of the world oil market in the 1950s by signing independent agreements with oil-producing countries that offered terms with more dignity and sovereign control than they were getting from the **major oil companies** (the **Seven Sisters**, a term coined by Mattei). ENI offered the producing countries an important role in the management of joint ventures. ENI also assumed all the risk of exploration. In a deal signed with **Iran** in August 1957, the National Iranian Oil Company (NIOC) would pay half of total costs only if oil were found. NIOC would also comanage the joint venture (Société Irano–Italienne des Pétroles, or SIRIP). ENI offered the government 50 percent of the venture's gross earnings and a 50–50 share in the net profits. Although this looked much better for a producing country than the **50–50 agreements** for profit sharing then offered by the multinationals, it did not necessarily yield more revenue once oil was discovered and produced, because of the different ways royalties and profits were calculated and applied. In practice, Iran would later earn less on a barrel of crude from SIRIP than on one from the **Iranian Oil Participants**, the Seven Sisters' consortium. Nine days after signing the agreement with Iran, ENI signed a similar one with **Libya**. Mattei's deals, vigorously opposed by the multinationals, had a broad impact on the psychology of the market and of producing countries. They helped usher in an era of greater assertiveness on the part of oil-rich nations. ENI took other unconventional actions for the time: In 1958, it began importing Soviet crude for processing in its refineries. By 1962, Italy was the Soviet Union's single largest export market for crude oil.

After Mattei's death in 1962, ENI continued to grow. Expansion into **petrochemicals**, begun in the 1950s, continued. ENI established a Norwegian subsidiary, AGIP Norge, in 1965 (later ENI Norge) and participated in the discovery and development of the **Ekofisk** field led by **Phillips Petroleum**. ENI subsidiaries won contracts to build **pipelines** and refineries and to conduct exploration in various parts of the world, including South America, **Australia**, and South Asia. However, various types of political unrest during the decade affected ENI's international operations, generally conducted under the AGIP name. In 1967, the Six-Day War interrupted a joint venture with **Egypt** that had begun in 1955. In 1969, the civil war over Biafra's secession disrupted AGIP refining activities in **Nigeria**.

In 1973, ENI reached a milestone agreement with **Algeria** to supply Italy with natural gas. It completed the Trans-Mediterranean Pipeline (later Enrico Mattei Pipeline), running from Algeria's Hassi R'Mel field across the Algerian desert, through Tunisia, under the Mediterranean (with a water depth of more than 2,100 ft) to Sicily, and on to the length of the Italian Peninsula, for a total distance of greater than 1,560 miles. The pipeline began operation in 1983. During the 1970s, the Italian government hobbled ENI by forcing it to acquire poorly performing companies, including 33 metallurgical companies in 1977. ENI also acquired troubled chemical companies. ENI emerged from these trials with some difficulty, entering the 1990s in a reasonable financial position. Through the 1990s, it won exploration contracts, including ones in **Kazakhstan**, **Russia**, **China**, and **offshore Azerbaijan**.

ENI became a joint stock company in 1992, and in 1995 the Italian government began privatizing the company in stages. It offered 15 percent of the company to private investors in 1995, 16 percent in 1996, 18 percent in 1997, and 14 percent in 1998. By the end of the 1998 round, the government had decisively relinquished its majority stake, and the company officially changed its name to ENI SpA. The government continued selling shares and by 2005 owned only 30 percent. In the late 1990s and early 2000s, ENI restructured, selling off more than $5 billion in assets and cutting about 42,000 jobs. It had diversified over the decades into various industries—even newspapers. It sold off such assets, and much of its petrochemicals business as well. In the 2000s, ENI decided to refocus on oil and gas. In 2001, it acquired Lasmo, establishing a presence in Asia and **Venezuela** and strengthening its positions in the **North Sea** and North Africa. ENI is also a member of the consortium developing Kazakhstan's offshore **Caspian** Kashagan field. Kashagan was discovered in 2000 and is one of the most important fields found anywhere in the world since the 1970s. In August 2015, ENI discovered Egypt's **supergiant** Zohr gas field.

In 2016, ENI controlled 3.398 Bb of oil **reserves** and 22.333 Tcf of gas. It produced 0.88 Mmbpd and 4.822 Bcfpd of oil and gas respectively. Ranked by total assets, it was among the eight largest multinational petroleum companies in the world.

ENTERPRISE DE RECHERCHES ET D'ACTIVITÉS PÉTROLIÈRES (ERAP). *See* ELF AQUITAINE.

ENVIRONMENTAL IMPACTS. The use of petroleum and its products has a number of environmental impacts. Land is cleared for infrastructure such as refineries, storage tanks, and **pipelines**. These may leak oil, gas, or harmful chemicals into the environment. **Tankers** and platforms can spill oil into the ocean, harming marine and bird life and polluting coastlines. Burning gasoline and diesel fuel in internal combustion engines releases particulates, nitrogen oxides, carbon monoxide, and other pollutants into the air. Burning any hydrocarbon releases carbon dioxide into the atmosphere. Carbon dioxide is a greenhouse gas, and its increasing emission into the atmosphere by human activity is accepted to be a cause of global warming. Global emissions of carbon dioxide from fossil fuels increased almost fivefold between 1950 and 2000. **Unconventional oil** sources can add some additional environmental impacts. The exploitation of the **Alberta oil sands**, for example, introduces environmental consequences similar to those of coal and other resource mining, and it consumes large quantities of freshwater (*see* FRACKING).

Oil is generally a cleaner energy source than coal, which on average produces more ash, smoke, and particulates and contributes more to acid rain. **Natural gas** is cleaner than oil. Methane produces only carbon dioxide and water when burned. Natural gas also adds a smaller amount of carbon dioxide to the atmosphere per unit of energy produced: about 74 percent of what is added by refined liquid fuel, and only 56 percent of what is added by bituminous coal. However, methane is a worse greenhouse gas than carbon dioxide, with a warming potential 23 times higher. Thus pipeline and storage leaks of natural gas are a considerable concern.

Government action and technology innovation, particularly in the developed world, has done much to mitigate the environmental impact of petroleum. New motor vehicles are much cleaner than those of 30 years ago, although the continuing increase in the total number of vehicles tends to offset this. Public anger has often focused on more immediate environmental consequences such as oil spills. The U.S. Oil Pollution Act of 1990, a response to the ***Exxon Valdez*** oil spill, created a spill cleanup fund, set penalties for spillers, and directed the federal government to respond quickly to spills. In 1993, the **United States** required double-hull construction for all tankers carrying oil to the United States. The petroleum industry claims to have spent about $17 billion complying with the terms of the Oil Pollution Act by the early 2000s.

On 11 December 1997, the Third Conference of the Parties of the United Nations Framework Convention on Climate Change (UNFCCC) adopted the Kyoto Protocol, whereby most industrialized countries would agree to legally binding restrictions on greenhouse gas emissions of an average of 6–8 percent below 1990 levels during the period between 2008 and 2012. The protocol was ratified by 180 parties by the end of 2007, but not by the United States. Among U.S. objections are the exemptions for developing and newly industrialized countries, including **China**. The petroleum industry has generally also opposed Kyoto.

A further agreement within the UNFCCC, the Paris Agreement, was adopted 12 December 2015 by representatives of 196 countries. Unlike Kyoto, the Paris Agreement does not have enforcement mechanisms. Greenhouse gas reduction targets are voluntary and determined at the national level. The United States was one of the countries agreeing to the consensus in 2015, but in 2017 President Donald Trump announced his intention to withdraw the country from the agreement.

Beginning in 2015 **ExxonMobil** became embroiled in a climate change controversy amid allegations that it made public statements downplaying the risks of anthropogenic climate change while internally doing research on the subject and acknowledging those risks. The allegations led to bad publicity and legal scrutiny. The company has denied any wrongdoing.

Overall, the petroleum industry has struggled with environmental issues. It has spent considerable amounts of money and management time complying with mountains of government regulations but is still popularly seen as an enemy of the environment. Part of the public resentment against the industry originates in the industry's historical behavior, such as the monopolistic practices of **Standard Oil** in the 19th and early 20th centuries and the perceived arrogance and secretiveness of the **Seven Sisters** in dealing with their own and producer countries' governments. Consumers angered by steep price increases during the 1970s and again more recently have tended to blame the petroleum companies for price gouging. These bad impressions doubtless contribute to anger and skepticism over the companies' environmental positions, and the industry really was more indifferent to environmental issues in the past. However, consuming more than 70 million barrels of crude oil a day, which must be extracted, stored, transported, and refined, cannot avoid leading to environmental consequences. Refer to appendix 9. *See also AMOCO CADIZ*; *DEEPWATER HORIZON* OIL SPILL; SANTA BARBARA OIL SPILL; *TORREY CANYON*.

EÖTVÖS, LORÁND (ROLAND). *See* GEOPHYSICAL PROSPECTING.

EQUATORIAL GUINEA. Oil production 2015: 0.248 Mmbpd (0.32 percent of world); **reserves** 1.1 Bb (0.07 percent of world); gas reserves 1.3 Tcf (0.02 percent of world). Oil production in Equatorial Guinea began in 1991 at about 100 bpd but increased rapidly over the decade. By 1995, it was about 0.052 Mmbpd, and by 2000 it was 0.167. By 2006, it had more than doubled from 2000 levels, to 0.343 Mmpd. After 2009 it began decreasing, reaching 0.248 Mmbpd. Equatorial Guinea is now sub-Saharan Africa's third-largest oil exporter, after **Angola** and **Nigeria**. Oil accounts for more than 90 percent of export earnings. The country joined the **Organization of the Petroleum Exporting Countries** (OPEC) in 2017; its Zafiro crude was added to the **OPEC Basket**. Equatorial Guinea is also a member of the **Gas Exporting Countries Forum**.

Equatorial Guinea's oil reserves are **offshore**, with the largest field being the **giant** Ceiba, discovered in 1999, with an **EUR** of about 500 Mmb. Although large in EUR, Ceiba had trouble meeting production expectations, and the leading producer has been the Zafiro field, discovered in 1995, with an EUR of about 400 Mmb. Zafiro is operated by a consortium led by

ExxonMobil, while Ceiba is operated by Hess (formerly **Amerada** Hess). Other foreign investors include Noble Energy and **Marathon**, which operates in the Alba field.

The petroleum sector is regulated by the Ministry of Mines, Industry, and Energy. The government also created a new national oil company, GEPetrol, which began operating in 2002. GEPetrol manages the government's interest in joint ventures and production-sharing agreements with foreign companies, which legislation set at a minimum of 35 percent. Equatorial Guinea has had a number of territorial disputes with neighboring countries that can affect petroleum development. Cameroon, Sao Tome and Principe, and Nigeria provisionally accepted an equidistant median line to delineate maritime boundaries unilaterally proposed by Equatorial Guinea. In July 2004, Equatorial Guinea and **Gabon** also reached an agreement allowing joint exploration around three disputed islands in the Gulf of Guinea. *See also* WONGA COUP.

ESSO. *See* EXXON.

ESTIMATED ULTIMATE RECOVERY. *See* EUR (ESTIMATED ULTIMATE RECOVERY).

ESTONIA. *See* OIL SHALE.

EUR (ESTIMATED ULTIMATE RECOVERY). An estimate of the total amount of oil or gas that will be recovered from a field over its entire lifetime. The EUR includes oil and gas that the field has already produced. Improvements in technology can often cause EUR to be revised upward (*see* RESERVE GROWTH). The same can happen if sustained price increases make more of a field's oil and gas economically recoverable. Sometimes the term EUR is erroneously applied on a global scale to mean **ultimately recoverable resource** (URR). *See also* RESERVES; TOTAL RESOURCE BASE.

EVELETH, JONATHAN C. *See* BISSELL, GEORGE HENRY (1821–1884).

EXXON. A **major oil company**; one of the **Seven Sisters**. The company that became Exxon (later **ExxonMobil**) originated as the Standard Oil Company of New Jersey (Jersey Standard), a product of the dissolution of **Standard Oil**. Before the breakup of Standard Oil, Jersey Standard was the holding company at the administrative center of the organization. At the dissolution of Standard Oil, Jersey Standard was by far the largest of the newly independent companies, with about $660 billion in assets, about 43 percent of the original Standard Oil's. Jersey's first president was **John D. Archbold**, a longtime associate of Standard Oil founder **John D. Rockefeller**. Archbold had effectively run Standard Oil after Rockefeller's retirement in 1896.

Jersey Standard retained interests in Carter Oil, **Imperial Oil (Canada)**, the Peoples Natural Gas Company, Standard Oil Company of Louisiana, and 28 other companies. It emerged with integrated operations in **Romania** and refineries in New Jersey, Louisiana, Canada, Cuba, Romania, and Germany, as well as marketing outlets in Europe and Latin America. During **World War I**, Jersey Standard was separated from many of its European properties and lost 23 **tankers** to German U-boats.

After the breakup of Standard Oil, Jersey Standard retained some small-producing properties, but like other new independent Standard Oils, it was short of crude supplies of its own. Its Carter Oil and Standard of Louisiana companies soon acquired some producing properties. However, the most important move was the acquisition, under the presidency of **Walter Teagle**, of 50 percent of the **Humble Oil and Refining Company** in 1919 for $17 million. This acquisition gave Jersey Standard sources of crude and an all-important presence in **Texas**. Humble operated with considerable independence from Jersey Standard but would have significant influence on its future and would feed talent into its top management ranks—*see* FARISH, WILLIAM STAMPS (1881–1942).

In 1920, Standard Oil Company of New Jersey was listed on the New York Stock Exchange as SNJ. In 1924, it joined with General Motors to establish the Ethyl Gasoline Corporation, and in 1926 it first registered the Esso trademark (derived from the initials for Standard Oil, "S. O.") in the **United States**. In 1928, Jersey Standard acquired a majority interest in **Creole Petroleum**'s **Venezuela** operations. Creole would dominate the Venezuelan industry by the 1950s. Jersey Standard also became a partner in the **Iraq Petroleum Company** (IPC) consortium, via the **Near East Development Corporation**. IPC membership would later give it interests in **Iraq**, **Abu Dhabi**, **Qatar**, and **Oman**. By the end of the 1920s, Standard Oil Company of New Jersey's net worth was about quadruple what it had been in 1911.

The company was hit hard by the Great Depression. In 1932, its net income fell to only $300,000 (less than Standard Oil's in 1873). That year, it acquired Pan American Petroleum and Transport from Standard Oil Company of Indiana (later **Amoco**). Pan American, founded originally by **Edward Doheny**, brought with it important properties in **Mexico**, but Standard Oil of New Jersey lost them to nationalization only six years later, in 1938. The company also saw its interests in **Bolivia** nationalized in 1937. On the positive side, the 1930s also saw significant technical progress. In 1937, Jersey

Standard chemists developed a process for making butyl rubber. In 1933, the company merged its Far East interests with Socony–Vacuum (later **Mobil**) to form Stanvac, which would prove to be a highly successful collaboration.

World War II brought difficulties and opportunities. Once again, Jersey Standard was separated from some overseas assets, and it lost 67 tankers, about a third of its fleet. However, it also managed to raise its crude production from 0.820 Mmbpd in 1940 to 1.12 Mmbpd in 1945, or 36.6 percent, while the industry average was 19 percent.

Humble doubled its production, and significant new supplies came from Venezuela. In 1941, Humble also completed the first commercial synthetic toluene plant, a significant contribution to the war effort because of toluene's role in the manufacture of TNT explosives. Jersey Standard developed fluid catalytic cracking for high-octane **gasoline** production, also essential during the war. The process also provided material for Buna synthetic rubber, which, along with the earlier Jersey process for butyl rubber, was invaluable in the production of tires, inner tubes, and self-sealing fuel tanks.

After the war, Standard Oil Company of New Jersey, already the world's largest producer, moved decisively into the Middle East in search of new supplies of crude. In 1948, it bought a 30 percent share in **Aramco**, giving it a position in **Saudi Arabia**, which would become the country with the world's largest **reserves**. In October 1954, along with the other Aramco partners, it joined the Iran consortium (**Iranian Oil Participants Ltd**.) with an 8 percent share, later reduced to 7 percent. Domestically, Jersey Standard continued to increase its stake in Humble, which had continued to operate with a large degree of autonomy. In 1948, Jersey Standard owned 72 percent of Humble. By 1954, it had increased its share to 88 percent. In 1958, it completed the acquisition and consolidated all its domestic operations with those of Humble.

In the postwar era, the major oil companies (the Seven Sisters) ruled the international petroleum market, and Standard Oil Company of New Jersey was arguably the most powerful of them all. But by the early 1960s, much of the power had begun to shift to producing countries. In 1960, Jersey Standard under **Monroe Rathbone** initiated a round of unilateral cuts in the **posted price** of oil that catalyzed the formation of the **Organization of the Petroleum Exporting Countries** (OPEC). The rising tide of nationalism in the Middle East prompted Jersey Standard to look for new sources of crude. In 1967, through Humble's partnership with **Arco**, it participated in the discovery of **Alaska**'s **Prudhoe Bay oil field**, the largest ever in North America. It also entered the **North Sea**.

In 1972, Standard Oil Company of New Jersey changed its name to Exxon Corporation. Domestic operations, of which Humble formed the backbone, were called Exxon USA. Exxon's revenue increased dramatically with the high oil prices of the 1970s (*see* FIRST OIL SHOCK; SECOND OIL

SHOCK). Revenue in 1972 was about $20 billion. In 1974, it was $40 billion, and by 1980 it had surpassed $100 billion. However, the decade of the 1970s also saw Aramco gradually nationalized by Saudi Arabia, beginning with the 1972 **participation agreement**. In 1979, Exxon lost its Iranian interests during the Iranian Revolution. Through the 1970s, it also lost to nationalization a number of other Middle Eastern interests that it held through IPC membership. Exxon embarked on a diversification program. It moved into coal, office equipment, and other areas. Most of these turned out to be failures, costing the company as much as $7 billion. Notably, it joined with Tosco in an attempt to create an **oil shale** industry in Colorado's **Green River Formation**. The Colony Oil Shale project, begun in 1980 when oil prices were near their peak, was canceled in 1982 when prices had dropped, without ever being fully tested.

In the 1980s, Exxon suffered decreased revenues along with the rest of the oil industry (*see* OIL COUNTERSHOCK). Like other companies, it responded by consolidating and refocusing on its core business of petroleum. It decreased its workforce and cut costs. Thus, although revenues went down, return on revenue increased. In 1986, Exxon consolidated all oil and gas subsidiaries outside North America into a single new division called Exxon Company International, headquartered in New Jersey. Exxon Company USA (headquartered in **Houston**) and Imperial Oil (headquartered in **Calgary**) continued to manage U.S. and Canadian operations respectively. In 1989, the company's reputation suffered greatly because of the ***Exxon Valdez*** oil spill in Prince William Sound, Alaska. During the continuing low oil prices of the 1990s, Exxon maintained profitability, largely by cutting jobs and slashing budgets. It eliminated 5,000 positions between 1990 and 1992 and continued cutting costs through the mid-1990s. In 1997, it had a net income of $8.46 billion on revenues of $120.28 billion. It was able to take some risks on exploration in **Russia**, **Indonesia**, and Africa.

In December 1998, Exxon agreed to acquire Mobil. Shareholders of both companies approved the merger in May 1999. In November 1999, Mobil became a wholly owned subsidiary of Exxon, which changed its name to ExxonMobil, in an $83 billion stock transaction. In 1998, the last full year before the merger, Exxon had a net income of $6.37 billion on revenues of $117.772 billion. It controlled 6.215 Bb in proved oil reserves and 42.294 Tcf of gas. It produced oil and gas at the rate of 1.515 Mmbpd and 6.984 Bcfpd respectively. That year it was the second-largest private sector petroleum company in the world measured by total assets, after **Royal Dutch Shell**. *See also* HOLMAN, EUGENE (1895–1962).

EXXONMOBIL. Largest private sector petroleum company in the **United States** by total assets in 2016, and second largest in the world behind **Royal Dutch Shell**. ExxonMobil was created by the 1999 merger of **Exxon** and

Mobil. The merger resulted in an estimated $4.6 billion of cost savings by 2001. In 2000, its first complete year as a merged company, ExxonMobil controlled 11.561 Bb in proved oil **reserves** and 55.866 Tcf of gas. It produced oil and gas at the rate of 2.501 Mmbpd and 11.447 Bcfpd respectively.

In the decade and a half since its creation as a new entity, ExxonMobil undertook exploration and development in **Angola**, **Equatorial Guinea**, **Chad**, **Brazil**, the **Caspian**, and far too many other places around the world to list here. In 2007 it became embroiled in a major dispute with **Venezuela** when that country decided to take majority control of the Orinoco heavy oil projects. Some companies accepted the new terms, but ExxonMobil sought arbitration with the International Center for Settlement of Investment Disputes (ICSID). Years later, On 9 October 2014, ICSID awarded ExxonMobil $1.6 billion. In 2017, press reports suggested that Infineum, a joint venture between ExxonMobil and Royal Dutch Shell, did business with Iran and other countries that were under U.S. **sanctions** at the time. ExxonMobil contended that the business did not violate sanctions because the joint venture was a European company and the transactions in question involved no U.S. employees. ExxonMobil also found itself the target of some hostile press coverage because of allegations that its public position on climate change differed markedly from what its internal research would suggest (*see* ENVIRONMENTAL IMPACTS). In 2017 ExxonMobil was denied a waiver by the Trump administration to continue certain exploration and development activities in **Russia** that are forbidden by U.S. sanctions.

In 2016 ExxonMobil's liquid reserves, at 8.737 Bb, were down compared to what they were in the year the new, combined company was formed. Its gas reserves were 33.434 Tcf. Oil and gas production occurred at the rate of 2 Mmbpd and 6.942 Bcfpd respectively. It should be noted that while ExxonMobil is the world's second-largest private sector petroleum company by some financial measures, its reserves are dwarfed by those of many national oil companies, particularly in the Middle East. As a final comment, it is worth noting that in February 2017, ExxonMobil's chief executive officer from 2006 to 2016 was sworn in as the 69th U.S. secretary of state in the administration of President Donald Trump.

EXXON VALDEZ**.** **Tanker** ship owned by **Exxon** that crashed into Bligh Reef in **Alaska**'s Prince William Sound on 24 March 1989, resulting in a major oil spill. The ship was outbound from the Alaska port of **Valdez**, terminus of the **Trans-Alaska Pipeline**, and was carrying a cargo of oil from **Prudhoe Bay**. The spill was the largest in the history of the **United States** up to that time (it has since been greatly exceeded by the ***Deepwater Horizon*** **oil spill**, although that was not a tanker spill) but only about the 35th-largest tanker spill in the world overall (refer to appendix 9). The total amount of oil spilled was about 257,000 barrels, making it about an eighth as large as the

Atlantic Empress spill and one-sixth as large as the ***Amoco Cadiz***. However, it is one of the most famous oil spills and received considerable media attention. The captain, John Hazelwood, had a history of alcohol abuse and admitted drinking alcohol the night the spill occurred, but he was not convicted of operating a vessel under the influence. He was found guilty of negligent discharge of oil and fined $50,000. This outraged many people, and a number of lawsuits were filed. In 1994, Exxon was ordered to pay $287 million in compensatory damages and $5 billion in punitive damages to a group of Alaska Native Americans and commercial fishers who had been adversely impacted. The $5 billion punitive award was halved on appeal in November 2001. In June 2008, the U.S. Supreme Court further lowered the award, concluding that punitive damages should roughly match actual damages, estimated at about $507 million.

The spill had a significant environmental impact. Part of the Prince William Sound shoreline was still contaminated in the early part of the 2000 decade. The spill killed at least 250,000 birds, 2,800 sea otters, 300 harbor seals, 250 bald eagles, and 22 killer whales. Billions of salmon eggs were destroyed. In 1990, 43 million pink salmon were caught in Prince William Sound; in 1993 only 3 million were caught. *See also* SANTA BARBARA OIL SPILL; *TORREY CANYON*.

F

FALL, ALBERT B. (1861–1944). U.S. secretary of the interior 1921–1923, under President Warren G. Harding; convicted in 1929 of taking bribes from oil executives in what became known as the **Teapot Dome scandal**.

FARISH, WILLIAM STAMPS (1881–1942). Organizer of the **Humble Oil and Refining Company**; president of Standard Oil Company of New Jersey (later **Exxon**), 1937–1942. William Stamps Farish was born in Mayersville, Mississippi, on 23 February 1881. He earned a law degree at the University of Mississippi in 1900 and practiced law for about three months before moving to Beaumont, **Texas**, after the **Spindletop** discoveries. He began working for his uncle there and learned the oil business, soon becoming an independent producer. He was one of the organizers of the Brown Farish Oil Company in 1902. The company went bankrupt after Brown's death. Farish then partnered in 1904 with Robert Lee Blaffer to do contract drilling and lease trading. They moved to **Houston** in 1905. Farish became a leading **independent** oil executive.

Farish joined with **Ross Sterling** and three other partners to form the Humble Oil Company in 1911. In March 1917, the company obtained a state charter as the Humble Oil and Refining Company. Farish became Humble's vice president, concentrating on production and technology. When Humble needed capital to expand, Farish negotiated with **Walter Teagle**, the head of Standard Oil Company of New Jersey, for that company to buy half of Humble's shares for $17 million in 1919. Jersey Standard was a guaranteed market for Humble's oil, and the capital infusion helped Humble build a large refinery in Baytown, Texas. In 1922, Farish became president of Humble, which operated independently from Jersey Standard. In 1933, he became chairman of the board of Jersey Standard while Teagle was still president. In 1937, the two switched positions.

Farish was an expert on new production technology and a prominent advocate for the petroleum industry in the United States. He became a founder of the **American Petroleum Institute** and served as its president in 1926. He served on the U.S. Federal Oil Conservation Board and was on the **Petroleum Industry War Council** at the beginning of **World War II**. He died 29 November 1942 in Milbrook, New York.

FESSENDEN, REGINALD (1866–1932). Inventor who developed the fundamental techniques of seismic prospecting. Reginald Aubrey Fessenden was born 6 October 1866 in Milton, Quebec. He worked for Thomas Edison and **Westinghouse** and was a professor at Purdue and Western University of **Pennsylvania** (now the University of Pittsburgh). In 1902, he formed the National Electric Signaling Company. He is best known as the first person to transmit voice and music over radio.

Fessenden's geophysical work began with his invention of the "fathometer," which used reflected sound waves in the ocean to determine the distance to objects such as icebergs. Fessenden was inspired to invent this device by the 1912 sinking of the *Titanic*. He applied similar principles to elastic waves in the earth. In 1913, he developed seismic instruments that detected reflected and refracted waves in tests near Framingham, Massachusetts. He applied for a patent on a "Method and Apparatus for Locating Ore Bodies" on 15 January 1917. The patent (U.S. Patent No. 1,240,328) was issued 18 September 1917. Fessenden's techniques were licensed by Geophysical Research Corporation, an **Amerada** subsidiary organized in 1925 by **Everette DeGolyer**, and were applied successfully to oil prospecting. Fessenden died on 22 July 1932 in Hamilton, Bermuda. *See also* GEOPHYSICAL PROSPECTING.

50–50 AGREEMENTS. Agreements by which profits from crude oil sales would be split evenly between oil companies and oil-producing countries; often referred to simply as "50–50." Before the late 1940s, oil-producing countries typically received a minority share of petroleum profits. **Venezuela** was the first major producer to get a 50–50 agreement, in 1948. Although this was done under the auspices of a military government, the groundwork had been laid by Minister of Production **Juan Pablo Pérez Alfonso** during the preceding democratic administration of **Rómulo Betancourt**. When **Ibn Saud**, the king of **Saudi Arabia**, learned of Venezuela's agreement, he asked for a similar formula for his country from the **Aramco** partners. Saudi Arabia got a 50–50 plan in late 1950. This was done at considerable expense to the American taxpayer, since oil royalties paid to producing country governments were made tax deductible by a bill originally drafted by **George McGhee**, the acting U.S. secretary of state for the Middle East and a negotiator of Saudi Arabia's deal. Once Saudi Arabia had a 50–50 agreement, 50–50 was extended to all the Arab oil producers.

The obstinacy shown by the **Anglo–Iranian Oil Company** (later **BP**) in refusing to grant 50–50 to **Iran** further damaged the already deteriorating relations between the company and the government of Iran, which would lead to **Mohammed Mossadegh**'s attempts to nationalize the oil industry and the subsequent coup against Prime Minister Mossadegh backed by the **United States** and the **United Kingdom**—*see* AJAX, OPERATION (1953). Since profit shares were calculated on the **posted price**, actual country shares based on market prices could differ. In the 1950s, the posted price was typically higher than the market price, so governments could receive more than half of the actual profits. On the other hand, companies were also accused of deliberately obfuscating profits in order to cheat on the agreements and pay lower shares. The 50–50 system lasted until the early 1970s. The **Tehran** agreement of 1971, which raised shares of **Persian Gulf** producers to 55 percent, signaled the end of it.

FINA. *See* PETROFINA.

FIRST OIL SHOCK. A large increase in oil prices in the early 1970s, mostly between October and December 1973; average prices in 1974 were more than four times their 1972 levels. The First Oil Shock was caused by a number of factors, not all entirely related. One was the oil embargo by the **Organization of Arab Petroleum Exporting Countries** (OAPEC) on the **United States** and the Netherlands resulting from the Arab–Israeli War of 6–23 October 1973 (Yom Kippur War). On 6 October, **Egypt** and Syria invaded Israel to take back territory they had lost in the Six-Day War of 1967. Israel was nearly overwhelmed. On 13 October, it begged for U.S. assistance and received an airlift of arms. On 16 October, six **Persian Gulf** members of the **Organization of the Petroleum Exporting Countries** (OPEC) raised the price of **Arab Light** crude from $2.90 to $5.11 a barrel. The next day, OAPEC announced a production cut of 5 percent, to be followed by further 5 percent cuts each month that Israel failed to withdraw to its pre-1967 borders. OAPEC also began a selective oil embargo on the United States and the Netherlands for supporting Israel. The embargo had an immense psychological effect on the markets, out of proportion to the actual loss of supply. Since not even all of OPEC participated, this amounted at its peak to less than 6 percent of world consumption. Countries unaffected by the embargo could also resell their crude. However, the embargo caused widespread fear of future supply losses, and this helped drive up prices more.

The embargo was not the only cause of the oil shock. Other factors that had been developing for some years paved the way. One was the 1970 **peaking** of oil production in the United States (this had been predicted by **M. King Hubbert** in 1956). In 1971, the **Railroad Commission of Texas** opened the taps and allowed all available production in **Texas**. However, there was little spare capacity left in Western countries. The first oil from **Prudhoe Bay** in **Alaska** would not be shipped until 1977. In the **North Sea**, **Ekofisk** and **Brent** had only been discovered in 1969 and 1971 and had not yet entered commercial production. **Mexico**'s **Cantarell** field would not be discovered until 1976, and the major deepwater discoveries in the U.S. **Gulf of Mexico** were still in the future. The early 1970s were thus a time of decreasing flexibility in the major Western consuming nations.

At the same time, OPEC was becoming more assertive, even before the Yom Kippur War. The **Tehran Agreement** of 1971 raised producer country profit shares. Individual OPEC countries began leapfrogging each other to gain additional concessions from the **major oil companies**. OPEC members nationalized their industries. **Algeria** expropriated 51 percent of foreign oil interests between 1968 and 1971. **Libya** began expropriation in 1971. **Iraq** completed its nationalization process in 1975. **Venezuela** had been squeezing foreign companies harder and harder for two decades and nationalized its industry in 1975–1976. Arab Gulf countries were beginning to take over their oil industries through **participation agreements**. Added to all this was a U.S. foreign policy that somewhat favored high oil prices because these helped strengthen the shah in **Iran**, seen as a major pillar of anticommunism.

This potent mix created the conditions for the First Oil Shock, which was an important turning point in the world oil market. It ended a relative stability that had existed since the end of **World War II** and undermined the Western presumption of an easy and guaranteed prosperity fueled by cheap petroleum. *See also* OIL COUNTERSHOCK; SECOND OIL SHOCK; THIRD OIL SHOCK.

FIRST OIL WELL. The question of who drilled the first modern oil well is not an easy one to answer and depends to a great extent on definitions. **Edwin Drake**'s 1859 well near **Titusville, Pennsylvania**, is popularly considered the first modern oil well. But there were contemporaneous and even earlier efforts elsewhere. Drake's well is most likely singled out because it had several key attributes: (1) it was drilled, not hand dug, and it was drilled deliberately to look for oil; (2) a steam engine was used to power the drill; (3) the well was cased with cast-iron pipe to prevent flooding by groundwater; and (4) the well resulted in a recognizably modern oil industry boom.

The notable earlier efforts around the world are not known to have had all these elements simultaneously. The wells drilled in **West Virginia** in 1859 just before Drake did not touch off an immediate boom of the magnitude seen in Pennsylvania. A group under Major Alexeev of the Bakinskii Corps of Mining Engineers drilled a well in the **Baku** region of **Azerbaijan** 13 years earlier than Drake—*see* VOSKOBOINIKOV, NIKOLAI (1801–C. 1861). However, their drilling equipment was not powered by an engine, and it is unlikely that the well was cased. **Ignacy Łukasiewicz**'s oil wells in **Galicia** in the 1850s fed a refining operation and were earlier than Drake's, but they were hand dug. **James Miller Williams** was working his oil discoveries in Ontario before Drake, but he left no record of his early work. It is not clear whether his early wells were drilled or hand dug. The wells drilled at **Pechelbronn** (Alsace, France) in 1813 were not engine powered, and they were not drilled directly to obtain oil, but rather to guide an **oil sands** mining operation. Overall, Drake's well and its consequences are regarded as more prototypical of the modern industry than the earlier efforts were.

FISCHER–TROPSCH PROCESS. *See* GAS TO LIQUIDS.

FISHER, JOHN ARBUTHNOT (1841–1920). Head of the British Admiralty 1904–1910 and 1914–1915; modernizer of the Royal Navy who began the conversion of ships from coal to oil. John Arbuthnot Fisher, popularly known as Jackie, was born in Ceylon (now Sri Lanka) on 25 January 1841. Joining the British navy at age 13, he served in the Crimean War and in **China**, became a captain in 1874, and by 1892 was third sea lord and controller of the Royal Navy. In October 1904, he became first sea lord and embarked on an ambitious program of modernization. He reformed the Royal Navy's personnel systems and revised its outdated and unduly harsh disciplinary procedures. He championed electric lights, large-caliber guns, long-range gunnery, torpedo-boat destroyers, and submarines. Fisher's name is now linked inextricably with the HMS *Dreadnought*, the revolutionary prototype of the fast "all-big-gun" ship.

Fisher also began the conversion of Royal Navy ships from coal propulsion to cleaner and more efficient oil propulsion, championing the internal combustion engine. This move was to have far-reaching consequences in the oil industry and oil geopolitics, ultimately leading to the direct engagement of the British government in the oil business in **Iran**. Fisher's passion for oil propulsion earned him the nickname "Oil Maniac." When **Winston Churchill** became first lord of the Admiralty, he took up the cause of converting the navy to oil and ensuring its oil supplies with enthusiasm. Fisher had retired from the navy in 1910, but Churchill used him as an adviser and in 1912 made him head of the **Royal Commission on Fuel and Engines**, which studied the issue and further championed oil.

By 1914, Fisher was again first sea lord, under Churchill. However, he disapproved of Churchill's plan for a naval expedition through the Dardanelles, which proved to be a disaster. He resigned on 15 May 1915 in protest against Churchill's leadership of the Admiralty. He published his memoirs in 1919 and died in London on 10 July 1920.

FISHER COMMISSION. *See* ROYAL COMMISSION ON FUEL AND ENGINES (1912–1913).

FLAGLER, HENRY M. (1830–1913). American entrepreneur who partnered with **John D. Rockefeller** to establish the **Standard Oil Company**. Henry Morrison Flagler was born 2 January 1830, in Hopewell, New York. He became a grain merchant in Ohio, where he met Rockefeller. Rockefeller was working as a commission merchant and wholesaler at the time, and Flagler sold grain through him. Flagler went to Michigan to begin a salt manufacturing venture, but he failed and returned to Ohio.

By 1867, Rockefeller was already successful in the oil business with his partner Samuel Andrews. Flagler joined them to form Rockefeller, Andrews, and Flagler. Three years later, Rockefeller and Flagler reorganized the partnership as an Ohio joint stock company, Standard Oil. Rockefeller was president and Flagler was secretary. Flagler played an active role in the development of Standard Oil, and he served as a director until 1911.

Apart from the oil business, Flagler spearheaded the development of Florida as a tourist destination. He first visited Florida in 1883. In 1886, he bought several railway lines and organized them as the Florida East Coast Railways. He then built a chain of luxury hotels to be served by the line. He died on 20 May 1913 in West Palm Beach, Florida.

FORT McMURRAY. City (pop. 64,500, 2007) in **Alberta**, **Canada**, 270 miles northeast of **Edmonton**; major center of oil production from the **Alberta oil sands**. Mines and upgrading facilities in the Athabasca oil sands are located about 25 miles north of Fort McMurray. In the summer, the city serves as a port. Lying at the confluence of the Athabasca and Clearwater Rivers, it is the southernmost node in the navigation network formed by the MacKenzie River and the Great Slave Lake.

FRACKING. (Fracturing.) The technique of stimulating oil and gas production by fracturing rock using, for example, a pressurized fluid, thus enhancing the flow of oil or gas. Fracking can make it possible to produce oil and gas that would be otherwise inaccessible because it is trapped in low-permeability formations and unable to flow satisfactorily.

The application of fracking technology, along with **horizontal drilling**, has enabled huge increases in the production of **shale oil** and **shale gas** in the years since 2007 in the **United States**, and dramatically reversed, at least temporarily, the long decline in U.S. hydrocarbon production. By 2015, U.S. production of **tight oil**, generally shale oil, represented nearly 49 percent of U.S. oil production and brought production levels to their highest point since 1972. U.S. production of **tight gas** and shale gas in 2015 accounted for nearly 69 percent of total gas production, bringing levels to their highest point in history. Fracking and horizontal drilling have rejuvenated the petroleum industry in several states, notably **Texas**, **Oklahoma**, **North Dakota**, and **Pennsylvania**.

Outside the United States and **Canada**, fracking has not yet been extensively used, but shale oil, shale gas, and tight gas resources are plentiful in various countries and have tremendous potential. **China** and **Argentina** are particularly well endowed with technically recoverable shale gas resources (estimated respectively at 1,115 and 802 Tcf, compared to the United States' 623). Meanwhile, **Russia** has an impressive endowment of tight oil, estimated at 74.6 Bb, nearly as much as the United States' 78.2 Bb.

During the first two decades of the 21st century, fracking appeared to shift the balance of power away from traditional oil-exporting countries such as those of the **Organization of the Petroleum Exporting Countries** (OPEC), although it must be noted that production using fracking also carries higher costs in both money and energy than typical conventional oil and gas production. In addition, there is considerable controversy about fracking's **environmental impacts**, including concerns about contamination of drinking water and even induced seismicity. These and other concerns have led to moratoria or outright bans on fracking in a number of jurisdictions. Examples include, but are not limited to, Germany, Scotland, Wales, Bulgaria, the **Australian** state of Victoria, the Canadian province of Quebec, the U.S. states of Vermont and New York, several counties in **California**, and even some jurisdictions in Texas.

Although fracking's impact has been significant in the early part of the 21st century and has garnered widespread media coverage, the ideas behind fracking are not new, and it has been practiced in some form or other for over a century. The use of explosives to fracture rock near a well bore and thereby stimulate production dates back at least to the 1860s, with the development of the "petroleum torpedo," a device containing tamped explosive that could be lowered down a well and detonated at depth. The torpedo was patented by Edward A. Roberts, who formed the Roberts Petroleum Torpedo Company in 1865. The first test was at the Ladies' Well in Pennsylvania in 1866, not far from the **Drake** Well. That test used black powder, although the black powder was soon replaced with nitroglycerin. The practice, known as "shooting a well," became a successful method of stimulating production. Other methods of creating fractures and channels in rock were developed, including the use of acids in limestone—*see* FRASCH, HERMAN (1851–1914)—and the use in the 1930s of shaped charges to shoot steel bullets into the rock surrounding a well.

A more recognizably modern method of fracking was developed by Riley "Floyd" Harris and Bob Fast, working for Stanolind Oil in Tulsa, Oklahoma, using a liquid to fracture limestone. The first test of the so-called Hydrafrac process was in the Klepper No. 1 well in the Hugoton natural gas field in southwestern **Kansas** in 1946. The process was eventually shown to be successful in rejuvenating partially depleted gas wells and showed potential for increasing the productivity of new ones. Fracking was also applied successfully to stimulate production of tight gas in **Ohio**'s Clinton Sandstone from 1951 to 1957.

Interest in the concept of fracking continued, even reaching the point of testing the use of nuclear explosions for the purpose, in both the United States and the USSR (*see* NUCLEAR EXPLOSION STIMULATION). In 1975 the test well drilled for Project Wagon Wheel, a nuclear test that was planned but never took place, was used by the El Paso Natural Gas Company to conduct tests of "massive hydraulic fracturing," involving the pumping of water into a well until the pressure built up to a point that it fractured the rock. At the time, El Paso concluded that the technique was not commercially viable. Large-scale, commercially viable fracking would have to await the groundwork laid by Mitchell Energy during the 1990s, with the help of the **Gas Research Institute**, in Texas's **Barnett Shale**—*see* HORIZONTAL DRILLING; MITCHELL, GEORGE P. (1919–2013).

FRANCE. *See* ALGERIA; COMPAGNIE FRANÇAISE DES PÉTROLES (CFP); ELF AFFAIR; ELF AQUITAINE; PECHELBRONN; SCHLUMBERGER LTD; TOTAL.

FRASCH, HERMAN (1851–1914). Chemist who developed a process, bearing his name, for the removal of sulfur from oil. The son of a pharmacist, Herman Frasch was born 25 December 1851 in Gaildorf, Wurttemberg, Germany. In 1868, he immigrated to the **United States** and taught at the Philadelphia College of Pharmacy while studying chemistry. He developed an improved process for refining wax and sold it to a subsidiary of **Standard Oil** in 1877. He moved to Cleveland and established a practice as a chemical consultant to the petroleum industry. Frasch developed a process for reducing the sulfur content of "sour" petroleum. This was an extremely important achievement, as a high sulfur content makes petroleum difficult to refine and degrades the ultimate products. At the time, the most important application was to produce better kerosene for illumination, free of foul odors. He sold the process to Canada's **Imperial Oil** Company in 1882.

Standard Oil became interested in Frasch's process when it discovered quantities of sour **Lima–Indiana** oil in Ohio and Indiana. Standard bought the process and hired Frasch as a consultant, putting him in charge of what was likely the first systematic scientific research program in the petroleum industry. Frasch's process entailed treating crude oil with metallic oxides to precipitate the sulfur. It was patented in 1887. Frasch was offered a high position in Standard Oil but refused, preferring to remain a consultant.

In addition to the Frasch process for sulfur removal, there is also a Frasch process for sulfur production. Frasch developed a completely different process, not directly related to the petroleum industry, for mining deep sulfur deposits by melting the sulfur and pumping it to the surface (almost like crude oil). He organized the Union Sulfur Company in 1892. This other Frasch process enabled the United States to become self-sufficient in sulfur for its chemical industry, and it broke the near monopoly on sulfur previously held by Sicily, whose deposits were, unlike America's, close to the surface and easily mined.

Yet another of Herman Frasch's contributions was the idea, patented in 1896, of using acid to dissolve limestone, thus stimulating oil production by opening channels through which oil could flow. Frasch received the Perkin Medal in 1912. He died in Paris on 1 May 1914.

FRASER, WILLIAM (1888–1970). Chairman of the **Anglo–Iranian Oil Company** (later **BP**) 1941–1956. William Fraser was born in Scotland on 3 November 1888. His father was the founder of Pumpherston Oil Company, a leading Scottish **oil shale** concern. Fraser joined Pumpherston in 1909 and became joint managing director in 1915. He helped the British government obtain oil supplies during **World War I** and was honored for his services in 1918 with an MBE. He joined the Anglo–Persian Oil Company (later Anglo–Iranian Oil Company) in 1923. In 1928, he became deputy chairman of the company, reporting to the chairman, **John Cadman**. Fraser contributed to the expansion of production in **Iran**, **Iraq**, and **Kuwait**. In 1941, he succeeded Cadman as chairman, a post he held until 1956.

Fraser led the company during perhaps the most tumultuous years in its history. Continuing disputes with the government of Iran culminated in the nationalization of the Iranian oil industry in 1951. Fraser was undoubtedly competent and was a valuable adviser to the British government on many aspects of the oil industry, but he did not manage the relationship with Iran well. He was reputed not to respect Iranians, viewing them as greedy and ungrateful. Fraser reportedly viewed the concession negotiated by Cadman and Shah **Reza Pahlavi** in 1933 as fair and believed the Iranians should live up to it and not make constant additional demands. This attitude did not facilitate compromise.

Fraser only began showing a small amount of flexibility when he heard that a **50–50 agreement** on profit sharing was about to be concluded for **Saudi Arabia** (the agreement was announced in December 1950). By then it was too late to stave off nationalization and the disasters that followed, including the overthrow of Prime Minister **Mohammed Mossadegh** in a coup backed by the **United States** and the **United Kingdom**—*see* AJAX, OPERATION (1953). When Anglo–Iranian was finally let back into Iran in 1954, it was as a member of a consortium (*see* IRANIAN OIL PARTICIPANTS LTD.) with no ownership of oil assets or facilities. Anglo–Iranian had a 40 percent interest in the consortium. Steadfast to the end, Fraser demanded compensation for Anglo–Iranian, now called British Petroleum (later BP), and actually received it. Fraser died on 1 April 1970.

FRIENDSHIP PIPELINE. *See* DRUZHBA PIPELINE.

FUEL ADMINISTRATION. A **World War I** agency in the **United States**, operating from 23 August 1917 to 30 June 1919. Its purpose was to control the allocation of coal and oil in order to meet the needs of the war effort, as well as routine domestic requirements. The agency prodded the energy industry to increase production. It also encouraged conservation, with appeals such as the one to residents of the eastern United States to observe "gasolineless Sundays." It restricted use of coal and oil by industries that were not critical for the war effort. It also set maximum prices for fuel in some cases.

The Fuel Administration set a pattern for close cooperation between the government and industry that would also occur in **World War II**—*see* PETROLEUM ADMINISTRATION FOR WAR (PAW) (1942–1946). The head of the Fuel Administration's Oil Division, Mark L. Requa, worked closely with the National Petroleum War Service Committee, composed of industry representatives and led by Alfred Bedford, the president of Standard Oil of New Jersey (later **Exxon**). The wartime demand for fuel led to an increase in the price of oil. By 1918, the price was about double what it had been four years earlier. Requa urged the industry to practice voluntary price restraint, threatening more direct government price controls and the withholding of iron and steel, which were critical materials for the industry, if it did not comply. The Oil Division and the National Petroleum War Service Committee arrived at a voluntary agreement known simply as "The Plan" on 9 August 1918 to assure the flow of oil and stabilize prices.

Despite Requa's imposition of the government's will over the industry, he was seen as a strong defender of its interests. He was also an advocate of employing U.S. military force in **Mexico** to ensure access to that country's **reserves**, even if it meant overthrowing the Mexican government. The administration of Woodrow Wilson did not act on his advice.

G

GABON. Oil production 2015: 0.23 Mmbpd (0.29 percent of world); **reserves** 2.0 Bb (0.12 percent of world); gas reserves 1 Tcf (0.01 percent of world). Gabon joined the **Organization of the Petroleum Exporting Countries** (OPEC) in 1975 but left the organization at the end of 1994 because of the high annual dues. It rejoined in 2016, and its Rabi Light crude was added to the **OPEC Basket**. Production of oil in Gabon reached a high point of about 0.370 Mmbpd in 1997 and has been declining since; 2015 production was near 1989 levels. Nevertheless, oil exports account for 51 percent of GDP and 63 percent of government revenues. A large part of Gabon's current oil is in the **giant** onshore Rabi-Kounga field operated by **Royal Dutch Shell**. The field, with **EUR** of about 1 Bb, was discovered in 1987. The **offshore** Tchatamba Marin and Etame fields are also important.

Gabon generally encourages exploration through its tax system. The oil ministry regulates the sector. The national oil company, Société Nationale Petrolière de Gabon (SNPG), was shut down in 1987. In 2011 the government formed a new national oil company, the Gabon Oil Company, to participate in new ventures. In 2014 a new law gave the Gabon Oil Company a 15 percent interest in all new projects. Companies operating in Gabon include Royal Dutch Shell, **Total**, **ENI**, and the **independent company** Addax Petroleum (Gabon's largest producer). In 2014 there were a number of promising **pre-salt** discoveries in the country's offshore: Total's Diaba, ENI's Nyonie Deep, and Royal Dutch Shell's Leopard-1. *See also* ELF AFFAIR.

GALBREATH, ROBERT. *See* GLENN POOL.

GALEY, JOHN H. *See* GUFFEY, JAMES McCLURG (1839–1930).

GALICIA. A province of the Austro-Hungarian Empire that was the locus of one of the world's earliest modern petroleum industries in the 19th and early 20th centuries. The territory of Austro-Hungarian Galicia passed to Poland after **World War I** and now lies in both Poland and Ukraine. Between 1880 and 1900, Galicia produced more total oil than any country in the world except the **United States** and **Russia**. Galician oil production peaked in 1909 at nearly 41,000 bpd, about 5 percent of world production. By 1925, its production was down to about 16,000 bpd, and it was 10th in the world, overtaken by **Mexico**, **Iran**, **Indonesia**, **Venezuela**, **Romania**, and others.

The modern industry in Galicia originated in 1853 when Jan Zeh, **Ignacy Łukasiewicz**, and others developed methods for **refining** crude oil into a usable illumination oil, probably close to kerosene, and a suitable lamp for burning the refined product. On 31 July 1853, the Lvov General Hospital became the first public building in the world lit by petroleum lamps, and in 1858–1859 the Emperor Ferdinand Northern Railway converted to petroleum illumination. In the early years, oil was obtained from hand-dug pits. Drilling equipment was introduced to the area in 1863.

GAS EXPORTING COUNTRIES FORUM (GECF). An organization of countries established in 2001 in Tehran to discuss **natural gas** and **liquefied natural gas** (LNG) issues and to foster "tangible cooperation among gas-producing and exporting countries." Member countries meet annually at the ministerial level. Topics discussed include natural gas projects, pricing, technology, and trade. Among the initial motivations for the creation of the GECF were the challenges posed by European Union competition rules, as well as the EU gas market liberalization following the First Gas Directive of 1998 and continuing after the Second Gas (Acceleration) Directive of 2003. In particular, members were concerned about rules on the illegality of destination clauses in contracts and the retroactive application of those rules.

The members of GECF are **Algeria**, **Bolivia**, **Egypt**, **Equatorial Guinea**, **Iran**, **Libya**, **Nigeria**, **Qatar**, **Russia**, Trinidad and Tobago, the **United Arab Emirates**, and **Venezuela**. Observer countries include **Azerbaijan**, **Iraq**, **Kazakhstan**, the Netherlands, **Norway**, **Oman**, and **Peru**. It should be noted that GECF membership has not been stable. Turkmenistan, for example, was a founding member but dropped out after the first meeting. Even some of the more regular members do not appear to participate in every meeting. The GECF began as a relatively informal organization, without statutes, charters, or membership fees, but more formal procedures were gradually put in place beginning in 2007. Regular GECF members accounted for roughly 70–75 percent of estimated global gas reserves in 2007, and about 40 percent of production.

Over half the GECF members (Algeria, Equatorial Guinea, Iran, Libya, Nigeria, Qatar, the United Arab Emirates, and Venezuela) are also members of the **Organization of the Petroleum Exporting Countries** (OPEC). Some observers have expressed concern that the GECF might become a cartel along OPEC lines, but the GECF does not appear to have been created specifically for this purpose and has not achieved OPEC levels of international power. However, one cannot completely exclude the possibility that the GECF might provide a platform for some subset of the countries to form such an alliance in the future. In the 2007 GECF meeting in Doha, Qatar, it was decided to form a high-level committee to study pricing policies. Russia has shown interest in leading a more cartel-like alliance, particularly one including the **Persian Gulf** members. Venezuela and Iran have also been reputed to be interested in forming an organization that acts more like a cartel.

It should be noted that a cartel along OPEC lines may not be easy to establish, because the gas market differs significantly from the oil market. Gas does not have a global market to the same extent that oil does. Most gas contracts are long term, and prices are set in advance. Gas is generally not traded on the open market as oil is. However, patterns of transportation (e.g., LNG versus **pipelines**) and pricing may change in the future. *See also* NATURAL GAS LIQUIDS (NGL).

GAS RESEARCH INSTITUTE (LATER GAS TECHNOLOGY INSTITUTE). Institute funded by the **United States** federal government that played a significant role in the development of **fracking** and **horizontal drilling** technologies. The Gas Research Institute was established in 1976 and funded by a tax on interstate shipments of **natural gas**. The institute worked with Mitchell Energy—*see* MITCHELL, GEORGE P. (1919–2013)—in the 1990s, subsidizing the company to apply fracking and horizontal drilling in the **Barnett Shale** of **Texas**. Mitchell performed the first such combination in the Barnett Shale in 1991.

The tax that funded the Gas Research Institute was discontinued in 2000. The Gas Research Institute merged with the Illinois Institute of Technology's Institute of Gas Technology to form the nonprofit Gas Technology Institute.

GAS TECHNOLOGY INSTITUTE. *See* GAS RESEARCH INSTITUTE (LATER GAS TECHNOLOGY INSTITUTE).

GAS TO LIQUIDS. A family of processes to convert **natural gas** to a liquid product, generally one resembling a refined fuel such as **gasoline** or diesel. Gas-to-liquids processes are based on variants of the Fischer–Tropsch process, a catalyzed reaction to make hydrocarbons from synthesis gas, a mixture of carbon monoxide and hydrogen. Coal or natural gas can be used as feedstocks for making synthesis gas. Franz Fischer and Hans Tropsch developed their process in the 1920s at the Kaiser Wilhelm Institute in Berlin. They filed patents, including a U.S. patent in 1926. Germany used gas-to-liquids processes with coal as a feedstock in **World War II** to alleviate fuel shortages. Sasol in South Africa did the same during the apartheid-era ban on oil imports. Sasol also opened plants in the 1990s. Gas-to-liquids processes are complex and energy intensive. When using natural gas as a feedstock, they can consume up to 45 percent of it just for energy. Gas-to-liquids installations are being developed in **Qatar** by a number of companies, including **ExxonMobil**, **Royal Dutch Shell**, **Marathon**, and Sasol, mainly to produce clean diesel fuel from natural gas in the immense **North Field**. An American company, Syntroleum, has built a demonstration plant near **Tulsa**, **Oklahoma**, and produced almost 10,000 barrels of diesel and jet fuel from natural gas.

GASBUGGY PROJECT. *See* NUCLEAR EXPLOSION STIMULATION.

GASOLINE. Arguably the most important **refined** product of crude oil and a major driver of the petroleum industry in the 20th century. Before the widespread use of automobiles, the most important refined product of crude oil was kerosene for illumination. In the 19th century, the gasoline fraction was generally regarded as undesirable waste. Gasoline increased steadily in quality during the first half of the 20th century.

An important measure of gasoline quality is the octane number, an indicator of the resistance of a motor fuel to knocking and hence the suitability of the fuel for use in high-power engines. Knocking refers to the uneven combustion that can occur in cylinders and damage engines, particularly at high compression ratios. The octane number was developed in 1927 by the American chemist Graham Edgar of the Ethyl Corporation, working with the Cooperative Fuel Research Committee. The committee, with representatives from the **American Petroleum Institute**, the American Manufacturers Association, the U.S. National Bureau of Standards, and the Society of Automotive Engineers, was established to test the knocking properties of motor gasolines. The committee performed its tests using a single-cylinder engine with a variable compression ratio built by John Campbell of General Motors.

Edgar proposed first testing mixtures of two hydrocarbons, normal heptane and iso-octane. Heptane is prone to knocking, whereas iso-octane, first synthesized by Edgar, is much less so. Once the heptane/iso-octane mixtures were calibrated, gasolines could be tested and compared with the performance of the mixtures. A gasoline with the same knocking propensity as pure heptane would have an octane number of 0, while one with the same propensity as pure iso-octane would have an octane number of 100. A gasoline with the same knocking properties as a mixture of 80 percent iso-octane and 20 percent heptane would have an octane number of 80, and so on. Heptane and iso-octane were good choices because they could both be produced in sufficient purity and quantity and had similar volatility properties.

The committee concluded that no one test could cover the whole range of operating conditions. Two methods were thus defined: the motor method and the research method. Both methods compared gasoline antiknock performance with that of heptane/iso-octane mixtures, but (simplifying) the motor method measures performance at high speeds and loads, and the research method measures it at low speeds. In the **United States**, the reported octane number is usually the average of the octane numbers obtained by the two methods. In Europe, it tends to be the research number.

In 1921, T. Midgely of General Motors discovered that the addition of tetraethyl lead could improve the antiknock properties of gasoline. Leaded gasoline was commonly used from the 1920s to 1970s, when environmental concerns dictated its gradual abandonment. Antiknock properties could also be improved via cracking processes that caused the rebranching of molecules. Hydrocarbons with branched chains of carbon atoms (like iso-octane) tend to burn more evenly than ones with straight chains (like normal heptane).

In the 1920s, gasolines with octane ratings of 50 were still common. By 1930, gasolines reached octane numbers between 60 and 70. This was insufficient for newer, high-performance engines. The process of thermal cracking—*see* BURTON, WILLIAM M. (1865–1954)—brought the achievable octane number to 82. Early catalytic cracking—*see* HOUDRY, EUGENE (1892–1962)—could produce octane numbers up to 95.

The matter of high-octane fuel had a particular significance in aviation, where performance was at a premium. Military airplanes in 1928 used 60-octane gasoline. By 1930, the U.S. Army aviation standard was 87-octane. Engines developed for 87-octane gasoline had a power-to-weight ratio that was one-third better than 60-octane engines. Engines developed for 100-octane fuel had a further 30 percent improvement. Airplanes with 100-octane engines had 20 percent lower takeoff distances and 40 percent greater climbing speeds than ones with 87-octane engines. The U.S. Army adopted 100-octane fuel as its aviation standard in 1936. During **World War II**, 100-octane fuel proved a decisive advantage for the Allies, as the Axis planes did not use it, and leadership in the production of the fuel rested with the United States.

GAZPROM. The world's largest **natural gas** production company, headquartered in Moscow, **Russia**. It is also a significant oil company through **Gazprom Neft**. Ranked by total assets, it was the fifth-largest oil and gas company in the world in 2016. The company has roots in the Soviet Ministry of the Gas Industry, which oversaw gas production in the Soviet Union. Gas production began in the Saratov region in 1941. By 1946, **pipelines** carried Saratov gas to Moscow. In 1989, two years before the breakup of the Soviet Union, the gas ministry was reorganized into a company and renamed Gazprom. After the breakup of the Soviet Union in 1991, the Russian government moved toward privatization of state monopolies. Russian president Boris Yeltsin signed a decree on 5 November 1992 transforming Gazprom into a Russian joint stock company.

The new Gazprom was initially owned about 15 percent by workers, 28 percent by the people of Russia's gas regions, and 40 percent by the Russian government. Top managers reportedly also retained substantial shares, including Viktor Chernomyrdin, the boss of Gazprom during the Soviet era, who became Russia's prime minister in 1992. Many people were dismayed by the potentially multibillion-dollar holdings of selected top managers and decried the privatization as corrupt. In 1996, Gazprom offered 1 percent of its stock to foreign investors. In 1998 and 1999, Ruhrgas acquired 4 percent of Gazprom. In 1997, Gazprom began building the Blue Stream Pipeline across the Black Sea to Turkey, gaining **ENI** as a backer and partner in 1999. In 1999, it began building its Yamal–Europe Pipeline for gas exports to Europe.

During the Yeltsin years (until the end of 1999), the Russian government did not take a very active role in Gazprom. Under chief executive officer (CEO) Rem Vyakhirev, Gazprom did not pay significant dividends and was not particularly diligent about paying its taxes. Vyakhirev also stripped Gazprom of some valuable assets and transferred them to holding companies controlled by himself, board chairman Chernomyrdin, senior managers, and various relatives. This changed after Vladimir Putin became Russia's president in 2000. Putin replaced Chernomyrdin with an ally, Dmitri Medvedev, who would succeed Putin as Russia's president in 2008. When Vyakhirev's term as CEO of Gazprom expired in 2001, he was succeeded by Aleksei Miller, another Putin ally. Gazprom then began to pay more dividends and taxes. In 2003, the Russian government increased its stake in Gazprom to 51 percent.

Gazprom under Putin became an instrument of state power. In 2000, Gazprom called for Media-Most to sell shares to it in order to settle debts. Media-Most had shares of NTV television, which regularly criticized Putin. This led to suspicions that the Kremlin was behind the deal in an attempt to muzzle media criticism. Eventually Gazprom sold its stake in NTV. In January 2006, when Gazprom suspended gas deliveries to Ukraine, many observers believed that the Kremlin was using gas to punish the Ukrainian government for an increasingly Western orientation. There was considerable unease in the European Union (EU), which purchases large quantities of gas from Russia through pipelines traversing the Ukraine and Belarus. The Russian government could use Gazprom and its gas as a political weapon in the future. The 2011–2012 completion of Nord Stream, the North European Gas Pipeline under the Baltic linking Russia directly to Germany, further cemented EU dependence on Gazprom. In 2012 the European Union launched an antitrust investigation on Gazprom, alleging among other things that the company overcharged customers and blocked rivals. In 2017 the EU and Gazprom were still in talks to settle the case.

In 2005, Gazprom acquired a 73 percent stake in oil company Sibneft for $13 billion. Sibneft changed its name to Gazprom Neft in 2006. In 2006 Gazprom took majority control of the **Sakhalin** II project, after an agreement with Royal Dutch Shell and Japanese partners, and in 2007 acquired Siberian gas assets from **TNK–BP** (later acquired by **Rosneft**). In 2014 Gazprom signed a contract with the China National Petroleum Corporation to deliver 1.342 Tcf of gas annually to China for 30 years beginning in 2018. In 2016 Gazprom controlled oil reserves of 17.431 Bb and gas reserves of 842.419 Tcf. It produced oil and gas at the rates of 1.257 Mmbpd and 40.645 Bcfpd respectively.

GAZPROM NEFT. Formerly Sibneft; a major Russian integrated oil company now mostly owned by **Gazprom**. (Gazprom Neft should not be confused with Gazpromneft, a former Gazprom subsidiary now called Invest Union.)

Sibneft was formed in 1995 during the privatization of former Soviet industries in **Russia** under President Boris Yeltsin. Sibneft was given the state's interest in the Soviet-era oil company Noyabrskneftegaz as well as the Omsk refinery, a distribution network, and other entities. Noyabrskneftegaz was the second most northern oil producer in western **Siberia**, with a large inventory of discovered but still undeveloped fields. Russian oligarchs **Boris Berezovsky** and **Roman Abramovich** acquired a controlling interest in Sibneft beginning in 1995 for about $100 million under the **Loans for Shares** scheme. In 2000, Berezovsky sold his interest to Abramovich.

In 1998, Sibneft attempted unsuccessfully to merge with **Yukos**, a large Russian oil company controlled by oligarch **Mikhail Khodorkovsky** that owned a 20 percent stake in Sibneft. In 1999, Sibneft became the first large Russian oil company to publish audited accounts according to U.S. Generally Accepted Accounting Principles. The company announced in 2000 that it would modernize Omsk, already Russia's largest and best refinery, increasing its capacity to produce unleaded **gasoline**. In 2001, it announced plans to explore for oil in the Chukotka autonomous district, of which Abramovich was the governor. The region has a geology similar to **Alaska**'s North Slope, causing some to believe it has good oil potential. In 2002, Sibneft bought 36 percent of a Moscow refinery from **Lukoil**. In 2005, Gazprom acquired a 73 percent stake in Sibneft from Abramovich's Millhouse Capital for $13 billion. Sibneft changed its name to Gazprom Neft in 2006.

GENEVA AGREEMENTS. Agreements in 1972 and 1973 between **major oil companies** and the **Organization of the Petroleum Exporting Countries** (OPEC) to correct oil prices in accordance with the value of the U.S. **dollar**. The Richard M. Nixon administration's abandonment of a fixed dollar price of gold on 11 August 1971 led to a devaluation of the dollar on 17 December 1971. Since internationally traded oil was priced in dollars, this led to an effective lowering of the value of OPEC countries' revenues. The first Geneva Agreement was concluded on 20 January 1972 and provided for an immediate 8.49 percent increase in the **posted price** of crude oil. This was close to what was required to compensate for the dollar's loss of value against gold. The agreement also included provisions to adjust the posted price until 1975 for changes in the value of the dollar.

Another major devaluation of the dollar on 12 February 1973 created pressure for a revision of the agreement. OPEC believed that the original agreement did not offer full compensation for devaluation and allowed some loopholes. The second Geneva Agreement was concluded 1 June 1973 between the major oil companies and eight OPEC member countries (**United Arab Emirates**, **Iran**, **Iraq**, **Kuwait**, **Libya**, **Nigeria**, **Qatar**, and **Saudi Arabia**). It provided for an immediate 6.1 percent increase in posted prices. Prices had already been increased after the 12 February devaluation according to the terms of the first Geneva Agreement. The new increase brought the total adjustment to 11.9 percent. Future increases were linked to changes in the value of the dollar against 11 major currencies.

GEOPHYSICAL PROSPECTING. The first commercial oil field to be discovered by geophysical methods was the Nash Dome in Fort Bend County, **Texas**, in 1924. The discovery was made by **Amerada Petroleum**, headed by **Everette DeGolyer**, using the torsion balance. The torsion balance was a gravimetric instrument invented near the end of the 19th century by Loránd (Roland) Eötvös (1848–1919) of Hungary. Before the 1924 discovery, the torsion balance had been tested and validated on some known oil fields, including the Czech Egbell field in 1915–1916 and **Spindletop** in 1922. The torsion balance was accurate but cumbersome and was eventually replaced in the industry by the field gravimeter. But gravity methods would not dominate in petroleum exploration. The industry quickly moved to seismic methods. The seismic method, specifically reflection seismology, with its ability to map subsurface geology, became by far the most important geophysical exploration technique for petroleum.

Although reflection seismology would dominate, the first petroleum discovery using seismic methods was made via refraction seismology. The refraction method was invented by Robert Mallet (1810–1881) in Ireland in 1845. During **World War I**, when both acoustic and seismic methods were developed for locating enemy artillery, a number of people filed for patents on improved refraction methods. A notable inventor was Ludger Mintrop (1880–1956), who filed for a patent in 1919 on a "method for the determination of rock structures." Mintrop established a company, Seismos Ltd., and began doing contract refraction seismic surveying for Marland Oil (later merged into **Conoco**) in 1921. The instrumentation was not sensitive enough, and the surveys were unsuccessful in finding oil. However, a similar survey conducted by Seismos for **Gulf Oil** discovered the Orchard Dome in Fort Bend County, Texas, in 1924. This was a salt dome structure near the **Gulf of Mexico** associated with commercial quantities of oil, and it was the first discovery using seismic methods. Seismos discovered several more salt domes over the next two years using refraction.

The reflection method replaced the refraction method during the 1930s and has been the almost exclusive method of the industry ever since. The reflection method is better able to generate interpretable observations over complex geologic structures. The method was invented by **Reginald Fessenden**, who applied for a patent on a "method and apparatus for locating ore bodies" on 15 January 1917. The patent (U.S. Patent No. 1,240,328) was issued 18 September 1917. The ideas grew out of Fessenden's earlier work on detecting icebergs by echo sounding (work that was inspired by the sinking of the *Titanic* in 1912). Early reflection experiments were carried out by John Clarence Karcher (1894–1978). In 1921, he mapped a shallow reflecting bed in central **Oklahoma**. Karcher is widely credited with reducing the reflection method to practice. In May 1925, Everette DeGolyer organized the Geophysical Research Corporation (GRC) as a subsidiary of Amerada, with Karcher as head. In 1926, GRC licensed Fessenden's patent. In 1927, it made the first commercial petroleum discovery using reflection seismology, the Maud field in Oklahoma. Other GRC reflection discoveries drilled between 1928 and 1933 included South Earlsboro, West Seminole, Carr City, and Sasakwa Townsite. GRC also conducted refraction surveys.

In the 1930s, as reflection seismology became established, it was used to discover 131 fields along the Gulf of Mexico, several with reserves of more than 100 Mmb. The introduction of geophysical techniques had an immediate and dramatic effect on discovery rates. Between 1900 and 1925, discoveries occurred at a rate of about 240 Mmb per year; between 1925 and 1935 the rate was about 1 Bb per year.

Subsequent decades saw the continual refinement of the reflection method. Advances were made in data recording, signal processing to facilitate the detection of seismic reflections in noise, and other areas. In July 1950, W. Harry Mayne of Petty Geophysical Engineering Company filed a patent for the common depth point (CDP) method. This was a data stacking technique that attenuated long wavelength noise (ground roll) without smearing subsurface detail. Although it represented a major advance, it did not see widespread use until the 1960s. There were also improvements in seismic sources.

The Vibroseis technique developed by John M. Crawford, William E. N. Doty, and Milford R. Lee of Conoco in the late 1950s gave the industry a well-characterized source that could also be used in areas where dynamite was prohibited.

In the 1970s and 1980s, three-dimensional (3-D) seismic reflection data acquisition and processing techniques were developed and made usable by advances in computer workstations and visualization software. The use of 3-D seismic greatly expanded in the 1990s. It enabled more accurate and useful mapping of subsurface structures. It saw considerable use in the Gulf of Mexico and in Texas's Laredo gas play, where it significantly reduced dry holes. In Texas's Midland basin, 3-D allowed imaging of small subsurface reef structures that could only be exploited if accurately mapped. **Royal Dutch Shell** announced in the early 1990s that it would henceforth only acquire 3-D data. **Aramco** also mandated 3-D over its large producing fields. As seismic methods began to be used in production and development as well as in exploration, four-dimensional (4D) methods were developed, where the fourth dimension is time. Several 3-D surveys taken over time can allow fluid movements to be tracked as production proceeds. *See also* GEOPHYSICAL SERVICE INCORPORATED (GSI); SCHLUMBERGER LTD.; WESTERNGECO; WESTERN GEOPHYSICAL.

GEOPHYSICAL RESEARCH CORPORATION. *See* AMERADA PETROLEUM CORPORATION; DEGOLYER, EVERETTE (1886–1956); GEOPHYSICAL PROSPECTING.

GEOPHYSICAL SERVICE INCORPORATED (GSI). Seismic services company founded in 1930 by John Clarence Karcher and Eugene McDermott, with the secret backing of **Everette DeGolyer**. DeGolyer, who was the general manager of **Amerada Petroleum**, had formed an Amerada subsidiary, Geophysical Research Corporation (GRC), headed by Karcher that had pioneered the application of reflection seismology to petroleum prospecting. DeGolyer wanted GRC to sell its seismic services to outside companies but had been prevented by Amerada's president. This led DeGolyer to support the formation of GSI, which would become the first geophysical contractor to offer reflection seismology services.

In 1939, the Coronado Corporation was formed as a parent company for GSI. On 6 December 1941, GSI was purchased from Coronado by McDermott and partners including Cecil Green. During **World War II**, the company used its expertise in wave propagation and seismic instrumentation to produce submarine detection devices. The company diversified into electronics unrelated to petroleum and created a Laboratory and Manufacturing (L&

M) Division in 1947. L&M grew faster than GSI, and in 1951 L&M became Texas Instruments (TI), with GSI as a subsidiary. From 1988 to 1991, TI sold GSI in stages to **Halliburton**.

Halliburton merged GSI with Geosource (formerly Petty-Ray Geophysical), another geophysical contractor it had acquired, to form Halliburton Geophysical Services (HGS). In 1992, Davey Einarsson, an executive from the original GSI, purchased GSI's data for **offshore Canada** and formed a new GSI in **Calgary** offering marine seismic, gravity, and magnetic services. In 1994, Halliburton sold HGS, which was unprofitable, to **Western Atlas**. Eventually the geophysical parts of Western Atlas, including what was left of the old GSI, were acquired by **Schlumberger**. *See also* GEOPHYSICAL PROSPECTING; WESTERNGECO; WESTERN GEOPHYSICAL.

GEORGE GETTY OIL COMPANY. *See* GETTY, J. PAUL (1892–1976).

GESNER, ABRAHAM (1797–1864). Canadian geologist who developed processes for extracting high-quality illuminating oil from asphalt and coal, and who coined the word "kerosene" to denote it. Abraham Pineo Gesner was born in Cornwallis, Nova Scotia, **Canada**, on 2 May 1797. He trained as a doctor in Nova Scotia and the **United Kingdom** but gave up practicing medicine to pursue scientific work in geology, chemistry, and other fields. In 1838, he was appointed as a geologist for the lower provinces of British North America (Canada). In 1850, he was appointed Indian commissioner in Nova Scotia, having acquired a knowledge of several native languages and dialects.

In 1851, Gesner developed a distillation process yielding an illuminating oil from Trinidad asphalt. He developed further processes for obtaining illuminating oil from coal and bituminous shale. This oil would prove to be a good substitute for the whale oil that had been used hitherto. He called the product "kerosene," from the Greek words *keros* and *elaion* ("wax" and "oil"). Gesner patented a number of his processes. He assigned them to the New York Kerosene company, for which he supervised the construction of facilities. The refineries occasionally extracted kerosene from crude oil found in surface seeps in **Pennsylvania** before **Edwin Drake** drilled his well in 1859. This was a portent of things to come, as petroleum would soon become the major source of kerosene. It is worth noting that in 1852, around the same time that Gesner was obtaining kerosene from asphalt, **Ignacy Łukasiewicz** of Poland developed a method for distilling it from crude oil, and **Samuel Kier** was also selling a kerosene-like product refined from petroleum. There was also similar activity in the state of **West Virginia**. Gesner died in Halifax, Nova Scotia, on 19 April 1864. He had returned there expecting to become professor of natural history at Dalhousie University but died before this could happen. *See also* WILLIAMS, JAMES MILLER (1818–1890).

GETTY, GEORGE FRANKLIN. *See* GETTY, J. PAUL (1892–1976).

GETTY, J. PAUL (1892–1976). Founder of **Getty Oil Company**. Jean Paul Getty was born in Minneapolis, Minnesota, on 15 December 1892. His father, George Franklin Getty, founded the Minnehoma Oil Company in **Oklahoma**. Getty worked in his father's company during the summers while he was a student (he attended the University of California at Berkeley, the University of Southern California, and Oxford). In 1914, Getty began an independent career in Oklahoma, with financial backing from his father. He struck oil in 1916 and became a millionaire. He lived a life of leisure for a time, but in 1919 he went back to the oil business. Getty gradually moved the focus of his operations to **California**. He acquired a one-third interest in Minnehoma, soon to become the George Getty Oil Company.

When his father died in 1930, Getty became president of the company. However, he did not have a controlling interest. His father, who disapproved of his reputation as a playboy, left the controlling interest to Getty's mother. Getty ran the company in the 1930s and also bought a controlling interest in the Pacific Western Oil Corporation, a large California company that had both oil **reserves** and cash. He negotiated with his mother and was eventually able to acquire a controlling interest in the George Getty Oil Company. He wanted to make George Getty Oil an integrated company, with downstream operations in addition to exploration and production. In pursuit of this goal, he acquired controlling interests, through Pacific Western, in the Tidewater Oil Company and the Skelly Oil Company, both controlled by the powerful Standard Oil Company of New Jersey (later **Exxon**). He did this gradually, and somewhat surreptitiously, in a series of complex maneuvers lasting from the mid-1930s to 1950. In 1934, Jersey Standard had set up the Mission Corporation to hold substantial blocks of shares in Tidewater and Skelly. Getty first gained control of Mission and a direct minority interest in Tidewater. He then used Mission to accumulate further shares in Tidewater while using his own direct stake in Tidewater to block other suitors. Getty merged his various holdings into the Getty Oil Company in 1967.

In 1949, Getty acquired a 60-year **concession** for **Saudi Arabia**'s interest in the **Neutral Zone** between that country and **Kuwait** for $9.5 million. Getty gave Saudi Arabia generous terms that drew the disapproval of the **major oil companies**. Getty saw himself as a strong **independent** who could stand up to the **Seven Sisters**. After three years and an investment of $30 million, Getty found large quantities of oil in the Neutral Zone, eventually producing more than 16 Mmb per year. In 1957, he topped the list of richest men in America published by *Fortune* magazine.

The public was fascinated not only by Getty's wealth but also by his reputed stinginess and the turbulence of his personal life. He was married five times and lost two of his five sons. In a case that garnered considerable news coverage, his grandson, J. Paul Getty III, was kidnapped in Italy and had part of his ear cut off before being returned for ransom. Getty was also a major art collector, opening the J. Paul Getty Museum in 1954 in his house in Malibu, California. The museum, which moved to a magnificent hilltop site in west Los Angeles, is widely visited. Getty died on 6 June 1976.

GETTY OIL COMPANY. Independent oil company founded by **J. Paul Getty** that discovered and produced oil in the **Neutral Zone** between **Saudi Arabia** and **Kuwait**. Getty obtained a **concession** from the Saudi side. Getty Oil Company was only known formally as such after 1967. Before that it consisted of various holdings, all controlled by J. Paul Getty, including the Pacific Western Oil Corporation, Tidewater Oil Company, and Skelly Oil Company. Getty Oil was acquired beginning in 1984 by **Texaco**, leading to a legal battle with prior corporate suitor **Pennzoil** (***Pennzoil v. Texaco***). A large Getty-branded marketing operation with 1,300 **gasoline** stations in the **United States** was acquired by **Lukoil** in 2001.

GHAWAR. World's largest oil field, located in **Saudi Arabia**. Ghawar is roughly 174 miles in length and 16 miles wide (about 31 miles at its widest point) and is the world's only **megagiant** oil field. In energy-equivalent terms, however, Ghawar is smaller than **Qatar**'s **North Field**, which is a **natural gas** field. The Ghawar field was discovered by **Aramco** in 1948 and first began producing oil in 1951. Production peaked in 1980 at approximately 5.6 Mmbpd. Production in the early 2000 decade still hovered around 5 Mmbpd. Ghawar typically accounts for 50–60 percent of total Saudi production. Total cumulative production by the end of 2005 was more than 60 Bb. The amount of remaining oil in Ghawar is a subject of some debate. **EUR** ranges from approximately 65 Bb, which would mean the field is nearly depleted, to 150 Bb.

Geologically, Ghawar is a north-trending anticlinal structure, expressed on the surface by outcrops of Tertiary rocks. The first discovery in 1948 was in the northern part of the structure. In 1949, a discovery was made in the southern portion. The northern and southern discoveries were initially considered separate, and Aramco believed that it had discovered several fields. In 1955, maps began to show it as a single entity. Aramco eventually defined six operating areas in the complex. These are, from north to south, Fazran,

Ain Dar, Shedgum, Uthmaniyah, Hawiyah, and Haradh. These areas are all considered part of Ghawar but are sometimes treated as different fields. They differ significantly in productivity. Ain Dar and Shedgum, toward the north, are the most productive.

GIANT. There are a number of definitions of a giant oil or **natural gas** field. One commonly accepted definition published by the **American Association of Petroleum Geologists** (AAPG) is an oil field with an estimated ultimate recovery (**EUR**) of 500 Mmb or more. In the case of a gas field, a giant field is one with 500 Mmb or more oil equivalent, where a **barrel** of oil is very roughly equivalent to 6,000 cubic feet of gas. If a field yields both oil and gas, it is considered a giant according to this definition if the combined amounts of oil and gas are at least 500 Mmboe. A **supergiant** field in this definition has an EUR of at least 5 Bboe. A **megagiant** has an EUR of at least 50 Bboe. Depending on which estimates one accepts, there may be only one oil megagiant in the world: **Saudi Arabia**'s **Ghawar** field. **Kuwait**'s **Burgan** and Saudi Arabia's **Safaniya** may possibly qualify if one accepts the upper range of their EURs. There are three gas megagiants: **Qatar**'s **North Field**, **Iran**'s Pars South, and **Russia**'s Urengoy.

The set of definitions outlined above is not universally accepted. In the **United States**, the original definition of a giant field was one with at least 100 Mmb of ultimately recoverable oil equivalent. Now some define a giant field as one with more than 1 Bb of ultimately recoverable oil equivalent. Often when a field yields both oil and gas, only oil might be considered in the definition, depending on the context of the discussion. There is also some inconsistency in how "ultimately recoverable oil" is estimated. Some fields might be characterized in terms of "oil in place," while others might be characterized according to the stricter criteria of various **reserves** definitions. A further inconsistency arises from whether condensate and **natural gas liquids** are included in the size estimate. Also, at least one investigator, Matthew Simmons, has argued that production rather than estimated ultimate recovery or reserves should be used as the criterion for defining a giant field. Production is much more easily and accurately measured. Simmons has proposed defining a giant oil field as one producing at least 0.1 Mmbpd. In the discussion about oil fields below, the 500 Mmb EUR definition is used, and only crude oil is counted, not natural gas, natural gas liquids, or condensate.

Giant fields are extremely important for the world's oil supply. Only about 1 percent of the world's oil fields are giants, yet they account for approximately two-thirds to over three-fourths of the world's estimated recoverable oil resource. The hundred largest oil fields accounted for about 45 percent of the world's oil production in the mid-2000s.

The geographic distribution of the known largest oil and gas fields around the globe is nonuniform. In 2016, there were 996 giant oil and gas fields, of which 590 were primarily oil fields and the rest primarily gas. Of the total number of oil and gas giants, nearly half were in the Middle East and the former Soviet Union (223 and 219 respectively). Considering only the 20 largest fields that are primarily oil, 15 are in the Middle East, with eight in Saudi Arabia (including Ghawar, by far the largest oil field in the world); two are in **Iraq**; three are in Iran; one is in Kuwait; and one is in the **United Arab Emirates**. Of the remaining five, Russia has two, and **Venezuela**, **Mexico**, and **China** have one each (provided we lump Venezuela's **Bolivar Coastal Complex** and China's **Daqing Complex** into one category each—not everyone does).

Geologically, the largest fields have tended to involve anticlinal traps, with sedimentary reservoirs (mostly sandstones). Because large anticlinal structures are easier to spot during exploration, this has led some people to assume that the largest and most obvious giants in the world have already been found. The discovery rate for giant fields has been declining since the 1960s, both in terms of the number of fields and their average size. In the 1940s, the decade when Ghawar was discovered, the average size of new giant oil fields was about 8.6 Bb, and around 30 were discovered. In the decade of the 1960s, about 120 giant oil fields were discovered, with an average size of about 4.4 Bb per field. In the 1990s, about 50 giant oil fields were discovered, with an average size of 1.2 Bb per field. In the first decade of the 21st century, through 2005, 14 giant oil fields were discovered. One of these was **Kazakhstan**'s **Kashagan** field, which was large enough to raise the average size to 1.5 Bb. Without Kashagan, the average size was around 0.5 Bb, the lower cutoff for a giant field. Newer giant discoveries have increasingly been made **offshore** and in deeper water.

Many of the largest fields are very old now. Ghawar was discovered in 1948. The second-largest field, Kuwait's Burgan, was discovered in 1938. Safaniya in Saudi Arabia was discovered in 1951. The fields of Venezuela's Bolivar Coastal Complex were discovered beginning in 1917. Not one of the world's top 20 oil fields was discovered after 1980. The most recent are Mexico's **Cantarell** (1976) and Iraq's East Baghdad (1979) (the latter may or may not be in the top 20, depending on definitions).

The decline in the rate of discovery of giant fields, the declining average size of new giant field discoveries, and the depletion of the older supergiants is a cause of some concern. Many advocates of the **peak oil** theory use these facts, along with others, to argue that the world is approaching a dangerous era of resource scarcity, with many dire economic and political ramifications. Refer to appendixes 11–13.

GIUSTI LOPEZ, LUIS EDUARDO (1944–). Head of **PdVSA** 1994–1999; a driving force behind the **Apertura Petrolera**, the opening of **Venezuela**'s petroleum industry to foreign investment. Born in an oil field camp, Luis E. Giusti Lopez graduated in 1966 from the University of Zulia in Maracaibo. In 1971, he earned a master's degree in petroleum engineering from the University of Tulsa. Giusti worked for **Royal Dutch Shell** in Venezuela until the nationalization of 1975–1976 transformed Shell's interests into **Maraven**, a PdVSA affiliate. He became Maraven's general manager of **refining** in 1985, and in 1987 he was appointed to the Maraven board. His career spanned all major areas of the industry. He worked to make the company more efficient and began developing ideas of reopening the Venezuelan industry to foreign investment. Giusti became president of PdVSA in 1994. Under his leadership, PdVSA proposed and lobbied for the Apertura Petrolera. After a difficult political process, the Venezuelan Congress approved it in July 1995.

The Apertura would prove both controversial and effective. Critics complained of a loss of sovereignty, and the government's revenue share from oil fell from about 66 percent to 33 percent. On the other hand, the Apertura reversed a significant decline in Venezuelan oil production that had been taking place since the early 1970s. Production in 1999 was nearly double the levels of a decade earlier. Major new infrastructure was also created for upgrading **heavy oil**, of which Venezuela has massive **reserves** in the **Orinoco Oil Belt**. PdVSA became a more modern and efficient oil company, although its increasing independence worried some in the government.

The election of **Hugo Chávez** as president of Venezuela in 1998 signaled the eventual end of the Apertura and of PdVSA's independence. Giusti left PdVSA in 1999. Since then, he has been a senior adviser at the Center for Strategic and International Studies in Washington, D.C., and has been involved in various energy-related ventures.

GLADYS CITY OIL, GAS, AND MANUFACTURING COMPANY. *See* HIGGINS, PATTILLO (1863–1955).

GLENN POOL. Early major oil field in **Oklahoma**, discovered 22 November 1905 near **Tulsa**. The field was named after Ida E. Glenn, a Creek Indian who owned the land. The Glenn Pool was discovered by Robert Galbreath and Frank Chelsey. Their initial strike, Ida E. Glenn No. 1, flowed at 75 bpd from a reservoir somewhat less than 1,500 feet deep. The crude was light and sweet, of a higher quality than that found at **Spindletop**. The Glenn Pool touched off a boom and transformed Oklahoma into a major oil-producing state, alternating with **California** for the top position in the **United States** over the next two decades. Among the entrepreneurs who made their first fortunes in the Glenn Pool were **Harry Sinclair** and **J. Paul Getty**. **Pipelines** were quickly built to the **Gulf** and **Texaco** refineries in **Texas** and to **Standard Oil**'s Whiting refinery in Indiana. Local refineries were also built. The Glenn Pool produced about 325 Mmb by 1986. In 1964, the original well was abandoned. In the early 2000s, the field was under waterflood and still producing small quantities from stripper wells.

GLUCKAUF. One of the first modern oil **tankers**. The *Gluckauf* was designed by engineer Henry Swan and built in 1885 at the Armstrong–Whitworth works in Newcastle-upon-Tyne, **United Kingdom**. The 2,700-ton steamer was not the first ship to carry oil, and its design owed much to **Ludvig Nobel**'s earlier tankers. At 318 feet long, with a 37-foot beam and a 19-foot draft, it was considerably larger than the earlier ***Zoroaster*** designed by Nobel. Unlike Nobel's early tankers, which plied the **Caspian**, the *Gluckauf* was an oceangoing tanker meant to carry oil from the **United States** to Europe. In July 1886, it delivered its first cargo of American oil at Geestemunde in Germany.

The *Gluckauf* used large, hull-integrated tanks filled and emptied by pipes leading ashore. Like Nobel's tankers, it obviated the costly need to place the oil in barrels before shipping. The oil storage tanks were separated from each other by a bulkhead along the centerline and a series of transverse bulkheads. There was a trunk above the tanks to allow for the expansion of the oil in warmer waters. A pump room lay between the tanks and the engine room, which was in the rear. On 25 March 1893, the *Gluckauf* ran aground on Fire Island in New York State.

GOLD, THOMAS (1920–2004). Astrophysicist who developed an **abiogenic theory** of the origins of petroleum. Gold was born in Vienna on 22 May 1920 and settled in the **United Kingdom** with his parents at age 17. During **World War II**, he did important work in radar and other areas. In 1957, he became a professor at Harvard University, later moving to Cornell. He made many widely recognized contributions in astronomy and astrophysics. He was also a passionate believer that petroleum did not come from the decomposition of dead organisms but had a primordial cosmic origin, and that hydrocarbons come from deep within the earth. He cited the ubiquity of methane in the solar system and the existence of methane hydrate deposits on earth, among many other pieces of evidence. He believed that deep flows of primordial **natural gas** of nonbiological origin feed thermophile bacteria living at great depths and that these produce crude oil. According to his hypothesis, these bacteria are responsible for the biomarkers found in oil.

Gold's theories were similar in some respects to those published by Soviet geologists in the 1950s, particularly **Nikolai Kudryavtsev**. Gold most likely came up with his theories independently. His efforts and theories eventually resulted in the drilling of a deep (nearly 22,000 feet) well in Sweden. The well was drilled in granite, in the area of an ancient meteorite impact. Most geologists would not expect oil to be found in such an environment. However, some oil was indeed found, along with microbes. Gold believed the well proved his hypotheses. Other scientists disputed this, claiming contamination. The debate over Gold's theories was often rancorous. If the ideas are correct, there may be much more oil in the earth than most people think, although it may not necessarily be accessible. Gold remained committed to his ideas until the end of his life. The majority of geological opinion is still against him. He died 22 June 2004.

GÓMEZ, JUAN VICENTE (1857?–1935). Dictator of **Venezuela** 1908–1935 who encouraged foreign oil companies to operate in his country. Juan Vicente Gómez was a follower of Cipriano Castro, who seized control of Venezuela in 1899. Gómez became Castro's vice president. Nine years later, he deposed Castro and became dictator. Until his death, Gómez ruled either directly or through puppet intermediaries.

In 1912, Gómez forced the country's supreme court to revoke landowners' rights to subsoil resources and place those rights in the hands of the central government. A similar move by **Mexico** in 1917 caused great consternation in the oil industry, but in the case of Venezuela, the **major oil companies** had much less to fear. The process of getting **concessions** in Venezuela was corrupt and fraught with cronyism, but once a concession was granted, Gómez guaranteed a certain predictability and stability in the political and administrative environment.

Gómez maintained good relations with foreign countries overall. He eliminated Venezuela's foreign debt and improved public works, using tax revenue from oil production. But he was thoroughly ruthless in maintaining his power. He established an oppressive police state, imprisoning and torturing opponents. He also used his position to amass a huge personal fortune, acquiring businesses and farms. He was supposedly the wealthiest individual in South America at one time. There were several attempts to overthrow his regime, but none was successful. A student uprising in 1928 was put down brutally, but among its leaders was **Rómulo Betancourt**, who would later become president of Venezuela and would ultimately improve the terms with which his country did business with the major oil companies.

GREAT CANADIAN OIL SANDS. *See* ALBERTA OIL SANDS (TAR SANDS); SUNCOR ENERGY; SUN OIL COMPANY.

GREEN RIVER FORMATION. Geological formation in the **United States** (Colorado, Utah, and Wyoming) estimated to hold about half the world's known resources of **oil shale** (not to be confused with **shale oil**, a type of **tight oil**). The Green River Formation is a lacustrine sedimentary formation of Eocene age. The formation is estimated to hold between 1,500 and 1,800 Bb of oil in deposits of adequate grade (capable of yielding 0.36 barrels per ton or more). About 1,000 Bb are located within the Piceance Creek basin of western Colorado. Of this, about 500 Bb are believed to be in relatively rich deposits that may yield at least 0.6 barrels per ton. The amount of oil that is actually recoverable in a practical sense, taking all technological, economic, and environmental factors into account, is still a matter of debate. Proponents point out that if even a fifth of the oil thought to reside in the Green River Formation is practically recoverable, the available oil would exceed the known conventional oil **reserves** of **Saudi Arabia**.

There have been some attempts to create an oil shale industry based on the Green River Formation. None have succeeded economically. In 1980, a consortium led by **Exxon** and Tosco began the Colony Oil Shale project on private land in the Piceance basin. The goal was to produce 47,000 bopd via mining and retorting (heating and separating out the resulting liquid). The project was canceled in May 1982 and never tested. Unocal (formerly **Union Oil Company of California**, later part of **Chevron**) began a project a few years later, also on private land in the Piceance basin. It built a plant with a projected output of 9,000 bopd. The system did produce oil, but at only half the projected rate. Fixing the plant proved too expensive, and Unocal canceled the project in 1991. However, both projects were terminated in an environment of relatively low oil prices. Their failure does not prove that oil production from oil shale in the United States is nonviable.

The U.S. government set aside three Naval Oil Shale Reserves (NOSRs) in the early 20th century on federal lands in the Green River Formation. These have not been developed and are no longer considered necessary for national defense. NOSR Nos. 1 and 3 are located near Rifle, Colorado. The government has made them available for commercial mineral leasing. Interest would not necessarily be limited to oil shale development. It would probably center on **natural gas** and **conventional crude oil** exploration. NOSR No. 2 is located in Utah and was transferred to the Ute Indian Tribe in 2000–2001.

GREENWAY, CHARLES (1857–1934). Early head of the Anglo–Persian Oil Company (APOC, later **Anglo–Iranian Oil Company**, eventually **BP**) who persuaded the British government to acquire a 51 percent stake in the company. Charles Greenway was born 13 June 1857 in Taunton, **United Kingdom**. He went to work for a cotton trader in London and then for the trading company that later became Shaw, Wallace, and Company. He left to work in **India** for another trading firm (Ralli Brothers) and returned to Shaw Wallace in 1893. There he concentrated on the part of the business that marketed kerosene in India for the **Burmah Oil Company**.

Burmah invested in **William Knox D'Arcy**'s **concession** in **Iran** in 1905. When D'Arcy's venture made a major oil find in 1908 at Masjid-e-Suleiman, Burmah decided to float a separate company. Charles William Wallace, Greenway's superior at Shaw Wallace, was also an executive director of Burmah and assumed the responsibility of setting up the new firm, the Anglo–Persian Oil Company. Greenway played a leading role in realizing the company's April 1909 launch. **Donald Smith** became chairman of APOC, and Wallace became its managing director. Greenway was Wallace's assistant and had a seat on the board; a year later he was managing director, with Wallace as chairman.

Greenway had ambitious plans for the company that clashed with Burmah's more conservative style. When Burmah cut off new money to APOC, Greenway entered into a long and complex series of negotiations with the British government that culminated in the government's acquisition of a 51 percent stake in the company for £2.2 million on 20 May 1914. Although the path was a tortuous one, Greenway had some fundamental forces on his side. The Royal Navy under First Sea Lord **John Arbuthnot Fisher** had begun conversion from coal to oil propulsion. The cause was later taken up by First Lord of the Admiralty **Winston Churchill**, who wanted to secure oil supplies for the British navy as war with Germany loomed on the horizon.

During and after the negotiations, Greenway clashed frequently with **Henri Deterding** of **Royal Dutch Shell**—an alternative supplier of oil to the Royal Navy—and never hesitated to appeal to jingoistic sentiments. Greenway raised the specter that if the British government did not provide financial assistance to APOC, the company would be acquired by foreign interests. The "foreign" interests in question were the **Shell Transport and Trading Company**, which was domiciled in London and had a majority of British directors on its board. However, Shell was part of the Royal Dutch Shell Group, which was only 40 percent British owned, the majority 60 percent stake being in Dutch hands. Greenway played up the Dutch ownership, implying that this would make the company susceptible to German influence. He also allegedly used the fact that **Marcus Samuel** (founder of Shell) was Jewish to further portray the company as "foreign."

Greenway became chairman of APOC in 1914 and continued as managing director until 1919. He turned APOC into a vertically integrated company. In 1917, he bought **British Petroleum**, the marketing subsidiary of a German company expropriated during the war, and began building more refineries and a **tanker** fleet. He expanded APOC's global exploration activities as well as its global downstream business. He retired in 1927. Greenway died on 17 December 1934 at his estate Stanbridge Earls.

GRONINGEN GAS FIELD. Supergiant gas field discovered in the Netherlands in 1959. At the time of its discovery, Groningen was the largest nonassociated **natural gas** field ever found. It was later significantly eclipsed by others. Groningen was important because of its proximity to Western European markets and because it stimulated interest in exploring the **North Sea**. The field is onshore, but petroleum companies speculated that the seabed might contain similar structures. Groningen has an **EUR** of about 43 Tcf of gas. By the mid-2000s, it was perhaps half exhausted.

Groningen was found in July 1959 near the village of Slochteren by Nederlandse Aardolie Maatschappij (NAM). NAM was a joint venture founded in 1947 and equally owned by **Royal Dutch Shell** and the Standard Oil Company of New Jersey (later **Exxon**). NAM realized that it had made a major find but kept it secret because it had only an exploration license, not a production license, for the province of Groningen. It wanted to explore the discovery further without competition. The "silence of Slochteren" lasted more than a year. It was broken on 14 October 1960, when the Belgian politician V. Leeman revealed the discovery during a discussion on energy policy in the European Parliament. NAM was allowed a production concession, with equity participation by the Dutch government. A new partnership was formed, with the government owning 40 percent of the shares through Energie Beheer Nederland (EBN). Production began in 1964. Groningen formed the basis of a significant new natural gas industry in the Netherlands. *See also* DUTCH DISEASE.

GUFFEY, JAMES McCLURG (1839–1930). Oil entrepreneur, cofounder of Guffey and Galey, Guffey Petroleum, and Gulf Refining. Guffey Petroleum and Gulf Refining became the **Gulf Oil Corporation**. James McClurg Guffey was born 19 January 1839 in Westmoreland County, **Pennsylvania**. He attended Iron City Commercial College in Pittsburgh and went to work as a clerk for the Louisville and Nashville Railroad. After another job in Nashville, Tennessee, he returned to Pennsylvania in 1872 and became a salesman of oil well machinery and supplies. He learned the oil business and began to explore successfully on his own. In 1880, he partnered with John H. Galey to establish the Guffey and Galey company, which became a leading **independent** producer. Guffey and Galey drilled the Matthews gusher that year, opening the McDonald field. They were involved in other famous fields, including Coalinga in **California**, Sand Fork and Kyle in **West Virginia**, and **Corsicana** and **Spindletop** in **Texas**. At one point, the Guffey and Galey firm was the largest producer in the world, at 40,000 bpd.

In 1894, a water well drilled in Corsicana, Texas, found the state's first significant oil. Ralph Beaton, one of the well's shareholders, managed to interest Guffey and Galey in the discovery. Both Guffey and Galey visited Corsicana to evaluate the prospects. Galey thought Texas had considerable potential, but Guffey was not impressed. Nevertheless, Guffey and Galey entered into a 50–50 venture with Beaton and his associates to test the field with three additional wells, two of which struck small amounts of oil. In 1897, Guffey and Galey offered to sell their interest to the locals or buy the locals' interest for $30,000. The local shareowners chose to buy the Guffey and Galey interest, which they then sold to a company that later formed part of the **Magnolia Petroleum Company** (eventually part of **Mobil**).

Guffey and Galey left Texas but returned in 1901 with financing from **Andrew Mellon** to back the 10 January 1901 Spindletop gusher drilled by **Anthony Lucas**. Guffey had a five-eighths interest, Galey had one-fourth, and Lucas one-eighth. In May 1901, Guffey organized the J. M. Guffey Petroleum Company, which acquired the Galey and Lucas shares. Guffey had a seven-fifteenths interest in the new company, the remainder belonging to Andrew Mellon, Richard Mellon, and several associates. Later in 1901, the same group established the Gulf Refining Company of Texas to refine the crude oil produced by the Guffey company. Gulf Refining built a refinery at Port Arthur, and Guffey was head of both companies. A 1902 reorganization, prompted by declining Spindletop production, placed **William L. Mellon** effectively in charge, although Guffey remained titular chief. In January 1907, the Gulf Oil Corporation was established. Guffey was bought out for $3 million.

Guffey had many other business interests. He developed **natural gas** fields around Pittsburgh and headed a number of gas firms. He had mining and real estate interests in Idaho, Colorado, Florida, and **Canada**. He also had coal interests in West Virginia. Despite his tremendous success, Guffey was not particularly wealthy when he died, mostly because he lost a large amount of money in the panic of 1907. Guffey died on 20 March 1930.

GUFFEY AND GALEY. *See* GUFFEY, JAMES McCLURG (1839–1930).

GUFFEY PETROLEUM COMPANY. *See* GUFFEY, JAMES McCLURG (1839–1930).

GULBENKIAN, CALOUSTE (1869–1955). Armenian entrepreneur who played an important role in opening the petroleum resources of the Middle East to Western development. Calouste Sarkis Gulbenkian was born on 29 March 1869 in Uskudar, near Istanbul, in the Ottoman Empire (now Turkey). He studied petroleum engineering at King's College in London and then went to **Baku** to learn about the industry. He returned to London and began his career as a deal maker in the oil business. He became a naturalized British citizen in 1902, mainly to facilitate his business dealings. Gulbenkian played a role in the 1907 merger of the **Royal Dutch Company** and the **Shell Transport and Trading Company** to form the **Royal Dutch Shell** Group, and he acquired substantial shares as a result.

Gulbenkian was a mastermind behind the 1912 creation of the Turkish Petroleum Company (TPC, renamed **Iraq Petroleum Company** in 1929). TPC was a consortium whose goal was to obtain a petroleum **concession** in Mesopotamia (**Iraq**), then a territory of the Ottoman Empire. TPC was half owned by the Anglo–Persian Oil Company (APOC), which became the **Anglo–Iranian Oil Company**, eventually **BP**. Deutsche Bank and Royal Dutch Shell each owned 22.5 percent of TPC, and Calouste Gulbenkian owned 5 percent. TPC got a promise of a concession, but its efforts were interrupted by the outbreak of **World War I**. It would not obtain a concession in Iraq until 1925.

After the war, the ownership of TPC was restructured at the **San Remo Conference**. The German share was transferred to France and would form the basis of the **Compagnie Française des Pétroles** (CFP, later **Total**). Italy and the **United States** were excluded from TPC at San Remo and demanded a share in the resources of the Middle East. After years of negotiations, U.S. oil companies were eventually allowed to buy into TPC. The agreement was not finalized until July 1928, nearly a year after TPC had struck oil at Baba Gurgur, near **Kirkuk**, on 15 October 1927. APOC, Royal Dutch Shell, CFP, and the **Near East Development Corporation** (a consortium of five U.S. companies) each got a 23.75 percent share. Through it all, Gulbenkian retained his 5 percent. His persistence in retaining his share earned him the nickname "Mr. Five Percent." The consortium members agreed not to compete with each other in the territory of the former Ottoman Empire. Gulbenkian drew a line on a map with a red pencil to delineate the territory, and the noncompetition agreement became known as the **Red Line Agreement**.

In 1938, Gulbenkian incorporated the Participations and Explorations Corporation (Partex) in Panama to hold his oil interests. When Gulbenkian died on 20 July 1955 in Lisbon, he had a net worth that might have been as high as $840 million. He had a famous art collection that was willed, along with part of his fortune, to the Calouste Gulbenkian Foundation. The foundation supports the arts, charity, education, and science, and it established the Calouste Gulbenkian Museum in Lisbon to display the Gulbenkian collection.

GULF COOPERATION COUNCIL (GCC). Officially, the Cooperation Council for the Arab States of the Gulf, an organization of six Arab states fostering economic, technical, military, and political cooperation. The GCC includes all the countries of the Arabian Peninsula except **Yemen**. The members are **Saudi Arabia**, the **United Arab Emirates**, **Kuwait**, **Bahrain**, **Qatar**, and **Oman**. Yemen is a potential future member, even though it does not have a coastline on the **Persian Gulf**.

The GCC was founded on 25 May 1981. On 11 November 1981, the member countries signed the Unified Economic Agreement eliminating customs duties and providing for free movement of people and capital. Some GCC members have territorial disputes with each other, and the GCC has helped mediate. The GCC was formed partly in response to the outbreak of the **Iran–Iraq War** as well as the 1979 Iranian Revolution and Soviet invasion of Afghanistan. In theory, an attack on one of the countries is deemed to be an attack on all of them, although it was not the GCC that repelled **Iraq** from Kuwait in 1991—*see* PERSIAN GULF WAR (1991). Still, the GCC has made progress in the security arena. It established the Peninsula Shield Force with 5,000 troops in 1984 and increased its size to 22,000 troops in 2000.

In 2011 Saudi Arabia, wanting to restrain the influence of Iran in the region, made a proposal to deepen the integration of member countries, making the GCC into a "Gulf Union" along the lines of the European Union. The proposal was backed by Bahrain and later Kuwait and Qatar, but at a GCC summit in 2013, Oman voiced strong objections, even threatening to withdraw from the group.

In 2017 the GCC entered a period of crisis as three members—Saudi Arabia, the UAE, and Bahrain—along with nonmember **Egypt**, cut diplomatic relations with member nation Qatar on 5 June. The stated reasons were Qatar's alleged support for groups such as the Muslim Brotherhood and Al Qaeda. Qatari individuals and businesses were then banned from various GCC member countries. Saudi Arabia and others have criticized Qatar for being too close to Iran. Iran and Qatar share the world's largest natural gas field complex. Iran's South Pars field is an extension of Qatar's **megagiant North Field**. Taken separately, North Field is the largest known hydrocarbon field on earth, and South Pars is the third largest.

GULF OF MEXICO. A roughly oval gulf of the Atlantic Ocean largely surrounded by the **United States** and **Mexico** and bounded on the southeast by Cuba. The Gulf of Mexico possesses significant **offshore** petroleum resources exploited by both the United States and Mexico; it is the top oil-producing area for Mexico and the second-highest producer in the United States after **Texas**. Much of the U.S. Gulf's resources are under federal jurisdiction, although some are under state control. Production from state areas is negligible compared to federal production. The Gulf has a surface area of about 600,000 square miles. Its greatest east–west and north–south extents are about 1,100 and 800 miles respectively. The Gulf's continental shelf varies from about 25 miles to 200 miles in width from the shore. Much of the Gulf is shallow, but it can reach depths of 17,000 feet.

The 23 November 1970 treaty between the United States and Mexico established maritime boundaries out to 12 nautical miles in the Pacific and from the mouth of the Rio Grande into the Gulf of Mexico. A second treaty of 4 May 1978 extended both boundaries seaward to a limit of 200 nautical miles. Both countries had passed legislation creating a 200-nautical-mile resource zone. In the middle of the Gulf of Mexico, the 200-nautical-mile arcs drawn from the Louisiana coast and from Mexico's islands north of the Yucatán created a gap of about 129 nautical miles. A further treaty of 9 June 2000 delimited the continental shelf in the gap, joining the two segments of the 1978 boundary. The treaty contained a provision to handle transboundary resources. It created a buffer zone of 1.4 nautical miles on each side of the boundary, in which the United States and Mexico agreed to forbid drilling for 10 years or more. The two countries also agreed to share geological and geophysical information and to exploit resources equitably.

Although earlier offshore wells had been drilled elsewhere, the truly modern offshore petroleum industry is generally acknowledged to have begun in the U.S. Gulf, with a shallow-water well drilled by Kerr McGee in 1947 off the Louisiana coast, out of sight of land. Most of the U.S. activity in the Gulf has been off the Louisiana or Texas coasts. Between 1930 and 1971, 12 shallow-water **giant** fields were discovered, mostly within 20 miles of the Louisiana coast. Nine were predominantly oil, and three were predominantly gas. The largest was Bay Marchand (**EUR** 0.736 Bb and 0.8 Tcf of gas), discovered in 1949.

Beginning in the 1970s, exploration and development moved to deeper waters. From 1993 to 2002, six giant deepwater oil fields were discovered, at depths from 2,900 to 6,500 feet. Thunder Horse and Thunder Horse North, discovered by **ExxonMobil** and **BP** in 1999 and 2000, together have an EUR of 1.23 Bb and 1.5 Tcf of gas. Others include Mars (**Royal Dutch Shell**, 1993), Mad Dog (BP and **BHP Billiton**, 1998), Atlantis (BP and BHP Billiton, 2001), and Tahiti (**Chevron**, 2002). In 2006, Chevron drilled the **Jack No. 2** well to a depth of 28,175 feet (including a 7,000-foot water column). Jack, Mad Dog, and Thunder Horse, as well as a number of others in the 21st century, are in the technically difficult and costly **deepwater** and **ultra-deepwater subsalt** plays.

Mexico's offshore resources in the Gulf have thus far all been in the Bay of Campeche (Bahía or Golfo or Sondo de Campeche). This inlet west of the Yucatán Peninsula, covering about 6,000 square miles, contains seven of Mexico's 18 giant fields, including **Cantarell**. The Mexican reforms of 2013 ending the monopoly of state-owned Pemex in the country's petroleum industry appeared to have a salutary effect. In 2017, Premier Oil, a **UK** concern, announced a discovery of 1 Bb. Its Zama No. 1 well was the first offshore well drilled by a private company in Mexico.

In 2010 the Gulf of Mexico experienced its worst environmental disaster, indeed one of the worst such disasters in history, with the ***Deepwater Horizon* oil spill** that released nearly 5 Bb into the Gulf, devastated beaches, and also resulted in a moratorium on drilling for several months.

GULF OIL CORPORATION. Later **Chevron**; **major petroleum company** and one of the **Seven Sisters**. The Gulf Oil Corporation originated at **Spindletop**, in **Texas**. **James Guffey** and John H. Galey, with financing from **Andrew Mellon**, backed the successful 10 January 1901 gusher drilled by **Anthony F. Lucas**. In May 1901, Guffey organized the J. M. Guffey Petroleum Company, which acquired the Galey and Lucas shares. Guffey had a seven-fifteenths interest in the new company, the remainder belonging to Andrew Mellon, Richard Mellon, and several associates. Later in 1901, the same group established the Gulf Refining Company of Texas to refine the crude oil produced by the Guffey company. Gulf Refining built a refinery at Port Arthur. Guffey was the head of both companies, but a 1902 reorganization, prompted by declining Spindletop production, placed **William L. Mellon** effectively in charge, although Guffey remained titular chief.

In January 1907, the Gulf Oil Corporation was established with Andrew Mellon as president. Guffey's interest was purchased for $3 million. William Mellon became president in 1909. With Texas production temporarily in abeyance, Gulf Oil began **refining Oklahoma** crude. It built a **pipeline** to bring oil to Port Arthur from the **Glenn Pool**, discovered in late 1905. Over the next two decades, under William Mellon's leadership, Gulf became a major oil company. By 1928, it was producing about 0.214 Mmbpd. It built refineries in Fort Worth (1911); Bayonne, New Jersey (1925); Philadelphia (1926); and Sweetwater, Texas (1928), and a pipeline network connecting them to its producing fields. In addition to Texas and Oklahoma, it produced in other areas of the **United States** and in **Mexico** and **Venezuela**.

In 1928, Gulf joined the **Iraq Petroleum Company** consortium via the **Near East Development Corporation** and became bound by the **Red Line Agreement**. This prevented it from exploiting a concession it had obtained in **Bahrain** through New Zealander **Frank Holmes** in 1927. As a result, it transferred its interest to the Standard Oil Company of California (Socal, later Chevron) on 21 December 1928 for $157,149 in expense reimbursements. This launched Socal on a path that led later to the founding of **Aramco** and the discovery of oil in **Saudi Arabia**. Ironically, Gulf sold its interest in the Near East Development Corporation in the mid-1930s.

Gulf obtained a significant interest in **Kuwait**, which was outside the red line. It formed the Kuwait Oil Company (KOC) as a joint venture with the Anglo–Persian Oil Company (later **Anglo–Iranian Oil Company**, eventually **BP**) in 1934. KOC obtained a concession in Kuwait on 23 December

1934. In 1938, KOC discovered Greater **Burgan**, the second-largest oil field ever found. By 1967, Gulf's Kuwait interest yielded a production of 1.3 Mmbpd.

By the end of the 1920s, Gulf had an extensive network of gasoline stations, mostly in the southern and eastern United States. It embarked on a major retail expansion program into the Midwest in 1929. Unfortunately, its expansion coincided with the Great Depression, and in 1931 it lost money for the first time. By the mid-1930s, the company was doing well again, a trend that only intensified with the increased demand for petroleum brought about by **World War II**. By 1950, Gulf was capitalized at more than $1 billion and had overseas production in Venezuela, Kuwait, and **Canada**. It had a 10,000-mile pipeline network, a **tanker** fleet, and 36,000 gas stations. During the 1950s, it expanded into **petrochemicals**, and it expanded oil and gas exploration in the Louisiana **Gulf of Mexico**, which would become one of its top regions. In the 1960s and early 1970s, it made discoveries in **Bolivia**, **Nigeria**, **Angola**, and **Ecuador**.

In 1975, Gulf lost its Kuwaiti interests to nationalization. Without Kuwait, Gulf's **reserves** position suffered. Despite significant investments in exploration, Gulf's reserves declined 40 percent between 1978 and 1982. Gulf's reputation also suffered in the early 1970s when some of its top managers were accused of making illegal political contributions. In August 1983, Mesa Petroleum, headed by corporate raider **T. Boone Pickens**, made a hostile attempt to take over Gulf, which was the sixth-largest U.S. oil company at the time. Mesa built up an 11 percent share. To deter Pickens, Gulf opened itself to takeover by other parties. On 5 March 1984, the Gulf board voted to be acquired by Socal for $13.2 billion. This was the largest corporate merger up to that time. Despite Gulf's relatively depleted reserves, the merger nearly doubled Socal's reserves and briefly made it the largest U.S. oil company by assets. After acquiring Gulf, Socal changed its name to Chevron Corporation.

GULF WAR. *See* PERSIAN GULF WAR (1991).

H

HALBOUTY, MICHEL THOMAS (1909–2004). Important American petroleum geologist working primarily in **Texas**. Halbouty was born in Beaumont, Texas, on 21 June 1909 and graduated from Texas A&M with degrees in geology and petroleum engineering in 1930. During the course of his career he discovered over 50 oil fields and was an outspoken advocate for continued vigorous petroleum exploration in the **United States** to reduce reliance on foreign sources. He also published widely in the field of petroleum geology. He died 6 November 2004 in **Houston**.

HALLIBURTON COMPANY. Petroleum services firm. Halliburton was established in 1919 by Erle Palmer Halliburton (1892–1957) as the New Method Oil Well Cementing Company. The firm was based in Wilson, **Oklahoma**, and offered cementing services to the petroleum industry in Oklahoma, Louisiana, and Arkansas. The company developed the jet mixer, a machine for mixing cement and water in controlled proportions that saved labor and reduced waste. Halliburton developed a patent portfolio on cementing that placed it in an almost monopoly position. The company incorporated in Delaware in 1924, jointly owned by the Halliburton family and a number of petroleum companies. It bought the Perkins Cementing Company (where Erle Halliburton had worked briefly in 1916) in 1940, and in 1949 it went public on the New York Stock Exchange.

Cementing accounted for about 70 percent of revenues, but the company diversified into other oil field services, including **well logging** and a variety of well stimulation processes to enhance oil recovery. In 1962, Halliburton acquired Brown and Root Inc., a firm specializing in the construction of **offshore** platforms and **petrochemical** plants as well as military bases. Despite this and numerous other acquisitions, well cementing continued to be a mainstay. Between 1988 and 1991, Halliburton acquired **Geophysical Service Inc.** in stages from Texas Instruments, merging it with Geosource, another acquisition, to form Halliburton Geophysical Services (HGS).

In the soft oil market of the early 1990s, Halliburton restructured. It merged 10 energy services subsidiaries into a single entity, Halliburton Energy Services, and placed its non-oil-field construction business in a separate Construction Engineering Group. It sold off several units that were unrelated to its core businesses, including a health-care cost management company sold in 1992 for $24 million and the Highlands Insurance unit spun out in 1996. It also shed several underperforming units, such as HGS, which was sold to **Western Atlas** for $190 million in 1994.

In 1995, Richard Cheney, who had been secretary of defense under President **George H. W. Bush**, became Halliburton's CEO. Under Cheney's leadership, Halliburton acquired several companies (including Landmark Graphics in 1996) and merged in 1998 with **Dresser Industries**, creating one of the two largest oil field services companies in the world, the other being **Schlumberger**. In merging with Dresser, it acquired M. W. Kellogg, which it combined with its Brown and Root construction subsidiary to form Kellogg Brown and Root (KBR). In the early 2000s, Halliburton came under considerable criticism for its business practices. Some but not all of this criticism was related to KBR's activities in the **Iraq War**—for example, its obtaining no-bid contracts at a time when its former chief executive Cheney was vice president under President **George W. Bush**. In April 2007, KBR became an independent company. Halliburton was involved in the 2010 ***Deepwater Horizon* oil spill**. A U.S. district court found Halliburton "negligent," as opposed to the main partner **BP**, which it deemed "reckless." In 2014 Halliburton and **Baker Hughes** announced an intent to merge, but the plans were abandoned in 2016.

HAMACA PROJECT. *See* ORINOCO OIL BELT.

HAMMER, ARMAND (1898–1990). Head of **Occidental Petroleum Company** 1957–1990. Armand Hammer was born in New York on 21 May 1898. He did not enter the oil industry until 1957 at the age of 59. He was already retired from several careers as a doctor and in business, with unique connections in the Soviet Union. Hammer had met Vladimir Lenin and had run a successful pencil factory in the Soviet Union. The **United States** used him on a number of occasions as an ad hoc trade ambassador to the Soviet Union.

In retirement, Hammer decided to seek a tax shelter to shield some of his wealth. He settled on a small, ailing oil firm named Occidental. He bought the company in 1957 and loaned it money. He was not expecting large returns on his investment, but the company somewhat unexpectedly struck new oil south of Los Angeles, **California**. Hammer came fully out of retirement. As Occidental's president, he used the company's windfall to buy other small petroleum-related companies.

In 1961, oil was discovered in **Libya**, and Hammer saw an opportunity. He went there and convinced King Idris to grant Occidental an oil **concession**. He argued that the **major oil companies** would exploit Libya as they had done with other countries in the region. Occidental succeeded in Libya, and by 1973 it had also made discoveries in **Nigeria**, Peru, and the **North Sea**. Hammer diversified into coal and chemicals, but Occidental remained primarily focused on petroleum. In 1970, Muammar Qaddafi deposed King Idris and took over rule of Libya. He threatened nationalization and expropriation. In 1973, Hammer was obliged to sell 51 percent of Occidental's Libyan interests to the government for a very low price of $136 million.

Armand Hammer was famous as an art buyer and seller and as a philanthropist. He has been criticized by some observers, however. The most serious allegation is that he may have initiated human rights abuses in Occidental's Latin American operations. Hammer died on 10 December 1990.

HARDING, WARREN GAMALIEL. *See* TEAPOT DOME SCANDAL (1922–1930).

HARDISON, WALLACE. *See* STEWART, LYMAN (1840–1923).

HARDISON AND STEWART OIL COMPANY. *See* STEWART, LYMAN (1840–1923).

HARDSTOFT OIL FIELD. The first oil field in the **United Kingdom** (excluding colonies). During **World War I**, Britain depended heavily on imports from imperial dependencies, especially **Burma** and **Iran**, as well as the **United States** and **Mexico**. In order to decrease its risks, the government decided to search for oil within the home islands. It contracted S. Pearson and Sons—*see* PEARSON, WEETMAN (1856–1927)—to explore for indigenous petroleum resources. The firm drilled 11 wells, the first of which was at Hardstoft in Derbyshire. Hardstoft No. 1 struck oil on 27 May 1919 at a depth of 3,070 feet in a sandy limestone layer. The well produced a total of 29,000 barrels between 1920 and 1946.

HARKEN ENERGY. *See* BUSH, GEORGE W. (1946–).

HEAVY OIL. A form of crude oil that is dense and viscous and more difficult to extract than **conventional crude**. Heavy oil may flow in the reservoir but does not do so easily, and it may require special and costly techniques to extract and process. Heavy oils tend to be sour (high in sulfur) and dark. Although there are no universally accepted definitions, heavy oil has an **API gravity** somewhat less than 20°API, generally in the range of 8–15°API. Conventional crude, on the other hand, ranges from 20°API to above 40°API, although sometimes its gravity can be lower. Heavy oil generally has a viscosity between 1,000 and 10,000 centipoise, while for many conventional crudes, viscosities range between 10 and 100 centipoise. Occasionally, the heavier conventional crudes are called heavy.

Sometimes the term "heavy oil" or "extra-heavy oil" is intended to include **bitumen**, the substance found in **oil sands**, but the two are different. Bitumen is more viscous, with viscosities of 100,000 cp or more, and is typically immobile in the reservoir. Bitumen also tends to have somewhat lower gravity, in the 5–10°API range. Both bitumen and heavy oil represent a degraded form of crude oil that has lost most of its light fractions, bitumen more so than heavy oil. Approximately 90 percent of the world's known heavy oil resource is in the **Orinoco Oil Belt** of **Venezuela**. The Orinoco Belt constitutes a potentially huge set of deposits, estimated to contain about 1,200 Bb of heavy oil in place, with around 270 Bb recoverable with current technology.

HENRY HUB. The world's most fully developed natural gas trading hub, located in Erath, **Louisiana**. Henry Hub is both a physical hub and a trading center. The physical facilities are owned by Sabine Pipe Line Company and connect to nine interstate and four intrastate gas pipelines. The **New York Mercantile Exchange** (NYMEX) chose it as a registered location to settle ultimate delivery for gas futures contracts. Of all the various hubs, it is the one at which the most fully developed price discovery occurs. Other hubs in the United States use Henry Hub prices as a benchmark.

HESS CORPORATION. *See* AMERADA PETROLEUM CORPORATION; HESS, LEON (1914–1999).

HESS, JOHN B. *See* HESS, LEON (1914–1999).

HESS, LEON (1914–1999). Founder of Hess Corporation. Leon Hess was born 13 March 1914 in Asbury Park, New Jersey, and began his career in his father's fuel oil business in the early 1930s. He specialized the company into a niche business of buying up residual oil left over from **refining** processes and selling it as a fuel to hotels and power plants. He also speculated successfully by buying oil at low prices in the summer and selling it high in the winter. He expanded into storage, eventually building up one of the largest storage networks in the **United States**. During **World War II**, Hess was the head of transportation logistics for General Patton, in charge of fuel. He left the army as a lieutenant colonel with valuable organizational experience. In the late 1940s, he expanded his residual oil business. He moved into refining and marketing and was a pioneer in low-cost, low-service **gasoline** stations.

Hess took his company public in 1962 via a merger with Cletrac Corporation, creating Hess Oil and Chemical, of which he was president. In 1966, Hess Oil and Chemical acquired 10 percent of **Amerada Petroleum Corporation** after the Bank of England sold the stake it had acquired in World War II. Hess wanted to move upstream and ensure his own crude supplies. He completed the hostile acquisition of Amerada in 1969 after a successful competition with **Phillips Petroleum**. The enlarged company became Amerada Hess. Amerada brought Hess proven oil **reserves** of more than 500 Mmb in the United States and 75 Mmb in **Libya**. The merger proved successful. Amerada Hess drilled a successful wildcat well in **Alaska**'s **Prudhoe Bay** in 1970 and would eventually become one of the most effective explorers in the **North Sea**. In 1967, before the acquisition of Amerada, Hess completed a refinery in St. Croix that by 1979 was one of the world's largest, with a capacity of 700,000 bpd.

Amerada Hess prospered during the **First Oil Shock** and the embargo imposed by the **Organization of Arab Petroleum Exporting Countries** (OAPEC). Because of its huge storage capacity (68 Mmb), it was one of the few operators that could effectively supply the U.S. market. The 1970s were also a time of legal problems. In 1970, Amerada Hess was charged with import rule violations in connection with its St. Croix refinery, and in 1978 it was convicted of fixing gasoline prices in the United States.

Hess struggled to fight off hostile takeover attempts. His company was also hit hard by the low oil prices of the 1980s **Oil Countershock**. Hess's revenues in 1986 were $4 billion, 60 percent lower than in 1981. Leon Hess had retired briefly from Amerada Hess in the 1980s but returned in 1986 to lead the company again and groom his son John B. Hess to take over. After the construction of a **pipeline** to supply the domestic market in the **United Kingdom** with North Sea **natural gas** in the early 1990s, the company's prospects looked temporarily brighter. However, the low oil prices of that decade took their toll. The company had large losses in 1993 and 1995. It expended more than $1 billion to upgrade its St. Croix refinery, damaged by Hurricane Hugo in 1989. Its large storage network and extensive inventories were liabilities in a low-price environment.

In May 1994, Leon Hess retired and John B. Hess took over. He sold marginal properties and streamlined operations. These actions and higher oil prices significantly strengthened the company in the 2000 decade. In 2006, Amerada Hess changed its name to Hess Corporation. Leon Hess died in New York City on 7 May 1999. Apart from his career in petroleum, he was also famous for his ownership of the New York Jets football team.

HIBERNIA OIL FIELD. *See* CANADA; PETRO-CANADA.

HIGGINS, PATTILLO (1863–1955). American oil entrepreneur who was a driving force behind the **Spindletop** discovery in 1901. Pattillo Higgins was born 5 December 1863 in Sabine Pass, **Texas**. He apprenticed as a gunsmith with his father. After a violent youth that resulted in a murder trial, in which he was acquitted, and the loss of an arm, he became dogmatically religious. He worked in lumber camps, real estate, and brick making. In the 1880s, he became convinced that petroleum could be found in the salt dome structure at Spindletop Hill, south of Beaumont, Texas. With George W. Carroll, he bought half the hill and then formed a partnership with George O'Brien to acquire the other half. In August 1892, he formed Gladys City Oil, Gas, and Manufacturing Company with his two partners to explore for and develop oil at Spindletop.

At the time, conventional wisdom did not consider the area promising for petroleum. Gladys City drilled unsuccessfully in 1893. It then leased land to the Savage Brothers of **Corsicana**, who also drilled unsuccessfully in 1895. Higgins resigned from the company in a dispute, after which the company drilled a third dry hole. Higgins remained convinced that oil was to be found. By placing advertisements in a number of publications, he found **Anthony F. Lucas**, an expert on salt domes. Higgins helped Lucas negotiate a lease with Gladys City on 20 June 1899 for a 10 percent finder's fee. Lucas began drilling but ran out of money and also had technical problems. Lucas made a deal with **James M. Guffey** and John H. Galey of Pittsburgh, **Pennsylvania**, who also brought backing from **Andrew Mellon** for five exploratory wells. The new agreement cut Higgins out and lowered Lucas's profit share to one-eighth. Lucas signed new leases with Gladys City and other parties, and he made the famous Spindletop discovery on 10 January 1901.

Higgins sued Lucas and Gladys City on grounds that the lease he had helped Lucas obtain was still valid when Lucas embarked on the new leases with Guffey and Galey. The parties settled out of court. Higgins then formed his own company, the Higgins Oil and Fuel Company, and drilled successfully at Spindletop on his own lease. He sold his shares for $3 million to John Henry Kirby in 1902, having made too many real estate investments. He then formed the Higgins Standard Oil Company. For the rest of his life, he explored for petroleum in the salt dome structures of Texas near the **Gulf of Mexico**, with moderate success. It was his fate to make discoveries that he was not able to exploit fully, leaving the riches mostly to others. Pattillo Higgins died on 5 June 1955 in San Antonio, Texas.

HOLMAN, EUGENE (1895–1962). President of Standard Oil Company of New Jersey (later **Exxon**) 1944–1960, bringing the company into **Aramco** in 1948. Eugene Holman was born on 2 May 1895 in San Angelo, **Texas.** He graduated from Simmons College in 1916 and received a master's degree in geology in 1917 from the University of Texas. He joined the **United States Geological Survey** (USGS) in 1918, moving soon after to the **Humble Oil and Refining Company** (later part of Exxon). By 1926, he was Humble's chief geologist. In 1929, he went to **Venezuela** to improve Jersey Standard's operations there (Jersey Standard was Humble's majority owner). Holman's success in Venezuela contributed to his election as a director of Jersey Standard in 1940. By 1942, he was vice president, and in 1944 he became president. During **World War II**, he served on the Petroleum Industry War Council—*see* PETROLEUM ADMINISTRATION FOR WAR (PAW) (1942–1946). In 1954, he became chairman of the board of Standard Oil Company of New Jersey.

Holman was instrumental in establishing Jersey Standard as a major producer in the Middle East. He helped bring the company into Aramco in December 1948, negotiating for 30 percent of the stock. He retired from the company in 1960. In the same year, he received the **American Petroleum Institute**'s Gold Medal for Distinguished Achievement. He was also noted for his emphasis on the development of employees and for the employment of local talent in the foreign countries where his company operated. He died 12 August 1962 in New York City.

HOLMES, FRANK (1874–1947). New Zealand mining engineer who believed in the petroleum potential of the Arabian Peninsula when others did not and obtained many early **concessions** from countries in the region. Holmes was undercapitalized and unable to pursue seriously the opportunities he identified and nurtured. By a circuitous route, however, he was responsible for bringing the Standard Oil Company of California (Socal, later **Chevron**) to the Middle East, ultimately leading to the discovery and development of **Saudi Arabia**'s huge petroleum resources. His vision also led to major discoveries in **Kuwait** and **Bahrain**.

Holmes was born on a sheep farm in New Zealand in 1874. During **World War I**, he was in the Middle East as a quartermaster in the British army, reaching the rank of major. He became aware of oil seepages and was soon convinced that the Arabian Peninsula had significant petroleum resources. He formed the Eastern and General Syndicate and pursued concessions in the Arabian Peninsula. Not everyone shared his optimistic view of the peninsula's resources. The Anglo–Persian Oil Company (APOC), later the **Anglo–Iranian Oil Company**, eventually **BP**, thought the prospects very unpromising. Nevertheless, the British government took a negative view of Holmes's pursuit of concessions in what it regarded as APOC's sphere of influence.

When Holmes tried to get a concession in the future Saudi Arabia from King **Ibn Saud** in the early 1920s, the **United Kingdom** (via Percy Cox) pressured Ibn Saud not to grant it. Ibn Saud assented at first, because he was receiving an annual subsidy of £60,000 from the British. When they canceled the subsidy in 1923, Ibn Saud gave Holmes a petroleum concession. The concession covered 30,000 square miles in the Hasa region (now the Eastern Province) and cost Holmes's company £2,000 per year, but nothing came of it. Holmes was unable to make the payments after the first two years, and Ibn Saud canceled the agreement in 1928.

In 1925, Holmes obtained a concession in Bahrain. He had difficulty stirring up interest among the oil companies but in 1927 finally convinced **Gulf Oil Corporation** to take over the concession. Gulf eventually sold its interest to Socal, because after July 1928 it was constrained by the **Red Line Agreement** from operating independently in Bahrain. Holmes became Socal's first local representative in Bahrain. After some difficulties, Socal struck oil on 31 May 1932 via its new subsidiary, the Bahrain Petroleum Company. Socal moved on to get a concession for what is now the Eastern Province of Saudi Arabia in 1933. This would ultimately lead to the discovery of Saudi Arabia's huge **reserves**.

Gulf Oil went on to bid for a concession in Kuwait, which was outside the area of the Red Line Agreement. In its pursuit of the concession, it faced competition from APOC, which had become more interested in the prospects of the Arabian Peninsula after the Bahrain discovery. After a period of intense competition, the two companies formed the Kuwait Oil Company as a joint venture. Frank Holmes played a major role in the ensuing negotiations with Kuwait's ruler, Sheikh Ahmad, whom he had befriended. The Kuwait Oil Company obtained its concession on 23 December 1934. Sheikh Ahmad appointed Frank Holmes as his London representative to the company. Holmes would hold that position until his death in 1947. Kuwait Oil Company discovered **Burgan** in 1938, still the second-largest oil field ever found.

HORIZONTAL DRILLING. The process of **drilling** a well down from the surface to a point just above a desired hydrocarbon-bearing rock stratum, then deviating to a near horizontal inclination and thereby moving inside the stratum in a direction parallel to it. This has the effect of exposing a much greater length of well bore to the producing formation. The basic technology of drilling in a curved direction can be traced back to 8 September 1891, when John S. Campbell obtained a U.S. patent for flexible drill shafts, initially targeted at dental applications. In the petroleum industry, the first horizontal well was drilled in Texon, **Texas**, in 1929. A successful horizontal well was drilled in 1944 in the Franklin heavy oil field of **Venango County, Pennsylvania**. Horizontal drilling was also tried in **China** in 1957. The practice did not begin to find significant application anywhere, however, until the 1980s, with the development of better drilling motors and downhole telemetry apparatus.

Between 1980 and 1983, **Elf Aquitaine** conducted the first tests that showed horizontal drilling could be more than a novelty. Elf drilled horizontal wells in three European oil fields. In one of these, the Rospo Mare field **offshore** Italy, Elf achieved significant increases in output. The practice spread during the 1990s, especially in Texas's Austin Chalk formation, where it helped increase oil production. Also during the 1990s, horizontal drilling began to be combined with well stimulation techniques, particularly **fracking**, to make feasible the production of oil and gas trapped in low-permeability formations (**tight oil** or **shale oil**, **shale gas**, and **tight gas**). Important preparatory work was done from 1976 to 1992 by the **Eastern Gas Shales Project**. In 1991 Mitchell Energy—*see* MITCHELL, GEORGE P. (1919–2013)—aided by the federally funded **Gas Research Institute** drilled the first horizontal well with fracking in the gas-bearing **Barnett Shale** in Texas. At first, horizontal fracked wells were not economically successful in the Barnett, although success was achieved with vertical wells. Eventually horizontal fracked wells succeeded, and by 2007 their number surpassed those of vertical wells in the Barnett. By 2010, 70 percent of Barnett gas production was from horizontal wells.

HOT OIL. Oil produced in contravention of regulations limiting output, particularly in the southwestern **United States** during the market glut of the 1930s. The discovery of the **supergiant East Texas oil field** in 1930 flooded the market with oil and caused prices to collapse from prevailing levels of over $1 per barrel in the 1920s to 10 cents a barrel or even lower. The **Railroad Commission of Texas** imposed production quotas in order to conserve the reservoir and stabilize prices. Many producers defied the regulations, producing beyond their quotas and smuggling away the "hot oil." Estimates of East Texas hot oil production in late 1934 and early 1935—when it is thought to have peaked—ranged from 50,000 to 150,000 bpd.

Hot oil was also produced in **Oklahoma**. Both states resorted to martial law in the oil fields at some point to try to control the problem. In **Texas**, the legislature made hot oil production a felony in 1932, and in 1935 it declared that hot oil was contraband that could be confiscated. The National Industrial Recovery Act of 1933 contained a section prohibiting the interstate transport of oil produced in excess of state quotas, but this provision was struck down by the U.S. Supreme Court in 1935. Congress responded with the **Connally Hot Oil Act** of 1935, which achieved the same purpose in a slightly different way. It gave the president power to regulate interstate flow of petroleum and petroleum products and to require certificates of clearance for petroleum products in interstate commerce.

HOUDRY, EUGENE (1892–1962). Engineer who devised a process for the catalytic cracking of petroleum, increasing the yield and quality of **gasoline**. Eugene Jules Houdry was born in Domont, France (near Paris), on 18 April 1892. He graduated first in his class in 1911 from the École des Arts et Métiers in Paris with a degree in mechanical engineering. He then began working in his father's steel business. He fought in **World War I**, becoming a lieutenant in the French army tank corps. His experience in the tank corps led to a fascination with motor vehicles and a desire to improve the quality of fuels. In 1922, he began work for the French government on creating **syncrude** from bituminous coal and lignite. He succeeded technically, but the product was not economically competitive.

Houdry turned his attention to producing better fuels for motor vehicles from crude oil. He found that the efficiency of the thermal cracking process (such as that developed by **William Burton**) could be increased through the use of a catalyst. Houdry's catalysts were solid pellets of aluminum and silicon oxides. The catalytically cracked gasoline had octane numbers as high as 88 (later 95), as opposed to a maximum of 72–82 for gasoline obtained by thermal cracking processes at the time. When Houdry's process was perfected, it would ultimately produce a 60–75 percent yield of gasoline from crude oil, versus 40 percent for thermal cracking. Houdry tried but failed to interest the French oil industry in the process. In 1930, however, the **Vacuum Oil Company** of New Jersey offered to finance Houdry's research.

Houdry moved to the **United States** and set up a pilot plant at the Vacuum research center in Paulsboro, New Jersey. When Vacuum merged with the **Standard Oil Company of New York** to become Socony–Vacuum (later **Mobil**), Houdry's work was interrupted. **Sun Oil Company** of Philadelphia offered to participate financially. Houdry formed the Houdry Process Corporation, of which he, Socony–Vacuum, and Sun each owned a third. In 1936, Houdry's plant opened in Paulsboro, producing 2,000 bpd of high-octane gasoline. During **World War II**, the Houdry process produced about 90 percent of the high-quality aviation fuel used by the Allies. Houdry also invented a catalytic converter for motor vehicle exhaust systems.

Houdry received several honors, including the Perkin Medal of the Society of the Chemical Industry in 1959 and the Industrial and Engineering Chemistry Award of the American Chemical Society in 1962. He died on 18 July 1962 in Upper Darby, **Pennsylvania**. *See also* REFINING.

HOUSTON. Largest city (pop. 2.3 million, 2016) in **Texas** and a major center of the oil and gas industry. Houston is considered by many to be the petroleum capital of the world. Almost all important entities in the industry have some kind of representation there. The city was founded on 30 August 1836 by Augustus Chapman Allen and John Kirby Allen. From 1837 to 1839, it served as the capital of the then independent Republic of Texas. It was connected to the **Gulf of Mexico** by Buffalo Bayou. Navigation was difficult, however, and only the opening of the **Houston Ship Channel** in 1914 turned the city into a true deepwater port.

Houston rapidly became a major center of the petroleum industry after the 1901 discovery of oil at **Spindletop**. Oil companies located refineries along the Houston Ship Channel, where they were safer from storms from the Gulf of Mexico than they would be on the coast. **Sinclair Oil Corporation** built the first large refinery in 1918. By 1929, 40 oil companies had offices in Houston, including the Texas Company (later **Texaco**), **Humble Oil and Refining Company** (later part of **Exxon**), and **Gulf**. **World War II** increased Houston's importance, not only because of the national need for fuel but also because of the demand for synthetic rubber and other **petrochemicals**, as well as raw materials for explosives. Many of the necessary ingredients could be found near Houston in coastal salt deposits. The city and its surrounding area developed one of the world's largest petrochemical complexes.

Houston's role as a world oil capital expanded with the high prices of the 1970s (*see* FIRST OIL SHOCK; SECOND OIL SHOCK). When oil prices fell in the 1980s (*see* OIL COUNTERSHOCK), the city lost population for the first time in its history. However, the city continues to be a major world energy capital. Its concentration of petroleum-related industry, skills, and knowledge is formidable. Houston is unique among America's major cities in its lack of zoning laws. Some observers believe this has made the city ugly and haphazard, while others insist that the absence of restrictions has encouraged business expansion.

HOUSTON SHIP CHANNEL. Artificially enhanced waterway connecting **Houston**, **Texas**, to the **Gulf of Mexico** and turning the city into a major deepwater port. From its earliest days, Houston could be reached by boat from the Gulf of Mexico via the Buffalo Bayou, but navigation was difficult. In 1869, a group of Houston business executives organized the Buffalo Bayou Ship Channel Company to dredge a more navigable channel. Shipowner Charles Morgan bought the company in 1874 and by 1876 had dredged a channel from Galveston Bay to the site of what is now Clinton, near Houston. Morgan became more interested in railroads than ships and sold his improvements to the U.S. government in 1890.

The city of Houston collaborated and shared costs with the federal government to enlarge the channel. The improvements were completed on 7 September 1914, making Houston ready to become a major deepwater port. Because of **World War I**, however, it was not until 1919 that the first oceangoing vessel loaded a shipment of cotton in Houston. With the buildup of the Texas oil industry, the channel soon became a crucial artery for shipping petroleum and petroleum products. Many companies built refineries along the channel. The channel was further widened and deepened later in the century. Currently it is 45 feet deep, 530 feet wide, and 50 miles long.

HUBBERT, M. KING (1903–1989). Geoscientist famous for his accurate mathematical prediction, made in 1956, that oil production in the **United States** would peak around 1970. Marion King Hubbert was born 5 October 1903 in San Baba, **Texas**. He attended Weatherford College (Texas) and the University of Chicago, from which he received his PhD in 1937. He taught geophysics at Columbia University from 1931 to 1940, partly while he worked on his doctorate. During **World War II**, he worked for the Board of Economic Warfare in Washington, D.C. In 1943, he joined **Royal Dutch Shell** as a research geophysicist, remaining there until 1964. He was working at Shell when he made his **peak oil** prediction at the 1956 meeting of the **American Petroleum Institute** in San Antonio, Texas.

Hubbert's prediction was valid for **conventional** oil in the lower 48 states, excluding production that was later added from new provinces (**Alaska** and the federal offshore areas, mostly in the **Gulf of Mexico**) and the development of **unconventional** resources, especially **shale oil**. Lower-48 U.S. oil production did peak in 1970, at 9.408 Mmbpd, and decline roughly symmetrically with its rise. Hubbert saw nuclear energy as a potential long-term replacement for petroleum. Although he is now heralded for his ideas, the industry was not ready for them at the time. Many ridiculed his prediction. His paper also caused some anger in Shell's management ranks. Hubbert's methodology has been widely used by scientists attempting to predict the peak of global oil production, although in that case it is more difficult to apply. "Hubbert peak" has become shorthand for resource exhaustion.

Among geologists, Hubbert is perhaps even more known for his other work, particularly on fluids and fluid migration in rock and on petroleum migration and entrapment. In 1959, he published two seminal papers with William Rubey, explaining the geological observation that huge rock slabs are sometimes displaced more or less intact, often over considerable distances. They posited that pore fluid pressure between the slab and the underlying rock can support the weight of the slab and thereby reduce the frictional force. Hubbert published the books *Theory of Ground Water Motion* in 1940 and *Energy and Resources* in 1962. After retiring from Shell, Hubbert joined the **United States Geological Survey** and taught at Stanford University. He died on 11 October 1989 in Bethesda, Maryland. *See also* PEAK OIL.

HUBBERT PEAK. *See* HUBBERT, M. KING (1903–1989); PEAK OIL

HUGHES, HOWARD SR. (1869–1924). Inventor of a superior drill bit to penetrate hard rock; founder of the **Hughes Tool Company** (later **Baker Hughes**). Howard Hughes Sr. is not to be confused with his son Howard Hughes Jr., the founder of Hughes Aircraft Company. Howard Robard Hughes was born in Lancaster, Missouri, on 9 September 1869. He studied at Harvard University and the University of Iowa without graduating from either institution. He briefly practiced law without a degree, but he soon became interested in mining. After the **Spindletop** discoveries in 1901, Hughes moved to Beaumont, **Texas**, and began working as a **drilling** contractor with partner Walter B. Sharp.

Hughes's experiences in the drilling business convinced him of the need for better drill bits that could penetrate hard rock. He proceeded to invent one. His bit, for use in rotary rigs, had cone-shaped cutters studded with steel teeth, and it ground the rock rather than chipping or scraping it. He filed two patent applications in 1908 and was granted them on 10 August 1909. The bit was able to drill through hard rock that other rotary equipment could not penetrate, and it was quickly adopted all over the world. In 1909, Hughes and Walter Sharp founded the Sharp-Hughes Tool Company. Hughes managed the company after Sharp's death in 1912 and a few years later purchased Sharp's stock from his widow. He renamed it the Hughes Tool Company in 1915. Hughes continued to improve his bit, obtaining 25 additional patents on it. Hughes died on 14 January 1924.

HUGHES TOOL COMPANY. Later **Baker Hughes**; an oil field services company. **Howard Hughes Sr.** and Walter B. Sharp founded the Sharp-Hughes Tool Company in 1909. The company marketed a superior rotary drill bit invented by Hughes that was able to drill through hard rock. Hughes bought Sharp's share after Sharp's death in 1912 and renamed the company the Hughes Tool Company in 1915. After Hughes died in 1924, ownership passed to his son Howard Hughes Jr. The younger Hughes had a number of interests unrelated to drill bits and used the Tool Company as a source of cash to fund them. Hughes Jr. founded the Hughes Aircraft Company, made substantial investments in the entertainment industry (RKO Pictures), and purchased 78 percent of TransWorld Airlines, among other things. Still,

Hughes Tool Company continued to be successful and dominated the drill-bit business even after its crucial patents expired. Howard Hughes Jr. finally took the company public in 1972.

The newly public concern began diversifying. In 1974, it bought the Byron Jackson division of Borg Warner, which manufactured oil field equipment. In 1978, it acquired Brown Oil Tools. By 1981, more than half its sales were unrelated to drill bits, although still related to oil field services. Hughes Tool Company was overextended and did not fare well in the soft oil market that followed. It lost $200 million between 1983 and 1986. In 1987, it merged with Baker International (formerly **Baker Oil Tools**) to form Baker Hughes.

HUMBLE OIL AND REFINING COMPANY. Important **Texas** oil-producing company; later **Exxon**. Oil was noticed seeping from the ground near Humble, Texas, in 1887 but was not produced until the early 1900s, shortly after the **Spindletop** find. Like Spindletop, the Humble fields are associated with a salt dome. **Ross Sterling**, a future governor of Texas, entered the oil business near Humble by purchasing two wells. In 1911, Sterling joined with **William Stamps Farish**, Walter William Fondre, and Robert L. Blaffer to form the Humble Oil Company with an initial capitalization of $150,000. The capital was doubled a year later. In June 1917, the company incorporated as the Humble Oil and Refining Company, with a capitalization of $1 million. Several oil companies were merged into Humble, including the Paraffine Oil Company, Flaffer and Farish, Schulz Oil, Ardmore Oil, and Globe Refining. Sterling served as Humble's first president, with Farish an influential vice president.

In 1919, needing capital for expansion, Humble sold a half interest to Standard Oil Company of New Jersey (later Exxon) for $17 million. Farish negotiated the deal with **Walter Teagle**. The 1911 dissolution of **Standard Oil** had stripped Jersey Standard of most of its crude oil production. Jersey now gained a supply of crude, and Humble gained a market for its oil. The capital from the deal enabled Humble to develop large new fields in Eastland County, Texas, and build a refinery in Baytown. In 1920–1921, the company built new headquarters in **Houston**. Farish, who became president in 1922, also promoted both vertical and horizontal integration. The company developed a larger **pipeline** system and built several more refineries in the 1920s, including a large one at Brownsville, Texas.

Humble operated with considerable independence from Jersey Standard but fed its top management ranks. In 1933, Farish became chairman of the board of Jersey Standard, while Teagle was president. In 1937, the two switched positions. Humble continued to make large additions to its **reserves** during the decade of the 1930s. During **World War II**, Humble became the largest oil producer in the **United States**, and it continued in that role for some years. In 1949, Humble was producing nearly 276,000 bopd (about 5.5

percent of U.S. production) from just under 10,000 wells. Apart from Texas, the company had fields in Louisiana, New Mexico, Mississippi, Alabama, and **California**.

Humble continued to operate autonomously from the Standard Oil Company of New Jersey, even though Jersey Standard kept increasing its stake. By 1948, Jersey Standard owned about 72 percent of Humble's stock, and by 1954 it owned 88 percent. In 1958, Jersey Standard increased the stake to 98 percent and in 1959 consolidated all its domestic operations with those of Humble. In 1967, Humble, partnered with **Arco**, participated in the discovery of **Alaska**'s **Prudhoe Bay oil field**, the largest ever in North America. Humble Oil and Refining Company thus formed the domestic backbone of the company that would adopt the Exxon name in 1972.

HUNT, H. L. (1889–1974). Founder of Hunt Oil Company and its president 1936–1974. Haroldson Lafayette Hunt was born in Carson, Illinois, on 17 February 1889. After running a cotton plantation in Arkansas, he became a lease broker in the petroleum industry that sprang up around El Dorado. He had quick success there followed by losses. On 26 November 1930, he bought **Dad Joiner**'s well and the surrounding leases in the **supergiant** new **East Texas oil field**. Two years later, he had 900 wells in the field. In 1935, he put his interests, collected in the Placid Oil Company, into trusts for six of his children. Hunt had also fathered four children in a bigamous marriage, and he adopted four more that were actually his own biological children. By 1948, Hunt was producing 65,000 bopd, and his petroleum assets were estimated to be worth $263 million. He may have been the richest person in the **United States** at the time. He became noted for his right-wing politics, expressed through two radio shows, and he endorsed Senator Joseph McCarthy. Hunt died 29 November 1974.

HUNT, NELSON BUNKER. *See* LIBYA.

HUNT, THOMAS STERRY (1826–1892). American chemist and mineralogist who made several contributions to petroleum geoscience in the 19th century, perhaps the most important of which was the elucidation of the anticlinal theory of trapping of hydrocarbons in 1861. Credit for that theory is also given to E. B. Andrews, who arrived at it independently at nearly the same time. Hunt spent the most productive part of his career at the Geological Survey of Canada, from 1845 to 1872.

HUSSEIN, SADDAM (1937–2006). President of **Iraq** 1979–2003; completed nationalization of Iraq's oil industry (1972–1975). Saddam Hussein Abd al-Majid al-Tikriti was born into a peasant family in Awja, Iraq, on 28

April 1937. He joined the Arab nationalist Ba'ath Socialist Party in 1957 and two years later had to flee Iraq after participating in an unsuccessful assassination attempt on Prime Minister Abd al-Karim Qasim. Saddam studied at Cairo Law School, returning to Iraq in 1963 when the Ba'ath party assumed power. The Ba'athists were overthrown that year and Saddam was imprisoned. He escaped and became leader of the party. In 1968, he played a key role in a coup that brought the Ba'ath Party to power. He served as a powerful vice president under Ahmed Hassan al-Bakr and created a repressive security force.

From 1972 to 1975, Saddam Hussein completed the nationalization of Iraq's oil industry, a process that had begun in 1961. He cemented his control over the government and became president in 1979. In 1980, he invaded **Iran**, beginning the bloody, costly, and unsuccessful **Iran–Iraq War** that would last eight years. There were a number of causes leading to the Iran–Iraq War, including Iraq's desire to broaden its access to the **Persian Gulf** for its oil exports. Another goal may have been gaining control of Iran's considerable petroleum **reserves** in the province of **Khuzestan**, which had a large ethnic Arab population.

In August 1990, with his country weakened and in a precarious financial condition after the war with Iran, Saddam invaded **Kuwait**. A desire to control Kuwait's considerable oil reserves was certainly one motivation for the invasion. Saddam Hussein also accused Kuwait of pumping Iraqi oil from the Rumaila field. Iraqi forces were driven out in the 1991 **Persian Gulf War** by a U.S.-led coalition. Saddam survived the defeat, largely because the **United States** and other Western countries did not want to risk deposing him and creating a power vacuum that might be exploited by Iran. The next decade was marked by continual tension with the West. Saddam Hussein's regime was subjected to severe sanctions that included a ban on oil exports. The sanctions were gradually relaxed for humanitarian reasons (*see* OIL FOR FOOD PROGRAM), and Iraq cheated continuously, often with tacit Western cognizance.

On 17 March 2003, U.S. forces invaded Iraq, initiating the **Iraq War**, and deposed Saddam Hussein. The major reason given by the **George W. Bush** administration was that his regime possessed weapons of mass destruction, although this was disputed from the beginning and turned out not to be true. Saddam went into hiding on 9 April 2003, when U.S. forces took Baghdad, but he was captured on 13 December. He was tried by the new Iraqi government beginning in October 2005, charged with the 1982 murder of 148 citizens in the Shiite town of Dujail. Convicted, he was executed in Baghdad on 30 December 2006.

I

IBN SAUD (1876?–1953). Founder of **Saudi Arabia** and king from 1932 to 1953. Ibn Saud (Abdul Aziz ibn Abdul Rahman ibn Faisal ibn Turki al-Saud) was born in Riyadh sometime between 1876 and 1880. He went with his family into exile after his father lost a conflict with the rival Rashid clan in 1891. The family wandered for two years and settled in **Kuwait**. The Sauds had ruled a sizable part of Arabia in the late 18th and much of the 19th century but had lost ground to the Rashids. In 1902, the 22-year-old Ibn Saud returned to Riyadh with 40 men and took the city from the Rashids.

Over the next 25 years, Ibn Saud brought all of the Arabian Peninsula under his rule except for **Yemen**, **Oman**, and the sheikhdoms that would eventually form the **United Arab Emirates**. On 2 December 1922, he agreed to the Protocol of Uqayr establishing British-drawn boundaries with **Iraq** and Kuwait (*see* NEUTRAL ZONE). The protocol contained safeguards protecting the rights of nomadic tribes in the vicinity of the border to cross as necessary. However, the border was still an alien concept to the Arab tribes.

Ibn Saud ran his country on fundamentalist Islamic principles (Wahhabism). He cultivated and used the Ikhwan, members of a militant Islamic movement drawn from Bedouin tribes. He was a skillful diplomat as well as a warrior. He made shrewd use of mercy to his defeated enemies and practiced strategic restraint when the situation called for it. In 1934, he defeated Yemen in a border war and perhaps could have conquered it, but he refrained and offered generous terms in the Treaty of Taif. After conquering the Hejaz region that contained the holy cities of Mecca and Medina from the Hashemite kingdom in 1925, he restrained the Ikhwan from further raids and concentrated instead on showing Muslims throughout the world that he could provide better stewardship than the previous rulers. In 1926, he banished the Ikhwan from the Hejaz after they attacked an **Egyptian** pilgrimage procession.

Ibn Saud's restraint of the Ikhwan and his failure to reward them with political appointments led to an Ikhwan revolt in 1927. They invaded Iraq without his authorization. They were repelled by the British, but their fight against Ibn Saud continued. He defeated them in 1929 at the Battle of Sibila.

This put a stop to the immediate threat from the Ikhwan, but their ideology was to have a continuing influence on Saudi society. Under the surface, Islamic extremists would resent Ibn Saud and his successors, considering the ruling Saud family corrupt and religiously impure. This mentality led to the attack on the Grand Mosque in Mecca in 1979 and continues today in the form of Islamic terrorism.

In 1932, Ibn Saud linked his kingdom explicitly to his family and named it Saudi Arabia. He designated his eldest son, Saud, as his successor. In 1923, Ibn Saud had granted an oil concession to **Frank Holmes**, who represented the Eastern and General, a British syndicate, but nothing came of it, and Ibn Saud canceled the agreement in 1928. On 29 May 1933, he granted Saudi Arabia's first successful oil **concession** to the Standard Oil Company of California (Socal, later **Chevron**). It was significant that he chose a company based in the **United States**. He had previously allied himself with the **United Kingdom** and in 1915 had signed a treaty that recognized his independence and guaranteed British protection. His rivals the Rashids were backed by the Turks at the time. However, he now wanted to cultivate a friendship with a powerful country that was not meddling in the Middle East (at least not yet). In this, he was influenced by his trusted adviser **Harry St. John Philby**. Although British himself, Philby strongly disapproved of Britain's domineering imperial role in the Middle East. Philby also played a key role in negotiating the concession. Another important reason Socal obtained the concession is that it was not party to the **Red Line Agreement** and could hence operate independently in Saudi Arabia. A further reason is that many other **major oil companies** were skeptical about Saudi Arabia's prospects.

Socal formed the Casoc (California Arabian Standard Oil Company) subsidiary, which made the first significant oil discovery in Saudi Arabia with its Dammam No. 7 well in the east near the **Persian Gulf** on 3 March 1938 under the leadership of chief geologist **Max Steineke**. Ibn Saud was present when the first Saudi crude was shipped from **Ras Tanura** on 1 May 1939. Casoc eventually gained additional partners and became **Aramco**. It went on to discover the world's largest oil deposits. Still, Ibn Saud did not begin to see significant oil revenue until the early 1950s. The lack of revenue severely constrained his ability to govern. But when the revenue finally arrived, it was too much too fast, and Ibn Saud was distressed by the effect it had on the traditional Saudi way of life.

Ibn Saud continued to cultivate U.S. friendship. He met with President Franklin D. Roosevelt on the USS *Quincy* at Great Bitter Lake in Egypt on 14 February 1945. The two had a wide-ranging discussion and developed a rapport. There is no record of whether oil was explicitly discussed. One topic that was discussed extensively was the potential future state of Israel. Ibn Saud insisted that a homeland be created for the Jews in Europe, not in Palestine. Roosevelt pledged that the United States would make no decision on Palestine without consulting the Arabs. President Harry S. Truman's later support for the creation of Israel in Palestine angered the Saudis, who thought the United States had reneged on Roosevelt's assurances to Ibn Saud. Ibn Saud's son Faisal urged him to sever relations with the United States, but taking a long-term view, he chose instead to get even closer to the United States in return for arms sales, training, and security cooperation.

Ibn Saud lived just long enough to see the beginnings of the massive oil revenues that would flow into Saudi Arabia for the next half century. When he learned of **Venezuela**'s **50–50 agreement** for sharing profits from crude oil sales, he demanded a similar formula for his country from the Aramco partners. Saudi Arabia got a 50–50 plan in late 1950.

Ibn Saud created Saudi Arabia and set it on a path of close linkage with the United States both militarily and in the petroleum industry. He died on 9 November 1953 at Taif and was buried in Riyadh.

ICE FUTURES. *See* INTERNATIONAL PETROLEUM EXCHANGE (IPE).

ICKES, HAROLD (1874–1952). U.S. secretary of the interior (1933–1946) and head of the **Petroleum Administration for War** (1942–1946). Harold LeClaire Ickes was born in Franklin Township, **Pennsylvania**, on 15 March 1874. He was educated at the University of Chicago and was admitted to the Illinois bar in 1907. A social activist with a reputation for honesty, he was a strong supporter of civil rights and civil liberties. He opposed the internment of Japanese Americans during **World War II**. President Franklin D. Roosevelt appointed him secretary of the interior in 1933, and he was also head of the Public Works Administration.

In 1941–1942, Ickes headed the Office of the Petroleum Coordinator for National Defense, and he was head of the Petroleum Administration for War from 1942 to 1946. He worked closely and cooperatively with the U.S. petroleum industry to meet the country's wartime needs for oil. Ickes recognized the central importance of oil to national security and the civilian economy. He tended to be very pessimistic about the future of U.S. domestic production and thought it imperative to obtain interests in **reserves** overseas. He championed the creation of the **Petroleum Reserves Corporation** and initially wanted the government to purchase interests in the new oil reserves of **Saudi Arabia** through that entity. He eventually gave up the idea of purchasing oil interests outright.

Ickes resigned from the cabinet in 1946 after President Harry S. Truman appointed **California** oil executive Edwin Pauley to the position of undersecretary of the navy. Until his death in 1952, Ickes wrote a regular column for the *New Republic* where, among other things, he forcefully spoke out against Senator Joseph McCarthy. He published five books in his lifetime, including a three-volume *Secret Diary* that appeared after his death.

IMPERIAL OIL. Canada's largest producer of crude oil, one of its leading producers of **natural gas**, and its largest refiner. The company is 69.6 percent owned by **ExxonMobil**. Imperial Oil was founded in 1880 by Frederick Fitzgerald of London, Ontario, with a capitalization of $500,000. The company opened refineries in Petrolia and Sarnia, Ontario. In 1882, Imperial Oil acquired **Herman Frasch**'s process for reducing the sulfur content of sour crude. In 1898, **Standard Oil** acquired a majority interest in Imperial Oil. When Standard Oil was dissolved in 1911, the interest in Imperial Oil remained with Standard Oil Company of New Jersey (later **Exxon**).

Imperial Oil played a leading role in the development of **Alberta**'s, and hence Canada's, petroleum industry. It developed Alberta's first commercial oil field in Turner Valley and made the giant Leduc and Redwater discoveries in 1946 and 1948. Leduc and Redwater established Canada as a major producer. Imperial formed and owned 49.9 percent of the Interprovincial Pipeline, which ran initially from Edmonton to Superior, Wisconsin, and enabled the export of Alberta oil. The **pipeline** was extended to Toronto by 1956 and Montreal by 1976. In 1973, Imperial joined the **Syncrude** consortium developing the **Alberta oil sands**. In February 1989, it bought **Texaco** Canada for $5 billion (Canadian). In 2016, Imperial Oil controlled 0.035 Bb of oil **reserves** and 0.495 Tcf of gas. It produced 0.011 Mmbpd and 0.123 Bcfpd of oil and gas respectively.

INDEPENDENT COMPANIES. The "independent" oil companies tend to be smaller than the **major oil companies** and tend to lack a global system of production, transportation, **refining**, and marketing. The term depends somewhat on the context. In the late 19th century, an independent oil company in the **United States** might have been any company that was not part of the **Standard Oil** empire. In the early **Russian** industry, it might have been any company other than **Branobel** or **Rothschild**. In the postwar era following **World War II**, an independent company might have been any company other than the **Seven Sisters** or the **Compagnie Française des Pétroles** (later **Total**).

The independent category is not generally thought to include national oil companies, although **ENI** has sometimes been classed as an independent because it pursued its own course in the Middle East in the 1950s, undercutting the majors there. Some independents can be very large companies indeed. For example, **Anadarko Petroleum** is generally regarded as an independent but in 2017 ranked fourth in total assets in the ***Oil and Gas Journal*** list of 150 largest companies (based on 2016 figures). **ConocoPhillips** is sometimes not considered a major, but it was third in total assets in the 2017 **OGJ150** list and could reasonably be argued to have a global reach. Other large and notable independents include **Occidental Petroleum** and **Marathon Oil**, fifth and sixth in total assets in the 2017 OGJ150.

INDIA. Oil production 2015: 0.76 Mmbpd (0.97 percent of world); **reserves** 4.621 Bb (0.28 percent of world); gas production: 1.022 Tcf (0.82 percent of world); reserves 43.34 Tcf (0.63 percent of world). India is more important to the petroleum market as a consumer than a producer. Increasing demand from India and **China** is one of the factors contributing to the **Third Oil Shock** in the 2000 decade. In 2015 India consumed almost five times as much oil as it produced. It was the world's third-largest consumer, after the **United States** and China.

Indian oil production has been systematically recorded since 1889. It remained at a low level for the first half of the 20th century, reaching only 9,000 bpd by 1961. It then increased rapidly, with particularly rapid growth after 1980. It peaked at 0.7 Mmbpd in 1989, after which it sank to a low of 0.534 Mmbpd in 1993. It peaked again in 1995 at 0.703 Mmbpd and stayed mostly in a band between 0.64 and 0.69 Mmbpd thereafter. The country's gas production, 0.051 Tcf in 1980, rose steadily to 1.067 Tcf in 2006. Of India's eight **giant** fields, five are predominantly oil, and three are predominantly gas. All except one were discovered before 1980.

Deliberate oil prospecting is known to have begun in India in the 1860s. An unsuccessful well was dug at Nahr Pong in Assam in 1866.

The first oil discovery was at Makum, Assam, a year later. Assam's Digboi field, India's first true commercial oil field, was discovered in 1889. The Assam Railway and the Oil Syndicate dug four successful wells between 1890 and 1893. By 1899, there were 15 wells, and the two entities merged to form the Assam Oil Company, which built a refinery at Digboi. By 1920, the field produced about 333 bpd.

The Assam Oil Company was acquired in 1920 by the **Burmah Oil Company**. In 1927, Burmah partnered with **Royal Dutch Shell** to form Burmah-Shell, a joint operating company for India. Burmah-Shell engaged in price wars against small Indian independents in the 1930s, eliciting protests in the national press. By the time India became independent, Burmah-Shell competed mainly with Stanvac (a joint venture between Standard Oil Company of New Jersey, later **Exxon**, and Socony–Vacuum, later **Mobil**) and **Caltex**.

In 1948, the government of newly independent India passed the Industrial Policy Resolution, which affirmed that the country's petroleum industry should be state owned and run. Nevertheless, the private companies continued to operate. Burmah-Shell and Stanvac both built refineries in the 1950s.

In 1953, Burmah-Shell discovered India's first giant oil field, Nahorkatiya (**EUR** about 0.5 Bb), in Assam. After a series of disputes, Burmah-Shell agreed to the Indian government's joining in the field's development.

In 1955, K. D. Malaviya, the minister for natural resources, led a delegation to Moscow to seek help from the Soviet Union in petroleum exploration. By January 1956, Russian technicians were reporting favorably on the prospects of the Cambay region of Gujarat. In May 1956, the government also established the Oil and Natural Gas Commission (ONGC) to execute upstream activities in the country's petroleum sector. The ONGC began prospecting with Soviet and **Romanian** assistance. It drilled its first well in the Cambay area in 1958, with success. In 1960, it discovered the giant Cambay Ankleshwar oil field (EUR 0.52 Bb). It also discovered oil in several **offshore** structures.

In 1959, the government established the Indian Oil Company as a vehicle for supplying refined products to Indian state enterprises. Its role expanded, and in 1964 it was merged with India's state refineries to form the Indian Oil Corporation (IOC). IOC would become the country's main downstream enterprise, complementing ONGC's upstream activities. ONGC, with Soviet and other foreign help, made several important offshore finds off the west coast of the country. These included India's largest oil field, the giant Bombay High Field (EUR 1.3 Bb and 0.8 Tcf of gas), discovered in 1974. Bassein North (EUR 0.55 Bb) and Bassein (7 Tcf and 0.2 Bb of condensate) were discovered in 1976 and 1977 respectively. The government nationalized the Burmah-Shell and Caltex refineries in 1976.

The 1980s saw no giant discoveries, and in 1991 the government began taking steps to deregulate the sector. ONGC became the Oil and Natural Gas Company Ltd. in 1993, although still wholly owned by the state. In 1999, the government forced IOC, ONGC, and GAIL (Gas Authority of India) to buy some of each other's shares from state holdings. Although this was ostensibly a step toward privatization and liberalization, critics pointed out that its main effect was to place $1.23 billion into the government treasury and deprive the companies of needed investment capital. In 2000, the Ministry of Petroleum and Natural Gas created the New Exploration License Policy (NELP). For the first time, foreign companies would be permitted to own 100 percent interest in oil and gas projects. Nevertheless, oil production in 2015 was only about 18 percent higher than in 2000.

INDIAN TERRITORY ILLUMINATING OIL COMPANY (ITIO). *See* CITIES SERVICE COMPANY.

INDONESIA. Oil production 2015: 0.69 Mmbpd (0.88 percent of world); **reserves** 3.23 Bb (0.20 percent of world); gas production: 2.665 Tcf (2.14 percent of world); reserves 98 Tcf (1.42 percent of world). Indonesia joined the **Organization of the Petroleum Exporting Countries** (OPEC) in 1962 and was a net exporter of oil for most of the next four decades. It became a net importer in 2004, and in 2009 it left OPEC. It rejoined in 2016 but left again later the same year because it would not comply with a 5 percent production cut demanded by the cartel. Indonesia remains an important net exporter of **natural gas**. In 2015 it was the world's fifth-largest exporter of **liquefied natural gas** (LNG).

The natives of Sumatra, in what is now Indonesia, knew of oil seeps and used the oil to make torches. **Aeilko Zijlker**, a Dutch plantation manager (Indonesia was then the Dutch East Indies), saw one of the torches and became interested in prospecting for oil. He obtained a **concession** for the area of Langkat and in 1885 struck the Telaga Tunggal No. 1 gusher. In 1890, he founded the **Royal Dutch Company**, one of the two companies that amalgamated in 1907 to form **Royal Dutch Shell**. The other company, **Shell Transport and Trading**, had operations in Borneo, in what is today **Malaysia**.

Other companies operating in Indonesia included **Caltex**, Stanvac (a joint venture between Standard Oil Company of New Jersey, later **Exxon**, and Socony–Vacuum, later **Mobil**), and **Gulf**. Standard Oil Company of California (Socal, later **Chevron**) began operating in Irian Jaya and Kalimantan in 1922 and in Northern Sumatra in 1929. Gulf also began working in Northern Sumatra the same year. In the 1930s, Socal's activities were placed under the Caltex name. Caltex was Socal's "east of Suez" collaboration with the Texas Corporation (later **Texaco**). In 1941, Caltex discovered the **giant** Duri field, Indonesia's largest (**EUR** about 1.7 Bb), in Sumatra; however, Duri would not enter production until 1958. Standard Oil Company of New Jersey operated in Java and Madura. After 1933, Jersey Standard's interests were handled by Stanvac. Stanvac supplied Jersey Standard's Indonesian oil to Socony–Vacuum's Asian and African markets.

Indonesia's oil production built up slowly from 1,600 bpd in 1893 to 0.17 Mmbpd in 1940. Indonesia was East Asia's largest producer between the world wars, and its petroleum resources were a significant Japanese strategic objective before and during **World War II**. One reason Japan attacked the **United States** at Pearl Harbor in 1941 was to protect its eastern flank while conquering Indonesia. The Dutch wanted to prevent the Japanese from gaining control of the country's petroleum reserves and destroyed many oil facilities, such as a Stanvac refinery at Palembang. They were unable to complete the process, and the Japanese did obtain control of some installations. The Japanese were able to complete the discovery of the Minas field, Indonesia's second largest, using a rig left behind by Caltex Pacific Indonesia, which had been prospecting in the area and intending to drill. Later, crude from the Minas field would form part of the **OPEC Basket** (until 2009).

In August 1945, Indonesia declared its independence from the Netherlands. Article 33 of the new Indonesian constitution established the first national oil company, Perusahaan Tambang Minyak Negara Republik Indonesia (PTMN-RI). The Indonesian government postponed granting new concessions to foreign companies. On 10 December 1957, the government established Exploitasi Tambang Minyak Sumatra Utara (PT ETMSU), which soon after changed its name to PT Permina. Indonesia made its first crude oil export on 24 March 1958. By 1965, Permina had discovered nine oil fields with a collective production of 21,000 bpd. Total Indonesian production that year was 0.48 Mmbpd.

In 1960, Law 44 prescribed that exploration and mining would only be done by state-owned companies. Foreign companies could no longer hold concessions. The three major foreign operators, Shell, Caltex, and Stanvac, would henceforth become contractors to state-owned companies. Stanvac would work with Permina, Shell would work with Permigan, and Caltex would work with Pertamin. The foreign companies would retain management of installations, but Indonesia would henceforth receive 60 percent of all profits. The government began purchasing Shell's assets in 1966.

In 1968, Indonesia formed a new national company, PT Pertamina, by merging Permina and Pertamin. In 1970, the **Union Oil Company of California** discovered the Attaka field, Indonesia's third largest, **offshore** Kalimantan (Indonesian Borneo). In 1971, Pertamina became the only company authorized to work everywhere in the country. In 1974, Pertamina renegotiated its profit share with foreign companies to take better advantage of rising oil prices in the wake of the **First Oil Shock**. It would take a base share of 60 percent as before, but if prices rose above a stipulated level, it would take 85 to 95 percent of the incremental profits. Indonesian oil production reached a peak of 1.696 Mmbpd in 1977. It varied mostly within a band between 1.3 and 1.6 Mmbpd for the next 20 years. After 1998, it declined rapidly. The 2006 level of 1.019 Mmbpd was under that of 1972.

In the 1990s, Pertamina was riddled with scandal. Corruption and inefficiency cost the company an estimated $6.1 billion in lost revenue during 1997 and 1998. The corruption took various forms, including embezzlement, illegal commissions, and unauthorized markups on procurement contracts. The company had also awarded 159 contracts to relatives and cronies of president Suharto. The government responded with reforms, including Oil and Gas Law No. 22 in 2001. The law moved Pertamina's regulatory functions and its management of **production-sharing agreements** to a new government agency, BP Migas, and ended Pertamina's downstream monopoly.

As a result, Shell opened Indonesia's first private **gasoline** station near Jakarta in 2005. The government reorganized Pertamina as a limited liability company in 2003. Pertamina was still state owned, but a possible candidate for future privatization. In 2006, the government introduced incentives to encourage upstream investment.

In 1971, Mobil discovered the Arun gas field (EUR 13.7 Tcf), ushering in the country's LNG industry. Two years later, **AGIP** (later **ENI**) discovered the giant Natuna offshore gas field. The field, later renamed East Natuna, did not go into development until much later because of technical difficulties. In 2017, partners including **ExxonMobil** were still trying to work out arrangements with the government. The Natuna field also acquired more significance in the 2010s because of the heating up of the **South China Sea** dispute. It lies within China's expansive claimed area of the sea.

Indonesia's gas production has been on a long-term growth trend for the last quarter century. In 1980, gas production was 0.63 Tcf. In 2015, it was 2.665 Tcf, nearly double the country's oil production in energy-equivalent terms.

INTERCONTINENTAL EXCHANGE FUTURES. *See* INTERNATIONAL PETROLEUM EXCHANGE (IPE).

INTERNATIONAL ENERGY AGENCY (IEA). Autonomous agency linked with the Organisation for Economic Co-operation and Development (OECD), with a core mission to maintain and implement systems for coping with oil supply disruptions. The agency also analyzes and disseminates information on oil markets, analyzes and makes recommendations for national energy policies, and fosters cooperative energy research and development. The International Energy Agency was conceived in the wake of the **First Oil Shock** in 1973. Oil prices nearly quadrupled in one year, and the **Organization of Arab Petroleum Exporting Countries** (OAPEC) embargoed the **United States** and the Netherlands because of their support for Israel during the Yom Kippur War. U.S. secretary of state Henry Kissinger proposed the creation of an energy agency within the OECD to mitigate the effects of future supply disruptions on Western nations. People referred to the idea as an "energy NATO." On 18 November 1974, 16 OECD countries concluded the Agreement on an International Energy Programme (IEP), the treaty that established the IEA. Membership in the OECD is a necessary but not sufficient condition for membership in the IEA.

The IEA is an autonomous agency within the OECD. It has its own executive director and governing board and its own budget. However, the budget is formally integrated into the OECD budget. IEA uses some OECD administrative services. The IEA's decision-making process differs from the

OECD's. The OECD makes decisions by consensus, but the IEA does so by majority voting. Votes are weighted by oil consumption. The IEA is thus capable of faster action than its parent organization.

The IEA's emergency response measures require member states to maintain oil stocks equivalent to at least 90 days' worth of net oil imports (**Canada** and **Norway**, which are net oil exporters, are not required to maintain stocks). In June 2007, total oil stocks in IEA countries were 4.1 Bb, or about 150 days' worth of net imports. If supply shortages occur for the whole IEA group, an oil-sharing plan may come into force. Surplus countries are required to provide oil for countries that need it. IEA member countries may be required to direct their oil industries to comply. Member countries are also required to implement measures to reduce demand.

The IEA may also decide to make oil available to the international market during supply disruptions or times when such disruptions are threatened. It did so in two notable cases. The first was during the 1991 **Persian Gulf War**. The IEA decided on 11 January 1991 to make 2.5 Mmbpd available to the international market. The IEA reversed the move on 6 March 1991, less than a week after cessation of hostilities. The second case was in 2005, in the aftermath of Hurricane Katrina, which devastated the U.S. city of New Orleans and damaged important oil facilities in the region. On 2 September 2005, IEA members unanimously agreed to make 60 Mmbo available to the market: 29 Mmb of oil and refined products were drawn from public stocks; the rest came from demand restraint and lowering industry stockholding obligations.

The IEA can have an important effect in a supply crisis simply by showing its willingness to provide oil to the market, even if the oil it makes available is not used. The presence of the IEA and its emergency response mechanisms can help prevent a panic. It is important to note that the IEA does not act, as a rule, in order to lower or stabilize prices. The IEA acts to avoid supply disruptions. Insofar as high or unstable prices may be correlated with supply disruptions, IEA actions may have an effect on prices, but high prices are not a trigger for IEA action. It should also be pointed out that the IEA is an agency of rich oil-consuming countries and is designed primarily to protect their interests. It performs some global functions, but it is not a true world energy agency.

Original 16 members of IEA: Austria, Belgium, Canada, Denmark, Germany, Ireland, Italy, Japan, Luxembourg, Netherlands, Spain, Sweden, Switzerland, Turkey, **United Kingdom**, United States. *Later additional members:* **Australia**, Czech Republic, Finland, France, Greece, Hungary, Korea, New Zealand, Poland, Portugal, Slovakia. Norway participates under a special agreement. The European Union also participates pursuant to Supplementary Protocol No. 1 of the OECD convention and Article 72 of the IEP. Refer to appendix 7. *See also* STRATEGIC PETROLEUM RESERVES.

INTERNATIONAL ENERGY PROGRAMME (IEP). *See* INTERNATIONAL ENERGY AGENCY (IEA).

INTERNATIONAL PETROLEUM EXCHANGE (IPE). Energy futures and options exchange based in London; later known as ICE Futures. The exchange was founded in 1980, and in 1981 it launched its first contract, for gas oil futures. In June 1988, the exchange began dealing in its most important contract, for Brent crude futures. **Brent Blend** from the **North Sea** is one of the three major **benchmark** crudes, the other two being **West Texas Intermediate** and **Dubai crude**. In 2001, the IPE was acquired by the Intercontinental Exchange (ICE) of Atlanta, Georgia. The IPE initially operated via a traditional open outcry mechanism. On 7 April 2005, it moved to an electronic trading system and changed its name to ICE Futures. On 22 April 2005, the exchange began trading carbon emission allowances with the European Climate Exchange (ECX).

INTERSTATE OIL COMPACT. An association of states of the **United States** formed in 1935 to make recommendations on oil conservation and stabilization of the oil market. The association was formed in the wake of an oil price collapse caused by overproduction in the **East Texas oil field** as well as the Seminole and Oklahoma City fields in **Oklahoma**. State commissions such as the Oklahoma Corporation Commission and the **Railroad Commission of Texas** realized that market regulation could be more effective if the various producing states cooperated with each other. Initially, representatives from **Texas**, Kansas, Oklahoma, and New Mexico began meeting with each other to discuss issues of mutual concern. The compact was officially formed in 1935 by authorization of the Oklahoma legislature. Membership eventually grew to 30 states, with an additional six associate members. The compact is headquartered in Oklahoma City.

IRAANSE AARDOLIE RAFFINAGE MAATSCHAPPJI. *See* IRANIAN OIL PARTICIPANTS LTD.

IRAANSE EXPLORATIE EN PRODUCTI MAATSCHAPPJI. *See* IRANIAN OIL PARTICIPANTS LTD.

IRAN. Oil production 2015: 2.856 Mmbpd (3.64 percent of world); **reserves** 158.4 Bb (9.62 percent of world); gas production: 6.68 Tcf (5.36 percent of world); reserves 1,183 Tcf (17.16 percent of world). Iran is a major oil-producing and exporting country, where petroleum has been known since ancient times. It is the oldest of the **Persian Gulf** producers, with significant commercial output dating back to 1913. Most of its oil reserves are in

Khuzestan and Fars Provinces, with significant **offshore** reserves in the Persian Gulf, where it also has huge gas reserves. Iran was a founding member of the **Organization of the Petroleum Exporting Countries** (OPEC) in 1960. Iran is a member of the **Gas Exporting Countries Forum**.

The modern petroleum industry in Iran began with the 1901 **concession** granted to **William Knox D'Arcy** by Shah Muzzafar al-Din. After several difficult years and an infusion of investment from the **Burmah Oil Company**, D'Arcy's prospectors, led by **George Reynolds**, discovered a **giant** oil field at Masjid-e-Suleiman in 1908. In 1909, Burmah formed the Anglo–Persian Oil Company (APOC), which later became the **Anglo–Iranian Oil Company** (AIOC) and eventually **BP**. APOC/AIOC would dominate Iran's industry completely until the 1950s. In 1912, the company completed a refinery at **Abadan**, which would become the world's largest. In 1914, the government of the **United Kingdom** acquired 51 percent of APOC, which would play an important role in supplying fuel during **World War I**—*see* CHURCHILL, WINSTON (1874–1965). APOC made a number of important discoveries in the 1920s and 1930s, including Haft Kel (1926, **EUR** 2 Bb), and the **supergiants** Gachsaran (1928, 11.8 Bb) and Agha Jari (1936, 5.76 Bb). Iran's oil production, which stood at just over 0.005 Mmbpd in 1913, was 0.214 Mmbpd in 1939.

In 1933, Shah **Reza Pahlavi** terminated the 1901 concession. There was an outcry from the British, who appealed to the League of Nations. Iranian representatives countered that APOC had falsified accounts to cheat Iran out of rightful payments. The shah negotiated a new agreement directly with **John Cadman**, head of APOC, with a reduced area and different financial terms. APOC became the Anglo–Iranian Oil Company. In 1941, the British and Soviets occupied Iran. They deposed Reza Pahlavi on 16 September, on the grounds that he was leaning toward the Nazis, and replaced him with his son, **Mohammed Reza Pahlavi**. Iranian production dropped during **World War II**; in 1941 it averaged 0.139 Mmbpd, 35 percent lower than in 1939. It recovered in the 1940s and reached 0.664 Mmbpd by 1950.

Iranians grew increasingly discontented with British domination of their petroleum industry. AIOC paid more in taxes to the British government than it did to Iran. As majority shareholder, the British government also collected substantial dividends. Iranian workers endured appalling conditions and had few opportunities for training and advancement. In 1947, there was a large strike of refinery workers in Abadan. The British countered by organizing mobs of paid provocateurs, giving AIOC a pretext to respond with force. AIOC arrogantly and stubbornly resisted reforms. Calls grew for renegotiation of the concession and then for outright nationalization of the industry. In May 1951, the government under Prime Minister **Mohammed Mossadegh** nationalized the petroleum industry. The new National Iranian Oil Company (NIOC) would own the assets and manage the sector.

The United Kingdom began an embargo of Iranian oil, with AIOC threatening legal action against anyone who tried to buy or transport Iranian crude, on the grounds that it belonged to AIOC. Iranian oil production fell to 0.021 Mmbpd, and the country neared economic collapse. Production would not surpass 1950 levels until 1957. Mossadegh nevertheless remained popular. The United Kingdom considered military action (*see* PLAN BUCCANEER) and urged Mossadegh's removal. The Harry S. Truman administration in the **United States** was reluctant to intervene, but the Dwight D. Eisenhower administration took a different stance and eventually engineered a coup that deposed Mossadegh in August 1953 (*see* AJAX, OPERATION (1953)). Shah Mohammed Reza Pahlavi, who had left the country, returned.

Iran's oil remained nationalized, but a new consortium of Western companies formed in 1954 would dominate the Iranian industry until the 1970s. The consortium, **Iranian Oil Participants** (IOP), consisted of the **Seven Sisters**—including AIOC, now renamed British Petroleum (BP)—and **Compagnie Française des Pétroles** (CFP, later **Total**). IOP ran the industry on behalf of NIOC, but in practice NIOC had a limited say. During the consortium years, Iran's oil production grew rapidly, as did government revenues. In 1958, Iran's largest oil field, Ahvaz (EUR 13.8 Bb and 23.3 Tcf gas), was discovered. Production surpassed 1 Mmbpd in 1960, 2 Mmbpd in 1966, 3 Mmbpd in 1969, and 4 Mmbpd in 1971. In 1974, it reached a peak of just over 6 Mmbpd.

Although this growth was impressive, the consortium was in fact restraining Iranian production somewhat to avoid flooding the world market. The consortium members had interlocking interests in other major consortia in the Middle East (**Aramco**, Kuwait Oil Company, **Iraq Petroleum Company**, and others; refer to appendix 2) and collaborated in secret to balance production across the region. Their methods became public in 1974 during investigations by the U.S. Senate Foreign Relations Subcommittee on Multinational Corporations. In July 1972, Iran began renegotiating its agreement with the consortium, and a year later a new agreement was signed, placing production policy in the hands of the government. Iran was also assertive on the international stage, advocating high prices within OPEC. On 28 February 1979, in the midst of the Iranian Revolution that established the Islamic Republic of Iran, NIOC unilaterally canceled the consortium agreement.

Although the consortium was dominant until the revolution, there were other Western players in Iran. A law passed in 1957 allowed NIOC to enter into joint ventures to explore and develop areas not leased to IOP. The first such venture was with **AGIP (ENI)**, then led by **Enrico Mattei**. Mattei enraged the **major oil companies** by offering Iran good terms and true joint management. The venture, Société Irano–Italienne des Pétroles (SIRIP), was formed on 27 August 1957. SIRIP explored both onshore and in the Persian

Gulf. It was a successful enterprise, although its production was relatively small. In the last five years before the Iranian Revolution, it produced 0.92 percent of Iran's oil.

Several joint ventures were created, but only four (including SIRIP) were successful. The largest producer was the venture created between NIOC and the Standard Oil Company of Indiana (later **Amoco**) in June 1958, named Iran Pan American Oil Company (IPAC). It was involved in developing the supergiant offshore Fereidoon field (EUR 10 Bb), as well as other fields—*see* KHARG ISLAND (KHARK). In the years leading up to the revolution, it produced almost 4 percent of Iran's oil. Another successful producer was Lavan Petroleum Company (LAPCO), created 13 February 1965. LAPCO was an offshore venture involving **Arco**, Murphy Oil, **Sun Oil Company**, and the **Union Oil Company of California**. From 1974 to 1978, it produced an average of 2.65 percent of Iran's oil. The Iranian Marine International Oil Company (IMINOCO), also formed 13 February 1965, involved AGIP (ENI), **Phillips Petroleum**, and the Indian Oil and Natural Gas Commission (*see* INDIA). IMINOCO produced 1.34 percent of Iran's oil in 1974–1978.

In 1966, Iran introduced a different type of arrangement, the service contract. The first, Société Française de Pétroles d'Iran (SOFIRAN), was with the French company Entreprise de Recherches et d'Activités Pétrolières (ERAP), which was later **Elf Aquitaine** and eventually Total. ERAP agreed to bear all exploration costs, to be refunded if oil was discovered. NIOC would be the sole owner of all produced oil. SOFIRAN would then act as a broker for NIOC for a 2 percent commission.

Together, the four successful joint ventures, IPAC, LAPCO, SIRIP, and IMINOCO, produced about 9 percent of Iran's oil in the late 1970s. After the Iranian Revolution, the Islamic Republic of Iran acquired the assets of all four. In August 1980, the government announced that the assets would be taken over by the new Iranian Oil Offshore Company (IOOC).

During the revolution, Iranian production plummeted. Average production between 1972 and 1978 was about 5.6 Mmbpd. In 1979, it was 3.168 Mmbpd. In 1980, it was 1.662 Mmbpd. The drastic cut in Iranian production was a factor in the **Second Oil Shock**. The outbreak of the **Iran–Iraq War** in 1980 further disrupted industry operations. The 1981 production was down to 1.38 Mmbpd. By 1988, production had recovered to 2.240 Mmbpd, still far lower than in the 1970s. Part of the reason for this was deliberate government policy. The government believed that Iran under the shah had overproduced and had spent its oil revenue wastefully. Nevertheless, production crept up. By 1996, production was 3.664 Mmbpd. By 2004, it had surpassed 4 Mmbpd again.

In 1987, Iran passed a new petroleum law. NIOC was allowed to make agreements with foreign oil companies under a buyback system. The foreign company would make all the investments, then turn over all operating rights to NIOC at the end of the contract. It would then receive a share of production at a set price. A number of companies signed such contracts. Total and Petronas (*see* MALAYSIA) developed the offshore Sirri A field, which came onstream in 1998. In 1999, Elf Aquitaine and ENI signed a contract to enhance oil recovery at the Doroud field. In 2000, **Statoil** undertook to perform exploration in the **Strait of Hormuz** area. In 1999, NIOC discovered Azadegan, one of the world's most important finds in two decades. Initial estimates of reserves ranged from 26 to 70 Bb, but 6–10 Bb is probably more realistic. Iran negotiated with Japanese companies, as well as Total, Statoil, and Lukoil, to develop Azadegan.

Iranian **natural gas** production has grown considerably since 1980. It was 0.25 Tcf that year. It surpassed 0.5 Tcf in 1985, 1.0 Tcf in 1994, and 2.0 Tcf in 1999. In 2015, it was 6.68 Tcf, or 3.24 Mmboepd. In energy-equivalent terms, 2015 gas production was about 13 percent higher than oil production. The **megagiant** offshore South Pars gas field (EUR 350 Tcf gas, 77.4 Bboe total, including oil and condensate) was discovered in 1991. South Pars is an extension of **Qatar**'s megagiant **North Field**. Taken separately, North Field is the largest known hydrocarbon field on earth, and South Pars is the third largest. **Saudi Arabia**'s **Ghawar** is second, although it is the largest oil field. Development of South Pars was complicated by the U.S. government's **Iran–Libya Sanctions Act** of 1996. However, in 1998 the Clinton administration, anxious to avoid a major confrontation with the European Union, waived the sanctions on a Total-led consortium to develop South Pars. Total's partners in the $2 billion project were **Russia**'s **Gazprom** and **Malaysia**'s Petronas.

In 2008 Iran opened the Iranian Oil Bourse, trading initially in petrochemicals but intending to move into crude oil and other petroleum-related commodities, and employing currencies other than the U.S. **dollar**. As of 2017, success was limited.

In the 2010s Iran's petroleum industry was crippled by international economic **sanctions**. Oil production fell about 20 percent between 2011 and 2013 but recovered after many sanctions were lifted in 2015. *See also* SANCTIONS AGAINST IRAN.

IRAN–IRAQ WAR (1980–1988). Major war between **Iran** and **Iraq** lasting from 1980 to 1988, during which production and exports of petroleum were disrupted. (The war is sometimes called the "first Gulf war," to distinguish it from the 1991 **Persian Gulf War** that followed Iraq's 1990 invasion of **Kuwait**.) The Iran–Iraq War was one of the bloodiest in the 20th century, with as many as 1.3 million estimated casualties.

The war began on 22 September 1980 when Iraq invaded Iran. The Iraqis made initial gains in Iran's oil-rich **Khuzestan** Province. They captured Khorramshahr, Iran's major port, and besieged **Abadan**, whose refinery was

Iran's largest and one of the largest in the world. However, the Iranians repelled the invasion by 1982. The Iranians then invaded Iraq, vowing to replace **Saddam Hussein**'s secular rule with an Islamic republic. The Iranians attempted and failed to take **Basra**. This was followed by years of brutal offensives and counteroffensives by both sides. The result was carnage and stalemate. Hostilities finally ceased on 20 August 1988.

The war had a number of causes. One was Iraq's desire to control Khuzestan, which is not only rich in petroleum resources but also has a substantial ethnic Arab population. Some in Iraq believed that Khuzestan belonged rightfully to their country and had been unjustly handed over to Iran by the Ottomans. In the 1960s and 1970s, Iraq encouraged ethnic Arabs in Khuzestan to revolt against the government of Iran. Each country also encouraged unrest in the other's Kurdish populations.

Another cause of tension was the Shatt al Arab waterway flowing into the **Persian Gulf**. The waterway is important for the oil exports of both countries. In 1975, Iran and Iraq had signed the Treaty of International Boundaries and Good Neighborliness, which took effect on 17 September of that year. The two countries agreed to recognize the land boundaries between them as set forth in the 1913 Protocol of Constantinople. They also agreed to mark boundaries according to the *thalweg*, the median line of the main navigable channel. This was a victory for Iran. The 1847 Treaty of Erzurum had given the Shatt al Arab entirely to the Ottoman Empire, of which Iraq was then a part, although Iran was granted right of free passage. Iraq wanted to keep full control of the Shatt al Arab and was unhappy with the 1975 treaty, claiming it had been signed under the pressure of a Kurdish revolt promoted by Iran. Iraq abrogated the treaty five years after it took effect, a few days before invading Iran. The Shatt al Arab, Khuzestan, and other boundary and nonboundary disputes were doubtless important, but the war also may have been a fundamental struggle for broad geopolitical dominance in the region.

One phase of the Iran–Iraq War of particular relevance to petroleum was the "Tanker War." Beginning in 1981, both countries attacked oil **tankers** in an effort to disrupt each other's trade. This included attacks on ships from third countries. Iran attacked a Kuwaiti tanker on 13 May 1984 and a Saudi tanker three days later. Iran accused noncombatant Arab countries in the Persian Gulf region of collaborating with Iraq and providing it with financial support. From 1984 onward, the attacks on tankers increased rapidly. The Tanker War damaged 546 commercial ships (not all of them oil tankers) and killed more than 400 civilian sailors.

On 1 November 1986, Kuwait asked the international community for protection. The Soviet Union began chartering tankers in response, and the **United States** began a program of reflagging Kuwaiti tankers with U.S. flags and protecting them. This drew the United States into direct conflict with Iran. When Iran attacked the reflagged Kuwaiti tanker *Sea Isle City*, the

United States retaliated in October 1987 by attacking Iranian oil platforms (Operation Nimble Archer). On 14 April 1988, the U.S. naval frigate *Samuel B. Roberts* was damaged by an Iranian mine. There were no casualties, but four days later the United States retaliated (Operation Praying Mantis), which led to a full-blown battle between the United States and Iran in or near the **Strait of Hormuz**. The U.S. Navy sank two Iranian warships and several armed speedboats. It also destroyed two Iranian oil platforms.

IRAN–LIBYA SANCTIONS ACT. A 1996 U.S. law punishing energy sector investments in **Iran** and **Libya**; became the Iran Sanctions Act in 2006. On 6 May 1995, U.S. president Bill Clinton issued Executive Order 12959, banning U.S. trade with and investment in Iran. The order was motivated by U.S. perceptions of a growing Iranian nuclear weapons development program and Iranian support for a number of what the **United States** considered terrorist organizations. The United States wanted to stifle the modernization of Iran's petroleum sector, a critical source of revenue for the Iranian regime.

Allies of the United States refused to participate in the sanctions. The United States responded with Public Law 104-172, signed on 5 August 1996, which sought to punish non-U.S. investment in Iran's energy sector and applied the rules to Libya as well. The law—the Iran–Libya Sanctions Act (ILSA)—required the president to impose at least two out of a list of seven possible sanctions on non-U.S. companies investing more than $20 million in Iran's or Libya's energy sector. The list of possible sanctions included denial of U.S. bank loans, denial of U.S. export licenses for militarily useful technologies, restriction on imports from the entity in violation, and other measures.

The European Union (EU) considered ILSA an extraterritorial application of U.S. law and threatened action in the World Trade Organization (WTO). The United States and EU agreed in 1997 to avoid a showdown over ILSA. Thus in 1998 the Clinton administration waived ILSA sanctions on a **Total**-led consortium to develop Iran's **megagiant** South Pars gas field. Total's partners in the $2 billion project were **Russia**'s **Gazprom** and **Malaysia**'s Petronas.

When ILSA was due to expire on 5 August 2001, the United States enacted renewal legislation on 3 August 2001 (Public Law 107-24). Relations between the United States and Libya improved in 2004, when the Libyan regime renounced development of weapons of mass destruction. In 2006, when ILSA was set to expire again, Libya was dropped from its coverage. A new law (Public Law 109-293), called the Iran Sanctions Act (ISA), was passed, with other modifications.

It does not appear that ILSA was particularly effective at slowing down foreign investment in Iran's energy sector. Non-U.S. companies, including European, Japanese, and Chinese firms, made more than $80 billion in new investments in Iran's oil and gas industry between 2000 and 2006. *See also* SANCTIONS AGAINST IRAN.

IRAN PAN AMERICAN OIL COMPANY. *See* IRAN.

IRAN SANCTIONS ACT. *See* IRAN–LIBYA SANCTIONS ACT.

IRANIAN MARINE INTERNATIONAL OIL COMPANY (IMINOCO). *See* IRAN.

IRANIAN OIL BOURSE. An international petroleum exchange located on the Iranian island of Kish in the **Persian Gulf**. The exchange opened on 17 February 2008. It began trading in petrochemicals with an intention to expand to a large range of petroleum-related commodities, including crude oil. It traded a shipload of crude oil in 2011 but has had trouble becoming a consistently operating oil exchange. The Iranian Oil Bourse is notable for its intent to deviate from the international practice of denominating international oil trades in U.S. dollars, favoring euros, Iranian rials, and combinations of other currencies. Iran began moving away from dollars for its own exports in 2007, often specifying payment in euros. Iran has talked of using the bourse to create a new **benchmark** crude blend denominated in a non-U.S. currency. Iran has suffered greatly from U.S. **sanctions** on its petroleum industry, and an alternative trading paradigm would reduce its exposure to such disruptions—although it has also faced sanctions from the EU. In January and February 2008, five sets of submarine communications cables were disrupted, leading some observers to allege sabotage as part of an attempt to interfere with the opening of the bourse; such allegations have never been proven.

IRANIAN OIL EXPLORATION AND PRODUCTION COMPANY. *See* IRANIAN OIL PARTICIPANTS LTD.

IRANIAN OIL PARTICIPANTS LTD. Consortium of companies formed in October 1954 to operate in **Iran**. The Iranian Oil Participants (IOP) involved all the **Seven Sisters** plus **Compagnie Française des Pétroles** (later **Total**) and a number of American **independent companies**. Iran, led by Prime Minister **Mohammed Mossadegh**, nationalized its oil industry in 1951 and formed the National Iranian Oil Company (NIOC). The British **Anglo–Iranian Oil Company** (AIOC, later **BP**), which had been running

Iran's oil industry, was expelled. Mossadegh was removed from power in 1953 by a coup engineered by the **United States** and the **United Kingdom**. Shah **Mohammed Reza Pahlavi** was reinstalled with increased power—*see* AJAX, OPERATION (1953).

The return of the shah did not mean that AIOC would be able to operate exactly as before. Under pressure from the United States, AIOC was reluctantly forced to accept membership in a consortium of companies that would include American **major oil companies**. All parties acknowledged that NIOC owned the oil and the facilities. The consortium's role was to operate and manage the industry, buy the oil output, and market it through the various companies' outlets. The initial members of the consortium included BP (formerly AIOC) with a 40 percent interest, **Gulf** with an 8 percent interest (later reduced to 7 percent), **Royal Dutch Shell** with a 14 percent interest, and Compagnie Française des Pétroles (later Total) with a 6 percent interest. The four **Aramco** partners—Standard Oil Company of New Jersey (later **Exxon**), Socony–Vacuum (later **Mobil**), the Texas Company (later **Texaco**), and Standard Oil Company of California (later **Chevron**)—each had an 8 percent interest (later reduced to 7 percent).

The consortium was constructed in the form of a holding company, the IOP, with two operating entities, both incorporated in the Netherlands. These were Iraanse Exploratie en Producti Maatschappji (Iranian Oil Exploration and Producing Company) and Iraanse Aardolie Raffinage Maatschappji (Iranian Oil Refining Company, which ran the **Abadan** refinery). All the member companies of the consortium were already involved in the Arab Middle East, and all had experience in joint ventures. The market would need to make room for Iranian oil to come back after the hiatus following nationalization, and only these companies could manage it.

It is worth noting that many of the companies were reluctant to join the consortium. BP was dissatisfied with only 40 percent of something it once owned completely. But the agreement provided for BP to be compensated. It was to receive £25 million from the government of Iran in 10 annual installments. It was also to receive an immediate £32.4 million from the other consortium members, plus a special royalty on oil produced, until a sum of £510 million was reached. The U.S. companies were wary of operating in Iran and of being seen to form a cartel. U.S. government persuasion was a key factor causing them to join. The government assured them that participation in the consortium would not lead to accusations of restraint of trade. The companies also realized that if they did not have a hand in controlling Iranian production, someone else would.

A few months later, the American members of the consortium each gave up one percentage point of their share to a new entity called Iricon. With a total share of 5 percent, Iricon was composed initially of nine American independent oil companies: **Aminoil**, Richfield Oil (later **Arco**), Atlantic

Refining (later Arco), **Sohio**, Pacific Western/**Getty Oil Company**, Tidewater Associated Oil Company (Getty), Signal Oil and Gas Company, Hancock Oil Company, and San Jacinto Petroleum Corporation. Iricon was set up largely for domestic political reasons in the United States.

The IOP members were given exploration, production, and **refining** rights in the agreement area, which comprised the southwestern part of Iran, and permission to operate the Abadan refinery. The rest of Iran was left to NIOC. The companies of the consortium explored for, developed, and produced the oil in the agreement area and then had to pay to buy it. Each company was obliged by the agreement to set up an independent trading company registered in Iran. The trading companies purchased the crude oil and gas at the wellhead from NIOC and then either resold it for export or refined it and sold the products. The companies were paid their operating costs and a fee for the oil they produced.

In practical terms, IOP ran the Iranian oil industry, despite the nationalization, and this led to tension with the government. The consortium members, which also had interlocking interests in other major consortia in the Middle East (Aramco, Kuwait Oil Company, **Iraq Petroleum Company**, and others; refer to appendix 2) deliberately restrained Iranian production in the 1950s and 1960s so as to avoid depressing world prices, and also so as not to offend the governments in their other operating areas in the Middle East. They applied a secret and complex procedure to arrive at an "aggregate programmed quantity" (APQ) of production for the consortium. Their methods became public in 1974, during investigations by the U.S. Senate Foreign Relations Subcommittee on Multinational Corporations.

In July 1972, Iran began renegotiating its agreement with the consortium, and a year later a new agreement was signed. Under the new agreement, NIOC effectively realized the original goal of the 1951 nationalization: to take over the operation of the Iranian petroleum industry. The new agreement took away the consortium's control of production policy and placed it in the hands of the government. The agreement also contained provisions guaranteeing NIOC a similar revenue per barrel of oil as that enjoyed by Iran's Arab neighbors that signed **participation agreements**. On 28 February 1979, in the midst of the Iranian Revolution, NIOC unilaterally canceled the consortium agreement.

IRANIAN OIL REFINING COMPANY. *See* IRANIAN OIL PARTICIPANTS LTD.

IRAQ. Oil production 2015: 3.978 Mmbpd (5.08 percent of world); **reserves** 142.5 Bb (8.66 percent of world); gas production: 0.037 Tcf (0.03 percent of world); reserves 111.5 Tcf (1.62 percent of world). Iraq is a major petroleum country that has been prevented from living up to its full potential first by colonial exploitation and then by decades of wars with its neighbors and the West. Iraq was a founding member of the **Organization of the Petroleum Exporting Countries** (OPEC) in 1960 and joined the **Organization of Arab Petroleum Exporting Countries** (OAPEC) in 1972.

The first commercial discovery of oil in Iraq was by the Anglo–Persian Oil Company (later **Anglo–Iranian Oil Company**, eventually **BP**) at Naftkhana in the border territories transferred from **Iran** to the Ottoman Empire by the 1913 Turco-Persian Frontier Commission. The production from this discovery was used to satisfy domestic demand. In 1925, the Turkish Petroleum Company (**Iraq Petroleum Company**, or IPC, after 1929) obtained a **concession** in the country and discovered the **supergiant Kirkuk** field (**EUR** at least 25 Bb and 8.2 Tcf gas) with its Baba Gurgur well on 15 October 1927. The IPC renegotiated the terms of its concession and signed a new one in 1931 with a significantly reduced area. But by 1941, through acquisition and other means, it had obtained rights to almost the whole country again. Oil production in Iraq was 926 bpd in 1927. By 1938, it had reached 89,400 bpd. Exports began in 1935 through IPC's **pipeline** from Kirkuk to Tripoli, Lebanon, and Haifa, Palestine. IPC went on to discover most of Iraq's supergiant fields, including (giving EUR figures for oil only) Nahr Umr (1948, 6.5 Bb), Zubair (1949, 8.2 Bb), and the great Rumaila field (at least 30 Bb). There were several other **giant** discoveries as well. Iraqi production dipped during **World War II** but increased rapidly after the late 1940s. It surpassed 0.5 Mmbpd in 1953, 1 Mmbpd in 1962, and 2 Mmbpd in 1973.

In 1952, Iraq obtained a **50–50 agreement** for profit sharing with IPC, following a trend begun by **Venezuela** that spread to the Middle East. Iraq's oil revenue quadrupled between 1952 and 1956. However, disputes still festered between IPC and the Iraqi government, which was largely excluded from the management and operation of its petroleum sector. A Nasser-inspired revolution overthrew King Faisal II in 1958 and ushered in a more assertive government under General Abd al-Karim Qasim. In 1961, the government passed Law No. 80, revoking 99.5 percent of IPC's concession area but allowing it to keep operating its producing interests. The government created Iraq National Oil Company (INOC) in 1964 and in 1967 gave it the exclusive right to exploit new resources as well as to develop Rumaila.

Iraq's output grew in the 1960s, but not as quickly as that of other Middle Eastern countries. The IPC partners introduced **Libyan** oil to the market and boosted output from **Saudi Arabia**, Iran, and **Kuwait**. The IPC continued to treat Iraq with a certain amount of arrogance. In 1966 and 1967, IPC had a dispute with Syria over transit fees for the Kirkuk pipeline. Syria shut down the pipeline. IPC simply increased output from other countries, while Iraq, which was not even party to the dispute, suffered the consequences. The IPC partners also cut the output of the Kirkuk oil fields in half in 1972 to put pressure on Iraq, which had been steadily trying to restrict the scope of their operations. Nevertheless, by 1975 Iraq had completed the nationalization of all IPC assets (including the assets of its subsidiaries, the Basra Petroleum Company and the Mosul Petroleum Company). Iraq managed its petroleum industry in the 1970s with both financial and technical assistance from the Soviet Union, East Germany, and Hungary. Iraq's production reached an all-time high of 3.477 Mmbpd in 1979. In the 1970s, the country also discovered some supergiant fields, including Majnoon (1977, 12 Bb) and Baghdad East (1979, 16 Bb). There would be no further supergiant discoveries after 1980, and only one giant (Subba, in 1988, with EUR of 2.2 Bb).

In 1980, Iraq invaded Iran, initiating the eight-year **Iran–Iraq War**. The war had a devastating effect on both nations. In 1981, Iraq's oil production was down to 1 Mmbpd. It recovered to about 2.9 Mmbpd by 1989. In 1990, Iraq invaded Kuwait, resulting in the 1991 **Persian Gulf War**. Oil production dropped precipitously in both countries. Iraq's 1991 production was 0.305 Mmbpd. Iraq then entered a crippling phase of economic sanctions levied by the international community. They were partially lifted in 1996 when the United Nations allowed Iraq to generate $4 billion worth of annual oil revenue to meet humanitarian needs. Iraq cheated on the terms of this **Oil for Food Program**, often with Western cognizance. Oil production increased, reaching 2.57 Mmbpd in 2000.

In 2003, the **United States** invaded Iraq, initiating the **Iraq War**. The United States toppled the regime of **Saddam Hussein** and occupied the country. The United States withdrew most forces in 2011 but partially reentered in 2014 to provide air support against the Islamic State. Iraqi oil production was 1.303 Mmbpd in 2003, recovering to 2.011 Mmbpd in 2004. Despite the country's troubles, oil production rose steadily beginning in 2005, more than doubling in 10 years to 4 Mmbpd.

A new Iraqi constitution was negotiated in 2005 and approved by a national referendum.

It entered into force in 2006. The constitution recognized northern Kurdistan as a special region. Article 112 grants the central government a conditional right to undertake management of oil and gas extracted from existing fields. The **Kurdistan Regional Government** (KRG) argues that Article 112 does not grant the central government the right to manage nonproducing, underproducing, or as yet undiscovered fields, and that such fields are under the control of regional governments. The KRG also argues that Article 110 of the constitution gives Kurdistan self-determination in petroleum issues. The KRG signed its own contracts with oil companies in 2007, but the central government threatened to cancel them. In mid-2008, the Iraqi government negotiated with **ExxonMobil**, BP, **Total**, **Royal Dutch Shell**, **Chevron**, and others for controversial no-bid service contracts for some of the country's largest oil fields. These contracts would bring the original members of the

IPC, and other Western companies, into Iraq for the first time since nationalization—although as service providers, not equity-holding concessionaires. In late 2008, however, Iraq turned from the Western companies and instead negotiated a deal with **China**.

Tensions between the KRG and the central government in Baghdad over oil and other issues continued unabated, especially after the KRG took control of **Kirkuk** in 2014 and the Kurdish area held an independence referendum in 2017. In 2016 the KRG and Kirkuk accounted for around a quarter of Iraqi oil production. Despite all this, there have been many international oil companies operating in Iraq and the KRG, in a sector that is now much more open than it was in the past. In late 2017 the Iraqi army took back control of Kirkuk.

Iraq's **natural gas** production is much less important than its oil production. It was 0.062 Tcf in 1980 and dipped to 0.018 Tcf in 1983 during the Iran–Iraq War. It reached a peak of 0.215 Tcf in 1989 and then dropped to 0.039 Tcf in 1991 because of the Persian Gulf War. It recovered to 0.112 Tcf in 1994 and then stayed nearly flat until 2000. It fell back again to 0.039 Tcf in 2003 because of the Iraq War. Although it rose afterward, in 2016 it was back down at 0.037 Tcf.

IRAQ PETROLEUM COMPANY. A consortium of Western companies established to operate in **Iraq**; formerly the Turkish Petroleum Company (TPC). The consortium later functioned as a cartel controlling the exploitation of petroleum resources in the territory of the former Ottoman Empire. The Turkish Petroleum Company was established in 1912 to obtain a petroleum **concession** in Mesopotamia (Iraq), then part of the Ottoman Empire. One of the main movers behind the creation of TPC was Armenian entrepreneur **Calouste Gulbenkian**. TPC was half owned by the Anglo–Persian Oil Company (APOC; later **Anglo–Iranian Oil Company**, eventually **BP**).

Deutsche Bank and **Royal Dutch Shell** each owned 22.5 percent, and Calouste Gulbenkian owned 5 percent. TPC got a promise of a concession, but its efforts were interrupted by the outbreak of **World War I**. It would not obtain a concession in Iraq until 14 March 1925. After the war, the ownership of TPC was restructured at the **San Remo Conference**. The German share was transferred to France and would form the basis of the **Compagnie Française des Pétroles** (CFP, later **Total**).

Both Italy and the **United States** were excluded from TPC at San Remo. American oil companies and the U.S. State Department argued for an "open door" principle, allowing companies of any nationality an equal chance at Middle Eastern petroleum resources. After years of negotiations, U.S. oil companies were eventually allowed into TPC. The agreement was not finalized until 31 July 1928, nearly a year after TPC had struck oil at Baba Gurgur on 15 October 1927, discovering the immense **Kirkuk** field. APOC, Royal

Dutch Shell, CFP, and the **Near East Development Corporation** (NEDC) each got a 23.75 percent share of the reorganized company. NEDC was a consortium of five U.S. companies.

All the members of the consortium pledged not to compete with each other anywhere in the territory of the former Ottoman Empire. The territory was reportedly delineated freehand by Calouste Gulbenkian on a map with a red pencil, and the noncompetition agreement was dubbed the **Red Line Agreement**. In other words, the door was opened a crack to let some American companies in, and then shut again. The TPC was renamed the Iraq Petroleum Company (IPC) in 1929. By 1934, all the American partners had dropped out except Socony–Vacuum (later **Mobil**) and the Standard Oil Company of New Jersey (later **Exxon**), which each ended up with a share of 11.875 percent.

TPC's 1925 concession in Iraq covered the entire country except for some formerly Iranian territories that had been transferred to the Ottoman Empire in 1913. However, the concession required TPC to select, within 32 months, 24 plots of eight square miles each, so that the remaining territory could be leased to the highest bidders. TPC felt constrained by this requirement, particularly since large areas of the country near Turkey were effectively impassable at the time. TPC received an extension of time but wanted more. In 1928, King Faisal refused to grant further extensions. After protracted negotiations, the Iraq Petroleum Company (IPC), as it was now known, obtained a new concession in 1931. The concession covered a reduced area: Baghdad and Mosul Provinces east of the Tigris River. IPC also agreed to build a **pipeline** from the Kirkuk field to the Mediterranean by 1935, a feat it was able to accomplish.

In 1932, the Iraqi government granted a concession for territory west of the Tigris to British Oil Developments Ltd. (BOD), a competitor who appeared on the scene in 1928 promising to build a railway to the Mediterranean at its own expense. BOD was given eight and a half years to reach export levels of more than 7 Mmb per year, with 20 percent of the oil going to the government. IPC created a holding company, Mosul Oil Fields Ltd., and used it to acquire BOD shares. By 1937, it had majority control. In 1941, it formed a wholly owned subsidiary, Mosul Petroleum Company, and took over the BOD concession.

In July 1938, the Iraqi government granted a third concession, covering **Basra** Province, to the Basra Petroleum Company, an IPC sister company. Thus, by the 1940s, IPC had concessions covering the entire country of Iraq, except for the territories acquired from Iran, mentioned above. IPC went on to discover a large share of Iraq's major oil fields. In the 1930s, the IPC partners also created consortia with the same ownership structure to explore for and develop petroleum in other countries within the Red Line: **Abu Dhabi** (**United Arab Emirates**), **Qatar**, and **Oman**.

In 1952, IPC negotiated a **50–50 agreement** on profit sharing with Iraq, following a trend begun by **Venezuela** that spread to the Middle East. Despite the increase in oil revenue that this represented, disputes festered between IPC and the Iraqi government, which was largely excluded from the management and operation of its petroleum sector. A Nasser-inspired revolution overthrew King Faisal II in 1958 and ushered in a more assertive government. In 1961, the government passed Law No. 80, revoking 99.5 percent of IPC's concession area but allowing it to keep operating its producing interests. The government created the Iraq National Oil Company (INOC) in 1964 and in 1967 gave it the exclusive right to exploit new resources. By 1975, Iraq completed the nationalization of IPC's assets, turning them over to INOC. In the 1970s, the IPC partners also saw all or part of their interests nationalized in Qatar, Oman, and the United Arab Emirates. The IPC was never a profit-making entity. It was a consortium that delivered crude to its partners, which were multinational oil companies. The IPC largely ceased to exist, but its partners continue on, as BP, **ExxonMobil**, Royal Dutch Shell, and Total.

IRAQ WAR (2003–2011). War in **Iraq** initiated in 2003 by the **United States**, with the aid of the **United Kingdom** and some minor coalition partners. A complete understanding of the causes of the Iraq War may have to await the verdict of history. Among the reasons stated by President **George W. Bush** were that Iraq possessed weapons of mass destruction, although none were ever found. Another stated goal was the liberation of the Iraqi people from **Saddam Hussein**'s oppressive regime and the promotion of democracy in the Middle East. These reasons were widely disbelieved, particularly outside the United States. It is likely that at least some of the motivation for the war was related to petroleum.

This does not mean the United States sought to confiscate Iraq's oil. It is more likely that the United States wanted to install a friendlier regime in order to open up Iraq's oil industry and relieve disproportionate Western dependence on the oil resources of **Saudi Arabia** and **Russia**. The United States was thought to be losing patience with the stalemate over the sanctions against Iraq resulting from the 1991 **Persian Gulf War** (*see* OIL FOR FOOD PROGRAM). Iraq was violating the sanctions, sometimes with tacit U.S. cognizance, and thus the sanctions were ineffective. At the same time, the sanctions did not allow complete development of Iraq's petroleum industry and resources. The political environment in the United States after the terrorist attacks of 11 September 2001 facilitated the justification of the war to the American public. Many U.S. government officials have denied that the war was about oil and claim that oil was rarely directly discussed. However, oil does not need to be discussed. It is fully embedded in the basic architecture of U.S. policy. If Iraq did not have nearly 9 percent of the world's oil **reserves**, it would hardly command even a fraction of the attention accorded to it.

The invasion of Iraq began on 20 March 2003. The Iraqi army did not prove to be a serious obstacle. By 7 April, U.S. forces were in Baghdad and British forces were in **Basra**. U.S. and British casualties were minimal at this point. Saddam Hussein went into hiding but was captured on 13 December 2003 and executed on 30 December 2006 after trial by the interim Iraqi government. After the initial military victories, U.S. occupation forces found themselves in the middle of sectarian conflicts, mostly between Shiite and Sunni Arabs. They also faced a protracted counterinsurgency campaign. At the end of 2008, the United States still had more than 130,000 combat troops in Iraq. The United States withdrew the bulk of its troops in 2011. In the aftermath of the war, Iraq suffered from sectarian violence. The country also suffered from attacks by the forces of the Islamic State group, which took the cities of Mosul and Tikrit in 2014, while Iraq's own **Kurdistan Regional Government** took **Kirkuk**. The United States reentered Iraq in 2014 to conduct air operations against the Islamic State. Fighting still continued in 2017. In 2017 the Iraqi army took back Kirkuk.

Any immediate negative effect of the invasion itself on the global oil market was relatively small. **Brent Blend** spiked up to $34.50 per barrel immediately before the invasion but fell to $24.20 by 7 April 2003. There was no large-scale destruction of oil installations as had taken place in **Kuwait** during the 1991 Persian Gulf War. The uncertainty introduced by the war may have contributed to the **Third Oil Shock** in 2004–2008, but other factors were also at work, such as strong global demand, including significant demand growth in **China** and **India**. Meanwhile, Iraqi oil production more than tripled between 2003 and 2015, and many international oil companies now operate in the country. If that was one of the goals, it was achieved.

IRAQI PIPELINE ACROSS SAUDI ARABIA (IPSA). Pipeline built to export Iraqi oil via **Saudi Arabia**. IPSA connects the oil fields of southern **Iraq** to Mu'ajiz near the Saudi port of **Yanbu** on the Red Sea. It was built during the **Iran–Iraq War** to offer Iraq an alternative to shipping its oil through the **Persian Gulf**, where its **tankers** were being attacked by **Iran**. IPSA runs south from Iraq through Saudi Arabia until it joins the path of the main Saudi **Petroline** around pump station no. 3. It runs parallel to Petroline thereafter, west to the Red Sea. It is a 48-inch line that had a 1.65 Mmbpd capacity when it was built. Saudi Arabia closed the pipeline soon after Iraq's August 1990 invasion of Kuwait. In June 2001, Saudi Arabia seized ownership. In 2012 Saudi Arabia reopened the pipeline to decrease dependence on the **Strait of Hormuz**.

ISRAEL. *See* LEVANT BASIN.

J

JABLONSKI, WANDA (1920–1992). Petroleum industry journalist who founded ***Petroleum Intelligence Weekly*** (*PIW*). As an investigative reporter and publisher, Wanda Jablonski was a powerful figure in the oil world from the 1940s to the late 1980s. She exposed and explained the secretive dealings of the international oil business. Through her private networking, she helped shape the debate between the **major oil companies** and the oil-producing countries that led to the creation of the **Organization of the Petroleum Exporting Countries** (OPEC) and to the oil shocks of the 1970s (*see* FIRST OIL SHOCK; SECOND OIL SHOCK).

Jablonski was born 23 August 1920 in what was then Czechoslovakia. Her father was a petroleum geologist. She graduated from Cornell University in 1942 and studied at Columbia University (1942–1943). The *New York Journal of Commerce* hired her as an office messenger, but she quickly became its oil reporter. In 1948, she unveiled plans by **Juan Pablo Pérez Alfonso**, **Venezuela**'s oil minister, for a gradual nationalization of his country's oil industry, and she wrote controversial commentaries from Tehran in 1953–1954 describing **Iran**'s perspective on its oil nationalization. At McGraw-Hill's *Petroleum Week* (1955–1961), Jablonski wrote extensively about the rise of oil nationalism. In 1961, she founded the *Petroleum Intelligence Weekly*, which became very influential in the industry. She managed it as publisher until selling it in 1988.

At the 1959 Cairo oil congress, Jablonski introduced Pérez Alfonso to **Saudi Arabia**'s **Abdullah al-Tariki**. The two then organized the secret **Mehdi Pact**, an alliance that presaged OPEC's founding in September 1960. Jablonski was also close to Tariki's successor as oil minister, **Ahmed Zaki Yamani**. In 1972, *PIW* achieved a major scoop when it published details of the **participation agreement** between five Arab oil-producing countries and leading oil companies. In 1986, Jablonski exposed the underlying reasons for King Fahd's firing Yamani after 24 years as oil minister. Wanda Jablonski died in New York on 28 January 1992.

JACK NO. 2. Gulf of Mexico well drilled by **Chevron** in 2006, significant for its depth and the potential for new hydrocarbon **reserves** at previously unexplored depths in the **offshore**. The Jack No. 2 well was drilled in 7,000 feet of water. Other wells have been drilled at that water depth and even greater. What made the Jack well remarkable at the time was that it penetrated more than 20,000 feet under the sea floor, for a total depth of 28,175 feet. The Jack No. 2 test flowed at 6,000 bpd. Chevron estimated that the 300-square-mile region around the Jack No. 2 well could hold from 3 to 15 Bb of oil and **natural gas liquids**. In 2015 Chevron's Jack/St. Malo unit had a production capacity of 0.177 Mmboepd.

JERSEY STANDARD. *See* EXXON.

JOINER, DAD (COLUMBUS MARION) (1860–1947). American petroleum entrepreneur who discovered the **supergiant East Texas oil field**. Columbus Marion Joiner was born on 12 March 1860 in Lauderdale County, Alabama. He practiced law in Tennessee beginning in 1883, and he served in the Tennessee legislature from 1889 to 1891. In 1897, Joiner entered the oil business in **Oklahoma**, where he experienced temporary success but ultimate failure. In 1926, he began exploring for oil in Rusk County in eastern **Texas**. He drilled two unsuccessful wells with primitive equipment. On 3 October 1930, he succeeded with his third well, Daisy Bradford No. 3, thus discovering the East Texas field. This was the largest field in the world at the time. Although he was the "father" ("Dad") of the East Texas field, many others would grow much richer from it. He faced continual financial difficulties, eventually selling his well and leases to **H. L. Hunt** for $1 million. Dad Joiner died 27 March 1947, leaving no substantial estate.

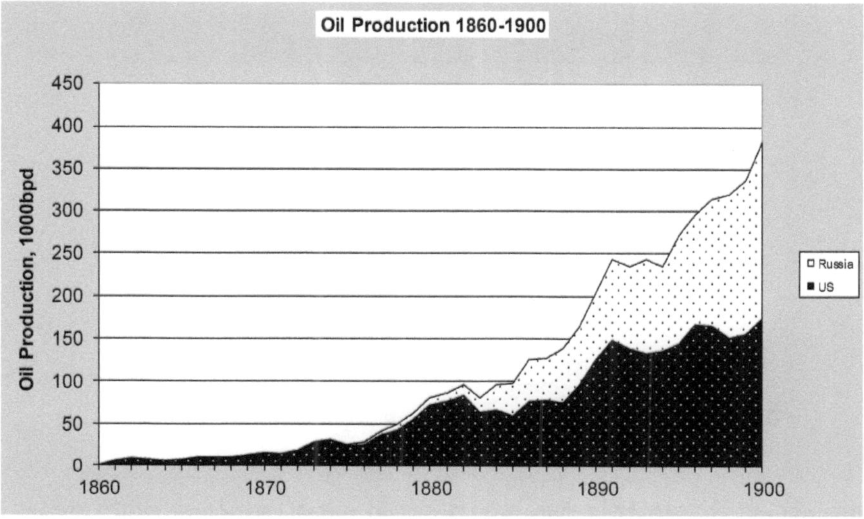

Oil production, 1860–1900. Data from Fanning (1950).

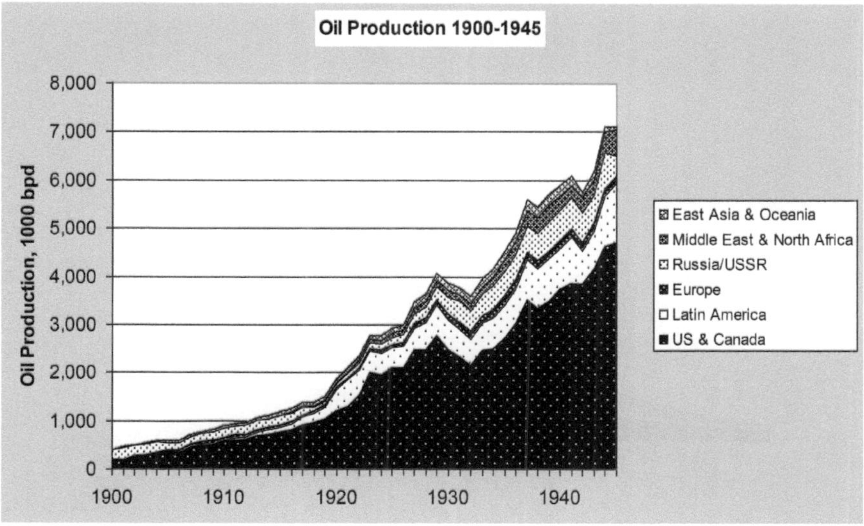

Oil production, 1900–1945. Data from Fanning (1950).

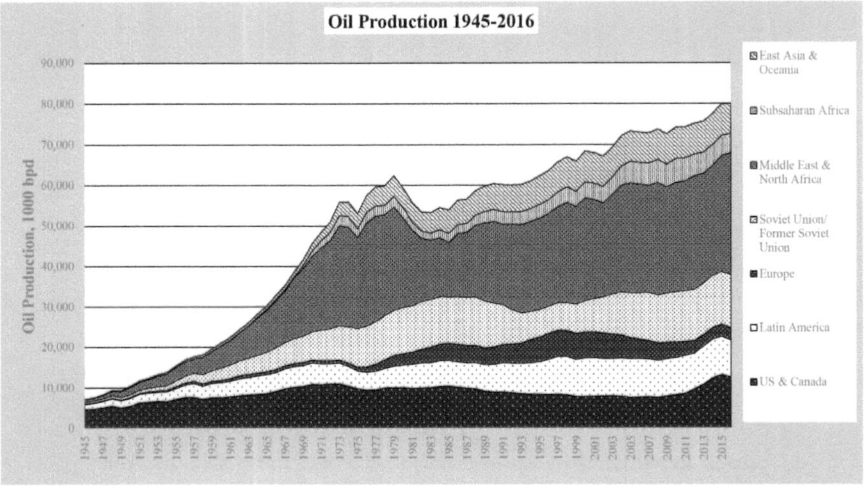

Oil production, 1945–2016. Data from DeGolyer and MacNaughton (2007), Salvador (2005), and U.S. Energy Information Administration.

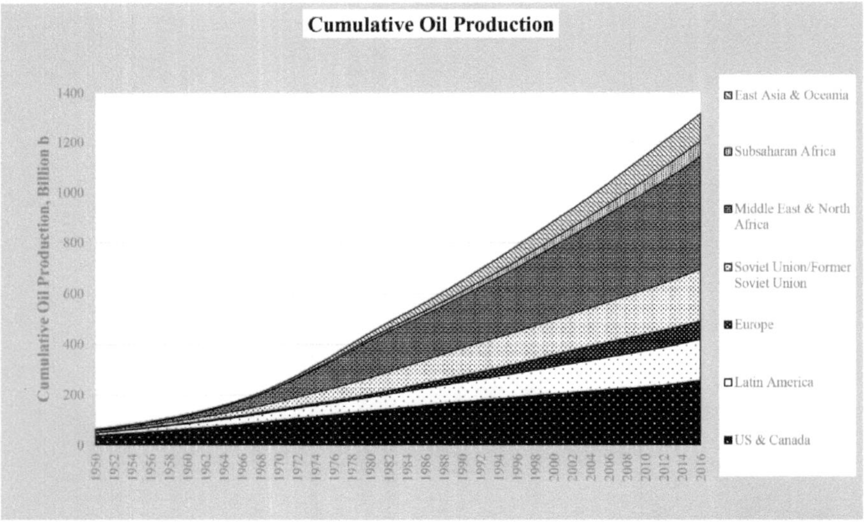

Cumulative oil production, 1950–2016. Data from Salvador (2005) and U.S. Energy Information Administration.

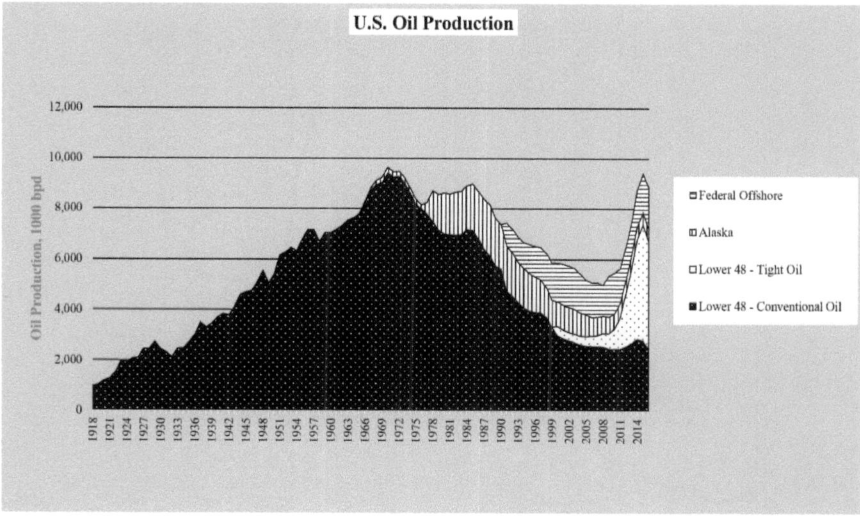

U.S. oil production, 1918–2016. Data from Salvador (2005) and U.S. Energy Information Administration.

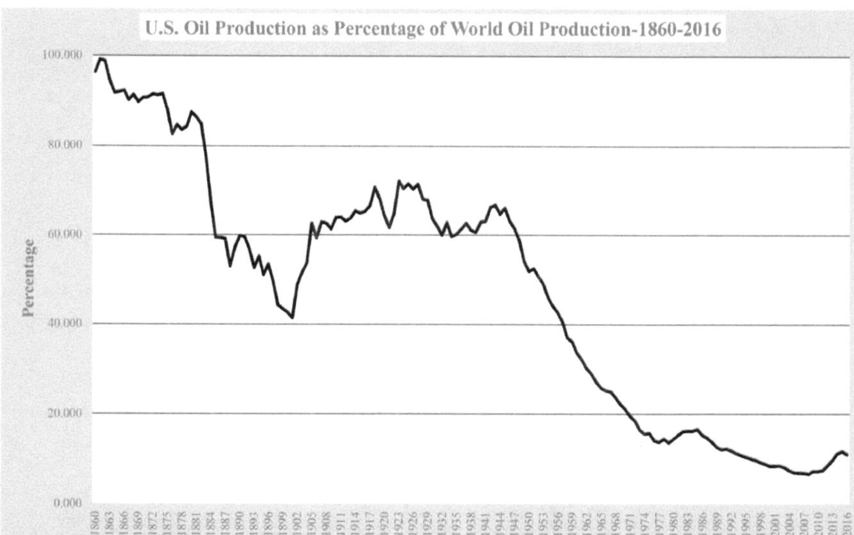

U.S. oil production as a percentage of world oil production, 1860–2016. Data from Fanning (1950), Salvador (2005), and U.S. Energy Information Administration.

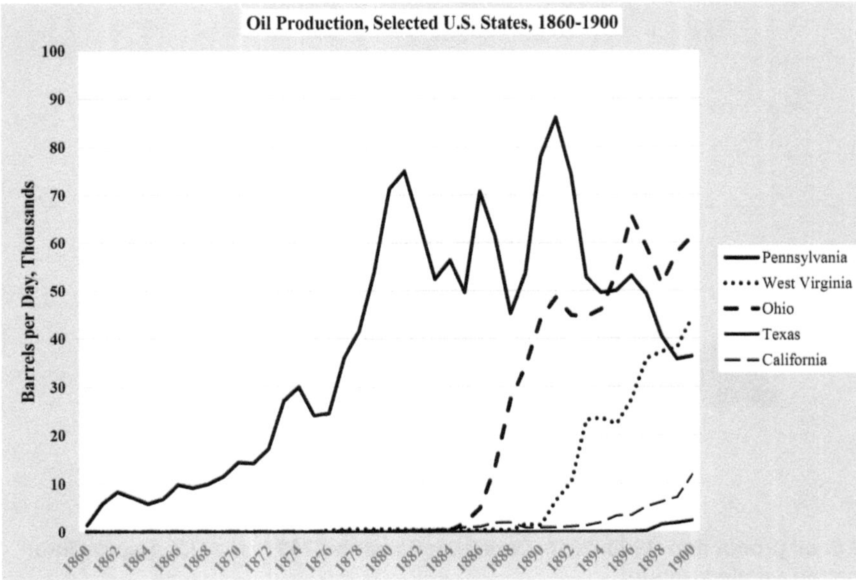

Oil production in selected U.S. states, 1860–1900. Data from Day (1922).

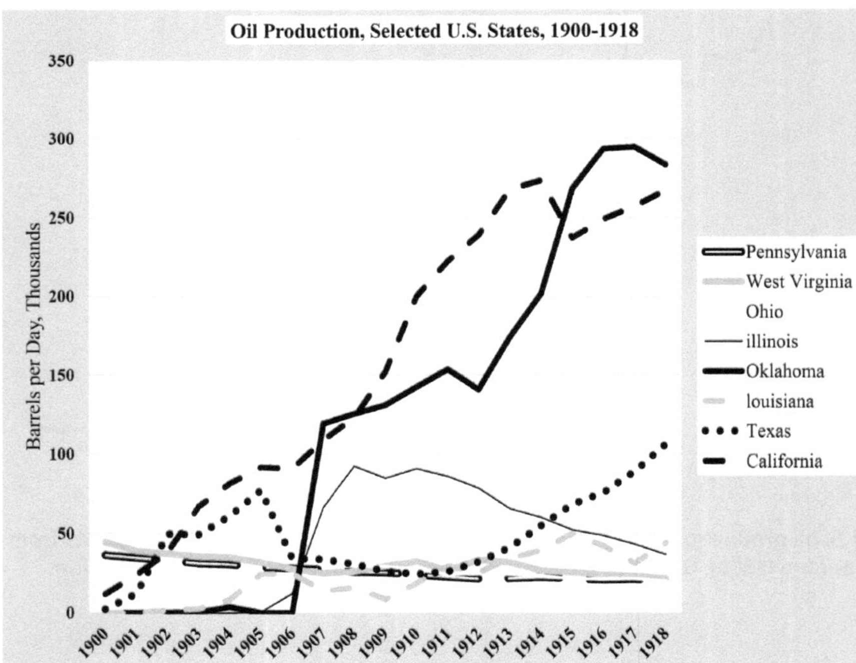

Oil production in selected U.S. states, 1900–1918. Data from Day (1922).

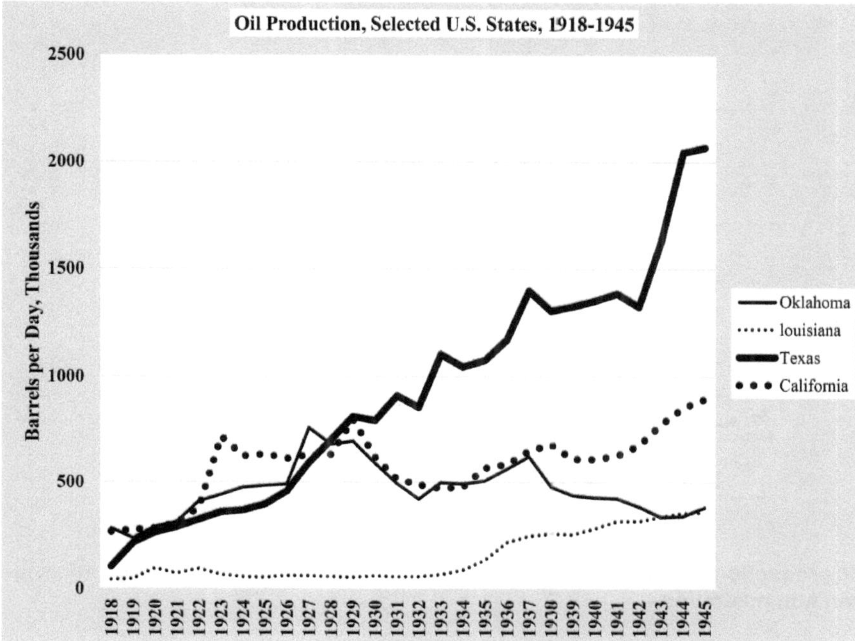

Oil production in selected U.S. states, 1918–1945. Data from DeGolyer and MacNaughton (2004).

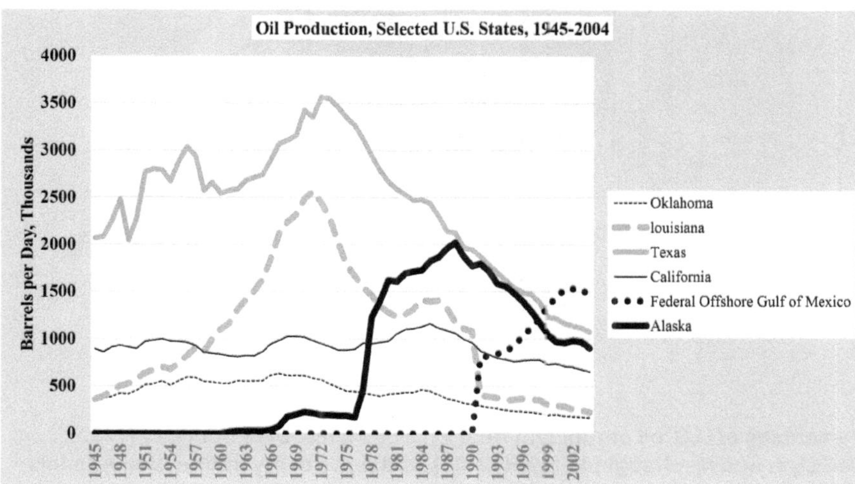

Oil production in selected U.S. states, 1945–2004. Data from DeGolyer and MacNaughton (2007) and U.S. Energy Information Administration.

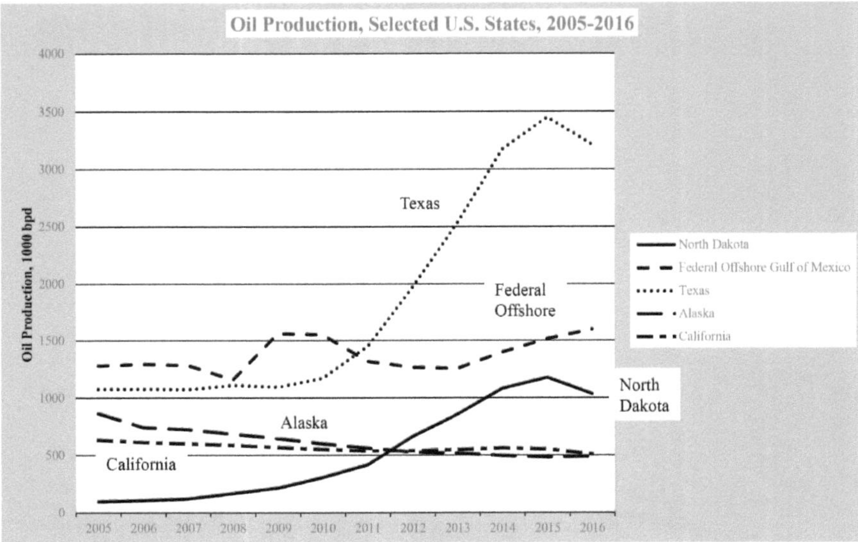

Oil production in selected U.S. states, 2005–2016. Data from U.S. Energy Information Administration.

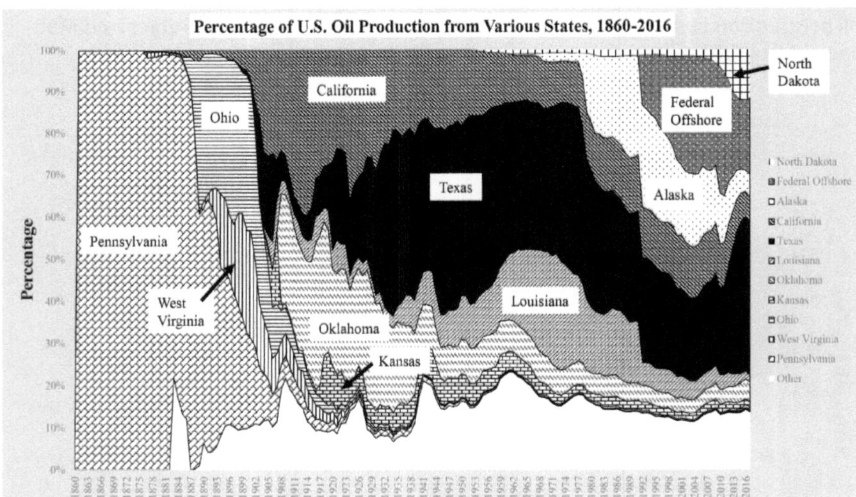

Percentage of U.S. oil production from various states. Data from Day (1922), DeGolyer and MacNaughton (2004, 2007), and U.S. Energy Information Administration.

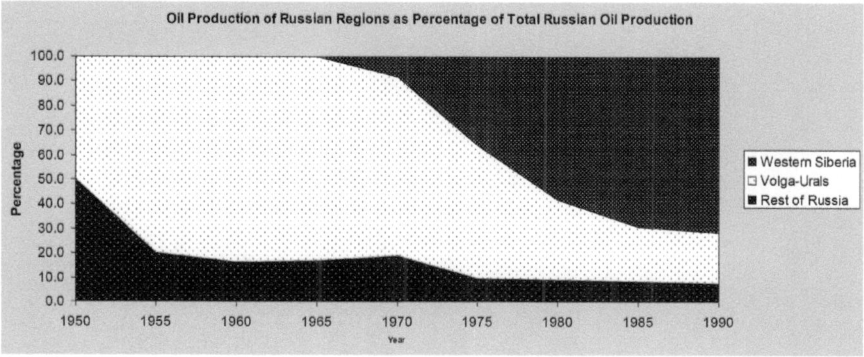

Oil production of Russian regions as a percentage of total Russian oil production, 1950–1990. Data from Krylov et al. (1998).

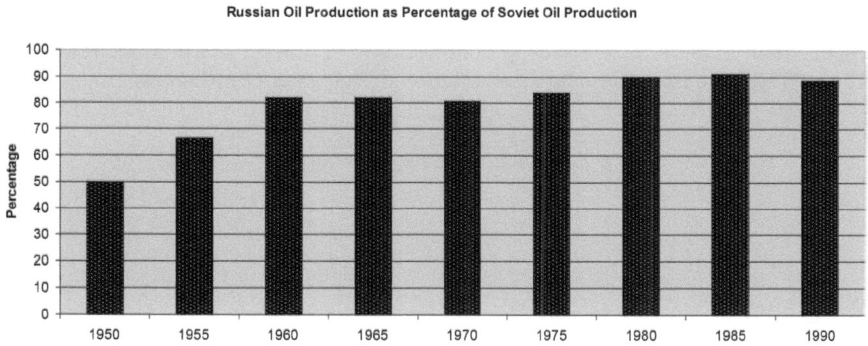

Russian oil production as a percentage of Soviet oil production, 1950–1990. Data from Krylov et al. (1998).

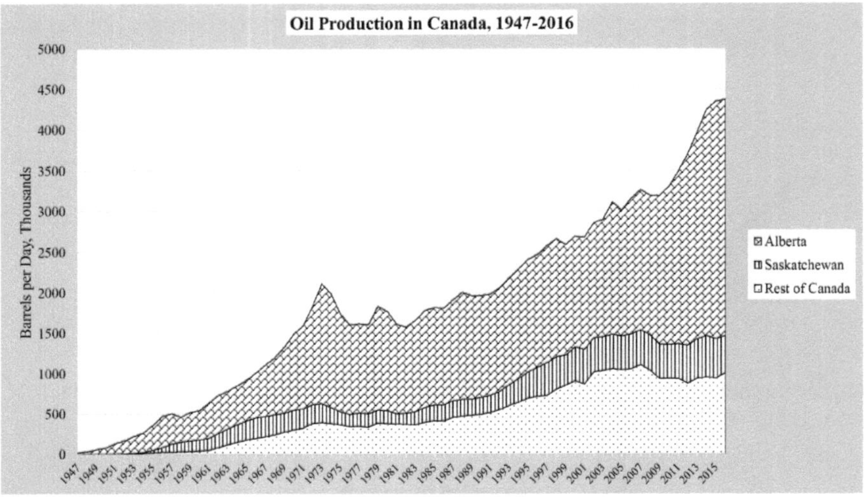

Oil production in Canada, showing contributions of Alberta and Saskatchewan, 1947–2016. Data from Canadian Association of Petroleum Producers.

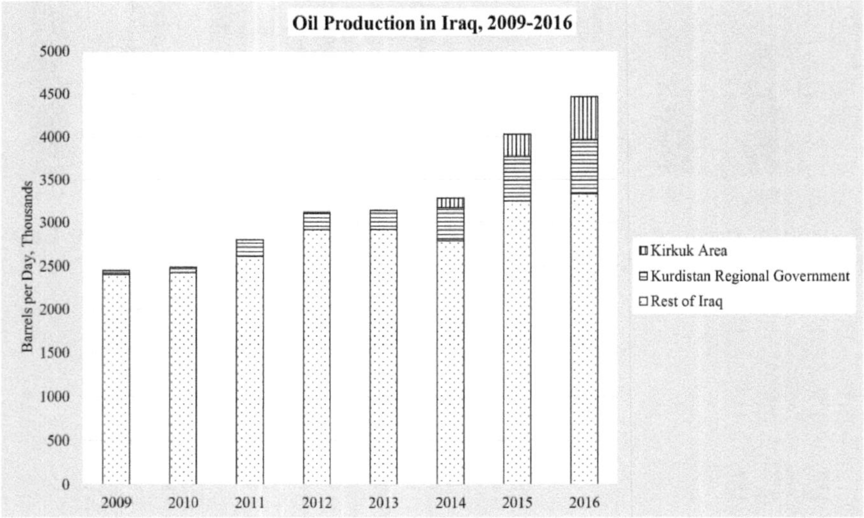

Oil production in Iraq, showing contributions of Kurdish Regional Government and Kirkuk area, 2009–2016. Data from Oxford Institute for Energy Studies and U.S. Energy Information Administration.

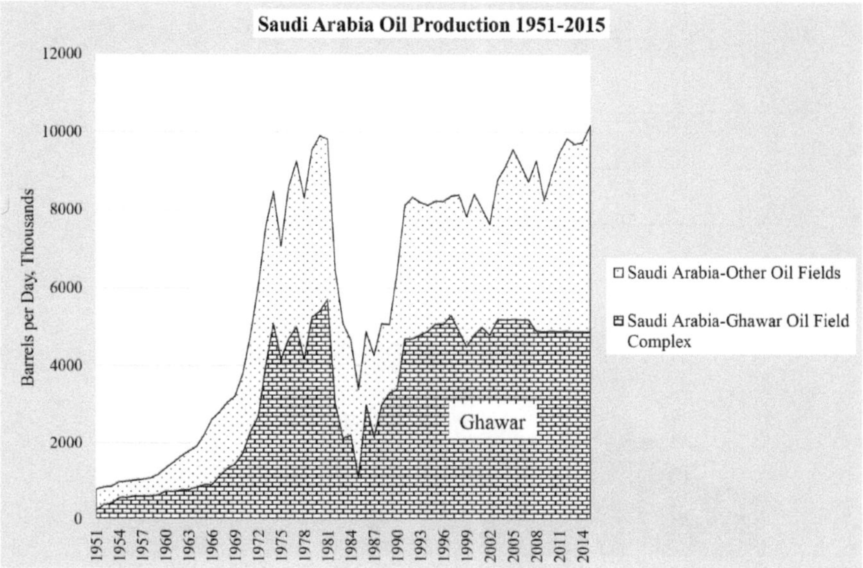

Oil production in Saudi Arabia, showing contribution of Ghawar oil field, 1951–2014. Data from Energy Consulting Group; DeGolyer and MacNaughton (2007); and U.S. Energy Information Administration.

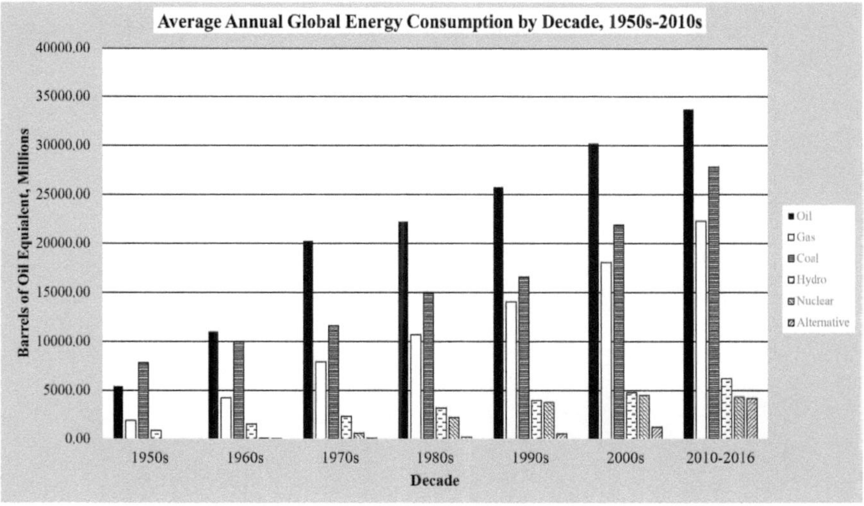

Average global energy consumption by fuel source, decades from 1950s to 2010s. Data from Salvador (2005) and *BP Statistical Review of World Energy 2017*.

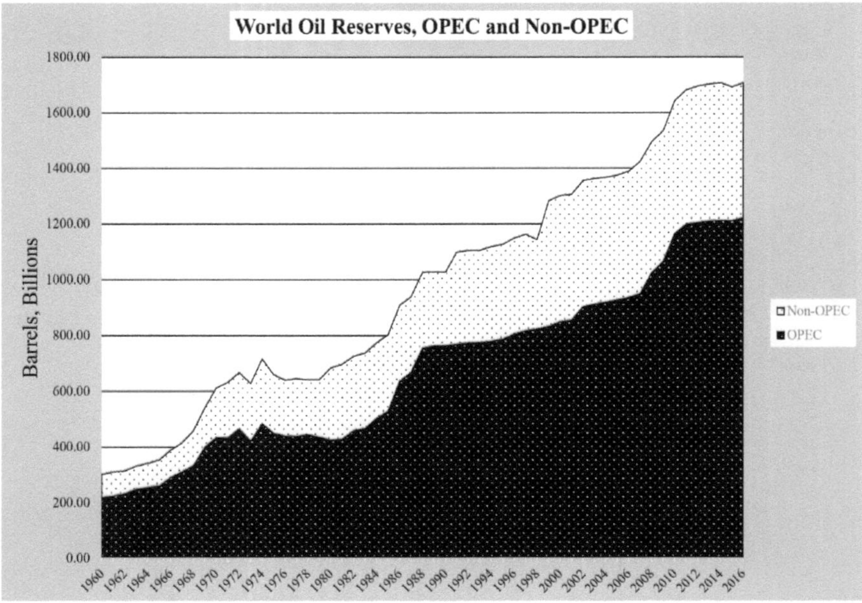

World oil reserves, OPEC and non-OPEC, 1960–2016. Data from Salvador (2005) and *BP Statistical Review of World Energy 2017*.

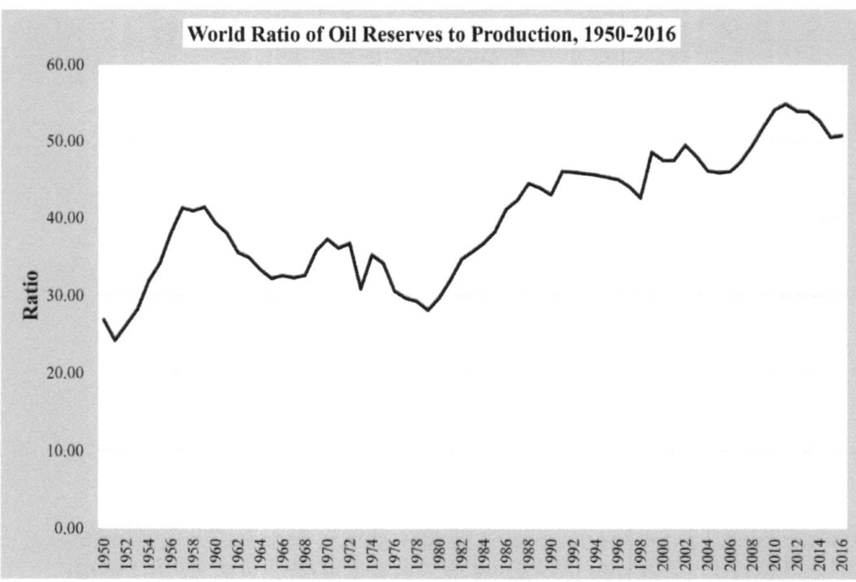

World ratio of oil reserves to production, 1950–2016. Data from Salvador (2005) and *BP Statistical Review of World Energy 2017*.

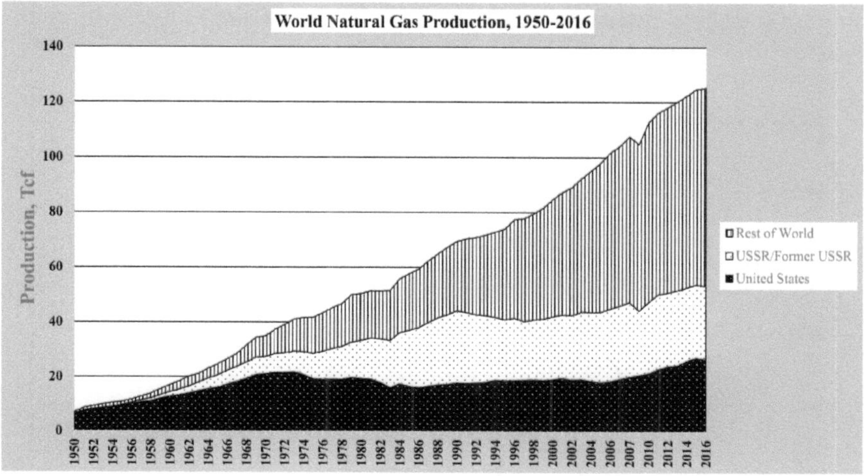

World natural gas production, 1950–2016. Data from Salvador (2005) and *BP Statistical Review of World Energy 2017*.

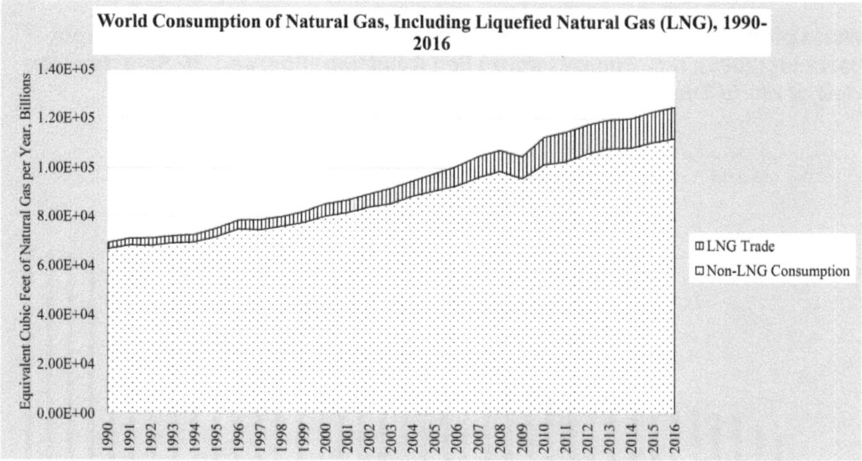

World consumption of natural gas, including liquefied natural gas, 1990–2016. Data from International Gas Union 2017 World LNG Report and *BP Statistical Review of World Energy 2017*.

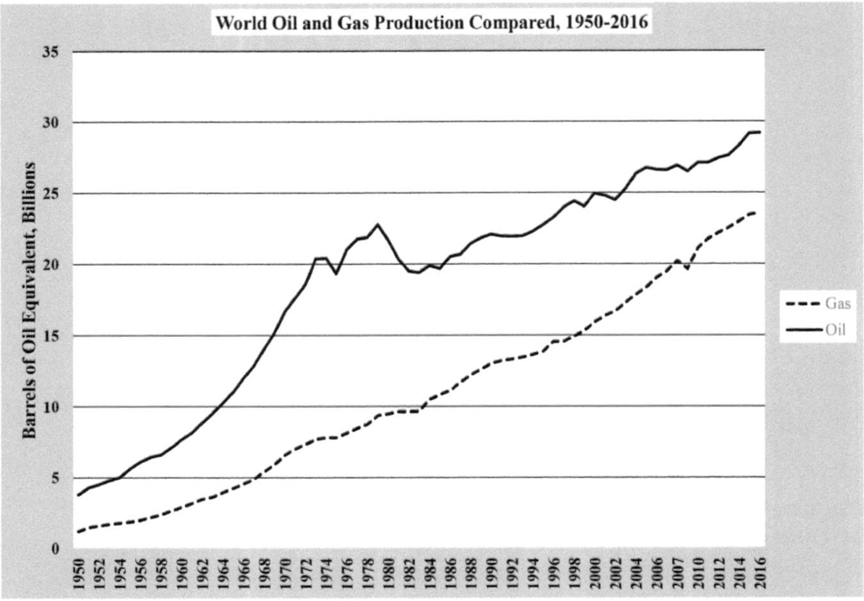

World production of oil and natural gas compared in energy terms. Data from Salvador (2005); U.S. Energy Information Administration; and *BP Statistical Review of World Energy 2017*.

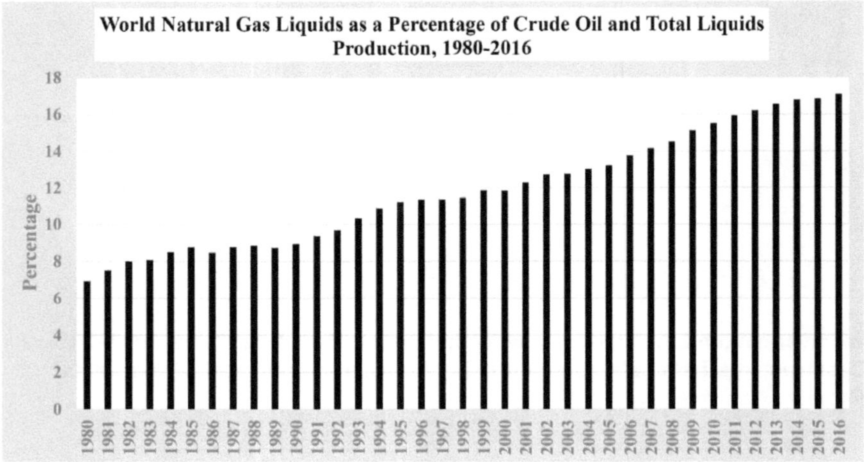

World natural gas liquids as a percentage of total production of crude oil and liquids. Data from U.S. Energy Information Administration.

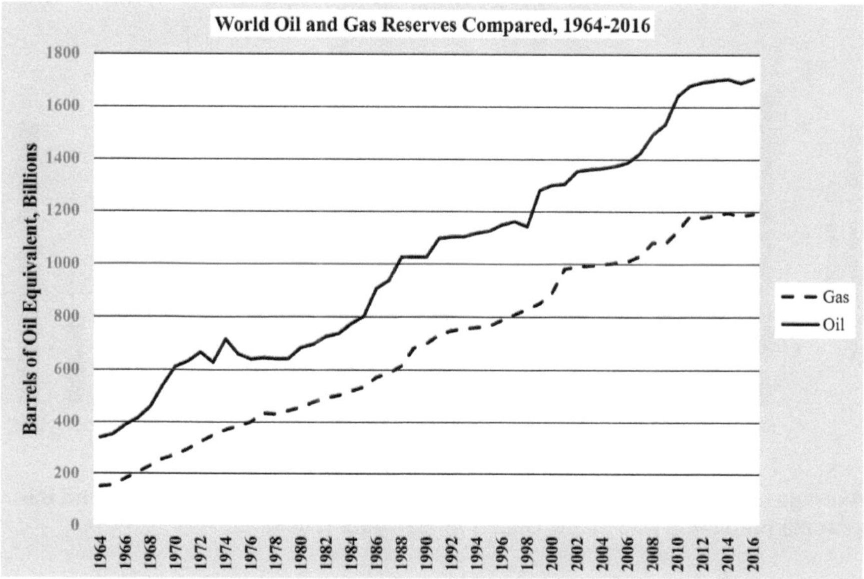

World reserves of oil and natural gas compared in energy terms. Data from Salvador (2005); U.S. Energy Information Administration; and *BP Statistical Review of World Energy 2017*.

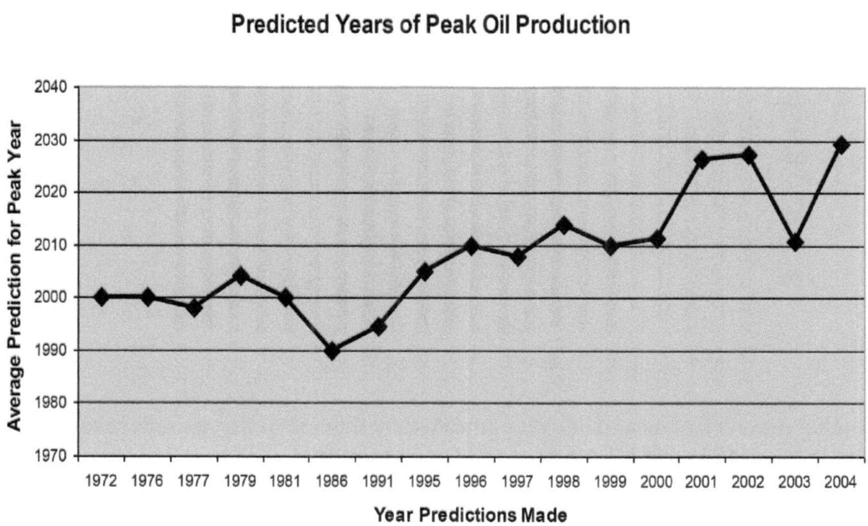

Predicted years of peak oil production made by various investigators, plotted against the year each prediction was made. Data from Salvador (2005).

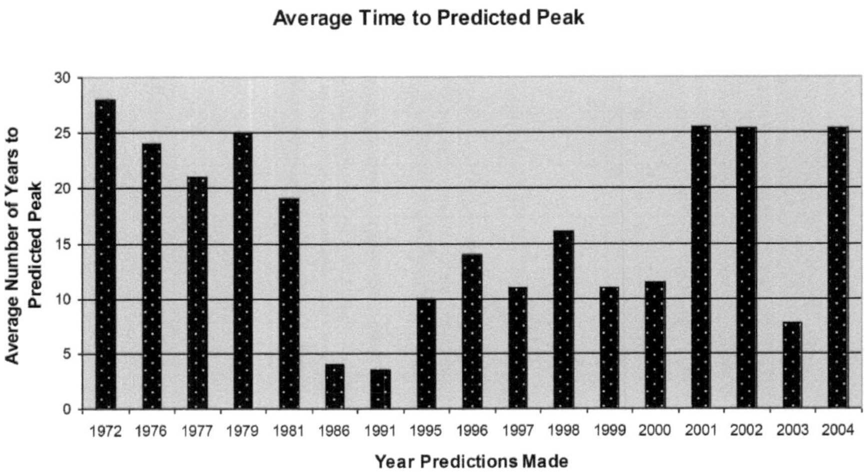

Average number of years between predicted year of peak oil production and the year the prediction was made. Data from Salvador (2005).

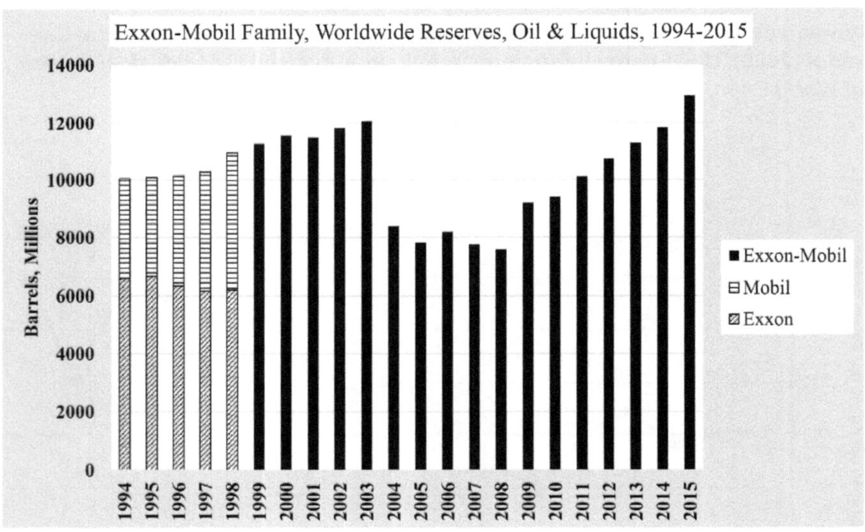

Global reserves of oil and liquids controlled by ExxonMobil and predecessor companies, 1994–2015. Data from *Oil and Gas Journal.*

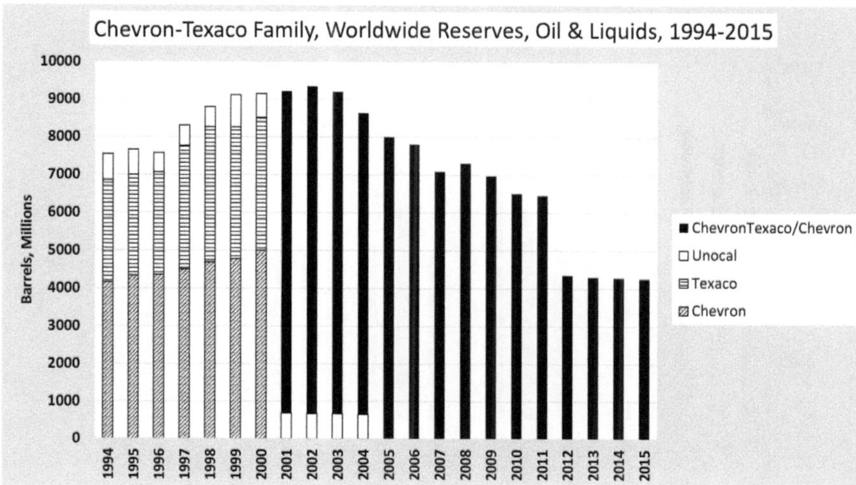

Global reserves of oil and liquids controlled by Chevron and predecessor companies, 1994–2015. Data from *Oil and Gas Journal*.

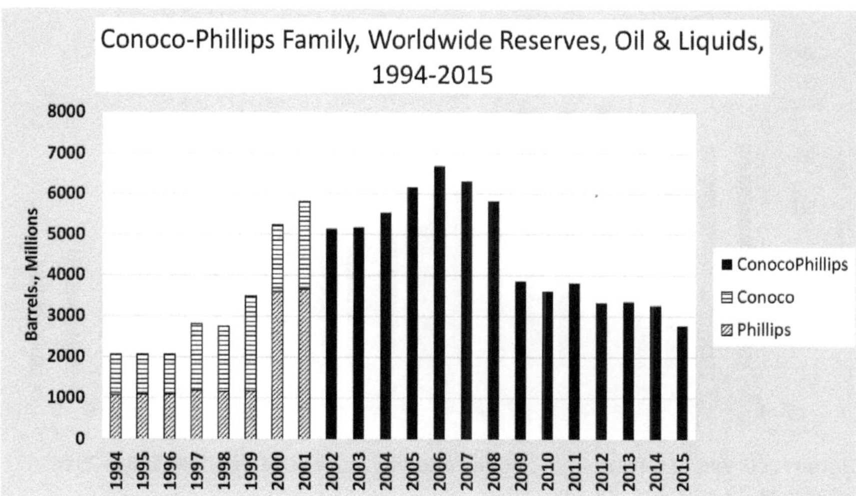

Global reserves of oil and liquids controlled by ConocoPhillips and predecessor companies, 1994–2015. Data from *Oil and Gas Journal*.

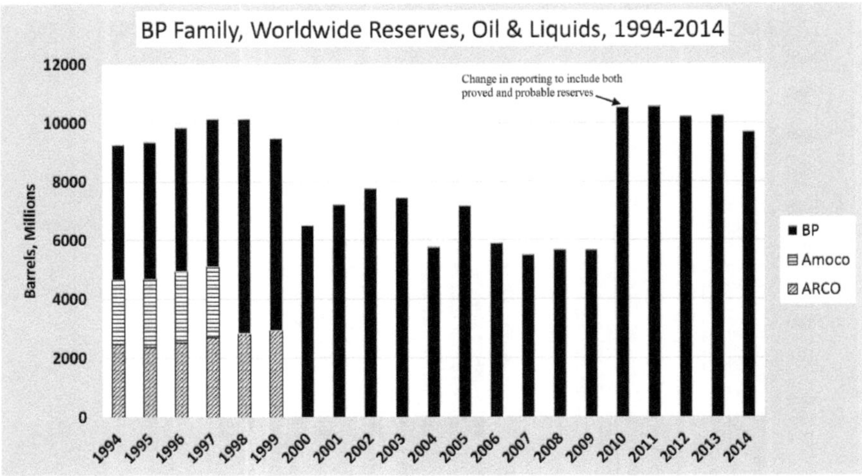

Global reserves of oil and liquids controlled by BP and predecessor companies, 1994–2014. Data from *Oil and Gas Journal*.

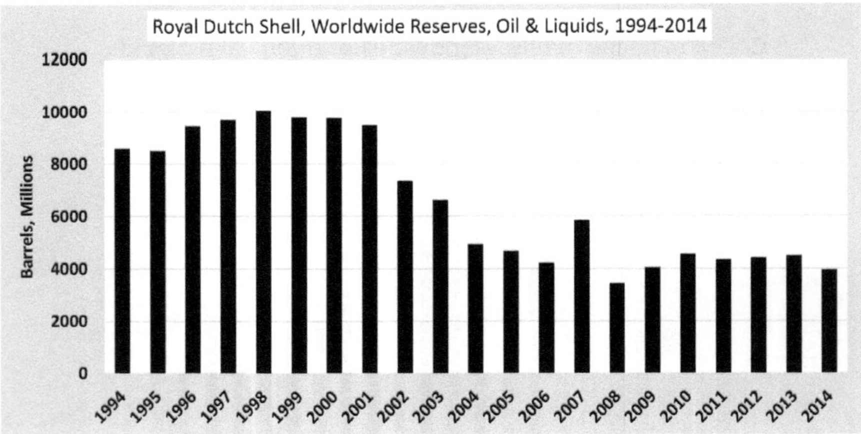

Global reserves of oil and liquids controlled by Royal Dutch Shell, 1994–2014. Data from *Oil and Gas Journal*.

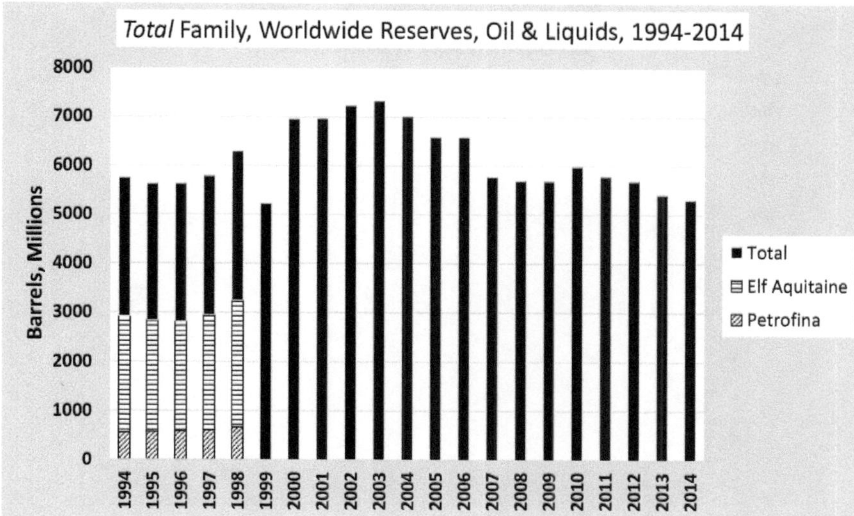

Global reserves of oil and liquids controlled by Total and predecessor companies, 1994–2014. Data from *Oil and Gas Journal*.

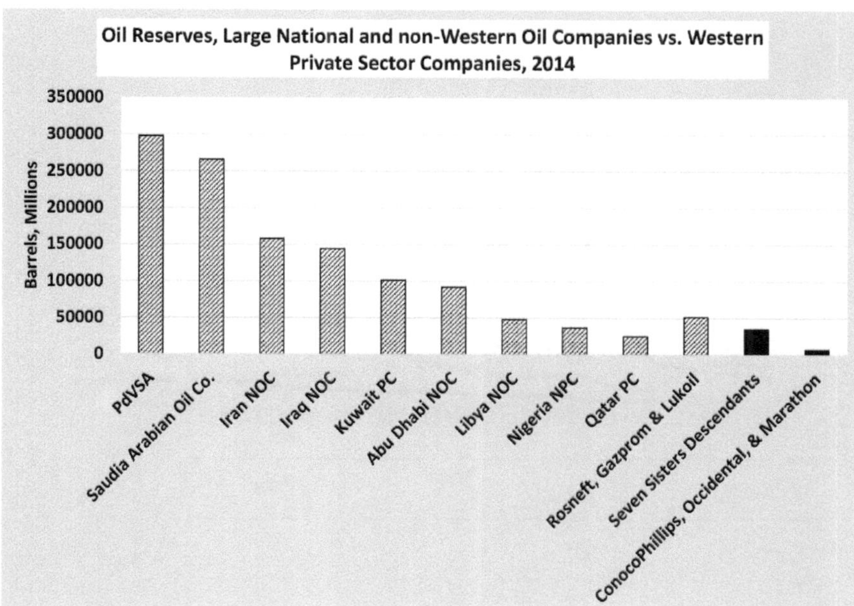

Global oil reserves controlled by large national oil companies and selected Western private sector companies, 2014. Data from *Oil and Gas Journal*.

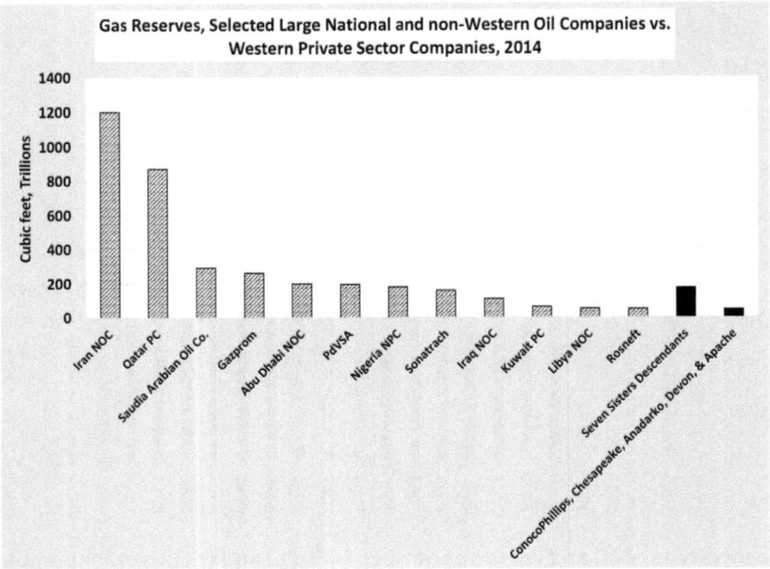

Global gas reserves controlled by large national oil companies and selected Western private sector companies, 2014. Data from *Oil and Gas Journal*.

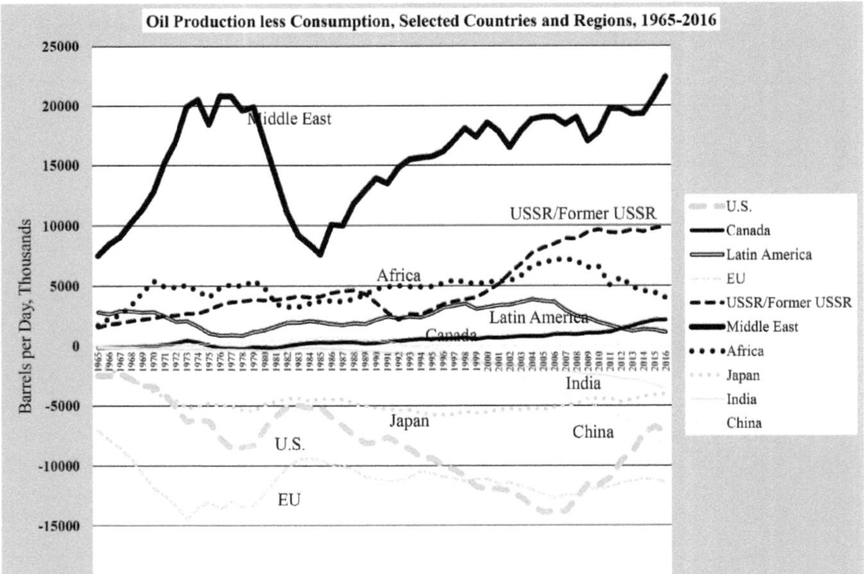

Oil production less consumption, selected countries and regions, 1965–2016. Data from Salvador (2005) and U.S. Energy Information Administration.

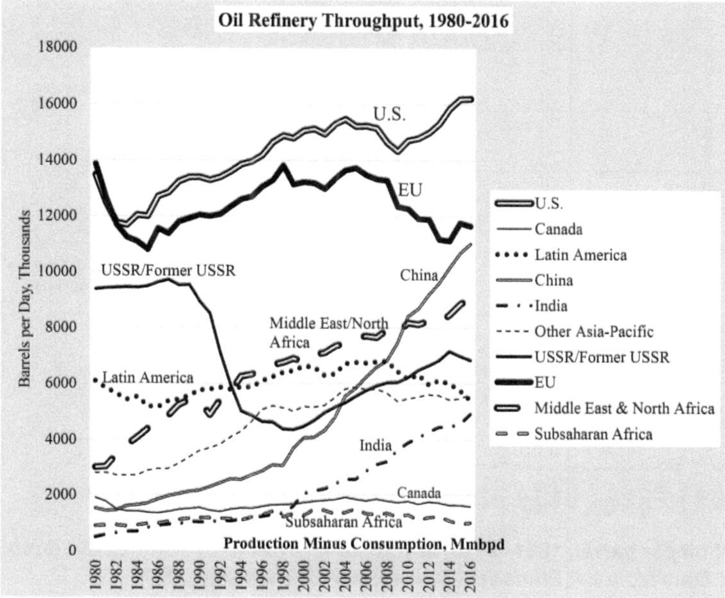

Oil refinery throughput, 1980–2016. Data from *BP Statistical Review of World Energy 2017*.

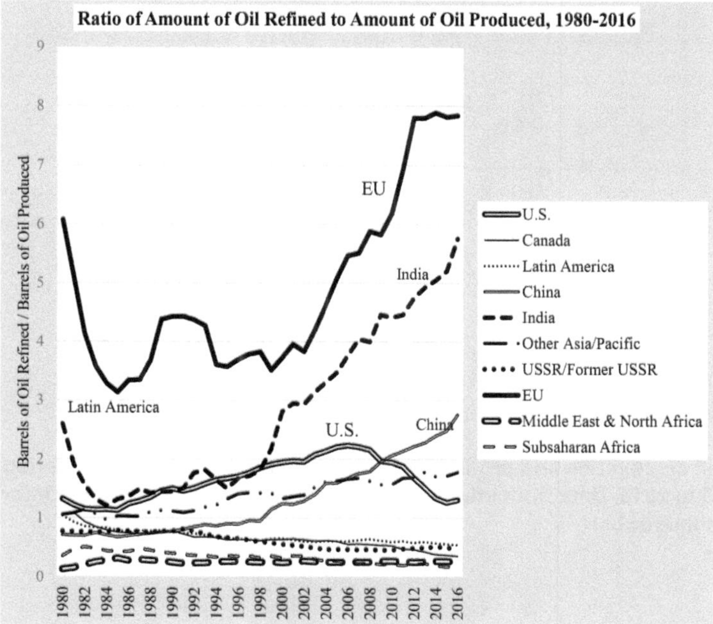

Ratio of amount of oil refined to amount of oil produced, 1980–2016. Data from *BP Statistical Review of World Energy 2017*.

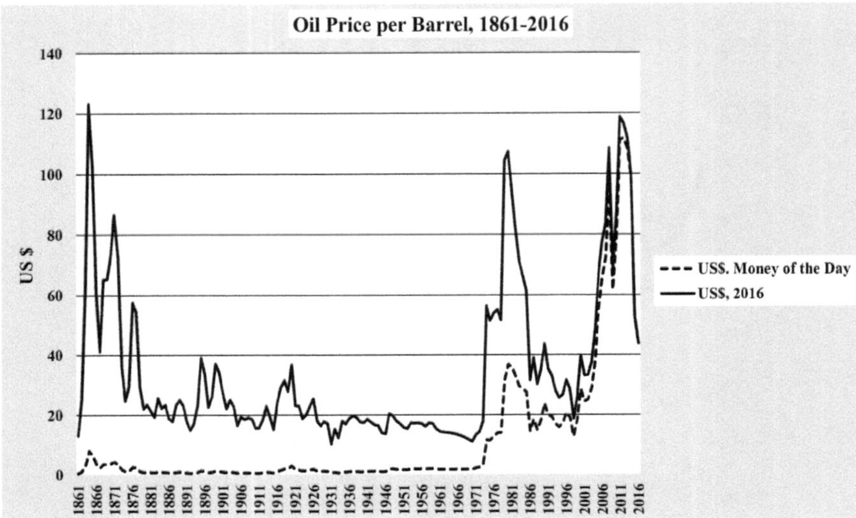

Price of oil per barrel, 1861–2016, in U.S. dollars of the day and 2016 constant dollars. Data from *BP Statistical Review of World Energy 2017*.

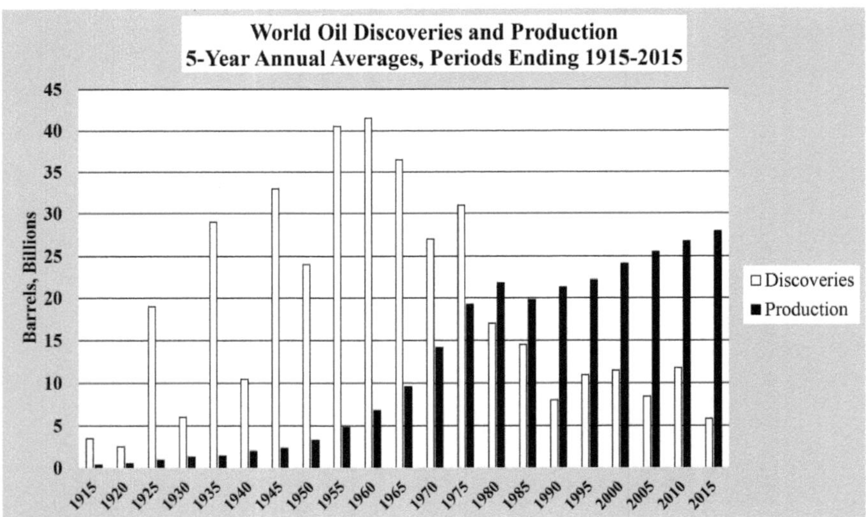

World oil discoveries and production, five-year annual averages for periods ending 1915 to 2015. Data from International Energy Agency; U.S. Energy Information Administration.

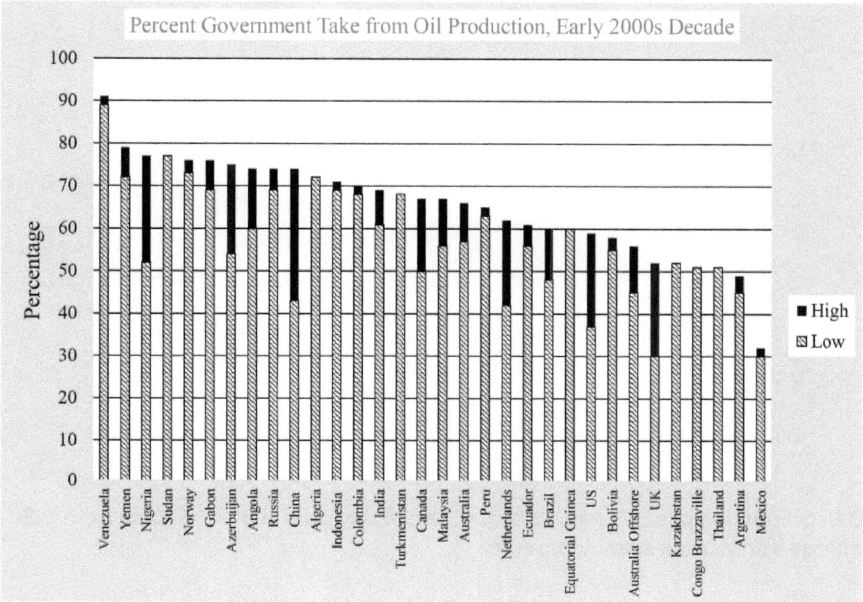

Percent government take from oil production, early 2000s decade.

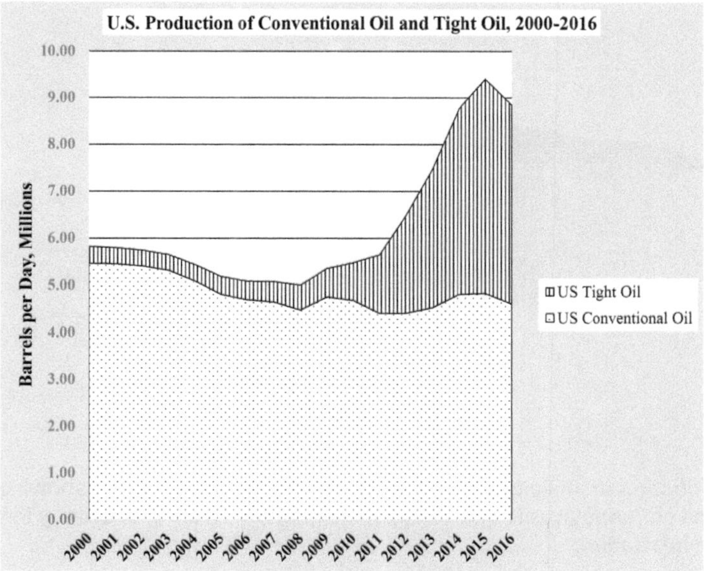

U.S. production of conventional oil and tight oil, 2000–2016. Data from U.S. Energy Information Administration.

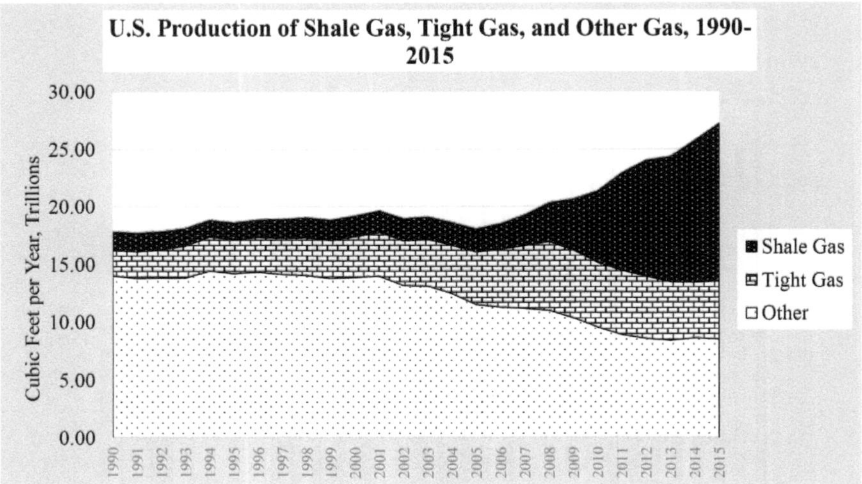

U.S. production of shale gas, tight gas, and other gas, 1990–2015. Data from U.S. Energy Information Administration.

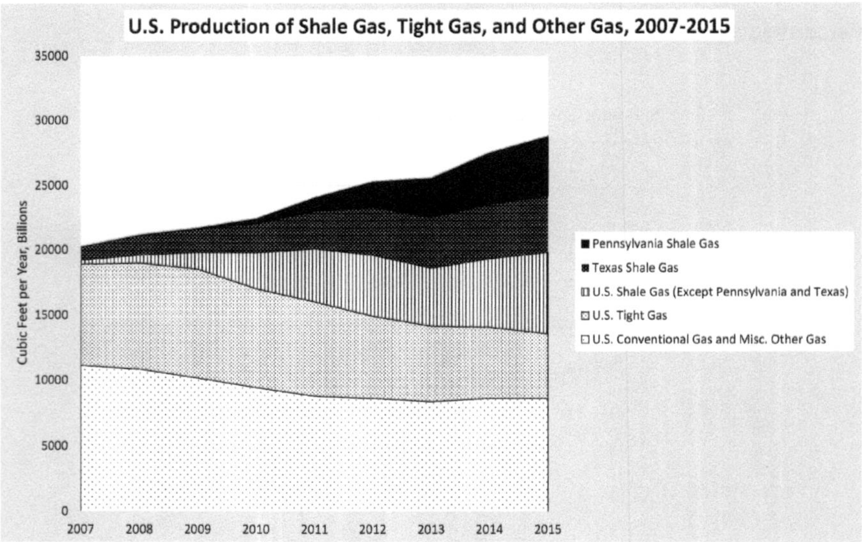

U.S. production of shale gas, tight gas, and other gas, 2007–2015, showing contributions of Pennsylvania and Texas shale gas. Data from U.S. Energy Information Administration.

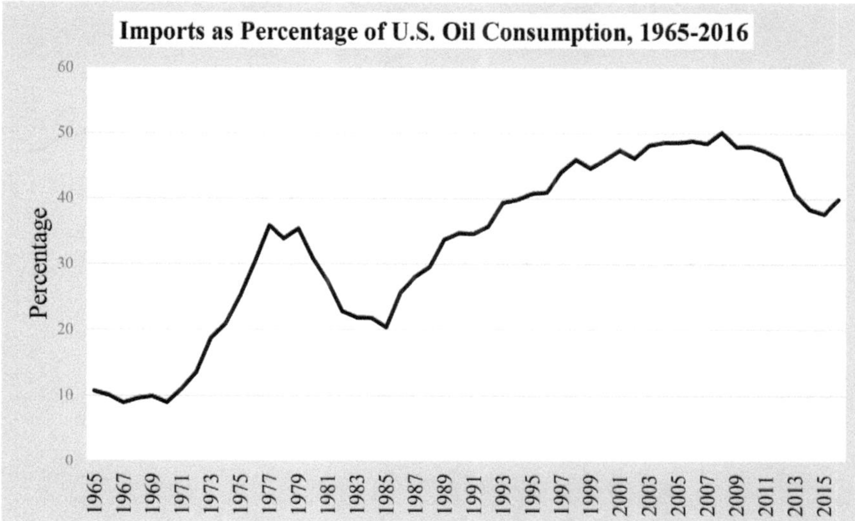

Imports as percentage of U.S. Oil Consumption, 1965–2016. Data from U.S. Energy Information Administration.

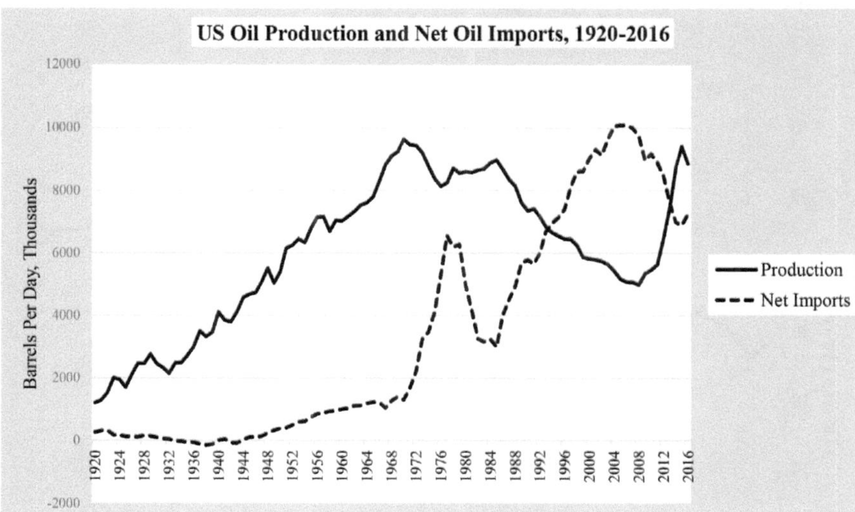

U.S. oil production and net imports, 1920–2016. Data from U.S. Energy Information Administration.

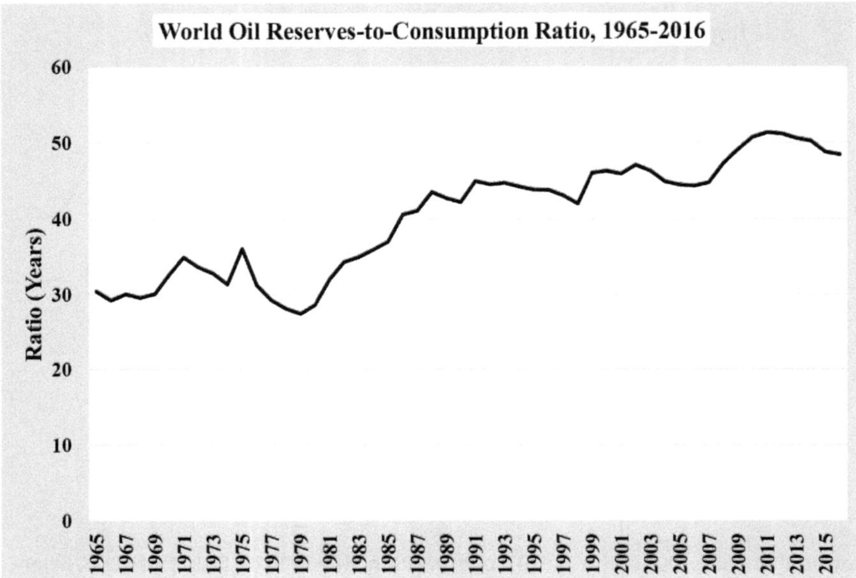

Ratio of world oil reserves to world oil consumption, 1965–2016. Data from *BP Statistical Review of World Energy 2017*.

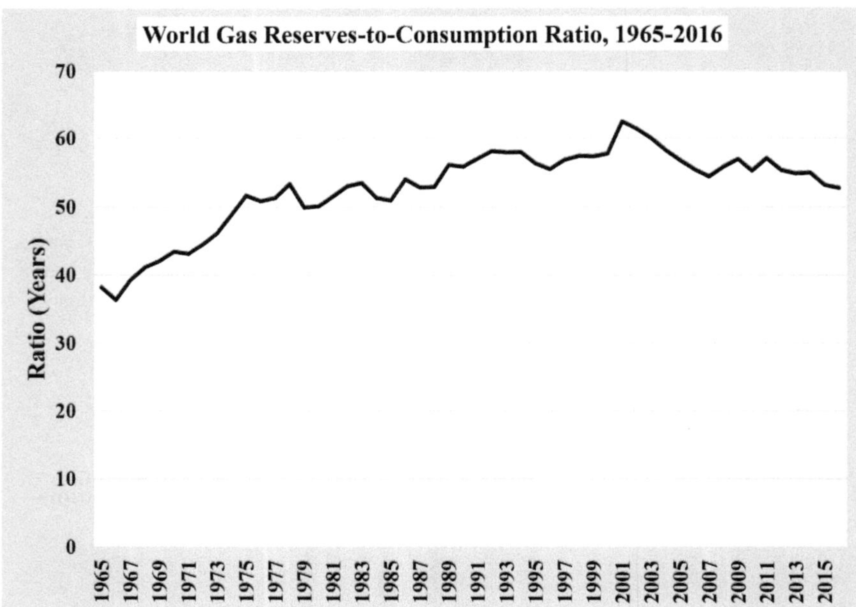

Ratio of world gas reserves to world gas consumption, 1965–2016. Data from *BP Statistical Review of World Energy 2017*.

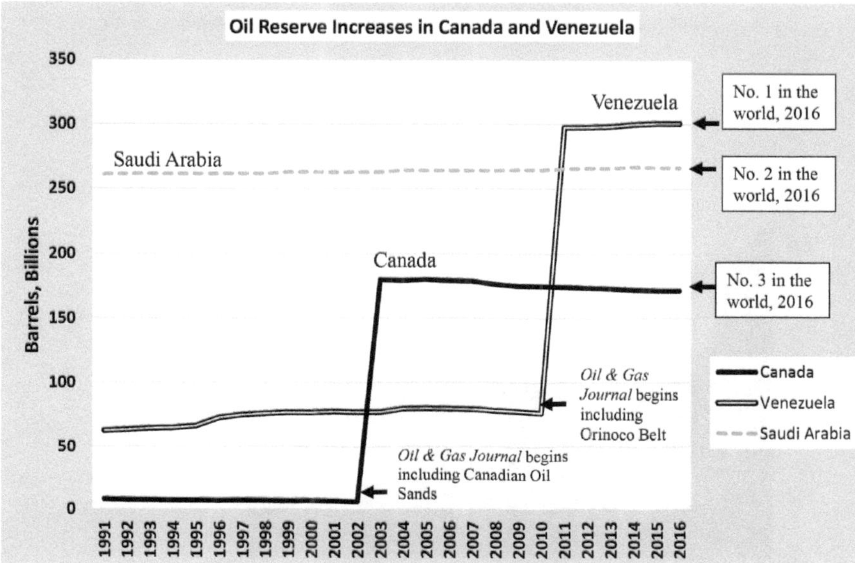

Reserve increases in Canada and Venezuela as Alberta oil sands and Orinoco Belt heavy and extra-heavy oil are included. Data from *Oil and Gas Journal*.

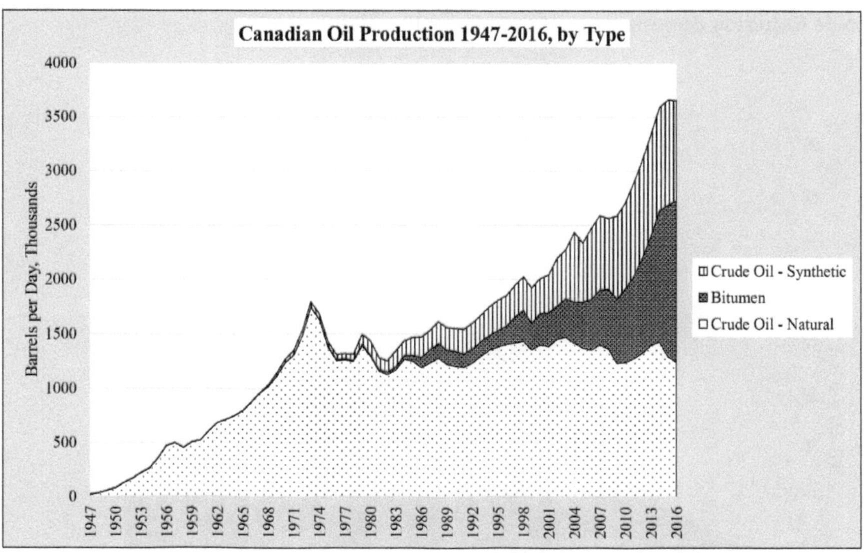

Canadian oil production, 1947–2016, showing contribution of bitumen and synthetic crude oil from the Alberta oil sands. Data from Canadian Association of Petroleum Producers.

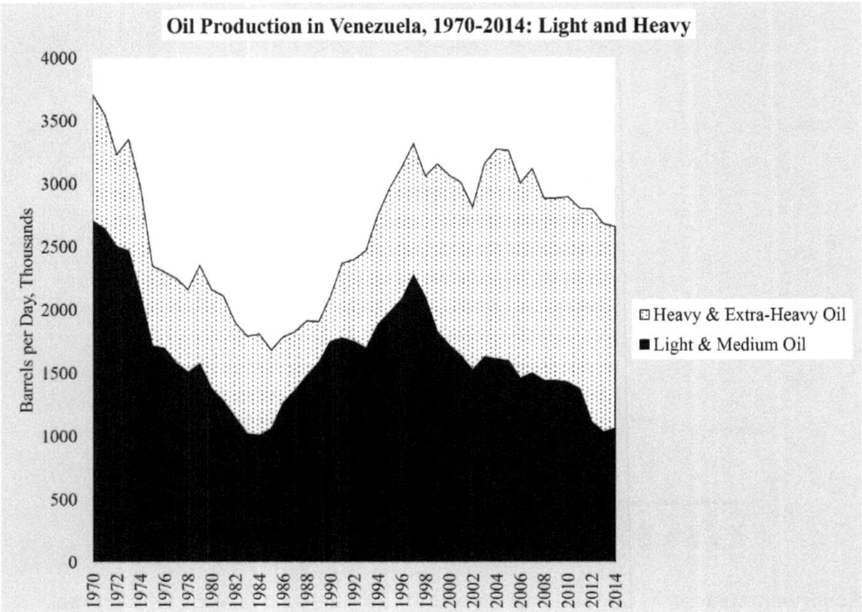

Venezuelan oil production, 1970–2014, showing contribution of heavy and extra-heavy oil. Data from International Center on Energy and the Environment, Instituto de Estudios Superiores de Administración, Venezuela.

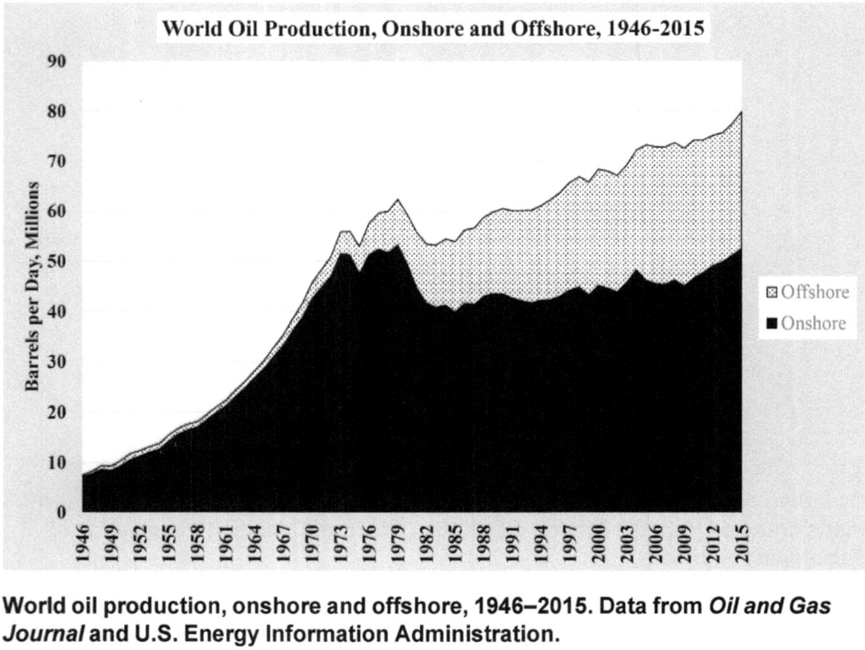

World oil production, onshore and offshore, 1946–2015. Data from *Oil and Gas Journal* and U.S. Energy Information Administration.

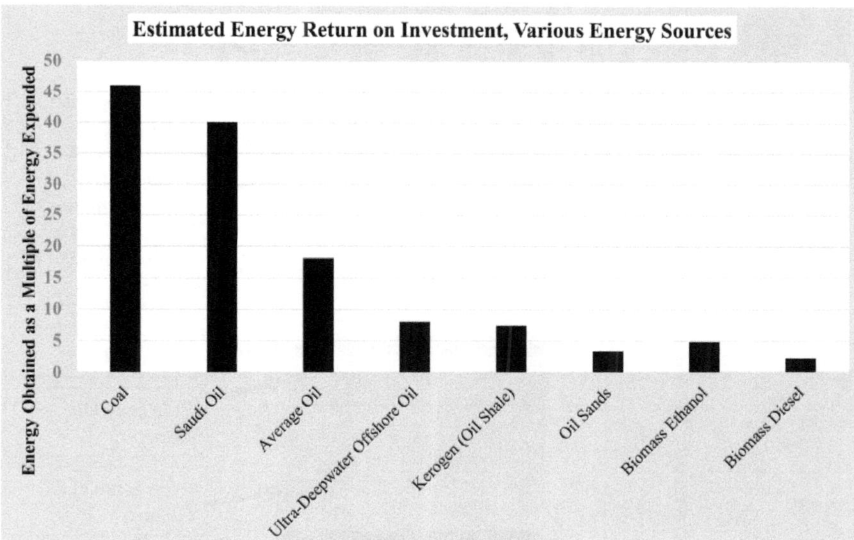

Estimated energy return on investment from various energy sources. Data from Hall and Klitgaard (2013) and Lambert et al. (2013).

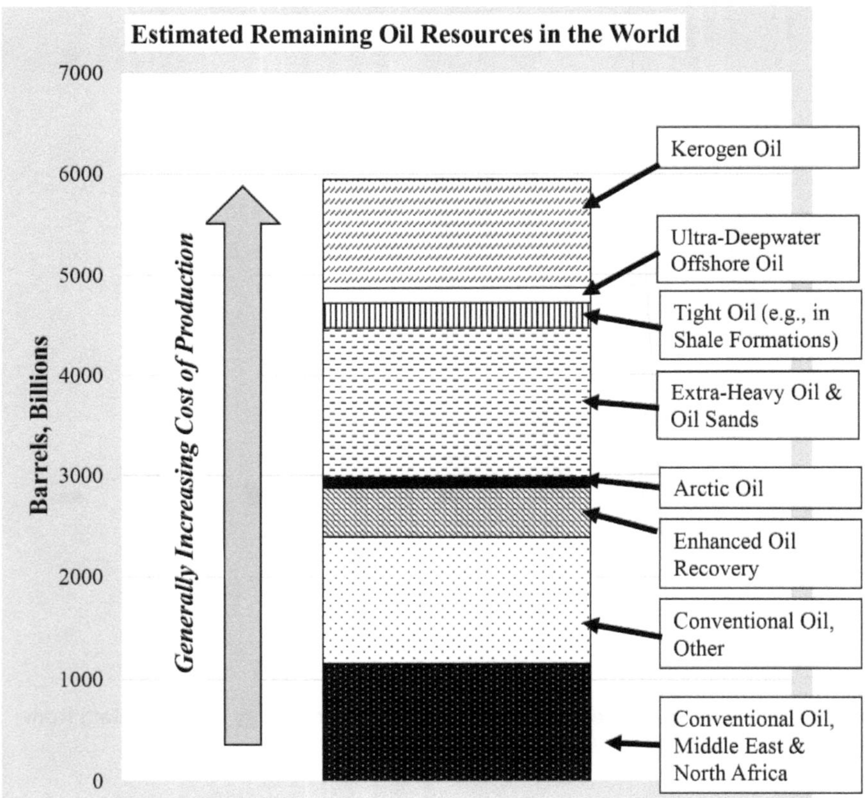

Estimated remaining oil resources (not reserves) of various types in the world. Difficulty and cost of extraction generally increases upward in the figures, but there are overlaps in cost and difficulty between various types. Data from International Energy Agency Resources to Reserves 2013.

K

KANSAS. Oil production 2015: 0.125 Mmbpd (1.3 percent of U.S.); **reserves** 0.337 Bb (1.04 percent of U.S.); gas production: 0.27 Tcf (1 percent of U.S.); reserves 3.18 Tcf (1.03 percent of U.S.). Kansas is a relatively minor producer today but is historically important. The first discovery of oil was the Norman No. 1 well near Neodesha, on 28 November 1892, by Albert P. McBride and Camden L. Bloom. This was also the first discovery in the vast **Mid-Continent Oil Region** spanning **Texas**, **Oklahoma**, **Louisiana**, Arkansas, and Nebraska. The Norman No. 1 well attracted attention from experienced oilmen, including **James M. Guffey** and John H. Galey, and by 1925 Kansas was the fifth-largest oil-producing state. On 5 October 1915 the Empire Gas Company (formerly Wichita Natural Gas, and acquired by **Cities Service Company**) drilled the Stapleton No. 1 well, discovering the **El Dorado Oil Field**. This was a highly significant event, because the timing of production from that field provided the United States with the margin of capacity it needed to supply the Entente forces and make a contribution to victory in **World War I**. The discovery of El Dorado was also significant because it was an early example of drilling being guided by the science of geology.

KAPELIUSHNIKOV, MATVEI (1886–1959). Russian/Soviet petroleum scientist and engineer. Matvei Alkumovich Kapeliushnikov was born 1 September 1886 in Adigeni, Georgia, which at the time was part of the **Russian** Empire. He graduated from the Tomsk Technological Institute in 1914. Working with S. M. Volokh and N. A. Kornev, he developed a turbodrill for **drilling** wells. With **Vladimir Shukhov**, he designed the first Soviet cracking plant, built in **Baku** in 1931. Among his many other innovations, he proposed in 1949 (with V. M. Fokeev) the use of high-pressure gas injection to increase **oil recovery**. He died in Moscow on 5 July 1959.

KARCHER, JOHN CLARENCE. *See* GEOPHYSICAL PROSPECTING.

KAZAKHSTAN. Oil production 2016: 1.581 Mmbpd (2.02 percent of world); **reserves** 30 Bb (1.82 percent of world); gas production: 0.657 Tcf (0.53 percent of world); reserves 85 Tcf (1.23 percent of world). The petroleum industry of Kazakhstan was first developed by the Soviet Union. Two **supergiant** fields, Tengiz and Karachaganak, were discovered in Soviet times. Tengiz, which lies on the northeast shore of the **Caspian** and has an **EUR** of 5.829 Bb (7.812 Bboe including gas), was discovered in 1980. Since 1993, Tengiz has been developed by the Tengizchevroil consortium, of which **Chevron** is the principal member and which also includes **ExxonMobil**, **KazMunaiGaz** (Kazakhstan's national petroleum company), and Lukarco. Lukarco, a former joint venture between **Lukoil** and **Arco** (later part of **BP**), is now owned by Lukoil and BP. In 2007, Tengiz produced about 280,000 bpd. Most of Tengiz's oil goes to the Russian Black Sea port of **Novorossiysk** via the Caspian Pipeline Consortium (CPC) **pipeline**.

Karachaganak, a gas and condensate field near the border with **Russia**, was discovered in 1979. It has an EUR of 5 Bb of condensate and 49.2 Tcf of gas, for a total of 13.2 Bboe. Karachaganak is now operated by **ENI**, British Gas, Lukoil, and Chevron. In 2007, Karachaganak produced about 250,000 bpd of condensate. The Kashagan field, discovered in 2000 in the Kazakhstan Caspian **offshore**, is the most important oil field discovered anywhere in the world since the late 1970s. It has an EUR of 10 Bb, 13.3 Bboe if gas is included. Kashagan began producing in 2016. The Kashagan project is operated by a consortium initially led by ENI, although in 2008 KazMunaiGaz became the largest shareholder, with a 16.81 percent interest. Other companies include ExxonMobil, **Total**, and **Royal Dutch Shell**, each with 16.66 percent (the same as ENI's new share), and **ConocoPhillips** and Inpex, each with 8.28 percent.

Oil production in Kazakhstan has risen markedly since the breakup of the Soviet Union and the country's independence. Oil production in 1992, at 0.444 Mmbpd, was less than a third of 2015 levels. **Natural gas** production also roughly tripled during the same period.

KAZMUNAIGAZ. National petroleum company of **Kazakhstan**.

KELLOGG, FRANK BILLINGS (1856–1937). Prosecutor in the case of the U.S. government against **Standard Oil** (1906–1909); U.S. secretary of state 1925–1928. Frank Billings Kellogg was born in Potsdam, New York, on 22 December 1856. He was admitted to the bar in 1877. Kellogg served as city and county attorney for St. Paul, Minnesota, and as special counsel to the U.S. government. In 1906, he led the Theodore Roosevelt administration's antitrust proceedings against Standard Oil in the federal circuit court in St. Louis (*see* SHERMAN ANTITRUST ACT). The case lasted three years and produced nearly 14,500 pages of transcripts. The court ordered the dissolution of Standard Oil in 1909 as a result. Standard Oil appealed to the U.S. Supreme Court and finally lost in 1911. Kellogg later became secretary of state under President Calvin Coolidge. He negotiated the Kellogg-Briand Pact for the maintenance of international peace and received the Nobel Peace Prize in 1929.

KESSLER, D. A. J. (1855–1939). Early exploration and production manager for **Royal Dutch Company** (later part of **Royal Dutch Shell**). Dominicus Antonius Josephus Kessler was born 21 August 1855 in Batavia, Dutch East Indies (Jakarta, **Indonesia**). He had a plantation near Garut, West Java, but in 1894 joined his older brother **Jean-Baptist Kessler** at the Royal Dutch Company as an exploration and later a production manager. After the death of his brother, he moved to the Netherlands and devoted his life to charity, building a permanent home for the homeless in The Hague. He died on 10 September 1939 at Kapelle op den Bosch, Belgium.

KESSLER, JEAN-BAPTIST (1853–1900). Managing director of **Royal Dutch Company** (later part of **Royal Dutch Shell**) who made the company viable. Jean-Baptist August Kessler was born 15 December 1853 in Batavia, Dutch East Indies (Jakarta, **Indonesia**). He worked as a trader with uneven success. In late 1891, he took over operations for the Royal Dutch Company, a bit less than a year after the death of its founder, **Aeilko Zijlker**. Zijlker had brought the venture to the brink of success, but after his death, the company slid backward and its production declined.

Kessler reversed the decline of the company and put it on a firm footing. He built an eight-mile **pipeline** through the jungle between the company's wells and Pangkalan Brandan on the Babalan Bay, where **tankers** would be loaded. He also constructed a refinery at the loading site where crude could be processed before shipment. He drilled more wells and increased the company's supply of crude. In April 1892, he began marketing kerosene under the brand name of Crown Oil. Under difficult conditions, with the considerable help of his engineer **Hugo Loudon** and his brother **D. A. J. Kessler**, he made the company profitable. Production increased fivefold between 1895 and 1897.

By the mid-1890s, Royal Dutch had become a potential takeover target. **Standard Oil** was already feeling threatened by **Russia**'s oil. The rise of Royal Dutch and **Marcus Samuel**'s **Shell Transport and Trading Company** in the East exacerbated these worries. Standard Oil had already initiated a price war in East Asia in 1894, sometimes giving oil lamps away for free. Kessler greatly feared Standard Oil, but apparently Standard Oil also feared him. Moreover, Shell and Royal Dutch were discussing the possibility of cooperating with each other. Such a combination would be even worse for Standard Oil. The discussions did not yield an immediate agreement. Kessler wanted some form of cooperative marketing arrangement for Asia, whereas Samuel wanted to acquire Royal Dutch. Standard Oil, which had already attempted unsuccessfully to acquire both Shell and **Branobel** in Russia, proposed to acquire Royal Dutch in 1897. Standard Oil offered to quadruple Royal Dutch's capital, taking a majority stake. It also offered not to interfere in operations. Kessler persuaded the Royal Dutch board to decline the offer.

In late 1897 and 1898, Royal Dutch began to have problems with its production. Its wells were in decline, flowing increasing proportions of saltwater. The value of the company's shares plummeted on the Amsterdam Stock Exchange. Ironically, neither Standard Oil nor Shell took the opportunity to buy up the shares at an attractive price. Thanks to Hugo Loudon, Kessler's company was able to find new sources of oil 80 miles to the north, in Perlak. The difficulties of turning Royal Dutch into a successful company and then maintaining it as such took their toll on Kessler. In November 1900, he cabled the Netherlands saying he was exhausted and wanted to go there. Unfortunately, he died en route in Naples on 14 December 1900.

KHARG ISLAND (KHARK). Iranian island in the northern **Persian Gulf** and **Iran**'s main oil export terminal. Kharg is located about 34 miles from Bushehr on the Iranian coast. It is about five miles long and 2.5 miles across at its widest point. Construction of petroleum installations on Kharg began in 1956. Within 20 years, Kharg had three large oil terminals: Kharg, Sea Island, and Darius. By the mid-1970s, it also had a major gas terminal, the Khemco Terminal. In 1975, Kharg Terminal could accommodate 10 tankers of up to 200,000 deadweight tons (DWT) each, and Sea Island could handle up to three vessels of 200,000–500,000 DWT. Darius could accommodate tankers up to 160,000 DWT.

In 1961, the **giant** Doroud (formerly Darius) field was discovered **offshore** Kharg by **Amoco** in partnership with the National Iranian Oil Company (IPAC venture—*see* IRAN). The field has an **EUR** of 1.655 Bb and 5 TCF of gas, for a total of 2.826 Bboe. It produces high-quality light oil (34.4°**API gravity**) and is linked to Kharg by **pipeline**. Other giant fields further offshore, and also linked to Kharg by pipeline, include Soroush (Cyrus), Foroozan (Fereidoon), and Abouzar (Ardeshir). They were discovered by Amoco in 1962, 1966, and 1969 respectively.

Kharg was bombed repeatedly during the **Iran–Iraq War**, and its petroleum infrastructure was severely damaged. Iraqi attacks on Kharg in August 1982 cut Iranian oil exports roughly in half. In 1986, the Kharg facilities were put completely out of commission for a time. Reconstruction began after the cease-fire of 1988 and proceeded with difficulty. Before the war, Kharg's offloading capacity was about 14 Mmbpd. In 1993 after partial reconstruction, that capacity was 2 Mmbpd. By 2007, it had reached 5 Mmbpd, with a storage capacity of 16 Mmbo. In the 2000s, Kharg was responsible for about 90 percent of Iran's crude exports.

KHAZ'AL KHAN (1861–1936). Sheikh in the oil-rich province of **Khuzestan, Iran**, who facilitated early British oil operations. Khaz'al Khan became ruler of Mohammerah (now known as Khorramshahr) and overlord of the Arab tribes of Khuzestan in June 1897 on the assassination of his brother, an event with which he is thought to have been connected. Khuzestan had a large Arab population and enjoyed considerable autonomy from the central government in Tehran. Nonetheless, Khaz'al Khan wanted more. He believed the central government was weak and that Iran would fragment. He wanted to create a completely independent state in Khuzestan in the event. He sought support from the **United Kingdom** beginning in 1898.

Once oil was discovered in 1908 in his province by the forerunner to the Anglo–Persian Oil Company (APOC), which later became the **Anglo–Iranian Oil Company**, eventually **BP**, his ties to Britain strengthened. In 1909, the British government asked Percy Cox, British resident at Bushehr, to negotiate an agreement with Khaz'al for APOC to obtain a site on **Abadan** Island for a refinery, depot, storage tanks, and other operations. The refinery was built and began operating in 1912. Khaz'al was knighted in 1910 and supported the United Kingdom in **World War I** while the central government in Tehran remained neutral. Khaz'al's fortunes fell when the British failed to establish a protectorate in Iran in 1921. He fought the forces of Reza Khan (later Shah **Reza Pahlavi**) and was defeated in 1924. He died under house arrest in Tehran on 25 May 1936. Khuzestan lost its autonomy.

KHODORKOVSKY, MIKHAIL (1963–). Russian business executive who headed **Yukos** Oil Company; imprisoned from 2003 to 2013. Mikhail Borisovich Khodorkovsky was born on 26 June 1963. In 1986, at the age of 23, he was the deputy head of the Komsomol at the Mendeleev Institute in Moscow. In that position, he was able to begin building a fortune in the waning years of the Soviet Union by exploiting certain oddities in the Soviet monetary system. The Soviet Union had a type of money designated as "noncash," used as an accounting device by state enterprises. Beginning in 1987, the Komsomol was allowed to turn noncash into cash. Khodorkovsky set up a clearinghouse for executives in state enterprises wishing to turn noncash into cash. He also began a cooperative that (possibly illegally) imported computers and liquor. He made enough money to establish his own bank, Menatep. He was well connected in the Soviet government and in **Russia**'s post-Soviet government under President Boris Yeltsin.

Khodorkovsky acquired a controlling interest in Yukos in 1995 via the **Loans for Shares** scheme for the very low price of $350 million. He increased his stake to 85 percent in 1996. Khodorkovsky's power continued to rise during the Yeltsin years, but the situation changed abruptly when Vladimir Putin became Russia's president in 2000. There were several points of conflict between Khodorkovsky and the new government. Putin wanted an increasing share of energy revenue for the state. Khodorkovsky resisted state interference in the private sector. The government tried to raise excise tax rates on oil exports in the 2003 Duma session. Khodorkovsky mobilized parliamentary opposition and killed the measure. He also led a private initiative to expand and privatize Russia's **pipeline** system. The government opposed this not only because of the revenue it gained from its pipeline monopoly, but also because of the information that the monopoly gave it on petroleum transactions. The government was also annoyed by his outspoken criticism of official corruption—not least because he himself had acquired Yukos under highly suspicious circumstances.

In October 2003, Russian state commandos seized Khodorkovsky at the Novosibirsk airport and imprisoned him in Moscow. He was accused of defrauding the state of privatization revenues, personal and corporate tax evasion, and other charges. Yukos was given a bill for more than $15 billion of back taxes and fines. On 31 May 2005, Khodorkovsky was found guilty and sentenced to nine years in prison. He was moved to a prison camp in Krasnokamensk. In 2010 the government leveled additional charges against him, and his prison sentence was extended to 2014. He was pardoned by Putin in 2013 and released. He moved to Switzerland. Before his arrest in 2003, Khodorkovsky was believed to be the richest person in Russia and the 16th richest in the world. His net worth after his release could still have been as high as $500 million.

KHOMEINI, RUHOLLAH MUSAVI (1902?–1989). Supreme leader of the Islamic Republic of **Iran** 1979–1989; instrumental in paralyzing the Iranian oil industry and precipitating the **Second Oil Shock**. Ruhollah Musavi Khomeini was born Ruhollah Musavi in Khomein, Iran; his exact birth date is unknown, but 17 May 1900 and 24 September 1902 are commonly mentioned. He took the name of his hometown as a surname when he was about 30 years old. Both his father and grandfather were mullahs, and he was educated in Islamic schools. He became an Islamic scholar and religious leader, rising to the rank of ayatollah in the 1950s and grand ayatollah in the 1960s.

Khomeini was a vocal opponent of Shah **Mohammed Reza Pahlavi** and his relatively secular rule. He protested vigorously against the shah's progressive treatment of women. He also opposed aspects of the shah's land reforms, because they had the effect of reducing religious properties. He was arrested on 5 June 1963 after delivering a sermon denouncing the shah. Khomeini's arrest led to several days of demonstrations across the country. He was imprisoned for one year and then exiled on 4 November 1964. He settled in **Iraq** and opposed the shah's regime from exile. **Saddam Hussein**, then vice president of Iraq, ejected him on 6 October 1978 under pressure from the shah's government. Khomeini moved to France, settling outside Paris.

Khomeini continued to oppose the shah and incited his supporters to demonstrate and strike, using recorded messages. Strikes paralyzed the country's oil industry and helped precipitate the Second Oil Shock of the 1970s. In 1978, Iranian production fell from 5.7 Mmbpd to 0.7 Mmbpd. Exports fell to zero by the end of the year, and global prices surged ($30.03 per barrel for 1979 vs. $13.60 the previous year). On 26 December 1978, Khomeini declared that as long as the shah remained in the country, there would be no oil exports. The shah left Iran on 16 January 1979, and Khomeini returned from exile on 1 February. In December 1979, Khomeini became the supreme leader of the new Islamic Republic of Iran.

Khomeini pursued a domestic policy firmly based on Islamic law; his foreign policy was adamantly anti-Western, leading him to a path of confrontation with the **United States**. He approved of the 4 November 1979 seizure of the U.S. embassy in Tehran and the holding of American personnel there as hostages. The state of tension between the two countries continued after his death in 1989 and even intensified in the early 2000s. During the **Iran–Iraq War**, Khomeini pushed implacably for victory against the secular Saddam Hussein, probably prolonging the war and seriously damaging his country's economy. Khomeini died 3 June 1989 in Tehran.

KHUZESTAN (KHUZISTAN). Oil-rich province of **Iran** at the head of the **Persian Gulf** and bordering **Iraq**. The province has a significant population of ethnic Arabs and was formerly known as Arabestan. It enjoyed considerable autonomy until the defeat of its ruler **Khaz'al Khan** by Shah **Reza Pahlavi** in 1924. Oil was discovered at Masjid-e-Suleiman in 1908 by the forerunner to the Anglo–Persian Oil Company (later **Anglo–Iranian Oil Company**, eventually **BP**). A refinery was built and began operating in **Abadan** in 1912, after negotiations by the **United Kingdom** with Khaz'al Khan. The Masjid-e-Suleiman field has an **EUR** of 1.13 Bboe. It is now possibly nearing exhaustion, with an estimated remaining recovery ranging from 2 to 362 Mmboe.

Over the course of the 20th century, 14 **giant** oil fields and seven **natural gas** giants have been discovered in Khuzestan. Early discoveries include the oil giant Haft Kel in 1926 (EUR 2 Bboe), the **supergiant** Agha Jari field in 1936 (EUR 5.76 Bboe), and the gas giant Pazanan (EUR 5 Bboe) in 1936. The Ahvaz oil supergiant was discovered in 1958 near the provincial capital and has an EUR of 13.8 Bb and 23.3 Tcf gas. Another huge field, Marun, with an EUR of 12.63 Bb and 75.3 Tcf gas, was discovered in 1964. Khuzestan is also the site of one of the world's most important recent supergiant discoveries, the Azadegan field, with an EUR of 6.4 Bboe, discovered in 1999.

The oil industry has made Khuzestan prosperous and a crucial part of Iran's economy. Khuzestan also benefited from the Trans-Iranian Railway, an 871-mile line completed in 1939 linking Tehran to both the Persian Gulf and the **Caspian Sea**. Abadan and Khorramshahr have become major ports. Because of its proximity to Iraq, Khuzestan suffered the most damage during the **Iran–Iraq War** of 1980–1988. Iraq under **Saddam Hussein** occupied western Khuzestan in 1980, including the city of Khorramshahr. The Iraqis also bombed the Abadan oil refinery. By 1982, however, Iran had taken back Khuzestan.

KIER, SAMUEL M. (1813–1874). Early producer and marketer of **Pennsylvania** oil; one of the first to establish a market value for petroleum. Samuel Martin Kier was born 19 July 1813 in Indiana County, Pennsylvania. His father manufactured salt from brine. Kier began his career as a forwarding merchant in Pittsburgh, eventually becoming a partner in the company. The business foundered in the panic of 1837. In 1838, he established a canal boat company plying the route between Pittsburgh and Philadelphia.

In the 1840s, Kier began to notice the oil that flowed as an undesirable by-product from his father's brine wells in Tarentum, Pennsylvania. In 1848, he began bottling it and marketing it as a patent medicine. He advertised it as "Kier's Petroleum or Rock Oil, nature's remedy celebrated for its wonderful curative powers, taken from a well four hundred feet below the earth's surface." The oil was claimed to ameliorate suffering due to rheumatism, gout, neuralgia, colds, sores, and a host of other problems, including diabetes. Kier also called the petroleum "Seneca Oil," after the local Native Americans who had reportedly long used the substance for medicinal purposes. The sales of patent medicine produced reasonable revenue but did not consume all the oil flowing out of the wells.

Kier sought other uses for the oil and in 1850 began experimenting with it as an illuminant. Crude oil produced too much smoke and an unpleasant odor. He sought help from Professor Booth, a Philadelphia chemist, who advised him to distill the oil. By doing so, he was eventually able to produce a fairly effective illuminant, probably similar to kerosene, that he marketed as "carbon oil." He built a single-barrel still on 7th Avenue in Pittsburgh, becoming one of America's first oil **refiners**. Kier's business records show that he was selling his "carbon oil" as early as 1851. By 1853, Kier's product was selling for $1.50 a gallon ($63 per barrel, an enormous sum at the time). Among the early producers and marketers of kerosene, Kier was roughly contemporaneous with **Ignacy Łukasiewicz** in Poland and a tad later than refiners in Parkersburg and the Kanawah Valley of **West Virginia**. He preceded **James Miller Williams** in Ontario. Kier bought crude oil from operators of brine wells and early hand-dug oil wells throughout the region, and he was one of the first people to establish a commercial market and value for petroleum. In 1853, he contracted to purchase the entire crude output of one well for five years at 66.5 cents a gallon ($27.72 per barrel).

After **Edwin Drake**'s well of 1859 produced an oil boom, Kier began to invest in the Pennsylvania oil regions. He helped develop **Oil City** at the mouth of Oil Creek into a marketing and shipping port for petroleum. He built storage tanks and operated a crude oil trading and shipping business in the 1860s. Kier died on 6 October 1874. *See also* GESNER, ABRAHAM (1797–1864).

KIRKUK. City (pop. ca. 850,000, 2009) in northeastern **Iraq**; major center of Iraq's petroleum industry. Kirkuk is considered a Kurdish city by many Kurds, but it is not de jure part of the **Kurdistan Regional Government** (KRG)'s zone. Nevertheless, it came under de facto KRG control in 2014 when the Iraqi army retreated in response to an offensive by forces of the Islamic State, and Kurdish forces took the city. The Iraqi army took back control of Kirkuk in 2017. Kirkuk is a continuing source of tension between the KRG and the rest of Iraq. The region around the city is replete with **giant** oil fields. Iraq's first major discovery was the **Kirkuk oil field**. This field, with an **EUR** of about 26 Bboe, was discovered about 50 miles south of the city in 1927. The Khabbaz field discovered in 1985 is only 10 miles west. Kirkuk is also the starting point for the **Kirkuk–Ceyhan Pipeline**.

KIRKUK–CEYHAN PIPELINE (KIRKUK–YUMURTALIK PIPELINE). Iraq's largest export **pipeline** for crude oil, completed in the 1970s. The pipeline is 600 miles long. It has two parallel lines, a larger one with a diameter of 46 inches and a smaller one of 40 inches. The combined maximum capacity is 1.6 Mmbpd, at least theoretically. Iraq relied heavily on the pipeline during the 1980s **Iran–Iraq War**, when shipping through the **Persian Gulf** was restricted by the fighting. The pipeline was closed in August 1990 under United Nations Security Council Resolution 661, which placed an embargo on Iraq because of its invasion of **Kuwait**. The embargo was partially lifted by Security Council Resolution 986 in 1995. Iraq was allowed to export a limited amount of crude oil in order to provide for the humanitarian needs of its people (*see* OIL FOR FOOD PROGRAM). These exports reached about 2 Mmbpd in 1999, with about half flowing through the pipeline. Iraq exported about 700,000–800,000 bopd through the pipeline in the years immediately preceding the **Iraq War** initiated by the **United States** in 2003.

During the Iraq War, the Kirkuk–Ceyhan Pipeline operated irregularly because of frequent sabotage and other damage. Major repairs were required after a bridge collapsed on the pipeline near Baiji as a result of bombing by American planes. These repairs were completed in 2005 and included construction of a new tunnel under the Tigris River. Pumping stations have also been significantly damaged by looters. In June 2006, exports to Ceyhan were set to resume at a significantly diminished capacity of 150,000–500,000 bopd, but on 9 July 2006 a sabotage attack closed down both lines. On 31 July 2006, another attack delayed the resumption of operations. The pipeline operated sporadically in 2007. In October 2007 it was moving about 400,000 bopd. However, sabotage continued, with attacks in 2009 and 2013.

KIRKUK OIL FIELD. First major oil discovery in **Iraq**; a **supergiant** that has produced for over seven decades. The field was discovered at Baba Gurgur about 50 miles south of the city of **Kirkuk** on 15 October 1927. It began producing in 1934 under the management of the **Iraq Petroleum Company**. The field has an **EUR** of 25 Bb and 8.2 TCF of **natural gas**, for a combined total of 26.37 Bboe. Estimates of remaining recoverable reserves range from 192 Mmboe to nearly 10 Bboe. There are some worries that the Kirkuk field may have been damaged by poor reservoir management practices under the government of **Saddam Hussein**. For example, too much fuel oil reinjection increased the viscosity of the oil. In 2014 the Kirkuk field came under the de facto control of the **Kurdistan Regional Government**. In 2017 the Iraqi army took back control of Kirkuk.

KOGALYMNEFTEGAZ. *See* LUKOIL.

KONINKLIJKE NEDERLANDSCHE MAATSCHAPPIJ TOT EXPLOITATIE VAN PETROLEUMBRONNEN IN NEDERLANDSCHE INDIE. *See* ROYAL DUTCH COMPANY.

KUDRYAVTSEV, NIKOLAI (1893–1971). Russian petroleum geologist and advocate of the **abiogenic theory** of petroleum formation. Nikolai Alexandrovich Kudryavtsev was born in Opochka, **Russia**, on 21 October 1893 and graduated from the Leningrad Mining Institute in 1922. He became a professor there in 1941 after obtaining a doctorate in geology in 1936. He began his geological career in 1920, before graduating from university. He would later spend several years in forced labor camps under Joseph Stalin.

The abiogenic theory postulates that petroleum is formed from nonbiological sources of hydrocarbons located deep in the earth's crust and mantle, and not from the decay of biological organisms as is commonly believed. Kudryavtsev cited examples of significant petroleum deposits being found in basement rocks, or in sediments directly overlying fractures in the basement. The **megagiant Ghawar** field of **Saudi Arabia** can be said to fall in the latter category. He also believed that the large deposits in the **Alberta oil sands** of **Canada** would have required huge quantities of source rock in the conventional theory, whereas none have been found.

He cited much other evidence for the abiogenic theory, and he also formulated **Kudryavtsev's rule**, which is consistent with the idea of deep sources of hydrocarbons. Kudryavtsev's notions were championed by American astrophysicist **Thomas Gold**, who independently arrived at many similar conclusions and added elements of his own. The great majority of petroleum geologists in the West discount the abiogenic theory. They argue that the presence of oil in basement rocks is the result of migration and that there are undeniable biological markers in oil.

Kudryavtsev's geologic work resulted in discoveries of petroleum in the Grozny district (Chechnya), **Timan–Pechora**, Central Asia, and many other areas of the Soviet Union. He also worked in western **Siberia**, which proved to be a prolific oil and gas province. He died in Leningrad (St. Petersburg) on 2 December 1971.

KUDRYAVTSEV'S RULE. Rule formulated by **Nikolai Kudryavtsev**, stating that a region in which hydrocarbons are found at one depth will also have hydrocarbons in some quantity and in some form at all levels down and into basement rock, with **natural gas** usually being the deepest.

KURDISTAN REGIONAL GOVERNMENT (KRG). Governing authority of an autonomous, primarily Kurdish area in **Iraq**, operating under the country's 2005 constitution. The KRG does not have de jure control of **Kirkuk** but took de facto control in 2014—although it lost that control in 2017. In 2014 the KRG's oil production accounted for nearly 12 percent of Iraq's total, and Kirkuk accounted for 3.4 percent. In 2016 the percentages were about 14 and 11 percent respectively. Petroleum is a large cause of tension between the KRG and the federal Iraqi government in Baghdad. In the 2000s and 2010s the KRG began independently signing contracts without Baghdad's approval, and this led to recriminations. As an example, in 2012 Iraq barred **Chevron** from oil agreements with the central government because the company had signed a contract with the KRG. Companies continued to sign contracts and agreements with the KRG anyway. **Rosneft** did so in early 2017. Further exacerbating tensions with Baghdad, the Kurds held an independence referendum on 25 September 2017, and a majority voted for independence—for both the KRG and other "Kurdish areas." Independence could greatly complicate the KRG's petroleum industry. Independence is opposed not only by Iraq, but also by Turkey; the latter country represents the KRG's only practical outlet for oil exports, via the **Kirkuk–Ceyhan Pipeline**.

KUWAIT. Oil production 2015: 2.779 Mmbpd (3.55 percent of world); **reserves** 101.5 Bb (6.17 percent of world); gas production: 0.584 Tcf (0.47 percent of world); reserves 63 Tcf (0.91 percent of world). An important petroleum exporter, Kuwait was a founding member of both the **Organization of the Petroleum Exporting Countries** (OPEC) in 1960 and the **Organization of Arab Petroleum Exporting Countries** (OAPEC) in 1968.

Observers from the **United Kingdom** noticed surface outcrops of asphalt in Kuwait early in the 20th century. The Anglo–Persian Oil Company (APOC), which later became the **Anglo–Iranian Oil Company**, eventually **BP**, first expressed interest in a **concession** there in 1911. In 1913, a member of a British Admiralty Commission visiting the **Persian Gulf** area conducted a promising geological survey of Kuwait. In that year, Shaykh Mubarak agreed to issue a concession only to an entity nominated and recommended by the British government. In 1914, an APOC geologist examined oil seeps at Burgan and Bahrah, and in 1917 APOC made a more comprehensive survey. Despite APOC's interest in Kuwait, it took until 1934 for a concession to be granted. The **Gulf Oil Corporation** became interested in the 1920s but was stymied by the fact that it was not a British company (*see* BRITISH NATIONALITY CLAUSE). The two eventually formed a joint venture, the Kuwait Oil Company (KOC), incorporated in London on 2 February 1934 with an initial capitalization of £50,000. In December 1934, KOC obtained a 74-year petroleum concession in Kuwait.

In May 1936, KOC drilled a hole to 7,950 feet at Bahrah, but it was dry. The company conducted a geophysical survey in the winter of 1936–1937 and selected a location in the Burgan area. On the night of 23–24 February 1938, KOC drilled the discovery well of the **Burgan oil field**, at the time the world's largest and still the second-largest oil field ever discovered (after **Ghawar** in **Saudi Arabia**). **EUR** for Burgan ranges as high as 70 Bb. **World War II** interrupted Burgan's development, and production did not begin until 1946. Production that year was 0.016 Mmbpd, but it grew explosively, surpassing 1 Mmbpd in 1955 and 2 Mmbpd in 1963.

KOC continued exploration in the 1950s and made a number of major discoveries. It found Ahmadi and Magwa, recognized as part of the Greater Burgan complex. Among others, KOC found the **giant** fields Bahrah, Sabriyah, Minagish, and Raudhatain. **Supergiant** Raudhatain, with an EUR of 7.2 Bb and 6.7 Tcf of gas, was the largest after Burgan. It was discovered in 1955.

Iran's 1951 nationalization of its petroleum industry and the resulting Western boycott removed significant supplies of crude from the world market, and Kuwait increased its exports to make up the difference. By 1956, Kuwait was the leading exporter in the region, a position it held for a decade. Of Kuwait's 10 giant oil fields (excluding the **Neutral Zone** shared with Saudi Arabia), six were discovered before 1960. All but the smallest one were discovered before 1980.

In 1972, Kuwaiti oil production peaked at 3.291 Mmbpd. It then declined, falling to a low of 0.823 Mmbpd in 1982. The Kuwaiti government made a deliberate decision to restrain production. The country was already very wealthy, and it made more sense for it to conserve its oil resources rather than trade them for more U.S. **dollars**. In the 1970s, Kuwait also nationalized its petroleum industry. It negotiated a **participation agreement** in the OPEC framework, which would have provided for an immediate 25 percent government ownership and a phased-in increase over a period of years. However, Kuwait's parliament failed to ratify the agreement. The government signed a new agreement with BP and Gulf, by which it acquired a 60 percent stake in KOC in 1974. The new agreement called for the government to acquire 100 percent by 1979, but Kuwait completed its acquisition in 1975.

In 1980, Kuwait created the Kuwait Petroleum Corporation (KPC) as an umbrella organization for its industry. The KPC held a number of companies, including KOC for domestic upstream operations and Kuwait National Petroleum Company (KNPC) for domestic downstream operations. KPC evolved into one of the most successful and best-integrated national oil companies. It made a number of international acquisitions in the 1980s and 1990s and joined in international upstream ventures. It was the first national oil company from the developing world to sell petroleum products under its own brand ("Q8," introduced in Europe in 1988) and from its own stations.

During the **Iran–Iraq War**, Kuwait helped **Iraq** meet its production obligations. Along with Saudi Arabia, it also provided aid to Iraq. Kuwait joined with Saudi Arabia in flooding the market in 1986, contributing to the **Oil Countershock**. The collapsing prices limited Iran's ability to prosecute its war, while the Kuwaiti and Saudi aid helped cushion Iraq somewhat. After the Iran–Iraq War, Kuwait angered Iraq by insisting on the repayment of $10–12 billion that Kuwait reckoned it was owed for having supplied Iraq's customers with oil. Iraq also accused Kuwait of stealing Iraqi oil from the Rumaila oil field, which is mostly in Iraq but extends into Kuwait.

These factors, as well as a long-standing belief that Kuwait properly belongs to Iraq, led Iraq to invade Kuwait in August 1990, precipitating the 1991 **Persian Gulf War**. Although Iraq was expelled from Kuwait by a coalition led by the **United States**, its retreating forces did a large amount of damage to Kuwait's oil infrastructure and environment. They dynamited 732 producing wells, and 650 of them caught fire, including 365 in the Greater Burgan field. The fires burned for more than eight months. Kuwaiti production collapsed in the wake of the war. In 1991, it was 0.19 Mmbpd, versus 1.783 Mmbpd in 1989. By 1994, it had recovered to 2.025 Mmbpd. It hovered around 2 Mmbpd until 2003, rising to 2.535 Mmbpd in 2006, close to the 2015 level of 2.779 Mmbpd.

Kuwait's **natural gas** production has varied in a manner similar to its oil production. In 1980, it was 0.244 Tcf. It dipped to a low of 0.1 Tcf in 1982 and rose to a peak of 0.288 Tcf in 1989. In 1991, in the wake of the Persian Gulf War, it was 0.018 Tcf, recovering to 0.328 Tcf in 1996. It hovered around 0.3 Tcf until 2003 and then rose. The 2015 level of 0.584 Tcf was equivalent in energy terms to about 0.29 Mmboepd. *See also* HOLMES, FRANK (1874–1947).

L

LA BREA OIL FIELD. Also called Negritos oil field. The world's first-discovered giant oil field (EUR about 1 Bb), located in the northern coastal region of **Peru**. It was discovered between 1868 and 1873.

LAGOVEN. Autonomous subsidiary of **PdVSA** formed in 1975 to take over the Venezuelan petroleum operations of **Creole Petroleum** (**Exxon**) after **Venezuela** nationalized its petroleum industry. The subsidiary was merged back into PdVSA in 1998. The subsidiary participated in joint ventures in the **Orinoco Oil Belt** during the **Apertura Petrolera**.

LANGELED PIPELINE. *See* PIPELINES; STATOIL.

LANGEPASNEFTEGAZ. *See* LUKOIL.

LANSDOWNE DECLARATION. Declaration made in May 1903 by British foreign secretary Henry Charles Keith Petty-Fitzmaurice, 5th Marquess of Lansdowne, affirming the **Persian Gulf** as a zone of vital British interest. The declaration was made in the House of Lords and stated that foreign powers attempting to establish a military presence in the region would be resisted by any means at Britain's disposal. Later in the century, the United States would adopt a similar stance in the **Carter Doctrine**. *See also* ANGLO–IRANIAN OIL COMPANY (AIOC); CHURCHILL, WINSTON (1874–1965); UNITED KINGDOM (UNITED KINGDOM OF GREAT BRITAIN AND NORTHERN IRELAND).

LAVAN PETROLEUM COMPANY (LAPCO). *See* IRAN.

LEACH, THOMAS WITT (1897–1966). Considered the father of the **North Dakota** petroleum industry. Leach first visited North Dakota in 1928 as chief geologist for the Transcontinental Oil Company and became convinced that the state, particularly the Williston basin, contained significant oil. He moved to North Dakota in 1936, opening his own offices in Bismarck. In 1950 he managed to convince the **Amerada Petroleum Corporation** (later Amerada Hess, eventually **Hess** Corporation) to drill at the Nesson anticline. Amerada struck oil on 4 April 1951 with the Clarence Iverson No. 1 well, the first producing oil well in North Dakota. Leach continued to play a major role in developing the Williston basin. His company, North American Royalties, went public in 1953. For a time it was the largest petroleum operator in the state. North Dakota continued to produce oil, but Thomas Leach did not live to see it move into the front rank of producing states, a development that occurred four decades after his death.

LEASE CONDENSATE. *See* NATURAL GAS LIQUIDS (NGL).

LEVANT BASIN. Subsea basin in the eastern Mediterranean estimated by the **United States Geological Survey** as holding technically recoverable resources of about 1.7 Bb of oil and 122 Tcf of gas. Israel's Noa gas field was found in June 1999, followed by Mari-B in February 2000 and some other successes. In June 2010 a team led by Noble Energy found the Leviathan gas field, with an estimated 16 Tcf of reserves. In 2011 a Noble-led team also found Cyprus's Aphrodite gas field, approximately 20 miles from Leviathan. The Cyprus exploration proceeded under the shadow of threats from Turkey, even though Aphrodite's reserves, which are between 3 and 6 Tcf of gas, are within Cyprus's exclusive economic zone (the **United Kingdom** also has some claims because of its sovereign base areas on the island). The hostile relationship with Turkey also complicates the potential transport routes for gas to the rest of the European Union. In August 2015, **ENI** discovered **Egypt**'s **supergiant** Zohr gas field.

LIBYA. Oil production 2015: 0.404 Mmbpd (0.515 percent of world); **reserves** 48.363 Bb (2.94 percent of world); gas production: 0.402 Tcf (0.32 percent of world); reserves 53.144 Tcf (0.77 percent of world). Libya is a significant petroleum exporter. Its crude tends to be sweet and is sought after by importers. Libya joined the **Organization of the Petroleum Exporting Countries** (OPEC) in 1962 and was a founding member of the **Organization of Arab Petroleum Exporting Countries** (OAPEC) in 1968. Libya is also a member of the **Gas Exporting Countries Forum**.

Libyan oil production began in 1961 at 18,000 bpd and grew explosively thereafter. It surpassed 0.5 Mmbpd in 1963, 1 Mmbpd in 1965, and 3 Mmbpd in 1969. It peaked at 3.3 Mmbpd in 1970. It then declined rapidly, to 1.48 Mmbpd in 1975. This was due to production restraint and OPEC quotas and not primarily to field exhaustion. In 1977, production was again above 2 Mmbpd, staying near that level until 1979. It declined to 1.14 Mmbpd in 1981 and stayed in a band between roughly 1 and 1.5 Mmbpd until 2004.

Natural gas production in Libya is less important than oil production. It was 0.18 Tcf in 1980. It hovered a bit above 0.2 Tcf from 1986 until 2003, increasing to 0.526 Tcf (about 0.240 Mmboepd) in 2006.

Attempts to find petroleum in Libya date back to the period of Italian colonization in the first half of the 20th century. In 1914, a water well at Sidi Mesri, near Tripoli, detected gas. In the 1920s and 1930s, the Italians drilled for water and often found petroleum, but not in commercial quantities. In 1935, Italian professor Ardito Desio systematically watched for oil and gas in water wells and produced a comprehensive geologic map. The Italians explored for oil on Desio's recommendation but were unsuccessful. **World War II** interrupted their efforts. After the war, two geologists with the Standard Oil Company of New Jersey (later **Exxon**) conducted a survey of Libya, but there was no **drilling**. In 1953, the Libyan government passed a new minerals law, and nine international companies obtained exploration permits. A 1955 oil discovery at Edjeleh, in **Algeria**, suggested that Libya might have oil as well.

On 18 July 1955, Libya issued its first petroleum law, Royal Decree No. 25. Prime Minister Mustafa Ben Halim had supervised the drafting of the law, consulting experts from **independent companies** as well as the **major oil companies**. The law declared that all subsurface minerals were the property of the state. It also created specific rules for **concessions**, limiting the number of concessions and the total acreage that could be obtained by any one company—provisions suggested by the independents on Ben Halim's panel. The government could also retract and reassign concessions that remained unexercised after a specified time. The law provided for 50 percent of net income from concessions to accrue to the government. Within a year of the law's passage, Libya issued 51 concessions to 17 companies. Libya was attractive for a number of reasons. For one thing, the **United States** and the **United Kingdom** had treaties with King Idris. For another, Libyan oil would not have to pass through the **Suez Canal** to reach markets in the United States and Europe.

The Standard Oil Company of New Jersey began drilling near the Algerian border. It found oil in 1958, but the quantity was too low and the location was too remote. In 1959, however, the company discovered the Zelten field, a major find. Numerous **giant** fields followed in the same year. The Oasis Group, a consortium of **Conoco**, **Amerada** (later Hess), **Royal Dutch Shell**, and **Marathon**, discovered what was to become the giant Waha field and followed that with the Dahra field. The Standard Oil Company of New Jersey discovered the Mabruk field. Amoseas (**Texaco** and California Asiatic) found the Beda field. Socony Mobil (later **Mobil**) and Gelsenberg of Germany discovered the giant field Amal, which has an **EUR** of 3.32 Bb and 3.5 Tcf of gas.

Nelson Bunker Hunt, son of **H. L. Hunt**, had two concessions in Libya. He was wealthy as an individual but was a minor operator by oil industry standards who might not be able to explore and develop his concessions. **BP**, which had bid on one of Hunt's concessions and lost, approached him in 1959 to set up a joint venture. In 1960, they reached an agreement whereby BP would pay all exploration costs in return for a 50 percent interest in the concession. In November 1961, after unsuccessful wells, BP discovered the Sarir field, another giant.

Major discoveries continued throughout the 1960s, but the end of the decade would bring a momentous change for the industry. On 1 September 1969, Muammar Qaddafi deposed King Idris in a military coup and became Libya's ruler. Qaddafi, an Arab nationalist, soon moved toward nationalization of foreign petroleum interests. After a series of negotiations between Qaddafi's government and the various companies, the state acquired a 51 percent share in the Libyan operations of **ENI**, Oasis, Exxon, Mobil, and **Occidental** in 1973. Qaddafi completely nationalized the Libyan interests of Amoseas/Texaco, Royal Dutch Shell, **Arco**, and the partnership of BP and Bunker Hunt. The government settled most compensation claims with the companies in 1977, following arbitration.

After nationalization, company concessions were converted to exploration and production-sharing agreements (EPSAs). The government issued three rounds of EPSAs from 1974 to 1988, generally improving the terms for the companies with each round. In EPSA-3 (1988), the government dropped its share to as low as 65 percent (it had been as high as 85 percent), depending on the projected quality of the acreage. Libya was attempting to attract investment in the wake of U.S. sanctions, instituted to punish Libya for its alleged support of international terrorism. President Ronald Reagan had ordered American companies out of Libya in 1986. Occidental, Conoco, Marathon, Amerada Hess, and others had ceased operations. Libya did succeed in obtaining investment throughout the EPSA rounds, but no significant new discoveries were made. During President Bill Clinton's administration, the **Iran–Libya Sanctions Act** (ILSA) also sought to punish non-U.S. companies making investments in the energy sectors of Libya and **Iran**.

When Qaddafi renounced weapons of mass destruction in 2004, the United States lifted sanctions against his government. (In 2006, ILSA would be reduced to ISA, the Iran Sanctions Act.) Libya launched EPSA-4 and received many winning bids from American companies, including Occidental, Hess, and **Chevron**. Other winners included Sonatrach (*see* ALGERIA) and Petrobrás (*see* BRAZIL). A further round in 2005 garnered **ExxonMobil**, ENI, and the BG Group among others. In 2005, it appeared that the new investment was yielding results. A consortium led by **Repsol** announced a significant discovery of light sweet crude in the Murzuq basin. Repsol announced another discovery in the basin in 2013. These discoveries followed on earlier discoveries in the 1980s by Rompetrol, known collectively as the El-Sharara development. Libya is in considerable need of such new discoveries. Excluding the Repsol find, all but three of Libya's 26 giant fields were discovered before 1980, with 13 discovered in the 1960s.

In the 2010s Libya plunged into chaos. A revolt broke out in February 2011, after the Arab Spring uprisings in Tunisia and **Egypt**, and sparked a civil war as Qaddafi attempted to hold on to power. The United Nations Security Council passed a resolution sanctioning a no-fly zone and allowing military intervention to protect civilians. Although it is unclear if civilians were protected, the United States and NATO mounted massive air strikes that all but ensured the overthrow of Qaddafi, who was killed by rebel forces on 20 October 2011. Afterward, the country was carved up between different opposing factions. To oversimplify, the two largest were the UN-backed Government of National Accord and the Libyan National Army. The presence of forces loyal to the Islamic State in Syria and Iraq complicated the situation. Oil production declined by almost 73 percent between 2010 and 2011, to 0.465 Mmbpd. It recovered partially in 2012 but was down again to 0.471 Mmbpd in 2014, and went even lower in 2015 and 2016. As of 2017, continuing political instability made the petroleum industry's future unclear. The El Sharara development, which had become Libya's largest producer, was shut down several times in the 2010s by armed militias, protesters, or other causes. *See also* LIBYAN SAFETY NET; PRODUCTION-SHARING AGREEMENT (PSA).

LIBYAN SAFETY NET. Secret agreement signed in 1971 between 22 oil companies active in **Libya** to support each other in the event of disputes with the government of Muammar Qaddafi. Specifically, the parties to the agreement agreed to supply each other with oil in the event one or more of them were cut off because of a dispute with the government. The agreement was not followed. In 1973, for example, BP Bunker Hunt, a venture between **BP** and Bunker Hunt that operated the giant Sarir field, had its interests nationalized, but the other parties to the agreement did not fulfill their pledges.

LIMA–INDIANA OIL FIELD. Early **giant** oil field with sour oil (refer to appendix 5), exploited by **Standard Oil** using the **Frasch** desulfurization process. The Lima–Indiana field, also called the Trenton field after its limestone reservoir rock, is located in northwestern **Ohio** and Indiana. It covers a large area that extends, with interruptions, from Lake Erie to Marion, Indiana. The field produced 514 Mmb of oil over its lifetime.

Gas seeps were long known in the area of Findlay, Ohio, and gas may have been used there as early as 1838. A well drilled in 1884 discovered gas at 1,100 feet. Another well drilled for gas in 1886 discovered oil. On the Indiana side, gas was discovered in 1885, and oil was discovered in 1889. The field made Ohio the top oil-producing state in the **United States** from 1895, when it surpassed **Pennsylvania**, to 1902, when it was surpassed in turn by **California**. The Ohio portion of the Lima–Indiana field peaked in 1897, and the smaller Indiana portion peaked in 1906. The field as a whole produced 68,500 bpd in 1896, amounting to 41.1 percent of the oil output of the United States. By 1910, 85 percent of its recoverable 514 Mmb had already been extracted. In 1914, its production was 13,700 bpd, or 1.9 percent of U.S. output (which had grown greatly from other fields).

Lima–Indiana oil was sour, that is, high in sulfur. This made it difficult to use in lamps, as its products emitted a foul smell when burned. Nevertheless, it came to represent a competitive advantage for Standard Oil. Standard, which had hitherto concentrated on downstream operations and shunned oil exploration and development, bought large amounts of Lima–Indiana producing property in the 1880s at attractive prices. Standard's head, **John D. Rockefeller**, had consulted with Herman Frasch, who developed a process to reduce the sulfur content of the oil. The company completed a large refinery at Whiting, Indiana, near Chicago in 1890 for **refining** Lima crude via the Frasch process, and it soon converted other refineries. Standard could now use cheap Lima crude to produce refined products equal in quality to those from more expensive oil.

LIQUEFIED NATURAL GAS. Natural gas cooled and stored in liquid form for long-distance transport by **tankers**. Liquefied natural gas (LNG) facilitates the international trading of gas. Gas is generally liquefied at a port-side facility in the producing country. It must be cooled to -162°C; cooling accounts for about half the total cost of delivery. Liquefaction also removes impurities. The LNG is transformed back to gas at a regasification terminal in the receiving country and is then introduced into that country's distribution network.

In 2016, LNG accounted for about 10.1 percent of total global gas consumption; this was almost triple the proportion in 1990. Over half of all LNG exports in 2016 went to just three countries: Japan, South Korea, and **China**, with Japan alone accounting for 32.3 percent. Top LNG exporters in 2016 were **Qatar**, **Australia**, **Malaysia**, **Nigeria**, **Indonesia**, **Algeria**, and **Russia**, in that order. Qatar accounted for about 30 percent of total LNG exports.

LNG infrastructure is expensive and poses considerable safety concerns. In 1944, an LNG explosion in Cleveland killed 128 people. An explosion in Algeria killed 27 people in 2003. LNG infrastructure and tankers are also vulnerable to terrorist attack, with potentially disastrous consequences. These and other considerations tend to drive construction of new terminals offshore, further adding to the cost.

LIQUEFIED PETROLEUM GAS. *See* NATURAL GAS LIQUIDS (NGL).

LITTLE BIG INCH. *See* BIG INCH AND LITTLE BIG INCH.

LITTON INDUSTRIES. *See* WESTERN GEOPHYSICAL.

LOANS FOR SHARES. A post-Soviet privatization scheme in **Russia** that resulted in state enterprises, including major petroleum companies, passing into the control of a small group of well-connected individuals usually called "oligarchs." The process is generally considered to have been corrupt. On the surface, Loans for Shares seemed a method by which the cash-starved government of President Boris Yeltsin could borrow money in order to meet its 1996 budget goals. The collateral was in the form of shares of state enterprises. If the government could not repay the loans, the lenders would buy the shares at auction and the government would repay the loans from the proceeds. However, the government was unable and perhaps never intended to repay the loans before the auctions. It has been alleged that some of the loans were completely staged and that the state borrowed its own money that it had already placed on deposit. Loans for Shares is generally acknowledged to have been a method by which state enterprises could be deliberately sold to privileged buyers at bargain prices.

The privatization auctions that began in 1996 are widely thought to have been rigged: the companies that had made the loans to the government were the organizers and sole bidders. The auctioned properties included large companies producing nickel, aluminum, and petroleum. The process was sometimes violent, particularly in the case of the aluminum industry. Some company managers resisting the privatization died under suspicious circumstances. The owners of the new companies were sometimes hidden, and profits were often shifted to offshore accounts. It is generally believed that Yeltsin was trying to shore up his power by spreading wealth to friends and supporters.

The oil company Sibneft (later **Gazprom Neft**) was created by special decree in order to be privatized in this fashion. The oligarchs **Roman Abramovich** and **Boris Berezovsky** acquired a controlling interest in Sibneft beginning in 1995 for about $100 million. In 2006, **Gazprom** bought the Sibneft stake for $13 billion. Similarly, oligarch **Mikhail Khodorkovsky** acquired a controlling interest in **Yukos** on 8 December 1995 for $159 million, only $9 million above the starting price in the auction. By 2004, Yukos was valued at about $30 billion.

Even the privatizations not related directly to Loans for Shares resulted in very cheap sales of valuable state property. Overall, Russian oil wells were sold for $0.04 per realized barrel at a time when the North American average price was $7.06 per realized barrel.

LOMONOSOV, MIKHAIL (1711–1765). Russian scientist and writer who put forward the theory of biogenic origin of petroleum and coal. Mikhail Vasilievich Lomonosov was born into a peasant family on 19 November 1711 in the far north of European **Russia**. As a boy, he walked to Moscow in search of education. He would eventually graduate from academies there and elsewhere, including Germany, and he later cofounded Moscow State University. In 1745, he became the first Russian professor of chemistry at the St. Petersburg Academy of Science. Lomonosov made many fundamental contributions to chemistry, physics, astronomy, metallurgy, geology, and literature. In the field of petroleum science, he concluded that oil was of biogenic origin, and he postulated the migration of petroleum from source rocks to reservoir rocks. He also conducted distillation experiments on crude oil a century before Western scientists such as **Benjamin Silliman Jr.**, although in the same general time frame as investigators at **Pechelbronn**. Lomonosov died 15 April 1765. *See also* ABIOGENIC THEORY.

LOS ANGELES OIL COMPANY. *See* STEWART, LYMAN (1840–1923).

LOUDON, HUGO (1860–1941). Engineer and petroleum explorer who was critical in finding new sources of oil for the **Royal Dutch Company** (later **Royal Dutch Shell**) in the 1890s. Born 18 June 1860 in The Hague, Hugo Loudon graduated from the Technical University in Delft in 1885. He worked in land reclamation in Hungary and in railway construction in the Transvaal, South Africa. In 1894, he joined **Jean-Baptist Kessler** in the Dutch East Indies (**Indonesia**) to work for the Royal Dutch Company, managing its Langkat operations. He developed a reputation for integrity and technical competence, and he became Kessler's deputy.

In 1898, when Royal Dutch's wells were depleted and the company was in danger of dissolution, Loudon led an expedition to Perlak (about 80 miles north of the company's existing operations) to search for new sources of oil. He negotiated skillfully with both the rajah of Perlak and the rebels who were fighting him. Loudon and his team of geologists and workers struck oil on 28 December 1899, just six days after starting to drill. Arguably, Loudon saved the company that would one day become Royal Dutch Shell. In 1902, Loudon became a director of Royal Dutch, and he later served as chairman of the board, a position he held until 1936. He died 6 September 1941 in Wassenaar, the Netherlands.

LOUISIANA. Oil production 2015: 0.172 Mmbpd (1.83 percent of U.S.); **reserves** 0.424 Bb (1.31 percent of U.S.); gas production: 1.74 Tcf (6.41 percent of U.S.); reserves 16.1 Tcf (5.23 percent of U.S.). According to generally accepted history, oil production in Louisiana began spectacularly in 1901 with the discovery of the Jennings oil field by W. Scott Heywood. Heywood was an oil entrepreneur who had already had some success at **Spindletop** in **Texas**. His 21 September well was a gusher initially flowing 7,000 barrels a day. However some scholars believe that the first commercial well in Louisiana was considerably earlier, in 1886 at Sulphur. It is also noteworthy that gas was discovered in 1870 in an ice plant in Shreveport by a night watchman who lit a match near a well drilled for water. That gas was used locally for illumination. In any case, after Jennings, exploration and drilling continued, with significant oil discoveries in 1913 (the Bull Bayou field) and gas discoveries in 1916 and 1921 (the Monroe and Haynesville fields respectively). On 14 November 1947, Kerr-McGee completed the first producing oil well out of sight of land in the **Gulf of Mexico** 45 miles south of Morgan City. This began Louisiana's **offshore** industry and, arguably, was the start of the modern offshore industry in the world at large. Two years later the offshore Bay Marchand field was discovered. It would go on to reach a cumulative production of 500 Mmb, making it one of the more prolific oil fields in U.S. history. In 1959 Louisiana overtook **California** to become the second-largest U.S. oil-producing state, behind Texas. It retained this position until 1979, when it was overtaken by **Alaska**. Louisiana oil production peaked in 1971 at 2.56 Mmbpd and never recovered. Gas production was remarkably steady between 1982 and 2000 at around 5 Tcf per year, then declined significantly. The Louisiana petroleum industry also suffered significant damage from Hurricanes Katrina and Rita (29 August and 24 September 2005). Louisiana enjoyed a resurgence in gas production beginning in 2009 through the application of **fracking** and **horizontal drilling** in the Haynesville Shale, although production levels in 2015 were still only a third of the high levels seen in the 1980s and 1990s.

LUCAS, ANTHONY F. (1855–1921). Engineer and geologist who discovered oil at **Spindletop** on 10 January 1901, initiating the **Texas** oil boom. Anthony Francis Lucas was born Antonio Francesco Luchich in the Austrian Empire and was an officer in the Austrian navy. He immigrated to the **United States** and became an American citizen in 1885. From 1893 to 1896, he worked in the salt mining industry in Louisiana and became a recognized expert on salt dome formations. This suited him for the task of exploring for oil at the Spindletop salt dome in Texas, an endeavor he undertook after answering an advertisement placed by **Pattillo Higgins**. Higgins, who had been exploring there but had fallen out with his business partners, helped Lucas obtain a lease.

Lucas drilled in 1900 but failed. He sought backing from the Guffey and Galey firm—*see* GUFFEY, JAMES McCLURG (1839–1930)—which also brought financing from **Andrew Mellon**. The new deal gave Guffey a five-eighths share in profits, with a quarter for Galey and an eighth for Lucas. This was less than Lucas had hoped for, but his bargaining position was not strong and he accepted. The deal also cut Higgins out completely, and this later led to recriminations and a lawsuit after Lucas's Spindletop discovery. Lucas's well was a gusher struck at 1,139 feet. It blew a column of oil 100 feet high and was not brought under control until nine days later. Lucas drilled further wells for Guffey and then became a consulting engineer in **Mexico** (for **Weetman Pearson**), **Russia**, **Romania**, and the United States.

Lucas was a respected engineer and industry leader. He was the first chairman of the Oil and Gas Committee of the American Institute of Mining and Metallurgical Engineers. This committee, later called the Petroleum Division, would evolve into the **Society of Petroleum Engineers** (SPE). Lucas died on 2 September 1921 in Washington, D.C.

LUFKIN INDUSTRIES. Diversified industrial company and developer of the familiar counterbalanced oil field pump. Lufkin was founded in 1902 as the Lufkin Foundry and Machine Company, focused on repairing sawmill machinery. The first president was Joseph Hubert Kurth. In 1905, the company hired **Walter C. Trout**, who developed the firm into more of a manufacturing concern for sawmill equipment. Trout developed a new counterbalanced oil field pump that he patented in 1925. The pump became an industry standard, recognizable the world over by its familiar horse-head shape. Sales of the pump to operators in the **East Texas oil field**, discovered in 1930, carried the company through the Great Depression. Trout served as president of Lufkin from 1930 to 1947. Lufkin still manufactures variations of its pump, as well as industrial gears and truck trailers.

ŁUKASIEWICZ, IGNACY (1822–1882). Early oil entrepreneur and a pioneer in **refining** kerosene from crude oil. Ignacy Łukasiewicz was born 8 March 1822 in Zaduszniki, Poland, which was then in Austro-Hungarian **Galicia**. He became a pharmacist's apprentice in Lancut and also began an involvement in anti-Austrian politics that landed him in prison on 19 February 1846. After his release on 27 December 1847, he began working in a pharmacy in Lvov. With help from his employer, he studied pharmaceutical science at the University of Krakow and in Vienna, graduating in 1852 from the University of Vienna.

Łukasiewicz had long been interested in the oil seeps of the Carpathians, and in the possibility of using petroleum instead of whale oil for illumination. Working with his senior associate Jan Zeh, he succeeded in distilling kerosene from crude oil in 1853. He and his coworkers also developed a lamp to burn the kerosene. On 31 July 1853, one of Łukasiewicz's lamps provided the light for an operation in Lvov's hospital.

In the fall of 1853, Łukasiewicz went to Gorlice, where he began to build distilleries. He formed the Łukasiewicz-Trzecieski company with Tytus Trzecieski, who got him interested in the oil seeps in the Bóbrka Forest. The company began to exploit the seeps, using hand-dug wells. In the years that followed, Łukasiewicz and his various partners dug several other wells (drilling equipment was not introduced to the area until 1863). In 1856, he built a facility for distilling crude oil in Ulaszowice, which was one of the world's first modern refineries.

Łukasiewicz continued to work as a pharmacist until the early 1860s, by which time he was wealthy. He continued to work in the local oil industry, entering into several ventures and actively promoting the industry's development. He built another refinery in Chorkówka, where he had settled on a large estate. In 1876, he was elected to parliament, where he worked on taxation, occupational safety, and other issues. He died of pneumonia on 7 January 1882. *See also* GESNER, ABRAHAM (1797–1864); KIER, SAMUEL M. (1813–1874); WEST VIRGINIA; WILLIAMS, JAMES MILLER (1818–1890).

LUKOIL. Large Russian integrated oil company. Lukoil's initial core resource base was in **Siberia**. Although Siberia is still important, Lukoil has expanded aggressively into other areas. Its main exploration and production activities have been in Russia, but it has also had upstream activities in a number of countries, including **Azerbaijan**, **Kazakhstan**, Uzbekistan, **Egypt**, **Iran**, **Iraq**, **Saudi Arabia**, **Colombia**, **Venezuela**, and Côte d'Ivoire. Lukoil was formed in 1991, before the breakup of the Soviet Union, by decree of the Council of Ministers. Three state-owned oil and gas companies, namely Langepasneftegaz, Uraineftegaz, and Kogalymneftegaz, were combined to form it. The "Luk" in Lukoil's name is formed from the initials of those three companies. All three companies sprang from oil discoveries in Siberia in 1964. In 1993, after the dissolution of the Soviet Union, the Russian government transformed Lukoil into a public joint stock company. Privatization began in 1994.

In 2000, the company began to make major hydrocarbon discoveries in the Russian part of the **Caspian Sea**. In 2006, it opened the V. Filanovsky field, with probable and possible **reserves** estimated at 600 Mbo and 1.2 Tcf gas. In 2006, Lukoil collaborated with **Norsk Hydro** to obtain exclusive rights to develop the Azar field in the Anaran block of western Iran. The field's resources may be as high as 2 Bb. In 1995, **Arco** (later part of **BP**) bought a 6.3 percent stake in Lukoil, and in a joint venture Lukarco was formed. This was the first strategic partnership between a major Russian company and a foreign one. Through Lukarco, Lukoil obtained a stake in the **supergiant** Tengiz field in Kazakhstan and in the Caspian Pipeline Consortium (CPC) connecting Tengiz with the port of **Novorossiysk** on the Black Sea. In 2002, Lukoil signed an agreement with Ecopetrol to conduct exploration and development in the Condor block of Colombia, discovering oil in commercial quantities by 2006.

In the late 1990s and the 2000s, Lukoil expanded downstream as well as upstream activities, and it entered into a number of additional agreements and alliances with Western and other companies. In 1995, Lukoil acquired petroleum companies including Astrakhanneft and Permneft. Three years later, it joined with **Conoco** (later **ConocoPhillips**) to develop oil and **natural gas** in northern Russia. In 1998 and 1999, Lukoil expanded its **refining** capacity by acquiring control of refineries in **Romania**, Bulgaria, and Ukraine. In 2001, Lukoil bought **Getty** Petroleum Marketing and acquired 1,300 gasoline stations in the **United States**. It also acquired Canadian oil company Bitech, which had operations in Russia, and the Nizhny Novgorod refinery. In 2004, Lukoil formed a strategic alliance with ConocoPhillips, which purchased 7.59 percent of its shares (increased to 20 percent in 2006). Lukoil also acquired 779 gas stations from ConocoPhillips in 2004. In 2010 ConocoPhillips sold its stake back. Lukoil made many significant acquisitions in the 2000s and 2010s. In 2005, it acquired Nelson Resources for $2 billion, gaining oil and gas assets in Kazakhstan. In 2006, it acquired the Khanty-Mansiysk Oil Corporation, a **Marathon Oil** asset in Siberia, for $847 million. In 2013, Lukoil bought Samara-Nafta, **Hess** Corporation's Russian subsidiary, for slightly over $2 billion. In 2016, Lukoil controlled 12.189 Bb of oil reserves and produced oil at the rate of 1.822 Mmbpd. It controlled 23.328 Tcf of gas reserves and produced 2.329 Bcfpd of gas.

M

MAADI PACT. *See* MEHDI PACT (MAADI PACT).

MAGNOLIA PETROLEUM COMPANY. Texas petroleum company organized in 1911 as a merger of older companies; later part of **Mobil**. John Sealy became the first president of Magnolia Petroleum in 1911 after his John Sealy Company purchased the Navarro Oil Company and the Security Oil Company. The Navarro Oil Company was the successor to the J. S. Cullinan Company. Cullinan began operating its own refinery in **Corsicana**, Texas, on 25 December 1898. It was founded by **Joseph S. Cullinan**, who also founded the Texas Company (**Texaco**). The Security Oil Company was the successor to the George A. Burts Refining Company, formed in 1901 to **refine** oil from **Spindletop**. Magnolia was originally an unincorporated joint stock association, with an issue of 24,500 shares at $25 each. On 21 November 1925, Magnolia incorporated formally in Texas. Magnolia also acquired the Texas interests of the Guffey and Galey company—*see* GUFFEY, JAMES McCLURG (1839–1930).

A month later, the company became a wholly owned subsidiary of the **Standard Oil Company of New York** (Socony, later Mobil), which had been acquiring its stock for some time. By 1949, the subsidiary was headquartered in Dallas and operating in 20 states. It had assets of about $700 million and more than 12,000 employees. It also had more than 8,500 miles of **pipelines** in a pipeline subsidiary. In 1955, Magnolia's parent, which was called Socony–Vacuum after the 1931 merger of Socony and **Vacuum Oil**, reorganized to become the Socony Mobil Oil Company. In 1959, there was a further reorganization. Magnolia's operations, along with those of General Petroleum Corporation and Mobil Producing Company, were folded into the Mobil Oil Company, a wholly owned subsidiary of Socony Mobil. The Magnolia Pipe Line Company, a pipeline subsidiary organized in 1925, remained a common carrier affiliate of Socony Mobil. In 1966, Socony Mobil became Mobil Oil Corporation. (In 1999, Mobil would merge with **Exxon** to form **ExxonMobil**).

MAJOR OIL COMPANIES. Large, multinational oil companies with a global system of production, transportation, **refining**, and marketing. In the period following **World War II**, the term was used nearly synonymously with the **Seven Sisters**: **Royal Dutch Shell**, **BP**, **Exxon**, **Mobil**, **Gulf**, **Chevron**, and **Texaco**, with the frequent addition of **Compagnie Française des Pétroles** (CFP, later **Total**). By 2007, the original Seven Sisters had merged into four: **ExxonMobil**, Chevron, BP, and Royal Dutch Shell, plus Total. The major oil companies are distinguished from **independents** not only by their larger size but also by their global reach and their historical geopolitical influence. Large national oil companies such as **Aramco** are usually placed in a separate category.

MALAYSIA. Oil production 2015: 0.66 Mmbpd (0.84 percent of world); **reserves** 3.6 Bb (0.22 percent of world); gas production: 2.519 Tcf (2.02 percent of world); reserves 41.765 Tcf (0.61 percent of world). Oil was produced at a low level in Malaysia before 1970. In that year, production was still only about 18,000 bpd. It increased more than 40-fold to a high of 0.755 Mmbpd in 2004. **Natural gas** production was about 45 times higher in 2015 than it was in 1980. In 2016 Malaysia was the world's third-largest exporter of **liquefied natural gas** (LNG).

Oil was first found in the territory that is now Malaysia at the end of the 19th century. In 1910, **Royal Dutch Shell** first drilled for oil in Sarawak, in the northern part of the island of Borneo. Sarawak was then a British colony. On 22 December of that year, the company struck oil on top of Canada Hill in Miri, Sarawak, with well Miri 1. Shell built the first refinery in what is now Malaysia in 1914, also in Miri. Sarawak and Sabah (also in northern Borneo) joined the Federation of Malaya to form Malaysia in 1963. Shell was still the only company operating in the country at the time. It began producing from the country's first **offshore** field, off Borneo, in 1968.

The Malaysian federal government wanted to encourage exploration in other areas of the country. It licensed exploration off the coast of the state of Trengganu on the Malay Peninsula. Standard Oil Company of New Jersey (soon to be **Exxon**), **Conoco**, and **Mobil** were all involved, although by 1974 only Exxon remained. Exxon discovered natural gas at first and then oil, and Trengganu became a larger producer than Sarawak and Sabah. In August 1974, the Malaysian government established Petroliam Nasional (Petronas) as its national oil company, modeled on **Indonesia**'s Pertamina. Petronas replaced the **concessions** held by Royal Dutch Shell in Borneo and Exxon on the Malay Peninsula with new **production-sharing agreements** that came into effect in 1976. The contracts provided for 70 percent of net income after federal and state royalties to accrue to the government. The government later offered better terms to companies, for example to **Elf Aquitaine** (later **Total**) in 1982. In 1976, Malaysia became a net oil exporter, although at too low a level to join the **Organization of the Petroleum Exporting Countries** (OPEC). Petronas began **refining** and distributing in 1983.

Between 1970 and 1975, Royal Dutch Shell discovered the offshore **giant** Central Luconia gas fields. Taken together, the three fields had an **EUR** of 9.4 Tcf of natural gas. In the late 1970s, Petronas began exporting liquefied natural gas (LNG). By the early 2000s, Malaysia was competing with **Algeria** to be the world's second-largest LNG exporter. In 1975, Exxon discovered the giant offshore Seligi oil field (EUR about 0.5 Bb). It followed up with the Guntong field (also about 0.5 Bb) in 1983. About 30 percent of Malaysia's 2000 oil production came from the Seligi, Guntong, and Tapis fields. Guntong was the highest single producer. By 2014 about a quarter of the country's oil output was coming from Tapis.

The Malaysian government has been worrying about oil depletion since the 1980s. It signed 22 new contracts with 31 companies in 1989. At first, the contracts were too restrictive, but the government has relaxed their terms over the years to encourage investment. There were two major discoveries after 1990: the K05-1 gas field (1992, with EUR of 5 Tcf) and Shell's deepwater Gumusut oil field (2003, with EUR 0.55 Bb), which began producing nearly 0.15 Mmbpd in 2014.

MALLET, ROBERT. *See* GEOPHYSICAL PROSPECTING.

MANTASHEV (MANTASHIAN), ALEXANDER (1842–1911). Russian oil entrepreneur of Armenian descent; founder of the **Mantashev Company** in 1899. Born in Tbilisi, Georgia, which at the time was part of the **Russian** Empire, Alexander Ivanovich Mantashev began his career in cotton and banking. In the 1890s, he began buying up marginally successful oil wells and making them profitable. He was also known to have an uncanny ability to choose successful drilling sites. Mantashev had a refinery in **Baku** to produce kerosene and a canister manufacturing factory in **Batumi**, among many other interests. He helped fund the **Baku–Batumi Pipeline**, was the leading oil entrepreneur in Russia after the **Nobels** and **Rothschilds**, and was a significant shareholder in several petroleum companies, including **Branobel**. Mantashev was also a prominent philanthropist. He died on 19 April 1911.

MANTASHEV COMPANY. Leading Russian **independent** petroleum company in the **Baku** area. Founded by **Alexander Mantashev**, the Mantashev Company was organized as a joint stock company in 1899 but had been doing business in the petroleum industry for several years. Its capitalization in 1899 was 22 million rubles, rising to 30 million by 1914. The company competed with **Branobel** and the **Rothschilds**. In 1900, the Rothschilds and the Nobels controlled about half of **Russia**'s crude production, two-thirds of its oil refineries, half the Russian domestic market, and three-quarters of Russia's kerosene exports. The Mantashev Company and some allied independents controlled about a third of the domestic market and about a quarter of kerosene exports.

MARACAIBO. City (pop. 2.7 million, 2013) in **Venezuela** and a major center of the Venezuelan petroleum industry. Before oil was discovered in 1914, Maracaibo was a small coffee port. It lies on the western shore of a channel connecting **Lake Maracaibo** with the Gulf of Venezuela. The city was founded in 1574 as Nueva Zamora de la Laguna de Maracaibo. It is the capital of Venezuela's Zulia state.

MARACAIBO, LAKE. Large lake in the state of Zulia, northwestern **Venezuela**, in a major petroleum-producing area. The lake is about 130 miles long from north to south and up to 75 miles wide. It reaches a maximum depth of nearly 200 feet in the south, although it is mostly shallower than this. It is connected to the Gulf of Venezuela by a narrow strait, making it slightly saline in the north. Before the 1930s, navigation was restricted by a large bar near the lake's mouth, but this problem has been alleviated by dredging.

Lake Maracaibo is one of the world's most prolific petroleum-producing regions. The **giant** Mene Grande field was discovered in 1914 to the east of the lake's widest point. Giant fields of the **Bolivar Coastal Complex** also line the land off its eastern shore. These represent some of the earliest significant discoveries in Venezuela and include Cabimas, discovered in 1917; Lagunillas, 1926; Tia Juana, 1928; and Bachaquero, 1930. There are also giant **offshore** fields in the lake itself, including Lama, Centro, and Lamar, all discovered in 1957; Urdaneta, discovered 1955; and Lago, discovered 1958. Giants on the western shore include La Paz, discovered in 1925 west of the city of **Maracaibo**; Mara, discovered 1945; and Boscan, discovered 1946.

MARATHON OIL CORPORATION. Large petroleum company headquartered in **Houston**, **Texas**. Marathon has exploration and production operations in the **United States**, the **United Kingdom**, Ireland, **Norway**, **Libya**, and West Africa (**Angola**, **Gabon**, and **Equatorial Guinea**). Marathon's lineage extends back to 1887, when the Ohio Oil Company was founded in Lima, Ohio (*see* LIMA–INDIANA OIL FIELD), by driller Henry Ernst and four partners. It was an **independent** company that competed with **Standard Oil**. As with many such companies, it was not long before Standard Oil acquired it. This happened in 1889, by which time Ohio Oil was already the largest producer in Ohio. In 1905, Ohio Oil moved its headquarters to Findlay, Ohio. In 1911, Standard Oil was broken up by order of the U.S. Supreme Court, and the Ohio Oil Company became independent again. It expanded exploration into **Wyoming**, **Kansas**, Texas, and **Louisiana**. In 1924, it discovered oil near the Pecos River in Texas and bought Lincoln Oil Refining, thus adding a downstream activity for the first time. It expanded downstream operations in the 1920s.

After **World War II**, the company formed Conorada Petroleum, later called Oasis. This was a joint venture with Continental Oil—the ancestor of **Conoco**—and **Amerada** (later Hess) to explore overseas. Conorada began activities in South and Central America and in Africa. In 1955, Conorada (Oasis) acquired **concessions** on more than 60 million acres in **Libya**. Oasis (which also included **Royal Dutch Shell**) discovered what was to become the giant Waha field, and it followed that with the Dahra field. In 1962, the Ohio Oil Company bought Plymouth Oil and changed its name to Marathon Oil Company. Marathon was a brand name used by Ohio Oil since the 1930s. In 1976, Marathon bought ECOL and gained 200,000 barrels a day of refinery capacity in Louisiana. In 1990, the company moved its headquarters to Houston.

Both **Mobil** and U.S. Steel became interested in Marathon as a takeover target. U.S. Steel ultimately won and acquired Marathon in 1982 for $6.2 billion. In 1986, U.S. Steel changed its name to USX and acquired Texas Oil and Gas. But 1986 was also the year that President Ronald Reagan's administration began economic sanctions on Libya. This was a blow to Marathon, which had considerable interests there. In 1990, USX combined Marathon and Texas Oil into one operation.

In 1991, USX created two separate stocks, one for USX-Marathon and one for USX-U.S. Steel. In 1992, it created another stock, USX-Delhi Group, for its **pipeline** division. In 1994, USX-Marathon was part of a consortium to develop petroleum in the **offshore** area of **Russia**'s **Sakhalin** Island but sold out its share in 2000. In 1998, Marathon merged downstream operations with Ashland, forming Marathon Ashland Petroleum (MAP). Marathon owned 62 percent of MAP. In 2005, Ashland sold its share of MAP to Marathon; MAP became Marathon Petroleum, a Marathon subsidiary. In 1998, Marathon also acquired the Canadian company Tarragon Oil and Gas. In 2000, Marathon intensified exploration in the **Gulf of Mexico**. It also began a significant move into West Africa, starting with exploration offshore **Congo Republic** (Brazzaville). In 2002, it bought CMS Energy's interests in Equatorial Guinea. It also acquired Globex Energy for $155 million, further increasing its West African interests.

In 2001, USX spun off U.S. Steel. Ironically, U.S. Steel was the initial basis of USX; now what was left of USX was not a steel company but an oil and gas company. USX renamed its remaining company the Marathon Oil

Corporation. In 2003, Marathon went back into Russia by acquiring Khanty Mansiysk Oil. By 2006, it was getting out again, selling Khanty-Mansiysk's interests to **Lukoil** for $787 million. In 2011 Marathon spun off its downstream operations into an independent company, Marathon Petroleum. During the next three years, Marathon sold off a number of operations, mostly to strengthen its balance sheet and help fund exploration and development elsewhere. In 2013 it sold its stake in the Angolan Block 31 field to Sinopec (China Petrochemical Corporation) for a bit over $1.5 billion, and followed up with a sale of its stake in Block 32 to Sonangol (*see* ANGOLA) for a bit under $600 million. The next year it sold its Norwegian operations Marathon Oil Norge to Det Norske Oljeselskap for about $2.1 billion. Marathon Oil Norge accounted for about 80,000 bpd in 2013. In 2016, Marathon was the sixth-largest petroleum company headquartered in the United States, ranked by total assets. It controlled liquid reserves of 0.996 Bb and 2.446 Tcf of **natural gas**. It produced 0.216 Mmbpd of liquids and 1.213 Bcfpd of gas.

MARAVEN. Autonomous subsidiary of **PdVSA** formed in 1975 to take over the Venezuelan petroleum operations of **Royal Dutch Shell** after **Venezuela** nationalized its petroleum industry. The subsidiary was merged back into PdVSA in 1998. The subsidiary participated in joint ventures in the **Orinoco Oil Belt** during the **Apertura Petrolera**.

MARCELLUS SHALE. Formation in the Appalachian basin of the Eastern **United States**, with estimated proved **reserves** of **natural gas** exceeding 77 Tcf, and unproven technically recoverable resources of 211 Tcf. Proven oil reserves are estimated at 143 Mmb. The hydrocarbon-rich portion of the Marcellus spans New York, **Pennsylvania**, **Ohio**, **West Virginia**, and Kentucky. The exploitation of the Marcellus has been made possible by **fracking** and **horizontal drilling** technology. The first well in the modern Marcellus play, Renz No. 1, was drilled by Range Resources in October 2004 in Washington County near Pittsburgh, Pennsylvania. The well was drilled and completed using conventional techniques and was productive. Range then began experimenting with fracking and horizontal drilling, following the model of the **Barnett Shale** in **Texas**. It succeeded by 2007, and many other companies followed suit, including Chesapeake Energy, EQT, Rice Energy, CONSOL Energy, and **Chevron**. Marcellus gas production in 2015 was 15.9 Bcfpd, an over eightfold increase over the level in 2010. In the second decade of the 21st century, Pennsylvania, the titan of oil in the 19th century whose petroleum glory had long faded, became the second-largest gas-producing state in the United States after Texas, producing 18 percent of the nation's gas in 2015.

MARKER CRUDES. *See* BENCHMARK CRUDE.

MARLAND OIL. *See* CONOCO.

MATTEI, ENRICO (1906–1962). Head of the Italian state energy company Ente Nazionale Idrocarburi (**ENI**) from its creation in 1953 until his death. Enrico Mattei played a role in dismantling the postwar oil market regime of fixed prices and tight control by **major oil companies**. He dubbed the cadre of multinational oil companies the **Seven Sisters**, a label that defined an era and has endured. Mattei was born 29 April 1906 in Acqualagna, Psaro, Italy. He was the son of a police officer and worked in a number of jobs as a young man. Despite a youthful flirtation with the Fascist Party, he fought in the Italian resistance against the Germans in **World War II** and became a prominent Christian Democrat.

After the war, Mattei was named provisional administrator of **AGIP**, the state petroleum company that had been established by Italy's Fascist government in 1926. His charter was to dismantle the company, but he moved instead in the opposite direction. In 1947–1949, AGIP discovered **natural gas** in the Po Valley, and Mattei argued forcefully for the establishment of a new integrated national petroleum company. He dreamed of making Italy self-sufficient in energy, using this as a basis for resuscitating the country's devastated economy and effecting a broad modernization. In 1953, Mattei got his wish when the Italian parliament passed a law creating ENI. Mattei was innovative and in many ways ahead of his time. He put Italy ahead of the rest of Europe in the use of natural gas, setting up an advanced distribution network. He moved ENI into **Egypt**, where it acquired a stake in the International Egypt Oil Company (IEOC).

Mattei's biggest impact on the world oil market was in the independent agreements he signed with oil-producing countries, offering them favorable terms and taking the unprecedented step of allowing the producing countries a significant voice in management of the ventures. In 1957, ENI and the National Iranian Oil Company (NIOC) signed a contract creating SIRIP (Société Irano–Italienne des Pétroles), a joint company to explore for and develop petroleum in **Iran**. ENI offered the Iranian government 50 percent of the company's gross earnings and a 50–50 share in the net profits. This looked better than the **50–50 agreements** then offered by the majors—and the majors had only offered 50–50 under protest. NIOC would manage SIRIP jointly with ENI. ENI assumed all the risk: NIOC would pay half the exploration cost only if petroleum was found. Deals such as Mattei's, however, did not necessarily offer producing countries more revenue per barrel once oil was discovered and produced. They could in fact offer less because of the different ways royalties and profits were calculated and applied. However, they did offer more sovereign control. Only nine days after signing this agreement, Mattei signed a similar one with **Libya**.

The multinational oil companies lobbied hard, with the help of the U.S. State Department, to prevent ratification of the ENI-NIOC deal by the Iranian and Italian parliaments. Elements within the U.S. intelligence agencies began to consider Mattei a threat and argued for stopping him in some way. President Dwight D. Eisenhower and Secretary of State John Foster Dulles rejected this advice on the grounds that what Mattei was doing was simply engaging in free market competition, and also on the grounds that ENI was too small to make much difference. On the latter point, they were correct in the sense that ENI was a relatively small company compared to the major multinationals. However, Mattei's deals had a broad impact on the psychology of the market and of producing countries, and they helped usher in an era of greater assertiveness on the part of oil-rich nations.

Mattei also made a deal to buy oil from the Soviet Union, another relatively unusual step for the time. He openly supported Nasser in Egypt, and he championed **Algeria**'s independence. Mattei died on 27 October 1962 when an airplane he was on exploded in midair. His death spawned a number of conspiracy theories. Some people think he was killed at the behest of the United States, although this is unlikely since the John F. Kennedy administration was already effecting a rapprochement with him. His death was fodder for a number of books and a 1970 feature film, *Il Caso Mattei*. The Trans-Mediterranean Pipeline that began transporting Algerian gas to Italy in 1983 is also known as the Enrico Mattei Pipeline in his honor.

MAZUT. Russian petroleum company established by the **Rothschilds** in 1897, operating in the **Caspian** area. The company concentrated on the transportation and marketing of crude oil and petroleum products. It had a fleet of **tankers**, with 13 just in the Caspian. It also had a large fleet of barges and tugs carrying petroleum products up the Volga. The Rothschilds sold their petroleum interests in **Russia** to **Royal Dutch Shell** in 1912. The interests were later expropriated by the Soviets.

McCLENDON, AUBREY (1959–2016). Early adopter of **fracking** and **horizontal drilling** for **unconventional natural gas**. Born in Oklahoma City, **Oklahoma**, on 14 July 1959, Aubrey McClendon founded Chesapeake Energy with Tom Ward in 1989 with $50,000. Working early in unconventional reservoirs in Oklahoma and **Texas**, the company was successful enough to go public in 1993. McClendon was an advocate for natural gas as a clean fuel, and in the 2000s Chesapeake produced **tight gas** and **shale gas**, the latter from plays in the **Barnett Shale**, **Marcellus Shale**, and Haynesville

Shale among others, becoming one of the largest gas producers in the United States. In 2013 McClendon stepped down as head of Chesapeake and formed American Energy Partners. At that time Chesapeake was the ninth-largest oil and gas company in the United States by total assets (nearly $42 billion in 2013). In 2013 it produced more than 1 Tcf of gas and controlled reserves of 11.73 Tcf. Throughout his career McClendon was noted for his bold and risky business tactics. On 1 March 2016 a federal grand jury indicted him on a number of charges including rigging the bidding process for oil and natural gas leases in a conspiracy to lower their prices. McClendon denied all charges. The next day, he died after driving his car into a concrete wall. He was the only occupant of the vehicle. Despite the circumstances, the death was not ruled a suicide.

McCLINTOCK NO. 1. The oldest continuously producing oil well in the world. The McClintock No. 1 well began producing oil for Brewer Watson & Company on 16 August 1861 on **Oil Creek**, a few miles from the **Drake** Well in **Pennsylvania**. It was still producing oil 150 years later when a small anniversary celebration was held at the site.

McGHEE, GEORGE (1912–2005). American geophysicist, petroleum entrepreneur, and diplomat. George Crews McGhee was born in Waco, **Texas**, on 10 March 1912. He showed an early interest in geology and graduated with a degree in that field from the University of **Oklahoma** in 1933. That year, he joined **Conoco** as a geophysicist, participating in the company's first discovery in the **Gulf of Mexico** made with the aid of reflection seismology (*see* GEOPHYSICAL PROSPECTING). Working with E. V. McCullom, he developed new ideas for estimating weathering corrections, which Conoco patented. He left Conoco to attend Oxford University, **United Kingdom**, on a Rhodes scholarship and received a doctorate in physical sciences in 1937. He then became a vice president of the National Geophysical Company in the **United States** and conducted the first reflection seismic survey in Cuba. In 1940, he joined **DeGolyer and MacNaughton**.

McGhee served in the U.S. Navy in **World War II**. After the war, he entered public service. He was coordinator for aid to Greece and Turkey, overseeing the disbursement of about $400 million. He served as ambassador to Turkey (1951–1953) and West Germany (1963–1968), as well as undersecretary of state for political affairs (1961–1963). During 1950, he was assistant secretary of state for Near Eastern affairs and played an important role in the complex negotiations leading to a **50–50 agreement** for profit sharing between **Aramco** and Saudi Arabia. The Aramco partners agreed to the final terms in McGhee's office.

McGhee maintained an interest and a presence in the petroleum industry throughout his life. He was a successful independent oil producer and from 1969 to 1982 was a director of **Mobil.** He had many other interests as well. He was chairman of the *Saturday Review* and published eight books, either as author, coauthor, or editor. He died 4 July 2005 of pneumonia in Leesburg, Virginia.

MEGAGIANT. A **giant** oil and/or **natural gas** field with an estimated ultimate recovery (**EUR**) of 50 Bboe or greater. There is one oil megagiant, **Ghawar** of **Saudi Arabia** (EUR 65–150 Bb depending on source of estimate). **Kuwait**'s **Burgan** and Saudi Arabia's **Safaniya** may also possibly qualify if one accepts the upper range of their EURs. There are three gas megagiants: **North Field** of **Qatar** (EUR about 176 Bboe), South Pars of **Iran** (EUR about 77 Bboe), and Urengoy of **Russia** (EUR about 59 Bboe). North Field and South Pars can be considered linked, making an even larger field. *See also* GIANT.

MEHDI PACT (MAADI PACT). Secret nonbinding 1959 agreement signed by representatives of a group of oil-producing countries, presaging the formation of the **Organization of the Petroleum Exporting Countries** (OPEC). During the First Arab Petroleum Congress in April 1959, a number of participants met separately and discreetly at a club in the Cairo suburb of Mehdi (Maadi). The movers behind the meeting were **Juan Pablo Pérez Alfonso** of **Venezuela** and **Abdullah al-Tariki** of **Saudi Arabia**. Also present were Ahmed Sayyid Omar of **Kuwait**, Manucher Farmanfarmaian of **Iran**, Seleh Nesim of the United Arab Republic (the brief Union of **Egypt** and Syria), and **Iraq**'s Mohammed Salman, the head of the Arab League's Petroleum Committee. The Egyptian was probably invited because his country was hosting the congress. Of all those present, only Pérez Alfonso and Tariki actually had authority to represent their governments, and it seems likely that of the two, only Pérez Alfonso would have had his government's approval.

The pact was a relatively modest list of recommendations. One recommendation was that the **posted price** structure should be maintained, but with no further unilateral price changes by oil companies. Another was that petroleum-producing countries should receive a greater share of petroleum revenues. Also, the pact recommended that national oil companies and agencies regulating the industry be established and refining capacity in producing countries be increased. The signatories were supposed to bring to the attention of their governments the idea of creating a joint consultative petroleum commission for producing countries that would meet annually to discuss issues of common interest. Despite its modest scope and lack of legal standing, the Mehdi Pact was important because it represented a first step in jointly formulating a set of common producer country policies. OPEC was formed about a year and a half later.

MELLON, ANDREW (1855–1937). Banker with diverse business interests, including petroleum; helped organize the **Gulf Oil Corporation**. Andrew William Mellon was born in Pittsburgh on 24 March 1855. He attended the University of Pittsburgh but left before graduation to start a lumber and building business in **Pennsylvania**. In 1874, he began working at his father's bank, helping rescue it from financial difficulty. By 1882, he owned the bank. He played a key role in the growth of various industries, including aluminum, steel, railroad cars, chemicals, abrasives, and power. He was able to obtain good positions for the bank in a number of key growth companies that became giants, including Alcoa, and companies that would be merged into Bethlehem Steel and U.S. Steel (which would later acquire **Marathon Oil Corporation**).

During the early 1890s, Mellon along with his brother Richard and nephew **William L. Mellon** began coordinating oil fields in western Pennsylvania with **pipelines** and with a refinery at Marcus Hook, Pennsylvania, on the Delaware River. Rather than pursue the inevitable fight with **Standard Oil**, he sold out to it very profitably in 1895. In 1901, Andrew and Richard Mellon financed **James McClurg Guffey** and John Galey, who in turn backed **Anthony F. Lucas** in his successful discovery at **Spindletop**. The Mellon family helped organize the J. M. Guffey Petroleum Company in 1901. Guffey was president, with seven-fifteenths of the shares. Andrew and Richard owned the rest. In 1901, they also organized the Gulf Refining Company, with a large refinery at Port Arthur, Texas. These two companies formed the core of the new Gulf Oil Corporation in 1907. Andrew Mellon died 26 August 1937.

MELLON, WILLIAM L. (1868–1949). Co-organizer and early president of the **Gulf Oil Corporation**; nephew of **Andrew Mellon**. William Larimer Mellon was born 1 June 1868 near Pittsburgh. He entered the **Pennsylvania** oil industry in 1889, with financing from his uncles Andrew and Richard Mellon. He produced petroleum and transported it by tank car and eventually **pipeline** to a refinery he had acquired at Marcus Hook, Pennsylvania, on the Delaware River. The Mellon operation, managed by William, was successful enough to arouse the fury of **Standard Oil**. Mellon's uncles decided to sell out profitably to Standard in 1895 rather than fight the monopoly.

In 1901, Mellon's uncles financed **James McClurg Guffey** and John Galey, who in turn backed **Anthony F. Lucas** in his successful discovery at **Spindletop**. The Mellon family helped organize the J. M. Guffey Petroleum

Company in 1901. Guffey was president, with seven-fifteenths of the shares. William L. Mellon oversaw the family's interests in the Guffey company and determined that Guffey was a better promoter and explorer than a manager. Mellon became executive vice president of Guffey Petroleum and Gulf Refining, the associated **refining** entity. He hired new executives and improved efficiency. As Spindletop production declined, the companies faced financial hardship. Mellon recommended that the Mellon family invest in developing new sources of oil in **Oklahoma**. The Mellons bought out Guffey's interest and in 1907 formed the Gulf Oil Corporation. In 1909, Mellon became its president. He served until 1930, at which point he became chairman of the board, a post he held until his retirement in 1948. Mellon turned Gulf into a **major oil company**, the fourth largest in the **United States**, with extensive upstream and downstream operations. The company also expanded internationally, particularly in **Mexico**.

Apart from his career in the petroleum industry, Mellon was both a railroad and a bank president. In 1949, he established a graduate school of industrial administration at the Carnegie Institute of Technology in Pittsburgh. The school is now known as the David A. Tepper School of Business at Carnegie Mellon University. Mellon died on 8 October 1949.

MENDELEEV, DMITRI (1834–1907). Russian chemist who developed the periodic table of the elements and had significant involvement in the early oil industry around **Baku**. Dmitri Ivanovich Mendeleev was born in Tobolsk, **Russia**, on 7 February 1834, the last of 17 children. He graduated first in his class from the University of St. Petersburg with a degree in chemistry in 1855 and received an advanced degree in 1856. After a few years abroad in France and Germany, he returned to teach in St. Petersburg. In 1869, he published the periodic table of the elements, which is a cornerstone of the modern science of chemistry.

In addition to pure science, Mendeleev was greatly interested in a broad range of technological problems, including those of the petroleum industry. He visited Baku in 1863 to consult on an early refinery operation, developed by Vasily Kokarev and others to produce kerosene from an asphalt-like substance called "kir." He was also a consultant for the **Nobel** brothers (Ludvig and Robert), major entrepreneurs in the Baku oil industry, who began distilling on a continuous basis using some of Mendeleev's ideas. Mendeleev was a major proponent of the idea of **refining** close to centers of consumption rather than centers of production. This brought him into conflict with some of Baku's smaller **independent** producers. He championed a **pipeline** between the oil fields of Baku and the Georgian Black Sea port of **Batumi**. After years of debate and a decade of construction, a products pipeline was completed in 1906 (*see* BAKU–BATUMI PIPELINES).

Mendeleev also argued for better economic incentives for oil production, including longer leases on government lands and low or no taxation on petroleum products. In 1876, he visited the **United States** and attended the Centennial Exposition in Philadelphia as a member of the Russian delegation. He took the opportunity to visit the **Pennsylvania** oil fields and subsequently published *The Oil Industry in the North American State of Pennsylvania and the Caucasus*. Mendeleev was an early proponent of the **abiogenic** theory of petroleum formation. He believed petroleum originated from the action of water on metallic carbides at depth. Mendeleev died of influenza in 1907 in St. Petersburg. The 101st element, Mendelevium, is named in his honor, as is the lunar crater Mendeleev.

MESA PETROLEUM. *See* PICKENS, T. BOONE (1928–).

MEXICAN EAGLE PETROLEUM COMPANY. *See* PEARSON, WEETMAN (1856–1927).

MEXICO. Oil production 2015: 2.308 Mmbpd (2.94 percent of world); **reserves** 7.64 Bb (0.46 percent of world); gas production: 1.898 Tcf (1.52 percent of world); reserves 12.651 Tcf (0.18 percent of world). Oil **concessions** were awarded in Mexico as early as 1865 but were unproductive. In the late 19th century, Henry Clay Pierce and William H. Waters brought their Waters–Pierce Oil Company into Mexico. Waters–Pierce was a **Standard Oil** affiliate. It concentrated on **refining** and on obtaining possible crude supplies for Standard. The dictator Porfirio Díaz wanted to encourage foreign investment to develop Mexico's oil resources. His government's 1884 petroleum law overturned a Spanish legal tradition that subsoil resources were the property of the nation, making such resources the property of the surface landowner.

The Mexican petroleum industry began in earnest with the involvement of two entrepreneurs, the American **Edward Doheny** and the British **Weetman Pearson** (later Lord Cowdray). Doheny's first discovery was the **giant** Panuco–Ebano field (**EUR** almost 1 Bb) at El Ebano near Tampico in 1901. The oil was heavy; he exported it to the **United States** for use in paving and was eventually able to refine fuel oil. In 1905, Doheny acquired new properties in the Tuxpan district farther south. He created the Huasteca Petroleum Company to hold these new properties and explored them intensively. Pearson, who had completed major engineering projects for the Díaz government and was well connected, formed the Compañía Mexicana de Petróleo el Águila, or Mexican Eagle Petroleum Company, in 1908. In 1910, Doheny drilled several producing wells at Casiano that yielded a lighter and better-quality oil than his properties at El Ebano. His gusher Casiano No. 7 released 402,000 barrels before it was brought under control. On 27 December 1910, Pearson's Potrero del Llano No. 4 well (sited by **Everette DeGolyer**) struck a gusher that flowed 1 Mmbo before it was brought under control. Casiano and Potrero del Llano were both in the rich oil area that would be known as the Faja de Oro, or Golden Lane, which was the principal source for Mexican oil for the next 15 years.

In 1911, the Mexican Revolution overthrew Porfirio Díaz and ushered in a protracted civil war lasting until 1920, with sporadic aftershocks lasting even longer. Still, oil production increased explosively. It increased 53-fold between 1910 and 1921, to 0.53 Mmbpd—a level it would not reach again until 1974. Mexican oil was important to the Western allies during **World War I**, and by 1920 Mexico supplied about a fifth of U.S. demand.

After 1915, the **major oil companies** began to dominate the Mexican industry. In 1917, Standard Oil Company of New Jersey (later **Exxon**) bought the **independent** Compañía Transcontinental for $2.47 million in gold. In 1919, **Royal Dutch Shell** bought Pearson's Mexican Eagle, which discovered the giant Poza Rica field (EUR 2 Bb) in 1930. In 1925, Standard Oil Company of Indiana (later **Amoco**) acquired Doheny's interests, although in 1932 it sold them to Standard Oil Company of New Jersey. Other oil companies with interests in Mexico included **Gulf**, the Texas Company (later **Texaco**), **Sinclair**, and **Cities Service Company**.

Mexico's 1917 constitutional convention adopted Article 27, reserving ownership of subsoil wealth to the nation. This set the legal basis for several future government policies, including the ultimate nationalization of the industry two decades later. The U.S. oil companies doing business in Mexico protested to the Mexican government and enlisted the U.S. State Department to protect their interests. The United States withheld recognition of the new Álvaro Obregón regime that came to power in 1920 until Mexico guaranteed U.S. property rights. The Bucareli Accords of 1923 affirmed that Article 27 should not be enforced retroactively on older concessions, provided that some positive effort had been made to extract oil before 1917. This gave the companies some of what they wanted, although they would have preferred Article 27 to be discarded altogether.

In 1925, the Plutarco Calles government passed a law to oblige companies to exchange their titles for 50-year concessions on claims that had already been worked, and 30-year terms on ones that had not. This led to a new round of disputes. Major companies began diverting investment to **Venezuela**, where the government of **Juan Vicente Gómez** was offering a better environment for them. Production in Mexico decreased drastically. In 1932, it was 0.09 Mmbpd, only about 17 percent of the peak level in 1921.

Relations between the foreign oil companies and the Mexican government continued to deteriorate in the 1930s. The government believed that the companies treated it as a subordinate. The companies also treated their workforce very poorly. This led to a major strike in 1937, causing President **Lázaro Cárdenas** to appoint a presidential commission to review the activities of foreign oil companies in Mexico. The commission recommended higher wages and better benefits for oil workers, and it called for foreign technical personnel to be replaced by Mexicans within two years. The companies protested to the Mexican government and to their own. Franklin D. Roosevelt's administration, anxious to preserve good relations between the United States and Mexico in a time of looming war, gave little support to the American companies.

On 18 March 1938, President Cárdenas signed the order to nationalize the oil industry and expropriate foreign assets, fully implementing Article 27. This affected 17 foreign oil companies. The Mexican government was willing to pay compensation for surface assets but not the high sums demanded by the companies for the value of subsoil resources. Mexico refused to discuss the fact of nationalization and would talk only about the amount and manner of compensation. The companies waged economic warfare against the country. They started a run on Mexican banks and on the peso. They also attempted a boycott of Mexican oil. American companies persuaded U.S. secretary of state Cordell Hull to stop silver purchases from Mexico. However, that policy hurt some U.S. interests as well and was reversed. Some oil company executives advocated the use of military force. The U.S. government persisted in maintaining good relations with Mexico and did nothing of the kind. Mexico signed various compensation agreements with companies between 1940 and 1943, involving a combination of up-front and installment payments. Mexico made its last payment in 1968.

When the government nationalized the industry, it created its national oil company, Petróleos Mexicanos (Pemex). It gave Pemex exclusive rights to all exploration and production in Mexico, including the **offshore**. Pemex's most important discovery between 1940 and 1960 was the **supergiant** Bermudez Complex (EUR 7 Bb and 17.5 Tcf gas) in 1958. In the 1970s, Pemex opened up two major new producing areas: the Reforma (Chiapas-Tabasco) onshore area and the offshore Bay of Campeche in the **Gulf of Mexico**. In 1976, Pemex made the first discovery in the supergiant **Cantarell** Complex in the Campeche Bay. The Cantarell complex is Mexico's largest oil accumulation. Mexico's production grew fairly steadily between 1950 and 1970 (from 0.198 to 0.487 Mmbpd), but after 1974 the increases were drastic. Production reached a peak of 2.78 Mmbpd in 1984, with exports at 1.7 Mmbpd. Mexican production played a role in breaking the power of the **Organization of the Petroleum Exporting Countries** (OPEC) during the 1980s (*see* OIL COUNTERSHOCK). After 1984, production declined and followed an irregular pattern, reaching a second peak of 3.07 Mmbpd in 1998 and a third of 3.383 Mmbpd in 2004.

Mexico's **natural gas** production has been less important than its oil production. It was 0.9 Tcf in 1980. It remained mostly in a band between 0.8 and 1.0 Tcf until 1995, after which it began to grow. In 2009 it reached a historic high of 2.080 Tcf. Still, its only slightly lower 2015 level of 1.898 Tcf was about 40 percent as high as oil production in energy-equivalent terms.

Both Pemex and Mexico have long been worried about declining oil reserves. The older giant fields are being exhausted, and not adequately replaced by new discoveries. Pemex has also had a reputation for inefficiency and a bloated payroll. The company has been heavily taxed by the government (about 60 percent in royalties and taxes). The company has also been dogged by charges of corruption that involve the government as well.

In 2013 Mexico began a sweeping reform of its petroleum sector, ending the 75-year Pemex monopoly and allowing foreign companies to compete. On 15 July 2015 Mexico held its first postreform auction, attracting a number of bidders despite low oil prices at the time. A consortium of companies including the **UK**'s Premier Oil, the U.S.'s Talos Energy, and Mexico's Sierra Oil and Gas won on some Gulf of Mexico blocks and two years later announced a discovery estimated at 1 Bb with its Zama No. 1 well. Subsequent bidding rounds garnered additional interest, and the Mexican petroleum sector was infused with new optimism.

MID-CONTINENT OIL REGION. A large oil province in the central **United States**, spanning portions of the states of Arkansas, **Kansas**, **Louisiana**, New Mexico, **Oklahoma**, and **Texas**. The region was opened up with the discovery at the Norman No. 1 well near Neodesha, Kansas, on 28 November 1892, and then the Nellie Johnstone No. 1 well near Bartlesville, Oklahoma, on 15 April 1897 that discovered the Bartlesville–Dewey field. The **Spindletop** discovery in Texas on 10 January 1901 was another major success. The Mid-Continent Oil Region also includes the **El Dorado** field, the **supergiant East Texas oil field**, and the **Glenn Pool** among many others. The Mid-Continent Oil Region played an important role in supplying the allies in **World War I** and **World War II**. Until the discovery of oil in the Middle East, the Mid-Continent was possibly the world's most important oil region.

MINNEHOMA OIL COMPANY. *See* GETTY, J. PAUL (1892–1976).

MINTROP, LUDGER. *See* GEOPHYSICAL PROSPECTING.

MISSION CORPORATION. *See* GETTY, J. PAUL (1892–1976).

MISSION TRANSFER COMPANY. *See* STEWART, LYMAN (1840–1923).

MITCHELL, GEORGE P. (1919–2013). Considered the "father of **shale gas**," George Phydias Mitchell was born in Galveston, Texas, on 21 May 1919 to Greek immigrant parents. He graduated in petroleum engineering in 1940 from Texas A&M University. In 1946 he bought a share of Roxoil Drilling, which changed its name to Oil Drilling Inc. Several successful developments, acquisitions, and name changes later, the company became Mitchell Energy and Development in 1971. Mitchell and his company are now most famous for their crucial experiments with **fracking** and **horizontal drilling** in the **Barnett Shale** during the 1990s in **Texas**, with the support of the **Gas Research Institute**. Mitchell's first horizontal fractured well in the Barnett was in 1991. The company pioneered the economic extraction of shale gas, which led to the 21st-century boom in hydrocarbon production in the **United States**. Mitchell Energy was acquired by Devon Energy in 2002 for $3.5 billion. In 2001, the last year before the takeover, Mitchell Energy controlled oil reserves of 241 Mmb and gas reserves of 1.96 Tcf. George Mitchell died 26 July 2013. His net worth was approximately $2 billion.

MITCHELL ENERGY. *See* FRACKING; HORIZONTAL DRILLING; MITCHELL, GEORGE P. (1919–2013).

MOBIL. A **major oil company** and one of the **Seven Sisters**; later **Exxon-Mobil**. Mobil originated in the 1931 merger of two former **Standard Oil** companies, the **Standard Oil Company of New York** (Socony) and the **Vacuum Oil Company**. It was first called Socony–Vacuum Corporation, changing in 1934 to Socony–Vacuum Oil Company. The merged company began using variations of Vacuum's "Mobil" trade name. In 1933, Socony–Vacuum joined with Standard Oil Company of New Jersey (later **Exxon**) to form Standard–Vacuum Oil Company (Stanvac). This joint venture would supply Socony–Vacuum's Asian and African markets with oil from fields in **Indonesia** controlled by Jersey Standard. Stanvac constituted an important share of Socony–Vacuum's earnings, 35 percent by 1941.

Socony–Vacuum's strong international presence, inherited from the large overseas role that Socony had played under Standard Oil, cost it dearly during **World War II**. A large Socony–Vacuum refinery in France was destroyed by the French army to keep it from falling into German hands. Similarly, a Stanvac refinery at Palembang, Indonesia, was destroyed to keep it from coming under Japanese control. Socony–Vacuum also lost 32 **tankers** during the war to German submarines.

After the war, Socony–Vacuum bought into **Aramco**, ensuring a supply of Saudi crude for decades. It was initially supposed to buy a 20 percent interest but balked at the magnitude of the investment and bought 10 percent instead. This turned out to be a major mistake, but even that small share proved to be an enormous benefit. In October 1954, along with the other Aramco partners, Socony–Vacuum joined the **Iran** consortium **Iranian Oil Participants Ltd.** with an 8 percent share, later reduced to 7 percent. By 1966, the Middle East, mostly **Saudi Arabia**, accounted for 43 percent of the company's oil production. In 1955, the company changed its name to Socony Mobil to take advantage of the considerable consumer recognition of its brand. In 1966, it became Mobil Oil Corporation and 10 years later shortened its name to Mobil Corporation.

During the 1950s and 1960s, Mobil's sales grew at a rapid rate as automobile use increased and the Interstate Highway System expanded in the **United States**. Sales were $2.8 billion in 1958, rising to $6.5 billion by 1967. Mobil also had a strong presence in **petrochemicals**. It formed Mobil Chemical in 1960. While petrochemicals accounted for a relatively small percentage of sales, they were considerably more profitable than Mobil's other products. To decrease its dependence on Middle Eastern crude, Mobil developed interests in both the **North Sea** and **Alaska** during the late 1960s. In the 1970s, Mobil benefited from high oil prices (*see* FIRST OIL SHOCK; SECOND OIL SHOCK). By 1977, its sales were $32 billion, about triple the level of four years earlier. It diversified away from petroleum, most notably by buying Montgomery Ward. On the negative side (from Mobil's point of view), the 1970s also saw Aramco gradually nationalized by Saudi Arabia, beginning with the 1972 **participation agreement**. Mobil also lost its Iranian interests in 1979 as a result of the Iranian Revolution. Through the 1970s, it also lost to nationalization a number of other Middle Eastern interests that it held through its membership in the **Iraq Petroleum Company** consortium, which Standard Oil Company of New York had joined in 1928.

The 1980s were a more difficult decade. Falling oil prices (*see* OIL COUNTERSHOCK) left Mobil particularly vulnerable because of expensive long-term contracts it had entered into for Saudi oil, and also because of debt from its $5.7 billion acquisition of Superior Oil in 1984. Mobil was forced to streamline. It scaled back its downstream operations and sold subsidiaries such as Montgomery Ward that were unrelated to petroleum. The difficult period continued through the early 1990s. Mobil continued restructuring and divesting itself of assets, including oil fields. Its reputation also suffered in 1992 when a jury awarded a former Mobil Chemical employee $1.75 million for having been wrongfully discharged after refusing to falsify environmental audits. The company rebounded in the mid-1990s, with strong positions in Saudi Arabia, **Nigeria**, and Indonesia. Its previous cost cutting made it one of the industry's most profitable companies.

In 1998, with oil prices at a three-decade (inflation-adjusted) low, cost cutting and consolidation again became urgent imperatives in the industry. Mobil and Exxon agreed in December 1998 to a merger. Shareholders of both companies approved the merger, and in November 1999 Mobil became a wholly owned subsidiary of Exxon, which changed its name to ExxonMobil, in an $83 billion stock transaction. In 1998, the last full year before the merger, Mobil had a net income of $1.704 billion on revenues of $53.531 billion. It controlled 4.738 Bb in proved oil **reserves** and 15.712 Tcf of **natural gas**. It produced oil and gas at the rate of 0.934 Mmbpd and 4.296 Bcfpd respectively.

MORALES, EVO (1959–). President of **Bolivia** (2006–) who nationalized its **natural gas** industry. Juan Evaristo Morales Ayma, more commonly known as Evo Morales, was born 26 October 1959 in Orinoca, Bolivia, into a poor rural family. He is of Aymara Indian descent and has been described as Bolivia's first full-blooded indigenous leader since before the Spanish conquest. He has a high school education and has worked as a bricklayer, baker, and farmer among other occupations. His courage and his organizational skills led to his election in 1985 as the general secretary of his coca farmers union. His opposition to government attempts to eradicate coca production led him to be jailed several times, and in 1989 he was severely beaten and left for dead by government forces.

Morales continued his political activities and was elected to parliament in 1997 as a member of a union of leftist parties. He was expelled from parliament on 24 January 2002 after saying openly that peasants had a right to resist government troops militarily. His popularity rose, and he ran for president that year. Although he lost, his showing was impressive enough to gain him considerable renown. He ran again in 2005 and won, despite or perhaps partly because of opposition by the **United States**. As president, Morales immediately began undertaking economic, educational, and other reforms, including ones to give more power to indigenous peoples.

On 1 May 2006, Morales signed a law nationalizing Bolivia's hydrocarbon reserves, which are mostly in the form of natural gas. He sent the Bolivian military and engineers from the state petroleum company Yacimientos Petrolíferos Fiscales Bolivianos (YPFB) to occupy natural gas facilities. He gave foreign companies six months to adjust to the new terms, according to which they would henceforth be regarded as service providers to YPFB. Royalties would also increase from 50 percent to 82 percent. On 29 October 2006, the key foreign firms operating in Bolivia, including Petrobrás and **Repsol YPF**, reluctantly accepted the new terms, although Petrobrás in particular continued to resist referring to itself as a "service provider." In May 2007, Morales issued a decree prohibiting foreign companies from exporting petroleum products from Bolivia, reserving that right for YPFB. In 2009 he was elected to a second term, and in 2014 to a third. In 2016 Bolivia had a referendum on whether or not Morales could run for a fourth term in 2020, and voters said no.

MOSSADEGH, MOHAMMED (1880?–1967). Prime minister of **Iran** 1951–1953 who nationalized its petroleum industry and was ousted in a coup backed by the **United States** and the **United Kingdom**. Mohammed Mossadegh was born Mirza Mohammed Khan to a wealthy and well-connected family in Tehran. There is some uncertainty about his date of birth, with some sources citing 1880 and others 1882 as the year. He obtained a degree in law at Neuchatel University in Switzerland, returning to Iran in 1914. At that point, he became known as Mohammed Khan Mossadegh al-Saltaneh.

Mossadegh was appointed governor of Fars Province in 1920. He resigned after the coup of 1921, led by Ziya Tabatabai and supported by Reza Khan (later Shah **Reza Pahlavi**). When Ahmad Ghavam became prime minister, Mossadegh was appointed Iran's minister of finance. In 1922, after his unsuccessful attempts to reform the ministry, he was appointed governor of Azerbaijan Province. A year later, he became foreign minister for about four months. When Mossadegh opposed the elevation of Reza Khan to the status of shah in 1925, he was forced out of public life, unable to return until the shah was forced by the United Kingdom and the Soviet Union to abdicate in favor of his son **Mohammed Reza Pahlavi** in 1941. During the intervening years, Mossadegh remained a critic, and he was briefly imprisoned in 1940.

In 1944, Mossadegh reentered politics via election to the Majlis, the Iranian parliament. He was a vocal nationalist and an advocate of nationalizing Iran's petroleum industry, then completely dominated by the British **Anglo–Iranian Oil Company**. He helped block the granting of a new oil **concession** to the Soviet Union in northern Iran and stepped up his calls for nationalization of British interests. In 1949, he led a crowd of supporters into the shah's palace to protest electoral fraud in the parliamentary elections. The demonstrators elected a committee of 20, of which Mossadegh was the leader. This became the seed of Mossadegh's National Front movement. The National Front formulated and promoted the oil nationalization legislation that was passed by the Majlis and took effect 1 May 1951. Mossadegh was elected prime minister by a large margin on 28 April 1951. His standing and popularity was such that Shah Mohammed Reza Pahlavi had to assent.

Mossadegh's nationalization of the oil industry led to a Western boycott of Iranian oil, with painful economic consequences for the country between 1951 and 1953 and a continuing struggle for control between Mossadegh and the shah. The United Kingdom wanted Mossadegh removed. The United States under President Harry S. Truman was opposed to taking any action against him, but this would change in 1953 during Dwight D. Eisenhower's administration. The shah dismissed Mossadegh in August 1953, but the move backfired as Mossadegh supporters took to the streets. The shah was forced to flee the country. By the end of the month, however, a coup engineered by the United States deposed Mossadegh—*see* AJAX, OPERATION (1953). The shah returned to Iran more powerful than ever. Mossadegh was convicted of treason and sentenced to three years in prison. After serving his sentence, he was placed under house arrest in Tehran until his death on 5 March 1967.

Despite Mossadegh's removal, the nationalization of Iran's petroleum was not reversed. However, Western companies—including Anglo–Iranian, now renamed British Petroleum (**BP**)—were allowed to form a consortium, the **Iranian Oil Participants**, to operate in Iran, and in practice they ran the country's petroleum sector for the next two decades. *See also* PLAN BUCCANEER.

MOSUL PETROLEUM COMPANY. *See* IRAQ PETROLEUM COMPANY.

N

NAIMI, ALI BIN IBRAHIM AL (1935–). First Saudi president and chief executive officer of **Aramco** (1984–1995); minister of petroleum and mineral resources for **Saudi Arabia** (1995–2016). Ali bin Ibrahim al-Naimi was born at Rakah in Saudi Arabia's Eastern Province. He began working for Aramco in 1947 when he was only 12 years old, beginning as a foreman and rising through the ranks. In 1953, he started working in the company's Exploration Department, first as a hydrologist and later as a geologist. In the 1950s, he studied at the American University in Beirut, Lebanon. He earned a bachelor's degree from Lehigh University in 1962 and a master's degree from Stanford University in 1963, both in geology.

Naimi returned to full-time duties at Aramco, working from 1967 to 1969 in the Economics and Public Relations Department and the **Abqaiq** Producing Division. He was promoted to assistant director of production in the Northern Province in 1972 and later became director. In 1975, he was appointed vice president of production affairs for Aramco, then senior vice president for petroleum affairs in 1978. In 1980, he became a director and in 1982 executive vice president of operations for Aramco (now called Saudi Aramco). He became Aramco's first Saudi president in 1984 and four years later was appointed CEO. In August 1995, he was appointed minister of petroleum and mineral resources for Saudi Arabia and chairman of the board of Aramco.

At the 2014 meeting of the **Organization of the Petroleum Exporting Countries** (OPEC), a number of oil ministers, especially those from **Venezuela**, **Algeria**, and **Iran**, wanted the cartel, especially Saudi Arabia, to cut production to prop up prices. Al-Naimi argued that it was better to keep prices low enough to avoid losing market share to higher-cost **unconventional** oil, particularly **shale oil** from the **United States**. He said Saudi Arabia would not cut production and urged the other members not to do so either. In 2016 al-Naimi ended his term as oil minister. Some media outlets reported the end of his tenure as a dismissal in the wake of continuing low oil prices.

NATIONAL PETROLEUM WAR SERVICE COMMITTEE. *See* FUEL ADMINISTRATION.

NATURAL GAS. Hydrocarbons found in natural accumulations in gaseous form, either alone or in association with crude oil. Associated gas was often wasted in the past, disposed of by burning, or "flaring." This was either because oil was the objective and gas was of no interest, or because markets for gas were too distant and no infrastructure existed for storage or distribution. Flaring is less common today but is still sometimes practiced on **offshore** platforms and in some parts of the Middle East.

Natural gas is usually 80–90 percent methane in composition. "Wet" natural gas has a higher proportion of **natural gas liquids** than "dry" natural gas. "Conventional" natural gas is that found in pressurized reservoirs, and it is recoverable with relative ease. Unconventional natural gas is more difficult to recover and comes in a number of forms. Coal-bed methane is found in coal deposits and must typically be recovered by fracturing. Tight gas reservoirs are ones in which the gas is held in low-permeability formations, and fracturing may be required. Another type of unconventional gas is geopressurized or ultra-deep gas, dissolved in deep aquifers. Methane hydrates are largely submarine deposits of ice-like solids with trapped gas.

Estimated global reserves of **conventional** natural gas in 2016 were 6,589 trillion cubic feet (Tcf), or about 1,167 billion barrels of oil equivalent (Bboe). This is of the same order of magnitude as estimated global conventional oil reserves of 1,319 Bb (including natural gas liquids). **Russia** had 17.3 percent of these reserves, and the **Persian Gulf** region had 42.3 percent, with 18 percent in **Iran** and 13 percent in **Qatar**. Beyond known reserves, global conventional gas resources could be an additional 2,800 Bboe, and **unconventional shale gas** resources are estimated at about 7,500 Tcf, or 1,330 Bboe, excluding methane hydrate.

Global natural gas production in 2016 was 125.1 Tcf or about 60.7 Mmboepd, around 65 percent of oil production in energy terms. In 1980, the figures were 24.4 Mmboepd and 41 percent. The **United States** accounts for about a fifth of total demand. Natural gas has constituted an increasing proportion of the world's total primary energy supply: in 1950 it accounted for 9.9 percent; in 2016 it accounted for 22.7 percent. Among natural gas's major uses are heating, cooking, and electricity generation. In the United States, almost all new power plants in the 2000s used natural gas. Natural gas is a relatively clean-burning fuel and adds the smallest amount of carbon dioxide to the atmosphere per unit of energy of any fossil fuel. Some vehicles are powered by natural gas. Former corporate raider **T. Boone Pickens** has been a staunch advocate of natural gas use in cars.

Natural gas was observed in premodern times. In **China** around 500 BCE, natural gas was transported in bamboo pipes and was used to make salt from seawater. Some seeps ignited by natural processes appeared to ancient peoples as "eternal flames," as on Mount Parnassus in ancient Greece. Such burning seeps were also known in the **Baku** area as well as a number of other places. In 1626, French explorers observed Native Americans lighting seeps around Lake Erie, and in 1770 George Washington noted such phenomena in what is now **West Virginia** and southeastern Ohio. An early modern use of natural gas was in Fredonia, New York. In 1821, a local gunsmith, William Aaron Hart, noticed gas bubbles and dug a 30-foot well. He was able to serve a number of residences in the area with gas for lighting. In 1827, Hart transported gas through hollowed-out pine logs to supply a nearby lighthouse on Lake Erie under a government contract. In 1858, the Fredonia Gas and Light Company was established, perhaps the first modern natural gas company.

The effective use of natural gas depended crucially on the development of **pipeline** technology. In 1872, a 2-inch, 51-mile wrought-iron pipeline was built to transport associated gas from western **Pennsylvania** oil fields to **Titusville**. Early pipelines suffered from wasteful and dangerous leaks. The problem was partially solved in 1887 by Solomon Dresser (*see* DRESSER INDUSTRIES), who patented a leakproof coupling. It was not fully effective at first, but he improved it, as well as the process of joining segments together. **George Westinghouse** developed a system of distributing natural gas to homes in Pittsburgh and by 1887 supplied 5,000 residential and 470 industrial customers in that city. New gas discoveries in the southwestern United States, far from major markets, necessitated longer pipelines. The development of improved welding techniques in the 1920s enabled them. In 1927, **Cities Service Company** built the first long-distance gas pipeline, running 250 miles from northern **Texas** to Kansas. A major milestone was the 980-mile, 24-inch line from Texas to Chicago, completed in 1931 by the Natural Gas Pipeline Company.

In 1978, the U.S. Natural Gas Policy Act created the Federal Energy Regulatory Commission (FERC). FERC issued a number of regulatory changes, culminating in Order 636 of 8 April 1992, which essentially transformed interstate pipelines into common carriers. Gas utilities and end users could contract directly with producers. In the European Union (EU), markets were liberalized by the First Gas Directive of August 1998 and the Second (Acceleration) Directive of July 2003. Among other things, these opened up access to pipeline networks. Gas also became an important political issue in the EU in the late 1990s and the 2000s, with fears of increasing energy dependence on Russia.

Because gas depends on a fixed distribution network, and for other reasons, it does not have as well developed a global market as oil does. Most gas contracts are long term, with prices set in advance. The relatively recent use of **liquefied natural gas** (LNG) has begun to change patterns somewhat, introducing more of a global market for gas. LNG is still a relatively small portion of total gas sold. *See also* BARNETT SHALE; FRACKING; GAS EXPORTING COUNTRIES FORUM (GECF); GAS TO LIQUIDS; GAZPROM; HORIZONTAL DRILLING; MARCELLUS SHALE; NORTH FIELD.

NATURAL GAS LIQUIDS (NGL). Hydrocarbon liquids recovered from **natural gas** reservoirs. The first process for capturing natural gas liquids was developed and patented by **Walter Snelling** in 1913. Liquids recovered at the well or in small field separators are called "lease condensate." Lease condensate consists primarily of pentanes and heavier molecules. Liquids recovered at natural gas processing plants are referred to as "natural gas plant liquids" or "liquefied petroleum gases" (LPG). These consist primarily of ethane, propane, butane, and isobutane. Gas reservoirs without natural gas liquids are referred to as "dry." "Wet gas" can have NGL contents of 200 barrels per Mmcf of gas. Production and **reserves** of natural gas liquids are not consistently reported. As a very rough estimate, the world's total resource of NGL is probably 20 percent or more of its total resource of **conventional crude oil**.

NATURAL GASOLINE. A **natural gas liquid** that somewhat resembles ordinary **gasoline** but is not suitable for use in automotive engines because of a low octane number and the presence of impurities such as sulfur. It can be refined and upgraded to the point of being usable in engines. It is used as a thinner for very **heavy oil** and **bitumen**.

NAVAL PETROLEUM AND OIL SHALE RESERVES. Oil-bearing lands set aside in the early 20th century by the **United States** for possible exploitation by the U.S. Navy in the event of war. The **reserves** were not actual stores of previously extracted crude oil but rather lands that contained exploitable deposits. The designation of the reserves was authorized by the Pickett Act of 25 June 1910, which allowed the federal government to classify public lands according to their best uses and either sell, lease, or manage them accordingly.

President William Howard Taft signed an executive order on 2 September 1912 establishing NPR-1, also known as Elk Hills, in Kern County, **California**; the **giant** Elk Hills field would be discovered there in 1919. President Taft also established NPR-2, located immediately south of NPR-1, via an executive order on 13 December 1912. NPR-2 is also known as Buena Vista Hills. NPR-3, or Teapot Dome, was established in Wyoming by President Woodrow Wilson in 1915. NPR-4, in the North Slope of **Alaska**, was established on 27 February 1923 by President Warren G. Harding. Both Elk Hills and Teapot Dome would gain a measure of infamy in the 1920s because of the **Teapot Dome scandal**, which arose from the discovery of their improper leasing to private parties in exchange for bribes.

In addition to the Naval Petroleum Reserves, the government also set aside Naval Oil Shale Reserves (NOSRs) in the area of the **Green River Formation** during the same time frame. NOSR-1 and NOSR-3 were located near Rifle, Colorado. NOSR-2 was located in Utah. Oil from **oil shale** is more difficult to extract than **conventional crude oil**. The U.S. government was farsighted in considering it and in recognizing the enormous potential resources in the Green River Formation. Although the NOSRs were never developed, oil shale may indeed become important.

The Naval Petroleum and Oil Shale Reserves changed administrative jurisdiction a number of times. They started out under the Department of the Interior. In 1920, control passed to the U.S. Navy's Fuel Oil Office. In 1921, President Harding put them back in the Department of the Interior. After the Teapot Dome scandal broke out, control shifted once again to the U.S. Navy. The Office of Naval Petroleum and Oil Shale Reserves was created in 1927. The reserves remained under the navy until 1977, when they were transferred to the new Department of Energy.

The Naval Petroleum Reserves remained largely idle until the 1970s, apart from some production during **World War II**. On 5 April 1976, President Gerald Ford signed the Naval Petroleum Reserves Production Act, allowing production of the reserves. Elk Hills (NPR-1) produced at a peak rate of 181,000 bpd in 1981. For a time, Elk Hills was the largest-producing oil field in the lower 48 states. It contributed more than $17 billion in profits to the U.S. Treasury. In September 1992, it produced its billionth barrel of oil. Its total estimated ultimate recovery (**EUR**), including gas, is projected at about 1.65 Bboe. On 5 February 1998, the Department of Energy sold Elk Hills to the **Occidental Petroleum Corporation** for $3.65 billion.

After the privatization of Elk Hills, the Department of Energy transferred the Oil Shale Reserves NOSR-1 and NOSR-3 to the Bureau of Land Management. They are available for commercial mineral leasing, not necessarily limited to oil shale development. In 2000–2001, the government transferred NOSR-2 to the Northern Ute Indian Tribe. After the enactment of the Energy Policy Act of 2005, the Department of Energy transferred Buena Vista Hills (NPR-2) to the Department of the Interior. The Department of Energy still retains NPR-3 (Teapot Dome), where it runs the Rocky Mountain Oil Field Testing Center (RMOTC). RMOTC tests new oil field technologies and production strategies.

NEAR EAST DEVELOPMENT CORPORATION (NEDC). A consortium of U.S. oil companies formed to participate in the Turkish Petroleum Company (later **Iraq Petroleum Company**). The Turkish Petroleum Company (TPC) was a European consortium formed in 1912 to obtain and develop a petroleum **concession** in Mesopotamia (**Iraq**), then part of the Ottoman Empire. **World War I** interrupted its progress, and its ownership was restructured at the postwar **San Remo Conference**, which granted its German shares to France. San Remo did not provide for participation of U.S. companies in TPC.

American oil companies and the U.S. State Department strongly lobbied for an "open door" principle, whereby the petroleum resources of the Middle East would be open to all companies equally, regardless of nationality. After several years of negotiations, U.S. companies were allowed to buy into TPC. The Near East Development Corporation was a vehicle created for a set of U.S. companies to do so.

NEDC consisted of the Standard Oil Company of New Jersey (later **Exxon**), the **Standard Oil Company of New York** (Socony, later **Mobil**), **Gulf**, Pan American Petroleum and Transport (majority controlled by Standard Oil Company of Indiana, later **Amoco**), and Atlantic Refining Company (later **Arco**). The agreement formalizing NEDC's 23.75 percent share in TPC was concluded in July 1928, nearly a year after TPC struck oil at Baba Gurgur near **Kirkuk** (on 15 October 1927). In 1929, TPC was renamed the Iraq Petroleum Company (IPC). During the 1930s, Gulf, Pan American, and Atlantic sold out their shares. By 1934, NEDC became jointly owned by Socony (by then, Socony–Vacuum) and Standard Oil Company of New Jersey. Much later, these two partners would merge as **ExxonMobil**.

NETBACK PRICING. A system whereby the price of a batch of crude is determined by deducting the cost of refining, marketing, transportation, and a fixed margin from the price of the products derived from it. Refiners purchasing oil under the terms of such a contract have very little to lose. They may sell products at whatever price it takes to move them, and they are guaranteed a profit.

In 1985, **Saudi Arabia**, frustrated as other members of the **Organization of the Petroleum Exporting Countries** (OPEC) cheated on their quotas, offered netback pricing in order to gain market share. In September 1985, it signed netback agreements with the four **Aramco** partners, **Exxon**, **Mobil**, **Texaco**, and **Chevron**, for 900,000 bpd. By early February 1986, it had netback agreements covering 3 Mmbpd. While Saudi Arabia's netback pricing was intended to gain back market share from the rest of OPEC, OPEC as a whole soon adopted netback pricing in order to gain back market share from non-OPEC producers, especially those operating in the **North Sea**. The inevitable result was the **Oil Countershock**—a price collapse. In 1986, prices averaged $14.32 a barrel, down from $27.53 a year earlier. At their low point during 1986, they were in the neighborhood of $8 a barrel. After this, OPEC tried in vain to reestablish fixed pricing, but the modern global free market in oil emerged instead.

NETHERLANDS. *See* DUTCH DISEASE; GRONINGEN GAS FIELD; ROYAL DUTCH COMPANY; ROYAL DUTCH SHELL.

NEUTRAL ZONE. An area shared by **Saudi Arabia** and **Kuwait**. The Neutral Zone was established by the Uqayr Conference of 1922. The conference did not address the issue of political sovereignty over the zone but allowed equal economic rights. In practice, each country had its own portion of the zone and administered it as if it were formally part of its territory. In 1965, Kuwait and Saudi Arabia agreed to an equal partition of the zone, with each effectively annexing its portion. However, the two countries share economic sovereignty in the zone and share equally in oil production. They renamed the area the Saudi–Kuwaiti Partitioned Neutral Zone (PNZ).

The Neutral Zone has a number of **giant** oil fields: Khafji, Wafra, Hout, and South Umm Gudair. Wafra (**EUR** about 1.7 Bb) was discovered in 1953 by **Getty Oil Company** and **Aminoil** (American Independent Oil, a consortium of oil companies from the **United States**). Aminoil had obtained Neutral Zone rights from Kuwait in 1946, and Getty (Pacific Western Company) had obtained rights from Saudi Arabia in 1949. Getty had won the Saudi Neutral Zone **concession** after **Aramco** had relinquished its rights to it, having viewed the terms Kuwait had granted to Aminoil as unfavorable. Getty and Aminoil entered into a joint operating agreement in 1960 to minimize duplication of operations in fields and improve overall efficiency. The agreement set the framework for future cooperation between Saudi Arabia and Kuwait in the zone. In 1962, Getty and Aminoil developed South Umm Gudair, the southerly extension of Kuwait's Umm Gudair field.

Khafji, the zone's largest oil field (EUR 8.65 Bb), is an **offshore supergiant** field discovered in 1959 by the Arabian Oil Company (AOC). AOC was a Japanese consortium that obtained offshore concessions for the Neutral Zone from both Saudi Arabia and Kuwait. AOC also discovered the Hout oil field in 1963 and the giant Dorra gas field in 1967.

The Kuwaiti government nationalized Aminoil's interests on 19 September 1977 and formed Kuwait Wafra Oil Company to take over its operations. Kuwait Wafra was dissolved less than a year later, on 16 April 1978. The Kuwait Oil Company took over its upstream activities, and the Kuwait National Petroleum Company took over its downstream operations. On the Saudi side, **Texaco** obtained Getty's Neutral Zone operations in 1984 when it acquired Getty Oil. In 1993, it renamed its operations there Saudi Arabian

Texaco. The interests passed to ChevronTexaco (later just **Chevron**) when Texaco and Chevron merged in 2001. The Neutral Zone's onshore operations are now managed by a Joint Operations Committee of the Kuwait Oil Company and Saudi Arabian Texaco. The offshore operations are managed by the Arabian Oil Company, in which the Kuwait and Saudi governments both have an interest.

The Saudi–Kuwait Neutral Zone was not the only such zone. There was also a neutral zone between Saudi Arabia and **Iraq**, which was effectively dissolved by a border agreement in 1981.

NEW YORK MERCANTILE EXCHANGE (NYMEX). The world's largest futures exchange for physical commodities; headquartered in New York City. The New York Mercantile Exchange began trading energy futures and options contracts in 1978. The petroleum-related commodities that NYMEX trades include crude oil, **natural gas**, **gasoline**, and heating oil. The major underlying commodity in the NYMEX futures price for crude oil is **West Texas Intermediate** crude delivered at **Cushing**, **Oklahoma**, although the contract allows other grades as well. The contract trades in lots of 1,000 barrels. West Texas Intermediate is one of the three most important international **benchmark crudes**, the other two being **Brent Blend** and **Dubai crude**. For natural gas, NYMEX uses **Henry Hub**.

NYMEX now includes both the New York Mercantile Exchange and the New York Commodities Exchange (COMEX), which were once separate companies. NYMEX traces its roots back to 1872, when a group of Manhattan dairy merchants established the Butter and Cheese Exchange of New York, soon to become the Butter, Cheese, and Egg Exchange. Trade was opened to other commodities, including dried fruits, canned goods, and poultry, and in 1882 the organization changed its name to the New York Mercantile Exchange.

The exchange's offices in the World Trade Center were destroyed in the terrorist attacks of 11 September 2001. On 26 February 2003, NYMEX moved into the New York Board of Trade's World Financial Center headquarters. NYMEX now maintains a backup trading facility outside New York City.

NIGERIA. Oil production 2015: 1.815 Mmbpd (2.32 percent of world); **reserves** 37.062 Bb (2.25 percent of world); gas production: 1.752 Tcf (1.41 percent of world); reserves 186.61 Tcf (2.71 percent of world). Nigeria is one of sub-Saharan Africa's most important oil producers and has large reserves of **natural gas**. Nigerian crude oil is generally light and low in sulfur, and it is well situated for export to the **United States** and the European Union. Nigeria joined the **Organization of the Petroleum Exporting Countries**

(OPEC) in 1971. Nigeria is also a member of the **Gas Exporting Countries Forum**. As recently as the 2000 decade, Nigeria was flaring about 40 percent of its annual gas production. The country's industry is still plagued by a lack of production infrastructure, but in 2016 Nigeria was nevertheless the world's fourth-largest exporter of **liquefied natural gas**.

Modern exploration in Nigeria dates back to 1903. Early companies included the Mineral Survey Company and a German concern called the Nigerian Bitumen Company. Following Nigeria's unification in 1914, Mineral Oil Ordinance No. 17 of 1914 was passed. One of its main provisions was that only British oil companies could obtain oil exploration licenses in Nigeria. The provision was repealed in 1958 (it had already been circumvented by non-British companies forming locally registered British subsidiaries, as was done by Socony Mobil, later **Mobil**, in 1955).

Serious exploration for oil did not begin until 1937, when the **Anglo–Iranian Oil Company** (later **BP**) and **Royal Dutch Shell** formed the joint venture Shell D'Arcy. Shell D'Arcy acquired a license covering about 370,000 square miles, and it carried out preliminary exploration, but **World War II** interrupted its efforts. In 1949, Shell D'Arcy reduced its license area to about 60,000 square miles, including the Niger Delta. In 1951, a new company, the Shell D'Arcy Petroleum Development Company of Nigeria, was formed to execute the activities of the joint venture. Conditions were primitive, and progress was slow. The company drilled a number of dry holes. In 1953, it discovered oil at Akata on the eastern edge of the Niger Delta, but not in commercial quantities. The first commercial discovery was the relatively small Oloibiri field in January 1956. That year Shell D'Arcy changed its name to Shell–BP. The company soon discovered a second field at Afam, and in February 1958 Nigeria exported its first cargo of oil from **Port Harcourt**. In 1958, the **giant** Bomu field was discovered southeast of Port Harcourt. It has an **EUR** of 0.311 Bb of oil and a total of 0.608 Bboe including gas.

Nigeria became an independent nation on 1 October 1960. Shell–BP began reducing its acreage, converting exploration licenses into prospecting licenses that allowed development and production. As it did so, other companies took up the relinquished licenses and began exploring. These included Socony Mobil (1959), Tennessee Nigeria (1962), **Gulf** (1963), and **ENI** (1964). Gulf was the first to discover oil and began producing in 1965. Shell–BP and its successors remained dominant in Nigeria's industry, however.

From 1967 to 1970, there was a civil war in Nigeria. The eastern, mainly Igbo, region seceded and formed the Republic of Biafra, taking a large part of the petroleum industry with it. In 1970, the eastern region rejoined the country. Nigeria was angered by the support some foreign oil companies had given to Biafra. This gave an additional impetus to an already growing desire to take more control over the country's petroleum industry. In 1971, Nigeria joined OPEC and created its national oil company, first called the Nigerian National Oil Corporation (NNOC); in 1977 it became the Nigerian National Petroleum Corporation (NNPC). NNOC acquired a 35 percent share in foreign operations in 1973. The equity share increased during the 1970s. In the case of Shell–BP, Nigeria nationalized BP's holding completely in 1979, and the company became Shell Petroleum Development Company of Nigeria (SPDC). SPDC is still the dominant company in the country, responsible for about half of production. Today more than 95 percent of Nigeria's oil production is controlled by joint ventures between foreign oil companies and NNPC.

Nigeria's petroleum industry continues to be plagued by violence and instability. In the 2000s, it has suffered from attacks by the Movement for the Emancipation of the Niger Delta (MEND), a militant group of the Ijaw tribe. Multinational oil companies face considerable grassroots opposition on environmental and human rights grounds. The government is often heavy-handed in its dealings with such opposition. In November 1995, the government executed author and activist Ken Sarowiwa, a vocal critic of multinational oil companies in his country, particularly Shell. His supporters accused Shell of collusion with the government. In 2009 the government announced an amnesty for MEND militants, and oil production increased about 12 percent over the next two years.

Nigeria has 31 giant oil and gas fields, 17 of which are **offshore**. The three largest fields are offshore in deep water, an increasingly important segment since 1990. Bonga and Bonga Southwest were discovered in 1996 and 2001 respectively and are operated by a consortium led by SPDC. The fields lie under an average water depth of about 3,300 feet. Bonga has an EUR of 1.171 Bb of oil, or 1.2 Bboe including gas. Bonga Southwest has an EUR of 0.75 Bb, 0.5 Bb of condensate, and 0.5 Tcf of gas, for a total of 1.333 Bboe. The third-largest field, Akpo, was discovered in 2000 and is operated by **Total**. It lies under up to 5,600 feet of water and has an EUR of 0.95 Bb, 1.143 Bboe when gas is included. **Bonny** has been the main port city for oil export.

NOBEL, EMANUEL (1859–1932). Head of **Branobel** from 1888 until 1918, when it was nationalized by the Bolsheviks. Emanuel Nobel took over the company after the death of his father, **Ludvig Nobel**, and continued to lead it in a similarly effective and innovative fashion. He was a director of the State Bank, and like his father he had good connections at the tsar's court and in the ministries. He was able to maintain Branobel's leadership in a difficult competitive environment and an increasingly difficult political one. By 1905, the Caucasus was racked by unrest and revolution. The oil industry around **Baku** was attacked by rampaging mobs, with about a thousand wells destroyed. Foreign companies were particularly targeted. Nobel was relatively unscathed; it had always behaved as a Russian company, and Emanuel Nobel was a Russian citizen.

In 1916, the Nobels controlled, or had substantial interest in companies controlling, about a third of **Russia**'s crude production and 40 percent of refined products. Branobel also had the largest private fleet in the world. Emanuel Nobel developed new oil fields and constructed new chemical plants. He also introduced American rotary **drilling** techniques, permitting exploration and development at greater depths. In 1898, he signed a license agreement in Berlin with Rudolf Diesel and then built the world's first diesel engine plant in St. Petersburg. He installed diesel engines in power plants, tugs and **tankers**, and pumping stations along the newly completed (1906) **Baku–Batumi Pipeline**.

Branobel's position became untenable after the Bolshevik Revolution of 1917. Nobel's assets and even his home were taken from him, and he was forced to flee Baku in the spring of 1918. Like a number of other industrialists and aristocrats, he settled first in Mineralye Vody on the north slope of the Caucasus. The area was not yet under Bolshevik control, but the safety was only temporary. Later in the year, he fled with a forged passport, dressed as a peasant. First he went to Stavropol, where he narrowly escaped the advance of the Red Army, and then Kiev, Warsaw, Berlin, and finally Stockholm. His brothers Gustav and Emil also managed to make it there, but they spent time first in the Cheka prison and narrowly escaped being executed. For more than a decade, Emanuel Nobel and his family fought to recover their assets from the Soviet government, but they were unsuccessful.

NOBEL, LUDVIG (1831–1888). Swedish entrepreneur and engineer who was a major force in creating the modern oil industry in **Russia**. Ludvig Nobel was a remarkable member of a remarkable family. His father, Immanuel, was a self-taught engineer who began a business in St. Petersburg and is credited with many inventions, including naval mines for the Russian military. Ludvig's brother Alfred invented dynamite and made a bequest establishing the Nobel Prize. Immanuel Nobel was brilliant but suffered financial difficulties. In 1833, he went bankrupt in Sweden and began a new business in St. Petersburg, making equipment for the Russian Army. The business did very well for a time but faltered after the end of the Crimean War (1853–1856). The company's creditors appointed Ludvig Nobel to run it. Immanuel returned to Stockholm with his wife and his sons Emil and Alfred, while his other son Robert stayed behind to work with Ludvig. In Stockholm, Immanuel began experiments with nitroglycerine that killed his son Emil in 1864 but formed the basis for the work that later made Alfred famous.

Meanwhile, through great effort, Ludvig and Robert put the St. Petersburg factory back on its feet. At one point, Ludvig asked Robert to tour southern Russia in search of wood from which to make rifle stocks. Robert made his way to the Caucasus and eventually visited **Baku**, where he became interested in the oil industry. Instead of buying wood, he purchased an interest in a refinery in 1876. Ludvig soon became involved, and in 1879 the two formed the Nobel Brothers Petroleum Production Company, or **Branobel**. The company was wildly successful. By 1883, it controlled about one-quarter of Russian oil production and around half the kerosene market. By 1916, under the leadership of Ludvig's son **Emanuel Nobel**, it would control about a third of Russian production and supply two-thirds of domestic consumption.

Ludvig Nobel was the driving force of Branobel. His innovations were legion and extremely important, not only for the Russian industry but for the world as a whole. He designed the world's first modern oil **tanker**, the ***Zoroaster***, and continued to improve on it. The company had a fleet of tankers plying the **Caspian Sea** in the 1880s. He also installed Europe's first **pipelines** and was the first to put tanker cars on rail anywhere in Europe. He built the world's first full-scale continuous distillation refinery in 1881, incorporating ideas from **Dmitri Mendeleev**. While patents for the idea or aspects of it existed as early as 1860 in the **United States** and were further developed there and elsewhere, Ludvig Nobel's system was the first functioning commercial unit. He also developed oil burners for industrial use and even powered his tankers with a type of fuel oil when the rest of the industrialized world was still using coal. All in all, he effectively created a modern petroleum production, storage, **refining**, transportation, and marketing infrastructure in what was still arguably a backward country.

He was also uniquely concerned with the welfare of his workers. This was true not only by prevailing Russian standards but by global ones, for labor practices were harsh throughout the world at that time. He abolished child labor and shortened the workday from 14 to 10 and a half hours. He built decent housing for workers and provided them with free medical care, technical training, and elementary education for their children. He also created a cooperative bank for workers. With his brother Robert, he established a large park in Baku that still exists. While the Nobel Prize is today associated with the award established by Ludvig's brother Alfred, there was another Nobel Prize for a time in Russia named after Ludvig. The Russian Imperial Technical Society awarded it beginning in 1896 for outstanding innovations in metallurgy or the oil industry. The prize consisted of a gold medal and the interest income on a principal of 6,000 rubles.

Today Ludvig Nobel and his family do not generally receive the credit they deserve for their achievements in the petroleum industry. Part of the reason doubtless lies in the fact that their company was nationalized by the Soviet government, which tended to obscure or obliterate recognition of industrial achievements in the tsarist era.

NOBEL, ROBERT. *See* NOBEL, LUDVIG (1831–1888).

NOBMAZUT. A cartel agreement concluded in May 1903 by the **Rothschilds** and the **Nobel** brothers to coordinate **Russia**'s domestic sales of petroleum products. At times, the cartel controlled up to 80 percent of the Russian domestic market.

NORSK HYDRO. Diversified Norwegian company that had significant operations in all phases of the petroleum industry. Its petroleum operations merged with **Statoil** in 2007 to form StatoilHydro. Norsk Hydro-Elektrisk Kvaelstofaktieselskap (Norwegian Hydro-Electric Nitrogen Corporation) was established in **Norway** in 1905 by Sam Eyde and Kristian Birkeland. The company's first product was fertilizer, using hydroelectric power to provide the energy for the manufacturing process. During **World War II**, the German occupying forces used the company's Rjukan plant to produce heavy water (a by-product of the plant's ammonia production) for nuclear research. The plant was destroyed by Allied saboteurs. After the war, the Norwegian government took a 48 percent interest in the company.

Norsk Hydro was primarily a fertilizer and chemical company but in the 1960s diversified into other areas, including petroleum. It partnered with **Phillips Petroleum**, which discovered the **Ekofisk** field in the Norwegian **North Sea** in 1969. It also partnered with ERAP (**Elf Aquitaine**), whose consortium discovered the Frigg field in 1971. That year, the Norwegian government increased its share of Norsk Hydro to 51 percent. In 1972, it established Statoil. Norsk Hydro began operating Norway's largest refinery, at Mongstad, in 1975. The company operated the Oseberg field, which began producing in 1988. Norsk Hydro became a leader in the technology of enhanced **oil recovery** in deepwater projects. In the early 1980s, it had the radical idea of producing oil from Oseberg by injecting gas from the giant Troll gas field. The technology worked well; when Oseberg opened, it had a record recovery factor of 72 percent. Estimates of recoverable **reserves** increased from about 0.994 Bb to 2.2 Bb.

In the 1990s, Norsk Hydro moved into retail operations. In 1990, it bought 330 Danish **gasoline** stations from UNO-X, and six years later acquired UNO-X's gas stations in Sweden. In 1992, it acquired **Mobil**'s marketing and distribution system in Norway. In 1994, it merged its oil and marketing operations in Norway and Denmark with **Texaco**'s. In 1996, it became a partner in **Petro-Canada**'s fields in the Canadian Atlantic. In 1999, Norsk Hydro acquired Saga Petroleum, Norway's largest **independent** petroleum producer. Saga operated several fields in the Norwegian North Sea, including Snorre, Vigdis, and Tordis. The Norwegian government's stake in Norsk Hydro was reduced from 51 percent to about 45 percent that year. In the 2000s, petroleum continued to be a major focus of the company's business, along with other energy and aluminum.

In December 2006, a merger was announced between Statoil and Norsk Hydro's oil and gas business. The Storting (the Norwegian parliament) approved the merger on 8 June 2007, and it was completed on 1 October 2007, forming StatoilHydro. The value of the deal was $30 billion. The Norwegian government owns 62.5 percent of StatoilHydro. In 2006, the last full year before the merger, Norsk Hydro had a net income of $2.685 billion on revenues of $30.593 billion. It controlled 0.748 Bb of oil reserves and 6.611 Tcf of gas. It produced oil and gas at the rate of 0.387 Mmbpd and 1.042 Bcfpd respectively.

NORTH DAKOTA. Oil production 2015: 1.177 Mmbpd (12.5 percent of U.S.); **reserves** 5.2 Bb (16.1 percent of U.S.); gas production: 0.382 Tcf (1.41 percent of U.S.); reserves 6.2 Tcf (2.02 percent of U.S.). The first producing well in North Dakota was the Clarence Iverson No. 1 in the Williston basin, struck 4 April 1951 by the **Amerada Petroleum Corporation** (later Amerada Hess, eventually **Hess** Corporation), under the persistent influence of **Thomas W. Leach**. North Dakota remained a relatively low producer of oil for decades, averaging 0.078 Mmbpd in the period between 1952 and 2005. Production grew by a factor of 12 in the decade after that, owing to the application of **fracking** and **horizontal drilling** in the **Bakken Shale**, placing North Dakota in the front rank of American oil-producing states, behind only **Texas** and the federal offshore areas in the **Gulf of Mexico**. Gas production increased eightfold in the same period. Hess is still a major producer in the state, along with other large companies such as **Marathon**, **ExxonMobil**, and **Statoil**. The two leading oil producers in 2013 were Continental Resources and Whiting Oil and Gas Corporation.

NORTH FIELD. The world's largest **natural gas** field. The field, which belongs to **Qatar**, has an **EUR** of 900 Tcf of gas and about 26.1 Bb of condensate, for a total of 176.1 Bboe. In energy-equivalent terms, it is larger than **Ghawar**, the world's largest oil field, and is the largest hydrocarbon complex ever discovered. The above numbers do not count **Iran**'s **mega-giant** South Pars gas field discovered in 1991, which is considered an extension of the North Field. The North Field is off the Qatar Peninsula in the

Persian Gulf, under about 165 feet of water. It was discovered in 1971 by **Royal Dutch Shell**. Production began on 3 September 1991, on the 20th anniversary of Qatar's independence. **Liquefied natural gas** exports began in 1997.

NORTH SEA. An Atlantic sea of 220,000 square miles with petroleum resources shared mostly by **Norway**, the **United Kingdom**, and the Netherlands, but also Denmark and Germany. In 2016, the North Sea region had **reserves** of about 12 Bb (some estimates are higher), nearly 90 percent of which belonged to Norway and the UK. It had about 100 Tcf of natural gas, mostly belonging to Norway and the Netherlands (again, some estimates are higher). North Sea oil production peaked in 1999 and has been declining since. The North Sea has 36 **giant** fields, 23 of them predominantly oil. It has four oil **supergiants** (the largest being Statfjord, EUR 9 Bb, discovered 1974) and one gas supergiant (Troll, EUR 22.8 Tcf, discovered 1979). All the supergiants were discovered before 1980. Production from the North Sea played an important role in reducing the power of the **Organization of the Petroleum Exporting Countries** (OPEC) in the 1980s and contributing to the **Oil Countershock**. North Sea oil production now appears to be in long-term decline.

The North Sea is mostly shallow, with few parts exceeding 300 feet in depth. In the southern part, the average depth is closer to 120 feet. One important exception is the Norwegian trench near the southern coast of Norway; it reaches a maximum depth of more than 2,300 feet. Other trenches include a 320-foot-deep trench off England's Wash and the 1,500-foot Devils Hole off Edinburgh. To the north of the North Sea is the Norwegian Sea, a section of the North Atlantic reaching a maximum depth of more than 13,000 feet, whose petroleum resources have been exploited by Norway. Farther north is the **Barents Sea**, shared between Norway and **Russia**.

Serious interest in the North Sea began after 1959, when the **Groningen gas field** was discovered in the Netherlands. Although there was considerable skepticism—with one expert famously offering to drink all the oil found in the North Sea—exploration proceeded in the 1960s. Development depended critically on two factors. One was the development of appropriate **offshore** technologies that could withstand the difficult environment. Another was the development of international law that could guide the equitable division of the sea's resources among the littoral states.

The **United Nations Convention on the Law of the Sea** did not complete its work until 1982. However, many of the principles of that convention, such as using median lines to divide the continental shelf, were agreed to at the 1958 Geneva Convention on the Continental Shelf, signed by 46 nations. Since the North Sea is mostly less than 200 meters (656 feet) deep—the standard set by the convention—it could be argued to represent one continuous continental shelf, on which median lines could be applied. However, there were still some difficulties. First, the existence of convex and concave coastlines could lead to unreasonable results using median lines. Second, the fact that the North Sea is deeper than 200 meters in some places could invalidate the median line principle. Germany raised the first objection. Denmark and the United Kingdom could have raised the second but did not do so.

Spurred on by oil companies anxious to ensure the security of their future claims, the North Sea littoral states divided the continental shelf according to the rules of the 1958 Geneva Convention. Bilateral treaties were concluded between the United Kingdom and Norway, the United Kingdom and Denmark, the United Kingdom and the Netherlands, and Norway and Denmark, establishing median lines as boundaries. The United Kingdom and Denmark could have argued that the Norwegian Channel formed a natural interruption in the continental shelf and could have attempted to exclude Norway from a large part of its current share, which would have gone to Denmark. However, they did not advance the argument.

Denmark and the Netherlands agreed on their boundary up to the median line with the United Kingdom but did not consult Germany, which protested. The shape of the German coast was such that applying a principle of equidistance yielded it a small share in proportion to the length of its coastline. The three countries negotiated unsuccessfully, ultimately submitting the case to the International Court of Justice. The court ruled in 1969 that the equidistant principle was not obligatory because Germany had never signed the 1958 convention and because it was not yet firmly established in international law. The court recommended a division guided by geology and coastal configuration. In 1971, Germany concluded treaties with Denmark and the United Kingdom, giving it an additional 2,734 square miles from the former and 1,953 square miles from the latter.

The division of the North Sea was quite amicable overall. Probably this was at least partly because no one yet had any idea how the resources were distributed between the sectors. As it happened, the United Kingdom and Norway ended up with almost all the known oil resources and about 80 percent of the **natural gas**. The Netherlands got a zone with good gas prospects. Denmark received a relatively large area that was nowhere near as productive as the United Kingdom's or Norway's. Germany received a relatively small and fallow share. Belgium received less than 1 percent of the total area and has not attempted to develop it. Some of the resources in the rich UK and Norwegian areas straddle the boundary between the sectors. In April 2005, the United Kingdom and Norway signed a bilateral treaty specifying the disposition of such resources, which had previously been handled on a case-by-case basis. *See also* BRENT OIL FIELD; EKOFISK OIL FIELD; NORSK HYDRO; STATOIL.

NORWAY. Oil production 2015: 1.604 Mmbpd (2.05 percent of world); **reserves** 6.611 Bb (0.4 percent of world); gas production: 4.125 Tcf (3.31 percent of world); reserves 65.543 Tcf (0.95 percent of world). Norway is a major oil exporter, playing a crucial role in supplying Europe; in 2015, 97 percent of its 1.3 Mmb of crude exports went to European countries. Norway has petroleum resources in the **North Sea**, the Norwegian Sea, and the **Barents Sea**. The Norwegian industry began in earnest after the discovery of **Ekofisk** in 1969. In 1972, Norway created its national oil company **Statoil** (Den Norske Stats Oljeselskap) and established the **Norwegian Petroleum Directorate** to manage the petroleum industry. Norway favored a discretionary licensing system, rather than a pure auction system as in the **United States**, in order to avoid complete foreign domination of its industry. Norway has also been a high-tax country for the industry, with a total government take of about 75 percent of revenue, sometimes higher. The country has a stabilization fund in which it has invested oil revenues to help provide for the day when its resources are exhausted.

Norway's oil production increased rapidly after 1982, when it was 0.492 Mmbpd. It peaked at 3.14 Mmbpd in 1997. It then dipped and reached a second peak of 3.226 Mmbpd. Between 2001 and 2006, production decreased about 23 percent. Production in 2006 was at roughly the same rate as in 1994; production in 2016 was at about 1990 levels. Reversing these declines will be difficult and will likely involve more intensive exploration in frontier areas such as the Barents Sea. A treaty with Russia delimiting each country's share of the sea, in force since 2011, has removed some of the political uncertainty. **Natural gas** production was 0.917 Tcf in 1980, and it stayed almost flat for over a decade. In 1993, it was 0.967 Tcf. It then increased rapidly, more than tripling by 2006. In 2015 it was higher still, at 4.125 Tcf. Norway discovered 15 **giant** oil and gas fields in the northern North Sea between 1969 and 1991 but only two after 1980. It discovered 11 giants in the Norwegian Sea between 1981 and 2000, more evenly spread across the time period. *See also* NORSK HYDRO.

NORWEGIAN PETROLEUM DIRECTORATE (NPD). Norwegian government agency created 14 July 1972 to manage **Norway**'s oil and gas resources. The Norwegian Petroleum Directorate (*Oljedirektoratet*) is located in **Stavanger**, Norway. In 2004, it had a staff of about 200. Before the creation of the NPD, Norway's hydrocarbon resources were managed by the Ministry of Industry, which began granting exploration and development licenses in the Norwegian sector of the **North Sea** in 1965. The Norwegian government left broad policy formulation to the Ministry of Industry, creating a separate Division of Petroleum and Mining within it. Detailed administration was allocated to the NPD. Meanwhile, the state oil company **Statoil**

(Den Norske Stats Oljeselskap) was created at the same time as the NPD in 1972 to handle state participation in production licenses and other business functions.

In 1978, a separate Ministry of Petroleum and Energy was created, and responsibility for oil and gas policy moved there from the Ministry of Industry. The NPD reported to the Ministry of Petroleum and Energy. After 1979, it reported simultaneously to the Ministry of Local Government and Labor, which had taken over responsibility for safety, working environment, and emergency preparedness in connection with oil and gas. In 2001, this second reporting relationship shifted to the Ministry of Labor and Government Administration after the relevant safety responsibilities were transferred to that ministry. On 1 January 2004, the NPD's safety responsibilities were spun off into a new Petroleum Safety Authority Norway. The NPD retained its core function of managing Norway's oil and gas resources and reported solely to the Ministry of Petroleum and Energy from that time on.

NORWEGIAN SEA. *See* NORWAY.

NOVOROSSIYSK. Russian Black Sea port (pop. 242,000, 2010); export terminal for **Caspian** oil. In 2003, about 52 Mt of oil and oil products were transshipped through Novorossiysk. The port can accommodate **tankers** up to 150,000 deadweight tons (DWT). Novorossiysk is the terminus of the Northern Early Oil **Pipeline** from **Baku**. It is also the terminus of the 944-mile Caspian Pipeline Consortium (CPC) pipeline from the **supergiant** Tengiz field in **Kazakhstan**. The CPC was opened in 2001 with a capacity of 350,000 bopd. In 2007, capacity was nearly double that. Novorossiysk's future importance may have decreased because of the completion of the **Baku–Tbilisi–Ceyhan** (BTC) Pipeline in 2005. This pipeline takes Caspian oil directly to the Mediterranean, bypassing the Black Sea and the **Bosporus**. From the point of view of the **United States** and some other Western countries, BTC also offers the advantage of avoiding Russian territory altogether.

NUCLEAR EXPLOSION STIMULATION. Both the **United States** and the **Soviet Union** conducted tests using nuclear explosions to stimulate hydrocarbon production by creating fractures to increase effective porosity and permeability. The United States conducted three tests in tight gas formations, and the Soviet Union conducted two in carbonate formations bearing oil. The U.S. tests were conducted under the rubric of the Plowshare Program, dedicated to investigating potential peaceful uses of nuclear explosions.

The U.S. tests were Project Gasbuggy, involving a 29-kiloton device in the San Juan basin of New Mexico near Farmington on 10 December 1967; Project Rulison, involving a 43-kiloton device in the Mesaverde Formation at Rifle, Colorado, on 10 September 1969; and Project Rio Blanco, involving three devices of 33 kilotons each, also at Rifle, Colorado, on 17 May 1973. A fourth test, planned for the Pinedale Anticline in Sublette County, **Wyoming**, and involving the El Paso Gas Company, never occurred. This fourth test, the Wagon Wheel Project, would have detonated five 100-kiloton nuclear explosions a few minutes apart. Vigorous local opposition scuttled the project.

The first Soviet test involved two 2.3-kiloton devices detonated separately on 30 March and 10 June 1965, probably in an oil field near Kuybyshev. The second test involved two devices of roughly 8 kilotons each, detonated separately on 2 and 8 September 1969 in an oil field about 150 miles southwest of Perm.

The U.S. tests succeeded in increasing the production of gas, but there were some important problems. First, the nuclear explosion caused volatilization of rock and fluids, releasing large amounts of carbon dioxide into the produced gas and lowering its quality. Second, the natural gas produced was radioactive. The increase in productivity was not significant enough to offset these two problems. The Soviet tests showed both increased oil production and increased estimated total recovery. Problems, if any, are not publicly known, but in any case the Soviets did not continue with their tests either.

O

OASIS GROUP. *See* LIBYA; MARATHON OIL CORPORATION.

OCCIDENTAL PETROLEUM CORPORATION. Large American **independent** petroleum company founded in 1920, often called OXY (its stock symbol). The Occidental Petroleum Corporation operated in **California** for decades without great success. It was acquired by **Armand Hammer** in 1957. Hammer was not an oil entrepreneur at the time but had achieved great financial success in the 1920s trading with the Soviet government, with which he had developed a close relationship, and engaging in various other ventures. Hammer's initial objective in acquiring the company was to use it as a tax shelter. However, the company struck new oil south of Los Angeles in 1957. In 1961, it made a large **natural gas** strike at Lathrop in northern California, and even more importantly, it obtained an oil concession from the government of **Libya**. The company discovered oil and prospered from its Libyan operations during the 1960s, especially in shipping oil to Europe.

When Muammar Qaddafi became the leader of Libya in 1970, Occidental found itself paying significantly increased taxes to the Libyan government. In 1973, it was forced to sell 51 percent of its Libyan interests to the government for a low price of $136 million. In 1985, Occidental sold 21 percent of its Libyan holdings to Österreichischen Mineralölverwaltung Aktiengesellschaft (OMV) of Austria. In 1986, the **United States** imposed economic sanctions on Libya and prohibited U.S. firms from doing business there. In 1992, the United Nations (UN) also imposed sanctions on Libya because of Libya's refusal to extradite two suspects in the 1988 bombing of Pan American Flight 103. UN sanctions were lifted in 1999, but U.S. sanctions continued until 2004 (*see* IRAN–LIBYA SANCTIONS ACT). In 2005, Occidental was able to resume Libyan production.

In 1982, to help offset the effects of Libyan nationalization, Occidental began to shift to domestic sources of energy and hydrocarbons. It acquired **Cities Service Company** in 1982 and Midcon Corporation in 1986. Midcon had one of the largest **natural gas pipelines** in the United States. Occidental also acquired the former Elk Hills **Naval Petroleum Reserve** in California in

1998 and Altura Energy Ltd. of **Texas** in 2000. The company also has businesses in coal mining and **petrochemicals**, among other interests. In the 2000s Occidental was embroiled in a dispute with **Ecuador**, which ejected it from the country in 2006 and expropriated its assets. In 2016 Ecuador agreed to pay the company $980 million after lengthy arbitration and negotiation processes. In 2016, Occidental was the fifth-largest petroleum company headquartered in the United States, by total assets. It controlled liquid **reserves** of 1.777 Bb and 3.774 Tcf of gas. It produced 0.474 Mmbpd of liquids and 0.956 Bcfpd of gas.

OCTANE NUMBER. *See* GASOLINE.

OFFICE OF THE PETROLEUM COORDINATOR FOR NATIONAL DEFENSE. *See* PETROLEUM ADMINISTRATION FOR WAR (PAW) (1942–1946).

OFFICE OF THE PETROLEUM COORDINATOR FOR WAR. *See* PETROLEUM ADMINISTRATION FOR WAR (PAW) (1942–1946).

OFFSHORE OIL. Offshore oil has become an increasingly important component of global oil production. As late as 1972, it accounted for only roughly 5 percent of crude oil produced in the world; by 1984 the proportion was about 27 percent, and in 2000 it was nearly 34 percent. The years between 2000 and 2015 saw similar percentages. Global production of onshore crude oil between 1982 and 2006 was essentially flat, hovering near 40 Mmbpd. The growth in global crude production, which stood at greater than 73 Mmbpd in 2006, was mostly from offshore oil. After 2006, increasing growth also came from **unconventional oil**.

Offshore oil exploration and production began in the 19th and early 20th centuries as a natural extension of onshore activity. That is, certain onshore fields seemed to extend offshore, and thus development moved there, in shallow waters. Overwater **drilling** and development took place in the **Bibi–Eibat** field in the **Baku** area as early as 1877.

Some exploitation of offshore oil seeps in Bibi–Eibat may have occurred as early as 1798. By 1924, the Bibi–Eibat offshore operation was relatively sophisticated, with drilling and production in waters as deep as about 250 feet. In the **United States**, the exploitation of the Summerland field in **California**'s Santa Barbara County was extended offshore in 1896. Operators had noticed in the previous two years that wells drilled closer to the beach were more prolific, and they began developing new ones offshore from piers. Water depth ranged up to about 30 feet. In the 1920s, **Creole Petroleum** (Standard Oil Company of New Jersey, later **Exxon**) began drilling shoreline wells in northeast Lake **Maracaibo**, in the **Bolivar Coastal Complex** of **Venezuela**. They used woodpiles, which were attacked by lacustrine termites. In 1928, they began using concrete piles and were drilling and producing in 60 feet of water by 1934. By 1960, there were about 4,000 wells in the lake, producing 1.4 Mmbpd.

The truly modern offshore industry began in the **Gulf of Mexico**. The first successful well was in October 1937 in 14 feet of water, a bit over a mile from the Louisiana coast. The well that heralded the modern industry, however, was completed by Kerr McGee on 14 November 1947. Although this well was in shallow water (about 20 feet), it was drilled out of sight of land. The rig was movable, and the major patterns of future offshore development were established. The world's largest offshore oil discovery, **Saudi Arabia**'s **Safaniya**, happened soon afterward in 1951, with first production in 1957. Again the water was relatively shallow.

As the industry matured, drilling occurred in progressively deeper waters. By the 1980s, "deepwater" meant about 800 feet of water. By 2008, anything less than 1,500 feet was commonly considered shallow. Definitions vary, but 1,500 feet and above is now considered deepwater, with depths more than 7,000 feet often called "ultra-deepwater." As water depths increased, there was a move away from fixed platforms to cheaper and more mobile submersible and semisubmersible platforms, as well as drill ships. The tallest conventional fixed steel platform was **Royal Dutch Shell**'s 1989 Bullwinkle in the Gulf of Mexico's Manatee field. The structure is 1,386 feet tall, in 1,326 feet of water, and used 44,500 tons of steel and 9,500 tons of piling. Bullwinkle represented a practical limit for this type of structure. Beginning in the 1980s, Petrobrás used a different approach in its deepwater plays offshore **Brazil**. It placed wellheads on the sea floor and produced oil through flexible risers connected to floating platforms. This type of approach became more common in deeper waters.

The period between 1973 and 1990 saw the discovery and development of **giant** offshore fields in the **North Sea**, **Mexico**, the **Caspian**, the **Arctic**, and Brazil, as well as Atlantic **Canada**. The period also witnessed the first deepwater discoveries in the U.S. part of the Gulf of Mexico. Since 1990, there have been giant offshore discoveries in West Africa, Brazil, and the U.S. Gulf of Mexico. Overall success rates increased during the 1980s, owing partly to improvements in seismic imaging. The overall global success rate since 1985 has been 30 percent, whereas in earlier years it was closer to 10 percent.

Early major discoveries in the North Sea were in water depths less than 500 feet. **Brent** is about 460 feet, and **Ekofisk** is about 250 feet. However, depths as much as 3,900 feet (Ormen Lange) have been reached in North Sea/Norwegian Sea projects. **Chevron**'s 2006 **Jack No. 2** well in the Gulf of Mexico was drilled not only in 7,000 feet of water but also more than 20,000 feet under the seafloor. The years between 2006 and 2015 have seen spectacular ultra-deepwater discoveries in the Brazilian **pre-salt** plays and off the shore of **Angola**.

OGJ100. A tabulation of the largest oil and gas companies outside the **United States**, published by the ***Oil and Gas Journal***. Like the **OGJ200** list of U.S. companies, the tabulation compiles data in 14 financial and operational categories, including total assets, revenue, and **reserves**. Unlike the OGJ200, the OGJ100 list has many blank spots because of the limited public availability of critical data, especially for national oil companies in the Middle East. The table thus does not rank companies the way the OGJ200 does but simply lists them by region.

OGJ150. *See* OGJ200.

OGJ200. A tabulation of the largest oil and gas companies headquartered in the **United States** and having oil or gas **reserves** there, published by the ***Oil and Gas Journal***. The tabulation ranks companies in 14 financial and operational categories, including total assets, revenue, and reserves. The OGJ200 also forms the basis of a stock tracking index. Originally, the list had many more members. It was first published on 17 October 1983 as the OGJ400. Mergers, consolidations, and liquidations pruned the list over the years. In 1991, it was renamed the OGJ300, and in 1996 it became the OGJ200. Although the list shrank in the 1990s, the total assets that it represented remained fairly steady until 1998, when oil prices decreased. In the 2000 decade, mergers and consolidations continued, and the list shrank further, although this did not reflect decreased performance for the group. In 2006, the list reached a low of 138 companies but continued to be called the OGJ200. In 2017, the list was called the OGJ150. *See also* OGJ100.

OGJ300. *See* OGJ200.

OGJ400. *See* OGJ200.

OHIO. Oil production 2015: 0.073 Mmbpd (0.78 percent of U.S.); **reserves** 0.062 Bb (0.19 percent of U.S.); gas production: 0.955 Tcf (3.53 percent of U.S.); reserves 12.1 Tcf (3.93 percent of U.S.). Ohio is a minor oil producer and somewhat less minor gas producer today but is historically important, not least because it was where **John D. Rockefeller** founded **Standard Oil** in 1870. The first commercial oil well in Ohio dates back to 1860, in the Macksburg Sand of Washington County, only months after the **Drake** Well in **Pennsylvania**. In 1884 the **giant Lima–Indiana oil field** was discovered in the Trenton Limestone of northwestern Ohio. The 1884 discovery was gas, but oil was discovered in 1886. Production from Lima–Indiana made Ohio the **United States**' largest oil-producing state in 1895, surpassing Pennsylvania. However, Ohio was soon exceeded by **California**, in 1903, and would never again be a dominant producer. There was to be another, much smaller oil boom in 1963, with a discovery in the Cambrian Trempealeau Dolomite in Morrow County.

In 1887 gas was discovered in the Clinton Sandstone in Fairfield County. Interestingly in the light of much later developments in the petroleum industry, the Clinton was a low-permeability formation, and thus the gas had the character of **tight gas**. Early **fracking** technology was applied in the Clinton to stimulate production beginning in 1951 and was successful for several years. A later phase of Clinton tight gas production occurred during the 1970s and 1980s. In recent years Ohio has begun producing shale gas, applying modern fracking and **horizontal drilling** technologies. Production began in 2012, growing from 0.014 Tcf that year to 0.955 Tcf in 2015. *See also* DONNELL, JAMES C. (1854–1927); DONNELL, OTTO D. (1883–1961); MARATHON OIL CORPORATION.

OHIO OIL COMPANY. *See* DONNELL, JAMES C. (1854–1927); DONNELL, OTTO D. (1883–1961); MARATHON OIL CORPORATION.

***OIL AND GAS JOURNAL* (*OGJ*).** Leading technical and business journal for the petroleum industry, reporting news and publishing statistics, analyses, and technical articles. The *Oil and Gas Journal* began publishing on 24 May 1902 in Beaumont, **Texas**, as the *Oil Investors Journal*. It was started by St. Louis journalist Holland Reavis. Its purpose was to report on drilling and business activity in the area surrounding the **Spindletop** oil discovery. Reavis sold the monthly magazine in 1910 to Patrick C. Boyle. Boyle changed the name to *Oil and Gas Journal*, moved the headquarters to **Tulsa**, **Oklahoma**, and began publishing weekly. After Boyle's death in 1920, his son-in-law Frank T. Lauinger took over; he was succeeded in 1931 by his son P. C. Lauinger, who would lead the magazine for 40 years. The publisher of *OGJ* evolved to become PennWell, a publisher of several industry journals and magazines. It also publishes technical books and runs a series of major industry conferences and exhibitions. PennWell is still based in Tulsa, although *OGJ* is now based in **Houston**. *OGJ* publishes important statistical features and reports for the industry. Among these is the annual *Worldwide Report*, listing estimates of oil and gas **reserves** worldwide and oil production field by field, as well as production from all the world's refineries. The journal is also known for its **OGJ100** and **OGJ200** summary statistics tables of the world's largest petroleum companies.

OIL AND NATURAL GAS COMMISSION. *See* INDIA.

OIL CITY. City (pop. 11,500, 2000) in **Venango County, Pennsylvania**; an early center of the petroleum industry in the **United States**, located at the mouth of Oil Creek, about 70 miles north of Pittsburgh and about 16 miles south of the site of the 1859 **Edwin Drake** well. Oil City served as a transshipment center for the early Venango County fields around Oil Creek, reaching its greatest importance in the 1860s. Riverboats plying the nearby Allegheny River transported oil to Pittsburgh. Oil City was also the site of the **Oil City Oil Exchange**. Petroleum-related industries remain important for Oil City to this day. **Pennzoil** has offices there.

OIL CITY OIL EXCHANGE. Early oil exchange in **Oil City**, **Pennsylvania**, that played a major role in setting the world price for crude oil in the 19th century. It was first established in 1870 but disbanded in 1873 and reorganized on 4 February 1874, and it progressively adopted more formal procedures. The Oil City Exchange was the first to apply clearinghouse processes to the oil business. It stopped trading oil certificates in 1904 and closed completely in 1908.

OIL CONTROL BOARD. Subcommittee of the **United Kingdom** War Cabinet during **World War II**, exercising control over the petroleum industry. The Oil Control Board was charged with implementing conservation measures and ensuring adequate supplies of petroleum. It also decided on the relative priorities of competing demands for petroleum. The leader was Secretary of Mines Geoffrey Lloyd. It included **John Cadman**, the chairman of the **Anglo–Iranian Oil Company**.

OIL COUNTERSHOCK. A collapse in oil prices between 1985 and 1986. In 1986, prices averaged $14.32 a barrel, down from $27.53 in 1985. At their low point during 1986, they were in the neighborhood of $8 a barrel. The **Second Oil Shock** had brought prices to record highs after the **Iranian** Revolution of 1979. The outbreak of the **Iran–Iraq War** helped sustain high prices for a time, but in the early 1980s a number of factors exerted downward pressure on prices. One was decreasing demand in the consuming countries and the cumulative effect of conservation measures. Another was an increased supply of crude from sources outside the **Organization of the Petroleum Exporting Countries** (OPEC), including the **North Sea** and **Mexico**. In addition, new oil from **Alaska** alleviated some of the **United States**' import requirement.

In 1980, the combined output from the North Sea's large producers, the **United Kingdom** and **Norway**, was about 0.5 Mmbpd; by 1985 it reached about 3.5 Mmbpd. Mexican production increased from 1.46 Mmbpd in 1979 to 2.745 Mmbpd in 1985. Alaska's oil production was 0.464 Mmbpd when oil first started flowing through the **Trans-Alaska Pipeline**; by 1985 it had reached 1.825 Mmbpd. The Soviet Union was producing at an average rate higher than 11 Mmbpd throughout the period from 1980 to 1985. In March 1982, OPEC moved to support prices by limiting its total production to 17.5 Mmbpd (44 percent lower than its 1979 production) and imposing quotas on its members. However, OPEC had difficulty maintaining unity and discipline among its members.

Saudi Arabia cut its production to support prices. Saudi production averaged 9.9 Mmbpd in 1980 and 3.388 by 1985. Deciding to gain back market share, Saudi Arabia in September 1985 began offering attractive **netback pricing** to customers and increasing its production. This was the immediate cause of the countershock. Prices decreased by more than 68 percent over the next six months.

After 1986, there was a partial recovery of prices. Non-OPEC exports declined. This was due to a multitude of reasons, including high taxes in the United Kingdom, underinvestment in Mexico because that country was struggling to pay its foreign debt, and the implosion of the Soviet Union. Still, many of the underlying factors that contributed to the countershock were felt through the 1990s. In that decade, there was also increased production from a variety of countries, such as **Venezuela**, that liberalized their energy sectors (*see* APERTURA PETROLERA). Prices dipped below $10 a barrel at some points during 1998 and 1999 before finally beginning what seemed like an inexorable increase in the **Third Oil Shock**. *See also* FIRST OIL SHOCK.

OIL CREEK. *See* BISSELL, GEORGE HENRY (1821–1884); KIER, SAMUEL M. (1813–1874); OIL CITY.

OIL CRISES. *See* FIRST OIL SHOCK; OIL COUNTERSHOCK; SECOND OIL SHOCK; THIRD OIL SHOCK.

OIL FOR FOOD PROGRAM. Plan allowing **Iraq** to sell some oil in order to meet basic humanitarian needs, thus loosening sanctions imposed by the United Nations (UN). After Iraq's invasion of **Kuwait** in 1990, the UN Security Council passed Resolution 661 imposing economic sanctions on Iraq. Iraqi oil exports were embargoed. The sanctions had a severe effect on Iraq's population. After the **Persian Gulf War** in 1991, a UN mission to Iraq reported the makings of "an imminent catastrophe." In 1991, the Security

Council offered Iraq the opportunity to export oil at a level sufficient to meet the basic needs of its people (Resolutions 706, August 1991, and 712, September 1991). The Iraqi government, under the dictator **Saddam Hussein**, refused. Widespread shortages of food, medicine, and other essentials continued for five years. In 1995, the Security Council adopted Resolution 986 establishing the framework for an "oil for food" program that would leave the overall sanctions regime in place but permit some oil exports in order to generate revenue for basic humanitarian needs. On 20 May 1996, the UN Secretariat and the Iraqi government reached agreement on the implementation of the resolution.

The first oil was shipped under the program in December 1996, and the first humanitarian supplies arrived in Iraq in March 1997. Iraq was allowed to sell $2 billion worth of oil every six months. Two-thirds of the revenue had to be used to meet basic humanitarian needs. In 1998, the limit was raised to $5.26 billion every six months, and in December 1999 the limit was abolished. Over the life of the program, the initial emphasis on food and medicine expanded to include infrastructure restoration and other activities. From the beginning, revenues from the program were apportioned to other uses as well, some of them self-serving for the victorious allies. Of the $65 billion in revenue between 1996 and 2003, $46 billion went to Iraqi aid, $18.2 billion went to Kuwait as reparations, and $1.9 billion went directly to the UN as authorized commissions. More than $500 million went to pay for ongoing coalition operations in Iraq, including weapons inspections.

On 17 March 2003, the UN evacuated its staff in Iraq in anticipation of the U.S. invasion that initiated the **Iraq War**, which began two days later with the American bombing of Baghdad. On 22 May 2003, the Security Council adopted Resolution 1483 lifting sanctions on Iraq. The resolution provided for the termination of the Oil for Food Program within six months. In 2004, accusations of corruption in the program surfaced and received considerable news coverage. It appears that Saddam Hussein evaded UN sanctions and exploited loopholes in the program to extract $21 billion in illegal revenue between 1990 and 2003. He derived about $13.6 billion from direct illegal sales to Syria, Turkey, Jordan, and some other countries. The rest came from complex pricing schemes, kickbacks, exploitation of sellers of humanitarian supplies, and other shady dealings.

The Oil for Food scandal damaged the reputation of UN secretary general Kofi Annan, particularly in the **United States**. It also provided U.S. critics of the UN with considerable propaganda fodder. However, the UN was not responsible for policing the Oil for Food Program. That responsibility lay with the Sanctions Committee, composed of member nations including the United States. Some observers believe the United States condoned Iraq's illegal oil trade because the main beneficiaries were U.S. allies such as Turkey, Jordan, and **Egypt**, and that trade was outside the scope of the Oil for

Food Program in any case. These observers further maintain that the United States knew of the other corruption and accepted it because it wanted to maintain the overall sanctions regime that hampered Iraq's weapons programs.

OIL INVESTORS JOURNAL. *See OIL AND GAS JOURNAL (OGJ).*

OIL RECOVERY. The processes by which crude oil is extracted from a reservoir. Primary recovery is the first stage of recovery, driven by natural pressure in the reservoir from gas or water. As oil is extracted, the pressure falls and secondary recovery techniques are applied. These include pumping, water flooding, **natural gas** reinjection, and gas lift (injecting air, carbon dioxide, or another gas). An early application of water flooding was in the **Bradford oil field** in **Pennsylvania**. Once secondary recovery can no longer extract significant quantities of oil, tertiary recovery techniques may be used. Tertiary recovery is also commonly termed enhanced oil recovery (EOR). EOR generally seeks to reduce the viscosity of the remaining oil. This can be done by heating, for example by steam injection. In some cases, in situ combustion is used: some of the oil in the reservoir is burned in order to heat the surrounding oil. Another way to decrease oil viscosity is to apply detergents or other chemicals.

Primary and secondary recovery methods together can typically extract about one-third of the original oil in the reservoir, although there is a large variation across reservoirs. The amount extracted using primary and secondary methods can be as low as 5 percent and as high as 80 percent. EOR can extract varying quantities of what remains, but at increasing costs of both energy and money.

OIL RESERVES SCANDAL. *See* TEAPOT DOME SCANDAL (1922–1930).

OIL SANDS. Deposits of **bitumen**, a degraded form of crude oil that has lost its lighter fractions and is both dense and too viscous to flow in the reservoir. Oil sands are also known as tar sands and bituminous sands. The most notable deposits are the **Alberta oil sands** in Western **Canada**. Oil sands can be processed to yield a liquid similar to crude oil, usually called synthetic crude or **syncrude**. *See also* PECHELBRONN.

OIL SHALE. Various types of fine-grained sedimentary rock that contain kerogen, an organic substance. Oil shale (not to be confused with **shale oil**) can be mined and subjected to destructive distillation and processing, yielding gas and a liquid similar to crude oil. This oil can be considered a type of synthetic crude, or **syncrude**. In the past this syncrude was sometimes called "shale oil," but that term is now more often used to refer to a type of **tight oil** held in shale formations. Commercial grades of oil shale yield between 100 to 200 liters (0.63 to 1.26 barrels) of syncrude for each metric ton of rock. Measurements are in liters or barrels per metric ton (l/mt or b/mt). The **United States Geological Survey** (USGS) has used a lower limit of 40 l/mt (0.25 b/mt) to define a viable oil shale. Others have used limits as low as 25 l/mt (0.16 b/mt).

While oil shale deposits are found throughout the world, the most substantial ones are located in the **United States**. The term "oil shale" comprises a variety of different rock types, with terrestrial, marine, and lacustrine origins. It is difficult to make a reliable resource estimate, but there are thought to be about 3.3 trillion barrels of potentially recoverable oil contained in the world's deposits. About half of these are found in the lacustrine **Green River Formation** in the U.S. states of Colorado, Wyoming, and Utah. The other half of the world's resource is distributed in a large number of smaller deposits—although a few of these are still immense, exceeding 200 Bb. There were some attempts to create an oil shale industry in the 1970s and 1980s, but they did not succeed in the long term (*see* EXXON). An oil shale industry could still be viable in the United States at some point in the future.

An oil shale industry arose in Scotland in the 19th century. James Young established a **refining** company to produce oil products from petroleum seeps associated with coal mines and from coal by-products. Later he switched to extracting oil from bituminous shales. There was also an oil shale industry in **Australia** in the 1800s, which ceased operations when it ran out of high-grade oil shale. The Australian industry has recently been resuscitated in Queensland. **Brazil** has operated an oil shale industry since the 1980s.

The longest continuously operating oil shale industry has been in Estonia. Production began during **World War I** and was still continuing in the 2000 decade. In 1918, 17,000 metric tons of oil shale were mined in Estonia. By 1940, annual production was 1.7 million metric tons. Estonian production peaked in 1980 at roughly 31 million metric tons. Estonia accounted for more than 60 percent of the world's production that year. In 2000, Estonia produced about 13 million metric tons of oil shale, or about 85 percent of the world's output. About 80 percent of the mined oil shale (marine kukersite) is burned directly as fuel in two large power plants. A small amount is used to manufacture cement. About 15 percent of the mined oil shale is distilled to produce syncrude, which is used mostly as a **petrochemical** feedstock.

OIL SHOCKS. *See* FIRST OIL SHOCK; OIL COUNTERSHOCK; SECOND OIL SHOCK; THIRD OIL SHOCK.

OIL SPILLS. Refer to appendix 9. *See AMOCO CADIZ; DEEPWATER HORIZON* OIL SPILL; *EXXON VALDEZ*; SANTA BARBARA OIL SPILL; *TORREY CANYON*.

OKLAHOMA. Oil production 2015: 0.447 Mmbpd (2.34 percent of U.S.); **reserves** 1.262 Bb (3.9 percent of U.S.); gas production: 2.34 Tcf (8.63 percent of U.S.); reserves 28.49 Tcf (9.26 percent of U.S.). Oklahoma became the top oil-producing state of the **United States** in 1907. It was overtaken by **California** in 1909, then took the lead again in 1915, until California overtook it again in 1919. Until 1928, Oklahoma traded places with California for the leading position, holding it in 1920–1922 and in 1927. In 1928, both Oklahoma and California were overtaken by **Texas**. Oklahoma oil production rose rapidly in the early 20th century to an all-time peak of 0.761 Mmbpd in 1927. It then declined unevenly to 0.337 Mmbpd in 1943 and rose unevenly to another peak of 0.632 Mmbpd in 1967. After 1967, oil production entered a phase of long-term decline. Oklahoma's 2006 oil production was comparable to its output in 1913. Oklahoma's nine **giant** fields (seven predominantly oil and two predominantly gas) were all discovered before 1953; seven were discovered before 1920.

Oklahoma's Native Americans long knew of oil seeps in the territory, and early European settlers mined them for asphalt. The first subsurface oil was discovered in 1859, 48 years before Oklahoma became a state. The well, near what is now Salina in Mayes County, had been drilled for salt. The oil was sold for use in lamps. In 1884, the Cherokee Nation passed a law authorizing the "organization of a company for the purpose of finding petroleum, or rock oil, and thus increasing the revenue of the Cherokee Nation." In 1889, Edward Byrd obtained Cherokee mineral leases and drilled Oklahoma's first intentional oil well near Chelsea. The well, flowing at 0.5 bpd, was not commercially successful.

On 15 April 1897, Albert P. McBride and Camden L. Bloom completed Oklahoma's first commercial oil well near the new small town of Bartlesville, on behalf of the Cudahy Oil Company. McBride and Bloom had earlier drilled Norman No. 1, Kansas's first commercial well. Their Oklahoma well, Nellie Johnstone No. 1, was named for the six-year-old daughter of William Johnstone, a Cudahy partner, and was the discovery well of the giant Bartlesville–Dewey field. The field would be the largest ever found in Oklahoma. It became profitable in 1900, when rail transportation became available in the area. Bartlesville–Dewey did not touch off an immediate boom, but by 2002 it had produced 1.5 Bb.

In 1905, the **Glenn Pool** was discovered. Although smaller than Bartlesville–Dewey, it was quicker to be exploited and initiated a major industry boom. When Oklahoma became a state in 1907, it was the top U.S. producer. Major discoveries included the Cushing field in 1912 and Sho-Vel-Tum in

1914. However, the most important finds after Bartlesville–Dewey were in the 1920s. On 7 March 1926, the Indian Territory Illuminating Oil Company (ITIO, later **Cities Service Company**) discovered the first pool in the greater Seminole field, a giant with an **EUR** of about 0.822 Bb. In 1928, ITIO discovered the even larger Oklahoma City field (EUR 0.829 Bb). It is worth noting that significant developments in modern **geophysical prospecting** occurred in Oklahoma during this period. In 1927, **Geophysical Research Corporation**, a subsidiary of **Amerada Petroleum**, discovered the Maud field. This had the distinction of being the first field ever discovered using reflection seismology.

Production from the Seminole and Oklahoma City fields flooded the market and caused prices to drop. This tendency was amplified by the discovery in Texas of the **East Texas oil field** in 1930. The Oklahoma Corporation Commission had more de jure power to regulate production to stabilize prices than did the **Railroad Commission of Texas** (at least at first), and attempted to do so. Producers largely ignored their quotas, however, and by mid-1931 prices fell to about 20 cents a barrel. In August 1931, Oklahoma governor William "Alfalfa Bill" Murray called out the National Guard, as was his penchant. The guard created 3,000 martial law areas, 50 feet around each oil well. Governor Murray pledged to keep the guard in place until the price of oil reached $1 a barrel. The guard stayed until 1933, paid by a tax levied on oil production.

Oklahoma is a major transportation hub for oil and **natural gas**. Major north–south and east–west **pipelines** meet there, at **Cushing**. Cushing is the price settlement point for **West Texas Intermediate** (WTI) crude oil on the **New York Mercantile Exchange**. During the later 2000s and the 2010s, Oklahoma experienced a resurgence as a gas producer, enabled by a 25-fold increase in **shale gas** output between 2007 and 2015.

OLJEDIREKTORATET. *See* NORWEGIAN PETROLEUM DIRECTORATE (NPD).

OMAN. Oil production 2015: 0.981 Mmbpd (1.25 percent of world); **reserves** 5.373 Bb (0.33 percent of world); gas production: 1.241 Tcf (1 percent of world); reserves 23 Tcf (0.33 percent of world). Oil production in the Arabian Peninsula country of Oman began in 1967 at 63,000 bpd. It rose rapidly, reaching 0.332 Mmbpd in 1970. It hovered about 0.3 Mmbpd until 1982, after which it entered a period of long-term growth, reaching 0.97 Mmbpd in 2000. It declined thereafter. Production in 2006 was near the level in 1992. However, Oman oil production turned around with the application of enhanced oil recovery techniques and reached nearly 1 Mmbpd in 2016.

Gas production stood at 0.028 Tcf in 1980. It grew slowly over the next 20 years, reaching 0.197 Tcf in 1999, then began growing rapidly, reaching 1.241 Tcf in 2015.

The **Anglo–Iranian Oil Company** (later **BP**) obtained a petroleum **concession** in Oman in 1925 but was unsuccessful in exploration. Oman was within the area of the **Red Line Agreement**, so members of the **Iraq Petroleum Company** (IPC) consortium could not explore there independently after 1928. In 1937, a company called Petroleum Concessions (Oman), whose ownership structure was the same as the IPC's, obtained a 75-year concession from Sultan Said bin Taimur, but it found no petroleum. The consortium abandoned part of its concession, covering Dhofar Province, in 1951. In May 1952, the sultan awarded a Dhofar concession to Wendell Phillips, an American archeologist who signed the concession over to Dhofar Cities Service Petroleum Corporation in January 1953. Dhofar Cities Service was jointly owned by **Cities Service Company** and Richfield Oil (later **Arco**). Dhofar Cities Service drilled an unsuccessful wildcat at Dauka No. 1 in 1955. Production from Dhofar Province would not begin until much later, in 1980.

In the 1950s, the IPC consortium began more serious exploration. Its first well, drilled in 1956 at Fahud, was unsuccessful. In 1957, it discovered the **giant** Marmul field (**EUR** 0.7 Bb) but judged the oil too viscous and heavy to produce commercially. All the owners withdrew except for **Royal Dutch Shell** and **Calouste Gulbenkian**'s Partex company, which were left with 85 percent and 15 percent shares respectively. The reduced partnership discovered Oman's largest field, Yibal (EUR 1.3 Bb and 4.5 Tcf gas) in 1962. Two other giant discoveries followed—Natih in 1963 and Fahud in 1964—very close to the first dry well of 1956.

In 1967, Partex sold some of its shares in the venture, now called Petroleum Development (Oman), to **Compagnie Française des Pétroles** (CFP, later **Total**), which as a member of the IPC consortium had been one of the original partners. The restructured partnership was 85 percent owned by Shell, 10 percent by CFP, and 5 percent by Partex. In 1975, the government of Sultan Qaboos acquired 60 percent of Petroleum Development (Oman), leaving 34 percent for Shell and 6 percent for CFP. Four more giant oil and gas fields were discovered in Oman, the largest new oil field being Nimr in 1980. Saih was discovered in 1990, and the Barik and Khazzan gas fields were found in 1991 and 2001 respectively. In 2001, crude from Oman was added to the **Dubai crude benchmark**. In 2009 a major discovery of **tight gas** was made in the area of the Khazzan field. The government awarded a development concession to BP. Production began ahead of schedule at 1 Bcfpd in 2017.

OPEC BASKET. A group of 11 crude oils used by the **Organization of the Petroleum Exporting Countries** (OPEC) to monitor the world oil market. Before 16 June 2005, the basket consisted of seven crudes; some but not all were included in the revised basket (refer to appendix 6). The OPEC Basket is heavier and sourer (refer to appendix 5) on average than **Brent Blend** or **West Texas Intermediate**, and more reflective of the average grade of oil sold in the world. OPEC adopted the basket in 1987 to replace **Arab Light** from **Saudi Arabia**, formerly the pricing reference crude. Initially the basket was intended to help establish fixed prices. However, by 1988 OPEC abandoned the fixed price system in response to market pressures caused by the flood of non-OPEC oil on the world market and other factors.

ORCUTT, WILLIAM W. (1869–1942). Petroleum geologist who organized the first oil company geological research department. William Warren Orcutt was born in Minnesota and moved with his family at the age of 13 to Santa Paula, **California**. He attended Stanford University and graduated in the pioneering geology class of 1895. In 1899, he went to work for the **Union Oil Company of California**, headquartered in Santa Paula and later in Los Angeles. He made a detailed geologic map of the Santa Maria basin, from which Union Oil discovered its Santa Maria fields.

Orcutt helped establish geology as a crucial science in the search for petroleum. At Union Oil in 1900, he organized the first known oil company geological research department. In 1902, he was the first to recognize that the bones buried in asphalt at Rancho La Brea in Los Angeles were not cow bones. This led to the famous excavations of the La Brea Tar Pits. Orcutt served on the executive board of Union Oil for 34 years.

ORGANIZATION OF ARAB PETROLEUM EXPORTING COUNTRIES (OAPEC). Intergovernmental organization formed 9 January 1968 by **Saudi Arabia**, **Kuwait**, and **Libya** and headquartered in Kuwait. The agreement establishing the Organization of Arab Petroleum Exporting Countries was signed in Beirut, Lebanon, in 1968. At first the organization was intended as a forum for conservative Arab states (at the time, Muammar Qaddafi had not yet taken power in Libya) with a primary reliance on petroleum revenue. It later admitted more states and became more activist. The membership criteria were relaxed to admit countries where petroleum was not the principal export but an important one. Membership increased to 11 countries by 1982 as the following were admitted: **Algeria** (1970), **Bahrain** (1970), **Qatar** (1970), **United Arab Emirates** (1970), **Iraq** (1972), Syria (1972), **Egypt** (1973), and Tunisia (1982). In 1979, OAPEC suspended Egypt's membership for signing the Camp David Accords and making peace with Israel. Egypt was readmitted in 1989. In 1986, Tunisia requested to withdraw from the organization, and its membership was suspended, but it is free to reactivate its membership.

The heyday of OAPEC's international power was in 1973, when the organization was spurred to action by the Yom Kippur War, which began when Egypt, Syria, and Iraq attacked Israel. On 16 October 1973, 10 days after the war began, OAPEC met in Kuwait. It resolved to cut its collective oil production by 5 percent every month until Israel withdrew to its pre-1967 borders. It also imposed a selective embargo on the **United States** and the Netherlands for supporting Israel. OAPEC's actions contributed to the **First Oil Shock**. OAPEC ended its action on 18 March 1974. The United States and the **United Kingdom** (which was not embargoed), working with Egyptian president Anwar Sadat, convinced Saudi king Faisal to stop the boycott. Faisal, a staunch anticommunist who did not want to excessively weaken the West, prevailed on other OAPEC member countries. In later years, OAPEC focused on regional integration and collective oil projects. All currently active members of OAPEC except Bahrain, Egypt, and Syria are also members of the **Organization of the Petroleum Exporting Countries** (OPEC).

ORGANIZATION OF THE PETROLEUM EXPORTING COUNTRIES (OPEC). International organization of petroleum exporting countries that coordinates the petroleum policies of its members and has often functioned as a cartel to support prices. The two visionaries behind OPEC were **Venezuela**'s **Juan Pablo Pérez Alfonso** and **Saudi Arabia**'s **Abdullah al-Tariki**. Pérez Alfonso had studied the workings of the **Railroad Commission of Texas**, which regulated production to stabilize the market, and he dreamed of a global analog. Tariki had done an internship at the Railroad Commission of Texas, and he was a nationalist who wanted his country to take control of its petroleum industry. Tariki and Pérez Alfonso were introduced to each other by petroleum journalist **Wanda Jablonski**.

The immediate catalyst for the formation of OPEC was a round of unilateral cuts in **posted prices** by **major oil companies** in 1959 and 1960, which had the effect of reducing producer country revenues. Tariki and Pérez Alfonso mobilized representatives from other oil exporting countries meeting at the 1959 Arab Oil Congress in Cairo to sign the secret **Mehdi Pact**. The pact was a nonbinding "gentleman's agreement" embodying recommendations the representatives would make to their governments. On 14 September 1960, after another unilateral cut in the posted price by the major oil companies, OPEC was founded. The original members of OPEC were Saudi Arabia, **Kuwait**, **Iran**, **Iraq**, and Venezuela. They were joined by **Qatar** (1961), **Libya** (1962), **Indonesia** (1962), **Abu Dhabi** (**United Arab Emirates**) (1967), **Algeria** (1969), **Nigeria** (1971), **Ecuador** (1973), **Gabon** (1975), **Angola** (2007), and **Equatorial Guinea** (2017). Ecuador suspended its membership in 1992 but reactivated it in 2007. Gabon suspended its membership in 1994, unwilling or unable to pay the membership fees, but resumed membership in 2016. Indonesia left OPEC in 2009, rejoined briefly in 2016, then left again later that same year.

OPEC sought to increase the control of its members over their petroleum industries, set reference prices, gradually reduce **concession** areas, and require the major oil companies operating in OPEC countries to maintain transparent accounts available to host governments for inspection. OPEC's specific initial demands, however, were rather modest: that posted prices be restored to their levels before the cuts. OPEC did not have much influence in the 1960s, but the 1970s were its heyday.

In 1971, OPEC countries of the **Persian Gulf** region signed the **Tehran Agreement** with major oil companies. The agreement raised prices and increased producing countries' profit shares. In the same year, OPEC members delivering oil to the Mediterranean area signed the even more aggressive **Tripoli Agreement**. These agreements signaled the beginning of OPEC's true pricing power. In 1971 and 1973, OPEC signed the **Geneva Agreements** with the major companies, correcting oil prices in accordance with the value of the U.S. **dollar** so that producing countries would not lose revenue as a result of dollar devaluations. The 1970s also saw a number of OPEC members nationalize their industries, according to OPEC's **participation** framework. The participation framework provided for gradual assumption of majority control, but the countries generally exceeded its terms and took control sooner. Saudi Arabia acquired all of Aramco by 1980, Kuwait acquired the Kuwait Oil Company by 1974, and Abu Dhabi acquired majority control of its industry in 1974. Other OPEC members, including Algeria, Libya, Iraq, and Venezuela, nationalized their industries in the early to mid-1970s outside the participation framework. Still others, such as Iran and Indonesia, had nationalized long before.

On 16 October 1973, 10 days after the start of the Yom Kippur War, which began when Egypt, Syria, and Iraq attacked Israel, six Persian Gulf OPEC countries raised the price of **Arab Light** crude from $2.90 to $5.11 a barrel. The next day, the **Organization of Arab Petroleum Exporting Countries** (OAPEC) announced a production cut of 5 percent, to be followed by a further 5 percent each month that Israel failed to withdraw to its pre-1967 borders. OAPEC also began a selective oil embargo on the **United States** and the Netherlands for supporting Israel. The embargo had an immense psychological effect on the markets, out of proportion to the actual loss of supply. These factors contributed to the **First Oil Shock** and to an image of invincibility for OPEC. This image was enhanced as the **Second Oil Shock**—which was proximately caused by the Iranian Revolution of 1979 and not, at first, deliberate OPEC action—drove prices up even further. Average prices for 1979 were $31.61, more than twice 1978 levels, and about 12 times as high as 1972 levels.

OPEC was able to maintain high prices through the early 1980s, aided by the outbreak of the **Iran–Iraq War** in 1980. However, the world oil situation had changed, and OPEC's power was soon broken. This was largely due to the large increases in non-OPEC production from the **North Sea** and **Mexico**, as well as increased production in **Alaska** that alleviated some of the United States' import requirement. OPEC had trouble maintaining unity, and countries cheated on their quotas. Saudi Arabia cut its production and supported prices alone for a time but then gave up on that strategy. In 1985, it offered attractive **netback pricing** to customers and boosted production. In 1986, prices collapsed in the **Oil Countershock**. OPEC was saved from 1985 to 1990 as production by the **United Kingdom** and Mexico declined and the Soviet Union imploded, but it did not achieve its former unity and power.

OPEC was relatively weak in the 1990s, but in the 2000s its hand was temporarily strengthened again. The large increase in prices from 2003 to 2008 (the **Third Oil Shock**) was not primarily the result of concerted OPEC action, but OPEC members benefited from it. Also, as certain important non-OPEC exporters, including Mexico, the United Kingdom, **Norway**, and others, appeared to be declining, the OPEC states temporarily found themselves more powerful once again. As the 21st century progressed, OPEC found itself challenged again by the explosive increases in production of **shale oil** and **shale gas** in the United States, which have the potential to spread globally. OPEC countries still have a cost advantage over such **unconventional** oil and gas, and this can act as an incentive to keep prices low rather than high. At the 2014 meeting, a number of oil ministers, especially those from Venezuela, Algeria, and Iran, wanted the cartel, especially Saudi Arabia, to cut production to prop up prices. Saudi oil minister **Ali al-Naimi** argued that it was better to keep prices low enough to avoid losing market share to the new unconventional sources. *See also* OPEC BASKET.

ORIMULSION. *See* ORINOCO OIL BELT.

ORINOCO OIL BELT. A region of **heavy oil** and **bitumen** deposits in eastern **Venezuela**, estimated to contain about 90 percent of the world's heavy oil. The Orinoco Oil Belt may have more than 1,200 Bb of heavy and extra-heavy oil in place, with perhaps 270 Bb recoverable with current technology. The deposits occur as a continuous belt about 435 miles long and 30–60 miles wide located north of the Orinoco River. Most of the deposit is in the form of heavy oil of 8–10°**API gravity**. The oil is viscous but generally flows into wells without thermal stimulation because of high reservoir temperatures. Thermal stimulation methods have also been used to increase the flow rate. The development of the Orinoco Oil Belt was a major goal of Venezuela's **Apertura Petrolera**. Four joint ventures were initiated in the mid-1990s, with **PdVSA** having an interest in all four: the Hamaca, Sincor, Petrozuata, and Cerro Negro projects.

The Hamaca Project was begun by PdVSA (Corpoven subsidiary), **Chevron**, and **Phillips Petroleum** (now **ConocoPhillips**). The group was known collectively as Ameriven. In 2001, it began extracting extra-heavy crude with gravity in the range of 8–10°API. By October 2004, the project had completed an upgrading facility in the Jose industrial complex on the country's northeast coast. The upgrading facility allows production of 180,000 bpd of 25.9°API **syncrude** from 190,000 bpd of extra-heavy oil. The Sincor Project was begun by PdVSA, **Statoil**, and TotalFinaElf (now **Total**) to upgrade 8.5°API natural bitumen, producing 180,000 bpd of 32°API syncrude. The Petrozuata Project was begun by PdVSA (Maraven subsidiary) and Conoco in 1995. An upgrading plant in Jose processes 120,000 bpd of 9°API heavy oil into 103,000 bpd of lighter syncrude. The Cerro Negro Project was begun by PdVSA (Lagoven subsidiary), Veba Oel, and **Mobil** (now **ExxonMobil**) and by 2007 was producing about 100,000 bpd of syncrude from 9°API heavy oil.

By 2007, all four of the projects were in turmoil as Venezuelan president **Hugo Chávez** moved to reverse the Apertura Petrolera and nationalize the projects. In May 2006, the Venezuelan National Assembly amended the Hydrocarbons Law of 2001 to levy a 33.33 percent royalty on the extra-heavy crude ventures in the Orinoco Oil Belt. The government also wanted PdVSA, hitherto a minority partner, to assume majority control. Some of the foreign oil companies, including Statoil, Chevron, and Total, agreed to reduce their stakes. However, ExxonMobil and ConocoPhillips resisted the government moves. They, along with **Petro-Canada**, withdrew from the projects. ExxonMobil sought arbitration with the International Center for Settlement of Investment Disputes (ICSID) and in early 2008 succeeded in getting courts in the **United States** and the **United Kingdom** to freeze about $12 billion in Venezuelan assets pending the outcome. The turmoil affected the market perception and credit rating of the projects. Years later, on 9 October 2014, ICSID awarded ExxonMobil $1.6 billion.

Another ongoing activity in the Orinoco Oil Belt is the production of Orimulsion, a substance created by the emulsion of natural bitumen (7.5–8.5°API) with water and surfactants. Orimulsion was developed by PdVSA in the early 1980s and can be pumped, stored, transported, and burned in boilers with minor equipment modifications. Exports began in

1988. Orimulsion is marketed by PdVSA's subsidiary Bitor (Bitúmenes del Orinoco). The product, which unfortunately has a high sulfur content and does not burn clean, is used mainly in power plants.

OXY. *See* OCCIDENTAL PETROLEUM CORPORATION.

P

PACIFIC WESTERN OIL CORPORATION. *See* GETTY, J. PAUL (1892–1976).

PAHLAVI, MOHAMMED REZA (1919–1980). Shah of **Iran** 1941–1979; a pro-Western autocratic leader who attempted to modernize his country using oil revenue and also aggressively sought oil price increases in the 1970s. Mohammed Reza Pahlavi was born 26 October 1919 in Tehran. He studied in Switzerland, returning to Iran in 1935. In 1941, he became shah after his father, **Reza Pahlavi**, abdicated under Allied pressure during **World War II**. In March 1951, nationalist leader **Mohammed Mossadegh** succeeded in getting a bill passed through the Majlis, the Iranian parliament, nationalizing the oil industry. The shah lost relative power and standing to Mossadegh and was forced to name him prime minister a month later. The **United Kingdom**, whose interests were most affected, embargoed Iranian oil, causing tremendous economic dislocations in Iran.

The shah attempted to dismiss Mossadegh in August 1953 but instead was forced into exile. By the end of the month, the shah was back in the country, and Mossadegh was under house arrest, following a coup engineered by the **United States** and supported by the United Kingdom—*see* AJAX, OPERATION (1953). Iran's oil remained nationalized, but Western companies—including the **Anglo–Iranian Oil Company**, renamed **BP**—were allowed to form a consortium to operate in Iran. The consortium, **Iranian Oil Participants Ltd.**, essentially ran the petroleum sector in Iran until the early 1970s, despite nationalization. With his power solidified and the backing of the United States, the shah embarked on a program of modernization and development. In January 1963 he launched the so-called White Revolution of reform and modernization. The country's infrastructure, including railways, roads, airports, and dams, was updated and expanded. The shah also sponsored programs to improve health care and literacy and undertook land reforms.

In all these endeavors the shah was helped by the considerable oil revenues flowing into his country. His desire for more revenue—both for his development programs and to support a strong military—made him an aggressive advocate for higher oil prices during the heyday of the **Organization of the Petroleum Exporting Countries** (OPEC) in the 1970s. He thinly camouflaged his motives, speaking of a "new discourse of oil," which he termed "a noble commodity" that should not be burned, but rather should be conserved for "higher, nobler purposes" such as **petrochemicals**. Like some other OPEC leaders, he did not fully appreciate what effect higher prices might have on demand. Although the shah's White Revolution enhanced his domestic support to an extent, there was also considerable dissatisfaction. Some segments of the population did not believe his reforms went far enough, or that oil wealth was distributed equitably. Other segments resented his secularism and Western orientation. The shah was also an ostentatious and autocratic ruler with a ruthless secret police apparatus (Savak) that did not balk at imprisoning and savagely torturing opponents of his regime.

The undercurrent of fearful opposition to the shah eventually exploded into a mass movement. In 1978, there were strikes and violent demonstrations. The strikes, some of which were instigated from exile by religious opposition leader Ayatollah **Ruhollah Khomeini**, paralyzed the country's oil industry and catalyzed the **Second Oil Shock** of the 1970s. In 1978, Iran's production fell from 5.7 Mmbpd to 700,000 bopd. Exports fell to zero by the end of the year, and global prices surged ($30.03 per barrel for 1979 versus $13.60 the previous year). On 26 December 1978, Khomeini declared that there would be no oil exports from Iran as long as the shah was still in the country. The shah left Iran on 16 January 1979, never to return. The country became an Islamic republic under Khomeini. Mohammed Reza Pahlavi died of cancer in Cairo on 27 July 1980.

PAHLAVI, REZA (1878–1944). Shah of **Iran** 1925–1941; terminated and then renegotiated the oil **concession** of the Anglo–Persian Oil Company (later **Anglo–Iranian Oil Company** and **BP**). Reza Khan, the son of a common foot soldier, was born in a village near the **Caspian Sea**. He joined the Iranian Cossack Brigade at the age of 15 and rose through the ranks, eventually becoming a brigadier general. In 1921, he supported a coup against the ruling government. The leader of the coup, Ziya Tabatabai, became prime minister, and Reza Khan became head of the army, still nominally under the reigning but weak Ahmad Shah of the Qajar dynasty. Reza Khan soon turned on Tabatabai and removed him from office. He pressured Ahmad Shah to make him minister of war and then prime minister.

In 1925, Reza Khan removed Ahmad Shah and assumed the throne as Shah Reza Pahlavi, founding the Pahlavi dynasty. He then embarked on a major reform and modernization program, rebuilding the army and the civil service and reforming and secularizing the legal system. He also modernized the country's infrastructure and education system. The Trans-Iranian Railway was completed during his reign, in 1938. Reza Pahlavi was a nationalist who wanted to build a modern industrial state that stood on its own. In 1933, he terminated the oil concession originally granted to **William Knox D'Arcy** in 1901 by Shah Muzaffar al-Din, and operated since 1909 by the Anglo–Persian Oil Company, which became the Anglo–Iranian Oil Company. He established a new concession with a quarter of the area, and he established a new royalty and payment schedule. The new concession was not favorable enough to Iran in the long term to stave off eventual nationalization of the industry.

Reza Pahlavi was deposed on 16 September 1941 after the **United Kingdom** and Soviet Union invaded Iran during **World War II**. Although Iran was officially neutral, he was suspected of leaning toward the Nazis and had refused to expel German citizens. The Allies did not want to risk the critical supply line to the Soviet Union that Iran represented. The shah was replaced on the throne by his 21-year-old son **Mohammed Reza Pahlavi**. Reza Pahlavi died under house arrest in Johannesburg, South Africa, on 26 July 1944. Overall, he was a modernizer who benefited his country in a number of ways, but he was also a ruthless autocrat who eliminated opponents and often turned against supporters.

PAN AMERICAN PETROLEUM AND TRANSPORT COMPANY. *See* AMOCO; DOHENY, EDWARD (1856–1935).

PARTEX. *See* GULBENKIAN, CALOUSTE (1869–1955).

PARTICIPATION AGREEMENTS. A framework by which some members of the **Organization of the Petroleum Exporting Countries** (OPEC) acquired portions of the foreign companies operating in their territory. In the early part of the 20th century, a producing country would typically grant **concessions** to a foreign company, giving it the right to explore for and produce oil in a defined area within the country's territory. As time went on, the terms of concession agreements were often viewed as unfair to the producing countries. Sometimes concessions were renegotiated. Sometimes the tensions led to outright nationalization of the oil industry, as happened in **Mexico** and **Iran**, for example.

In some of the **Persian Gulf** oil producers, such as **Saudi Arabia**, the relations between the governments and the foreign companies were not quite so contentious, but the countries still eventually wanted more control over their resources. However, some of these countries did not consider a sudden, outright nationalization of their oil industry to be a reasonable alternative.

Suddenly removing **major oil companies** from upstream operations and effectively making them detached buyers of crude to refine and sell—buyers who would then shop for the best deal—could create downward pressure on prices. This led to the concept of participation, championed by Saudi oil minister **Ahmad Zaki Yamani**, by which oil-producing countries could gain gradual control.

In October 1972, a participation agreement was concluded between some OPEC nations in the Persian Gulf region and the major oil companies. It created the option for an immediate 25 percent participation, rising to 51 percent by 1983. Participation was not supposed to be confiscation—the oil companies were to be compensated on net book value. The oil companies reluctantly agreed because they feared the alternative might be outright nationalization. The numbers were not followed precisely. The **United Arab Emirates** established the **Abu Dhabi** National Oil Company (ADNOC), and it acquired 25 percent of the two foreign consortia operating there (Abu Dhabi Petroleum Company and Abu Dhabi Marine Associates) on 1 January 1973. In December 1974, ADNOC increased its participation to 60 percent. Saudi Arabia acquired its immediate 25 percent of **Aramco** in 1973 and increased its stake to 60 percent in 1974. In 1980, it acquired 100 percent. **Kuwait**'s parliament failed to ratify the participation agreement, but in January 1974 Kuwait signed a new agreement providing for an immediate 60 percent and by 1975 acquired 100 percent.

Other OPEC members rejected the framework. Iran had already nationalized its industry in 1951. By the early 1970s, Iran had improved the terms of its agreements with the Western consortium **Iranian Oil Participants** that essentially still ran its petroleum sector. The country reclaimed sovereign control and ensured it received contract terms no worse than those of its Arab neighbors who were signing participation agreements. **Indonesia** had nationalized its sector in 1960. **Iraq** had begun taking steps toward nationalization in 1961 and completed the process in 1975. **Algeria** nationalized in steps from 1965 to 1971, not always taking a 100 percent share but establishing control. **Libya** did the same in 1973. **Venezuela** nationalized its industry in January 1976.

PDVSA. National petroleum company of **Venezuela**. Petróleos de Venezuela, Sociedad Anónima (PdVSA) was formed in 1975 when President **Carlos Andrés Pérez** nationalized all hydrocarbon assets in Venezuela, paying the relatively small sum of $1 billion for all foreign-owned petroleum holdings. Henceforth the properties would be held by PdVSA. Various independent subsidiaries were formed. Maraven took over **Royal Dutch Shell** operations. Lagoven took over operations of **Creole Petroleum** (owned by Standard Oil

Company of New Jersey, later **Exxon**). Corpoven took over the holdings of other companies. All the subsidiaries were eventually merged into the broader PdVSA in 1998.

In 1986, PdVSA bought 50 percent of Citgo from Southland Corporation, acquiring the other 50 percent in 1990. Citgo had once been the **refining** and marketing arm of **Cities Service Company**. Beginning around 1995, the **Apertura Petrolera** opened the Venezuelan industry to foreign investment again. Under the leadership of **Luis Giusti**, PdVSA entered into 33 operating service agreements, eight risk/profit-sharing agreements, and four strategic associations, these last to develop the **heavy oil** resources of the **Orinoco Oil Belt**. Venezuelan oil production increased by about 1 Mmbpd between 1994 and 1997, to 3.6 Mmbpd.

In 1998, while PdVSA profits suffered because of record low oil prices, **Hugo Chávez** was elected president of Venezuela. Chávez wanted tighter controls on both PdVSA and foreign partners. He accused PdVSA of acting as a "state within a state." He replaced Giusti and made other changes that created discontent at the company. In December 2001, PdVSA managers and professionals initiated a work stoppage that lasted until early 2002. Chávez's further attempts to increase his grip on the company created more discontent and there was a serious strike from December 2002 to February 2003 that severely disrupted the company's operations. Chávez retaliated by firing upper management and dismissing 18,000 employees, largely skilled workers. This disrupted operations even more, resulting in domestic shortages and significant decreases in Venezuelan oil exports. In 2004, Venezuelan oil production was down to about 2.5–2.6 Mmbpd, and by 2016 it had not recovered.

After a number of years acting as an arguably more efficient and market-oriented company during the Apertura, PdVSA became once again a firmly national oil company controlled by political factors. Also, the loss of experienced workers after the firings of 2003 caused problems in maintenance and reservoir management, among other areas. PdVSA had an impressive portfolio as of 2016. It controlled 302.25 Bb of proven oil **reserves** and 202.7 Tcf of **natural gas**, and it produced oil and gas at the rates of 2.373 Mmbpd and 2.682 Bcfpd. However, the company has been hampered by inadequate reinvestment of earnings into exploration and development, as the government has siphoned profits into its various social and other programs. The unfavorable climate for foreign investment has also done damage. PdVSA is still an important, but declining, exporter of oil to the **United States** and has refining and marketing operations in Europe, the Caribbean, and the United States (Citgo).

PEACE RIVER OIL SANDS. *See* ALBERTA OIL SANDS (TAR SANDS).

PEAK OIL. The point at which peak production of oil is reached, after which production declines. In the petroleum industry, many are aware of what is called the peak oil theory, which argues that the world is nearing the peak of its production of **conventional crude oil**, and that after the peak, production will decline irreversibly, with potentially dire economic, social, and political consequences. Since oil is believed to be a finite and nonrenewable resource, it stands to reason that continuing consumption will eventually exhaust it, and some sort of peak in production will be encountered along the way. The main points of contention are how much oil is really left and when the peak will be reached.

M. King Hubbert accurately predicted in 1956 that oil production in the lower 48 states of the **United States** would peak around 1970. Hubbert applied a mathematical model of resource exhaustion and fit bell-shaped logistic curves to the data available to him. Despite the rigor of his analysis, many found his prediction astonishing, as oil production was still rising. However, U.S. conventional oil production in the lower 48 states did peak in 1970, at 9.408 Mmbpd. **Alaska** production was still small in 1970; including it raises the figure to 9.636 Mmbpd, still a peak for the United States overall. The oil production of the lower 48 states after 1970 followed a decline that was roughly symmetric with its rise, as Hubbert predicted. In 2004, 34 years after the peak, lower-48 production was 2.991 Mmbpd—very close to the value of 3.013 Mmbpd in 1936, 34 years before the peak. The overall production of the United States also declined, but not in such a symmetric fashion, because of the addition of production from two new provinces: Alaska and the federal **offshore** areas, mostly in the **Gulf of Mexico**. The explosion in tight oil production beginning in the late 2000s served to reverse the decline altogether, and by 2015 production was at roughly 1972 levels again. Hubbert was right, but only about conventional oil in the lower 48 states; to be fair, that is the only oil he knew about.

Others have attempted to apply Hubbert's method to world oil production and have issued predictions of the peak year of world oil production, but the years have come and gone without a peak. In 1977, Hubbert himself predicted inaccurately that world oil production would peak in the late 1990s. The pattern of world oil production does not follow the rising portion of a bell-shaped curve exactly. Fitting such a curve to the data is not a reliable method for inferring a peak year and the value of the world's **ultimately recoverable resource** (URR) of oil. If one makes an assumption about the value of the URR based on geological data (for example, using estimates such as those prepared by the **United States Geological Survey**), one can further constrain the problem. However, the inferred peak year is then quite sensitive to errors in the URR—which must, of necessity, be rather large. One must also bear in mind that the URR is a function of price, technology, and other variables. Improved technology can increase the URR. Higher prices can also do so, by making currently marginal resources economical to extract. Also, the lower 48 states of the United States, on which Hubbert based his prediction, constituted a province subject to extensive and relatively unconstrained exploration. This is not true of the world as a whole.

Peak oil theorists do have some other arguments on their side, however. The most serious are the depletion of the world's **giant** oil fields, the decreasing rate of discovery of giant fields, and the failure of discoveries as a whole to replace consumed **reserves**. Giant fields (**EUR** greater than or equal to 500 Mmb) are crucial for the world's oil supply. Only about 1 percent of the world's fields are giants, but they account for about two-thirds of the world's estimated recoverable oil. The hundred largest oil fields accounted for about 45 percent of the world's oil production in 2005. Geologically, the largest fields have tended to involve anticlinal traps, with sedimentary reservoirs (mostly sandstones). Since large anticlinal structures are relatively easier to spot during exploration, this has led some people to assume that the largest and most obvious giants in the world have already been found.

Giant field discoveries have been declining since the 1940s, when they peaked. In the 1940s, 30 giant fields were discovered, with an average EUR of 8.6 Bb. That is, on average, they were **supergiants**. In the 1990s, 50 giant fields were discovered, but their average size was 1.2 Bb. The very largest fields, where a disproportionate share of the oil resides, are very old and have been depleted. **Saudi Arabia**'s **Ghawar**, the largest oil field ever known, was discovered in 1948. The second-largest field, **Kuwait**'s **Burgan**, was discovered in 1938. The fields of **Venezuela**'s **Bolivar Coastal Complex** were discovered beginning in 1917. Not one of the world's top 20 fields has been discovered since 1980. Of course, this can change at any time; still, the trends are worrying.

Overall, oil discoveries have not been replacing consumed reserves since about 1981. Although some years since then have seen new discoveries exceeding production by a small margin, most years have seen the opposite. Global oil production was about 28.6 Bb in 2016. Replacing that would entail discovering the equivalent of a new **Kirkuk** field. Replacing five years of production at that level would entail discovering the equivalent of a new Ghawar (if one accepts high estimates of Ghawar's oil). It will not be easy to achieve such feats. One should note, however, that oil exploration is a costly undertaking. It is difficult to predict what might be accomplished if prices rise high enough, and the need (and potential profit) is great enough. *See also* TOTAL RESOURCE BASE.

PEARSON, WEETMAN (1856–1927). (Later Lord Cowdray.) Pioneer of the petroleum industry in **Mexico**. Weetman Dickinson Pearson was born on 15 July 1856 near Huddersfield, England. Before he entered the oil business, he was a prominent engineering contractor. After a formal education lasting to the age of 16, he became an apprentice in the firm of S. Pearson and Son, founded by his grandfather Samuel Pearson. He rose rapidly in the firm and made it into one of the world's largest and most successful engineering contractors. Among the projects Pearson completed were the Hudson River Tunnel connecting New York and Jersey City (1889–1895) and the Blackwall Tunnel under the Thames (1891–1897). He also did a large amount of infrastructure work in Mexico. He built a 200-mile railroad connecting the Atlantic and Pacific oceans across the Isthmus of Tehuantepec (1898–1906). He also constructed a large-scale drainage system for Mexico City (completed 1896) and a modern harbor at Veracruz (1895–1902), among many other achievements.

While working on his various Mexican projects, Pearson and his employees noticed signs of petroleum deposits. A missed train connection in April 1901 put Pearson in Laredo, **Texas**, for a night. This was three months after oil had been struck at **Spindletop**, and oil was the talk of the town. Pearson's brief stay in Texas stimulated his thinking on the potential for an oil industry in Mexico. He hired **Anthony F. Lucas**, who had drilled the first gusher at Spindletop, as a consultant. In 1902, Pearson began to obtain large, 50-year, tax-exempt oil **concessions** in Mexico. His infrastructure projects had gained him excellent government connections, including a friendship with the dictatorial president Porfirio Díaz. Díaz wanted to offset the influence of the **United States** in his country and was motivated to give Pearson more favorable terms than those he granted to Pearson's American competitors.

By 1907, the firm of S. Pearson and Son owned 600,000 acres of land and subsoil leases for more than 1 million acres. Taking a risk, Pearson built a refinery (at Minatitlán, near the Atlantic terminus of the Tehuantepec railway), bought **tankers**, and signed supply contracts before finding any oil. By 1908, he had invested nearly his entire personal fortune in the venture and had still found no substantial quantities of oil. He would be forced to buy crude oil from Texas in order to feed his refinery and meet his commitments. In 1908, fires destroyed large quantities of Pearson's oil. That same year, a ruinous price war began with American competitor Henry Clay Pierce.

Pearson had been operating his oil business under the umbrella of his engineering firm, S. Pearson and Son, but in 1908 he decided to establish a Mexican company to handle the oil business. He formed the Compañía Mexicana de Petróleo el Águila, the Mexican Eagle Petroleum Company. Mexican Eagle acquired all the oil business interests of S. Pearson and Son except for the tankers and the refinery. Pearson placed a number of prominent and well-connected Mexicans on the company's board, including Porfirio Díaz's son. In 1909, Pearson hired new geologists, including a young **Everette DeGolyer**. DeGolyer selected the location of the Potrero del Llano No. 4 well in Veracruz. On 27 December 1910, Mexican Eagle struck a gusher there that flowed 1 Mmbo before it was brought under control. The well would produce 140 Mmbo over its lifetime and would help open up the Golden Lane (Faja de Oro) fields. DeGolyer also sited the first successful well in the **giant** Los Naranjos field, which has produced more than 1.2 Bb.

Mexican production increased greatly after these discoveries. In 1908, it was 3.933 Mmbo. By 1914, it was 26.235 Mmbo, and Pearson's company controlled 60 percent of it. In 1914, Mexico was the world's third-largest producer, after the United States and **Russia**. Pearson formed two new British companies in 1912 for downstream business: the Eagle Oil Transportation Company handled transportation and distribution, and the Anglo–Mexican Corporation marketed Mexican Eagle's oil outside Mexico. Pearson's 25-year-old son was chairman of the second company. Pearson's companies won a contract to supply the Royal Navy with fuel oil. Pearson also began selling oil to Standard Oil Company of New Jersey (later **Exxon**), newly independent after the dissolution of **Standard Oil** and hungry for crude supplies.

In 1911 the Mexican Revolution began, and Porfirio Díaz was overthrown. The new president, Francisco Madero, assured Pearson that Mexican Eagle's concessions were still valid. In 1913, Madero was murdered by Victoriano Huerta. There was a period of civil unrest during which Pearson's refinery in Tampico was captured. U.S. president Woodrow Wilson was against Huerta and was angered when the **United Kingdom** recognized his regime. Wilson blamed Pearson for instigating the recognition. In 1914, Huerta was overthrown and Mexico was plunged into civil war. The continuing unrest made Pearson look for ways to reduce his exposure in Mexico. As early as 1913, he entertained the possibility of a buyout by Standard Oil Company of New Jersey, but it did not come to pass. He had similar discussions with **Royal Dutch Shell**. He also made repeated attempts to get the British government involved in his company in exchange for guaranteed supplies of oil to the Royal Navy. The British government instead bought a stake in the Anglo–Persian Oil Company (later the **Anglo–Iranian Oil Company**, eventually **BP**) in 1914. In 1917, the British government blocked Pearson from selling his company to Standard Oil Company of New Jersey on the grounds of national security in wartime. Matters became even more uncertain for Pearson in 1917 as Article 27 of the new Mexican constitution affirmed subsoil resources as the property of the nation.

In October 1918, Royal Dutch Shell approached Pearson again, and the British government did not object. By 1919, Pearson had sold Royal Dutch Shell a controlling interest in his Mexican oil business. Throughout his career, a number of contemporaries had commented on the good "Pearson luck." This was certainly an example. A few years after Pearson sold, Royal Dutch Shell had problems with saltwater flooding in its fields, and by 1938 it lost all its business in Mexico when the industry was nationalized.

Pearson continued his involvement in the oil business. In 1919, he organized the **Amerada Petroleum Corporation** (later Amerada Hess and Hess), again tapping the talents of Everette DeGolyer. He also continued his engineering contracting business. Pearson served in 1917 as president of the Air Board, significantly increasing British national aircraft production. He was a liberal member of Parliament for Colchester from 1895 to 1910, although he was so often absent that critics dubbed him the "member from Mexico." He died in Scotland on 1 May 1927. *See also* DOHENY, EDWARD (1856–1935).

PECHELBRONN. Early **oil sand** mining operation and oil field in Alsace-Lorraine, France. Oil sand mining and processing at Pechelbronn dates back to the middle of the 18th century. On 10 October 1734, Jean Théophile Hoeffel completed a thesis in medicine at Strasbourg. He described the thick, tarry bituminous substance floating to the top of an oil spring in the area and the nearby oil sand. He extracted some petroleum products, including something similar to kerosene, by laboratory distillation of the oil sand and the surface seep. In 1746, King Louis XV granted an exploration and mining concession for Baechel-Brunn (Pechelbronn) to Louis Pierre Ancillon de la Sablonnière and Jean d'Amascéne Eyrénis. They formed an oil company, issuing 40 shares. This might be regarded as the first modern oil company.

Sablonnière had been an interpreter with the French ambassador in Switzerland. Eyrénis was the son of a doctor of Greek descent, Eyrén Eyrénis, who had developed a **bitumen** mine in Switzerland. Sablonnière began large-scale mining of the oil sands. The first pit was about 30 feet deep. Later shafts were 100–1,000 feet. Underground tunnels reached a length of 1,500 feet by 1765. Oil sands were brought to the surface and washed with boiling water to release the bitumen. In 1866, it was discovered that oil could be obtained from deeper sands by letting it drain in place. In 1889, recovery of **conventional crude oil** began via **drilling** and pumping. Mining also continued into the 20th century, with shafts as deep as 13,000 feet. Between 1745 and 1977, when the operations stopped, about 43 percent of the oil in place was recovered by mining, and 17 percent was recovered by drilling. The remaining 40 percent was unrecoverable.

Pechelbronn was an incubator for new technology. As early as 1813, there was manual research drilling, to guide further mining operations. About 140 wells were sunk to depths as great as 640 feet. A school of petroleum technology was created that later evolved into the Institut Français du Pétrole. In 1879, the Flauvelle water injection system was put in place. Pechelbronn was also the birthplace of **well logging**. On 5 September 1927, Henri Doll, working for Conrad and Marcel **Schlumberger**, recorded the first electrical resistivity well log, at the Dieffenbach les Woerth site.

PEMEX. *See* MEXICO.

PENNSYLVANIA. Oil production 2015: 0.019 Mmbpd (0.2 percent of U.S.); **reserves** 0.013 Bb (0.04 percent of U.S.); gas production: 4.759 Tcf (17.59 percent of U.S.); 55.894 Tcf (18.16 percent of U.S.). Pennsylvania, a state in the northeastern **United States**, is now a minor oil producer, but it was one of the cradles of the modern petroleum industry, along with **West Virginia**, **Baku**, **Romania**, and Austro-Hungarian **Galicia**. **Edwin Drake**'s discovery of oil near **Titusville** in 1859 initiated a major exploration and production boom in the western part of the state, particularly in **Venango County**. Oil was transported by riverboats, which reached Pittsburgh via the Allegheny River. The 1871 discovery of the **Bradford oil field** shifted the locus of the industry from Venango County to northern Pennsylvania, near the New York border. In 1879, the world's first long-distance crude oil **pipeline** was completed between the Bradford field and Allentown, a distance of 109 miles.

Pennsylvania was the top oil-producing state in the United States until 1894, when it was overtaken by **Ohio** (*see* LIMA–INDIANA OIL FIELD). Ohio was overtaken in turn by **California** in 1902. From 1860 until 1874, Pennsylvania produced more than 90 percent of the oil in the world every year. By 1885, this proportion was down to 49 percent. Pennsylvania's oil production reached 75,000 bpd in 1881. Over the next few years, output moved up and down, dipping to 45,000 bpd in 1888. It then rose to an all-time high of 86,000 bpd in 1891. It declined to 21,000 bpd in 1912 and stayed near that level until 1924, after which waterflooding in the Bradford field led to another increase, with a secondary peak of 53,000 bpd in 1937. Thereafter, production entered a phase of apparently irreversible decline. Pennsylvania's oil production in 2006 was lower than in 1866.

Yet this state, the energy titan of the 19th century, would not yet exit the hydrocarbon business. In the 2000s and 2010s, **shale gas** production (*see* MARCELLUS SHALE) grew explosively in Pennsylvania, turning it into the second-largest gas-producing state in the United States after **Texas**. In 2015 Pennsylvania accounted for nearly a fifth of U.S. gas production. *See also* BISSELL, GEORGE HENRY (1821–1884); OIL CITY; SILLIMAN, BENJAMIN, JR. (1816–1885).

PENNSYLVANIA ROCK OIL COMPANY. *See* BISSELL, GEORGE HENRY (1821–1884).

PENNZOIL. Petroleum and automotive consumer products company formed in 1963 from the merger of **South Penn Oil Company** and **Zapata Petroleum** (a portion later became Pennzoil–Quaker State, eventually absorbed by **Royal Dutch Shell**; another portion became PennzEnergy, later Devon Energy). Pennzoil had headquarters in **Houston**, **Texas**, with offices in Los Angeles and **Oil City**, **Pennsylvania**. The Pennzoil name is famous, particularly for lubricating oils. "Pennzoil" began as a brand name of the Pennsylvania Refining Company early in the 20th century and then became the name of that company in 1924. In 1925, the Pennzoil company came under the partial ownership and control of South Penn. The name was retained as both a brand name and a company name until South Penn completed its acquisition of the company 30 years later. At that time, Pennzoil continued as a brand name but ceased to be a company name. The Pennzoil name reemerged in 1962 as the name of the enlarged corporation formed when South Penn merged with Zapata Petroleum.

The newly formed Pennzoil corporation was small by the standards of **major oil companies**: in 1963 it earned profits of about $7 million on sales of about $77 million. However, under the aggressive management of J. Hugh Liedtke, who had come from Zapata, it made some bold moves. In 1965, Pennzoil took control of United Gas Corporation, a company eight times its size, foreshadowing the type of corporate raid that would become commonplace two decades later—*see* PICKENS, T. BOONE (1928–). United Gas was a **natural gas** producer and one of the largest distributors in the **United States**. It also had a nonpetroleum mining subsidiary. By 1970, Pennzoil sales were about $700 million, up ninefold in seven years. Pennzoil created Pennzoil Offshore Natural Gas Operators (POGO) and Pennzoil Louisiana and Texas Offshore (PLATO) Inc. to expand exploration. By 1980, sales exceeded $2 billion. Despite Pennzoil's diversification, a large part of its business continued to come from its traditional competence in the production and sale of motor oil.

In 1984, Pennzoil embarked on another acquisition, reaching an agreement to purchase three-sevenths of **Getty Oil Company**. **Texaco** stepped in and bought all of Getty instead. Pennzoil sued Texaco as a result. In ***Pennzoil v. Texaco***, a jury found for Pennzoil, awarding it an astounding $10.5 billion in damages, later reduced on appeal to about $8.5 billion. Pennzoil settled with Texaco in 1987 for about $3 billion, approximately $2 billion of which it used to purchase shares in **Chevron**. Chevron initially mistook Pennzoil's share purchase as a hostile takeover attempt, although Pennzoil apparently was seeking to shelter the settlement money from taxes. This led to problems with the U.S. Internal Revenue Service in the years to come. In October 1994, Pennzoil paid $556 million in back taxes and accrued interest.

In 1998, Pennzoil spun off its Pennzoil Products Group, which produced motor oil and other products, and its Exploration and Production Company into separate entities. Pennzoil Products Group acquired Quaker State, which also produced motor oil, for around $1 billion. The combined company became Pennzoil–Quaker State, the world's largest automotive consumer products company. The Exploration and Production Company became PennzEnergy and was acquired by **Oklahoma**-based Devon Energy in 1999. Pennzoil–Quaker State suffered in the next few years from weak demand for its automotive consumer products and agreed in August 2002 to be taken over by Royal Dutch Shell, giving its motor oils a far broader distribution network, among other things.

PENNZOIL V. TEXACO. Lawsuit filed by **Pennzoil** against **Texaco** for tortuous interference in Pennzoil's 1984 agreement to acquire a portion of **Getty Oil Company**, resulting in a record jury award. In the early 1980s, Pennzoil considered Getty stock to be undervalued and began buying it. In January 1984, Pennzoil president J. Hugh Liedtke concluded an agreement with Gordon Getty (son of **J. Paul Getty**) for Pennzoil to buy three-sevenths of Getty Oil's outstanding shares for $112.50 a share, or about $3.9 billion. This represented a significant premium over the prevailing price of Getty stock. Getty's board approved the agreement, and it was publicly announced. However, Getty's investment bankers had not stopped soliciting higher offers. Texaco stepped in and offered $128 a share for all of Getty—a total of about $10 billion.

In *Pennzoil v. Texaco*, Pennzoil sued Texaco for tortuous interference. The jury found for Pennzoil and awarded it about $10.5 billion in compensatory and punitive damages, the highest award in history up to that time. Texaco appealed, but the verdict stood, although the award was reduced to about $8.5 billion. Texaco professed a desire to settle with Pennzoil on some smaller amount, threatening to file for bankruptcy if an accommodation could not be reached. It offered $2 billion, but in April 1987 Pennzoil turned it down. True to its word, Texaco filed for protection under Chapter 11 of the U.S. bankruptcy code. Pennzoil's capitalization immediately dropped by $631 million as investors feared Pennzoil might not get the cash it was expecting, at least not in the near future. Others did not believe Texaco was in any real danger, even in the face of such a large cash payout; they saw Texaco's bankruptcy as a stunt to pressure Pennzoil into accepting a lower figure. Pennzoil eventually did. At the end of 1987, Texaco agreed to pay around $3 billion—a far cry from $8.5 billion but still a huge number. The settlement enabled Pennzoil to purchase a substantial position (nearly 9 percent) in **Chevron** and to acquire Facet, manufacturer of Purolator oil filters.

PÉREZ ALFONSO, JUAN PABLO (1903–1979). A founder of the **Organization of the Petroleum Exporting Countries** (OPEC); **Venezuela**'s minister of production 1945–1948 and minister of mines and hydrocarbons 1959–1963. Juan Pablo Pérez Alfonso was born in Caracas, Venezuela, on 13 December 1903. He became politically active in 1936 after the death of dictator **Juan Vicente Gómez**. Pérez Alfonso helped found the democratic party Acción Democratica and formed a close friendship with **Rómulo Betancourt**, who named him minister of production in 1945 when he became president. Petroleum fell within Pérez Alfonso's purview. He formulated a policy of "No mas concesiones petroleras" (no more petroleum **concessions**) and was the principal author of the petroleum reforms enacted on 12 November 1948, even though by then he was no longer in the government.

The reforms established the **50–50 agreement** formula, by which international oil companies would pay 50 percent of their profits derived from Venezuelan operations to the government. This 50–50 plan set a new symbolic standard, and the important oil producers of the Middle East, including **Saudi Arabia**, soon demanded and received equally favorable terms from the **major oil companies**. During the years of military rule in Venezuela following the overthrow of President Gallegos in 1948, Pérez Alfonso spent seven months in prison and was then exiled. He lived in the **United States** and **Mexico**. During his years in the United States, he studied the history of the **Railroad Commission of Texas**, which sought to control prices by controlling production. This planted the seed of an idea for a global organization of oil-producing countries that would do the same thing. He returned to Venezuela when democracy was restored in 1958. Rómulo Betancourt became president once again and in 1959 named Pérez Alfonso his minister of mines and hydrocarbons.

In April 1959, Pérez Alfonso headed the Venezuelan delegation to the Arab Petroleum Congress in Cairo. There, he and Saudi Arabian oil minister **Abdullah al-Tariki** organized a secret meeting of representatives from other oil exporting countries. This group signed the **Mehdi Pact**, embodying recommendations the representatives would make to their governments. In cooperation with Tariki, Pérez Alfonso formulated the basis of a "Petroleum compact" in May 1960. By 14 September—after unilateral cuts in the **posted price** of crude by the major oil companies—Venezuela, Saudi Arabia, **Iraq**, **Iran**, and **Kuwait** joined together to form OPEC.

In 1961, Pérez Alfonso published the book *Petróleo: Jugo de la Tierra* (*Petroleum: Juice of the Earth*). He retired from public life in 1963 but continued to study the politics and economics of petroleum, further developing theories about the negative consequences of national wealth founded on petroleum, which he dubbed the "Venezuela effect." This led to his 1976 book *Hundiéndos en el Excremento del Diablo*, or "Sinking in the Devil's Excrement." Pérez Alfonso held that one can become ill not only from hunger, but also from indigestion. His goal had always been to ensure the orderly development of the petroleum industry in producing countries, not to generate windfalls and wasteful spending. In his later years, observing the negative effects of sudden wealth on oil-producing countries, he reportedly said that he wished he had never heard of OPEC. He died of cancer in Washington, D.C., on 3 September 1979.

PÉREZ RODRÍGUEZ, CARLOS ANDRÉS (1922–2010). President of **Venezuela** 1974–1979, 1989–1993. Carlos Andrés Pérez Rodríguez was born on 27 October 1922 in Rubio, Táchira state, Venezuela. His first presidential term was notable for the nationalization of Venezuela's oil industry that took effect 1 January 1976, and for a booming economy fueled by rising oil prices. His regime was dubbed the "Saudi Venezuela." Critics pointed out that the prosperity was also marked by extravagant and wasteful spending. His second term, by contrast, was dominated by low oil prices and austerity. He brutally crushed the resulting social protests, called the "Caracazo," using the national guard. Pérez survived two unsuccessful coup attempts against his leadership during 1992. The first was the most serious. It was led on 4 September 1992 by Lieutenant Colonel **Hugo Chávez**, who would be elected president of Venezuela in 1998. Pérez finally lost his office not because of a coup but because of corruption charges. He was found guilty of misappropriating 250 million bolivars from a presidential discretionary fund and on 20 May 1993 was removed from office by the Venezuelan Congress. In May 1994, he was placed under house arrest. He was released September 1996. He died in Miami on Christmas Day 2010.

PERMIAN BASIN. Oil-producing area in western **Texas** and southeastern New Mexico, and an early locus of **geophysical prospecting**. The Permian basin, which is about 300 miles long and 250 miles wide, is a subsurface basin covered by thick strata of Permian sedimentary rocks. Oil was first noticed by ranchers and farmers who drilled wells to obtain water. The first commercial oil discovery was in the Westbrook field in 1921, in Mitchell County, Texas. Several more fields were discovered in the 1920s, including World, McCamey, and Yates. The Permian basin was an early proving ground for geophysical prospecting techniques. The Hobbs field discovery of 1928 was made with the aid of magnetometer and torsion balance data. Initial discoveries were relatively shallow, but in 1928 a deep test in the Big Lake field struck significant oil at 8,525 feet. Because of the Permian basin's remoteness from infrastructure and major markets, deeper exploration did not become economical until **World War II**. Midland and Odessa emerged as the regional headquarters of the Permian basin petroleum industry, and the area's infrastructure was built up. By 1993, the Texas counties of the Permian basin had produced nearly 15 Bb. Since the advent of **shale gas** and **tight oil** extraction enabled by **fracking** and **horizontal drilling** technology, the Permian basin has experienced renewed attention and importance. The

Wolfcamp Formation in the Midland basin part of the Permian basin has been estimated by the **United States Geological Survey** to hold a potential 20 Bb of oil and 16 Tcf of gas.

PERSIAN GULF. An extension of the Indian Ocean, with a surface area of about 93,000 square miles, located between **Iran** and the Arabian Peninsula. The Persian Gulf is connected to the Gulf of Oman in the east via the **Strait of Hormuz**. It is shallow, with an average depth of about 65 feet. The term "Persian Gulf" is often used to denote not only the gulf itself but the entire region, including the bordering nations **Saudi Arabia**, **Kuwait**, **Oman**, **Qatar**, **Bahrain**, the **United Arab Emirates** (UAE), Iran, and **Iraq**. Iraq does not have a significant coastline but does have an outlet to the gulf. In 2016, the region had about 60 percent of the world's **conventional** oil **reserves**—47.5 percent if **Canada**'s **Alberta oil sands** and **Venezuela**'s **Orinoco Oil Belt** are included in the world total. The Persian Gulf countries also had 42.3 percent of the world's **natural gas** reserves.

There are significant **offshore** reserves in the Persian Gulf, including Saudi Arabia's **Safaniya**, the world's largest offshore oil field. There are 45 offshore **giants** in the Persian Gulf, 34 predominantly oil and 11 predominantly gas. Of these, Iran has 14, discovered between 1960 and 1993; Saudi Arabia has 13, discovered between 1951 and 1978; the UAE has 13, discovered between 1964 and 1980; and Qatar has five, discovered between 1960 and 1976.

The division of the Persian Gulf continental shelf was conducted via a series of bilateral treaties concluded between 1958 and 1974. The first was between Bahrain and Saudi Arabia in 1958. This was followed by treaties between Iran and Saudi Arabia (1968), Qatar and the UAE (more accurately, Qatar and **Abu Dhabi**, in 1969), Iran and Qatar (1970), Iran and Bahrain (1971), and Iran and Oman (1974). There are also understandings (based on the principle of median lines) between Iran and Kuwait, and between Iran and the UAE emirate of **Sharjah**, even though these understandings are not codified in bilateral treaties. There are continuing disputes between Iran and Sharjah over Abu Musa Island in the gulf. There was also a 1974 treaty between Iran and the UAE emirate of **Dubai** that was ratified by Iran but not the UAE. *See also* GULF COOPERATION COUNCIL (GCC).

PERSIAN GULF WAR (1991). War fought by a U.S.-led coalition in 1991 to repel **Iraq**'s 1990 invasion of **Kuwait**. The 1991 Persian Gulf War is sometimes called the Second Gulf War, with the 1980s **Iran–Iraq War** being regarded as the first. In the aftermath of the Iran–Iraq War, which ended in 1988, Iraq was significantly weakened and about $100 billion in debt. It owed much of the money to Kuwait, **Saudi Arabia**, and the **United**

Arab Emirates (UAE). Iraq resented having to pay back Arab countries when—in its view—it had fought a war with their natural enemy. Also, overproduction by the **Organization of the Petroleum Exporting Countries** (OPEC), fed at least partly by Kuwait and the UAE, was keeping prices low and making it difficult for Iraq to earn revenue to pay the debt. Iraq also had a dispute with Kuwait arising out of the fact that the **supergiant** Rumaila oil field, which lies mostly in southern Iraq, extends into Kuwait. Iraqi leader **Saddam Hussein** accused Kuwait of drawing oil from the Iraqi share of the field. Not far in the background was the Iraqi belief that Kuwait belonged to Iraq and had been artificially separated from it when boundary lines were drawn by the British after **World War I**.

At a summit meeting in Jedda on 11–12 July 1990, Kuwait and the UAE agreed to observe their OPEC production quotas. This only addressed some of Iraq's grievances. Iraq continued to be belligerent, making provocative troop movements. In an episode that has never been adequately explained, U.S. ambassador April Glaspie told Saddam Hussein on 25 July 1990 that the **United States** had no opinion on an "Arab–Arab conflict like your border dispute with Kuwait," which sounds almost like permission for an invasion. Although the United States apparently did not expect Saddam Hussein to invade, it had tilted toward Iraq during the Iran–Iraq War and had turned a blind eye to Iraq's 1988 gassing of Kurdish civilians at Halabja during that war. (Ironically, in 2002–2003 during the run-up to the **Iraq War**, the United States would cite the gassing of the Kurds as proof of Hussein's barbarism.)

On 2 August 1990, Iraq invaded Kuwait and quickly conquered it. The Kuwaiti emir, Shaikh Jabbar al-Ahmed al-Sabah, escaped to **Saudi Arabia**. At a stroke, Saddam Hussein added about a 12th of the world's proven oil **reserves** to his country's already vast endowment and now posed a threat to the most important reserves in the world, those of Saudi Arabia (in addition to those of the UAE). U.S. president **George H. W. Bush** worked through the United Nations (UN) to apply diplomatic pressure and economic sanctions on Iraq. He also assembled a multinational coalition and began a military buildup, Operation Desert Shield, in preparation for war. In November 1990, after nonmilitary measures failed to dislodge Iraq from Kuwait, the UN Security Council adopted a resolution authorizing the use of force if Iraq did not withdraw by 15 January 1991. The U.S. Congress did the same. The U.S.-led coalition went to war with Iraq, in Operation Desert Storm. A bombing campaign began on 16–17 January 1991 and a ground invasion on 24 February. Iraq's infrastructure was severely damaged, and as many as 200,000 Iraqis may have been killed, both civilians and soldiers in retreat. Coalition casualties numbered about 150. President Bush declared Kuwait liberated on 27 February 1991. The United States did not attempt to remove Saddam Hussein from power, most probably because it did not want to create a power vacuum that might be exploited by **Iran**.

Retreating Iraqi forces did immense damage to the oil infrastructure of Kuwait and to the environment. On 26 January 1991, they deliberately spilled 5.7 Mmb of oil into the **Persian Gulf**—the largest oil spill in history (refer to appendix 9). On 22 February, they dynamited 732 producing oil wells, and more than 650 caught fire. Of these, 365 were in the Greater **Burgan** field, Kuwait's most prolific. The fires burned for more than eight months. At their peak, they consumed more than 5 Mmbo and nearly 4 Bcf of natural gas per day and generated immense amounts of gaseous and particulate pollution. It took more than 10,000 workers to put the fires out. The dynamiting of the wells also created large pools of unburned oil. It took several additional months to recover the roughly 25–50 Mmbo in such pools.

The Iraqi invasion had immediately suspended Kuwaiti oil exports. Oil prices more than doubled between July and September 1990. Saudi Arabia and other OPEC members quickly increased production to make up the shortfall. Saudi production increased by about 40 percent to 8 Mmbpd between July and September 1990. By September 1990, OPEC production was nearly at its level before the invasion of Kuwait. The **International Energy Agency** also helped to avert panic before Desert Storm by making 2.5 Mmbpd available to the international market from January to March 1991.

PERTAMINA. *See* INDONESIA.

PERU. Oil production 2015: 0.058 Mmbpd (0.074 percent of world); **reserves** 0.473 Bb (0.03 percent of world); gas production: 0.438 Tcf (0.35 percent of world); reserves 14.09 Tcf (0.2 percent of world). Peru's peak oil production was only 0.195 Mmbpd, reached three out of four years from 1979 to 1982, and is now much lower. Peru's gas production is becoming more significant, though, and Peru became an exporter of **liquefied natural gas** (LNG) in 2010. Peru joined the **Gas Exporting Countries Forum** in 2017.

Peru's petroleum industry has one of the longest histories of any country's. Oil seeps were long known to the indigenous population before the arrival of the Spaniards. They used oil and tar to waterproof pottery and embalm mummies. During the colonial period the Spaniards used it to caulk ships. In 1863 a landowner named Diego de Lama enlisted a Scotsman named Farrier to dig for oil on his property near Piura, in the northwestern coastal region. Farrier found and extracted oil from hand-dug pits. In 1865 George Corey, who had experience in the oil fields of **Pennsylvania**, began working for the newly formed U.S. company Peruvian Petroleum Syndicate. He drilled several wells that year in the northwestern coastal region and struck oil. One well produced 25 barrels per day, and another produced 480 barrels per day before it was damaged in an accident involving tools at the bottom. Despite the engineering success, business success did not follow. However, these wells are among the first, and possibly the first, drilled in Latin America. They can be regarded as the discovery wells of the Zorritos oil field, which was further explored and developed by others beginning in the 1870s, especially the Genoese immigrant businessman Faustino Piaggio. Piaggio formed the Establecimiento Industrial de Petróleo de Zorritos in 1884. Piaggio had taken over the concessions of Captain Henry Smith, an important petroleum explorer and developer in the region. The Zorritos field produced continuously until 1965. The northwestern coastal region was also the area in which the world's first **giant** oil field was discovered. This field, the La Brea or Negritos field, had an EUR of a billion barrels of oil. Some sources say it was discovered in 1868. Other sources say it was discovered in 1873 by an engineer named Fowkes, who worked for businessman Henry Meiggs. War with Chile broke out in 1879, delaying development.

Peru has four large oil-producing regions. One is the North Coast region, where the earliest developments outlined above occurred. The region includes the Talara and Tumbes basins. The other regions are the Central Jungle, opened up in 1939 with the Agua Caliente field; the North Jungle, opened in 1971 with the Corrientes field; and the South Jungle, where significant activity has occurred since 2004 when production began in the Camisea gas field, discovered 20 years earlier.

The petroleum industry operating in Peru has sometimes had a rocky relationship with the national government. In the late 1960s the government accused the International Petroleum Company (IPC), a subsidiary of the **Standard Oil** Company of New Jersey (later **Exxon** and **ExxonMobil**), which held assets in the northern coastal region including the giant La Brea field, of political meddling and tax evasion. The company came to an agreement with the government of president Fernando Belaúnde Terry in 1968, but Belaúnde Terry was then overthrown in a military coup that same year. The military government expropriated IPC's oil. The **United States** suspended economic and military aid. In 1974 the United States and Peru reached a settlement and aid was restored.

Peru undertook significant reforms of its petroleum sector in the 1990s, when the country privatized its state-owned oil company Petroperú. The company has concentrated only on downstream operations since then. With the promulgation of Law No. 26221 in 1993, Peru created another private/state-owned company called Perupetro. Perupetro acts on behalf of the state to negotiate, sign, and supervise hydrocarbons contracts and technical evaluation agreements with petroleum companies from around the world. Companies operating in Peru in the 21st century include Petrobrás (*see* BRAZIL), **Repsol**, Pluspetrol, Hunt Oil—*see* HUNT, H. L. (1889–1974), Talisman (now Repsol), Perenco, **ConocoPhilips**, SK Corp., CNPC (*see* CHINA), and Sonatrach (*see* ALGERIA).

PETER I (PETER THE GREAT) (1672–1725). Tsar of **Russia** 1682–1725; turned Russia into a major modern European power. Pyotr Alexeyevich Romanov was born on 9 June 1672. He became tsar under a regent because of his young age but eventually gained full control and began efforts to modernize Russia. He took an active role in promoting an early petroleum industry in Russia, ordering a systematic search for petroleum and other mineral resources. He sent Gottlieb Schober, his physician, to the area between the Terek and Sunzha Rivers (**Volga–Urals** region) for the specific purpose of prospecting for oil. Schober found surface seeps. In 1702, Peter ordered oil samples to be sent to him from **Siberia** and ordered further scientific exploration of the vast region. Peter was also interested in the oil fields of **Baku** and the **Abseron Peninsula**, then part of the Persian Empire. He ordered the western shore of the **Caspian** mapped in 1719. One of the officers charged with the mission, Fyodor Soimonov, described the Baku fields. Oil may not have been the primary reason Peter conquered Baku in 1723, but he had voiced clear intentions to exploit the oil fields. A regular flow of oil began from Baku to St. Petersburg. Peter the Great died on 8 February 1725. In 1732, seven years after his death, Baku reverted to Persian control. Russia would gain it back in 1806.

PETROBRÁS (PETRÓLEO BRASILEIRO). *See* BRAZIL.

PETROBRÁS SCANDAL. **Brazil**'s state-majority-owned oil company Petrobrás was rocked by scandal beginning in 2014 with the beginning of the money laundering and corruption investigation known as Operação Lava Jato (Operation Car Wash). At the core of the scandal were allegations that large sums of money were kicked back to Petrobrás officials and to politicians by major Brazilian corporations in return for Petrobrás contracts. The ruling Partido dos Trabalhadores (PT) was heavily implicated. The scandal was highly complex, but testimony given by former senior managers at Petrobrás and companies doing business with it suggests that Petrobrás executives and those from contracting companies inflated prices of contracts and skimmed the difference. Supposedly, over $2 billion was stolen in this way, and some of it was used to finance the campaigns of PT and other politicians. The scandal exposed widespread corruption in large sectors of the country's economy. Many high-level businessmen and public figures were arrested in the course of the investigation into the scandal and corruption more generally, including former president Lula (Luiz Inácio Lula da Silva) himself. Lula was convicted of money laundering and corruption on 12 July 2017 and sentenced to nine and a half years in prison but remained free on appeal. The scandal and its associated damage brought down Brazilian president Dilma Rousseff. Despite no alleged involvement in the scandal itself, Rousseff had lost public confidence and was impeached for unrelated matters. She claimed that the real reason she was being impeached was that she had allowed investigations into the scandal to move forward. She was removed from office on 31 August 2016 and succeeded by Vice President Michel Temer for the remainder of her term.

PETRO-CANADA. **Canada**'s former national integrated petroleum company, based in **Calgary**, **Alberta**; now part of **Suncor Energy**. Petro-Canada was established in 1975 by the Petro-Canada Act in the wake of the **First Oil Shock**. It was capitalized directly by the Canadian federal treasury. It acquired the Canadian subsidiaries of large international oil companies, including **Arco** (1976), Pacific Petroleum (1979), **Petrofina** (1981), and **BP** (1983). In 1979, it shared in the discovery of the **giant** Hibernia oil field **offshore** Newfoundland by Standard Oil Company of California (later **Chevron**). In 1980, under the terms of the new National Energy Program, the government granted Petro-Canada a share of all oil and **natural gas** prospects on federal lands and offshore, and a share of output from the **Alberta oil sands**. In 1982, the company made its largest oil discovery to that date, at Valhalla, Alberta.

After the 1984 election brought the government of Brian Mulroney to power, Canadian energy policy had a more neoliberal orientation. The government told Petro-Canada to stop thinking of itself as an arm of the state and to conduct business as a commercial company. In 1984, Petro-Canada made its first large offshore discovery as an operator, the Terra Nova oil field offshore Newfoundland (first production 2002). The next year, it acquired Gulf Canada's retail operations. In 1991, the government passed privatization legislation and sold 20 percent of Petro-Canada to private investors. By 1995, the company was more than 80 percent privatized. In 1997, Petro-Canada began producing oil from Hibernia. During the 1990s and 2000s, it expanded internationally. In 1994, it made an oil discovery in Tamadenet, **Algeria**. In the 2000s, Petro-Canada was involved in **North Sea** discoveries in the Dutch and **United Kingdom** sectors, including the Buzzard field. In 2009 it merged with Suncor Energy.

PETROCHEMICALS. Chemical products, other than fuels, made from hydrocarbon raw material obtained from petroleum. Both crude oil and **natural gas** serve as sources. Globally, about 6 percent of produced crude oil is used for petrochemicals. Petrochemicals are used in detergents, fertilizers, medicines, paints, plastics, synthetic fibers, synthetic rubber, and explosives. It would be difficult to imagine the modern world without them. The two broad classes of raw materials for petrochemicals are olefins and aromatics. Olefins include ethylene and propylene, used in making plastics products, and butadiene, used in making synthetic rubber. Aromatics include benzene, used in making dyes and detergents; toluene, used in making explosives; and xylenes, used in making plastics and synthetic fibers. Olefins are produced at oil refineries by cracking and reforming processes. Aromatics are produced at refineries mostly by catalytic reforming or related processes.

Many substances similar to petrochemicals were known in the 19th and early 20th centuries. Celluloid was patented in 1870, and the thermoplastic Bakelite was invented in 1907 by Leo H. Baekland. Polyvinyl chloride was also known in the 19th century, as were polyethylene and polystyrene. However, it was only in the 1920s and later that processes for the large-scale production of new plastics and other synthetic materials from petroleum-derived raw materials began to be discovered. The development of thermal cracking by **William Burton** and his team had opened the door. The original purpose of the process was to increase **gasoline** yields, but it and related processes enabled petrochemistry as well.

In the 1930s, Imperial Chemical Industries (ICI) devised an industrial process for producing low-density polyethylene (LDPE). Manufacturing was ready to begin in 1939. In 1951, J. Paul Hogan and Robert Banks of **Phillips Petroleum** discovered a process for promoting polymerization at lower temperatures and pressures, producing high-density polyethylene (HDPE). Phillips marketed it under the trade name Marlex. A process for making polypropylene was also discovered by a number of investigators in the 1950s, including Hogan and Banks and Professor Giulio Natta. Commercial production began in 1957, although litigation over the invention continued until 1989. In 1934, Wallace Carothers's laboratory at DuPont created the first polyamide fiber, which the company patented as nylon.

The first nylon product was a toothbrush filament, introduced in 1938. The 1939 San Francisco Exposition famously featured nylon stockings. Carothers's team at DuPont also discovered neoprene, the first synthetic rubber, in 1930. In 1937, chemists at the Standard Oil Company of New Jersey (later **Exxon**) developed a process for making butyl synthetic rubber. **World War II** provided a major stimulus to the industry, with the demands it placed on the production of explosives and all manner of materials. Nylon was used in making parachutes. Plexiglas was introduced for airplane windows. Synthetic rubber was critical for making tires, inner tubes, and self-sealing fuel tanks. In the postwar years, the world at large was transformed by plastics and other petrochemicals.

Even if petroleum were not necessary as an energy source, it would still be needed for petrochemicals. Shah **Mohammed Reza Pahlavi** of **Iran** stated in the 1970s that oil should not be burned, because it was too valuable as a source of petrochemicals. Although he made his comments in the political context of pushing for higher oil prices, they are perhaps still worth noting.

PETROCHINA. *See* CHINA.

PETRODOLLAR RECYCLING. The process by which **petrodollars** earned by oil exporting countries find their way back into the U.S. financial system. From 1970 to 1980, the price of oil increased by nearly a factor of 20 (*see* FIRST OIL SHOCK; SECOND OIL SHOCK). The international oil trade was priced in U.S. **dollars** (and continues to be as of 2017). Oil exporting countries found themselves awash in dollars, often far exceeding their domestic investment needs. The oil exporting nations invested the "surplus petrodollars" with Western banks, typically in the **United States** or the **United Kingdom**. The banks would then lend the petrodollars out, often to governments of developing countries that needed the dollars to pay for oil. This paved the way for the Third World debt crisis of the early 1980s. Petrodollar recycling continued in later years, although recently it has taken a somewhat different form. The oil exporting nations of the Middle East have increasingly become importers of goods from Asia, particularly Japan and **China**, as well as the European Union. The petrodollars earned from Japan and China are recycled back to those countries, which then typically invest them in U.S. treasury securities. It is estimated that in 2006, 45 percent of the massive U.S. current account deficit was financed by recycled petrodollars.

PETRODOLLARS. U.S. **dollars** earned by countries through the exportation of petroleum. International oil sales have been priced for decades in U.S. dollars. The term "petrodollar" was coined in 1973 by Ibrahim Oweiss, a Georgetown University economics professor. *See also* PETRODOLLAR RECYCLING

PETROFINA. Former Belgian oil company headquartered in Brussels and once Belgium's largest corporation; now part of **Total**. Petrofina was founded in 1920 under the name Compagnie Financière Belge des Pétroles. The company acquired oil properties in **Romania** that were surrendered by the Germans after **World War I** and began downstream operations in the 1920s and 1930s. Business was disrupted by the German invasion of Belgium in **World War II**.

After the war, the company's Romanian oil properties were nationalized. Petrofina turned its attention to North America, the **North Sea**, and Africa. In 1955, Petrofina made the first commercial discovery of oil in **Angola**, in the onshore Kwanza basin. In the **United States**, it formed American Petrofina in 1956 (which became Fina in 1991). American Petrofina bought **Tenneco**'s operations in the U.S. Southwest in 1988. Petrofina made a number of other acquisitions in the 1980s as well, including Charterhouse Petroleum, based in the **United Kingdom**, in 1985. Petrofina also had a 30 percent share in **Norway**'s **Ekofisk** field in the North Sea. In 1996, Petrofina announced important oil discoveries **offshore** Angola.

In 1997, Petrofina became the first Belgian company listed on the New York Stock Exchange. Two years later, the company was acquired by Total for about $11 billion, and the combined company became known briefly as TotalFina. When it acquired **Elf Aquitaine** shortly thereafter, it became TotalFinaElf. In 2003, it changed its name to Total SA. In 1998, the last full year before it was acquired by Total, Petrofina had a net income of $0.519 billion on total revenues of $19.023 billion. It controlled 0.670 Bb of oil **reserves** and 1.336 Tcf of **natural gas**. It produced 0.140 Mmbpd of oil and 0.513 Bcfpd of gas.

PETRÓLEOS DE VENEZUELA. *See* PDVSA.

PETROLEUM ADMINISTRATION FOR DEFENSE (1950–1954). U.S. government agency administering policies for meeting military, government, industrial, and civilian requirements for oil and gas during the Korean War. The agency was created within the Department of the Interior on 3 October 1950 and abolished 30 April 1954. *See also* PETROLEUM ADMINISTRATION FOR DEFENSE DISTRICTS (PADD); PETROLEUM ADMINISTRATION FOR WAR (PAW) (1942–1946).

PETROLEUM ADMINISTRATION FOR DEFENSE DISTRICTS (PADD). Petroleum Administration for Defense Districts were established in **World War II** by the **Petroleum Administration for War** (PAW) of the **United States**. They were initially known as PAW districts. The purpose of the districts was to help the PAW organize the allocation of **gasoline**, diesel, and other petroleum-based fuels. The five districts were later renamed for the **Petroleum Administration for Defense** (PAD), an analogous agency operating during the Korean War. The PADDs are still used for data collection and reporting purposes. PADD 1, the East Coast District, is divided into three subdistricts. Subdistrict 1A (New England) includes Connecticut, Maine, Massachusetts, New Hampshire, Rhode Island, and Vermont. Subdistrict 1B (Central Atlantic) includes Delaware, the District of Columbia, Maryland, New Jersey, New York, and **Pennsylvania**. Subdistrict 1C (Lower Atlantic) includes Florida, Georgia, North Carolina, South Carolina, Virginia, and **West Virginia**. PADD 2, the Midwest District, includes Illinois, Indiana, Iowa, Kansas, Kentucky, Michigan, Minnesota, Missouri, Nebraska, North Dakota, South Dakota, Ohio, **Oklahoma**, Tennessee, and Wisconsin. PADD 3, the Gulf Coast District, includes Alabama, Arkansas, Louisiana, Mississippi, New Mexico, and **Texas**. PADD 4, the Rocky Mountain District, includes Colorado, Idaho, Montana, Utah, and Wyoming. PADD 5, the West Coast District, includes **Alaska**, Arizona, **California**, Hawaii, Nevada, Oregon, and Washington.

PETROLEUM ADMINISTRATION FOR WAR (PAW) (1942–1946). Agency of the U.S. government during **World War II** responsible for the development and utilization of petroleum resources to meet the nation's wartime needs. The Petroleum Administration for War was established by President Franklin D. Roosevelt on 2 December 1942 and was headed by Secretary of the Interior **Harold Ickes**. Before the establishment of the PAW, Ickes headed the Office of the Petroleum Coordinator for National Defense (1941–1942) and the Office of the Petroleum Coordinator for War (1942).

The PAW was a successful venture that involved a highly effective collaboration between the government and the petroleum industry. Although Ickes, with his progressive background and his abiding faith in the public sector, was considered by many to be biased against the petroleum industry, he made very pragmatic and effective use of industry talent and advice. Ickes's deputy administrator, Ralph Davies, was a vice president and director of the Standard Oil Company of California (later **Chevron**), and Ickes relied considerably on him for daily management and operational decisions. Other important PAW administrators were also from the petroleum industry. For example, George Walden had been an executive both with Standard Oil of New Jersey (later **Exxon**) and Stanvac (*see* EXXON; MOBIL). Another highly prominent administrator, Terry Duce, was a vice president and director of Casoc (later **Aramco**).

Ickes and Davies established several important industry committees. Two major ones were the Foreign Operations Committee and the Petroleum Industry War Council. The Foreign Operations Committee was composed of nine executives from the U.S. petroleum industry and two British observers. The Petroleum Industry War Council was composed of about 70 industry leaders and headed by William Boyd Jr., who was president of the **American Petroleum Institute** (API). The Petroleum Industry War Council was later succeeded in peacetime by the National Petroleum Council. Industry representatives voluntarily agreed to fund the committees by contributing $0.001018 per barrel of oil (1943 budget). The industry committees worked closely with the PAW and sometimes opposed Ickes's initiatives. The Petroleum Industry War Council was against Ickes's plans to form the **Petroleum Reserves Corporation** and his ideas to have that government entity purchase interests in **reserves** overseas. The council was very wary of having the government actually enter the oil business. The council did agree with Ickes that U.S. reserves were declining, but it favored solving the problem with higher oil prices to encourage domestic drilling. Despite such disagreements, the PAW and its industry partners were able to work together to meet military and civilian needs for petroleum. They managed to effect an increase in average crude oil production by 27 percent (well over a million barrels per day) compared to pre-1941 levels. Having fulfilled its purpose, the PAW was abolished on 3 May 1946 by President Harry S. Truman. *See also* PETROLEUM ADMINISTRATION FOR DEFENSE DISTRICTS (PADD).

PETROLEUM BOARD. Organization pooling and combining the operations of the domestic downstream **United Kingdom** petroleum industry during **World War II**. Planning for the Petroleum Board began in the summer of 1938, before the war began. On 8 March 1939, the **Anglo–Iranian Oil Company** (later **BP**) and the British operations of **Royal Dutch Shell** and Standard Oil Company of New Jersey (later **Exxon**) signed the Petroleum Board Agreement, along with some smaller companies. The companies agreed to pool their distribution networks, storage facilities, and transport infrastructure. Fuel was actually sold under the single brand name of "Pool." Each company's share of the Petroleum Board's business would be based on its British sales in the last full year of peace. The board chairman was Andrew Agnew, recently retired from Royal Dutch Shell. The Petroleum Board had a monopoly over petroleum and petroleum products distribution in Britain. It was supplemented in April 1940 with a parallel body, the Trade Control Committee, also with Andrew Agnew as chairman. The Trade Control Committee had a more international focus, encompassing Allies and the British Empire. On the government side, the petroleum industry was controlled by the **Oil Control Board**.

PETROLEUM HISTORY INSTITUTE. American professional society devoted to the study of the history of petroleum. The institute publishes the annual journal *Oil Industry History*, the only professional, peer-reviewed journal exclusively devoted to the history of the petroleum industry. The institute began as the **Drake** Well Foundation, established by the **American Petroleum Institute** in 1951 to manage the Drake Well Museum and Memorial Park in **Titusville**, **Pennsylvania**. In 2003 the Drake Well Foundation changed its name to the Petroleum History Institute. *Oil Industry History* began publication before then, in 2000.

PETROLEUM INDUSTRY WAR COUNCIL. *See* PETROLEUM ADMINISTRATION FOR WAR (PAW) (1942–1946).

PETROLEUM INTELLIGENCE WEEKLY (PIW). Newsletter covering the petroleum industry, founded in November 1961 by journalist **Wanda Jablonski** and noted for its investigative reporting and access to high-level sources. It is particularly praised for its coverage of **Saudi Arabia** and **Aramco**, and the **First Oil Shock** of 1973–1974.

In 1972, *Petroleum Intelligence Weekly* published the details of the **participation agreement** between five Arab oil-producing countries and **major oil companies**. Saudi oil minister **Ahmad Zaki Yamani** and the companies were angered by this and blamed each other for the leak. It is believed that *PIW* obtained the details not from the negotiating partners but from **Compagnie Française des Pétroles** (CFP, later **Total**). CFP had the information under the terms of the old **Red Line Agreement** of 1928. Although the agreement no longer governed the major actions of the signatories, it was still true that any change in one company's **concession** had to be disclosed to the others. *PIW* was sold in 1988 to the Petroleum Finance Company, a consulting firm in Washington, D.C. In 1996, it was sold to the Oil Daily Company. *Oil Daily* and *PIW* operations were merged to create the Energy Intelligence Group, which runs several other publications and provides other information services.

PETROLEUM RESERVES CORPORATION (1943–1944). U.S. government agency created during **World War II** to acquire foreign petroleum **reserves**. **Harold Ickes**, the secretary of the interior and head of the **Petroleum Administration for War**, authorized the formation of the Petroleum Reserves Corporation (PRC) under the Reconstruction Finance Corporation, with himself as head. The prominent geologist and oil executive **Everette DeGolyer** traveled to **Saudi Arabia** under the auspices of the PRC to assess that country's oil potential. He reported in early 1944 that here was likely the greatest accumulation the world had ever known, and that the center of gravity of the entire global oil industry would eventually move to that region.

Ickes, who had a pessimistic view of future domestic oil production, wanted the PRC to obtain a "participating and managerial interest" for the **United States** in Casoc's (**Aramco** after January 1944) **concession** in Saudi Arabia. This was a rare example of the U.S. government seeking to nationalize oil interests. In addition, Ickes wanted the PRC to oversee construction of a refinery and a major **pipeline** in Saudi Arabia. The PRC did not achieve any of these goals in its short lifetime, but they were realized later by other parties, in forms different from those originally envisioned. Military representatives on the PRC were in favor of purchasing a 51 percent interest in the Casoc concession. However, Ickes moved away from the idea and toward a more contractual model. He was convinced by other parties in the government that owning a foreign oil producer might complicate foreign policy and adversely affect relationships with other producing countries. In the end, the initiative was killed.

In February 1944, Ickes announced a tentative agreement with Standard Oil Company of California (later **Chevron**) and the Texas Company (**Texaco**), the two partners in Aramco, to construct a pipeline across Saudi Arabia from the **Persian Gulf** to the eastern Mediterranean. Other companies, led by Standard Oil Company of New Jersey (later **Exxon**) and Socony–Vacuum (later **Mobil**), lobbied against it. They couched their opposition as anti-imperialism, but the more probable reason is that those two companies were constrained by the **Red Line Agreement** and viewed the pipeline as a threat. This initiative too was killed. Both Jersey Standard and Socony–Vacuum would later become partners in Aramco. The PRC was abolished in 1944. Its function had been stymied both by industry intransigence and intragovernment disagreement.

PETROLINE (EAST–WEST PIPELINE). Saudi Arabia's main oil **pipeline**, running 745 miles from **Abqaiq** to **Yanbu** on the Red Sea. Petroline was built in 1981 as a single 48-inch line with a capacity of 1.8 Mmbpd. It was expanded to 3.2 Mmbpd in 1987 by constructing a new 56-inch pipeline parallel to the original. In 1993, capacity was increased to 5 Mmbpd by adding more pumping capabilities. The pipeline now has 11 pumping stations using on-site gas turbine electric generators. As of 2007, Petroline had significant unused capacity (about half).

PETROMIN. Saudi Arabian petroleum company; later part of **Aramco**. The General Petroleum and Mineral Organization, or Petromin, was founded by **Saudi Arabia** in 1962 and began marketing petroleum products in Saudi Arabia in 1963. It developed into an integrated oil company and by 1980 was marketing about 20 percent of Saudi production. Petromin, by proving that the Saudi government could control a large and complex oil operation on its own, may have smoothed the acquisition of Aramco by the Saudi government. Petromin was merged into Saudi Aramco on 24 October 2005.

PETRONAS. *See* MALAYSIA.

PETROZUATA PROJECT. *See* ORINOCO OIL BELT.

PEW, JOSEPH NEWTON. *See* SUN OIL COMPANY.

PHILBY, HARRY ST. JOHN (JACK) (1885–1960). British explorer and adviser to **Ibn Saud**, who helped negotiate the first successful oil **concession** in **Saudi Arabia**. Harry St. John Bridger Philby, more familiarly known as Jack, was born in Ceylon (Sri Lanka) on 3 April 1885. He graduated from Trinity College, Cambridge, and in 1907 joined the Indian Civil Service. In **World War I**, he served as a political officer with the Mesopotamian Expeditionary Force and was sent in 1917 to negotiate with Ibn Saud, the future founder of Saudi Arabia, whom the British considered an unreliable ally at the time. Philby established a rapport with Ibn Saud. After serving as chief British representative in Jordan from 1921 to 1924, he returned to start a business in Jedda, Saudi Arabia. Under Ibn Saud's tutelage, he converted to Islam. Ibn Saud gave him his Muslim name, Abdullah. Philby became a trusted adviser to King Ibn Saud and a member of his privy council.

Philby helped convince Ibn Saud to exploit his country's potential oil wealth. In the 1920s and early 1930s, that oil wealth was still theoretical, and many did not believe it existed. However, Philby held that Saudi Arabia was like someone "sleeping atop buried treasure." An early concession granted to **Frank Holmes** in 1923 had come to nought. In May 1932, Standard Oil of California (Socal, later **Chevron**), operating on a different concession once held by Holmes, discovered oil in **Bahrain**. Socal became interested in exploring in the Hasa region (now Eastern Province) in Saudi Arabia. Philby signed on with Socal as a consultant without informing the other interested party, the **Iraq Petroleum Company**, which was led in the negotiations by the Anglo–Persian Oil Company (APOC, later **Anglo–Iranian Petroleum Company**, still later **BP**). APOC was not as serious a bidder as Socal. It did not believe oil was to be found in Saudi Arabia but wanted to keep competitors out. Philby favored the American company in any case. Although British himself, he disapproved of British policy in the Middle East, which he saw as unwelcome Western dominance. At the time, the **United States** was not yet significantly engaged in the Middle East. Philby's primary loyalty was to Ibn Saud. He helped Socal negotiate a 60-year concession covering about 360,000 square miles in the Eastern Province in exchange for modest royalties and up-front advances in gold to Ibn Saud's ailing treasury. The concession agreement was signed on 29 May 1933. Other than Holmes's concession from the 1920s that had been canceled, this was the only concession agreement signed in Saudi Arabia (apart from the **Neutral Zone**). It would lead eventually to the formation of **Aramco** and the discovery of the world's richest oil deposits.

Philby returned to England in 1939 and tried unsuccessfully to enter politics. He was briefly imprisoned in 1940 because of his antiwar views. He returned to Saudi Arabia. In 1955, after Ibn Saud's death, Philby was expelled for criticizing the extravagance of his successor's regime. Apart from his political activities, Philby was an explorer and a linguist who undertook many expeditions in Saudi Arabia and contributed greatly to its cartography, archeology, and descriptions of its fauna and flora. His son, Kim Philby, was later revealed as a Soviet mole in British intelligence. Jack Philby died on 30 September 1960 in Beirut, Lebanon.

PHILLIPS, FRANK (1873–1950). Cofounder of **Phillips Petroleum** (later part of **ConocoPhillips**). Frank Phillips was born on 28 November 1873 in Nebraska. He began his career as a barber in Creston, Iowa, where he heard stories of an oil boom in the Indian Territory (**Oklahoma**) from a friend who had done missionary work near the town of Bartlesville. In 1903, he founded Anchor Oil and Gas Company and began exploring for petroleum in Oklahoma. After considerable difficulties, the company discovered oil in September 1905. Working with his younger brother Lee Eldas Phillips, he went on to drill 81 successful wells in a row. In 1917, the Phillips brothers incorporated under the name **Phillips Petroleum**, with Frank as president and Lee as vice president. At the time, they produced 384 bopd. By 1937, under Frank Phillips's leadership, production stood at 67,000 bopd. Frank Phillips was an innovative thinker and believed in the value of scientific and technological research. His company was a technological leader. For example, it developed a thermal polymerization process in 1930 that greatly reduced wastage of **natural gas** at oil wells. Phillips was a pioneer in extracting value from natural gas when many considered it a waste product of oil production. Phillips became chairman of the board of Phillips Petroleum in 1938 and retired in 1949. He died on 23 August 1950 in Atlantic City, New Jersey.

PHILLIPS PETROLEUM. Large **independent** petroleum company founded by **Frank Phillips** and his brother Lee Phillips in 1917 in Bartlesville, **Oklahoma**; eventually part of **ConocoPhillips**. The antecedent to Phillips Petroleum was Anchor Oil, established by Frank Phillips in 1903. Frank and his younger brother Lee Phillips made several successful discoveries on Native American lands in Oklahoma. The initial Anchor discovery was in September 1905, following a number of unsuccessful attempts. After that, the Phillips brothers drilled a remarkable 81 successful wells in a row and founded Phillips Petroleum. The company was an early leader in extracting value from **natural gas** at a time when many considered it a waste product of oil production.

The company moved into **refining** and marketing, acquiring a refinery in **Texas** in 1927 and selling **gasoline** under the iconic "Phillips 66" brand (a tribute to the fact that a test car reached 66 miles per hour on the famous Route 66). By 1930, Phillips had 6,750 retail gasoline outlets in 12 states. Phillips also made a major contribution to plastics and **petrochemicals**. In 1951, company chemists Paul Hogan and Robert Banks discovered methods of manufacturing crystalline polypropylene and high-density polyethylene (HDPE). These substances are regarded as important foundations for modern plastics. Phillips marketed HDPE under the trade name Marlex. Petrochemicals became an important part of Phillips's business, which included activities in every phase of the petroleum industry.

In 1969, Phillips led the consortium that discovered the **Ekofisk** field in the **North Sea**, one of its biggest successes. Soon after, the company's reputation suffered when it was found to have made illegal contributions to Richard M. Nixon's presidential campaign. In 1984, Phillips became a takeover target for corporate raider **T. Boone Pickens**, and a year later it was targeted by Carl Icahn. It defended itself successfully by repurchasing stock but took on huge debts (about $9 billion). It suffered under the debt burden and was forced to sell assets and eliminate jobs. In 1997, it entered into a joint refining venture in **Qatar** with the Qatar General Petroleum Company. In 1999, Phillips acquired **Arco**'s **Alaska** operations for about $7 billion. These properties had to be sold in order for **BP** to obtain regulatory approval for its 2000 acquisition of Arco. In 2001, Phillips acquired the refining and marketing company Tosco for $9.3 billion. In 2002, Phillips merged with **Conoco**, forming ConocoPhillips in a deal valued at $15.12 billion. In 2001, the last full year before the merger, Phillips was the third-largest petroleum company headquartered in the **United States** measured by total assets. It had a net income of $1.661 billion on revenues of $26.868 billion and controlled 3.660 Bb in proved liquid **reserves** and 8.829 Tcf of gas. It produced 0.619 Mmbpd of oil and 1.562 Bcfpd of gas. *See also* ADAMS, KENNETH S. (1899–1975).

PICKENS, T. BOONE (1928–). Oil entrepreneur and corporate raider known for his ability to identify undervalued companies, buy substantial stakes, and resell them at a large profit. Thomas Boone Pickens was born 22 May 1928 in Holdenville, **Oklahoma**. He graduated with a degree in geology from Oklahoma A&M University in 1951 and then joined **Phillips Petroleum**. In 1956, he struck out on his own, founding Petroleum Exploration Inc. (PEI). He explored the **mid-continent** and Rocky Mountain regions of the **United States**. In 1959, he acquired Altair Oil and Gas. He took his firm public in 1964, changing its name to Mesa Petroleum. In 1968, he had revenues of about $6 billion and decided to acquire the Hugoton Production Company, which owned a large **natural gas** field in **Kansas** and which he considered inefficiently managed. Pickens offered Hugoton Mesa preferred stock as payment. Hugoton refused and looked for a "white knight" (a potential acquiring company more agreeable to management), finding one in Reserve Oil and Gas. In what was to become a pattern, Pickens pursued a hostile takeover and by the end of 1968 owned a third of Hugoton's shares. Mesa acquired Hugoton on 7 April 1969. It tried and failed to acquire Southland Royalty, but it obtained a large interest in Pubco Petroleum.

Mesa did not limit itself to stock-market maneuvering. It was a serious oil company as well. In 1976, it discovered the Beatrice field in the **North Sea**, the largest field in the company's history. It sold the field, named for Pickens's wife, to the **British National Oil Company** for a $31.2 million profit. In 1979, Pickens restructured Mesa, creating a royalty trust that held the company's petroleum **reserves** and paid earnings directly to shareholders, thereby avoiding double taxation at the corporate and shareholder levels. The trust still existed in the 2000 decade.

The 1980s were the heyday of Pickens's corporate raiding activity in the petroleum industry. In the spring of 1982, Mesa made an offer to acquire **Cities Service Company**, which was more than 20 times its size. Cities Service responded by making a counterbid to acquire Mesa. Cities Service—ironically, given subsequent events—enlisted **Gulf Oil Corporation** as a white knight. Mesa failed to acquire Cities Service but sold its shares back to the company for $30 million in profit, a type of profit often labeled "greenmail." Pickens then turned his sights on Gulf. In 1983, Mesa acquired 5 percent of Gulf's stock. Pickens soon raised the stake to 11 percent and launched a proxy fight. Gulf offered him greenmail, but he turned it down. Gulf found its own white knight in the Standard Oil Company of California (Socal). Socal acquired Gulf and changed the name of its enlarged company to the familiar **Chevron**. Pickens's group made a $760 million profit on the deal.

Pickens then attempted an almost simultaneous takeover of Phillips Petroleum and Unocal (formerly the **Union Oil Company of California**). He failed in both attempts. He did make greenmail on the Phillips deal but not on the Unocal. Unocal had previously incorporated as a holding company in Delaware, where laws made it more difficult for outsiders to gain control of a company without the approval of its directors. By the end of 1984, Pickens had acquired 13.6 percent of Unocal, but Unocal stymied him with a buyback offer that specifically excluded shares owned by Mesa. Mesa sued Unocal, but the Delaware Supreme Court ruled its actions legal.

In 1985, Mesa merged with the Pioneer Petroleum Company of Amarillo, **Texas**, to form the Mesa Limited Partnership, then the largest independent oil company in the world. In 1991, it changed its name to Mesa Inc. In 1987, Pickens published *Boone*, his autobiography. In later years, Pickens ran hedge funds investing primarily in oil and natural gas. He also became an outspoken advocate of natural gas as a vehicle fuel. In 1997, he formed the Pickens Fuel Corporation, reincorporated in 2001 as Clean Energy, which owns and operates natural gas vehicle fueling stations on the western coasts of the United States and **Canada**. In 2007, he traveled to **China** to promote the idea there. Pickens expanded into other forms of **alternative energy**. In June 2007, he announced that he would build the world's largest wind farm, in the Texas panhandle. In 2008 Pickens formalized many of his ideas on energy policy in the *Pickens Plan*, which advocated the adoption of various forms of alternative energy and the conversion of commercial transportation from liquid fuels to natural gas.

PIPELINES. Pipelines constitute an economical method for transporting petroleum and petroleum products over long distances. Pipeline transport can be up to 15 times cheaper than truck, 12 times cheaper than rail, and cheaper than barge, though on average it is somewhat more expensive than **tanker**. It is also less energy intensive. A typical oil pipeline might consume about 450 Btu per ton-mile of transport, versus 540 for barges, 680 for rail, and 2,300 for truck (these are gross averages and there can be wide variation).

Petroleum pipelines have a long history. Bamboo pipes wrapped with waxed cloth were used to transport **natural gas** for lighting in **China** as early as 400 BCE. Cast-iron pipes were used in the 18th century to transport water, sewage, and gas. The introduction of steel pipe in the 19th century was a major innovation, greatly increasing the strength of pipes. The first long-distance crude oil pipeline was built in **Pennsylvania** in 1879, 20 years after the **Edwin Drake** well. It was 6 inches in diameter and 109 miles long, running from the **Bradford** oil fields to Allentown. In 1886, an 8-inch, 87-mile natural gas pipeline was built from Kane, Pennsylvania, to Buffalo, New York. In tsarist **Russia**, the 522-mile, 8-inch **Baku–Batumi** kerosene pipeline was completed in 1906. In 1930, the Soviet Union completed a crude oil pipeline between the two cities. The 1930 pipeline incorporated a number of ideas from the 1870s developed by Russian engineer Sigismund Wojslaw. For example, it eliminated intermediate tanks at pumping stations, a source of loss of light hydrocarbons.

Pipelines built before 1920 had to be created by threading together pipe segments, which was difficult to accomplish for large structures. Such threaded pipes also had a tendency to leak under pressure. The introduction of oxyacetylene welding in 1920 and electric arc welding later in the decade made it possible to build large, high-pressure pipelines that did not leak. The 1920s also saw the introduction of large, seamless steel pipe segments. After 1950, other major innovations included cathodic protection to reduce corrosion and the use of large side booms to lay pipe. Nondestructive inspection techniques, such as using X-rays to detect flaws in welding, were also developed. Another development was the use of mechanical "pigs" to clean pipes on the inside and to perform inspections. **Drilling** technology also advanced. Directional drilling allowed pipelines to be laid under rivers and other obstacles without the need for digging long trenches.

Pipeline construction in the **United States** accelerated in the 1920s. A 250-mile pipeline was built between the **Texas** panhandle and Wichita, Kansas, in 1927. It was later extended to Kansas City, Missouri. In 1928, a pipeline was built from **Oklahoma** to Chicago. During **World War II**, the fear of enemy attacks on tankers motivated the construction of the **Big Inch and Little Big Inch** Pipelines to move crude oil and petroleum products respectively from the petroleum regions of the southwestern United States to the East Coast.

The construction of Big Inch and Little Big Inch, which were later privatized, heralded the modern era of long-distance pipelines in the United States. A notable project of the 1960s was the **Colonial Pipeline**, a large product pipeline from Houston to New York built largely to counter the power of the maritime union. Perhaps the most famous modern American pipeline is the 800-mile **Trans-Alaska Pipeline**. Its construction in **Alaska**'s hostile arctic environment represented a significant engineering achievement. The pipeline was completed in 1977 and runs from the oil fields of the North Slope to the ice-free port of **Valdez** in southern Alaska. The United States has the most developed petroleum pipeline network in the world, with about 85,000 miles of oil pipeline, 92,000 miles of product pipeline, and 1.4 million miles of gas pipeline in 2000.

Pipeline construction proceeded rapidly in the Soviet Union in the postwar era. Motivated by the need to exploit the distant gas and oil fields of **Siberia**, the Soviets began building a large gas pipeline system in the 1960s. As in Alaska, hostile conditions were encountered. Within three decades, the system consisted of 126,000 miles of trunklines with diameters of 40–55 inches. The former Soviet Union also has about 30,000 miles of oil pipelines. The **Druzhba Pipeline** from the Urals to Eastern Europe, first operated in 1964 and extended in 1970, is the longest pipeline in the world, at 2,500 miles, with a capacity of 1.2–1.4 Mmbpd.

Canada has a number of major pipelines, such as the Interprovincial Pipeline, transporting oil 2,000 miles from Manitoba and Saskatchewan to Ontario. The Interprovincial has a diameter of up to 34 inches. The 825-mile Intermountain Pipeline links oil from the area of **Edmonton**, **Alberta**, to refineries in British Columbia and neighboring Washington State in the United States. The 36-inch Trans-Canada Pipeline transports gas 2,300 miles from Alberta to Montreal.

Middle Eastern oil producers transport all their inland oil destined for export via pipeline to seaports. The **Trans-Arabian Pipeline** (Tapline) is an important early example of a major pipeline in the region. It was completed in 1950 and connects oil fields of the Eastern Province of **Saudi Arabia** to the Lebanese Mediterranean port of Sidon, over a distance of more than

1,000 miles. **Iraq** has a 1,055-mile pipeline to the Turkish port of Iskenderun and another pipeline to the Turkish port of Yurmurtalik (the **Kirkuk–Ceyhan Pipeline**).

A notable recent major pipeline is the **Baku–Tbilisi–Ceyhan (BTC) Pipeline** completed in 2005. The pipeline was built to transport **Caspian** oil to Western markets without transiting Iranian or Russian territory and avoiding the Black Sea and the Bosporus. It is 1,099 miles long and has a maximum capacity of about 1 Mmbpd.

There are also major submarine pipelines, which are generally more expensive and difficult to construct than overland pipelines. An important early one was **PLUTO** (Pipeline Under the Ocean), which ran under the English Channel to supply Allied armies in continental Europe during World War II. The **North Sea** has major submarine pipelines. The Brent System Pipeline, connecting British North Sea oil fields (including **Brent** and Cormorant) to **Sullom Voe** in the Shetland Islands, is 92 miles long and 36 inches in diameter, lies under 525 feet of water, and transports about 1 Mmbpd. The 740-mile Langeled Pipeline began supplying the **United Kingdom** with gas from **Norway**'s Ormen Lange gas field in October 2007. It has a diameter of 43 inches and is the world's longest subsea petroleum pipeline.

PITCH LAKE. Natural asphalt lake in La Brea, Trinidad and Tobago; the world's largest commercial asphalt deposit. Pitch Lake covers an area of about 100 acres and contains an estimated 10 million tons of pitch in solid emulsified form. The refined product, Trinidad Lake asphalt, is exported and used for road construction, insulation, and other applications. The deposit was known to the indigenous Amerindians. The first known European discovery was made by Walter Raleigh in February 1595. The asphalt was not widely used commercially until after 1851, when British admiral Thomas Cochrane patented a process for using the asphalt in construction and other applications. The asphalt is extracted and refined by the Trinidad Lake Asphalt Company.

PLAN BUCCANEER. A plan conceived in the **United Kingdom** for military intervention in **Iran** in the wake of that country's 1951 nationalization of its petroleum industry. The **Anglo–Iranian Oil Company** (AIOC), of which the British government was majority owner, stood to lose most of its assets and future income. The plan involved seizing and holding **Abadan**, site of the world's largest refinery. The land, sea, and air forces needed were deployed in a state of readiness. The **United States** was alarmed, fearing that such intervention might drive Iran into the arms of the Soviet Union. In the end, the plan was not executed. The forces stood down on 4 October 1951. The United States sent Averell Harriman to speak with Iran's prime minister,

Mohammed Mossadegh, and mediate in the situation. This proved unsuccessful. The crisis continued, and the United States eventually engineered a coup to overthrow Mossadegh. *See also* AJAX, OPERATION (1953).

PLOIEŞTI (PLOEŞTI). City (pop. 230,000, 2002) in **Romania** and a group of nearby oil fields and installations that played a key role in supplying Germany with oil in both world wars. Ploieşti, situated about 35 miles north of Bucharest, was one of the world's major petroleum processing centers in the late 19th and early 20th centuries. The Ploieşti oil fields tend to occur along lines of uplift related to salt movements. The major fields are Moreni, Boldesti, Ceptura, and Ochiuri. The most important of these, Moreni, was discovered in 1907 by Romăno–Americană, a wholly owned subsidiary of **Standard Oil**. It has an **EUR** of 0.8 Bb, of which most has now been consumed. The Ploieşti fields accounted for 99 percent of Romanian output before 1938, although by the end of the 1970s the proportion had dropped to around 30 percent.

Germany invaded Romania in 1916, during **World War I**, to gain control of its oil resources and processing facilities, most of which were located around Ploieşti. The Allies managed to sabotage the oil fields and facilities, seriously crippling production for a time. In **World War II**, **Winston Churchill** labeled Ploieşti the "taproot of German might." The area supplied Germany with about 60 percent of its crude oil. Ploieşti boasted some 40 refineries, producing nearly 3.5 Mmb of refined fuel annually. The Soviet Union bombed Ploieşti on 23 June 1941, and the **United States** did so on 12 June 1942.

The major air raid, however, took place on 1 August 1943. Code-named "Tidal Wave," it involved 178 bombers of the U.S. Army Air Force (USAAF) taking off from Benghazi, **Libya**. The bombers flew 1,350 miles to reach their target, the farthest air raid ever attempted up to that time. The attackers used new low-flying tactics. But the Germans, who had been intercepting and decoding USAAF communications, were ready. The United States lost 54 bombers, and more than 500 crew members were killed, wounded, or captured. About 42 percent of Ploieşti's production capacity was destroyed, but not all capacity was being used. Within a few weeks, Ploieşti was producing more than it had before the raid. In April 1944, the USAAF began raiding Ploieşti from Italy, ultimately destroying most of its facilities.

After the war, the communist regime in Romania nationalized the oil industry and invested heavily in it, repairing the war damage in Ploieşti and elsewhere. The postcommunist government began allowing private investment in the 1990s. Ploieşti continues to be the hub of Romania's petroleum industry and a major downstream center for southeastern Europe.

PLOWSHARE PROGRAM. *See* NUCLEAR EXPLOSION STIMULATION.

PLUTO (PIPELINE UNDER THE OCEAN). Submarine **pipeline** laid under the English Channel to supply Allied forces in continental Europe with fuel during **World War II**. Two types of pipeline were created for PLUTO: HAIS (Hartley/**Anglo–Iranian**/Siemens) and HAMEL (Hammick and Ellis). HAIS was a flexible pipe with a 3-inch diameter lead core, and HAMEL was a 3-inch steel pipe. Both types were tested in May 1942 and passed. The first HAIS pipe was laid on 12 August 1944 from the Isle of Wight in the **United Kingdom** to Cherbourg in France, to be followed by another HAIS and two HAMELs. The early pipelines mostly failed, the first one being broken by a ship's anchor, but eventually 17 pipes were put into successful operation (11 HAIS and 6 HAMEL). These ran from Dungeness to Ambleteuse. The pipelines were fed by an extensive pipeline network in the United Kingdom and were linked to camouflaged pumping stations on the southern coast. They were extended overland in continental Europe as Allied forces moved forward. The PLUTO pipelines delivered nearly 24,000 bpd of fuel. PLUTO helped pave the way for the development of the **offshore** industry after the war.

POLAND. Oil production 2015: 0.018 Mmbpd (0.023 percent of world); **reserves** 0.138 Bb (0.01 percent of world); gas production: 0.146 Tcf (0.12 percent of world); reserves 2.884 Tcf (0.04 percent of world). Poland is estimated to have 145.8 Tcf in **unconventional** technically recoverable **shale gas** resources, as yet unexploited.

The **Galicia** region, the locus of one of the world's earliest modern petroleum industries in the 19th and early 20th centuries, was once part of the Austro-Hungarian Empire; it passed to Poland after **World War I** and now lies in both Poland and Ukraine. The modern industry originated in 1853, when Jan Zeh, **Ignacy Łukasiewicz**, and others developed methods for **refining** crude oil into an illumination oil similar to kerosene.

PORT HARCOURT. City in **Nigeria** (pop. 1 million, 2006) and a center of the Nigerian petroleum industry. Port Harcourt, capital of the Nigerian state of Rivers, is located about 40 miles from the sea, on the Bonny River, with many of Nigeria's giant oil and **natural gas** fields lying within 50 miles. Nigeria's first oil refinery, completed in 1965, is at Alesa-Eleme, about 12 miles to the southeast. Oil and natural gas are piped to Port Harcourt and to nearby **Bonny** for export. There is also a refinery in Port Harcourt. Refined products are sent to Makurdi in the Nigerian state of Benue.

POSTED PRICE. Official reference oil price, controlled by the **major oil companies** until the 1960s. The posted price was the price from which producer countries' taxes and royalties were computed. Cuts in the posted price in 1959 and 1960 by the major oil companies directly catalyzed the formation of the **Organization of the Petroleum Exporting Countries** (OPEC). After 1960, OPEC assumed increasing control of posted prices. Posted prices continued to play a role until the mid- to late 1980s, by which time market prices were the only ones that mattered. In theory, the posted price would reflect market prices, but in practice, gaps often arose. By the late 1950s, the posted price was higher than the market price, largely because of the appearance of Soviet oil on the international market. Since the producer countries' profit share was calculated on the posted price, their actual share based on the market price was higher. Instead of 50 percent, which had by then become typical because of the widespread adoption of **50–50 agreements**, it was closer to 60 percent or even 70 percent. The major international oil companies were absorbing the difference and wanted to cut the posted price. **BP** was the first to do so, cutting the posted price unilaterally by about 10 percent. The oil-producing countries were furious. On 19 August 1960, Standard Oil Company of New Jersey (later **Exxon**) cut its posted price unilaterally and without warning by about 7 percent, and other companies followed. OPEC was established on 14 September 1960.

PRE-SALT. A rock layer found in continental shelves in continents (such as South America and Africa) once joined together but broken apart by the extensional processes of plate tectonics. In particular the margins of what are now **Angola** and **Brazil** were joined when what is today the South Atlantic began to open in the Cretaceous period, about 140 million years ago. Brazil's Campos and Santos basins were once adjacent to Angola's Benguela and Kwanza basins. The "pre-salt" rock is overlain by the deposition of thick layers of evaporites, mostly salt. The salt layer acts as a seal trapping petroleum formed in the underlying sedimentary reservoirs. The pre-salt layer in the **deepwater** Brazilian **offshore**, generally carbonate, may contain technically recoverable resources of 50 Bb of oil. Exploration has been carried out in the Santos basin since 2006, with possibly 12 Bb of oil resources recoverable in existing finds. The deepwater pre-salt layer in Angola has been explored since 2011.

In the **Gulf of Mexico**, interesting new plays in the 21st century have tended to be subsalt rather than pre-salt. The distinction is that pre-salt layers are overlain by autochthonous salt, while subsalt reservoirs are overlain by allochthonous salt—that is, salt that has moved significantly from its locus of origin. Salt diapirs (salt domes) constitute an example.

Pre-salt exploration and development is costly. In the Brazilian case, for example, accessing the resources can mean drilling through more than 6,500 feet of rock and a similar thickness of salt, in water depths up to 10,000 feet.

PRICES. *See* FIRST OIL SHOCK; GENEVA AGREEMENTS; NETBACK PRICING; OIL COUNTERSHOCK; ORGANIZATION OF ARAB PETROLEUM EXPORTING COUNTRIES (OAPEC); ORGANIZATION OF THE PETROLEUM EXPORTING COUNTRIES (OPEC); POSTED PRICE; RAILROAD COMMISSION OF TEXAS; SECOND OIL SHOCK; TEHRAN AGREEMENT (1971); THIRD OIL SHOCK.

PRIMARY ENERGY. Energy obtained directly from a fundamental energy source, without conversion. For example, the direct burning of **natural gas** produces primary energy. Electricity generated by burning natural gas at a power plant represents secondary energy.

PRODUCTION-SHARING AGREEMENT (PSA). A type of agreement, with widely varying terms, between a petroleum company and a host government. The agreement often takes the form of a joint venture with a national oil company. The petroleum company typically pays costs and recoups those costs by selling "cost oil," a pre-agreed portion of the production. The sale of the remaining production, or "profit oil," generates profits that are split between the company and the government according to pre-agreed terms. *See also* CONCESSION.

PRUDHOE BAY OIL FIELD. Supergiant oil field discovered in 1967 on **Alaska**'s North Slope. Prudhoe Bay has an **EUR** of about 15 Bb and is the largest oil field discovered to date in North America. By the end of 2002, it had already produced 10.7 Bb. Prudhoe Bay produced at a rate of just over 1.5 Mmbpd between 1980 and 1986. Its oil, by reducing the import requirements of the **United States**, contributed to the diminished power of the **Organization of the Petroleum Exporting Countries** (OPEC) in the 1980s (*see* OIL COUNTERSHOCK). Prudhoe Bay production peaked at about 1.6 Mmbpd in 1987 and then began to decline. By 2002, production was about 0.415 Mmbpd, not very far above 1977's level and below 1978's.

Exploration in the North Slope of Alaska began in the early 1960s. The first company to drill was Colorado Oil and Gas in 1963, using a small rig left behind by the **United States Geological Survey**. **BP** and **Sinclair** were early believers in the North Slope. They formed a partnership and explored intensively. BP identified a large coastal arch from Colville in the west to Prudhoe Bay in the east and considered it very promising for a large oil field or even an entire oil province. Alaska opened the western half of the area for bidding in December 1964, and BP/Sinclair acquired a major part of it. In July 1965, the state opened a second round covering the Prudhoe Bay structure in the eastern part. BP thought this part even more promising, but Sinclair was discouraged by a series of dry holes drilled in the foothills to the south and pulled out, leaving BP to bid alone.

BP lost much of the acreage to a team of the **Humble Oil and Refining Company** (later **Exxon**) and Richfield Oil (which became part of **Arco** in 1966). BP drilled in the Colville structure to the west in 1965 and 1966 and found only small, uncommercial quantities. Arco and Humble drilled a dry hole in January 1967. At this point, BP was ready to give up. However, Arco decided to drill one more well, and it struck oil. A second well, drilled seven miles away in 1968, confirmed a huge find. BP drilled three miles south of the discovery well and realized that it had a major share in the field. It used its Prudhoe Bay acreage as currency to buy a controlling interest in **Sohio**, which had a downstream infrastructure for the oil in the lower 48 states. A consortium of companies built the **Trans-Alaska Pipeline** to transport Prudhoe Bay oil to the ice-free port of **Valdez** on Alaska's south coast for loading onto **tankers** bound mostly for the lower 48 states. *See also* ANDERSON, ROBERT O. (1917–2007).

PRYADUNOV, FYODOR. *See* TIMAN–PECHORA.

PULIDO PULIDO, MANUEL ANTONIO. *See* VENEZUELA.

PURE OIL COMPANY. Independent petroleum company founded in 1891; later part of **Union Oil Company of California** and eventually **Chevron**. Operating mostly in **Texas**, the **Gulf of Mexico**, and the U.S. Midwest, the Pure Oil Company originated in the Producers' Protective Association, formed in 1887 by independent producers in **Pennsylvania** to fight the overwhelming market power of **Standard Oil**. In 1891, the parties incorporated the Producers Oil Company Ltd., which was renamed the Pure Oil Company in 1895. To prevent Standard Oil from gaining control of their company, the founders created bylaws stating that no new shareholders could vote their shares without first being elected members of the Producers' Protective Association.

The company organized the U.S. Pipeline Company to build a **pipeline** from the oil fields of western Pennsylvania to the Delaware River. The pipeline was finished in 1901, overcoming opposition from Standard Oil. That year, the first Pure Oil refinery was built at Marcus Hook, Pennsylvania, with a 3,000 bpd capacity. In 1917, Pure Oil was acquired by the Ohio Cities Gas Company headed by Beman Dawes. Dawes had struck oil in **West Virginia** in 1914 while drilling for **natural gas** and believed Pure Oil could complement his operations. He adopted the Pure Oil name for his entire company in 1920. In 1924, Pure Oil bought oil fields in Texas developed by Albert Humphreys and became that state's third-largest producer. That year, Beman Dawes sold off his gas interests, making Pure Oil an oil company only, and retired. His younger brother Henry became president. In 1926, Pure Oil moved from the old Ohio Cities headquarters in Columbus, Ohio, to the Pure Oil building in Chicago.

In 1929, Pure Oil discovered the **giant** Van field in Van Zandt County, Texas, about 60 miles east of Dallas. Van had an **EUR** of 550 Mmbo, most of which has since been produced. Van's peak production was about 31,780 bopd in 1944; 50 years later, it produced about 1,700 bopd. Pure Oil owned 76.73 percent of the field, with the remaining interest owned by **Humble Oil and Refining Company** (later **Exxon**), **Royal Dutch Shell**, **Sun Oil**, and the Texas Corporation (the Texas Company, later **Texaco**). Pure Oil signed the Van Field Unit Plan of Operation with the other four companies. Under the agreement, Pure Oil would operate the entire field, charging each party a proportionate share of the costs and paying a proportionate share of the proceeds.

Pure Oil went through a difficult period in the Great Depression, including its first and only loss of $885,000 in 1934. It recovered quickly, earning $8.1 million the next year. In 1937, it struck oil at the Bunyan Travis field in southern Illinois, touching off a local boom. A year later, it moved **offshore** and discovered the Creole field in the Gulf of Mexico. This was one of the first offshore wells in the Gulf and was followed quickly by the discovery of the Cumberland field in **Oklahoma** in 1940. During **World War II**, Pure Oil manufactured 100-octane aviation fuel (*see* GASOLINE), toluene, and bases for synthetic rubber for the war effort. It lost three **tankers** to German submarine attacks.

After the war, Pure Oil expanded its exploration efforts in the Gulf of Mexico, discovering the Eugene Island gas field in 1949. The next year, it opened a new research and development laboratory in Illinois. In 1957, it discovered oil in Lake **Maracaibo**, **Venezuela**, with Signal Oil and **Sohio** as partners. In 1957, it ventured into Paraguay with **Sinclair**, **Gulf Oil**, and other partners; it discovered gas but abandoned the effort because the remote location made it commercially impractical. In the early 1960s, Pure Oil continued to discover gas in Texas and New Mexico, and it acquired Woodley Petroleum (1960). In 1965, after fighting off hostile takeovers, Pure Oil Company merged into the Union Oil Company of California via an exchange of stock.

Q

QADDAFI, MUAMMAR. *See* LIBYA.

QATAR. Oil production 2015: 0.66 Mmbpd (0.842 percent of world); **reserves** 25.244 Bb (1.53 percent of world); gas production: 6.315 Tcf (5.06 percent of world); reserves 858.1 Tcf (12.44 percent of world). Qatar joined the **Organization of the Petroleum Exporting Countries** (OPEC) in 1961 and the **Organization of Arab Petroleum Exporting Countries** (OAPEC) in 1970. Qatar is an oil exporter and the world's largest exporter of **liquefied natural gas**. It is a member of the **Gas Exporting Countries Forum**.

The Anglo–Persian Oil Company (later **Anglo–Iranian Oil Company**, still later **BP**) obtained a petroleum **concession** in Qatar in 1925 but was unsuccessful in exploiting it. The concession was transferred in 1935 to Petroleum Development (Qatar), a joint venture with the same ownership structure as the **Iraq Petroleum Company** (IPC). APOC was a partner in IPC, and Qatar was within the territory covered by the **Red Line Agreement**. In 1940, Petroleum Development (Qatar) discovered the **giant** Dukhan oil field. Dukhan is Qatar's largest oil field, with an **EUR** of about 1.6 Bb and 9 Tcf of **natural gas**. **World War II** interrupted operations. Production did not begin until 1949, at a level of about 2,000 bpd. It grew rapidly, surpassing 0.1 Mmbpd by 1955.

The concession held by Petroleum Development (Qatar), renamed Qatar Petroleum Company in 1963, covered the country's onshore area. In 1952, **Royal Dutch Shell** obtained an **offshore** concession. The offshore was to prove very fruitful, particularly for natural gas. Shell discovered the Idd al-Shargi field in 1960, Mydan-Mahzam in 1963, and Bu al-Hanin in 1970. All three are giants. Idd al-Shargi is an oil giant with an EUR of about 0.6 Bb. The other two are predominantly gas giants with significant amounts of oil. Their combined EUR is 8 Tcf gas and 0.635 Bb. Shell's most important discovery by far was in 1971. The **North Field**, with an EUR of 900 Tcf of gas and about 26.1 Bb of condensate, for a total of about 176.1 Bboe, is the world's largest gas field and the largest single hydrocarbon deposit ever discovered, larger than **Ghawar** in energy-equivalent terms.

Between 1974 and 1976, the government of Qatar purchased the Qatar Petroleum Company in two phases. It formed its national oil company, the Qatar General Petroleum Company (QGPC), in 1974. QGPC acquired Shell's interests in 1977, completing the nationalization of the sector. In 1984, the government officially decided to develop the North Field and established Qatargas. Production began on 3 September 1991, on the 20th anniversary of Qatar's independence. Liquefied natural gas exports began in 1997.

Qatar's oil production peaked in 1973 at 0.57 Mmbpd. It hovered between 0.43 and 0.5 Mmbpd until 1979, then dipped to a low of 0.295 Mmbpd in 1983. It hovered around 0.3 and 0.4 Mmbpd until 1990, then grew unevenly but firmly, reaching an average level of 1.545 Mmbpd for the years 2011–2016. Gas production was 0.184 Tcf in 1980 and stayed almost flat until 1989, after which it grew rapidly for the most part to its present high levels. *See also* GULF COOPERATION COUNCIL (GCC).

R

RAILROAD COMMISSION OF TEXAS. Texas agency responsible for regulating petroleum production in the state. The Railroad Commission of Texas strongly influenced oil supply and prices in the **United States** during the 20th century and formed a model for the **Organization of the Petroleum Exporting Countries** (OPEC). The original charter of the commission was to regulate railroads in Texas. It was created in 1891 during the administration of Governor James S. Hogg. In 1917, the Texas legislature expanded the commission's duties to the oversight of petroleum **pipelines** to ensure that they remained common carriers. This gave the commission a toehold in the petroleum industry. In 1919, it acquired responsibility for regulating the spacing of oil wells, and in the 1920s it began to oversee **natural gas** utilities.

The Railroad Commission's crucial petroleum oversight role was established in the early 1930s, when unrestrained production from the new **East Texas oil field** caused a serious market glut and a collapse in prices. At first, the commission had difficulty in attempting to regulate output because its authority was limited. It was permitted to regulate output in order to prevent physical wastage but had no authority to do so in order to stabilize prices. Overproduction can damage a field, but it was not always easy for the commission to prove that it was regulating to prevent premature exhaustion of a giant field rather than to stabilize the market.

Nevertheless, the commission issued its first proration order for the East Texas **supergiant** on 4 April 1931, reducing overall production from its prevailing level of about 200,000 bpd to 50,000 bpd. Producers largely disregarded the order, and by that summer total output was around 900,000 bpd. By 14 July 1931, crude prices were 13 cents a barrel; they had averaged $1.27 in 1929. The reservoir pressure in the field was also reduced. In a special session that month called by Governor **Ross Sterling**, the legislature introduced a bill to limit production in the field in order to prevent damage to the reservoir and stabilize the price of oil. This initiative was stopped on 28 July, when a federal court struck down the Railroad Commission's original 4 April proration order on the grounds that it had been an attempt to interfere in the market for the purpose of stabilizing prices. The court order prompted the Texas legislature to reverse itself and pass a law specifically prohibiting the Railroad Commission from regulating production for economic reasons.

In the Texas oil industry, there was a split between those who wanted to restrain production and stabilize prices, and those—mostly smaller **independents**—who wanted to keep producing and let the market sort itself out. Responding to the appeals of the former group, Governor Sterling ordered the National Guard and the Texas Rangers into the field on 17 August 1931 to shut it down and impose martial law. On 5 September 1931, production resumed under a new order with a limit of 400,000 bpd and a 225 bpd limit per well. Some producers sued and others ignored the order, smuggling out **hot oil** (illegally produced oil) despite the military presence. In 1932, Governor Sterling appointed **Ernest O. Thompson** to the commission. Thompson favored regulation for orderly production and price stability, and he became the leading member of the commission in the turbulent years that were to follow.

On 2 February 1932, a federal court ordered the rescission of martial law in the East Texas field. The Railroad Commission continued to issue proration orders, and federal courts continued to invalidate them. On 12 November 1932, Governor Sterling again called a special session of the legislature, which again passed a bill allowing the Railroad Commission to regulate production to stabilize the market. The legislature also made hot oil production a felony. Nevertheless, production for 1932 still averaged more than 400,000 bpd. On 6 April 1933, the commission shut down the field again. It reopened on 24 April with an allowable production set at 750,000 bpd. This was based on reservoir pressure and was too high to stabilize prices, especially given that hot oil was also still being produced. Prices collapsed to 10 cents a barrel and even lower.

In 1933, the federal government stepped in. The National Industrial Recovery Act (NIRA) of 16 June 1933 contained a provision prohibiting the interstate transportation of petroleum produced in excess of state quotas. On 14 July 1933, President Franklin D. Roosevelt signed an executive order enforcing the regulation of petroleum production. Secretary of the Interior **Harold Ickes** sent in teams of federal inspectors to enforce the order. By the end of November 1934, some progress had been made. A setback for the regulators emerged in January 1935, when the U.S. Supreme Court struck down the NIRA provision outlawing hot oil. The U.S. Congress responded by passing the **Connally Hot Oil Act** on 22 February 1935. This achieved the same end in a slightly different way. It gave the president power to regulate interstate flow of petroleum and petroleum products, and to require certificates of clearance for petroleum products in interstate commerce. The Texas legislature enacted the Texas Hot Oil Statute, declaring hot oil as contraband and authorizing its confiscation. The Railroad Commission went on to regulate production both for the purposes of reservoir conservation and to maintain a stable market. It regularly shut the field down for as many as 14 days a month.

The many small independent oil producers who owned a substantial portion of the East Texas wells tended to oppose the philosophy of prorationing output based on production potential, because it favored major producers. However, it eventually became apparent that the Railroad Commission was adopting policies that were advantageous to small independents. For example, the commission was permitting a narrow spacing of wells and had also set prorationing allocation on a per-well basis. The small producers eventually elected to work within the system. The concessions to the small producers, as well as the commission's practice of allowing flaring of gas, would now be considered wasteful, although at the time this was not as obvious. In any case, the Railroad Commission had to balance many competing considerations and could not always make what might seem to be the least wasteful policy choices. The major producers and those worried about preserving reservoir health still got most of what they wanted, which was relative price stability and more orderly production.

The Railroad Commission successfully regulated Texas petroleum production in the mid-20th century. Because of Texas's importance to U.S. production, and the importance of the United States in the global market, the Railroad Commission was an agency with international influence. **Venezuela**'s oil minister **Juan Pablo Pérez Alfonso**, one of the founding figures of OPEC, had spent considerable time studying the Texas Railroad Commission's activities, which formed a model for OPEC's operation. By the 1970s, the rise of OPEC and the decline of Texas oil **reserves** had eliminated the commission's influence over the broader oil market, although this has perhaps increased again somewhat in recent years with the turnaround in Texas production in the 2000s and 2010s. The commission continues to enforce environmental standards in Texas oil fields and maintain equity among Texas producers. It also continues to play an important role in regulating the state's natural gas utilities.

RANGOON OIL COMPANY. *See* BURMAH OIL COMPANY.

RAS AL-KHAIMA. One of the constituent emirates of the **United Arab Emirates** (UAE). Ras al-Khaima has only about 100 Mmb, or roughly 0.1 percent, of the UAE's 97.8 Bb of oil **reserves** (2015). Oil was first struck in Ras al-Khaima in 1984. By the late 1980s, production was still under 12,000 bpd. In 1983, a small **natural gas** field, Saleh, was discovered about 26 miles **offshore** in the **Persian Gulf**. There is a long-standing dispute between Ras al-Khaima and **Iran** over the sovereignty of the Tunb Islands, now held by

Iran. The Tunbs are important because of their location near the **Strait of Hormuz**, the only maritime outlet for Persian Gulf oil. The dispute is a matter of national pride in Iran and throughout the Arab world. When **Libya** nationalized **BP**'s interests in 1973, it cited as a cause the "complicity" of the **United Kingdom**, the former colonial power, in delivering the islands to Iran. *See also* SHARJAH.

RAS TANURA. Important Saudi Arabian oil port on the **Persian Gulf**, developed by **Aramco** beginning in the 1940s. Ras Tanura is in **Saudi Arabia**'s Eastern Province, north of the Dammam/**Dhahran**/Khobar metropolitan area. It has storage tanks and a major refinery and can accommodate the largest **tankers**. It is a major **pipeline** terminal. Ras Tanura's oil refinery was once supposedly the largest in the world, although it was overtaken by Iran's **Abadan** refinery complex and later by several others, including **Venezuela**'s Centro de Refinación de Paraguaná (CRP) complex, South Korea's Uslan, and **India**'s Jamnagar.

RATHBONE, MONROE J. (1900–1976). President of the Standard Oil Company of New Jersey (later **Exxon**) 1960–1965; initiated cuts in the **posted price** of oil in 1960 that helped precipitate the creation of the **Organization of the Petroleum Exporting Countries** (OPEC). Monroe Rathbone was born in **West Virginia**, the son and nephew of refiners for **Standard Oil**. He studied chemical engineering at Lehigh University and began working at Standard Oil Company of New Jersey's Baton Rouge refinery after **World War I**. He rose rapidly through the ranks of the company, becoming the chief executive in 1960. He was a technically competent and decisive leader, but he did not fully understand the motivations and psychology of the governments of Middle Eastern–producing countries. Rathbone made a fateful decision, executed on 9 August 1960, to cut the posted price of his company's crude oil by 7 percent without consulting the producing countries. There was a glut in the global oil market, and he wanted to gain market share. A cut in the posted price also lowered his company's payments to producing countries, aligning those payments more with market prices. Other **major oil companies** reluctantly followed suit.

Rathbone had seriously underestimated the vehement adverse and nationalistic reaction in producing countries. Their revenues decreased when posted prices were cut. The price cuts initiated by Rathbone accelerated the formal creation of OPEC just over a month later, on 14 September 1960. The new assertiveness of Middle Eastern oil-producing countries spurred Rathbone to make Standard Oil Company of New Jersey less dependent on the region. He diversified the company's sources and is credited with making the company truly international.

RED LINE AGREEMENT (1928–1948). Part of the 31 July 1928 reorganization of the Turkish Petroleum Company (TPC, renamed **Iraq Petroleum Company** in 1929). The agreement was meant to govern the activities of the TPC partners in the former Ottoman Empire. It supposedly received its name because during the negotiations no one seemed entirely sure of where the actual borders lay, and so the entrepreneur **Calouste Gulbenkian** took a red pencil and drew a line representing them. The area enclosed by the Red Line included what are now **Iraq** and **Saudi Arabia**, but not **Iran** or **Kuwait**.

The agreement included a clause prohibiting the principals of TPC from independently seeking additional **concessions** within the Red Line. Significantly, the clause prevented the partners from seeking independent concessions in Saudi Arabia. Such a concession was then obtained by a non–Red Line company, Standard Oil of California (Socal, later **Chevron**). Socal made significant discoveries in Saudi Arabia through its Casoc subsidiary (later **Aramco**), half of which was purchased in 1936 by the Texas Corporation (later **Texaco**). In 2001, Chevron and Texaco would merge to form ChevronTexaco, later known simply as Chevron.

The American members of the Iraq Petroleum Company (IPC) consortium that were bound by the Red Line Agreement became progressively displeased with its terms. However, it was not until November 1948 that the anticompetitive provisions of the Red Line Agreement were abrogated, after a long and complex process of legal maneuvering and negotiation. The way was cleared for Standard Oil Company of New Jersey (later **Exxon**) and Socony–Vacuum (later **Mobil**) to become partners in Aramco.

REFINING. The processes of extracting useful products from crude oil, including various types of fuel, lubricating oil, and raw materials for **petrochemicals**. Refining in the 19th century focused on distilling crude oil to obtain kerosene for illumination applications. Early refining operations began in **Russia**, the **United States**, **Canada**, **Romania**, and Austro-Hungarian **Galicia** in the early and middle part of the 19th century. **Standard Oil** had a virtual monopoly on refining in the United States in the latter part of the century. Standard Oil also adopted and applied the **Frasch** process for removing sulfur from crude oil, enabling it to make competitive products from the relatively cheap, low-quality oil in the **Lima–Indiana oil field**.

The invention and diffusion of the automobile changed the focus of refining operations. **Gasoline**, once considered a waste product of the fractional distillation of crude oil, became the most desirable product. A major leap in refining technology occurred with the development of thermal cracking to increase the yield of gasoline from a given quantity of crude oil. **William Burton** of the Standard Oil Company of Indiana (later **Amoco**) patented a thermal cracking process in 1913 that saw widespread use. **Vladimir Shukhov** had developed a thermal cracking process in Russia two decades earlier, but his process was not immediately applied to increase gasoline yields. Another important milestone in refining came in 1936, when **Eugene Houdry**'s catalytic cracking process, which increased both the yield and quality of gasoline, began commercial production.

In 2017, there were about 700 refineries operating worldwide, with a total throughput of more than 80 Mmbpd. The largest single refineries were Reliance Jamnagar refinery in **India** (1.240 Mmbpd), the Paraguana Refining Center in **Venezuela** (0.940 Mmbpd), the Uslan refinery in South Korea (0.820 Mmbpd), the Ruwais Refinery in the United Arab Emirates (0.817 Mmbpd), and the GS **Caltex** refinery in South Korea (0.73 Mmbpd). **Iran**'s **Abadan** refinery, operating since 1912 and once the world's largest, was 13th. The largest refining companies in 2017 included **ExxonMobil**, **Royal Dutch Shell**, **BP**, **ConocoPhillips**, **PdVSA**, **Total**, PetroChina, and **Saudi Aramco**. Profit margins from refining operations are often lower than those from upstream operations, especially during periods of high oil prices.

RÉGIE AUTONOME DES PÉTROLES (RAP). *See* ELF AQUITAINE.

REPSOL. Spanish petroleum company formed in 1987 by the merger of several state-controlled companies; later Repsol YPF, then again Repsol. In 1981, Spain created the Instituto Nacional de Hidrocarburos (INH) as a holding company for the state's various activities in the petroleum sector, which were mostly downstream, given Spain's relatively small endowment of hydrocarbon resources. When Repsol was formed in 1987, it took over INH's operating divisions; INH then owned Repsol for the Spanish government. However, within two years of forming it, Spain began to privatize Repsol. In 1989, it sold off about a third of the company in a successful public offering. Three further share issues in 1993, 1995, and 1996 reduced the government's stake to about 10 percent. By 1997, the company was fully privatized.

Repsol was Spain's largest industrial enterprise. It operated five refineries in Spain and had 2,900 **gasoline** stations in Spain, France, and Portugal. It also held 91 percent of Enagas, Spain's former national **natural gas** company. In the mid-1990s, it expanded significantly in Latin America, spending about $3 billion. It acquired gasoline stations in Peru and **Ecuador**. In 1996, it acquired 38 percent of Astra CAPSA, an oil and gas company in **Argentina** with both upstream and downstream operations. In 1997, it acquired a controlling interest in PlusPetrol, which had service stations and gas assets in Argentina.

In January 1999, for $2 billion, Repsol acquired the Argentine government's remaining 15 percent share in Yacimientos Petrolíferos Fiscales (YPF), the national oil company whose privatization had begun in 1992. It proceeded to buy up the remainder of the company, and by June 1999 it had acquired 97 percent of the stock for $13.4 billion. Repsol became Repsol YPF, an international integrated petroleum company. The acquisition of YPF increased Repsol's oil **reserves** by 236 percent (to 2.15 Bb) and its gas reserves by 613 percent (to 14.31 Tcf). Oil production increased 122 percent (to 0.451 Mmbpd), and gas production increased 386 percent (to 1.299 Bcfpd). The company invested aggressively around the world, particularly in Latin America. However, oil and gas reserves showed a significant net decline over the next seven years. In 2016, Repsol YPF controlled 0.584 Bb in proved oil reserves and 10.094 Tcf of gas. It produced oil and gas at the rate of 0.243 Mmbpd and 2.51 Bcfpd respectively.

In 2012 Argentina under President Cristina Fernández de Kirchner expropriated 51 percent of Repsol's shares of YPF. This action was condemned by both the United States and the European Union. Repsol sued for $10.5 billion. In 2013 Argentina offered as compensation $5 billion and a small share in the Vaca Muerta **tight oil** and **shale gas** field that Repsol had discovered in 2010. Repsol rejected the offer. However on 25 February 2014 Repsol accepted an offer of $5 billion in Argentine government bonds.

In 2015, Repsol acquired Talisman Energy of **Canada** for $8.3 billion. Talisman, once known as **BP** Canada, had operations all over the world, including Canada, the **United States**, **Colombia**, **Algeria**, the **UK**, **Norway**, East Asia, and the Middle East. By acquiring Talisman, Repsol gained 142 Mmb of oil reserves and 2.09 Tcf of gas.

Beginning in 2007, Repsol became a significant player in the exploration and development of **Brazil**'s **pre-salt** resources, through Repsol-Sinopec, its joint venture with Addax Petroleum UK Limited (which is a subsidiary of Sinopec, the China Petrochemical Corporation). In 2017 Repsol announced Horseshoe, a 1.2 Bb oil discovery in **Alaska**'s North Slope, possibly the largest onshore find in the United States in three decades.

REPSOL YPF. *See* ARGENTINA; REPSOL.

REQUA, MARK L. *See* FUEL ADMINISTRATION.

RESERVE GROWTH. Upward revisions of **reserve** estimates. There are a number of causes of reserve growth. One is that when a new field is discovered, its full extent may not be known. Another is that there is a natural tendency to be conservative in initial reporting, especially for publicly traded petroleum companies that are subject to the oversight of securities agencies. A third reason is that improvements in **drilling** and extraction technologies may make a greater portion of a reservoir recoverable. In an environment of rising prices, a greater portion of a reservoir may become economically recoverable as well.

Reserve growth is a phenomenon that has been observed in oil fields all over the world. Most fields produce considerably more oil over their lifetime than initial estimates predict. In a number of mature provinces, more reserves are added through reserve growth than through new field discoveries. In the **United States**, from 1978 to 1990, reserve growth accounted for 90 percent of additions to reserves. In the **United States Geological Survey** World Petroleum Assessment of 2000, future global reserve growth to the year 2025 was projected at 688 Bb, nearly as high as the 732 Bb of projected new field discoveries. *See also* OIL RECOVERY.

RESERVES. Petroleum reserves are distinct from petroleum resources. Reserves are identified quantities that are commercially recoverable using current technology, under current economic conditions and government regulations. Resources are quantities that might not currently be recoverable but may become so if conditions change. Improvements in technology and changes in regulations can turn resources into reserves as well as allow more resources to be discovered. Price is also a very important factor. Higher prices can increase estimates of reserves as previously uneconomic deposits become worth extracting. **Canada**'s **Alberta oil sands** are a case in point. As processing technology improves and prices rise, they become more and more competitive. In 2003, the ***Oil and Gas Journal*** began including the oil sands in its reserves estimates for Canada (increasing them by a factor of 37).

Methods of classifying and estimating resources and reserves have been developed at different times by different entities in different countries, for different purposes. One purpose of reserve estimation is for securities disclosure by petroleum companies, so that investors may adequately judge their asset positions. The U.S. Securities and Exchange Commission (SEC) has developed rules for this purpose. Documents associated with the Canadian Security Administrators (CSA) and the **United Kingdom** Statement of Recommended Practices (UK-SORP) also contain rules and practices aimed at this objective.

The SEC rules and guidelines established in 1978 are generally considered quite restrictive. They allow disclosure only of reserves and not of resources. They also allow only the disclosure of proved reserves (i.e., 1P—see below). No credit is given for probable or possible reserves. In addition, the SEC definition of proved reserves is somewhat more restrictive than that of the **Society of Petroleum Engineers** (SPE). The strictness of the SEC rules may mean that estimates of reserves by multinational oil companies are overly conservative. In 2004, **Royal Dutch Shell** caused a major stir in the world petroleum industry and beyond by decreasing its proven oil and **natural gas** reserves estimates by 3.9 Bboe, or about 20 percent. Many observers feared that this was an indication that oil companies had been inflating their reserves. Others countered that SEC regulations forced petroleum companies to be overly cautious and conservative, and Shell was merely striving for strict compliance. Shell made another reduction in estimates later in the same year.

Another purpose of reserve classification and estimation is for government assessment and reporting of national resources. The **Russian** Ministry of Resources, the **China** Petroleum Reserves Office, the **Norwegian Petroleum Directorate**, and the **United States Geological Survey** have all developed systems useful for this purpose. The reporting of reserves by countries often has a very strong political component. Production quotas set by the **Organization of the Petroleum Exporting Countries** (OPEC) are based on reserves, and an OPEC country wishing to justify an increase in its production quota may have an incentive to inflate its reported reserves. Between 1986 and 1989, for example, the reported reserves of many OPEC members showed large sudden increases. In 1987, **Iran**'s reported oil reserves increased from 47 Bb to 93 Bb, **Iraq**'s from 47 Bb to 100 Bb, **Abu Dhabi**'s from 31 Bb to 92 Bb, and **Venezuela**'s from 25 Bb to 56 Bb. In 1989, **Saudi Arabia** increased its reported reserves from 170 Bb to 255 Bb. Such increases were not justified by new oil discoveries. However, it is difficult to say with certainty if the increases were justified revisions based on better estimates or politically motivated. The nearly simultaneous timing is certainly suspicious.

Even the reporting of reserves by impartial authorities is subject to variation depending on differing assumptions and definitions in various institutions. The reserves estimates published by the *Oil and Gas Journal* are generally regarded as a good benchmark, but the reader should never forget that the determination of reserves is not an exact science and is subject to a great deal of interpretation.

There have also been a number of international efforts to standardize the way reserves and resources are classified and estimated. The Society of Petroleum Engineers (SPE) and the **World Petroleum Council** (WPC) independently published definitions and classifications for reserves in 1987. The two sets of definitions were similar, and in 1997 the SPE and the WPC jointly published a single set of classifications. In 2000, a classification system for all petroleum resources was released by the SPE, the WPC, and the **American Association of Petroleum Geologists** (AAPG). The SPE classification is regarded as a de facto international standard. The United Nations Framework Committee (UNFC) has developed a classification framework intended to be an international standard for all energy minerals, including coal and uranium. The UNFC incorporates the SPE classifications for petroleum.

In the SPE system, Proved (or Proven) Reserves of petroleum, denoted 1P, are those that are commercially recoverable from known reservoirs using current technology, under current economic conditions and government regulations. If probabilistic methods are used, there should be at least a 90 percent estimated probability that the quantities recovered will equal or exceed the reserve estimate. The area of a petroleum reservoir considered Proved includes that delineated by **drilling** and adjacent undrilled portions that appear with reasonable confidence to be continuous extensions based on geological, geophysical, and engineering data. Probable Reserves are additional reserves that are less certain to be recovered than Proved Reserves. Confidence in reservoir continuity or recovery efficiency for Probable Reserves is lower than in the case of Proved Reserves. In a probabilistic context, there should be at least a 50 percent probability that the quantities recovered will equal or exceed the estimate (denoted 2P) of Proved plus Probable Reserves. Possible Reserves are even less certain than Probable Reserves. They may be located in areas adjacent to those of Probable Reserves, where data and interpretations thereof are less certain. There should be at least a 10 percent probability that total recovered quantities will exceed the estimate (denoted 3P) of Proved plus Probable plus Possible Reserves. Thus, 1P is a conservative estimate of reserves, while 3P is a high estimate and 2P is intended to represent a best estimate.

In the case of resources, the SPE defines two broad classes: Contingent and Prospective. Contingent Resources are those quantities estimated to be potentially recoverable from known accumulations but which are not currently considered commercially recoverable. Contingent Resources may include those for which commercial recovery is dependent on technology that is still being developed. Or they may be resources where evaluation of the accumulation is still at an early stage. Contingent Resources may, like reserves, also be divided according to the confidence of the estimate of their magnitude. In decreasing order of certainty, these are termed Measured, Indicated, and Inferred. Low, best, and high estimates are respectively denoted 1C (Measured), 2C (Measured plus Indicated), and 3C (Measured plus Indicated plus Inferred).

Prospective Resources are those quantities of petroleum estimated to be potentially recoverable from as yet undiscovered accumulations. Prospective Resources are further subdivided according to the level of certainty associated with the estimate of their magnitude. For Prospective Resources, the low, best, and high estimates are simply described as such without any special names or alphanumeric designations. *See also* RESERVE GROWTH.

RESOURCES. *See* RESERVES; TECHNICALLY RECOVERABLE RESOURCES; TOTAL RESOURCE BASE; ULTIMATELY RECOVERABLE RESOURCE (URR).

REYNOLDS, GEORGE B. (1853–1925). British engineer and geologist who made the first major oil discovery in **Iran**, at Masjid-e-Suleiman in 1908, and the first commercial discovery in **Venezuela** in 1914 near Lake **Maracaibo**. George Bernard Reynolds was born 5 April 1853 in Sussex, England. He graduated from the Royal Indian Engineering College and worked for the Indian Public Works Department from 1875 to 1897, reaching the level of executive director and certified mine manager. He then worked for **Royal Dutch Company** (later **Royal Dutch Shell**) in the oil fields of Sumatra, **Indonesia**. In 1901, he began working for **William Knox D'Arcy**, prospecting for oil in Iran. Financial problems led to a 1905 investment and reorganization by the **Burmah Oil Company**. Reynolds's strong independent streak and sometimes sarcastic demeanor led to continual friction with the new management.

Reynolds persevered without much success for years, enduring extremely difficult field conditions. In January 1908, he finally began drilling at Masjid-e-Suleiman, his preferred location. Initially he had no success there either. The directors of Burmah Oil, short of cash, instructed him to abandon drilling if he could not find oil by the time he reached a depth of 1,500 feet. On 26 May 1908, he made a major find, effectively initiating the oil industry in the Middle East. In 1909, Burmah Oil formed the Anglo–Persian Oil Company (later **Anglo–Iranian Oil Company**, eventually **BP**). Reynolds's relations with Burmah and Anglo–Persian continued to be tense. Despite his astounding achievements, he was dismissed in 1911, with £1,000 and reportedly not a single word of praise or thanks.

After his dismissal from Anglo–Persian, Reynolds went to work for Royal Dutch Shell in Venezuela. He surveyed the area near Lake Maracaibo and in 1914 produced oil on a modest scale. On 14 December 1922, he discovered the large La Rosa field with the Barroso No. 2 well, also in the area of Lake Maracaibo. The discovery well blew out with a flow rate of about 100,000 bpd. Reynolds died in Seville, Spain, on 23 February 1925, his achievements never properly recognized in his lifetime.

REZA KHAN. *See* PAHLAVI, REZA (1878–1944).

RICHFIELD OIL COMPANY. *See* ARCO.

RIO BLANCO PROJECT. *See* NUCLEAR EXPLOSION STIMULATION.

ROBERTS, EDWARD A. *See* FRACKING.

ROCKEFELLER, JOHN D. (1839–1937). American industrialist who founded **Standard Oil**. John Davison Rockefeller was born 8 July 1839 in Richford, New York. His father, William Avery Rockefeller, was a traveling salesman of novelties and supposed cures for ailments, who allegedly became a bigamist in 1855. Rockefeller's mother, Eliza Davison Rockefeller, was a devout Baptist, as Rockefeller would be for the rest of his life. The family moved to Owego, New York, in 1849, and to Strongsville, Ohio, near Cleveland, in 1853. Rockefeller enrolled in Folsom Mercantile College in Cleveland to study bookkeeping and other practical topics. On 26 September 1855, at the age of 16, he got his first job as an assistant bookkeeper at Hewitt and Tuttle, commission merchants and produce shippers. He lived frugally and in 1859 used $1,000 in savings to enter into a produce wholesaling partnership with Maurice B. Clark (future grandfather of oil executive **Walter Teagle**).

Rockefeller was only 20 when **Edwin Drake** drilled his well near **Titusville, Pennsylvania**. As the industry boomed, he became interested in the possibilities and began investing in kerosene production. Rockefeller saw oil exploration as risky and concentrated instead on the **refining** of petroleum and the marketing of its products. In 1863, he formed Andrews, Clark and Company with Maurice Clark and Samuel Andrews. The company built a refinery in Cleveland. Two years later, Rockefeller bought out his partners for $72,500 and formed a new company with Samuel Andrews, Rockefeller and Andrews. By the end of the year, Rockefeller and Andrews had the largest refinery in Cleveland, with $1.2 million of gross annual revenue and a capacity of 505 bpd. Rockefeller enlisted his brother William as a partner, and they built a second refinery, the Standard Works. In 1865 and 1866, Rockefeller bought out 50 small refineries in Cleveland and 80 in Pittsburgh.

In 1867, Rockefeller brought **Henry Flagler** in as a partner, forming Rockefeller, Andrews, and Flagler. He also obtained additional capital from Stephen Harkess, his wife's cousin. He began negotiating aggressively with railroad officials for lower rates. He instituted a cost accounting system that kept track down to the third decimal. He also began the process of vertical integration. The company began to manufacture its own barrels and its own sulfuric acid for use in refining. It also provided for its own storage. Rockefeller was also interested in the "waste" products from manufacturing kerosene, ranging from lubricants to paraffin.

In 1870, Rockefeller and Flagler reorganized the partnership as an Ohio joint stock company, Standard Oil. This first Standard Oil company was not the same entity as Standard Oil of Ohio (**Sohio**), which was an independent successor company formed after the breakup of the Standard Oil trust in 1911. Standard Oil was capitalized initially at $1 million. Rockefeller was president, and Flagler was secretary. Rockefeller owned 27 percent of the stock.

Rockefeller sought to mitigate the chaos, inefficiency, and cutthroat competition characteristic of the early oil industry. He achieved this by creating a colossus of a company that was both horizontally and vertically integrated. Horizontal integration meant buying out or eliminating competitors. In this, Rockefeller was ruthless but also often generous. As his company became more and more dominant, he was able to coerce special rebates from the railroads transporting his oil (*see* SOUTH IMPROVEMENT COMPANY). He also gained control of oil **pipelines**. Since shipping costs were a very important factor in the business, Rockefeller was able to undercut his competitors, eventually forcing them to choose between bankruptcy or selling out to Standard Oil. However, when they did sell to him, he often offered a fair price. He also continued the process of vertical integration, creating a company that controlled every stage of the business from raw material to marketing. In marketing he was also ruthless, coercing local stores to sell only Standard kerosene (in a distinctive red five-gallon can) and lubricants.

By 1880, Standard Oil controlled an estimated 90 percent of refining capacity in the **United States**. Although Rockefeller was wary of upstream operations, he needed new supplies of crude, and he took a risk on the **Lima–Indiana** fields discovered in Indiana and Ohio in 1885. The Lima oil was high in sulfur, and its refined kerosene was unusable in lamps owing to its foul smell. There was no guarantee that this oil would ever be usable, but Rockefeller consulted with **Herman Frasch**, who developed a process to desulfurize the oil. Rockefeller was able to use cheap crude oil as an input to produce refined products that competed with those derived from more expensive Pennsylvania oil.

In 1882, Rockefeller created the Standard Oil Trust, headquartered in New York City. The trust united Standard Oil (Ohio) and various affiliated companies under one umbrella. There were nine trustees, including Rockefeller. Eventually the trust controlled 40 companies, of which 14 were wholly owned. The trust was deliberately confusing and opaque to discourage public scrutiny.

The Ohio Supreme Court outlawed the arrangement in 1892 and ordered the Standard Oil Trust dissolved. The trust nevertheless continued to operate from its New York City headquarters. Eventually Standard took advantage of favorable New Jersey laws allowing holding companies. In 1899, it renamed its New Jersey firm the Standard Oil Company (New Jersey) and placed it at the hub of the organization. Assets that were formerly in the Standard Oil Trust were put into the holding company. By this time, Rockefeller had retired from Standard Oil. He had begun reducing his involvement gradually since 1887 after suffering some sort of nervous breakdown from overwork. By 1897, he was no longer actively involved in the management of the company, which was carried out principally by **John Archbold**.

In 1906, the U.S. government brought suit against Standard Oil Company (New Jersey) under the **Sherman Antitrust Act**, beginning a process that ultimately led to the dissolution of the company into 34 different entities in 1911 by order of the U.S. Supreme Court. The dissolution of Standard Oil only made John D. Rockefeller richer, as he owned stock in the new and valuable companies. In 1902, an audit put his net worth at about $200 million. By some estimates, it may have been as high as $1.4 billion at the time of his death. Compared to the total wealth of the country at the time, he was probably the richest American that has ever lived.

John D. Rockefeller was a controversial figure and was much reviled by the press and the public at various times. His tactics were new to a nation unused to giant corporations and accustomed to doing business on a local or regional level. The Standard Oil Trust inspired the creation of many other similar structures in various industries, resulting in considerable public concern and a political backlash, including the 1890 passage of the Sherman Antitrust Act outlawing industrial monopolies. Rockefeller was attacked by progressive reformers and journalists. One of these, **Ida Tarbell**, wrote a famous and scathing history of the Standard Oil Company for *McClure's Magazine*. It was released in book form in 1904.

Despite these attacks, it should be noted that Rockefeller significantly reduced the price of kerosene to the American consumer, from 30 cents a gallon in 1866 to 5 cents in 1894. Unlike other industrialists, he lived a relatively frugal and unostentatious lifestyle. Finally, he was a philanthropist on a scale never before seen, commenting, "I have always regarded it as a religious duty to get all I could honorably and to give all I could." Rockefeller gave away hundreds of millions of dollars. Beneficiaries included the Rockefeller Institute for Medical Research (now Rockefeller University), the University of Chicago, Spelman College in Atlanta, and the Rockefeller Sanitary Commission, devoted to eradicating hookworm in the southern states. He established the Rockefeller Foundation in 1913 to "promote the well-being" of humanity. It is still operating today. Rockefeller died in his sleep at age 97 on 23 May 1937 while vacationing in Ormond Beach, Florida. *See also* KELLOGG, FRANK BILLINGS (1856–1937).

ROMANIA. Oil production 2015: 0.08 Mmbpd (0.11 percent of world); **reserves** 0.6 Bb (0.04 percent of world); gas production: 0.329 Tcf (0.26 percent of world); reserves 3.725 Tcf (0.05 percent of world). Romania, a minor producer today, was one of the cradles of the modern petroleum industry. Petroleum has been known in Romania since premodern times. It was long used as a medicine and to grease axles. Documents dating from the 15th and 18th centuries mention oil in Bacău and in the region around **Ploiești**. Oil was extracted via hand-dug pits. Romania exported small quantities of oil to Austria–Hungary and the Ottoman Empire beginning in the early 19th century.

Romania may be the first country with regularly and continuously recorded oil production. Records begin in 1857, when production occurred at the rate of 4.7 bpd. The **United States** and Italy produced statistics beginning in 1860, **Canada** in 1862, and **Russia** in 1863. The first refinery was in 1840 in Lukăceşti, Bacău. Refineries sprang up around the oil fields, with main centers evolving in Ploieşti and Cámpina (both in Prahova). In 1857, Theodor Mehedinţeanu also set up a large-scale (for the time) refining operation to provide oil for street lamps in Bucharest. In 1860, Romanian oil production was about 130 bpd, around a third of the U.S. level.

In 1861, the first well was drilled in Romania (as opposed to being dug by hand). The first foreign company, the Valachia Petroleum Company run by Jackson Braun, began exploring in the area of Ploieşti. Foreign investment in the latter 19th century was hampered by the large numbers of small **concessions** with an uncertain legal basis. C. M. Pleyte of the **Royal Dutch Company** said the country was an "El Dorado for lawyers." However, Romanians made attempts at creating larger oil companies. In 1889, the Societatea Română pentru Industria şi Comertul Petrolului (Romanian Company for the Manufacture and Marketing of Petroleum) was founded by Romanians with Austrian backing. It built a number of refineries and became prominent in the industry. After five years, it suffered some financial setbacks but managed to procure Hungarian financing for restructuring. It reorganized on 17 September 1895 as Steau Română SA and became the country's first integrated oil company, with both upstream and downstream operations. In 1896, it produced 29 percent of Romanian output, and 61 percent in 1899.

On 20 April 1895, the government issued a new mining law that encouraged foreign investment. The law established a distinction between surface and subsoil rights and affirmed that anyone of any nationality could explore for minerals if authorized to do so by the minister of domains. Further, such authorization was not necessary for exploration on one's own land. The government reserved the right to exploit resources if a surface landowner would or could not. In 1895, Romanian oil production was just under 1,600 bpd, with about 15 percent of output exported.

In 1904, after about a decade of assessing Romania's prospects, **Standard Oil** entered the country. Standard was opposed by many elements in the Romanian government, but in the end the government did not want to appear to be repudiating the spirit of the 1895 mining law. Standard formed the Româno–Americană Societatea Anonima pentra Industria, Comertul, şi Exportul Petrolui (Romanian American Company for the Manufacturing, Marketing, and Export of Petroleum). Româno–Americană was owned entirely by Standard Oil and capitalized initially at 2 million French francs. In 1905,

Standard tripled the investment and completed a refinery at Ploieşti. In 1907, Româno–Americană drilled the first well in the **giant** Moreni–Gura Ocnitei oil field (**EUR** about 0.8 Bb), about 20 miles west of Ploieşti. After the dissolution of Standard Oil in 1911, Româno–Americană passed to the Standard Oil Company of New Jersey (later **Exxon**). By 1928, it had assets of $18 million and produced 8,414 bpd, almost 9 percent of Romanian production.

In 1910, **Royal Dutch Shell** formed the Astra Romana Company to counteract a possible Standard Oil monopoly. In 1913, Romanian production peaked at more than 37,000 bpd. It decreased during **World War I** and by 1917 was about 11,000 bpd. Romanian oil was a strategic objective for both sides during the war, and the subject of espionage and sabotage. In 1924, the liberal government of Ion I. C. Brătianu passed a new mining law restricting foreign investment, but it was reversed in 1929 by another law giving foreign investment equal treatment. The Romanian industry modernized and expanded in the 1930s, and production reached a new peak of more than 174,000 bpd in 1936. During **World War II**, Romanian oil again became a strategic objective for both sides. Production decreased during the war, reaching a low point of about 72,000 bpd in 1944.

In 1948, the Romanian communist regime nationalized all private enterprises and their assets in the country. Oil production languished until 1951, when it took a steep turn upward. This was during the time of Soviet occupation, which lasted until 1958. The Soviets pushed production and took large quantities of oil for themselves. By 1958, production was more than 231,000 bpd. It continued to grow, but at a slower rate. It reached its all-time peak of 303,000 bpd in 1977. By the 1970s, Romania, with aging fields and a lack of investment, became a net oil importer. In 2005, it produced about 115,000 bpd (about the same as in 1930) and consumed a total of 241,000 bpd.

Petrom, founded in 1991 after the fall of communism, is Romania's largest oil company and is responsible for about 68 percent of production. Petrom was partially privatized by the government in 2004. Österreichischen Mineralölverwaltung Aktiengesellschaft (OMV) of Austria bought a 51 percent share; the government still owns about 41 percent. The Romanian press criticized the privatization because Petrom was assigned large reserves just before the deal was signed, and as a result OMV obtained a near monopoly on Romanian oil production. Another company, Rompetrol, was established in 1974 as the international operator of the communist Romanian petroleum industry. It was privatized to management and employees in 1993. Dinu Patriciu acquired control in 1998. In August 2007, **Kazakhstan**'s KazMunaiGaz acquired a controlling interest in Rompetrol.

ROMPETROL. *See* ROMANIA.

ROOSEVELT, KERMIT, JR. (1916–2000). Operational chief of the U.S.-orchestrated coup removing **Iran**'s prime minister **Mohammed Mossadegh** from power in 1953. Kermit Roosevelt Jr., a grandson of Theodore Roosevelt, was born in Buenos Aires, **Argentina**, on 16 February 1916. He was educated at Harvard University and joined the U.S. Office of Strategic Services, the forerunner of the Central Intelligence Agency (CIA), during **World War II**. He became director of the CIA's Near East and Africa Division and in that capacity engineered the coup that overthrew Mohammed Mossadegh following Iran's 1951 nationalization of its oil industry—*see* AJAX, OPERATION (1953).

The coup removed a popular and democratic leader and solidified the position of Shah **Mohammed Reza Pahlavi**, who became increasingly despotic. Iran's oil remained nationalized despite Mossadegh's removal, although a Western consortium (**Iranian Oil Participants**) essentially ran the country's petroleum sector until the 1970s. The reaction to the shah's autocratic rule and his eventual overthrow in 1979 helped precipitate the **Second Oil Shock**. It also placed Iran on a path of continuing confrontation with the **United States** and the West. Roosevelt defended his actions as necessary to prevent a communist takeover in Iran. In 1979, he published the book *Countercoup: The Struggle for the Control of Iran*. Roosevelt left the CIA in 1958. He worked for a number of American companies doing business in the Middle East, and he was also a lobbyist representing foreign governments in Washington, D.C. He died in Cockeysville, Maryland, on 8 June 2000.

ROSNEFT. Petroleum company largely owned by the government of **Russia**. Rosneft was established in 1993 as a state enterprise from assets previously held by Rosneftegaz, the successor to the Soviet Ministry of Oil and Gas. In 1995, it was transformed into a joint stock company. From 1995 to 1998, Rosneft's management changed frequently and its oil production decreased markedly. New management in 1998 improved the situation, and the company was profitable by 2000, increasing its production rapidly thereafter. In 2000, Rosneft's oil production was 0.27 Mmbpd. By 2004, it was up over 50 percent, to 0.406 Mmbpd.

In 2001, Rosneft began representing Russia's interest in **production-sharing arrangements**, including those to develop the **offshore** Kurmangazy structure in **Kazakhstan**'s **Caspian** shelf and to develop the resources of **Sakhalin** Island. Rosneft expanded internationally in the 2000 decade, for example in **Algeria**. In late 2004, Rosneft acquired **Yukos**'s main oil subsidiary, Yuganskneftegaz, for $9.4 billion. Rosneft had agreed to merge with **Gazprom**, but the merger was derailed by the Yuganskneftegaz acquisition. Rosneft wanted a renegotiation of the merger terms to account for its ownership of Yuganskneftegaz, but Gazprom was unwilling. There were also complications arising from loans secured by Yugansknefteg az assets and back taxes owed by Yukos in relation to Yuganskneftegaz. The Gazprom merger did not happen. Rosneft acquired most of Yukos's remaining assets in 2007.

In 2006, Rosneft was partially privatized. It sold off 15 percent of its shares, raising $10.7 billion. **BP** acquired a $1 billion stake in the company. In 2011 Rosneft and BP agreed to joint exploration and development in the South Kara Sea, which may contain 36 Bbo and 353 Tcf by Russian estimates. In the same year it formed an alliance for Black Sea exploration and development with **ExxonMobil**. On 21 March 2013 Rosneft acquired **TNK–BP**. In 2016, Rosneft controlled 27.335 Bb of oil **reserves** and 60.528 Tcf of **natural gas**. It produced 4.183 Mmbpd of oil and 6.492 Bcfpd of gas.

ROTHSCHILD COMPANY. Petroleum company in **Russia** founded by the Paris **Rothschild family**, operating in the **Baku** area and the Caucasus. The name of the company in English was the Caspian and Black Sea Oil Production and Trading Company. In French, it was the Société Commerciale et Industrielle de Naphte Caspienne et de la Mer Noire; in Russian, it was Kaspiysko Chernjmjrskoe Neftepromushlenoe i Torgovoe Obchshestvo (KChNTO). The company has widely been referred to as BNITO, based on the Russian acronym of a predecessor company, the **Batumi Oil Production and Trading Company**, which the Rothschilds took over in 1885. From that toehold, they formed a larger company on 30 July 1885. In 1887, the Rothschilds gained control of the **Mazut** Company transporting and marketing oil and petroleum products. The Rothschilds presented stiff competition for the **Nobel** brothers' company **Branobel**, accounting for more than half of Russian kerosene exports by 1888. In 1899, the Rothschilds produced about 533,000 metric tons of crude oil (about 5.8 percent of Russian production), about 13,000 tons of kerosene, and about 280,000 tons of fuel oil.

The Rothschild operation never achieved the same levels of profitability as the Nobels. The Nobels invested more consistently and were more vertically integrated than the Rothschilds, who concentrated on downstream operations and trading, especially in the early years. The Nobels also benefited from having acquired control of land earlier, when it was substantially cheaper. In 1903, the Rothschilds teamed with the Nobels to form a cartel, **Nobmazut**, to coordinate their efforts in the Russian domestic market. The Rothschilds sold their petroleum interests to **Royal Dutch Shell** in 1912. The interests were later nationalized by the Soviets.

ROTHSCHILD FAMILY. European banking dynasty with interests in **Russia**'s petroleum industry in the **Caspian** and Caucasus regions in the late 19th and early 20th centuries. The house of Rothschild was founded by Mayer Amschel Rothschild (1744–1812) and his five sons. By the 1820s, the

Rothschilds had branches in London, Paris, Vienna, and Naples. It was the Paris branch, de Rothschild Frères, founded by Jakob (or James, 1792–1868), that became involved in petroleum. Jakob's son Alphonse de Rothschild (1827–1905) headed the Paris house when it bought up the **Batumi Oil Production and Trading Company** in 1884. The Rothschilds formed an augmented company (Caspian and Black Sea Oil Production and Trading Company) in 1885 and the **Mazut** company in 1897. In 1905, Alphonse de Rothschild died, and his younger brother Edmond de Rothschild (1845–1934) took over management of the family petroleum interests. Edmond took over during a time of significant turmoil, including the unrest in **Baku** that led to large-scale arson in the oil fields. He reevaluated the family's position and sold the Rothschild interests to **Royal Dutch Shell** in 1912. The interests were ultimately nationalized by the Soviets.

ROYAL COMMISSION ON FUEL AND ENGINES (1912–1913). British government commission created by **Winston Churchill** and headed by Admiral **John Arbuthnot Fisher** to study the conversion of Royal Navy ships from coal to oil propulsion. The Royal Navy under Fisher had begun the conversion in 1905 and was already partly reliant on oil. Churchill wanted to complete the process and ensure the navy's oil supplies on a long-term basis. The commission was chartered officially on 31 July 1912 and began proceedings on 24 September. It studied the potential of oil use in the navy, as well as issues of oil supply and oil storage. It also studied trials of oil-fired marine and railway engines. It held 41 meetings and submitted reports on 27 November 1912, 27 February 1913, and 22 April 1913.

The commission reached four important conclusions. First, using oil for propulsion was of vital importance to the Royal Navy, as oil was a more powerful fuel, cleaner burning, and less personnel intensive than coal. Second, the ideal use of oil was via the internal combustion engine, as opposed to conventional boilers or as an adjunct to coal. Third, it was necessary to store oil in large quantities in Britain. Fourth, the Admiralty should be a large-scale buyer and storer of petroleum. The commission's findings, along with those of several other bodies, led to the complete adoption of oil by the Royal Navy and ultimately to the British government's momentous 1914 purchase of a majority interest in the Anglo–Persian Oil Company (later **Anglo–Iranian Oil Company**, eventually **BP**).

ROYAL DUTCH COMPANY. Early petroleum company that amalgamated in 1907 with the **Shell Transport and Trading Company** to form **Royal Dutch Shell**. *See also* DETERDING, HENRI (1866–1939); KESSLER, JEAN-BAPTIST (1853–1900); SAMUEL, MARCUS (1853–1927); ZIJLKER, AEILKO (1830–1890).

ROYAL DUTCH SHELL. A **major oil company** (one of the **Seven Sisters**), formed in 1907 by the amalgamation of the **Royal Dutch Company** and the **Shell Transport and Trading Company**. Although Royal Dutch under **Henri Deterding** was the dominant partner in the amalgamation, the company came to be universally known as "Shell." When Royal Dutch and Shell combined to form the Royal Dutch Shell Group in 1907 with Henri Deterding as managing director, they created a complex and somewhat curious capital structure, consisting of two holding companies. These were the Anglo–Saxon Petroleum Company and the Bataafsche Petroleum Maatschappij, or Batavian (Dutch) Petroleum Company. The former was incorporated under English law and the latter under Dutch law. Shell owned 40 percent of each, and Royal Dutch owned 60 percent of each. This basic capital structure would continue for almost a century until 2005, when the group was fused into a single Royal Dutch Shell—which nearly everyone still calls simply "Shell."

Deterding's main goal was to control the world oil market and give it stability. This gave him, and Shell, a penchant for trying to create cartels. He first tried to reach some sort of agreement with **Standard Oil**, and its descendant Standard Oil Company of New Jersey (later **Exxon**), to split the global market outside the **United States**. Standard Oil refused and started price wars instead. Shell decided to move into Standard's home turf in the Western Hemisphere. In 1913, Shell acquired operations in **California** and **Venezuela** and also joined the Turkish Petroleum Company (TPC) consortium in 1912 with a 22.5 percent interest (later to become a 23.75 percent interest in the **Iraq Petroleum Company**, or IPC). In 1910, Shell formed the Astra Romana Company in **Romania** to counteract a possible Standard Oil monopoly in that country. Not all Deterding's actions were ultimately successful. Two of his moves proved disastrous. In 1912, he acquired the **Rothschild** petroleum interests in **Russia**, and in 1918 he acquired control of **Weetman Pearson**'s Mexican Eagle Petroleum Company in **Mexico**. Both of these interests would ultimately be lost, nationalized by the Soviets in 1918 and Mexico in 1938. Shell also joined with BP in 1937 to form a joint venture in **Nigeria** that eventually came to dominate that country's industry. In 1979, the Nigerian government nationalized BP's holdings, leaving Shell as the primary foreign operator.

During **World War I**, Shell wanted to do business with the government of the **United Kingdom**, supplying the Royal Navy with fuel. The balance tilted away from Shell's favor in 1914, when the British government bought a 51 percent stake in the Anglo–Persian Oil Company (APOC; later the **Anglo–Iranian Oil Company**, still later **BP**). Deterding and **Marcus Samuel**, cofounder of Royal Dutch Shell, clashed with Anglo–Persian's **Charles Greenway**, who attempted to paint Royal Dutch Shell as foreign and unreliable, despite the fact that Shell, the 40 percent partner, was British. However, Shell did supply the British government with considerable fuel during the war.

After the war, the petroleum industry entered another phase of oversupply (due partly to Soviet oil) and competition, of the type that Deterding abhorred. In an attempt to control the market, he hosted a meeting at Achnacarry Castle in Scotland attended by **John Cadman** and **Walter Teagle**, heads respectively of APOC and Standard Oil Company of New Jersey. Cadman had already approached Teagle and had a draft agreement produced internally. The three magnates concluded the secret **As-Is Agreement** to limit competition and control the global oil market. The agreement would not prove successful in the long run. Too many **independent** oil companies were operating outside the agreement, and too many new sources of oil were soon developed, including the **East Texas oil field** discovered two years later.

During **World War II**, Germany invaded the Netherlands. Royal Dutch Shell's Dutch headquarters moved to Curaçao. The company's refineries in the United States produced large quantities of high-octane aviation fuel (*see* GASOLINE) for the war effort, and the Shell Chemical company manufactured butadiene for synthetic rubber. All of Royal Dutch Shell's **tankers** were placed under the control of the British government. The company lost 87 ships during the war to enemy attacks.

In the 1950s and 1960s, Royal Dutch Shell and the other Seven Sisters effectively controlled the world oil market. Royal Dutch Shell supplied about one-seventh of the world's oil products during this period. It expanded into **natural gas** and **offshore** oil. It did not operate in **Saudi Arabia** or **Kuwait**, but it had interests in **Iraq**, **Abu Dhabi**, **Qatar**, and **Oman** through its 23.75 percent interest in the Iraq Petroleum Company. In 1954, it also entered **Iran**, with a 14 percent interest in **Iranian Oil Participants Ltd**. (IOP). In 1959, working with Standard Oil Company of New Jersey, it discovered the **Groningen** gas field in the Netherlands. By the 1970s, the **supergiant** Groningen was supplying about half of Europe's gas. Shell also became active in the **North Sea**. In July 1971, again working with Standard Oil Company of New Jersey (renamed Exxon the next year), it discovered the **Brent oil field**. Crude from Brent and related fields later evolved into a major pricing **benchmark**, known as **Brent Blend**. Shell also made a major gas discovery off **Australia**'s northwest coast in 1971.

The 1970s, saw a decided shift in power away from multinational oil companies and toward producing countries. Royal Dutch Shell had many of its assets nationalized, for example in **Libya** (1973) and Venezuela (1976). In 1979, IOP was ejected from Iran. Through the 1970s, the interests of the Iraq Petroleum Company in various Middle Eastern countries were gradually nationalized as well (*see* PARTICIPATION AGREEMENTS).

Royal Dutch Shell suffered along with other oil companies because of low oil prices during the **Oil Countershock** of the mid-1980s. By the early 1990s, many of the major oil companies were retrenching and cutting investment. Shell did the opposite and in 1991 overtook Exxon as the world's largest private sector petroleum company. In 1993, Cornelius Herkstroter became chairman of Royal Dutch Shell. He was convinced that the company's highly decentralized management structure, which had afforded it resilience and agility for nearly a century, was becoming a liability with too much duplication, inefficiency, and chaos. He restructured, eliminating about a third of the jobs at the group's headquarters. During the 1990s, the company also came under considerable criticism from environmental and human rights groups. It wanted to decommission its Brent Spar offshore platform in the North Sea by sinking it, but protests and boycotts forced it to dismantle the platform on land, a far more expensive solution. In November 1995, the Nigerian government executed author and activist Ken Sarowiwa, a vocal critic of multinational oil companies in his country, particularly Shell. His supporters accused Shell of collusion with the government.

In 1998, with real oil prices at their lowest levels in over two decades, Royal Dutch Shell posted its largest annual loss: $2.47 billion. The mergers of BP and **Amoco**, and especially Exxon and **Mobil**, also knocked Shell from its top position among the multinational oil companies. The company restructured, selling off almost half of its chemical businesses. In 2002, it acquired **Pennzoil**–Quaker State, thereby becoming a leader in lubrication. In 2004, Royal Dutch Shell caused a major stir by decreasing its proven oil and gas **reserves** estimates by 3.9 Bboe (about 20 percent). Many observers feared that this was an indication that oil companies had been inflating their reserves. Others countered that U.S. Securities and Exchange Commission regulations forced petroleum companies to be overly cautious and conservative, and Shell was merely striving for strict compliance. Shell made another reduction in estimates later in the same year.

Shell also ran into trouble on **Sakhalin** Island. The company had led the first-ever **production-sharing agreement** (PSA) with Russia, issued in May 1996. In 2007, **Gazprom** took majority control over the project (Sakhalin II) for $7.45 billion. The Shell-led consortium had already spent $13 billion. Pressure from Russia clearly played a role in Shell's acquiescence, and many observers decried this. However, others point out that the original PSA was very unfair to Russia and that Shell's consortium had failed to meet local expectations for social development. In the 2000s and 2010s, Royal Dutch Shell has been involved in exploration and development projects in all parts of the world.

In 2016, Royal Dutch Shell controlled 6.258 Bb of oil reserves and 40.541 Tcf of gas. It produced 1.849 Mmbpd and 11.301 Bcfpd of oil and gas respectively. Ranked by total assets, it was the largest private sector petroleum company in the world.

RULISON PROJECT. *See* NUCLEAR EXPLOSION STIMULATION.

RUSSIA. Oil production 2015: 10.662 Mmbpd (13.6 percent of world); **reserves** 80 Bb (4.86 percent of world); gas production: 20.294 Tcf (16.28 percent of world); reserves 1,688.2 Tcf (24.48 percent of world). Russia has been one of the world's most important petroleum-producing countries since the 19th century. Along with the **United States**, it has been a prime mover in the development of the petroleum industry. Russia has the world's second-largest reserves of **conventional natural gas** after **Iran** and is a member of the **Gas Exporting Countries Forum**. Russia is also estimated to have 284.5 Tcf in technically recoverable resources of **unconventional shale gas** and 74.6 Bb of technically recoverable resources of tight oil. These are as yet unexplored and unexploited. In its estimated **tight oil** endowment, Russia is second only to the United States.

Petroleum has been known in Russia since premodern times, with evidence of oil being gathered and used going back to at least the 10th century. Russian tsar **Peter I** (the Great) showed considerable interest in oil and encouraged systematic exploration and exploitation in the early 18th century. The cradle of the modern Russian oil industry was in **Baku**, **Azerbaijan**, part of the Russian Empire and then the Soviet Union. The **Bibi–Eibat** area around Baku was the site of one of the world's first deliberately drilled modern oil wells—*see* VOSKOBOINIKOV, NIKOLAI (1801–C. 1861)—in 1846. Later in the 19th century, the Baku industry, led by **Branobel**, the **Rothschilds**, and others, propelled Russia to the status of the second- and sometimes the first-largest oil producer in the world in pre-Bolshevik times. The **Baku Oil Society**, formed in 1874, was possibly the world's first vertically integrated oil company. The early modern Russian/Azerbaijani industry was also a highly innovative one, spearheading the use of **tankers** and making great progress in storage and **refining**, as well as **offshore** drilling and production in the **Caspian Sea**. The early Russian industry also developed in the North Caucasus, with Russia's first gusher being struck on the **Taman Peninsula** in 1866.

The Baku and North Caucasus regions dominated Russian production until the formation of the Soviet Union. The Soviets actively explored other provinces. Over the course of the 20th century, huge deposits of oil and gas were discovered in the **Volga–Urals** region and in western **Siberia**, as well as in the **Timan–Pechora** area. By 1955, Volga–Urals dominated Russian production. It accounted for about 80 percent of Russian oil production and more than half that of the whole Soviet Union. By 1960, the numbers were nearly 85 percent and 70 percent respectively. Western Siberia began its ascent in the mid-1970s. By 1980, it had surpassed Volga–Urals. Near the peak of Soviet production in the 1980s, western Siberia accounted for about 70 percent of Russian oil production and about 63 percent of the production of the whole Soviet Union.

Soviet production peaked in 1986 at nearly 12 Mmbpd, of which more than 90 percent was accounted for by Russia. It declined by about 16 percent by 1991. Among other factors, there was reservoir damage in many fields caused by overdrilling and overproduction. The communist system had both good and bad effects on petroleum industry practice. On the one hand, it allowed a fairly patient and scientific approach to exploration, without the constant need to justify every move to nervous investors. On the other hand, it did not adequately account for total costs in such a way as to incentivize operators to conserve. Operators had an incentive to exhaust fields quickly and then look for new ones.

Production continued declining in the early 1990s in the turbulent years following the dissolution of the Soviet Union. In 1993, production in the countries of the former Soviet Union was 7.6 Mmbpd, and it hovered between 6.8 and 6.9 Mmbpd between 1994 and 1998. Russian production was about 6.7 Mmbpd in 1993 and fell to about 5.9 Mmbpd in 1998. In 1999, it began to rise again. In 2006, it was 9.247 Mmbpd, the highest level since the breakup of the Soviet Union. Russia's share of the production of former Soviet countries was about 80 percent in 2006, versus almost 90 percent in 1992, owing largely to production increases in **Kazakhstan**.

During the 1990s, Russia's petroleum industry was privatized, restructured, and opened to foreign cooperation, for example in **Sakhalin**. Although some of the changes often occurred via questionable practices, such as the **Loans for Shares** scheme, they ultimately made the industry considerably healthier and more self-confident. However, the industry might have become too confident and too aggressive, often with backing from the Russian government. In 2007, **Gazprom** wrested control of the Sakhalin II project from **Royal Dutch Shell**, and in 2008 **BP** alleged that its Russian partners were using questionable tactics to take control of the **TNK–BP** joint venture.

The Russian industry was helped greatly by the large increase in oil prices during the **Third Oil Shock** of the 2000 decade. Some observers do not believe Russia can sustain oil production at current levels over the long term. Large fields are not being discovered at the same rates as in the past. In western Siberia, for example, **Samotlor** and Urengoy (the largest oil field and the largest gas field) were discovered in the 1960s; 77 of western Siberia's 98 **giant** fields that were discovered between 1961 and 2000 were found before 1980. In the 2000s and 2010s Russia explored aggressively in frontier areas such as the **arctic**. It concluded a treaty with Norway delimiting the two countries' areas in the **Barents Sea**. Russia has been willing to cooperate with Western energy companies, but such collaboration has been increasingly hampered since Russia's 2014 annexation of Crimea and the imposition of successive rounds of Western sanctions.

Soviet gas production did not reach a high point until 1991, at about 28.6 Tcf for the year. The total gas production of former Soviet countries declined to a low of 23.9 Tcf in 1995 (of which 20.2 Tcf was from Russia) and rose to 28.9 Tcf in 2006 (of which 22.6 Tcf was from Russia), exceeding the previous peak. Russia's percentage of gas production of former Soviet countries was about 78 percent in 2006, versus nearly 86 percent in 1998. *See also* ABSERON PENINSULA (ABSHERON, APSHERON); BAKU–BATUMI PIPELINES; BAKU–TBILISI–CEYHAN (BTC) PIPELINE; BARENTS SEA; BATUMI (BATUM); DRUZHBA PIPELINE; GAZPROM NEFT; LOMONOSOV, MIKHAIL (1711–1765); LUKOIL; MANTASHEV (MANTASHIAN), ALEXANDER (1842–1911); MANTASHEV COMPANY; MAZUT; MENDELEEV, DMITRI (1834–1907); NOBEL, LUDVIG (1831–1888); NOVOROSSIYSK; ROSNEFT; SAKHALIN; STRIZHOV, IVAN (1872–1953); TAGIYEV, ZEINALABDIN (1823–1924); TNK–BP; YUKOS.

S

SAFANIYA. World's largest **offshore** oil field and one of the largest fields overall, possibly third largest. Safaniya, located just off the coast of **Saudi Arabia** in the **Persian Gulf**, was discovered in 1951. First production was in 1957. It is believed that the field has never been tapped to capacity because it produces relatively heavy (**API gravity** about 27°) and less desirable oil. In 1980 and 1981, a period of peak Saudi production, Safaniya produced about 1.5 Mbd. During the 2000 decade, it produced about 600,000 bpd. The **EUR** was 55.6 Bbo. The structure is an anticline about 40 miles long and 9 miles wide. It is possible that the Safaniya field and the Khafji field in the **Neutral Zone** are part of the same complex, which may even extend to the **Burgan** field in **Kuwait**. When the Burgan field was set on fire during the 1991 **Persian Gulf War**, a pressure drop might have been observed in Khafji and Safaniya, but this has not been verified.

SAGA PETROLEUM. *See* NORSK HYDRO.

SAKHALIN. Significant petroleum resources were discovered around Sakhalin Island during Soviet times, and since the 1990s some have been developed by consortia including foreign oil companies. Sakhalin's **reserves** are estimated at 12 Bb and 90 Tcf of **natural gas**. The first onshore field was discovered in 1922, followed by a series of smaller finds. Onshore production peaked in 1968. The first large **offshore** oil and gas complex, Odoptu, was discovered in 1977 by the Soviets. Others include Lunskoye (1984) and Piltun–Astokhskoye (1986). Piltun–Astokhskoye is primarily an oil field, about 10 miles offshore in the Sea of Okhotsk. It may have 1.2 Bb and 18 Tcf of gas. Lunskoye is primarily a gas field, with about 10 Tcf of gas.

After the breakup of the Soviet Union, **Russia** did not have the capital resources to develop the Sakhalin fields. Two foreign-led consortia entered into **production-sharing agreements** (PSAs) to do so. The projects are known as Sakhalin I and Sakhalin II. Sakhalin I, led by **ExxonMobil** (Exxon Neftegaz), is developing the Odoptu field and the Arkutun Dagi field. The consortium spent $3.4 billion in 2005 and 2006. Commercial production from the Chayvo field (part of the Odoptu complex) began in October 2005. By February 2007, 250,000 bpd were being exported, mostly to East Asian markets. Gas from the field was piped into the Russian distribution system.

Sakhalin II, led before 2007 by **Royal Dutch Shell**, is developing Piltun–Astokhskoye and Lunskoye. Sakhalin II resulted from the first production-sharing agreement ever signed with the Russian government. The agreement was signed on 22 June 1994, and a license was issued in May 1996. The consortium first included the McDermott and **Marathon** companies, but they sold out their interests in 1997 and 2000 respectively. Mitsui and Mitsubishi are also members, along with **Gazprom**. The consortium spent about $13 billion by 2007. In February of that year, Gazprom announced that it would take majority control of the project for $7.45 billion. Shell reluctantly buckled under pressure from the Kremlin, which had been applied in various ways. For example, the Russian Ministry of Natural Resources suddenly began backing Sakhalin environmentalists, revoking permits, and delaying work on **pipelines**.

The Western business press bemoaned Gazprom's takeover. However, most observers agree that the original PSA was unfair to Russia. The resources were already identified, so the Shell consortium did not face undue exploration risk. The agreement gave the Shell group (Sakhalin Energy Investment Company) the right to recoup all costs plus a 17.5 percent rate of return before the Russian government would see any of its 10 percent share. Ballooning costs further delayed when Russia would see a payoff. In addition, the consortium apparently failed to meet local expectations for constructing roads and schools. Safety and environmental problems, as well as the extreme disparity in wealth between expatriate employees and local residents, also contributed to Russian dissatisfaction. For its part, the consortium argued that it had invested $13 billion of its shareholders' money and the Russian government had risked nothing.

In July 2005, Shell estimated that the Sakhalin II project had about 17 Tcf and 1 Bb of recoverable gas and liquid reserves. In July 1999, the project began producing 80,000 bpd of oil in the summer months. In 2016 production of oil stood at 0.395 Mmbpd, and gas at 1.9 Bcfpd. Four newer projects, Sakhalin III–VI, are still in development.

SAMOTLOR OIL FIELD. Russia's largest oil field and one of the largest in the world, possibly seventh largest. Samotlor was discovered on 29 May 1965 in western **Siberia**. In the mid-1980s, the field produced about 3.5 Mmbpd, its highest level. It declined rapidly thereafter, producing at only about a 10th of that rate in the early 2000s. **Halliburton** became involved in projects to restore production and succeeded in preventing significant further declines for the time being. Samotlor has an **EUR** of about 5 to 20 Bb of oil, depending on whom one asks, and 5.3 Tcf of gas. During the first part of the 21st century, Samotlor was operated by the **TNK–BP** joint venture, until TNK–BP was acquired by **Rosneft** in 2013.

SAMUEL, MARCUS (1853–1927). Founder of **Shell Transport and Trading Company**; cofounder of **Royal Dutch Shell**. Marcus Samuel was born in London on 5 November 1853. He began his business career trading in curios, principally painted seashells and boxes made from them. On a business trip to Japan, he became interested in the petroleum industry and saw an opportunity in the transportation of oil. He acquired a **tanker** fleet and in 1892 managed to obtain permission for his tankers to transit the **Suez Canal**. His were the first oil tankers to satisfy the Suez Canal company of their safety and suitability. He was able to ship the **Rothschilds**' petroleum products from **Russia** to Bangkok and Singapore, where they could be sold for less than those of **Standard Oil**.

Wanting a source of crude oil closer to his markets, Samuel moved into upstream operations, exploring for oil in Borneo. He discovered commercial quantities in 1897 and in October of that year founded the Shell Transport and Trading Company, with an initial capitalization of £1.1 million. When Samuel died in 1927, the company was worth £27 million. Shell's activities in Borneo brought it into competition with another relatively new company operating in the area, the **Royal Dutch Company**. Samuel developed an interest in acquiring Royal Dutch, but that company only wanted a possible joint marketing arrangement. Meanwhile, Standard Oil was interested in acquiring either or both companies at different times, although that eventuality never came to pass. In 1900, Royal Dutch came under the management of **Henri Deterding**, who had a keen interest in an alliance with Shell. Unlike Deterding, who was focused exclusively on business, Samuel had several political and social interests vying for his attention. For example, in 1902–1903, he was lord mayor of London. Samuel made some mistakes that weakened Shell's position. In 1901, he contracted to buy oil from the **Spindletop** field in **Texas** and ship it to Europe. When fervent overproduction caused Spindletop to become prematurely exhausted, Samuel ran short of crude to ship in his costly tanker fleet.

In 1903, Samuel agreed to combine Asian marketing activities with Deterding and the Rothschilds. Shell, Royal Dutch, and the Rothschilds formed the Asiatic Petroleum Company for that purpose. Deterding became managing director. This was a forerunner of events to come. Shell's position continued to weaken, and in 1907 Samuel was forced to accept an amalgamation proposal from Deterding. Shell and Royal Dutch joined together in the Royal Dutch Shell Group, 40 percent controlled by Shell and 60 percent by Royal Dutch. (The capital structure was more complex, involving holding companies—*see* DETERDING, HENRI (1866–1939); ROYAL DUTCH SHELL. Deterding became the true head of the group, while Samuel was relegated to the role of a nonexecutive chairman.

Samuel was one of the first people to fully understand the potential of oil as a transport fuel in place of coal. He was important in influencing First Sea Lord **John Arbuthnot Fisher** to begin the conversion of Royal Navy ships from coal to oil in 1905. The cause was later championed by **Winston Churchill**. In many cases, the conversion took the form of simply using oil instead of coal to heat boilers in steam engines. Samuel was not satisfied with this and presaged the future by advocating the use of internal combustion engines in vessels—*see* ROYAL COMMISSION ON FUEL AND ENGINES (1912–1913). Samuel was made Baron Bearsted in 1921, and in 1925 was made 1st Viscount Bearsted. He died in London on 17 January 1927.

SANCTIONS AGAINST IRAN. There has been a varying and complex set of economic sanctions against **Iran** imposed by the **United States** and the international community since the overthrow of Shah **Mohammed Reza Pahlavi** and the beginning of the Islamic Republic in 1979. Some of these sanctions have been targeted at the Iranian petroleum industry, a major source of the country's revenue. One such set of sanctions from the United States was the **Iran–Libya Sanctions Act** (later the Iran Sanctions Act). Since 2006 there have been international sanctions against Iran related to the desire to suppress its nuclear program. Six United Nations Security Council Resolutions from 2006 to 2010 were aimed specifically at nuclear and weapons-related matters and did not target the petroleum industry. However, sanctions imposed by the United States and the European Union did. For example, in the United States, Section 1245 of the National Defense Authorization Act of 2011 restricted exports of Iranian oil. Executive Order 13590 of the same year sanctioned contributing to the maintenance or expansion of Iranian petroleum resources. The Iran Threat Reduction and Syria Human Rights Act of 2012 locked Iranian petroleum revenue in special escrow accounts and sanctioned support of Iran's petroleum sector. There were also petroleum-targeted provisions in Executive Order 13622 of 2012 and the Iran Freedom and Counter-proliferation Act of 2012. The U.S. sanctions permitted only a few countries to purchase Iranian oil at a fixed level. These were **China**, **India**, Japan, South Korea, Taiwan, and Turkey. Meanwhile the 2010 Council Decision 2010/413/CFSP in the European Union banned export to Iran of key equipment and technology related to the petroleum industry. In 2012 the European Union followed up with even more comprehensive petroleum-targeted sanctions with Council Decision 2012/35/CFSP, including a ban on import, purchase, or transport of Iranian crude oil and petrochemical products. Council Decision 2012/635/FSP added a ban on import, transport, and purchase of Iranian natural gas. The various U.S. and EU sanctions had a devastating effect on Iranian oil production, cutting it about 20 percent between 2011 and 2013 to around 3.615 Mmbpd (oil and liquids). In November 2013 Iran agreed to freeze or reverse various aspects of its nuclear program, leading to some sanctions relief from the United States and EU. In 2015 Iran and the so-called P5+1 group of world powers—the five permanent members of the UN Security Council plus Germany—came to an agreement to rein in Iran's nuclear program, bringing fairly comprehensive relief from many sanctions, including ones targeted at the petroleum industry. In 2016 Iranian production of oil and liquids stood at 4.6 Mmbpd, above the baseline level immediately preceding the strict sanctions of 2011–2012.

SANCTIONS AGAINST IRAQ. *See* OIL FOR FOOD PROGRAM.

SAN REMO CONFERENCE (1920). Meeting to determine the future of Ottoman Empire territories following **World War I** that restructured the ownership of the Turkish Petroleum Company (TPC, later **Iraq Petroleum Company**). The conference was held in San Remo, Italy, on 19–26 April 1920 and was attended by the leaders of the **United Kingdom**, France, and Italy, with representatives from Japan, Greece, and Belgium. The conference set the framework for the Treaty of Sèvres signed 10 August 1920, abolishing the Ottoman Empire. That treaty was modified by the 1923 Treaty of Lausanne, but the provisions eliminating Turkish control over the Arab Middle East and North Africa remained. The San Remo Conference gave France a mandate over Syria and Lebanon, and Britain a mandate over southern Palestine and Mesopotamia (**Iraq**).

Before World War I, the Turkish Petroleum Company was owned 50 percent by the Anglo–Persian Oil Company (later **Anglo–Iranian Oil Company**), 22.5 percent by **Royal Dutch Shell**, 22.5 percent by Deutsche Bank, and 5 percent by **Calouste Gulbenkian**. At San Remo, the German Deutsche Bank shares were given to France and would form the basis for the formation of the **Compagnie Française des Pétroles** (CFP, later **Total**). In return, the French agreed that Mosul—which the secret **Sykes–Picot Agreement** of 1916 had placed in a French zone of influence—would be joined to the British mandate in Iraq. The **United States** was not at San Remo, and its companies received no participation rights. American oil companies and the U.S. State Department lobbied vigorously for an "open door" policy to the resources of the Middle East, allowing all companies to compete regardless of nationality. An American consortium would later be allowed to buy into the TPC. *See also* NEAR EAST DEVELOPMENT CORPORATION (NEDC).

SANTA BARBARA OIL SPILL. Oil spill that occurred off the coast of Santa Barbara, **California**, on 28 January 1969 following a blowout in a **Union Oil Company of California** drilling platform about six miles offshore. The spill released about 71,430 barrels of crude oil and did considerable environmental damage, killing thousands of birds, seals, and fish. It was not a huge spill by global standards; the **Kuwait** spills during the 1991 **Persian Gulf War** were 80 times larger, for example. However, it was notable because it served as a major catalyst for organizing and motivating the **environmental** movement in the **United States**. Refer to appendix 9. *See also AMOCO CADIZ*; *EXXON VALDEZ*; *TORREY CANYON*.

SAUD, ABDUL AZIZ IBN SAUD AL-. *See* IBN SAUD (1876?–1953).

SAUDI ARABIA. Oil production 2015: 10.16 Mmbpd (12.96 percent of world); **reserves** 266.46 Bb (16.19 percent of world); gas production: 3.687 Tcf (2.96 percent of world); reserves 303.28 Tcf (4.4 percent of world). Saudi Arabia, a **Persian Gulf** country, has the world's largest reserves of **conventional crude oil** and is the world's most important oil-producing country. Its reserves are third in the world, behind **Venezuela** and **Canada**, when the **Orinoco** Belt and the **Alberta oil sands** are included in those countries' totals. Saudi Arabia has often played the role of swing producer in the world market. It was a founding member of the **Organization of the Petroleum Exporting Countries** (OPEC) in 1960 and the **Organization of Arab Petroleum Exporting Countries** (OAPEC) in 1968.

King **Ibn Saud** granted an exclusive petroleum **concession** to New Zealander **Frank Holmes** in 1923, covering a large part of the Hasa region (now Eastern Province). Holmes was unable to exploit the concession and fell behind in his payments. Ibn Saud canceled the concession in 1928. In 1933, Ibn Saud granted a concession to the Standard Oil Company of California (Socal, later **Chevron**). Unlike many of the **major oil companies**, Socal had not signed the **Red Line Agreement** and was not constrained from operating independently in Saudi Arabia. Many of the majors were skeptical of Saudi Arabia's oil potential. Socal formed the California Arabian Standard Oil Company (Casoc) and began exploring in September 1933. With conditions harsh and success elusive, Socal partnered with another company that had not signed the Red Line Agreement, the Texas Corporation, which was known before 1926 and after 1941 as the Texas Company and after 1959 as **Texaco**. The Texas Corporation acquired 50 percent of Casoc, and the two formed their own **Blue Line Agreement**, collaborating closely with each other in the area "east of Suez"—*see* CALTEX (CALIFORNIA–TEXAS COMPANY).

On 3 March 1938, Casoc made its first significant strike with its Dammam No. 7 well, under the leadership of chief geologist **Max Steineke**. The first Saudi crude was exported from the shipping terminal at **Ras Tanura** on 1 May 1939. Four months later, **World War II** broke out, interrupting operations. **Harold Ickes**, the U.S. secretary of the interior and head of the **Petroleum Administration for War**, pessimistic about future petroleum production in the **United States**, sent a team to Saudi Arabia that included **Everette DeGolyer** to assess Saudi Arabia's potential. DeGolyer concluded that the potential was huge. The U.S. government considered acquiring a controlling interest in Casoc through the **Petroleum Reserves Corporation**, but this did not take place. In January 1944, Socal and the Texas Company (as it was by then again known) changed Casoc's name to the Arabian American Oil Company (**Aramco**). Saudi oil production, which began in 1936 at the low level of 550 bpd, reached more than 58,000 bpd by the end of World War II. It then grew very rapidly, surpassing 1 Mmbpd in 1956 and reaching 8.481 Mmbpd in 1974.

In 1948, Aramco discovered **Ghawar**, the largest oil field complex in the world and the only oil **megagiant**. In December 1948, two new partner companies bought into Aramco: Standard Oil Company of New Jersey (later **Exxon**) and Socony–Vacuum (later **Mobil**), with respectively 30 percent and 10 percent interest. This left Socal and the Texas Company with 30 percent each. The two new companies had been parties to the Red Line Agreement and were finally able to join after lengthy negotiations with other signatories. In 1950, Aramco completed the 1,063-mile **Trans-Arabian Pipeline** (Tapline) between the oil fields of the Eastern Province and the Mediterranean Sea via Lebanon. Ras Tanura was also significantly developed, and exploration continued. In 1951, Aramco discovered **Safaniya**, the world's largest **offshore** oil field and possibly the third-largest oil field overall. In 1961, Aramco would begin processing and shipping liquefied petroleum gas from Ras Tanura.

In 1949, **independent** oil executive **J. Paul Getty** acquired a 60-year concession for Saudi Arabia's interest in the **Neutral Zone** between Saudi Arabia and **Kuwait** for $9.5 million. Getty gave Saudi Arabia generous terms that drew the disapproval of the major oil companies. His company went on to make substantial discoveries in the Neutral Zone. The concession would pass later to Texaco, which acquired Getty Oil Company between 1984 and 1987, and ultimately to Chevron, which acquired Texaco.

In 1950, Saudi Arabia negotiated a **50–50 agreement** for profit sharing with Aramco, following the lead of Venezuela. This was one signal of an impending shift in power between the major oil companies and producing countries. As time went on, some Saudis suspected that Aramco was cheating on its payments to the government by deliberately obfuscating costs and earnings. Saudi Arabia, like other oil-producing countries, was also angered by unilateral cuts in the **posted price** of oil by the major oil companies in 1959 and 1960; the cuts translated directly to lower government revenues. The cuts were an important catalyst for the formation of OPEC in 1960. Saudi oil minister **Abdullah al-Tariki**, along with Venezuela's **Juan Pablo Pérez Alfonso**, played an important role in establishing the organization.

On 9 January 1968, Saudi Arabia joined with Kuwait and **Libya** to form OAPEC, the Organization of Arab Petroleum Exporting Countries. At first, OAPEC was a conservative and moderate organization (at the time of its formation, Muammar Qaddafi had not yet taken power in Libya). However, OAPEC admitted more radical Arab states and changed to a more activist stance. The Yom Kippur War of October 1973 led to an almost immediate selective OAPEC embargo on countries supporting Israel, including the United States. The Aramco partners complied. OAPEC also pledged to cut its collective oil production by 5 percent every month until Israel withdrew to its pre-1967 borders. OAPEC's action, along with an OPEC price rise, contributed to the **First Oil Shock**. But the Western powers were able to convince Saudi king Faisal that the embargo was against his long-term interests and that weakening the West would strengthen communism, an ideology Faisal opposed. Faisal persuaded the other OAPEC countries to lift the embargo in early 1974.

The high prices of the 1970s led to a huge influx of **petrodollars** into Saudi Arabia, more than the country's economy could productively absorb. The Saudis reinvested their petrodollars in the Western financial system, particularly that of the United States. This **petrodollar recycling** constituted a major coup for the United States (*see* DOLLAR, U.S). Despite continuing U.S. support for Israel, Saudi Arabia became an important U.S. ally and was able to purchase advanced U.S. arms.

In 1973, Saudi Arabia acquired 25 percent of Aramco, following OPEC's **participation agreement** framework. This framework, championed by the Saudi oil minister, **Ahmad Zaki Yamani**, was concluded between the major oil companies and some OPEC countries in the Persian Gulf region. It was intended to lead to a gradual transition of control to the countries involved. It created the option for an immediate 25 percent participation, rising to 51 percent by 1983. However, the countries significantly accelerated the schedule and exceeded the envisioned shares. Saudi Arabia increased its stake in Aramco to 60 percent in 1974 and took complete control in 1980. Links continued between the now Saudi-owned company and the Aramco partners. They continued to provide services to Saudi Arabia and were slated to receive 21 cents a barrel. They also received the right to market 80 percent of Saudi production.

In 1981, Aramco constructed the **Petroline** (east–west **pipelines**) for crude oil and **natural gas liquids**, from the Eastern Province fields to **Yanbu** on the Red Sea. In 1983, **Ali al-Naimi** became the first Saudi president of Aramco. In 1988, Aramco officially became the Saudi Arabian Oil Company, or Saudi Aramco; it is still commonly called Aramco, however.

During the early 1980s, Saudi Arabia tried alone to maintain high oil prices against downward market pressures. While other OPEC countries cheated on their quotas, Saudi Arabia cut production from a high of nearly 10 Mmbpd in 1980 and 1981 to 3.383 Mmbpd in 1985. In late 1985, Saudi Arabia gave up and began increasing production, making attractive **netback pricing** deals with customers to gain back market share. Oil prices collapsed (the **Oil Countershock**). Saudi production exceeded 8 Mmbpd in 1991 and stayed between 7.6 and 8.4 Mmbpd for the next decade. In 2005, it was 9.55 Mmbpd. Saudi Arabia was alarmed by **Iraq**'s 1990 invasion of Kuwait and allied itself with the United States and the West during the 1991 **Persian Gulf War**. The resulting long-term U.S. military presence in Saudi Arabia became a focus of anger for Islamic extremists, who often cited it as one of the reasons for their attacks on the West. In the 2000s and 2010s, Saudi Arabia continued to be a major determining factor in world oil production, although the growth of **tight oil** production in the United States, and the prospect that such growth might spread elsewhere, began to weaken the Saudi role. Nevertheless, Saudi Arabia continues to have a cost advantage over such unconventional crude production. For the moment, Saudi Arabia still has immense power.

Saudi Arabia has 59 **giant** fields (excluding the Neutral Zone), of which 51 are primarily oil and eight are primarily gas. Of the 51 oil fields, nine are **supergiants**, and one (Ghawar) is a megagiant. By some rankings, Saudi Arabia has eight of the world's 20 largest oil fields: Ghawar (discovered 1948), Safaniya (1951), Berri (1964), Manifa (1957), Shaybah (1968), Zuluf (1965), Abqaiq (1940), and Khurais (1957). Of Saudi Arabia's 10 largest oil fields, six were discovered before 1960, and none were found after 1970. Of the 50 oil giants, 39 were discovered before 1980, and 18 were discovered before 1970.

Saudi **natural gas** production is not as important as oil production, and Saudi gas reserves are not as extensive as those of oil. Still, gas production has been rising almost continuously for the last quarter century. Gas production in 1980 was 0.334 Tcf. In 2015, it was 3.687 Tcf, or about 1.789 Mmboepd. *See also* DHAHRAN (ZAHRAN); McGHEE, GEORGE (1912–2005); PETROMIN; PHILBY, HARRY ST. JOHN (JACK) (1885–1960).

SAUDI ARABIAN OIL COMPANY. *See* ARAMCO.

SAUDI ARAMCO. *See* ARAMCO.

SCHLAET, ARNOLD. *See* CULLINAN, JOSEPH S. (1860–1937); TEXACO.

SCHLUMBERGER, CONRAD AND MARCEL. *See* SCHLUMBERGER LTD.

SCHLUMBERGER LTD. Oil field services company; a pioneer in **well logging**. Schlumberger owes its origins to two brothers, **Conrad and Marcel Schlumberger**. Conrad was born 2 October 1878 and Marcel 21 June 1894 in Guebwiller, Alsace. Conrad received a physics degree in 1900 from the École Polytechnique de Paris and in 1907 became a professor at the École Supérieur des Mines. Marcel studied civil engineering at the École Centrale des Arts et Manufactures, graduating in 1907.

In 1914, Conrad located a deposit of copper ore for a customer in Serbia using surface electrical measurements. This was the first commercial application of the technique. In 1919, the brothers formed a company with the support of their father, Paul. In 1923, they mapped an oil-bearing salt dome in **Romania** for a petroleum company. In 1926, they named their company the Société de Prospection Electrique, the precursor of Schlumberger Ltd.

Beginning in 1921, the company experimented with downhole electrical resistivity measurements, initially to validate surface measurements. It soon became apparent to the brothers that the downhole measurements had intrinsic value. On 5 September 1927, the Schlumberger brothers and Henri Doll, Conrad's son-in-law and also an engineer, recorded the first petroleum industry well log at the **Pechelbronn** oil field in Alsace. The technique, resistivity logging, exhibited tremendous potential to reveal the layering of the subsurface rocks. Soon the company discovered spontaneous potential (SP) logging, although this would not be introduced commercially until 1931. The two techniques proved useful in helping to locate permeable hydrocarbon-bearing layers.

In 1929, the company signed an agreement to work in the Soviet Union and also began working for **Royal Dutch Shell** in **Venezuela**. In 1932, they expanded into the **United States**, recording logs for Shell in **California** and the **Texas** coast along the **Gulf of Mexico**. They obtained additional business from **independents**. In 1934, the brothers established a new company, Schlumberger Well Surveying Corporation, in **Houston**, with Conrad as president. Conrad died two years later, and Marcel took over. During **World War II**, Henri Doll established his own firm in Ridgefield, Connecticut, the Electro-Mechanical Research company. In 1945, Schlumberger bought Doll's firm and made him the head of research and technology for the whole company. This was to prove a very profitable decision. Doll introduced numerous advances, including induction logging, that would account for about 40 percent of company revenues by the time he retired in 1967.

Marcel Schlumberger died in 1953. The firm remained somewhat rudderless for a time, with four divisions headed by family members, unable to decide on an overall leader. Marcel's son Pierre ran North American Wireline Operations, a role he had performed since 1946. Doll headed research and technology. Jean de Ménil, another Conrad son-in-law, ran operations in South America and the Middle East, while René Seydoux, Marcel's son-in-law, headed European operations. In 1956, Schlumberger Ltd. was formed as a parent company to unify the four divisions. Pierre Schlumberger became president, and Henri Doll became chairman of the board. The parent company was headquartered in Houston and went public in 1962. Although the family had supplied some excellent talent, the company henceforth would no longer grant preference to family members for top management posts.

In 1965, Jean Riboud was named president. He was not a family member but had worked for the company since 1951. Riboud's tenure was one of nearly uninterrupted growth, helped along by the high oil prices of the 1970s (*see* FIRST OIL SHOCK; SECOND OIL SHOCK). It was also one of unwise diversification, including the 1979 purchase of Fairchild Camera and Instrument Corporation. The low oil prices of the mid-1980s (*see* OIL COUNTERSHOCK) brought an abrupt decline in revenues, and Riboud resigned in 1985. The new president, Michel Vaillaud, was in office only one year. In 1986, Schlumberger posted a large loss of $1.6 billion, its first loss since incorporation. Vaillaud was replaced by D. Euan Baird, the first non-French president of the company.

Baird initially refocused the company on oil field services. In 1986, he bought 50 percent of Geco, a Norwegian geophysical company, which in 2000 would become part of a joint venture with **Baker Hughes** called WesternGeco (*see* WESTERN GEOPHYSICAL). In 2006, Schlumberger would acquire Baker Hughes's share. In 1987, Baird sold Fairchild for $220 million. In 1992, Schlumberger acquired Geoquest, another geophysical company. But in 2001, Baird ventured into information technology, acquiring Sema PLC and forming SchlumbergerSema. In 2002, a large write down in the value of SchlumbergerSema led to a loss of $2.32 billion. A year later, Baird resigned and was replaced by Andrew Gould, who would again refocus the company on oil field services. Gould sold most of SchlumbergerSema to Atos Origin in 2003. In 2015 Schlumberger agreed to plead guilty to violating sanctions against Iran and Sudan and to pay $232.7 million in fines and forfeited profits.

Schlumberger, from its inception, has been a technology leader. Its founders were gifted scientists, and the culture of research and technology has persisted. One estimate in the 1980s claimed that Schlumberger's well logging services saved the upstream oil industry $35 billion to $80 billion annually.

SECOND OIL SHOCK. Large increase in oil prices in 1979. Average prices (**Arab Light**) for 1979 were $31.61 a barrel, over twice 1978 levels, and about 12 times as high as 1972 levels, before the **First Oil Shock**. In percentage terms, the Second Oil Shock was not as dramatic as the first: prices roughly doubled rather than quadrupling. However, they doubled from an already high level.

The proximate cause of the Second Oil Shock was the Iranian Revolution. The popular uprising in **Iran** against Shah **Mohammed Reza Pahlavi** involved major dislocations in the country's petroleum industry, instigated to an important extent by Ayatollah **Ruhollah Khomeini**. On 13 October 1978, workers at the **Abadan** refinery, then the world's largest, went on strike. Within a week, the strike spread to include most of the country's oil industry. By early November 1978, Iranian production had decreased from about 5.7 Mmbpd to 0.7 Mmbpd. This brought Iranian exports to a halt. While Iran was the second-largest exporter of oil after **Saudi Arabia**, the 5 Mmbpd shortfall was still not a significant percentage of world demand, especially given that other members of the **Organization of the Petroleum Exporting Countries** (OPEC) made up about 3 Mmbpd of it. The net gap of about 2 Mmbpd was only about 3 percent of world demand at the time.

The large price rises resulted not just from the Iranian shortfall but also from the panic that resulted. Fearful oil companies, anxious to increase inventories, pushed demand up an extra 3 Mmbpd, effectively wiping out the increased OPEC production that partially filled the Iranian gap. Another factor was the Rotterdam spot market. When the new Iranian regime began selling more oil again in early 1979, it chose to do so on the spot market. Japan, which was dependent on Iranian oil, began buying in a panic and drove spot prices up. Although most oil at this time was still sold on term contracts and the spot market accounted for only 3 or 4 percent of the international oil trade, it nevertheless had a large influence. Spot prices could diverge considerably from official ones, and in this period the pressure on spot prices was upward. This made OPEC countries think that their official prices were too low. Many of the deals on the spot market were for small quantities, by independent operators willing to pay a premium. They were not directly translatable to the larger-volume, longer-term market, but they still influenced the psychology of OPEC producers. OPEC countries also began selling more of their oil on the spot market, causing it to grow to about a 10th of the overall market by late 1980.

The higher prices were not sustainable, particularly as oil from the **North Sea**, **Mexico**, and **Alaska** began to make its presence felt. However in 1980, the **Iran–Iraq War** broke out and resulted in an immediate shortfall of about 3 Mmbpd. Spot prices again rose and influenced OPEC official prices. Arab Light reached a historic high of $42 per barrel on the spot market and averaged $36.83 overall for 1980. The Second Oil Shock was followed in 1985–1986 by the **Oil Countershock**, when prices collapsed. *See also* THIRD OIL SHOCK.

SEISMIC EXPLORATION. *See* GEOPHYSICAL PROSPECTING.

SEMYONOV, VASILI (1801–1863). Russian government official who helped promote the ideas of **Nikolai Voskoboinikov** on drilling for oil instead of hand digging large pits, as was the practice in early 19th-century **Baku**. Vasili Nikolaevich Semyonov, a member of the Caucasus executive council, is sometimes wrongly credited with drilling the **first oil well**. The well in question was drilled in 1846 in the **Bibi–Eibat** area near Baku, under the command of Major Alexeev of the Bakinskii Corps of Mining Engineers, based on ideas conceived by Alexeev's predecessor, Nikolai Voskoboinikov. Semyonov's role was not in supervising the drilling of the well but in helping Voskoboinikov present his proposals to the government.

SENECA OIL COMPANY. Company that employed **Edwin Drake**, who drilled an important early oil well in **Titusville**, **Pennsylvania**, in 1859. *See also* BISSELL, GEORGE HENRY (1821–1884).

SESPE OIL COMPANY. *See* STEWART, LYMAN (1840–1923).

SEVEN SISTERS. The seven **major oil companies** that effectively controlled the global oil market in the quarter-century after **World War II**. The origin of the name has been attributed to **Enrico Mattei** of **ENI**. The Seven Sisters included the original signatories of the **As-Is Agreement**. The Seven Sisters were **Royal Dutch Shell** ("Shell"); **BP** (Anglo–Persian Oil Company 1909–1935, **Anglo–Iranian Oil Company** 1935–1954); **Exxon** (Standard Oil Company of New Jersey until 1972, Exxon 1972–1999, **ExxonMobil** 1999–); **Mobil** (Socony–Vacuum 1931–1955, Socony Mobil 1955–1966, Mobil 1966–1999, ExxonMobil 1999–); **Chevron** (Standard Oil Company of California, or Socal, until 1984, then Chevron 1984–2001, ChevronTexaco 2001–2005, Chevron 2005–); **Texaco** (Texas Company 1902–1926, Texas Corporation 1926–1941, Texas Company 1941–1959, Texaco 1959–2001, ChevronTexaco 2001–2005, Chevron 2005–); **Gulf Oil Corporation** (acquired by Socal 1984, enlarged to become Chevron).

To the Seven Sisters, an eighth is often added: **Compagnie Française des Pétroles** (CFP), which became Total CFP in 1985 and **Total** in 1991.

These companies shared interests throughout the Middle Eastern oil-producing nations, and they used their interlocking memberships in consortia to balance production and control prices. The **Iraq Petroleum Company** consortium, which controlled the industry in **Iraq**, **Abu Dhabi**, **Qatar**, and **Oman**, had BP, Shell, Exxon, Mobil, and CFP as members. **Aramco**, which controlled the industry in **Saudi Arabia**, was composed of Chevron, Texaco, Exxon, and Mobil. The Kuwait Oil Company, which controlled **Kuwait**'s industry, was composed of BP and Gulf. **Iranian Oil Participants**, which dominated **Iran**'s industry from 1954 to 1979, had all the Seven Sisters and CFP as members, plus some American **independents**. *The Seven Sisters* was the title of a book on the oil industry by Anthony Sampson published in 1975. Refer to appendix 2.

SHALE GAS. Unconventional natural gas contained in shale formations of low permeability, generally trapped between layers. Shale gas does not flow easily and is generally produced by applying hydraulic fracturing (**fracking**) and **horizontal drilling**. Shale gas is sometimes subsumed under the blanket term "**tight gas**," although that term is more commonly used to denote gas in low-permeability formations other than shale.

Shale gas production has grown explosively in the **United States** in the last decade. In 1990 it comprised under 9 percent of total gas production, a percentage that did not change greatly for 15 years. In 2007 it was still under 14 percent; by 2015 it was over 50 percent. Natural gas production from the **Marcellus Shale** propelled a hydrocarbon renaissance in **Pennsylvania**. The state has been transformed from an aging 19th-century American oil superpower well past its prime to an American gas superpower in the early part of the 21st century. In 2015 Pennsylvania accounted for about 18 percent of all U.S. gas production, behind only **Texas** with 26 percent. In Texas, large-scale gas production from the **Barnett Shale** using fracking was pioneered by **Mitchell** Energy Corporation in the 1980s and 1990s, with definitive commercial success by 2000.

Shale gas production in other countries has not yet taken off the way it has in the United States, but many countries around the world have shale gas potential, and the countries with the most potential are not necessarily the ones with the most reserves of **conventional** gas. Estimates are highly uncertain, but the world is thought to have about 7,577 Tcf of unproven technically recoverable shale gas resources. Of these, 1,115 Tcf are thought to be in **China**, 802 Tcf in **Argentina**, 707 Tcf in **Algeria**, 623 Tcf in the United States, 573 Tcf in **Canada**, 545 Tcf in **Mexico**, 429 Tcf in **Australia**, and 285 Tcf in **Russia**.

SHALE OIL. Tight Oil contained in shale formations, generally produced by applying technologies such as hydraulic fracturing (**fracking**) and **horizontal drilling**. The term was sometimes previously used to refer to a liquid, similar to crude oil, extracted from **oil shale** by destructive distillation.

SHARJAH. One of the constituent emirates of the **United Arab Emirates** (UAE). Sharjah has about 1.5 Bb, or roughly 1.5 percent, of the UAE's 97.8 Bb of oil **reserves** (2015). In 1972, the Crescent Petroleum Company, working with Buttes Oil and Gas, discovered the Mubarak oil field in the **Persian Gulf** about 45 miles from the Sharjah coastline. Mubarak production peaked at 38,300 bpd, declining to less than 2,000 bpd by the early 2000s. The discovery was in an area disputed between Sharjah and another UAE emirate, Umm al-Qaiwain. **Occidental Petroleum** claimed to have an overlapping **concession** in Umm al-Qaiwain that had been violated, and some complex legal battles ensued. In June 1973, Sharjah and Umm al-Qaiwain decided on a 70–30 split in Sharjah's favor, and Umm al-Qaiwain terminated Occidental. In 1980, the Standard Oil Company of Indiana (later **Amoco**) and the Sharjah Petroleum Department discovered the giant onshore Sajaa gas field, with an **EUR** of 3.3 Tcf of gas and 0.25 Bb of condensate.

Sharjah has a long-standing dispute with **Iran** over Abu Musa Island in the Persian Gulf, not far from the entrance to the **Strait of Hormuz**, the only maritime outlet for Persian Gulf oil. In 1971, when the **United Kingdom** left the area, it was agreed that Iran would share sovereignty with Sharjah over Abu Musa. Oil revenues from fields surrounding Abu Musa would also be shared. However, none of the parties were satisfied by the situation, and tensions continued to flare up. Both parties consider it a matter of national sovereignty.

SHELL. *See* ROYAL DUTCH SHELL.

SHELL–BP. *See* NIGERIA.

SHELL D'ARCY. *See* NIGERIA.

SHELL TRANSPORT AND TRADING COMPANY. Early petroleum and trading company that amalgamated in 1907 with the **Royal Dutch Company** to form **Royal Dutch Shell**. *See also* SAMUEL, MARCUS (1853–1927).

SHERMAN ANTITRUST ACT. Law passed by the U.S. Congress in 1890 to limit monopolistic and anticompetitive practices by large American corporations such as **Standard Oil**. The law was named for Senator John Sherman of Ohio, one of the original proposers. The act contained two main provisions. The first prohibited a group of companies from entering into contracts or conspiracies in restraint of trade. Such agreements included price fixing, market division agreements, and boycotts. The second provision prohibited a single company from monopolizing a particular market. Some observers believe the law was essentially written because of Standard Oil and its anticompetitive practices. The administration of President Theodore Roosevelt prosecuted Standard Oil for violations of the act beginning in November 1906, winning the case in 1909, when the company was ordered dissolved. Standard Oil appealed to the U.S. Supreme Court but ultimately lost on 15 May 1911. The Court, led by Chief Justice Edward White, found that Standard Oil's "unreasonable" restraints of trade included a number of practices. One was maneuvering smaller **independent** companies into a position where they had no choice but to be bought out, often on unfavorable terms. Another was selectively cutting prices in some markets in order to force rivals into bankruptcy while maintaining high prices elsewhere. As a result of the Court's decision, Standard Oil was broken up into 34 different companies. *See also* KELLOGG, FRANK BILLINGS (1856–1937).

SHUKHOV, VLADIMIR (1853–1939). Russian engineer and architect who invented an early thermal cracking process for petroleum. Vladimir Grigorievich Shukhov, born 28 August 1853 in Graivoron, **Russia**, graduated from the Imperial Moscow Technical School with a gold medal in 1876. His primary career was as a structural engineer and architect. He is famous for his hyperboloid towers, for example the Shukhov Tower, a broadcasting tower in Moscow. Among the industries where he applied his talents was the Russian petroleum industry. He designed the first oil **pipeline** in Russia for the **Baku** oil fields. He also designed high-performance barges, strong and inexpensive oil tanks, an oil pump, and one of the first furnaces to process residual oil.

One of Shukhov's most important contributions to the petroleum industry was for a thermal cracking process to break larger molecules down into smaller ones and thus increase the proportion of light fractions (such as **gasoline**) recovered in petroleum **refining**. Shukhov devised his process with the help of his associate Sergei Gavrilov. He also built a plant to implement it. The Shukhov Process was patented on 27 November 1891 (Russian Patent No. 12926), more than two decades before **William Burton**'s thermal cracking process in the **United States**. Shukhov's patent was used in U.S. litigation by companies trying to invalidate the Burton patent. However, it should be noted that Burton's process was the one that first achieved large-scale practical application to produce increased yields of gasoline from petroleum.

Shukhov was designated a "Hero of Labor" by the Soviet government in 1928 and won the Lenin Prize in 1929. He died on 2 February 1939 in Moscow.

SIBERIA. Russia's most prolific oil and gas producing region. Northwestern Siberia to the northeast of Tyumen, between the Urals and the Central Siberian Plateau, accounted for about three-quarters of Russian production in the 2000 decade, up from almost nothing in 1965. Eastern Siberia has not been as intensively explored. Russian colonists noticed evidence of petroleum in Siberia since at least the 17th century, in and around Lake Baikal, on the banks of the Yenisei River, and in the northern Urals. In 1685, Leonty Kislyansky, head of the fortress at Irkutsk, was instructed by the government to search for minerals. He found oil seepages near Irkutsk. In 1702, Tsar **Peter I** (Peter the Great) ordered Siberian oil samples sent to him and ordered further scientific exploration of the vast region. **Mikhail Lomonosov** mentioned Siberian oil from the Yenisei in 1741.

The modern systematic search for petroleum in Siberia began in the 20th century. Vladimir Ryazonov directed **drilling** in and near Lake Baikal in 1902–1905 and 1907–1909 and struck oil. Modern prospecting in what was to become the prolific western region did not begin until the 1930s. It continued even during **World War II**, intensifying after the war was over. The Tyumen oil prospecting expedition began its work in January 1948. On 21 September 1953, foreman V. Melnikov and his team found a gas field near Berezovo. On 25 September 1959, foreman Semyon Urusov and his group struck oil at about 4,600 feet near the Konda River's confluence with the Mulymya, but not quite in commercial quantities. They struck larger quantities in the same area on 25 April 1960. The Soviet Union Council of Ministers adopted resolution 241 on 20 March 1960 that urgent priority be given to the development of western Siberia's petroleum resources. On 21 June 1960, there was a major oil discovery near Shaim, heralding the beginning of western Siberian production on a large scale. On 23 May 1964, the first cargo of Shaim Oil was shipped via the Irtysh River to the Omsk refinery.

Several major fields were discovered in rapid succession, including Tkurskoye (September 1960), Megionskoye (March 1961), and Ust Balykskoye (October 1961). In May 1962, the Soviet Union formed Tyumenneftegaz to develop the region's resources. On 29 May 1965, the **Samotlor** field, Russia's largest oil field and one of the largest in the world, was discovered. The development of the significant resources found in western Siberia was hampered by the lack of an effective and economical transport infrastructure. The Soviet government ordered **pipelines** constructed, and this was accomplished in relatively short order. On 1 December 1965, first oil flowed in the 256-mile Shaim–Tyumen Pipeline, and in 1967 in the Ust-Balyk–Omsk Pipeline.

Between 1961 and 2000, 98 **giant** oil and gas fields were discovered in western Siberia, 77 of these before 1980. The region gave Russia the world's largest **reserves** of **natural gas**. The **megagiant** Urengoy gas field, with an **EUR** of about 335 Tcf, was discovered in 1966. Western Siberian oil production reached a high point in 1986 at more than 8 Mmbpd. By that time, overproduction and too much drilling had caused reservoir damage in many fields. Oil production in 2006 was about 14 percent lower than peak levels.

SIBNEFT. *See* GAZPROM NEFT.

SIDOROV, MIKHAIL. *See* TIMAN–PECHORA.

SILLIMAN, BENJAMIN, JR. (1816–1885). American chemist who distilled and analyzed crude oil samples and reported on their potential products and uses, leading to the drilling of the **Edwin Drake** well near **Titusville, Pennsylvania**, and helping launch the modern petroleum industry in the **United States**. Benjamin Silliman Jr., son of the famous geologist, chemist, and Yale University professor Benjamin Silliman (1779–1864), was born on 4 December 1816 in New Haven, Connecticut. He followed in his father's footsteps, assisting him in establishing a chemical laboratory at Yale, and taught chemistry and toxicology at the University of Louisville in Kentucky from 1849 to 1854. In 1854, he assumed his father's position of chairman of chemistry at Yale. He published *First Principles of Chemistry* in 1847 and *First Principles of Physics* in 1858, and he conducted research on gases, meteorites, and photographic processes. He also prepared several reports on mines and manufacturing processes.

In 1855, Silliman issued his *Report on the Rock Oil, or Petroleum from Venango Co., Pennsylvania, with Special Reference to Its Use for Illumination and Other Purposes*. In the report, he noted Pennsylvania oil's desirable properties and that it did not harden from continued exposure. He subjected the petroleum to fractional distillation and noted that the kerosene fraction produced an excellent flame for illumination. However, he added that "these [illumination] experiments were not continued, because it was assumed that other products now known and in use for gas making might be employed at less expense for this purpose than your oil." Perhaps he was influenced by the fact that he was a director in the New Haven Gas Light Company, or perhaps he thought that petroleum might be rarer and more expensive than it ultimately proved to be. Silliman also noted the paraffin fraction but did not recommend its use for candles. He sent some of the distilled oil to Boston to be tested for lubricating properties, but he did not have the results of the tests at the time he issued his report. Overall, Silliman concluded that the petroleum he tested was a raw material that could be used to manufacture very valuable products, using simple and inexpensive processes, and without waste.

Silliman also became involved with the embryonic **California** oil industry, with considerable damage to his reputation. He went to southern California in 1864 at the behest of Tom Scott of the Pennsylvania Railroad to assess the resource potential there. He examined the potential for precious metals and also for petroleum. On the basis of a hasty and superficial examination of potential prospects in the areas of Ojai, Los Angeles, and the Santa Clara Valley, he concluded that southern California possessed immense resources of high-quality oil. His reports contributed to the California oil bubble of 1865, during which a number of companies failed. They found that the oil was neither as easy to obtain nor of as high a quality as Silliman had predicted. One entrepreneur committed suicide and another sued Silliman. Silliman was also accused of adulterating the California oil samples he analyzed with higher-quality Pennsylvania crude and even refined kerosene. Such accusations were fodder for rival professors at Yale who already disapproved of Silliman's industrial consulting activities. At his peak year, Silliman may have made as much as $75,000 from consulting, nearly 30 times his professorial salary. Silliman was forced to resign from Yale College, although he retained his faculty position in the medical school. He also successfully resisted attempts to eject him from the National Academy of Sciences.

His reputation recovered somewhat when large quantities of oil were indeed discovered in southern California. Silliman also predicted that California's oil would be useful as a fuel, at a time when illumination was the only widely recognized application. Benjamin Silliman Jr. died 14 January 1885 in New Haven.

SINCLAIR, HARRY (1876–1956). Founder of the Sinclair Consolidated Oil Corporation and head of that corporation or its antecedents from 1916 to 1949. Harry Ford Sinclair was born in Wheeling, **West Virginia**, on 6 July 1876. He entered the oil business in Kansas in 1901, expanding into **Oklahoma** and other states. He founded Sinclair Oil and Refining Corporation in 1916 to consolidate many of the various interests he had built up. At the time, Sinclair's operations were producing about 9,000 bpd and had a **refining** capacity of more than 15,000 bpd. In 1917, Sinclair consolidated other properties into the Sinclair Gulf Oil Corporation. In 1919, he created the Sinclair Consolidated Oil Corporation as a holding company for his two major corporations and other interests he had acquired.

By 1929, Sinclair had upstream operations in Kansas, Oklahoma, **Texas**, Louisiana, and Wyoming. The company had numerous refineries, a **tanker** fleet, and 1,000 **gasoline** stations. From 1932 to 1943, the holding company dropped Sinclair from its name, becoming simply the Consolidated Oil Company, but in 1943 the Sinclair name was back; the company became **Sinclair Oil Corporation**.

Harry Sinclair played a major role in the **Teapot Dome scandal**, which led to his indictment on charges of bribery and conspiracy to defraud the government. He was acquitted but served six and a half months in prison for contempt of court and contempt of the U.S. Senate. However, during **World War II** he served on the Petroleum Industry War Council—*see* PETROLEUM ADMINISTRATION FOR WAR (PAW) (1942–1946). When Sinclair retired from his company in 1949, Sinclair Oil had assets of $1.2 billion, with earnings of $68 million. Apart from his career in petroleum, Sinclair was also one of the organizers of baseball's Federal League in 1914–1915. He died on 15 June 1953.

SINCLAIR OIL CORPORATION. Independent integrated petroleum company founded in 1916 by **Harry Sinclair**. The company was acquired by **Arco** in 1969; Arco was later acquired by **BP**.

SINCOR PROJECT. *See* ORINOCO OIL BELT.

SKELLY OIL COMPANY. *See* GETTY, J. PAUL (1892–1976).

SLADE COMMISSION (1913–1914). British government commission created by **Winston Churchill** and headed by Rear Admiral Edmund Slade to study the new oil fields of Persia (**Iran**) operated by the Anglo–Persian Oil Company (APOC, later **Anglo–Iranian Oil Company**, eventually **BP**) and report on their potential for ensuring supplies of oil to the Royal Navy. The Royal Navy under Admiral **John Arbuthnot Fisher** had begun the conversion away from coal in 1905 and was already partly reliant on oil. Churchill wanted to complete the process and ensure the navy's oil supplies on a long-term basis. The commission examined the geology of the oil fields and APOC's equipment and operations, issuing its final report on 6 April 1914. Its conclusions were favorable.

The commission advised the Admiralty to enter into a contract with APOC and to give it the financial support it needed in order to expand. The commission also advised the government to maintain some influence over the company's strategic direction, making sure new capital was used wisely and oil fields were developed intelligently. The government bought a majority stake in APOC on 20 May 1914, based on the findings of the commission and several other factors. Admiral Slade also investigated potential oil-bearing areas in **Bahrain** and **Kuwait** and recommended that actions be taken to ensure that the development of those areas would be British; this was partly achieved.

SMITH, DONALD (1820–1914). First chairman of the Anglo–Persian Oil Company (later **Anglo–Iranian Oil Company**, eventually **BP**). Donald Alexander Smith was born 6 August 1820 in Forres, Scotland, and relocated to **Canada**, where he worked for 30 years at the Hudson's Bay Company. He also served in the Canadian government. Smith made a fortune investing in stocks, real estate, railroads, and other areas unrelated to petroleum. He was later named the first Baron Strathcona. Toward the end of his life, he became chairman of the **Burmah Oil Company**, and in 1909 he became the first chairman of the Anglo–Persian Oil Company. He died on 21 January 1914.

SNELLING, WALTER (1880–1965). Father of **natural gas liquids**. Born 13 December 1880, Walter Otheman Snelling was the elder son of Walter Comonfort Snelling (1859–1893), noted for patenting an adding machine. He earned a bachelor's degree from Harvard (1904) and a PhD from George Washington University (1907). In 1910 while working for the U.S. Bureau of Mines he identified propane as a volatile component in gasoline. He built his own still using an old water heater and developed a method for capturing the various volatile components as natural gas liquids. A patent for Snelling's process was issued 25 March 1913. Snelling formed the American Gasol Company and in 1912 installed a propane tank at a commercial home, thus beginning the **Liquefied Petroleum Gas** industry. In 1913 Snelling sold his patent to **Frank Phillips**, the future founder of **Phillips Petroleum** (ultimately part of **ConocoPhillips**) for $50,000. Snelling died 10 September 1965.

SOCAL. *See* CHEVRON.

SOCAR. *See* AZERBAIJAN.

SOCIÉTÉ COMMERCIALE ET INDUSTRIELLE DE NAPHTE CASPIENNE ET DE LA MER NOIRE. *See* ROTHSCHILD COMPANY.

SOCIÉTÉ IRANO–ITALIENNE DES PÉTROLES (SIRIP). *See* ENI; IRAN; MATTEI, ENRICO (1906–1962).

SOCIÉTÉ NATIONALE DES PÉTROLES D'AQUITAINE (SNPA). *See* ELF AQUITAINE.

SOCIETY OF EXPLORATION GEOPHYSICISTS. Professional society founded 1930. The Society of Exploration Geophysicists (SEG) is headquartered in **Tulsa, Oklahoma**. SEG states that it "promotes the science of geophysics and the education of applied geophysicists" and "fosters the expert and ethical practice of geophysics in the exploration and development of natural resources, in characterizing the near surface, and in mitigating earth hazards." When the SEG began on 11 March 1930, it was known as the Society of Economic Geophysicists and had 30 members. The first president was Donald C. Barton. In 1931, the organization changed its name to the Society of Petroleum Geophysicists. A year later, it became affiliated with the **American Association of Petroleum Geologists** (AAPG) as the Division of Geophysics of AAPG. In 1937, the organization changed its name yet again to the Society of Exploration Geophysicists and formally resigned from the AAPG. It was established as an independent society, retaining affiliation with AAPG. SEG continued to hold its annual meeting jointly with AAPG until 1955.

In 1936, SEG began publication of its flagship journal, *Geophysics*. By 1950, the society had more than 2,500 members. Membership peaked at just under 20,000 in 1985 and then began to decrease in response to the low oil prices of the 1980s (*see* OIL COUNTERSHOCK) and other factors. Membership numbers recovered slowly and in 2003 exceeded the previous peak. In 2006, membership reached an all-time high of more than 27,000. In addition to its journal publications, SEG also publishes high-quality monographs in applied geophysics. These are mostly but not exclusively related to oil and gas exploration. SEG also helps to set a number of industry technical standards. *See also* GEOPHYSICAL PROSPECTING.

SOCIETY OF PETROLEUM ENGINEERS (SPE). International professional association founded in 1957 with a mission to "collect, disseminate, and exchange technical knowledge concerning the exploration, development and production of oil and gas resources, and related technologies for the public benefit." The SPE is headquartered in **Texas**, with offices in **Houston** and Richardson. It also has international branch offices in London, Kuala Lumpur, and **Dubai**. The SPE organizes more than 50 events a year related to the petroleum industry and publishes the highly respected *Journal of Petroleum Technology*. When the SPE was founded in 1957, it had 12,500 members. It had almost 18,000 members by 1970, almost 39,000 by 1980, and more than 51,000 by 1990. In 2017, it had more than 168,000 members.

The SPE grew out of the American Institute of Mining Engineers (AIME) Standing Committee on Oil and Gas, which was founded in 1913. In 1919, the AIME joined with the American Institute of Metals to form the American Institute of Mining and Metallurgical Engineers; the combined society kept the initials AIME. In 1922, the Standing Committee on Oil and Gas became the AIME Petroleum Division (one of 10 full divisions), under the influence of early industry leaders including **Anthony F. Lucas** and **Everette DeGolyer**. The Petroleum Division began with about 900 members and nearly tripled by 1948 when it became the Petroleum Branch, one of three major units within the AIME. In 1957, after a decade of explosive membership growth, the Petroleum Branch became the Society of Petroleum Engineers of AIME. In 1985, SPE incorporated separately from AIME.

SOCONY. *See* MOBIL; STANDARD OIL COMPANY OF NEW YORK (SOCONY).

SOCONY–VACUUM. *See* MOBIL.

SOHIO. Standard Oil Company of Ohio, later **BP**. The original **Standard Oil** was incorporated in Ohio in 1870 as Standard Oil (Ohio). It grew into a large empire that dominated the oil industry in the **United States** and had vast international operations, and it was ultimately dissolved in 1911 by order of the U.S. Supreme Court into 34 independent companies. One of the new independent companies was Standard Oil of Ohio, which was close in many ways to the original Standard Oil. It emerged from the breakup as a regional oil company centered in Ohio, with a refinery in Cleveland. The company began retail **gasoline** operations and in 1921 opened a new refinery in Toledo. By the 1930s, it had adopted the Sohio trade name. Over the course of the next several decades, it concentrated largely on downstream operations. In the late 1970s, however, events in **Alaska** transformed it into a more integrated petroleum company.

In 1967, **Arco** and the **Humble Oil and Refining Company** (later **Exxon** USA) discovered the **Prudhoe Bay oil field**. BP had also been exploring in Alaska and would eventually control a considerable portion of the field. However, BP did not have sufficient retail infrastructure in the United States to dispose of the crude once it was produced. Sohio had the infrastructure but not enough supplies of crude. The two formed a partnership, signing a memorandum of intent in June 1969. BP would acquire an initial 25 percent of Sohio. Sohio would take over BP's leases in Prudhoe Bay, as well as U.S. East Coast downstream interests that BP had acquired from Arco in 1968; the assets had once belonged to **Sinclair Oil Corporation**, and Arco had to divest them as part of its acquisition of that company.

The structure of the partnership was such that BP would gain an increasing stake in Sohio as Prudhoe Bay production increased. BP initially acquired 1,000 special shares in Sohio, each worth 4,466 ordinary shares, to give BP an initial 25 percent. The special shares would increase in value as a function of the company's production from Prudhoe Bay until that production reached

0.6 Mmbpd. At that point, BP would have a maximum 54 percent stake in Sohio. Beyond 0.6 Mmbpd, BP would not receive any additional stake in Sohio but would receive special royalty payments in cash or oil. If production exceeded 1.05 Mmbpd, Sohio would be the sole beneficiary.

The U.S. Department of Justice raised antitrust objections to the partnership. To satisfy the government, Sohio had to sell off about 30 percent of its retail outlets in Ohio as well as some of its service stations in western **Pennsylvania**. The deal was approved, and Sohio and BP worked together to develop Sohio's portion of Prudhoe Bay. Sohio, which did not have much upstream experience, needed BP's expertise. BP's stake in Sohio increased as forecast, and in June 1978 BP became Sohio's majority shareholder.

Despite BP's controlling interest, Sohio remained very independent in operations and attitude, sometimes to the point of noncooperation. When BP's crude supplies were curtailed by the Iranian Revolution of 1979 and by the takeover of its interests in **Nigeria**, BP sought to purchase crude from Sohio. It was not seeking especially favorable terms from a subsidiary, but simply a source of supply. Sohio raised so many legal and business objections that 18 months and 10 meetings later there was still no sale. One senior BP executive reportedly said that it was "easier dealing with Exxon." In 1987, BP acquired the rest of Sohio and merged it with other U.S. interests, including BP Alaska, to form BP America.

SONATRACH. *See* ALGERIA.

SOUTH CHINA SEA. Sea in the western Pacific region that is the focus of competing territorial claims between several countries, notably **China**, the Philippines, Vietnam, **Malaysia**, and **Indonesia**. Potential oil and gas resources constitute part of the various disputes, to a debatable extent. Some analysts argue that the disputes are more about fundamental sovereignty rights and that hydrocarbon exploration and development are important to the extent that they represent the exercise of such sovereign rights.

China claimed nearly the entirety of the South China Sea with the "Nine-Dash Line" (originally containing 11 dashes), drawn in 1949 after the formation of the People's Republic of China. The line was originally drawn by the predecessor government of China in 1947, based on a previous 1935 map. The Nine-Dash Line overlaps with the exclusive economic zones of other countries around the South China Sea, as defined by the 1982 **United Nations Convention on the Law of the Sea** (UNCLOS). In particular, there are several claimants to disputed archipelagos, islands, and their surrounding areas. These include the Paracel Islands (Chinese: Xisha), the Spratly Islands (Nansha), and Scarborough Shoal (Huangyan Dao). The Paracels are claimed by both China and Vietnam. The two countries fought a battle over them on

19–20 January 1974, and China has controlled them since then. The Spratlys have also been the site of skirmishes and are occupied by China, Malaysia, the Philippines, Taiwan, and Vietnam. In 2017 China completed the construction of artificial islands with military facilities on Fiery Cross, Mischief, and Subi reefs in the Spratlys. Elsewhere in the sea, Scarborough Shoal has been the cause of various standoffs between China and the Philippines since 2012. On 12 July 2016 an international tribunal in The Hague ruled that China's claims based on the Nine-Dash Line have no force in international law; the tribunal upheld the exclusive economic zones according to UNCLOS. China rejected the tribunal's ruling.

Most of the known hydrocarbon resources of the South China Sea are in undisputed areas relatively close to coastlines. Examples include Vietnam's Bạch Hổ (White Tiger), Dai Hung (Big Bear), and Hac Long (Black Dragon) fields. Other examples are China's gas fields in the Pearl River Mouth basin, including the Wushi field. However, Indonesia's **giant** East Natuna gas field is technically within China's Nine-Dash Line and is thus a cause for concern. The field was discovered in 1973 by **AGIP** (part of **ENI**), although development began much later. The proven and probable reserves of the South China Sea are estimated at 11 Bb oil and 190 Tcf gas. About 90 percent of those proven and probable reserves of both oil and gas are in undisputed areas. What is more contentious, and less certain, is the endowment of hydrocarbons in the lightly explored disputed areas. China National Offshore Oil Corporation has issued estimates of 125 Bb oil and 500 Tcf gas awaiting discovery in the South China Sea. Other estimates are as low as 3 Bb of oil.

Whatever the actual amount of oil and gas that may ultimately be discovered, the exploration process will be politically dangerous. China has already stopped Philippine exploration a number of times in the Reed Bank area since 2011. When China deployed its *Hai Yang Shi You* mobile offshore oil platform in disputed waters near the Paracel Islands in 2014, it sparked riots in Vietnam. Apart from oil and gas resource exploration and development, another factor making the South China Sea contentious is that it is a very important transit area for Middle Eastern and African oil destined for East Asian markets, as well as other world trade, entering from the **Strait of Malacca**.

SOUTH IMPROVEMENT COMPANY. An 1872 scheme whereby American **refining** companies and railroads attempted to control the market for oil and oil shipping. **Standard Oil** obtained preferential rates from freight companies, thus lowering its costs relative to competitors. The South Improvement Company was organized by Tom Scott, head of the Pennsylvania Railroad, as a supposedly independent **Pennsylvania** corporation in 1872. South Improvement's shareholders were large refiners such as Standard Oil, and three railroads—the Pennsylvania, the Erie, and the New York Central and Hudson River. Refiners that were members of South Improvement—mainly meaning Standard Oil—would obtain preferential shipping rates from South Improvement's railroad members. The agreement was administered as a system of rebates paid to Standard Oil, but not to other customers who shipped petroleum products. The rebates could be substantial. In one locality, South Improvement paid an effective rate of $0.44 per barrel after the rebate, while Standard Oil's competitors paid $1.50.

In February 1872, the railroads doubled the freight rates for refiners that were not members of South Improvement. Standard Oil exploited the panic that ensued over the next two months and bought out 22 of its 26 competitors in the Cleveland refining industry, in what became known as the "Cleveland Massacre." The scheme enraged the upstream producers in the Pennsylvania oil regions, and they boycotted refiners and railroads. South Improvement lost its charter in April 1872, about three months after it had been granted. The U.S. Congress launched an investigation, and **John D. Rockefeller**, the head of Standard Oil, was publicly identified as a mastermind behind the scheme. One newspaper dubbed him the "Mephistopheles of Cleveland." Despite the collapse of the scheme, he had achieved his primary purpose, which was to buy out his competition. Standard Oil and the railroads also continued to negotiate in secret, and Standard Oil continued to obtain preferential rates.

SOUTH PENN OIL COMPANY. One of the original companies of **Standard Oil**; became independent in 1911 and part of **Pennzoil** in 1962 (acquired in part by **Royal Dutch Shell** in 2002). The South Penn Oil Company was formed on 27 May 1889 by Standard Oil. Standard controlled about 90 percent of oil **refining** in the **United States** but was not as dominant in exploration and production of crude oil. It created South Penn to increase its sources of crude, beginning in the **Pennsylvania** oil regions. South Penn was successful in exploiting Pennsylvania oil, including the **Bradford** field, and it soon expanded into **West Virginia**. The West Virginia fields proved fertile, making South Penn Standard's leading upstream producer by 1898. However, the unit suffered from the exhaustion of the Appalachian resource province. Production increased about 10-fold during the 1890s but dropped by about half over the subsequent decade.

When Standard Oil was dissolved in 1911 by order of the U.S. Supreme Court, South Penn found itself independent. Because it operated in a declining province, it needed new avenues for growth. In 1925, it bought a controlling interest in Pennzoil. Pennzoil had grown out of the Pennsylvania Refining Company (PRC), formed in 1886 by Standard Oil. Unable to compete as a **gasoline** refiner, the company had developed and marketed high-quality lubricants under the Pennzoil brand name. The popularity of the brand name caused PRC to adopt it as a company name in 1924. South Penn completed its purchase of Pennzoil in 1955 and, while keeping the valuable brand name, changed the company's name to South Penn. In 1963, the enlarged South Penn Oil Company merged with the **Zapata Petroleum Corporation** to form the Pennzoil Company. Pennzoil was thus once again both a brand name and a company name. The lubrication part of Pennzoil was acquired by Royal Dutch Shell in 2002.

SOVIET UNION. *See* AZERBAIJAN; KAZAKHSTAN; RUSSIA.

SPECTRUM 7 ENERGY CORPORATION. *See* BUSH, GEORGE W. (1946–).

SPINDLETOP. Oil field discovered 10 January 1901 near Beaumont, **Texas**, by **Anthony F. Lucas**. The Spindletop oil discovery spawned or nurtured several major petroleum companies and initiated a petroleum boom in Texas. Spindletop is often considered to mark the transition from the petroleum industry's focus on illumination to its modern focus on transportation fuel.

By the end of 1902, more than 600 companies had been formed or had major operations at Spindletop; some would become giants of the modern international petroleum industry. The Guffey Petroleum Company—*see* GUFFEY, JAMES McCLURG (1839–1930)—and the Gulf Refining Company of Texas were both born at Spindletop. They later became **Gulf Oil** (eventually merged with **Chevron**). The Texas Company (later **Texaco**, eventually joined with Chevron) was also established at Spindletop. The **Humble Oil and Refining Company** (acquired by Standard Oil Company of New Jersey, later **Exxon**, eventually **ExxonMobil**) was founded by families connected with Spindletop. Another company that became successful through Spindletop was the **Magnolia Petroleum Company** (later part of **Mobil**, eventually ExxonMobil). **Sun Oil Company** was yet another. **Edward Doheny**'s Pan American Petroleum and Transport Company (later part of Standard Oil of Indiana, or **Amoco**) was also involved. The **Shell Transport and Trading Company** (later **Royal Dutch Shell**) played a role in transporting Spindletop oil to the **United Kingdom**.

The Spindletop field was discovered in the formations surrounding a salt dome structure. The first unsuccessful wells at Spindletop were **drilled** using cable tool equipment. Anthony F. Lucas's successful discovery well was drilled using a more modern rotary tool. Lucas's well was a gusher struck at 1,139 feet. It blew a column of oil 100 feet high and was not brought under control until nine days later. Spindletop as a whole produced nearly 48,000 bpd in 1902. By 1904, production was down to 10,000 bopd. Principles of reservoir management and conservation were not yet well developed, and too many wells had been drilled in the shallow caprock. Lucas reportedly said that "the cow was milked too hard; moreover, she was not milked intelligently."

A second Spindletop boom occurred when prospectors began drilling to deeper depths and exploring the flanks of the salt dome. On 13 November 1925, the Yount-Lee Oil Company discovered oil on the flanks at 5,400 feet, and this was followed by many other discoveries. The second boom was somewhat better controlled than the first. Spindletop production peaked in 1927 at about 57,500 bopd. By 1985, the Spindletop field had produced more than 153 Mmbo. In the 1990s, the field continued to produce a limited amount of oil from stripper wells. *See also* HIGGINS, PATTILLO (1863–1955).

STANDARD OIL. Dominant oil company in the United States in the late 19th and early 20th centuries. Standard Oil was the progenitor of several of the **major oil companies** that controlled the global oil market for a large part of the 20th century and still exert considerable influence today. Standard Oil was founded in 1870 by **John D. Rockefeller** and **Henry Flagler** as an Ohio joint stock company. The company had roots in a number of previous Rockefeller companies: Andrews, Clark and Company (1863), Rockefeller and Andrews (1865), and Rockefeller, Andrews, and Flagler (1867). This first Standard Oil company of 1870 was not identical to the Standard Oil Company of Ohio (**Sohio**), an independent successor company formed after the breakup of the Standard Oil trust in 1911. Standard Oil was capitalized initially at $1 million. Rockefeller was president, and Flagler was secretary. Rockefeller owned 27 percent of the stock.

Standard Oil was already a powerful company when it was formed, and it continued to grow through a combination of shrewd management and the application of a number of anticompetitive practices. As the company grew, it was able to negotiate secret rebates from freight companies, thus lowering its costs and further increasing its advantage against competitors (*see* SOUTH IMPROVEMENT COMPANY). It selectively lowered prices in markets where competitors operated, while keeping prices high elsewhere. It thus forced competitors to choose between going bankrupt or being acquired by Standard Oil. It also made some friendly acquisitions of companies eager to be affiliated with it. Standard was thus able to achieve significant horizontal integration of its operations. It also worked toward vertical integration, controlling **refining**, transportation, and marketing.

At first, Standard did not move upstream into oil exploration and production, which Rockefeller considered risky endeavors. However, the company was not immune to the vagaries of crude oil supply, especially as the initial discoveries in Pennsylvania began to run low. Rockefeller took a risk on the new **Lima–Indiana oil field** discovered in 1886 in northwestern Ohio and later in Indiana. The oil had a high sulfur content, making its refined products foul smelling and unsuitable for use in lamps. Standard Oil marketed the oil as fuel oil, but there was no guarantee that it would ever be usable for illumination. Rockefeller consulted with **Herman Frasch**, who developed a process to desulfurize the oil. Standard completed a large refinery at Whiting, Indiana, near Chicago in 1890 to refine Lima crude, and it soon converted other refineries to do so. Lima crude represented a competitive advantage for Standard Oil once it had the Frasch process. It could now produce refined products competitive with those derived from **Pennsylvania**–Appalachian oil, but using cheaper, lower-quality Lima oil as an input. Once Standard had begun upstream integration with its purchase of Lima properties, it continued the process with Appalachian properties. By 1891, Standard held properties producing about a quarter of the U.S. national output of crude.

In foreign markets, Standard Oil faced considerable competition from **Branobel** and the **Rothschilds** beginning in the 1880s, and from **Royal Dutch** and **Shell** beginning in the 1890s. It waged price wars against all these major foreign competitors, but it was ultimately unable to bankrupt them or buy them out. Royal Dutch and Shell instead later combined with each other to form the **Royal Dutch Shell** group in 1907.

By 1880, Standard Oil controlled about 90 percent of refining capacity in the United States. In 1882, Rockefeller created the Standard Oil Trust, headquartered in New York City. The trust united Standard Oil (Ohio) and various affiliated companies under one umbrella. It held "in trust" all the assets of the various Standard Oil companies. The structure, which was the first major implementation of the trust-type arrangement in the United States, was deliberately confusing and opaque to discourage public scrutiny.

Standard Oil's size and power led to public hostility and helped prompt the passage of the **Sherman Antitrust Act** in 1890. That year, the state of Ohio sued Standard Oil for illegal monopolization of the petroleum industry. On 2 March 1892, the Ohio Supreme Court convicted Standard Oil of violating the Sherman Antitrust Act by forming a holding company, interpreted as being a conspiracy in restraint of trade. Despite the Ohio court ruling, and continuing legal attempts to stop it, the trust continued to operate for several years.

Eventually Standard Oil took advantage of favorable New Jersey laws allowing corporations to hold stock in other corporations. In 1899, it renamed its New Jersey firm Standard Oil Company (New Jersey) and placed it at the hub of the organization. Assets that were formerly in the Standard Oil Trust were put into the holding company. The various Standard Oil companies continued to function effectively as a single entity, through interlocking directorates and various other complex arrangements. Despite the New Jersey charter, Standard Oil continued to be run from New York. By this time, Rockefeller was secretly retired and the company was being run largely by **John Archbold**.

The discovery at **Spindletop** in 1901 ushered in new competitors for Standard, including the future **Texaco** and **Gulf**, as petroleum's age of illumination gave way to its age of energy, and **gasoline** replaced kerosene as the most important refined product. In addition to the increased competition, Standard also faced continuing legal challenges. In November 1906, U.S. attorney general George Wickersham brought charges against the company for monopoly and restraint of trade in the Federal Circuit Court of the Eastern District of Missouri. The case, prosecuted by **Frank Billings Kellogg**, lasted three years. In 1909, the court found Standard Oil (New Jersey) guilty of violating the Sherman Antitrust Act by forming a holding company and by restraining competition among merged firms by fixing transportation costs, supply costs, and prices. Standard Oil appealed to the U.S. Supreme Court, which upheld the ruling on 15 May 1911. A short time later, the Supreme Court ordered Standard Oil dissolved into 34 independent companies.

Among the companies formed by the dissolution of Standard Oil were many that would continue to be giants in the global petroleum industry: Standard Oil of New Jersey, or Jersey Standard (later **Exxon** and **ExxonMobil**); Standard Oil Company of New York (Socony, later **Mobil** and ExxonMobil); Vacuum Oil Company (later Mobil and ExxonMobil); Standard Oil of Indiana (later **Amoco**, eventually merged into **BP**); Standard Oil of California (later **Chevron**); Atlantic Refining Company (later Atlantic Richfield, or **Arco**, eventually acquired by BP); Continental Oil Company (later **Conoco** and **ConocoPhillips**); South Penn Oil Company (later **Pennzoil**, eventually acquired in part by Royal Dutch Shell); Standard Oil Company of Ohio (Sohio, eventually acquired by BP); and Ohio Oil Company (later **Marathon**).

Of these, Jersey Standard was by far the largest, with about 43 percent of the assets of the original company. The second largest, Socony, had less than 5 percent. The newly independent companies were not functionally integrated, and many found themselves with refining and marketing capacity but no sources of crude. In the aftermath of the breakup, Jersey Standard, Socony, and others embarked on acquisition programs to gain control of crude **reserves**. Refer to appendix 3.

STANDARD OIL COMPANY OF CALIFORNIA. *See* CHEVRON.

STANDARD OIL COMPANY OF INDIANA. *See* AMOCO.

STANDARD OIL COMPANY OF NEW JERSEY. *See* EXXON; EXXONMOBIL.

STANDARD OIL COMPANY OF NEW YORK (SOCONY). One of the companies that formed **Mobil** (later **ExxonMobil**). Socony was one of the independent companies formed from the 1911 dissolution of **Standard Oil**. Its first president was Henry Clay Folger. At the breakup, Socony had assets of about $60 million, less than 5 percent of the former trust's assets. Although it was the second largest of the newly independent companies, it was only a 10th of the size of the largest, Standard Oil Company of New Jersey (later **Exxon**). As part of Standard Oil, Socony had handled the bulk of the trust's foreign sales and was relatively well known throughout the world. But the newly independent Socony was in a difficult position. It had **refining** and marketing operations but essentially no oil or gas production, no **pipelines**, and no transport capability. Its early years were spent recovering from this situation and building an integrated business.

As part of this thrust, Socony bought a 45 percent share in the **Magnolia Petroleum Company**, an integrated producer and refiner, in 1918. It completed the acquisition of Magnolia in 1925, and it purchased General Petroleum in **California** and White Eagle within a year. In 1926, it began trading on the New York Stock Exchange with the symbol NY. In 1928, it joined the **Iraq Petroleum Company** (IPC) consortium, via the **Near East Development Corporation**. IPC membership would later give it interests in **Iraq**, **Abu Dhabi**, **Qatar**, and **Oman**. In 1931, Socony merged with the **Vacuum Oil Company**, another Standard Oil successor, to form Socony–Vacuum, the company that would become Mobil.

STANDARD OIL COMPANY OF OHIO. *See* SOHIO.

STANDARD–VACUUM OIL COMPANY (STANVAC). *See* EXXON; MOBIL.

STATOIL. National oil company of **Norway**, partially privatized in 2001. In 2007, Statoil merged with **Norsk Hydro**'s oil and gas business to form StatoilHydro. Statoil (Den Norske Stats Oljeselskap) was established on 14 June 1972 as a vehicle for Norwegian government participation in and control over the development of Norway's petroleum. The government also established the **Norwegian Petroleum Directorate** to manage the petroleum industry and created the Division of Petroleum and Mining within the Ministry of Industry, which formulated broad policy and strategy.

Norway asserted sovereignty over its continental shelf in 1963, and in 1965 it began allowing exploration. In 1969, **Phillips Petroleum** (later **ConocoPhillips**) led the discovery of the **Ekofisk** field, ushering in Norway's era as a petroleum power. Several major discoveries followed in the 1970s. In February 1974, **Mobil** discovered the **supergiant** Statfjord field, in which

Statoil had a 50 percent interest. Statfjord, with an **EUR** of 9 Bb, is the **North Sea**'s largest oil field and was Statoil's and Norway's mainstay over the next two decades. Statoil took over operation of Statfjord in 1987. In 1979, **Royal Dutch Shell** discovered the supergiant Troll **natural gas** field, which would become a key gas supplier to mainland Europe. Shell operated the field in development, although it only owned an 8.3 percent interest; Statoil's share was 74.6 percent. Statoil took over operations in 1996.

In 1981, Statoil began operating the supergiant Gullfaks field (discovered 1978) with Norsk Hydro and Saga Petroleum, Norway's only **independent** oil company. Gullfaks, the first Norwegian North Sea field that was completely Norwegian owned, had an EUR of 5.3 Bb and began producing in 1987. In 1975, Statoil began Norpipe, its first undersea **pipeline**, extending to the **United Kingdom**. In 1985, Statoil began transporting North Sea gas to the Norwegian mainland via Statpipe, its natural gas pipeline system. Statpipe had about 520 miles of pipeline on the sea floor at depths down to about 1,080 feet.

In 1985 and 1986, Statoil moved downstream all the way to retail by acquiring Esso (**Exxon**) **gasoline** stations in Denmark and Sweden. In 1992, it acquired more stations from **BP** in Ireland, and it moved into **petrochemicals** in 1994. In 1990, Statoil made a major decision to internationalize operations. It bought into existing BP exploration and development licenses in **China**, Vietnam, **Congo**, and **Angola**. Through the 1990s, the company's reserves and assets grew, but revenues stayed flat or fell as they did in most of the industry. In 1999, Statoil helped Norsk Hydro acquire Saga Petroleum in return for some Saga assets.

In 2000, Statoil decided that it needed to act less like a national oil company guided by Norwegian politics and more like a major international petroleum concern. On 25 May 2001, it underwent a partial privatization and changed its name to Statoil ASA. The Norwegian government retained an interest of 81.7 percent, but the new company was listed on both the Oslo and New York stock exchanges. Statoil moved ahead with both expansion and internationalization. In 2002, it began the Snøhvit gas project in the **Barents Sea**, with partners Norsk Hydro, TotalFinaElf (later **Total**), and Gaz de France. Snøhvit is a complex project, with all development installations undersea. Statoil also entered into an agreement with the government of **Iran** to develop some phases of South Pars, the world's second-largest gas field (with EUR of 77.433 Bboe). Unfortunately the Iranian business also brought scandal with it, as Statoil was accused in 2003 of making inappropriate payments to secure deals in Iran. The scandal led to the resignation of CEO Olav Fjell.

Beginning in October 2007, Statoil began supplying up to 20 percent of British gas demand from the Ormen Lange gas field through the 740-mile Langeled Pipeline, the world's longest undersea pipeline. In October 2007,

Statoil merged with the oil and gas business of Norsk Hydro in a $30 billion deal and became StatoilHydro. Even though the Norwegian government continued to own 67.3 percent of the company, Statoil acted less like a national oil company in the 2000s and 2010s and more like a multinational major. It was involved in projects all over the world, including Angola, **Azerbaijan**, **Canada**, Iran, and **Venezuela**. Closer to home, it continued to make discoveries in the North Sea and the Barents Sea. In 2011 it acquired Brigham Resources for over $4 billion, thereby gaining entry into the **tight oil** of the **Bakken** Shale in **North Dakota**. In 2016, Statoil controlled 2.405 Bb of oil reserves and 14.637 Tcf of gas. It produced oil and gas at the rate of 1.024 Mmbpd and 4.658 Bcfpd respectively.

STATOILHYDRO. *See* NORSK HYDRO; STATOIL.

STAVANGER. City (pop. 128,000, 2012) in **Norway**; major center of the Norwegian petroleum industry. Stavanger is on the **North Sea** coast and well situated with respect to Norway's **offshore** fields. It hosts refineries and a large number of oil and gas service companies. Stavanger has an ice-free harbor and is Norway's closest major port to the **United Kingdom**. The **Norwegian Petroleum Directorate** is located in Stavanger. StatoilHydro (formed from the 2007 merger of **Statoil** with the oil and gas division of **Norsk Hydro**) is headquartered at Forus, near Stavanger.

STEINEKE, MAX (1898–1952). Important American petroleum geologist who played a major role in early oil discoveries in **Saudi Arabia**. Steineke, who graduated in geology from Stanford in 1921, was chief geologist at the California-Arabian Standard Oil Company (Casoc, later **Aramco**) from 1936 to 1950. He was responsible, after persisting through considerable adversity, for the discovery of the first commercial quantities of oil in Saudi Arabia with the Dammam No. 7 well. He was also instrumental in discovering the **Abqaiq** field. Steineke developed the technique of sinking relatively shallow wells for the purpose of mapping subsurface structures. This structural drilling technique ultimately contributed to the discovery of huge oil fields in Saudi Arabia, including **Ghawar**.

STERLING, ROSS (1875–1949). American oil executive and governor of **Texas** (1931–1933) during the collapse in oil prices following the discovery of the **East Texas oil field**. Ross Shaw Sterling was born 11 or 22 February 1875 in Chambers County, Texas. He was a farmer, shopkeeper, and banker before entering the oil business in 1910. He bought two wells and in 1911 cofounded the Humble Oil Company, which he served as president. Humble Oil formed the basis of the **Humble Oil and Refining Company**, which ultimately became part of Standard Oil Company of New Jersey (later **Exxon**). Sterling sold his shares of Humble in 1925 and invested in **Houston** real estate and newspapers.

In 1931, Sterling became governor of Texas. His term was during the Great Depression and was a difficult one. Among the crises he faced was the collapse in oil prices because of unrestrained production in the recently discovered East Texas field. When the East Texas producers ignored the production quotas imposed by the **Railroad Commission of Texas**, Sterling declared martial law in four producing counties and sent the Texas Rangers and the National Guard into the East Texas field on 17 August 1931 to shut down production. On 2 February 1932, a federal court ruled that he had exceeded his authority and ordered the rescission of martial law. Sterling ran for a second term as governor in 1932 but lost to Miriam A. Ferguson, who assumed power on 17 January 1933. Sterling returned to Houston and formed the Sterling Oil and Refining Company, which he led from 1933 to 1946. He died on 25 March 1949 in Fort Worth, Texas.

STEWART, LYMAN (1840–1923). Cofounder of **Union Oil Company of California**. Lyman Stewart was born in **Venango County**, **Pennsylvania**, on 22 July 1840, in an emerging center of the modern oil industry. His initial forays into the oil industry came early in life. In 1859, when Stewart was 19 years old, **Edwin Drake** drilled a successful well near **Titusville**. Shortly thereafter, Stewart invested $125 for one-eighth of an oil lease. Lacking money to drill, he and his partners lost the lease, Stewart lost all his savings, and others later discovered oil on the land. When he tried again in 1861, he and his partners discovered oil, but ambient prices were too low to justify development. He lost his savings again, and once again the field was later developed by others.

After serving in the Civil War and briefly studying bookkeeping, Stewart returned to Pennsylvania. In 1868, he joined with his brother Milton to form the Claremont Oil Company. The company was successful in the oil business, but an unrelated investment in agricultural machinery wiped it out. By 1872, Stewart had lost his savings once more. In 1877, he tried the oil business yet again. He formed a partnership with Wallace Hardison and invested in the area of the **Bradford oil field**. The partners found it difficult to operate in a market so dominated by **Standard Oil** and soon sold their holdings.

In 1883, Stewart and Hardison decided to try their luck in **California**. After seven dry holes in Pico Canyon in southern California and almost out of money, they struck a small amount of oil in 1884. They sold their interest to Pacific Coast Oil Company, the future Standard Oil Company of California (later **Chevron**). They explored and drilled constantly, and by 1886 they were producing about 10 percent of California's output. They allied themselves with Thomas R. Bard, who had made money in California oil and real estate. The partners organized their activities in three companies. One was the Hardison and Stewart Oil Company, with Stewart as president. Another was Bard's company Mission Transfer, which owned large tracts of land in Ventura County and also a refinery, storage tanks, and **pipelines**. Hardison and Stewart bought a half interest in Mission Transfer. A third was the Sespe Oil Company, formed to drill in the Sespe Canyon. Bard was president, while Stewart and Hardison were majority owners. In the late 1880s, Stewart designed a tanker ship carrying oil in steel tanks. Some wrongly claim this to have been the first oil tanker. However, that honor properly belongs to **Ludvig Nobel**'s ***Zoroaster*** that plied the Caspian as early as 1878.

In 1890, Hardison, Stewart, and Bard consolidated their three companies into the new Union Oil Company of California, headquartered in Santa Paula, Ventura County. That year, Union produced about 84,000 barrels—approximately a quarter of California's output. Bard served as the company's first president. Union discovered new oil in Torrey Canyon and in 1892 purchased the Los Angeles Oil Company. That company had no oil but held promising land in Los Angeles. Union began drilling immediately and discovered California's largest find to that time, Adams No. 28. It produced 15,000 barrels in its first year and initiated a California oil boom that flooded the market.

In 1894, Stewart replaced Bard as president. The two had disputed constantly since the company's formation. Stewart viewed Bard as a short-term profit taker, and Bard viewed Stewart as a poor manager. Bard continued to challenge Stewart for control of the company through the 1890s. In 1899, he attempted a hostile takeover but failed. A year later, he sold his shares and entered politics. He was elected in 1900 to complete an unexpired term in the U.S. Senate.

Stewart expanded his operations all over southern California. In 1901, he moved headquarters from Santa Paula to Los Angeles. He built a refinery in Bakersfield. He sold new issues of stock to raise capital but created a complicated structure of holding companies in order to retain control with a minority of shares. Stewart expanded relentlessly. By 1908, Union's capitalization was $50 million, 10 times its original level. The company became overextended. When it needed credit in 1914, Stewart was forced to resign as president as a condition of obtaining it. He was succeeded by his son **William Lyman Stewart** and became chairman. In 1916, the California courts ruled that the system of holding companies devised by Lyman Stewart to maintain control of Union Oil was illegal, and Union passed from the Stewart family's control. Lyman Stewart died in California on 29 September 1923.

STEWART, WILLIAM LYMAN (1868–1930). Son of **Lyman Stewart**; succeeded his father as president of the **Union Oil Company of California** (1914–1930). William Lyman Stewart was born in **Pennsylvania** in 1868. He graduated from the University of California, working in his father's oil fields in the summer and learning the business from the ground up. In 1898, he became general manager of Union Oil and served as his father's right-hand man. He succeeded his father as president in 1914. Stewart presided over Union Oil during **World War I** and the 1920s, when automobile use exploded. The company's growth reflected the strong demand for petroleum products. Union's production, which had been 4.5 Mmb in 1915, peaked at 18 Mmb in 1923. It was still over 13 Mmb in 1930. William Lyman Stewart died while still president of the company, on 21 June 1930. His sons Arthur C. Stewart and William Lyman Stewart Jr. became vice presidents of Union Oil.

STRAIT OF HORMUZ. Strait linking the **Persian Gulf** with the Gulf of Oman and the Arabian Sea; the only maritime passage by which Persian Gulf petroleum can be shipped to the rest of the world. In 2016, 18 Mmbpd—nearly a quarter of world oil production—flowed through the Strait of Hormuz, destined for Europe, East Asia, and the **United States**. On the north coast of the strait lies **Iran**, and on the south lie the **United Arab Emirates** (UAE) and Musandam, which is an exclave of **Oman**. The northern part of the strait, which is too shallow for large **tankers**, is under Iranian sovereignty. The southern part, which is more important for shipping, is under the joint sovereignty of Iran and Oman. The strait is about 21 miles wide at its narrowest point. Channels for inbound and outbound traffic are two miles wide, with a two-mile buffer zone.

There is considerable tension between the countries around the strait. For example, Abu Musa Island, which lies near the entrance to the strait on the Persian Gulf side, has long been disputed between Iran and the UAE emirate of **Sharjah**. The Tunb Islands, even closer to the strait, are disputed between Iran and the UAE emirate of **Ras al-Khaima**. In addition to such significant local problems, there has also been tension at varying levels between the United States and Iran ever since the Iranian Revolution of 1979. These tensions have threatened shipping through the strait at various times.

The problems between the United States and Iran were exacerbated during the 1980s **Iran–Iraq War**, when the United States tilted openly toward **Iraq**. Iran's leader, Ayatollah **Ruhollah Khomeini**, threatened to close the strait in 1983 if Iraq destroyed Iran's **Kharg Island** oil terminal in the Persian Gulf. U.S. president Ronald Reagan made it clear that the United States would intervene militarily to keep the strait open. Both Iraq and Iran attacked oil tankers during the war, including those of neutral countries. Iran placed mines in the Persian Gulf. On 14 April 1988, the U.S. naval frigate *Samuel B.*

Roberts was damaged by an Iranian mine. There were no casualties, but four days later the United States retaliated with Operation Praying Mantis, which led to a full-blown battle between the United States and Iran in or near the strait. The U.S. Navy sank two Iranian warships and several armed speedboats. It also destroyed two Iranian oil platforms. The battle lasted only a day but highlighted the vulnerability and importance of the strait.

More recently, tensions have flared between Iran and the United States over Iran's nuclear program and suspicions that Iran supported insurgents in Iraq fighting against U.S. troops. At various times after the turn of the 21st century, the Iranian leadership has made implicit threats that it might close the strait. However, it should be noted that closing the strait for a significant period of time may not be as easy to accomplish as it sounds. Legally, Iran would need Oman's agreement to close the strait peacefully. Closing the strait by force would be an act of war against Oman, which might mean a war with the entire **Gulf Cooperation Council**. Iranian mines in the southern channel could be countered by U.S. and British minesweepers already in the area. If Iran decided to fire on tankers, the United States and others would be in a position to neutralize the firing positions on the Iranian coast. It must also be noted that the Arab countries and the West are not the only ones who depend on the strait. Iran relies heavily on the strait as well, both for its own exports and for imports. Although a major oil-producing country, Iran relies heavily on imports of refined petroleum products such as **gasoline**.

An alternative route bypassing the Strait of Hormuz is Saudi Arabia's **Petroline** pipeline ending at **Yanbu** on the Red Sea, with a capacity of about 5 Mmbpd.

STRAIT OF MALACCA. A narrow seaway linking the Indian Ocean to the **South China Sea**, through which roughly 16 Mmbpd flowed in 2016. The Strait of Malacca is 1.7 miles wide at its narrowest point and carries almost all **Persian Gulf** oil destined for East Asian customers, including Japan and **China**. Ships transiting the strait, located between **Indonesia**, **Malaysia**, and Singapore, have been subject to significant threats of piracy and hijacking. Closure of the strait would force ships to take longer routes around the Indonesian archipelago. On 29 January 2015, China opened a 1,500-mile pipeline from Maday Island in **Burma** (Myanmar) to Kunming in China, with a capacity of roughly 0.26 Mmbpd, to reduce China's dependence on the strait.

STRATEGIC PETROLEUM RESERVES. Stocks of petroleum set aside as a cushion against supply disruptions. Strategic petroleum reserves may be classified into two broad types. The first are stores of petroleum that have already been extracted from the earth and perhaps subjected to preliminary processing and have then been stored for later use. These are perhaps more properly called "strategic petroleum stocks," as they are not **reserves** in the normal sense of the word. Strategic petroleum reserves of the second type really are reserves: they are known oil-bearing lands set aside for later exploitation.

The **Naval Petroleum and Oil Shale Reserves** of the **United States** were of the second type. In situ reserves of the second type have a number of problems that make them an unsuitable choice. For one thing, they are subject to commercial encroachment from drillers on adjacent lands, and they may not be located near the necessary processing and transportation infrastructure. Also, it is difficult to keep such reserves in a state of readiness so that they can rapidly achieve full production. A large amount of proved reserves must be set aside in order to be certain of rapidly achieving surge capacity, since only a small portion of a proved reservoir can be extracted in any one year. For these reasons, modern strategic petroleum reserves such as those mandated by the International Energy Programme—*see* INTERNATIONAL ENERGY AGENCY (IEA)—are generally of the first type, that is, stocks.

STRATHCONA, LORD. *See* SMITH, DONALD (1820–1914).

STRIZHOV, IVAN (1872–1953). Major organizer of the Soviet petroleum industry; explorer who discovered a number of important fields. Ivan Strizhov was born on 8 October 1872 in the area of Yekaterinburg in **Russia**. He graduated from Moscow State University with a gold medal in natural sciences in 1894 and began working in the mining industry. In 1898, he became involved with petroleum, exploring for the Cheleken–Dagestan Oil Company. He was responsible for some major finds in the North Caucasus and became manager of the company's fields. In June 1913, **Branobel** acquired 60 percent of the company, and Strizhov became the head of Branobel's North Caucasus exploration.

In 1920, the Soviet government placed Strizhov in charge of the Grozneft trust to restore the oil production of the Grozny fields. Within two months, he was transferred to Moscow as the deputy head of the Production and Technical Department of the Main Oil Committee of the Soviet Union, charged with restoring the country's oil industry after years of turbulence. He soon became head of the department. On 15 February 1924, Strizhov was put in charge of the **Baku** petroleum industry, and in 1926 he became the senior director for the oil industry on the Supreme Council for the National Economy. He succeeded in raising Soviet oil production 30 percent between 1926 and 1928, to about 0.234 Mmbpd.

In 1927, Strizhov made an official visit to the **United States** to examine American petroleum industry operations and technology. In doing so, he passed through Paris. Even though he made this trip at the direction of the Supreme Council, it was used against him in 1929. That year he was arrested for "consorting with capitalists" such as **Emanuel Nobel** and **Alexander Mantashev**, who resided in Paris at the time he visited. He spent time in prison in Moscow and then, escaping a death sentence, served in a labor camp for 10 years. He arrived at the Ukhta region's camp on 13 June 1931. There he was classified as a "free prisoner" because of his expertise, and he went to work on oil exploration in the area. He discovered oil and gas fields, including the **giant** Yaregskoye oil field in 1932, opening up the important **Timan–Pechora** oil province. Later in his life, Strizhov became a professor at the Moscow Oil Institute and performed important research in **natural gas**. He died in August 1953.

SUBSALT. *See* PRE-SALT.

SUDAN. (In 2011 South Sudan became an independent country. Sudan and South Sudan are covered together here.) Oil production 2015: 0.26 Mmbpd (0.33 percent of world); **reserves** 5 Bb (0.3 percent of world); gas reserves 3 Tcf (0.04 percent of world). Most of the oil deposits are in South Sudan, but South Sudan is a landlocked country and depends on (north) Sudan for pipeline and port access.

Combined Sudan is a relatively new oil producer of potentially increasing importance. Oil production began in 1992 at an almost negligible 300 bpd. However, by 1996 it reached 2,000 bpd, and then grew almost a hundredfold to 0.186 Mmbpd in 2000. Between 2000 and 2006, it more than doubled. Sudan has two **giant** oil fields, the Unity field (**EUR** 0.9 Bb, discovered in 1980) and the Palogue field (EUR 0.6 Bb, discovered in 2003). A 1,000-mile **pipeline** connecting fields in the Unity neighborhood with Port Sudan on the Red Sea was completed in 1999. A pipeline connecting Palogue to Port Sudan began operating in 2006.

The **United Kingdom** administered Sudan as two separate regions, northern and southern. The two regions differ in culture and religion. The north is dominated by Islam and Arabic culture; the south is Christian or Animist, and sub-Saharan in orientation. When the two regions were merged by the British in 1946, civil war ultimately resulted. The first civil war lasted from 1955 to 1972. Tensions remained unresolved, leading to a second civil war between 1983 and 2005. Much of Combined Sudan's known oil reserves lay in a border area disputed between the north and the south. Oil inflamed the tensions after the discovery of the Bentiu oil field in 1978. The central government raised southern suspicions by replacing the southern soldiers at

Bentiu with northerners. In 1980, a second field discovered near Bentiu was given the Arabic name of Heglig, further raising southern suspicions. The central government also tried to redraw provincial boundaries to transfer oil and other assets to northern provinces, leading to protests in the south. After the end of the second civil war in 2005, oil revenues were to be shared equally between the central government (representing the north) and the (then regional) government of south Sudan, under the oversight of the National Petroleum Commission (NPC) established that year. The NPC also issued new contracts and resolved problems arising from duplicate contracts previously allocated by the northern and southern regions.

South Sudan finally seceded and became fully independent in 2011, but violence continued and the country suffered from famine. Oil production nearly collapsed after secession, from 0.452 Mmbod in 2011 to 0.092 Mmbpd in 2012. There have been many disruptions. South Sudan shut down its oil production in 2012 because of disputes with (north) Sudan over pipeline fees. In 2013 oil production was shut down again because of civil conflict. In 2012 South Sudan established its own National Petroleum and Gas Corporation (NGPC) to supervise the petroleum sector. In 2011 it nationalized the assets of (north) Sudan's national oil company Sudapet and gave them to its own national oil company, Nilepet (Nile Petroleum Company).

In 1993, the **United States** declared (combined) Sudan a country believed to be sponsoring terrorism, and four years later American companies were prohibited from doing business there. This left the country's assets open to development by other interests. In 1996, the **Canadian** company Arakis formed the Greater Nile Petroleum Operating Company (GNPOC) with the **China** National Petroleum Corporation (CNPC), **Malaysia**'s Petronas, **India**'s Oil and Natural Gas Commission, and the Sudan National Petroleum Corporation (Sudapet). GNPOC operated in the Unity and Heglig oil fields near the border between northern and southern Sudan. **Qatar** General Petroleum Company (QGPC) began operating in the Adar Yal fields in 1997. In the 2010s the main foreign operators were still CNPC, Petronas, and ONGC.

SUEZ CANAL. Roughly 100-mile-long canal in **Egypt** linking the Red Sea with the Mediterranean. The canal saves a trip around the southern tip of Africa for oil **tankers** traveling from the Middle East to Europe. Although the largest tankers are unable to use it, the Suez Canal remains a major route and a serious choke point for **Persian Gulf** oil. In 2016, around 3.9 Mmbpd passed through the canal in both directions. The two main endpoints are Port Said in the north and Port Suez in the south.

French engineer and entrepreneur Ferdinand de Lesseps obtained a concession from Said Pasha in 1854 to construct the canal. The Suez Canal Company (Compagnie Universelle du Canal Maritime de Suez) was formed on 15 December 1858. The French financed most of the construction, which cost more than 450 million francs (about $100 million at the time) and took more than a decade. The project involved extremely hard and uncompensated or poorly compensated labor by Egyptian workers. The canal was opened in November 1869 under a 99-year concession. At the time, it was about 33 feet deep and 108 feet wide, and it was controlled by France and Egypt. Egypt was forced by external debts to sell its share to the **United Kingdom** in 1875 for £4 million. On 29 October 1888, the United Kingdom, **Russia**, Germany, Austria–Hungary, Spain, France, Italy, the Netherlands, and the Ottoman Empire signed the Convention of Constantinople, which declared the canal neutral. France remained the majority shareholder in the Suez Canal Company, although the United Kingdom had acquired de facto physical control over the canal and Egypt in the early 1880s.

The Suez Canal remained under French and British control until 1956, when Egypt nationalized it under the leadership of Gamal Abdel Nasser. The nationalization led to an invasion of Egypt by France, the United Kingdom, and Israel, with hostilities beginning on 29 October 1956. Egypt responded by sinking the 40 ships in the canal at the time, blocking the canal. The attack on Egypt was condemned by the international community, including the **United States** and the Soviet Union. Under intense international pressure, the aggressors began withdrawing on 22 December. The "Suez crisis," as it is commonly known, led to a significant loss of standing for both Britain and France as great powers. The canal was closed to navigation during the crisis, reopening in 1957. Since nationalization, it has been run by the Suez Canal Authority. The canal was closed again between 1967 and 1975 because of hostilities between Israel and neighboring Arab countries, beginning with the Six-Day War.

The canal was enlarged between 1960 and 1964 and again between 1975 and 1980 so it could accommodate larger oil tankers. Before the enlargements, the maximum size for tankers was about 65,000 deadweight tons (DWT). After enlargement, it reached about 150,000–200,000 DWT. The canal is now at least 590 feet wide and up to 53 feet deep. Future enlargements may eventually allow the canal to accommodate tankers of 200,000–300,000 DWT or more. The Suez Canal Authority earned over $5 billion in transit fees. *See also* SUMED PIPELINE (SUEZ–MEDITERRANEAN PIPELINE).

SUEZ–MEDITERRANEAN PIPELINE. *See* SUMED PIPELINE (SUEZ–MEDITERRANEAN PIPELINE).

SULLOM VOE. Major **United Kingdom** petroleum terminal and port on Mainland Island, part of Scotland's Shetland Islands. In the late 1990s, it handled about a quarter of the United Kingdom's petroleum output. Oil flows to Sullom Voe through the Brent System Pipeline from a number of important **North Sea** oil fields, including **Brent** and Cormorant. The **pipeline** is 92 miles long and is one of the deepest underwater oil pipelines in the world, lying under 525 feet of water. Sullom Voe has large storage facilities and a processing plant to separate **natural gas** from oil.

SUMED PIPELINE (SUEZ–MEDITERRANEAN PIPELINE). A **pipeline** linking the Red Sea with the Mediterranean. The Sumed Pipeline is an alternative to the **Suez Canal** for transporting oil. Some **tankers** too large to cross the canal fully laden may partially unload at one end, traverse the canal, and reload at the other end. The pipeline began operating in 1977 and has a capacity of 2.34 Mmbpd. It is about 200 miles long, running from Ain Sukhna terminal on the Gulf of Suez (Red Sea) to Sidi Kerir on the Mediterranean. Both terminals can accommodate ultralarge crude carriers (ULCCs). Sumed consists of two parallel lines, each of 42-inch diameter. It is owned by the Arab Petroleum Pipeline Company, which is a joint venture half owned by Egyptian General Petroleum Corporation, 15 percent by **Aramco** (originally **Petromin**), 15 percent by Abu Dhabi National Oil Company, 5 percent by Qatar General Petroleum Company, and the rest by three Kuwaiti companies with 5 percent each. Refer to appendix 8.

SUNCOR ENERGY. Integrated energy company with a primary focus on developing **Canada**'s **Alberta oil sands**. Suncor Energy began as a unit of the **Sun Oil Company**. In 1953, Sun incorporated Great Canadian Oil Sands Ltd. (GCOS) and acquired patents and leases around **Fort McMurray** for the purpose of developing and **refining** the Athabasca oil sands. Sun invested about $240 million to construct an oil sands processing plant, beginning 1964. This was the largest single private investment in Canadian history up to that time. The plant opened 30 September 1967, with an initial planned production capacity of 45,000 bpd of **syncrude** from oil sands. First oil flowed from Fort McMurray to Sarnia, Ontario, in 1968.

In 1979, Sun Oil merged all its Canadian operations with Great Canadian Oil Sands Ltd. to form Suncor Inc. A $185 million expansion project brought theoretical production capacity to 58,000 bpd in 1981, although the plant did not exceed 55,000 bpd until 1986. In 1981, the Sun Company, as Sun Oil was called since 1976, also sold 25 percent of its interest in Suncor to the province of Ontario. The company remained largely unprofitable during the **Oil Countershock** and the low oil prices in the 1980s.

On 18 March 1992, Suncor became a publicly traded company. The next year, the province of Ontario divested its interest, and the Sun Company reduced its interest to 55 percent. In 1995, the Sun Company sold its entire interest. The next year, the company reached a cumulative syncrude production of 500 Mmb. In 1997, Suncor changed its name to Suncor Energy and began trading on the New York Stock Exchange. In 1998, it began Plant 25, a $200 million expansion of its oil sands upgrading facility, and it also opened its Steepbank mine on a new lease across the Athabasca River from its existing operations. The next year, Suncor began Project Millennium to increase its output of syncrude. By 2002, total syncrude production was 205,800 bpd. In 2005, production reached 260,000 bpd, and by 2006 Suncor had shipped its billionth barrel. In November 2006, the company obtained regulatory approval to build a third oil sands upgrading facility. In 2009 Suncor acquired **Petro-Canada**. In 2016 it controlled oil reserves of 0.172 Bb and gas reserves of 0.031 Tcf. It produced 0.118 Mmbpd of oil and 0.027 Bcfpd of gas.

SUNOCO. *See* SUN OIL COMPANY.

SUN OIL COMPANY. Integrated petroleum company that eventually focused on downstream operations; became Sun Company and later Sunoco. The roots of the Sun Oil Company date back to 27 March 1886, when Joseph Newton Pew and Edward Octavius Emerson acquired two oil leases near Lima, Ohio (*see* LIMA–INDIANA OIL FIELD) for $4,500. Pew and Emerson were already partners in the Peoples Natural Gas Company of Pittsburgh. They discovered oil on their leases, and they acquired storage tanks and **pipelines**. By 17 March 1890, when they organized the Sun Oil Company of Ohio, they were already one of Ohio's largest producers of crude oil. In 1894, Sun joined Merriam and Morgan Paraffine Company to incorporate the Diamond Oil Company for the purpose of purchasing a refinery outside Toledo, Ohio, from the Crystal Oil Company. The Diamond Oil Company was the origin of Sun's and later Sunoco's familiar trademark of a diamond pierced by an arrow. In 1899, Pew bought out Emerson's interest in Sun Oil.

When oil was discovered at **Spindletop** in **Texas** in 1901, Pew sent his nephew J. Edgar Pew there. In 1902, Edgar Pew bought the oil properties of the bankrupt Lone Star and Crescent Oil Company. Sun built a refinery in Marcus Hook, **Pennsylvania**, to process its Spindletop crude. Marcus Hook was near major East Coast markets and infrastructure, and the oil could be shipped there by **tanker**, avoiding prohibitively expensive overland transport.

Joseph Newton Pew died in 1912 and was succeeded as president by his son J. Howard Pew. In 1920, Sun opened its first **gasoline** station in Ardmore, Pennsylvania. It sold gasoline under the Sunoco name and used essentially the same trademark seen today. In 1922, it shortened its name to Sun Oil Company. On 12 November 1925, Sun went public and was listed on the New York Stock Exchange. In 1926, the company introduced "Blue Sunoco" antiknock gasoline, which was deliberately dyed blue. It was advertised as "knockless fuel at no extra price." Although the gasoline was antiknock, it had no added lead. In 1931, Sun built the first long-distance product pipeline in the **United States**, running 730 miles from Twin Oaks, Pennsylvania, to Syracuse, New York, and Cleveland, Ohio.

During the 1930s, Sun helped finance **Eugene Houdry**'s research in catalytic cracking and owned a third of the Houdry process. The other two-thirds were owned by Houdry and Socony–Vacuum (later **Mobil**), in equal shares. Sun Oil opened a Houdry catalytic cracking plant at its Marcus Hook refinery, where Houdry had conducted much of his research and testing. Toward the end of **World War II**, Marcus Hook was producing more than 36,000 bpd of crucial 100-octane aviation fuel. In 1957, Sun Oil made a large oil discovery in Lake **Maracaibo**, **Venezuela**, which ultimately produced about 200 Mmbo for the company, until the 1975 nationalization of the Venezuelan oil industry. In 1965, Sun was a partner in the discovery of the Hewett **natural gas** field in the **United Kingdom** sector of the **North Sea**.

In 1967, Sun completed a refinery to process **Alberta oil sands** into **syncrude**. The plant, first known as the Great Canadian Oil Sands Plant and later as **Suncor**, produced 260,000 bpd of syncrude in 2005. Suncor became an independent company in 1995, after Sun sold all its interest. In 1968, Sun merged with Sunray DX, a company with **refining** and marketing operations in the Midwest that complemented Sun's in the East. In 1976, Sun Oil changed its name to Sun Company Inc., reflecting the company's diversity. Since 1916, for example, it had been involved in shipbuilding. In 1976, the company acquired a large interest in a medical supply firm, and in 1979 it acquired Elk River Resources, a coal-mining concern. It also acquired petroleum assets. In 1980, Sun acquired the U.S. petroleum properties of Texas Pacific Oil Company, and in 1984 it acquired Exeter Oil Company. In 1982, Sun became one of the first companies to explore for oil in **China**'s **offshore** under a new opening from the Chinese government. It also began reversing its diversification. It sold its shipbuilding business that year. In 1985, it sold its medical supply business, and in 1986 its trucking business, among other actions.

In 1988, after the **Oil Countershock** and the collapse in oil prices, Sun began divesting itself of upstream petroleum interests. It spun off its Sun Exploration and Production Company subsidiary. Meanwhile it expanded its downstream interests, acquiring refineries and gasoline stations and rebranding them as Sunoco. In 1992, it began reducing its ownership of Suncor, selling it completely by 1995. In 1996, it sold its international upstream interests to **ENI** and became a completely downstream company. In November 1998, it followed in the footsteps of many other petroleum companies by adopting its well-known brand name as its company name, changing its name to Sunoco Inc.

SUPERGIANT. A **giant** oil and/or **natural gas** field with an estimated ultimate recovery (**EUR**) of 5 Bboe or greater. *See also* MEGAGIANT.

SUPSA. Port (pop. 5,000, 2002) on the Black Sea coast of Georgia. Terminus of the Western Early Oil Pipeline bringing **Caspian** oil from **Baku**. This **pipeline** began operating in 1999. In 2005, the **Baku–Tbilisi–Ceyhan (BTC) Pipeline** was completed. BTC takes Caspian oil directly to the Mediterranean, avoiding the Black Sea and the **Bosporus**. The BTC Pipeline makes Supsa and other Black Sea ports relatively less important. In 2003, about 45 Mmb were shipped through Supsa.

SYKES–PICOT AGREEMENT (1916). A secret agreement negotiated between the **United Kingdom** and France during **World War I** to determine the postwar fate of various portions of the Ottoman Empire. The agreement was named for the British negotiator Mark Sykes and the French negotiator François Georges Picot. According to the agreement, the Middle East would be divided between French and British spheres of influence. Syria, Lebanon, and Cilicia would come under direct French control, while the Mesopotamian provinces (later **Iraq**) of Baghdad and **Basra** would come under direct British control. France would also get a zone of influence including Damascus and Aleppo and extending east through Mosul. Britain would get a zone of influence extending from Basra to Palestine. Palestine would be placed under international administration.

Russia supported the Sykes–Picot Agreement in return for direct control of a large part of eastern Anatolia. After the Russian Revolution, however, the Bolshevik government repudiated the agreement and made it public in 1918. This elicited a very negative reaction among the Arabs, who expected independence after the war. Although the Sykes–Picot Agreement was eventually repudiated, many of its principles remained influential in the postwar disposition of the Ottoman territories in the Middle East at the **San Remo Conference**.

SYNCRUDE. A liquid similar to **conventional crude oil**, produced by processing coal or unconventional hydrocarbon resources such as **oil sands**, **oil shale**, or **heavy oil**. Syncrude generally requires further **refining** to produce **gasoline** and other petroleum products. Syncrudes vary widely in their properties but can be light and sweet (refer to appendix 5). For example, some syncrude from the **Alberta oil sands** has an **API gravity** of 35° and a very low sulfur content. Sincor synthetic crude from the projects in the **Orinoco Oil Belt** in **Venezuela** has a gravity of 32° and also has a low sulfur content. A syncrude from the Stuart oil shale proposed for commercial development in **Australia** has a gravity of 48° and very low sulfur content.

SYNCRUDE (JOINT VENTURE). *See* ALBERTA OIL SANDS (TAR SANDS).

T

T2 TANKER. Oil **tanker** ship produced by the **United States** during **World War II**. About 600 were made between 1940 and 1946. The standard T2 was 501.5 feet long and 68 feet wide, with a deadweight of 15,800 tons. It had a single propeller driven by steam turbines and a top speed of 15 knots. A variant was the T2-SE-A1 built for Standard Oil Company of New Jersey (later **Exxon**) in 1940. The T2-SE-A1 was 523 feet long and had a deadweight of 16,613 tons. After World War II, many T2 tankers were sold for commercial use and formed an important part of many fleets.

TAGIYEV, ZEINALABDIN (1823–1924). Early entrepreneur in the **Russian**/Azeri oil industry in **Baku**. Zeinalabdin Tagiyev began life as a stonemason, apprenticed to a bricklayer at the age of 10. He gradually became a successful building contractor with diversified business interests, including small-scale oil **refining**. In December 1872, he partnered with an Armenian, Sarkissov, to purchase an oil property in a government auction. The property, **Bibi–Eibat**, was considered one of the least promising at the auction, and for more than a decade it produced only small amounts of oil with considerable difficulty. Sarkissov eventually gave up and sold his interest to Tagiyev.

On 27 September 1886, Tagiyev struck a gusher that completely drenched the surrounding area. It blew for 15 days and wasted an estimated 1.4 Mmbo. Tagiyev proceeded to build a refinery for what turned out to be very high quality oil. He also built a dock and acquired a fleet of **tankers** for the **Caspian** and barges for the Volga to transport his products. He joined forces with his competitors the **Nobel** brothers to help build the **Baku–Batumi** kerosene **pipeline**. Tagiyev eventually sold his oil company to British interests for 5 million rubles in gold.

TAMAN PENINSULA. The site of **Russia**'s first oil gusher. The Taman Peninsula juts into the Black Sea northwest of **Novorossiysk**, at the gateway to the Sea of Azov. Although the Russian oil industry started at **Baku** on the **Caspian**, the first gusher was not there but here in the North Caucasus, in 1866. Evidence of oil being gathered and used in the area goes back to at least the 10th century. In 1792, Catherine II granted the Black Sea Cossacks land on the peninsula and an oil monopoly. But it was not until the 1860s that systematic and modern prospecting occurred. Colonel Ardalion Novosiltsev explored the area around the Kudako, Psif, and Psebeps Rivers, all tributaries of the Kuban. On 15 February 1866, his team struck Russia's first gusher at about 180 feet. The well blew for 24 days. Another gusher on 26 April lasted 28 days. These wells opened up the Kuban oil fields, which **Dmitri Mendeleev** called a second **California**. Novosiltsev built a refinery and exploited the oil.

TANKER WAR. *See* IRAN–IRAQ WAR (1980–1988).

TANKERS. Ships used to transport petroleum and petroleum products. The **major oil companies** once directly controlled most of their tanker transportation needs, maintaining substantial fleets. After **World War II**, they shifted away from this direct control, depending more on the spot market for chartering ships from specialized companies. One of the first recorded voyages of a ship carrying a commercial quantity of oil was that of the brig *Elizabeth Watts*, which carried 224 tons from the **United States** to England in 1861. The *Elizabeth Watts* was not a specialized tanker in the modern sense. It did not carry the oil in bulk in its hull but rather carried barrels of oil.

The ***Zoroaster***, a steamer designed by **Ludvig Nobel** and built in 1877, could carry 242 tons of kerosene across the **Caspian** in 21 vertical cylindrical tanks. The *Zoroaster* was ahead of its time in that it burned a type of fuel oil to power its steam engine. In 1880, Ludvig Nobel completed the design of a newer tanker, the *Moses*, that used the hull of the vessel as the wall of a bulk cargo hold. In 1881, a similar Nobel tanker, the *Nordenskjold*, exploded while being loaded with kerosene in **Baku**, killing half the crew. This prompted Nobel to design a flexible loading pipe between the tanker and the shore. The pipe's couplings could move with the ship and remain fastened to the loading hole, reducing the risk of leaks and explosions. The Nobel brothers had a fleet of early modern tankers plying the Caspian by the mid-1880s.

Many consider Nobel's *Zoroaster* to have been the first modern tanker. Others accord that honor to the 2,700-ton ***Gluckauf*** built in 1885. Even though its design owed much to the Nobel tankers, it was, unlike them, an explicitly oceangoing vessel meant to cross the Atlantic. It carried bulk oil in large, hull-integrated tanks that were filled and emptied by pipes leading ashore. Carrying oil in bulk created stability problems for ships, which have been solved (beginning with Nobel's tankers and the *Gluckauf*) by subdividing the hull. By 1891, the **Standard Oil Company** and affiliated companies had a fleet of nearly 70 coal-powered tanker steamships carrying oil in bulk between the United States, Europe, and other areas at speeds of 8 to 10 knots. The tankers carried 20,000–30,000 barrels of oil.

After the invention of the diesel engine in 1897, tankers and other merchant ships began converting to oil power. By the mid-20th century, tankers were generally about the size of a **T2 tanker**, about 15,000 deadweight tons (DWT), built in large quantities during **World War II**. A 25,000 DWT tanker was considered a large one. Until 1956, an effective limit was imposed on tankers by the dimensions of the **Suez Canal**. The tonnage limit was around 65,000 DWT in the 1950s, although the canal has been progressively deepened and widened since then, raising the limit to about 150,000–200,000 DWT.

The Suez Canal was closed during the Suez crisis of 1956 and again during the Six-Day War of 1967. The first closure was brief, but the second lasted for nearly eight years. Suez closures forced tankers carrying oil between the Middle East and Europe to sail around South Africa. Also, demand was growing in distant markets such as Japan. These factors, combined with a desire to reduce unit transportation costs, led to a rapid increase in tanker sizes. In 1959, the 104,500 DWT ***Universe Apollo*** became the first tanker to surpass 100,000 DWT. The first 200,000 DWT tanker was delivered in 1966. This was a midrange very large crude carrier (VLCC). By July 1969, the first ultralarge crude carrier (ULCC) was introduced: the 332,178 DWT *Universe Iran*. By 1973, there were 366 VLCCs and ULCCs in operation, with 525 under construction.

Large tankers became a matter of global concern when they began having accidents and spilling oil (*see AMOCO CADIZ*; *EXXON VALDEZ*; *TORREY CANYON*). Many engineering innovations were introduced to help mitigate the problem, although spills still occur. Safe tank venting, satellite navigation, and more sophisticated engine control have all helped. The most important innovation is perhaps the double-hull design, already mandatory in some markets. Single-hulled tankers are supposed to be phased out by the middle of the 2010 decade by international agreement. Refer to appendix 8.

TAPLINE. *See* TRANS-ARABIAN PIPELINE (TAPLINE).

TAR SANDS. *See* ALBERTA OIL SANDS (TAR SANDS); OIL SANDS.

TARBELL, IDA (1857–1944). Investigative journalist who exposed the business practices of **Standard Oil** in the early 20th century. Ida Minerva Tarbell was born in Hatch Hollow, **Pennsylvania**, on 5 November 1857. She grew up in the oil regions, the daughter of an **independent** producer and refiner. The 1872 **South Improvement Company** scheme, by which Standard Oil bullied independent oil companies, had a significant negative effect on her father's business and on the prosperity of her family. A graduate of Allegheny College in 1880, Tarbell spent some time as a schoolteacher and began writing for magazines. S. S. McClure, founder of *McClure's Magazine*, hired her in 1894. When she began investigating Standard Oil, her father feared retaliation and advised her to desist. She pressed on nonetheless, and after two years of research, she was able to analyze **John D. Rockefeller**'s various complex maneuvers.

What was originally intended to be a three-part series was published in 19 installments between 1902 and 1904. While acknowledging Rockefeller's genius, Tarbell presented a detailed account of his questionable business tactics and their effects on independent producers. She became a leading exemplar of a crusading trend in the reporting of the effects of big business on society that President Theodore Roosevelt dubbed "muckraking" in 1906. Tarbell also did a two-part study of Rockefeller himself that was highly unflattering and perhaps somewhat unfair. Rockefeller was personally wounded by the exposé and refused to rebut it publicly. Tarbell's series in *McClure's* was published in book form as *The History of the Standard Oil Company*. She died of pneumonia in Bridgeport, Connecticut, on 6 January 1944.

TARIKI, ABDULLAH AL (1918–1997). First oil minister of **Saudi Arabia** and considered a cofounder of the **Organization of the Petroleum Exporting Countries** (OPEC); the first Saudi to become a member of the **Aramco** board. Abdullah al-Tariki was born in Zilfi, Saudi Arabia, on 9 January 1918, the son of a camel owner. He studied in **Kuwait** and obtained a degree in geology and chemistry from Fouad University in Cairo in 1945. He obtained a master's degree in petroleum geology from the University of **Texas** in 1946. Tariki did an internship at the **Railroad Commission of Texas**, where he would have learned the principles of controlling production to stabilize prices. He worked for a while at the Texas Company (later **Texaco**) in Midland, Texas (*see* PERMIAN BASIN), before returning to Saudi Arabia in 1948 to work in the finance ministry.

In 1955, Tariki was appointed head of the new Saudi Directorate of Oil and Mining Affairs. In 1960, he was promoted to the rank of minister by King Saud. Tariki was a fervent Arab nationalist who wanted Saudi Arabia to take control of its oil industry and obtain higher royalties. He confronted Aramco over its treatment of workers and accused it of obfuscating profits in order to avoid paying the full amounts it owed the Saudi government. He wanted to create an integrated Saudi company with full downstream operations and even floated the idea of nationalization of Aramco. He viewed

concession contracts as negotiable, not sacrosanct. The **major oil companies** disliked and feared Tariki. They dubbed him the "Red Sheikh," thereby insinuating that he was a communist.

In February 1959, **BP** made a 10 percent cut in the **posted price** of crude. Tariki and **Venezuela**'s **Juan Pablo Pérez Alfonso** joined forces to harness the anger that resulted from the cut. The two mobilized representatives from other oil exporting countries to sign the secret **Mehdi Pact** at the April 1959 Arab Oil Congress in Cairo. The pact was a nonbinding "gentleman's agreement" embodying recommendations the representatives would make to their governments, but it presaged the formation of OPEC. On 14 September 1960, after another unilateral cut in the posted price by the major oil companies, OPEC was born.

Tariki's career did not last much longer. His Nasserist pan-Arabism and his willingness to confront the West aggressively were too radical for the Saudi royal family at the time. They feared Tariki would cause a rupture between Saudi Arabia and the **United States**. In 1961, Tariki joined Prince Talal in accusing Crown Prince Faisal of corruption. Tariki offered evidence that 2 percent of the profits from the Arabian Oil Company, a joint venture with the Japanese, were assigned in perpetuity to Faisal's brother-in-law, Kamal Adham, who would later become Saudi Arabia's intelligence chief. However, Faisal was too powerful to be brought down in this way. In March 1962, he removed Tariki from the position of oil minister, replacing him with **Ahmad Zaki Yamani**. Yamani had Tariki removed from Aramco's board. Unemployed, Tariki settled in Beirut, Lebanon, and started a consulting business. He believed, wrongly, that there would be a revolution in Saudi Arabia that would overthrow the royal family, with possible **Egyptian** help. He remained in exile for much of the rest of his life. It was not until Faisal's murder in 1975 that he was able to visit home. Abdullah al-Tariki died in Cairo on 7 September 1997.

TEAGLE, WALTER (1878–1962). President (1917–1937) and chairman (1937–1942) of the Standard Oil Company of New Jersey (later **Exxon**); during his tenure, the company became the world's largest oil producer, with a global market share growing from 2 percent to 11.5 percent. Walter Clark Teagle came from a family with a tradition in the oil industry. Teagle's grandfather Maurice B. Clark was **John D. Rockefeller**'s first partner, in an agricultural produce wholesaling business in 1859. Thereafter, Clark both competed against and cooperated with Rockefeller in the oil business, ultimately merging his interests into **Standard Oil**. Teagle's father, John Teagle, was the president of an oil company in Cleveland, Ohio, that was also purchased by Standard Oil.

Teagle graduated from Cornell University in 1899 with a degree in chemistry. Two years later, Standard Oil bought the family company's refinery, and Teagle began to run it for the new owners. He rose to become a director of Standard Oil in 1910. After the breakup of Standard Oil in 1911, Teagle stayed on with the Standard Oil Company of New Jersey, or Jersey Standard, the most important of the newly independent companies. He managed aggressively, including expansion of the company's operations into **Venezuela**. In 1917, he became president of Jersey Standard.

In 1919, Teagle bought half of **Humble Oil and Refining Company** for Jersey Standard for $17 million. Humble became Jersey Standard's largest operating unit and would feed its top management ranks in the years to come—*see* FARISH, WILLIAM STAMPS (1881–1942). In 1920, Teagle made a less successful acquisition: Jersey Standard bought half of the **Nobel** family's interest in **Branobel** for $11.5 million. With the communists already in power, Teagle knew it was a gamble, but the prospect of access to **Russia**'s market was too enticing. As things turned out, he lost the gamble. The Soviets expropriated the company without compensation.

In 1928, Teagle met in Achnacarry Castle, Scotland, with **Henri Deterding** of **Royal Dutch Shell** and **John Cadman** of the Anglo–Persian Oil Company (later the **Anglo–Iranian Oil Company**, eventually **BP**) and signed the **As-Is Agreement** forming an international oil cartel that would limit production, fix prices, and divide markets.

The net effect of Teagle's various moves was that Jersey Standard became the world's largest oil producer. In addition, Teagle fostered significant new developments in petroleum products and **petrochemicals**. He showed some progressive tendencies in labor relations, pushing such reforms as an eight-hour workday and worker representation in some company councils. He served on the National Labor Board from 1933 to 1934 in the administration of President Herbert Hoover and on the National War Labor Board among other committees in the administration of President Franklin D. Roosevelt.

Teagle's reputation was tainted by his connection to the German chemical company IG Farben; he was a director of its U.S. subsidiary, and Jersey Standard conducted joint research with IG Farben. As a result, Germany was able to manufacture some critical materials that helped it in **World War II**. Teagle even sold the patent rights for synthetic rubber to IG Farben, which resulted in delays in American rubber production. This might have resulted in severe penalties for Teagle and his company, but in the end he paid a fine out of court and resigned from Jersey Standard in 1942. Teagle died on 9 January 1962.

TEAPOT DOME SCANDAL (1922–1930). A major oil-related scandal in President Warren G. Harding's administration, with an aftermath stretching across the whole decade of the 1920s. Harding's secretary of the interior,

Albert B. Fall, granted exclusive leases to Teapot Dome, a **Naval Petroleum Reserve** in Wyoming, to **Harry Sinclair**, head of **Sinclair Oil Corporation**, without competitive bidding (Sinclair had set up a company named Mammoth Oil for the purpose). It was discovered that Fall had received at least $300,000 from Sinclair. Fall had also received a $100,000 "loan" from oil entrepreneur **Edward Doheny** for leases to the Elk Hills and Buena Vista Hills Naval Petroleum Reserves in **California**. Elk Hills, Buena Vista Hills, and Teapot Dome were proven oil lands, awarded to the U.S. Navy during the presidencies of William Howard Taft and Woodrow Wilson to provide a source of fuel during wartime. For about $100,000, Doheny acquired control of oil **reserves** estimated at up to 250 Mmb, allegedly saying, "We will be in bad luck if we do not get $100 million in profit. But that will depend on the price of gasoline." The affair was investigated by the U.S. Senate in 1924. Both criminal and civil proceedings ensued. The leases appeared legal on the surface and were at first upheld but later canceled. Sinclair and Doheny were indicted for bribery and conspiracy to defraud the government. Both were acquitted, but Sinclair served six and a half months in prison for contempt of court and contempt of the U.S. Senate. Ironically, it was Doheny who had received the most valuable properties, and for a lower price than what Sinclair had paid. Fall was convicted in 1929 of accepting bribes, fined $100,000, and served nine months of a one-year prison sentence. The scandal severely tarnished the administration of President Harding, although the president was not personally implicated.

TECHNICALLY RECOVERABLE RESOURCES. The subset of **ultimately recoverable resources** that has not yet been produced and can be extracted and produced with currently available technology and practice. Technically recoverable resources may not be economically recoverable at prevailing price levels.

TEHRAN AGREEMENT (1971). Agreement between **major oil companies** and **Persian Gulf** members of the **Organization of the Petroleum Exporting Countries** (OPEC), raising prices and increasing producer countries' profit shares. The Tehran Agreement is regarded as an important turning point in OPEC's relations with the oil companies, with the power and the initiative passing decisively to OPEC; it also signaled the death of the **50–50 agreements** for profit sharing that had been common since the 1950s.

The Tehran Agreement was concluded on 14 February 1971 between the major international oil companies and the governments of **Saudi Arabia**, **Iran**, **Iraq**, **Kuwait**, **Qatar**, and the **United Arab Emirates**. It established 55 percent as a minimum profit share for producer countries, and it raised the **posted price** of oil by 35 cents a barrel (about 15 percent). The agreement also provided for subsequent price increases of 2.5 percent a year and an additional annual increase of 5 cents a barrel to compensate for the fact that crude oil is a nonrenewable resource. The terms of the agreement were to be applied retroactively to 1 January 1971. The agreement was meant to last five years.

The oil companies expressed a preference for negotiating with OPEC as a bloc, to avoid multiple agreements and constant leapfrogging of prices as members one-upped each other. However, the various OPEC countries differed in their approaches and their degree of radicalism, preferring to form smaller blocs at this time. Thus, in the same year, the **Tripoli Agreement** was separately concluded by OPEC members delivering oil in the Mediterranean region.

TEXACO. A **major oil company** and one of the **Seven Sisters**; later part of **Chevron**. Texaco's origins date back to **Spindletop**. **Joseph Cullinan** founded the Texas Fuel Company there in 1901 to purchase crude and resell it to refineries. The initial capitalization was $50,000. Cullinan and New York investment manager Arnold Schlaet set about soliciting additional investments and in 1902 reorganized the Texas Fuel Company as the Texas Company with a capitalization of $3 million. The Texas Company built a refinery at Port Arthur on the **Texas** coast and a **pipeline** connecting it to the Spindletop field. As Spindletop became prematurely exhausted in 1902, the company urgently needed new sources of crude. Over Schlaet's objections, Cullinan risked the company to drill three wells at Sour Lake, about 20 miles from Spindletop. The third well struck oil in January 1903, and the company survived.

The company continued to explore successfully and build more pipelines to connect the new fields to Port Arthur. By 1913, it had assets of $60 million. The Texaco name originated from the telegraph address of the company's New York offices. The Texas Company registered it as a trademark in 1906. The name gained brand recognition throughout the world, and in 1959 the Texas Company formally adopted it, becoming Texaco Inc. The Texas Company introduced its first consumer product in 1907: Familylite kerosene for illumination. However, it did not take long for **gasoline** to become its primary retail product. In 1911, it opened its first station in Brooklyn, New York, and by 1916 had 57 stations in the **United States**.

The demand for petroleum in **World War I**, coupled with increased civilian demand for gasoline, led to explosive growth for the company. It roughly quadrupled in size between 1914 and 1920. In 1920, it developed a continuous thermal cracking process for increasing gasoline yields from crude oil. During the 1920s, the company introduced premium gasoline and aviation fuel into its product line and continued to prosper. In 1928, the Texas Company acquired the California Petroleum Company and broke into West Coast markets. That year, the company controlled more than 4,000 gasoline stations nationwide.

The company changed its corporate structure in 1926 and then again in 1941. In 1926, the Texas Company reincorporated in Delaware to take advantage of more favorable laws. It formed the Texas Corporation, which functioned as a holding company for the Texas Company of Delaware. After the California Petroleum Company acquisition, the corporation also held the Texas Company of California. In 1941, the holding company merged with both operating companies and formed a single entity known once again as the Texas Company.

Like other petroleum companies, the Texas Corporation suffered during the Great Depression. It introduced a number of new retail gasoline and motor oil products to woo customers and was partially successful. During the 1930s, it also took some important actions abroad. Most notably, it entered into a joint venture in 1936 with the Standard Oil Company of California (Socal, later Chevron). The two companies pooled their resources "east of Suez." The collaboration made sense because the Texas Corporation had marketing outlets there but no crude; Socal had crude in **Bahrain**, but no marketing outlets. Together they formed the California–Texas Company (**Caltex**). As part of the arrangement, the Texas Corporation acquired a 50 percent interest in the California Arabian Standard Oil Company (Casoc)—the predecessor to **Aramco**. This would give the Texas Corporation a position in what would become the world's largest oil **reserves**. Casoc discovered oil in **Saudi Arabia** two years later.

During **World War II**, the Texas Company, as it was known again after 1941, produced aviation fuel, gasoline, and **petrochemicals** to support the Allied effort, working closely with the government—*see* PETROLEUM ADMINISTRATION FOR WAR (PAW) (1942–1946). The Texas Pipeline Company, a subsidiary, aided in the construction of the **Big Inch and Little Big Inch** Pipelines for the government. These pipelines ran from Texas to the northeastern coast of the United States.

After the war, the Texas Company and Socal, needing capital, sold a 30 percent interest in Aramco to Standard Oil Company of New Jersey (later **Exxon**) and a 10 percent interest to Socony–Vacuum (later **Mobil**). In 1954, the Texas Company acquired an 8 percent interest (later reduced to 7 percent) in the **Iran** consortium **Iranian Oil Participants Ltd**. During the 1950s and 1960s, Texaco, as it was formally known after 1959, acquired acreage in western Texas and bought a number of regional companies in the United States. It significantly expanded its **refining** capacity and its international operations, especially downstream. It acquired the Trinidad Oil Company in 1956 and the Seaboard Oil Company in 1958 to obtain reserves in

South America. Between 1969 and 1972, it helped build the Trans-Andean Pipeline and one of the pipelines across **Ecuador**. It increased refining, marketing, and petrochemicals production in Europe and Japan.

During the 1970s, Texaco suffered from the petroleum industry nationalizations in the Middle East. Saudi Arabia began acquiring Aramco under the 1972 **participation agreement** framework and by 1980 had acquired it all. In 1973–1974, **Libya** nationalized Texaco's Texas Overseas Petroleum Company. In 1979, Texaco lost its interests in Iran due to the Iranian Revolution. In response to these pressures, Texaco increased its exploration efforts in the U.S. Southwest and the **Gulf of Mexico** and ventured into the **North Sea**. It also consolidated downstream operations and pulled out of unprofitable markets at the retail level. In 1980, Texaco reorganized into three large divisions covering the United States, West Africa, and Europe and Latin America, plus one petrochemical division (Texaco Chemical Company).

In 1984, Texaco made what turned out to be a huge mistake when it attempted to acquire the **Getty Oil Company**. **Pennzoil** had already begun a partial acquisition process and sued Texaco for tortuous interference. The jury in ***Pennzoil v. Texaco*** found for Pennzoil, initially awarding it a stunning $10.5 billion. The case resulted in Texaco declaring bankruptcy in 1987. It settled with Pennzoil that year, paying about $3 billion. Texaco lost $4.4 billion in 1987. Corporate raider Carl Icahn took advantage of Texaco's distress and its depressed stock price to make a takeover attempt in 1988. He lost the proxy fight and was bought out. Texaco began selling assets to raise capital for new investment and expanded exploration in the North Sea and **offshore California**. In the continuing soft market for oil during the early 1990s, Texaco's revenues decreased. It responded by reducing its workforce and shedding noncore operations. In 1994, it sold the Texaco Chemical Company for $850 million to the Huntsman Corporation. It also sold several petroleum-producing properties that it considered marginal to Apache Corporation. It used the money to invest in exploration in **Russia**, **Colombia**, and **China**.

On 9 October 2001, Texaco was acquired in a friendly takeover by Chevron, its longtime partner in Aramco and Caltex. The motivations were cost savings and increased efficiency for both companies. The deal was completed for $38.3 billion in stock and the assumption of $6.7 billion in Texaco debt. Texaco became a subsidiary of Chevron, which changed its name to ChevronTexaco. Caltex also became a subsidiary. In 2000, the last full year before the merger, Texaco was the fifth-largest petroleum company headquartered in the United States, measured by total assets. It had a net income of $2.542 billion on revenues of $51.130 billion. It controlled 3.518 Bb in proved oil reserves and 8.292 Tcf of gas. It produced 0.8 Mmbpd of oil and

1.867 Bcfpd of gas. In 2005, ChevronTexaco dropped Texaco from its corporate name and became simply Chevron Corporation. Texaco continued to exist as a subsidiary and a brand. *See also* BEATY, AMOS (1870–1939).

TEXAS. Oil production 2015: 3.448 Mmbpd (36.65 percent of U.S.); **reserves** 11.759 Bb (36.39 percent of U.S.); gas production: 7.07 Tcf (26.13 percent of U.S.); reserves 78.866 Tcf (25.63 percent of U.S.). Texas was the **United States**' most important oil-producing state for most of the 20th century, and a significant force in world production for a large part of the century. Texas became the largest oil-producing state in the United States in 1928 and held that position until 1988, when it was overtaken by the federal **offshore** areas, mostly in the **Gulf of Mexico**. From 1988 to 1998, it was nearly tied with **Alaska**. In 2011 it overtook the federal offshore and became top U.S. producer again.

Texas oil production was about 0.001 Mmbpd in 1898. It rose to 0.050 Mmbpd in 1902 after the **Spindletop** discovery. By 1918, it was 0.106 Mmbpd and then began a very rapid increase. Although it went through some peaks and valleys, the long-term trend was up, for decades. Between 1931 and 1978, its production fluctuated between 35 percent and 45 percent of U.S. production. In every year from 1929 to 1953, Texas produced more than a fifth of the world's oil. In 1944 and 1945, its production was 29 percent of the world's total. By 1956, Texas produced 3.035 Mmbpd, representing 42.3 percent of U.S. output. It declined to 2.541 Mmbpd in 1960 but then increased to its all-time high of 3.566 Mmbpd in 1972. It seemed to most people that output in 1972 represented Texas's **Hubbert** peak (*see* PEAK OIL). After 1972, Texas production entered a phase of long-term decline. Its production in 2006 was slightly lower than it was in 1933. Yet by 2015, Texas production had rebounded to almost 1972 levels because of the explosive increase in **unconventional tight oil** production in the late 2000s and 2010s.

Signs of oil have long been noticed in Texas. The first recording of an observation was made by Luis de Moscoso, a survivor of the De Soto expedition. In July 1543, members of the expedition noticed oil floating on the surface of the water between Sabine Pass and High Island. They gathered some and used it to caulk their ships. The first well known to have been purposely drilled for oil was by Lyne T. Barret. He leased 279 acres near Oil Springs in Nacogdoches County on 15 December 1859. The Civil War interrupted his work for several years, but on 12 September 1866 he made a modest strike at 106 feet, producing about 10 bpd. He obtained the services of **John F. Carll** to test and develop the field. Although indications were positive, a combination of low oil prices and the uncertainties associated with Reconstruction made it impossible to obtain financing. The field would remain undeveloped until 1887, when others would exploit it.

The **Corsicana** oil field, found in 1894, is generally considered the first truly commercial discovery in Texas. However, it was not until the 1901 Spindletop discovery that a major boom ensued. Spindletop resulted in the formation of 600 companies. An astonishing number of **major oil companies** that would have a huge impact on the world had at least some roots at Spindletop: **Texaco**, **Gulf**, **Exxon**, **Mobil**, **Amoco**, and **Sun**. Discoveries continued, the industry expanded, and the opening of the **Houston Ship Channel** in 1914 attracted a large number of refineries after **World War I**.

By the end of the 1920s, exploration had expanded beyond the Gulf Coast (where Spindletop was) to encompass many other parts of the state, including the Panhandle, Central Texas, North Texas, and the **Permian basin**. On 28 October 1926, the Transcontinental Oil Company and Mid-Kansas Oil and Gas, a subsidiary of the Ohio Oil Company (later **Marathon**), discovered the **giant** Yates field in the Permian basin just west of the Pecos River. The field, with an **EUR** of nearly 2 Bb, produced its billionth barrel on 11 January 1985. Its production peak of 0.112 Mmbpd occurred early, in the third year after its discovery. Beginning in the 1940s, various secondary recovery methods were used to maintain production. In 1929, the **Pure Oil Company** discovered the Van field east of Dallas.

On 5 October 1930, **Dad Joiner** discovered the **East Texas oil field**, the "Black Giant," with an EUR of 5.4 Bb, of which about 5.2 Bb have already been produced. It was the largest field ever discovered in the United States outside Alaska. Uncontrolled overproduction in the East Texas field led to a collapse of oil prices in the 1930s, prompting action by the **Railroad Commission of Texas** and the U.S. government to stabilize prices.

Texas's second-largest field after East Texas is Wasson, in the Permian basin. This field, with an EUR of 2.1 Bb, was discovered between 1936 and 1939. It has such a large areal extent that the discovery wells were thought to be in distinct fields. The first, Bennet No. 1, was drilled by the Honolulu Oil Corporation and the Davidson Drilling Company (later Cascade Petroleum Company), striking oil on 6 April 1936. Another important well, Wasson No. 1, struck oil on 19 June 1937 and was drilled by Continental Oil (**Conoco**). The Railroad Commission of Texas ruled on 1 December 1939 that the various discoveries were in the same field and named it Wasson. Wasson produced about 1.82 Bb by 1992. Another of Texas's largest fields, Kelly–Snyder, is also in the Permian basin and has an EUR of 1.7 Bb. Gulf Oil in 1938 and **Humble Oil and Refining** (later Exxon) in 1948 drilled wells in the area but abandoned them. On 15 November 1948, **Magnolia Petroleum** completed the discovery well for the Kelly part of the field, and about two weeks later Standard Oil of Texas (a subsidiary of Standard Oil Company of California, later **Chevron**) discovered the Snyder portion. The field produced its billionth barrel in 1979.

Of Texas's 32 giant oil and gas fields, 26 were discovered before 1940. Of the 32 giants, 28 are primarily oil fields, and all but three of them were discovered before 1940. Texas's three largest oil fields (East Texas, Wasson, and Yates) were also all discovered before 1940.

Despite the aging of its **conventional** petroleum industry, in the 21st century Texas had a major resurgence in oil and gas production because of **shale oil** and **shale gas** production. This was driven by fracking and horizontal drilling, first in the **Barnett Shale** and then elsewhere. In 2015 Texas produced over a quarter of the United States' gas. *See also* CONNALLY HOT OIL ACT (1935); GUFFEY, JAMES McCLURG (1839–1930); HOT OIL; LUCAS, ANTHONY F. (1855–1921); MITCHELL, GEORGE P. (1919–2013); STERLING, ROSS (1875–1949); THOMPSON, ERNEST (1892–1966).

TEXAS COMPANY. *See* TEXACO.

TEXAS CORPORATION. *See* TEXACO.

TEXAS FUEL COMPANY. *See* TEXACO.

TEXAS INSTRUMENTS. *See* GEOPHYSICAL SERVICE INCORPORATED (GSI).

TEXAS RAILROAD COMMISSION. *See* RAILROAD COMMISSION OF TEXAS.

THIRD OIL SHOCK. Large increase in oil prices that began in 2003 and lasted over a period of years. Average annual prices more than doubled between 2003 and 2006. There was a particularly rapid increase between 2007 and early 2008. Average world prices for all crudes (FOB, weighted by export volume) were $54.63 a barrel in the first week of 2007; by early July 2008, they exceeded $135.

The causes for the rapid increase are not yet fully understood. Some observers believed the increase indicated that the world was approaching **peak oil** in 2008. However, this conclusion was not justified, given known **reserves**, likely undiscovered potential, and recent advancements in technology. The demand from a newly industrialized and increasingly affluent Asia was one factor. **India** and **China** experienced tremendous demand growth after 2000. In 2006 alone, China's demand grew by more than 450,000 bpd, and India's grew by 100,000 bpd. Another factor was the weakness of the U.S. **dollar**. An additional possible explanation is that with interest rates low and the stock market not offering attractive returns in 2007 and 2008, speculators bought up oil as a store of wealth and drove up the price; increasingly, oil became a financial asset rather than a commodity. Oil prices began to decrease after their July 2008 highs. In the fall of 2008, the world began to experience a global financial crisis, with lower economic activity and lower demand for oil. By December 2008, oil prices dropped below $50 a barrel. *See also* FIRST OIL SHOCK; OIL COUNTERSHOCK; SECOND OIL SHOCK.

THOMPSON, ERNEST (1892–1966). Leading member of the **Railroad Commission of Texas** who spearheaded efforts to regulate production and stabilize prices in the 1930s. Ernest Othmer Thompson was born on 24 March 1892 in Alvord, **Texas**. He graduated from the University of Texas in 1917 with a degree in law and served in **World War I**, rising to the rank of lieutenant colonel. After the war, he practiced law and owned a hotel. Elected mayor of Amarillo, Texas, in 1928, he fulfilled a campaign promise of increasing competition in the utility industry and cutting rates.

In 1932, Governor **Ross Sterling** appointed him to the Railroad Commission of Texas. Thompson became the commission's leading member and saw it through a difficult period of collapsing oil prices caused by unrestrained overproduction in the newly discovered **East Texas oil field**. He faced constant challenges by elements of the federal and state governments, as well as many private producers, to the commission's ability to act. Eventually the commission was able to bring order to Texas petroleum production, thereby having a major influence on national and international markets.

Thompson ran unsuccessfully for governor of Texas in 1938 and 1940. In his second campaign, he observed that Texas oil was an exhaustible resource that should be taxed at higher levels, and he proposed a tax of 5 cents a barrel to help pay for Social Security pensions. Thompson remained a member of the Railroad Commission until January 1965. He died on 28 June 1966.

TIDEWATER OIL COMPANY. *See* GETTY, J. PAUL (1892–1976).

TIGHT GAS. Unconventional natural gas contained in sandstone and carbonate formations of low permeability. In some cases, the term is used to encompass **shale gas** as well. Tight gas does not flow easily through the reservoir. It can be produced by applying hydraulic fracturing (**fracking**) and **horizontal drilling**. There is no universally accepted precise definition of a tight gas reservoir. One commonly accepted criterion, used by the Natural Gas Policy Act of 1978 in the **United States**, is that the average permeability of the reservoir be lower than 0.1 millidarcies (mD). The German Society for

Petroleum and Coal Science and Technology (Deutsche Wissenschaftliche Gesellschaft für Erdöl, Erdgas und Kohle, or DGMK) has established a higher cutoff, of 0.6 mD.

Tight gas was first established as a regulatory category of natural gas by the 1978 act. The tight gas category enjoyed favorable treatment as an incentive for increased production of new types of gas resources to offset a perceived growing scarcity of conventional gas. When the pricing of natural gas as a whole was deregulated in 1989, tight gas no longer constituted a separate regulatory category, but it continued to be one for geological and engineering purposes.

Tight gas production is important in the United States, where it occurs in a number of regions. It has grown in importance in the 21st century, though its growth has not been as explosive as that of shale gas. Tight gas production accounted for over 18 percent of total U.S. gas production in 2015, compared to about 12 percent in 1990.

TIGHT OIL. Crude oil contained in rock formations of low permeability. Such rocks do not allow easy movement of fluid; oil contained within them is consequently difficult to extract and produce. Tight oil can occur in shale, sandstone, and carbonate reservoirs, but the most important deposits are in shales. This sometimes leads the terms "tight oil" and "**shale oil**" to be used synonymously, although the latter is more precisely a subset of the former. Tight oil is considered **unconventional oil**, although its physical properties may be similar to those of conventional oil.

Production of tight oil is facilitated by the use of hydraulic fracturing (**fracking**) and **horizontal drilling**. Tight oil production has been particularly prolific in the **United States** during the past decade, thoroughly reversing long-term production declines, at least temporarily. In 2007 the United States produced 360,000 barrels of tight oil per day, amounting to about 6.2 percent of total production; by 2016 tight oil production stood at 4.25 million barrels per day, or nearly 48 percent of the total. Important oil-producing formations in the United States include the **Bakken** and Eagle Ford; the **Wolfcamp** shale has also been assessed to have huge future potential.

Tight oil production in other countries has not yet taken off the way it has in the United States, but many countries around the world have tight oil potential. Estimates are highly uncertain, but the world is thought to have about 419 Bb of unproven technically recoverable tight oil resources. Of these, 78.2 Bb are thought to be in the United States, 74.6 Bb in **Russia**, 32.2 Bb in **China**, 27 Bb in **Argentina**, 26.1 Bb in **Libya**, and 22.6 Bb in the **United Arab Emirates**.

TIMAN–PECHORA. The third-largest oil-producing region of **Russia** (after western **Siberia** and the **Volga–Urals**). Timan–Pechora is located south of the shores of the Pechora Sea and west of the Urals, in northwestern Russia. Signs of oil have long been known in the region. The 15th-century Dvina chronicle mentions oil seeping along the Ukhta River. In 1721, Grigory Cherepanov found an oil spring on the river. This was reported to Tsar **Peter I** (Peter the Great), who ordered the area surveyed and awarded Cherepanov 6 rubles. In 1724, a Captain Bosargin explored the spring again, and Peter I ordered samples sent to Europe for analysis. In 1746, Fyodor Pryadunov built a rudimentary installation to collect oil from the surface of the Ukhta River. The oil was analyzed in Hamburg and sold in Moscow. The enterprise continued for nearly the rest of the 18th century under various operators. **Mikhail Lomonosov** also mentioned Ukhta oil in his work on Russia's minerals.

The first geological survey of the region, in 1843, described oil seeps. In 1868, Mikhail Sidorov drilled near the confluence of the Ukhta and Neft-Yol Rivers and struck oil at about 40 feet. In 1872, he struck a weak gusher. The quantities were not commercial, but Sidorov believed in the field and wanted to develop it. He persisted despite the difficult terrain and bureaucratic barriers. He published a book in 1882 about oil in northern Russia, but he died 12 July 1887 without realizing his ambition.

Modern exploration of Timan–Pechora began in the 1930s. In 1932, **Ivan Strizhov** discovered the Yarega (Yaregskoye) field, with an **EUR** of about 0.8 Bb. Exploration intensified after **World War II**. There were several discoveries in the 1960s and 1970s, most of them relatively small. The **giant** Usinskoye field, discovered in 1961, has an EUR of about 3 Bb. Other large fields include Vozey (1972), Khar'yaga (1977), and Yuzhno Khilchuyu (1990). Some **offshore** fields have been discovered in the Pechora Sea. Prirazlomnaye, with an EUR of 0.6 Bb, is the most important. A field was also discovered on the island of Kolguev. Timan–Pechora oil production peaked in 1986 at roughly 0.370 Mmbpd but may increase in the future. *See also* BARENTS SEA.

TITUSVILLE. City (pop. 5,400, 2016) in Crawford County, **Pennsylvania**; early center of the petroleum industry in the **United States**. Titusville was founded in 1796 by surveyors Jonathan Titus and Samuel Kerr. In August 1859, an important early well was drilled by **Edwin Drake** outside Titusville, initiating the Pennsylvania oil boom. The well was just inside **Venango County**, where most of the oil exploration and development activity immediately ensued. An early refinery was built in Titusville. The last refinery closed there in 1950. The city's economy is now based on steel products, aluminum products, plastics, and lumber. Titusville has a regional campus of the University of Pittsburgh founded in 1963 and a large museum devoted to the Drake Well and the early Pennsylvania oil industry.

TNK–BP. Joint venture in **Russia** between **BP** and the Tyumen Oil Company (Tyumenskaya Neftyanaya Kompaniya, or TNK), acquired by **Rosneft** in 2013. The venture was created on 1 September 2003 to control the Russian and Ukrainian petroleum interests of BP and the Alfa Access Renova (AAR) group, which held, among other things, assets of the Tyumen Oil Company. Among TNK–BP's extensive interests was the **Samotlor** oil field, Russia's largest. Taken as a single entity, TNK–BP was the third-largest oil producer in Russia in 2007, with 1.6 Mmboepd (up from 1.3 Mmboepd four years earlier). There was considerable controversy surrounding the joint venture in 2008. BP alleged that the oligarchs behind TNK were attempting to take control, with help behind the scenes from the Russian government. For example, the government suddenly terminated visas and expelled BP executives and technical personnel. TNK–BP continued to operate, acquiring BP assets in Vietnam and **Venezuela** and a stake in some **Brazilian** exploration blocks. On 21 March 2013, TNK–BP was acquired by **Rosneft**.

TORREY CANYON. Oil **tanker** that ran aground on granite rocks near the Scilly Isles off the southwest coast of England on 18 March 1967, resulting in a major oil spill. The total amount of oil spilled was about 910,000 barrels, making it the fifth-largest tanker spill in history. It was the first of the truly major supertanker spills and essentially introduced the world to a new phenomenon. The spill also taught some important lessons. For one thing, it pointed out the inadequacy of ship design at the time. The **United Kingdom**'s bombing of the wreck to set the remaining oil on fire did not succeed. It was learned that oil spreads and thins rapidly, and also quickly loses its volatile parts. It was also learned that dispersants need to be applied quickly and in greater quantity than were done in this case. The use of detergents was shown to be environmentally damaging.

The *Torrey Canyon* spill showed that some of the best treatment for oil spills is provided by nature, in the form of storm action and metabolic breakdown by microorganisms. The *Torrey Canyon* incident also pointed out a number of legal complexities that could arise in oil spills. The ship was American owned, registered in Liberia, and had a multinational crew. It ran aground in British waters but also contaminated the coast of Brittany in France (which would suffer again nine years later in the ***Amoco Cadiz*** spill). Refer to appendix 9. *See also EXXON VALDEZ*; SANTA BARBARA OIL SPILL.

TOTAL. Petroleum company headquartered in France, descended from the **Compagnie Française des Pétroles** (CFP), **Elf Aquitaine**, and **Petrofina**. CFP changed its name to Total Compagnie Française des Pétroles in 1985, and in 1991 simply to Total. In 1991, it was listed on the New York Stock Exchange. The French government began to reduce its minority stake in the company in 1992. Total expanded the former CFP's hydrocarbon assets, to include shares in fields in the **Caspian** area, **Colombia**, and **Argentina**.

In 1995, Total began to develop important oil and gas fields in **Iran**, resisting pressure from the **United States** not to do so. It sold its share of American subsidiary Total Petroleum to Diamond Shamrock to protect itself from the U.S. government and then invested further in Iran with Petronas, **Gazprom**, and other partners. Even though Total had divested itself of its American subsidiary, the **Iran–Libya Sanctions Act** (ILSA) of 1996 provided for the punishment of non-U.S. companies doing business with Iran's energy sector. However, President Bill Clinton's administration, wanting to avoid a showdown with the European Union (EU), granted the Total consortium a waiver in 1998. The consortium was able to move ahead with development of the **megagiant** South Pars gas field.

In 1999, Total bought Petrofina for $11 billion, acquiring, among other things, interests in **Angola**. The new company briefly became TotalFina. TotalFina then made a hostile acquisition of Elf Aquitaine in 1999–2000 and became TotalFinaElf. Total paid $48.7 billion for 95 percent of Elf and acquired the remaining 5 percent soon after. In 2003, the company renamed itself yet again, becoming Total S.A.

In 2011 Total discovered the 10.6 Tcf Abseron gas field in the offshore area of **Azerbaijan**. In 2017 the company announced its intention to invest half a billion dollars over four years to develop **shale gas** in the Vaca Muerta Formation of Argentina. Argentina is estimated to have the world's second-largest endowment of technically recoverable shale gas resources. In 2016, Total controlled 5.414 Bb of proven oil **reserves** and 32.984 Tcf of **natural gas**. It produced 1.271 Mmbpd of oil and 6.447 Bcfpd of gas.

TOTAL RESOURCE BASE. As used by the **Society of Petroleum Engineers**, total resource base includes "all quantities of petroleum naturally occurring on or within the earth's crust (recoverable and unrecoverable) plus those quantities already produced. Further, it includes all types of petroleum whether currently considered '**conventional**' or '**unconventional**.'" The total resource base includes the **ultimately recoverable resource** (URR) and adds to it the unrecoverable portion of the resources.

TOWNSEND, JAMES M. *See* BISSELL, GEORGE HENRY (1821–1884).

TRANS-ALASKA PIPELINE SYSTEM (TAPS). An 800-mile **pipeline** delivering oil from the **Prudhoe Bay oil field** on **Alaska**'s North Slope to the ice-free port of **Valdez** in southern Alaska. The pipeline is an impressive feat of engineering in a hostile environment. It has a diameter of 48 inches. It has 78,000 support columns and 65,000 welds. It has 12 pump stations, with flow rates controlled by computer. As many as 20,000 workers were involved in its construction at any one time. The northern half of the pipeline is elevated to protect the permafrost. It is buried at river crossings and in areas threatened by avalanche. The pipeline also has a system to keep the oil warm for 21 days in the event of a shutdown.

Planning for the pipeline began almost immediately after the 1967 discovery of oil at Prudhoe Bay by the Atlantic Richfield Company (**Arco**) and the **Humble Oil and Refining Company** (later **Exxon**). After some initial studies, the pipeline subsidiaries of Arco, Humble, and **BP** announced plans on 10 February 1969 to build the pipeline. On 22 October 1969, they were joined by **Amerada Hess**, Home Pipe Line Company, **Mobil** Pipeline Company, **Phillips Petroleum**, and the **Union Oil Company of California** in a joint venture. The partners incorporated as the Alyeska Pipeline Service Company on 14 August 1970.

The concept faced considerable opposition. In April 1970, environmental groups and other parties began filing lawsuits to block construction. The creation of the U.S. Environmental Protection Agency and the requirement for environmental impact statements added new levels of scrutiny. However, the Trans-Alaska Pipeline Authorization Act was approved on 16 November 1973 during the **First Oil Shock**. On 3 January 1974, a U.S. federal right-of-way grant was issued. The first pipe was laid on 27 March 1975, and construction was completed on 31 May 1977. The final cost of the pipeline was $7.7 billion, much greater than the initial estimates of $1.5 billion.

On 20 June 1977, the first oil began to flow, reaching the terminal at Valdez on 28 July. On 1 August 1977, the Arco tanker *Juneau* left Valdez with the first oil shipment. The billionth barrel of oil arrived at Valdez on 22 January 1980. By 27 April 2000, 13 Bb had flowed through the pipeline. Partly because of the requirements imposed by the authorization act, the pipeline has had a very good environmental and safety record. The worst problem was a 16,000-barrel oil spill in 1978 caused by sabotage.

TRANS-ARABIAN PIPELINE (TAPLINE). Major oil **pipeline** running from **Abqaiq** in **Saudi Arabia** to the Lebanese Mediterranean port of Sidon. The 1,063-mile pipeline was completed in 1950 by the **Aramco** partners: Standard Oil Company of New Jersey (**Exxon**), Standard Oil Company of California (**Chevron**), Socony–Vacuum (**Mobil**), and the Texas Company (**Texaco**). The construction of a large pipeline in Saudi Arabia was advocated by U.S. secretary of the interior **Harold Ickes** and the **Petroleum Re-**

serves Corporation during **World War II**. Opposition from some **major oil companies** in the **United States** helped stop the project. The Aramco partners were able to achieve the goal later with their own capital. The strongest opposition to the pipeline during the war had come from the very same Standard Oil Company of New Jersey and Socony–Vacuum that would become Aramco partners later in the decade. At the time, the two companies were shut out of Saudi Arabia by the **Red Line Agreement** and viewed the pipeline as a threat.

Construction began in 1947, managed largely by Bechtel. The construction effort enjoyed the support of the Harry S. Truman administration, and this helped ensure the availability of steel. Steel was a critical material in short supply after the war and was still allocated by the U.S. government. The Tapline had an initial capacity of 320,000 bopd, which was raised to 450,000 in 1957 with the addition of more pumping stations. The original four pumping stations spawned new towns at Qaysumah, Rafha, Badana, and Turayf.

Aramco argued that the **50–50 agreement** for profit sharing with the government of Saudi Arabia covered only oil production and not transportation. However, Saudi Arabia insisted that it was entitled to 50 percent of Tapline's profits. In 1963, Aramco finally agreed to pay the government half the difference between the price of petroleum at **Ras Tanura** and the price at Sidon, after deducting costs.

Part of the pipeline ran through the Golan Heights of Syria. After the Six-Day War in 1967, this area came under the control of Israel. The Israelis permitted the pipeline to continue working, but in 1976 Saudi Arabia stopped operating the pipeline beyond Jordan. The cessation was caused by a number of factors, including disputes with Syria and Lebanon over transit fees and the availability of very large and ultralarge crude carriers (refer to appendix 8; *see* TANKERS) as an alternative mode of transportation. Saudi Arabia has also developed other pipelines, including the high-capacity **Petroline**. Saudi Arabia closed the pipeline because of disruptions surrounding **Iraq**'s 1990 invasion of **Kuwait** and the resulting 1991 **Persian Gulf War**. Future reactivation of the Tapline remains a possibility.

TRANS-CASPIAN TRADE PARTNERSHIP. *See* BAKU OIL SOCIETY.

TRINIDAD LAKE ASPHALT CO. *See* PITCH LAKE.

TRIPOLI AGREEMENT (1971). Agreement between **major oil companies** and members of the **Organization of the Petroleum Exporting Countries** (OPEC) doing significant business in the Mediterranean region. The Tripoli Agreement, concluded 2 April 1971, raised prices and increased producer countries' profit shares. It was negotiated by **Libya** on behalf of itself,

Algeria, **Saudi Arabia**, and **Iraq**, and it was more aggressive than the **Tehran Agreement** concluded a few weeks earlier. The **posted price** was increased immediately by 90 cents a barrel, or about 35 percent. The agreement also provided for a 2.5 percent annual increase thereafter, plus an inflation adjustment. Producer country profit shares were increased to 60 percent.

TROUT, WALTER C. (1874–1947). Developer of the familiar counterbalanced oil field pump, still widely used in the industry; president of **Lufkin Industries** (1930–1947). Canadian-born Walter Charles Trout joined Lufkin Foundry and Machine Company in 1905, three years after its founding. At the time, the firm specialized in repairing sawmill equipment. Trout led the diversification of the company into manufacturing, beginning with sawmill equipment. In 1923, Trout was invited to lunch by **Ross Sterling**, a future governor of **Texas** who had recently been president of the **Humble Oil and Refining Company** (later **Exxon**). Sterling articulated the need for an effective pump to lift oil from low-pressure wells. As a result, Trout began working on a design and prototypes. He patented his counterbalanced pump in 1925. The pump became an industry standard, and its familiar horse-head shape can be seen the world over. Trout served as president of Lufkin from 1930 until his death in 1947.

TRUCIAL STATES. *See* ABU DHABI; DUBAI; RAS AL-KHAIMA; SHARJAH; UNITED ARAB EMIRATES (UAE).

TULSA. City (pop. 400,000, 2015) in **Oklahoma** that became a boomtown after the 1905 discovery of the nearby **Glenn Pool** oil field. For much of the 20th century, Tulsa was a major center of the petroleum industry, although it is now less prominent in that role and has a more diversified economy. The **American Association of Petroleum Geologists** and the **Society of Exploration Geophysicists** are headquartered in Tulsa. *See also* GAS TO LIQUIDS; *OIL AND GAS JOURNAL* (*OGJ*).

TURKISH PETROLEUM COMPANY (TPC). *See* IRAQ PETROLEUM COMPANY.

TURKISH STRAITS. *See* BOSPORUS AND DARDANELLES.

U

ULTIMATELY RECOVERABLE RESOURCE (URR). All quantities of a resource occurring in the earth that have already been extracted and are projected to be extractable in the future. The ultimately recoverable resource is a very different quantity from **reserves**. The URR includes cumulative production to date, discovered reserves, and both known and undiscovered **technically recoverable resources**. The URR is thus a function of technology, price, and other economic conditions; higher prices can turn currently unrecoverable resources into recoverable ones, as can more advanced technology. One could speak of a URR for **conventional crude oil**, a URR for **natural gas**, a URR for heavy oil and **bitumen**, or a URR for all types of hydrocarbons. The URR is the ultimately recoverable subset of the **total resource base**, which also includes the unrecoverable portion of the resources.

ULTRA-DEEPWATER. As with "**deepwater**," there is no universally accepted definition of "ultra-deepwater" **offshore** oil and gas, and the definition changes with time and the progress of technology. A common definition of an ultra-deepwater play is one that lies beneath more than 7,000 feet of water.

UNCONVENTIONAL OIL AND GAS. Oil and gas that may have been overlooked in the earlier history of the industry because it is difficult to extract, unusually expensive in cost or energy, or inaccessible compared to **conventional oil and gas**. Unconventional oil includes very **heavy oil** and **bitumen**, **tight oil** (including **shale oil**), and **oil shale**. The term is sometimes used to refer to oil that might be otherwise conventional but is very difficult to access, for example **Arctic** oil and **ultra-deepwater** oil, especially in **pre-salt** and **subsalt** plays. Unconventional gas includes **tight gas** and **shale gas**.

UNION OIL COMPANY OF CALIFORNIA. Large **independent** petroleum company later known as Unocal; eventually part of **Chevron**. The Union Oil Company of California was founded in 1890 by **Lyman Stewart**, Wallace Hardison, and Thomas R. Bard as the "union" of three oil companies in southern California unaffiliated with **Standard Oil**: the Sespe Oil Company, the Hardison and Stewart Oil Company, and the Mission Transfer Company. Union Oil was first headquartered in Santa Paula, California, in a building that now houses a petroleum museum. Bard was the company's first president, but Stewart replaced him in 1894. In 1900, the company moved its headquarters to Los Angeles. That same year, the first known oil company geological research department was created at Union Oil by petroleum geologist **William Orcutt**.

Union Oil's predecessor companies were among the earliest commercial discoverers and producers of oil in California, beginning in Pico Canyon in 1884. By 1886, they were producing about a tenth of California's output. Union discovered new oil in Torrey Canyon and in 1892 purchased the Los Angeles Oil Company, which held promising land in southern California. Union began drilling immediately and discovered California's largest find to that time, Adams No. 28. It produced 15,000 barrels in its first year and initiated a California oil boom that flooded the market. In 1910, Union Oil, in partnership with the Lakeview Company, drilled the Lakeview gusher in the Midway–Sunset field, the second-largest field ever found in California and the third largest in the lower 48 states. Lakeview was the longest lasting (18 months) and the most productive (9 Mmb) gusher in the history of the **United States**.

As president, Stewart expanded the company too fast and in 1914 had to resign as a condition of obtaining needed credit. He was replaced by his son **William Lyman Stewart** and became chairman. In 1917, Union acquired Pinal-Dome Oil. It sold **gasoline** under the "76" brand name, which became a 20th-century icon. In 1922, Union thwarted a takeover attempt by **Royal Dutch Shell**. The company remained heavily concentrated in California well into the 1950s. In 1965, Union acquired the **Pure Oil Company**, which operated mostly in **Texas** and **offshore** in the **Gulf of Mexico**. The Pure acquisition doubled Union's size and made it a national oil company. A few years later, the company's reputation suffered heavily because of the 1969 **Santa Barbara oil spill**.

In the 1970s, Union invested in developing alternative energy sources and also began expanding overseas in earnest, particularly in **Indonesia** and Thailand. In the 1980s, it began an unsuccessful **oil shale** project in Colorado's **Green River Formation**. The company changed its name to Unocal in 1983. Corporate raider **T. Boone Pickens** attempted a takeover of Unocal in 1984. The company borrowed more than $5.5 billion to defend itself through stock repurchases. The heavy debt caused difficulties for the company, and over the next two decades its story became one of sell-offs and divestitures. In 1989, Unocal entered into a joint venture with **PdVSA** to run its U.S. Midwest refining and marketing. It continued to shift away from downstream operations in the 1990s, eventually even selling its venerable "76" brand to Tosco, which was acquired by **Phillips Petroleum** in 2001. On the environmental front, its reputation again suffered when it was found responsible for contaminating groundwater in California. It settled the case for $80 million in 1997.

During the 1990s, the company, which had been so intimately associated with California, began to leave the state. In 1996, it sold its California oil fields to Nuevo Energy. In 1999, it combined its operations in West Texas with those of Titan Exploration, forming Pure Resources a year later, which it acquired in toto two years after that. In 2003, it sold properties in Louisiana and the Gulf of Mexico to Forest Oil.

In April 2005, the company agreed to a merger with Chevron. Soon after, the China National Offshore Oil Corporation (CNOOC) made a rival bid. There was an uproar in the American press, and the U.S. House of Representatives passed a nonbinding resolution urging President **George W. Bush** to review the bid in the light of potential national security concerns. CNOOC withdrew its bid and Unocal was acquired by Chevron in August 2005 for $18.4 billion in cash and stock. In 2004, the last full year before its acquisition by Chevron, Unocal was the 13th-largest U.S. petroleum company by total assets. It had a net income of $1.208 billion on revenues of $8.204 billion and controlled 0.659 Bb of proved oil reserves and 6.568 Tcf of gas. It produced 0.159 Mmbpd of oil and 1.597 Bcfpd of gas.

UNITED ARAB EMIRATES (UAE). Oil production 2015: 2.88 Mmbpd (3.67 percent of world); **reserves** 97.8 Bb (5.94 percent of world); gas production: 2.117 Tcf (1.7 percent of world); reserves 215.1 Tcf (3.12 percent of world). An important oil producer and exporter, the United Arab Emirates is a federation of largely independent states of the eastern Arabian Peninsula that were previously known collectively as the Trucial States. In January 1968, the **United Kingdom**, long installed in the area, announced its intention to withdraw all military forces by 1971. In March 1971, the Trucial States joined **Bahrain** and **Qatar** to form the Federation of Arab Emirates. However, Bahrain and Qatar seceded in August 1971 and became independent states. On 2 December 1971, the six emirates of **Abu Dhabi**, **Dubai**, **Sharjah**, Umm al-Qaiwain, Ajman, and Fujairah formed the United Arab Emirates. In February 1972, they were joined by a seventh emirate, **Ras al-Khaima**.

According to the UAE constitution, each emirate is responsible for its own oil and gas affairs. The main oil-producing emirate is Abu Dhabi, with more than 94 percent of the federation's oil reserves. Dubai is a distant second, with about 4 percent. Sharjah and Ras al-Khaima are minor oil states, with about 1.5 percent and 0.1 percent of the reserves respectively. In 1967, Abu Dhabi joined the **Organization of the Petroleum Exporting Countries**

(OPEC). In 1970, both Abu Dhabi and Dubai joined the **Organization of Arab Petroleum Exporting Countries** (OAPEC). The UAE is a member of the **Gas Exporting Countries Forum**.

Oil production in what is now the UAE began in 1962 and increased rapidly to a peak of 2 Mmbpd in 1977. It then decreased to a low of 1.146 Mmbpd in 1984, after which it increased again to a second peak of 2.386 Mmbpd in 1991. The 1991 peak reflected the UAE's policy of raising output to help make up for lost production from **Iraq** and **Kuwait** because of the 1991 **Persian Gulf War**. After 1991, production was roughly 2–2.3 Mmbpd until 2003. In 2006, production was 2.636 Mmbpd, the highest level up to that time. UAE **natural gas** production was 0.2 Tcf in 1980. It grew fairly steadily, with some oscillations, to its 2006 level of 1.672 Tcf, or about 0.763 Mmboepd. In 2015 it was higher, at 2.117 Tcf.

UNITED KINGDOM (UNITED KINGDOM OF GREAT BRITAIN AND NORTHERN IRELAND). Oil production 2015: 0.894 Mmbpd (1.14 percent of world); reserves 2.564 Bb (0.16 percent of world); gas production: 1.387 Tcf (1.1 percent of world); reserves 7.319 Tcf (0.11 percent of world). In his 1188 *Itinerary through Wales*, Giraklus Cambrensis mentioned rock outcrops on the river Severn that appeared to bear petroleum, although there is no record of the outcrop ever being exploited. In 1848, James Young established a refinery near an oil spring in a Derbyshire coal works. When the supply was exhausted, he applied some of his techniques to bituminous **oil shales** in Scotland. Despite such early indications, the United Kingdom (UK) did not develop a significant indigenous petroleum industry until the late 20th century. Earlier in the century, the United Kingdom's main role was as a major power often in dire need of oil, and as a majority shareholder in the **Anglo–Iranian Oil Company** (later **BP**), which controlled **Iran**'s oil industry until 1951 and developed petroleum resources elsewhere as well.

The situation changed drastically with the discovery of petroleum in the **North Sea**. The United Kingdom became a major oil producer and a net exporter. The discovery of the **sugergiant** onshore **Groningen gas field** in the Netherlands in 1959 sparked UK interest in possible **offshore** resources. The first British discovery of North Sea **natural gas** was the West Sole field, off the coast of East Anglia, in 1965. This was followed by a number of other gas discoveries in the southern part of the North Sea, including the UK's largest gas field, Leman (**EUR** 11 Tcf), in 1966.

Oil proved more elusive at first. The Standard Oil Company of Indiana (later **Amoco**, eventually part of BP) discovered some shows of oil in the northern North Sea in 1969 but was unable to exploit them commercially. On 7 October 1970, however, BP announced the discovery of the **giant** Forties field, the North Sea's first major oil find and the UK's largest oil field. Forties (EUR up to 2.7 Bb) is located about 110 miles east of Aberdeen under

350 feet of water. Its production peaked at about 520,000 bpd in 1980. By 2003, production was down more than 10-fold, to about 45,000 bpd. In July 1971, **Royal Dutch Shell** and the Standard Oil Company of New Jersey (renamed **Exxon** the next year) discovered the **Brent oil field**, about 110 miles northeast of the Shetland Islands under 460 feet of water. Crude from Brent and related fields later evolved into a major pricing **benchmark** known as **Brent Blend**. Twenty giant fields, 12 primarily oil and the rest primarily gas, would be discovered in the UK North Sea, almost all before 1980. The oil giants are in the northern North Sea, while the gas giants are largely in the south. The most recent giant oil field was Buzzard (EUR about 0.5 Bb), discovered in 2001.

UK oil production built up rapidly during the 1970s, aided by the high oil prices of the period, which serendipitously made the relatively high-cost North Sea oil viable (*see* FIRST OIL SHOCK; SECOND OIL SHOCK). In 1976, as the sector was growing, the government also created the short-lived **British National Oil Company** (BNOC). UK production reached 1.62 Mmbd in 1980 and peaked six years later at 2.539 Mmbpd. It then declined rapidly to 1.8 Mmbpd in 1989. The decline was caused by a number of factors. One was the overall softness in the world market (*see* OIL COUNTERSHOCK). The 1987 fires at the Piper field were another factor. A third factor was probably taxation, which had been increasing for some years. The Petroleum Revenue Tax (PRT), introduced in 1975 on profits of individual fields, reached 75 percent by the mid-1980s, despite the Conservatives' control of the government. At one point, taking all levies together, the government's marginal take on oil income reached 88 percent.

The reversal of these taxation trends helped spur production in the 1990s. The PRT came down to 50 percent in 1993 and was abolished for new fields. By the late 1990s, the total government marginal take was down to about 65 percent, and 30 percent on fields developed after 1983. Oil production, which had stayed roughly flat near its 1989 low, began to increase after 1993 and reached a second peak of 2.68 Mmbpd in 1998 as new fields were developed. After that, it declined again, owing mainly to exhaustion of the major fields (which were 35 percent off peak by 1998). The new discoveries of the 1990s were all smaller than the giants discovered in the 1970s. The UK's 2006 oil production level of 1.49 Mmbpd was below its production in 1980. The UK was a net exporter of oil from 1981 to 2005, with most of its crude grades light and sweet, hence desirable. However, in 2006 it became a net importer and remained so. BP accounted for about a fourth of UK oil production in 2004. Other major operators include Royal Dutch Shell, **Chevron**, and **Total**. In the 2000 decade, **major oil companies** began to sell their UK assets as the focus of the UK industry shifted to smaller fields and to the improvement of existing fields.

Although the vast majority of the UK's petroleum resources are offshore, the country has one major onshore oil field, Wytch Farm, which has an EUR of 0.47 Bb and is the largest onshore oil field in Western Europe. The main complex was discovered in 1973 by British Gas, although nearby fields were discovered as early as 1959 and were producing in the 1960s. BP acquired Wytch Farm in 1984. It peaked in 1997 at 110,000 bpd. The UK's first onshore oil field in the home islands was the **Hardstoft** field, discovered in 1919.

The United Kingdom was the world's fourth-largest producer of gas for much of the 1990s, behind **Russia**, the **United States**, and **Canada**, and just ahead of **Algeria**. But by 2005, it had begun to fall behind. In 2006, it was in seventh place. It also became a net gas importer in 2004. UK gas production peaked in 2000 at 3.826 Tcf (about 1.747 Mmboepd). In 2015 it was 1.387 Tcf.

UNITED NATIONS CONVENTION ON THE LAW OF THE SEA (UNCLOS). Set of international rules for the use of the oceans and their resources. The work of the convention was completed on 10 December 1982 when it was opened for signature. It came into force 16 November 1994. Among other things it defined the exclusive economic zone (EEZ) 200 nautical miles from a country's coastline. *See* ARCTIC REGIONS; CASPIAN SEA; NORTH SEA; SOUTH CHINA SEA.

UNITED STATES (UNITED STATES OF AMERICA). Oil production 2015: 9.415 Mmbpd (12.01 percent of world); **reserves** 35.272 Bb (2.14 percent of world); gas production: 27.047 Tcf (21.69 percent of world); reserves 293.12 Tcf (4.25 percent of world). The United States has been a major force in the global petroleum industry since 1860. The age of oil has been an American age, and the American age has been an age of oil. As a nation, the United States has been a leader in both production and consumption, and its companies have dominated the global industry. **Standard Oil,** which would leave a legacy lasting to the present day, was an American company. Five of the **Seven Sisters** originated in the United States: **Exxon, Mobil, Gulf, Chevron**, and **Texaco**. The American oil industry contributed to the Allied victories in **World War I** and **World War II**. The United States has not only been a driving force in the technology and business processes of the petroleum industry; it has also been a driving force in developing and diffusing technologies and applications that consume oil. For example, it gave birth to an automotive culture that spread in varying degrees throughout the world. The United States has played the leading role in creating the age of oil, and has in turn been strongly shaped by it. The price of oil is denominated in U.S. **dollars**, and oil has been central to U.S. politics and foreign policy for decades.

The United States was one of the birthplaces of the modern petroleum industry. The early U.S. industry began in the mid-19th century in the states of **Pennsylvania**, **West Virginia**, and southeastern **Ohio**. **Edwin Drake**'s important 1859 well near **Titusville**, Pennsylvania, is popularly regarded as signaling the beginning of the modern industry. Drake's well was undoubtedly significant and initiated a major boom, but there were contemporaneous and even earlier efforts (*see* FIRST OIL WELL). Pennsylvania produced more than 90 percent of the world's oil until the 1870s, and it dominated U.S. production until 1894, when it was overtaken by Ohio (*see* LIMA–INDIANA OIL FIELD). In 1903, **California** became the United States' largest oil-producing state. It was overtaken in 1907 by **Oklahoma** but regained the top spot again in 1909 and held it until 1914. **Kansas** and **Louisiana** were also important during this period. California and Oklahoma traded places with each other until 1928, when both were overtaken by **Texas**. Texas was the United States' top oil-producing state for the next seven decades. In 1999, it was overtaken by the federal **offshore** area, mostly in the **Gulf of Mexico**. Texas regained its top spot in 2011 and was still on top in 2016. **Alaska** was also an important oil-producing state beginning in the late 1970s. From 1988 to 1996, Alaska's production was nearly tied with that of Texas.

U.S. oil production peaked in 1970 at 9.636 Mmbpd. The pattern of U.S. oil production in the lower 48 states (excluding Alaska and the federal offshore area) roughly followed a bell-shaped curve, as envisioned in 1956 by **M. King Hubbert**, who predicted the 1970 peak in production (*see* PEAK OIL). The production of the lower 48 in 2004, 34 years after the peak, was very close to the production of the lower 48 in 1936, 34 years before the peak. Overall U.S. production also declined after the peak, but not in such a symmetrical fashion because of the entrance of new producing provinces (Alaska and the federal offshore area) that were not active before the peak. In the 2010s, U.S. production not only declined at a slower rate, but it actually turned around. By 2015, U.S. production of **tight oil**, generally **shale oil** (especially in Texas and **North Dakota**), represented nearly 49 percent of U.S. oil production and brought production levels to their highest point since 1972. (U.S. production of **tight gas** and **shale gas**, especially in Texas and Pennsylvania, accounted for nearly 69 percent of total gas production in 2015, bringing levels to their highest point in history.) The United States, long a technological leader and innovator in the oil industry, perfected **fracking** and **horizontal drilling** to extract previously inaccessible oil and gas in impermeable formations.

The United States produced more than 90 percent of the world's oil until the late 1870s. By 1901, it was producing about 40 percent, owing mainly to the rise of **Russia**'s output. Between 1905 and 1945, the United States produced roughly 60–70 percent of the world's oil every year. After 1945, the percentage declined continuously even though U.S. production was still rising until 1970, because of increasing production elsewhere. In 2016, the United States produced about 13 percent of the word's oil and liquids but was still the largest single producer.

Despite being a major producer of oil, the United States is nevertheless the world's largest net importer. In 2016, it imported 60 percent of its total consumption. This percentage has been generally climbing since 1983, when it was 29 percent. The United States, with less than 5 percent of the world's population, consumes about 20 percent of the world's oil. The United States became a net importer in 1948. Domestic producers then were concerned about competition from low-cost oil from the **Persian Gulf** region, mainly imported by the multinational **major oil companies**. U.S. **independent** oil companies stood to lose in two ways: through lower prices and through lost sales to the major oil companies. In 1954, the Trade Agreements Extension Act provided for increased restrictions on imports on the grounds of national security. However, national security arguments could as easily be made in favor of imports as against them. For example, one could argue that imports help to conserve precious domestic resources. On the other side, one could argue that imports compromised the domestic industry, so that it might not be able to respond adequately should a war or other emergency interrupt those imports. In the end, the arguments against imports won the day. The United States adopted voluntary quotas on imports in 1957, followed by mandatory quotas in 1959. These quotas annoyed a number of exporting countries, particularly **Venezuela**, which was left out of a system of hemispheric preferences and exceptions. The U.S. import quotas, along with many other factors, contributed to the formation of the **Organization of the Petroleum Exporting Countries** (OPEC) in 1960. The quotas remained in effect until 1973.

The United States has often wrung its hands about its dependence on oil imports but has done little to mitigate its appetite for oil. The oil shocks of the 1970s (*see* FIRST OIL SHOCK; SECOND OIL SHOCK) prompted some conservation measures. However, after prices collapsed in the 1980s (*see* OIL COUNTERSHOCK), the United States relaxed. It continued and even expanded the production and use of inefficient vehicles and has not explored alternatives to oil in a serious and sustained manner. At the same time, it has often hobbled its domestic petroleum industry, limiting its domestic exploration capability for environmental and other reasons. Many jurisdictions in the United States have adopted bans on fracking, thus cutting off large potential sources of unconventional oil and gas. Historically, the country's main response to oil insecurity seems to have been a long and unfortunate record of meddling in the affairs of the Middle East and more recently Central Asia. *See also* AMERADA PETROLEUM CORPORATION; AMERICAN PETROLEUM INSTITUTE (API); AMINOIL (AMERICAN INDEPENDENT OIL COMPANY); AMOCO; ANADARKO PETROLEUM CORPORATION; ARCO; BAKKEN FORMATION; BARNETT SHALE; CARTER DOCTRINE; CONOCO; FUEL ADMINISTRATION; GREEN RIVER FORMATION; IRAQ WAR (2003–2011); MARATHON OIL CORPORATION; MARCELLUS SHALE; NEAR EAST DEVELOPMENT CORPORATION (NEDC); PENNZOIL; PERSIAN GULF WAR (1991); PETROLEUM ADMINISTRATION FOR DEFENSE DISTRICTS (PADD); PETROLEUM ADMINISTRATION FOR WAR (PAW) (1942–1946); PETROLEUM RESERVES CORPORATION (1943–1944); PHILLIPS PETROLEUM; UNION OIL COMPANY OF CALIFORNIA; UNITED STATES GEOLOGICAL SURVEY.

UNITED STATES GEOLOGICAL SURVEY. An agency of the U.S. Department of the Interior charged with studying the landscape, geology, and natural resources of the **United States**, as well as natural hazards. The United States Geological Survey (USGS) is headquartered in Reston, Virginia. In 2016, it had a budget of about $1 billion and over 8,000 employees. The USGS's Energy Resources Program produces a "World Petroleum Assessment" estimating the world's remaining endowment of oil and **natural gas** based on geologic analysis. The USGS is a scientific agency and does not have regulatory responsibility.

The USGS was established on 3 March 1879 during the administration of President Rutherford B. Hayes. It was charged with the "classification of the public lands, and examination of the geological structure, mineral resources, and products of the national domain." The agency's first director was Clarence King. He was succeeded two years later by John Wesley Powell, who had led a noted expedition through the Grand Canyon.

The USGS has been publishing assessments of world petroleum resources since at least the mid-1960s. The latest assessment was issued in 2012 and estimated a mean of 565.3 Bb of undiscovered technically recoverable conventional oil resources, 5,605 Tcf of technically recoverable conventional gas, and 166.7 Bb of technically recoverable natural gas resources.

UNITED STATES STEEL CORPORATION (U.S. STEEL). *See* MARATHON OIL CORPORATION.

UNIVERSE APOLLO. The first petroleum **tanker** to surpass 100,000 deadweight tons (DWT). The *Universe Apollo* was built at the former Japanese Imperial Naval Yard in Kure, Japan, by Daniel K. Ludwig's National Bulk Carriers Inc. The *Universe Apollo* was launched in 1959. It was 950 feet long, with a 135-foot beam and a tonnage of 104,500 DWT.

UNOCAL. *See* UNION OIL COMPANY OF CALIFORNIA.

URAINEFTEGAZ. *See* LUKOIL.

USX. *See* MARATHON OIL CORPORATION.

V

VACUUM OIL COMPANY. One of the companies that formed **Mobil** (later **ExxonMobil**). The Vacuum Oil Company was organized in 1866 by Hiram Bond Everest in Rochester, New York. The company had a patented vacuum distillation process developed by Everest's partner, Matthew Ewing. The heavy residue from the process formed a good lubricating oil, which became Vacuum's specialty. In 1879, after several years of successful expansion, Vacuum was purchased by **Standard Oil** for $200,000. Vacuum was the originator of the "Mobil" trademark, first as "Mobiloil" in 1899 and later as "Mobilgas."

When Standard Oil was dissolved in 1911, Vacuum became an independent company. Charles Everest, who had been president since 1906, continued in the post. Vacuum built a refinery in New Jersey in 1917. After **World War I**, it acquired Lubrite Refining, Wadhams Oil, and White Star Refining. It continued to be noted for its lubricating oils. In 1927, Charles Lindbergh used Vacuum's "Gargoyle Mobiloil" on his solo flight across the Atlantic. Vacuum scientists drained the oil and analyzed it after the flight, publishing the results. In 1929, Vacuum introduced Mobilgas and Mobilgas Ethyl **gasolines**. In 1931, Vacuum merged with the **Standard Oil Company of New York** (Socony) to form Socony–Vacuum, the company that later became Mobil.

VALDEZ. City (pop. 4,000, 2010) and important ice-free port in southern Alaska; terminus of the **Trans-Alaska Pipeline**. Alaska North Slope oil is loaded onto **tankers** at the Valdez oil terminal and shipped (today, mostly to the western **United States**). Valdez was named for the Spanish naval officer Antonio Valdez y Basan in 1790 by explorer Salvador Fidalgo. The modern town developed beginning in 1898. The city was destroyed in the great Alaska earthquake of 1964 and rebuilt on a new site nearby. The ***Exxon Valdez*** oil spill occurred in 1989 on Bligh Reef about 40 miles from Valdez. The spilled oil did not reach the town but caused massive environmental damage in the surrounding area.

VENANGO COUNTY. Western **Pennsylvania** county (pop. 53,000, 2015) considered one of the birthplaces of the modern petroleum industry. **Edwin Drake**'s 1859 well near **Titusville** touched off a **drilling** and production boom in the 1860s. Titusville is just across the county line in Crawford County, but Drake's well was within Venango County, as was most of the exploration and development activity that followed. Venango County remained the center of the oil industry in Pennsylvania and the **United States** until the 1871 discovery of the **Bradford** field in McKean County, Pennsylvania, extending slightly into the state of New York.

VENEZUELA. Oil production 2015: 2.4025 Mmbpd (3.07 percent of world); **reserves** 300.88 Bb (18.28 percent of world); gas production: 1.132 Tcf (0.91 percent of world); reserves 201.343 Tcf (2.92 percent of world). In 2011 the ***Oil and Gas Journal*** began including very **heavy oil** in the **Orinoco Oil Belt** in its reserves estimate for Venezuela, tripling the estimate in a short period of time. Venezuela is one of the world's most important oil-producing and oil-exporting countries. It was a founding member of the **Organization of the Petroleum Exporting Countries** (OPEC) in 1960 and is a member of the **Gas Exporting Countries Forum**.

Oil was likely known in premodern times by the native population in what is now Venezuela. Spanish explorers used seepages of asphalt along the eastern shore of Lake **Maracaibo** to caulk their ships. The historian Gonzalo Fernández de Oviedo Valdés mentioned oil specifically in his writings in 1535. The Spanish ship *Santa Cruz* is said to have transported a barrel of Venezuelan oil to Spain in 1539, to be used in treating King Charles I's gout.

The modern oil industry in Venezuela began on a small scale in the late 19th century. Camillo Ferrand obtained the first **concession** on 24 August 1865 to prospect for petroleum in the state of Zulia (in which lies Lake Maracaibo). He lost the concession the next year for noncompliance. On 3 September 1878, Manuel Antonio Pulido Pulido obtained a concession to exploit oil on his hacienda, "La Alquitrana," in Táchira state, south of Lake Maracaibo. He formed the Compañia Hullera del Táchira, later called the Compañia Minera Petrolia del Táchira. In 1883, he struck a small amount oil at about 200 feet in his Eureka well, which produced about 1.4 barrels a day. The next year, he drilled his Salvador well, which he had to abandon because of flooding. Nevertheless, the company was producing about 6 bpd by this time. The company refined the crude oil to obtain kerosene and marketed it to surrounding towns. Although Pulido died in 1892, the company continued its modest operations until the 1930s. Despite the small quantities involved, this was Venezuela's first integrated oil company.

In 1905, the government of Cipriano Castro formalized the legal basis for long-term petroleum concessions. The dictator **Juan Vicente Gómez**, who deposed Castro in 1908 and ruled until 1935, granted several concessions to foreign companies that then proceeded to create the basis for the Venezuelan petroleum economy. The Caribbean Petroleum Company obtained an extensive concession in 1912. **Royal Dutch Shell** acquired a majority interest in Caribbean Petroleum the same year. In 1913, the company made the first major discovery in Venezuela, the Guanoco field. In 1914, it made a more important one on the eastern shore of Lake Maracaibo. It struck oil with its Zumaque No. 1 well. Although this well flowed at a modest 200 bpd, it opened up the **giant** Mene Grande field (**EUR** 0.7 Bb). In 1922, Caribbean Petroleum's Barroso No. 2 well in Zulia came in as a gusher, flowing 100,000 bpd for days before it was brought under control. This confirmed the presence of vast reserves in the country.

From 1926 to 1930, the bulk of the **Bolivar Coastal Complex**, Venezuela's largest aggregation of **conventional crude oil** and one of the five largest in the world, was discovered (part of the complex was actually discovered in 1917, and another part was discovered in 1957). **Creole Petroleum**, acquired in 1928 by Standard Oil Company of New Jersey (later **Exxon**), also made significant discoveries in eastern Venezuela. Shell, Jersey Standard, and **Gulf** established themselves as the major players in the Venezuelan industry. **Edward Doheny**'s Pan American Petroleum and Transport Company also had Venezuelan interests, acquired in 1925 by Standard Oil Company of Indiana (later **Amoco**). Standard Oil of Indiana sold the interests to Jersey Standard in 1932. In 1928, Venezuela became the world's largest oil exporter, a position it held until 1970.

After the death of Juan Vicente Gómez, Venezuela moved to increase its share of revenue from the exploitation of its petroleum. From 1938 to 1942, fiscal revenue was about 11 percent of gross income from petroleum. In 1943, the country instituted a new hydrocarbons law extending the time period of company concessions, in return for higher royalties (16.67 percent) and the construction of refineries in the country. Following the passage of the new law, fiscal revenue rose to about 30 percent of gross income and stayed around there until 1957. The achievement of a **50–50 agreement** for profit sharing in November 1948 had tremendous symbolic value but in practice had little effect on the fiscal revenue share. Fiscal revenues peaked at about 86 percent of gross income in 1974.

Venezuela, under the leadership of Mines and Hydrocarbons Minister **Juan Pablo Pérez Alfonso**, was a motive force for the formation of OPEC in 1960. Pérez Alfonso's vision was to create a cartel that would stabilize prices by regulating production, as the **Railroad Commission of Texas** had done since the 1930s in the **United States**. Although a founding member of the cartel and often an aggressive voice for higher prices, Venezuela has often bristled at OPEC production quotas. The same year it joined OPEC, Venezuela formed its first national oil company, the Corporación Venezolana de Petróleos. In 1975, the government of **Carlos Andrés Pérez** passed a law nationalizing its petroleum industry, to take effect on 1 January 1976. The country also created a new national oil company, **PdVSA**, to be subject to the same fiscal structure as other companies had been. Foreign companies received total compensation of about $1.28 billion, a relatively small sum. The companies had already been progressively squeezed by the government, which had been taking higher proportions of petroleum earnings for itself. The nationalization law reduced taxes somewhat for PdVSA. The government's fiscal revenues fell from 86 percent of gross petroleum income in 1974 to about 66 percent between 1976 and 1992.

Venezuelan oil production peaked at 3.71 Mmbpd in 1970. It declined almost continuously until 1985, when it reached a low of 1.68 Mmbpd. There were many reasons for the decline, including government rapaciousness in the 1970s and lower demand for OPEC oil in the 1980s as production increased from nonmember countries such as **Norway**, the **United Kingdom**, and **Mexico**. After 1985, Venezuelan production began to rise again. The increase was further stimulated by the **Apertura Petrolera**, the reopening of the petroleum sector to foreign investment in the 1990s. In 1997, the country produced 3.28 Mmbpd, almost as high as the 1970 peak. However, the 1998 election of **Hugo Chávez** as president, and the subsequent reversal of the Apertura, caused another decline in production. Because of a series of severe strikes at PdVSA between 2001 and 2003 and Chávez's resulting mass firing of managerial and professional workers, 2003 production was 2.34 Mmbpd, around the same level as 1991. It stayed at a similarly low level for the next 12 years. The Chávez government's nationalization of the Orinoco heavy oil projects in 2007 further destroyed international confidence in Venezuela as a place to do business. ExxonMobil sued and was awarded $1.3 billion in damages by the International Center for Settlement of Investment Disputes in 2014. Chávez's successor Nicolás Maduro, who assumed office in 2013, continued the same sorts of policies.

Venezuela's **natural gas** industry is less developed than its oil industry. Most of Venezuela's gas has been associated with oil fields and not efficiently captured. To stimulate development of the country's natural gas industry, Chávez's government initiated a new "Apertura" for gas. In 1999, Venezuela instituted a gas hydrocarbons law to open all aspects of the Venezuelan gas industry to private investment, even allowing 100 percent private ownership. In 2005, the Ministry of Energy and Mines began awarding licenses for offshore gas exploration. Gas production did grow for a while, to a peak of about 1.28 Tcf in 2007, but then decreased again, although the rate in 2015 was still almost 15 percent higher than in 1999. There has been some recent progress. For example, in 2015 Repsol and ENI began producing gas from the giant offshore Perla gas field discovered in 2009.

Despite all the many problems induced by government policy and by the civil chaos that intensified in the 2010s, Venezuela is still a major oil country. It is true that its very largest fields of **conventional** oil were discovered in the 1920s, and of its 30 giant fields overall, only five have been discovered since 1980. However, the country is still an important oil producer and has the potential to be an even more important one. The heavy oil of the Orinoco Belt still represents a tremendous resource as the world inevitably moves away from "easy oil." *See also* BETANCOURT, RÓMULO (1908–1981); CALDERA RODRÍGUEZ, RAFAEL (1916–2009); GIUSTI LOPEZ, LUIS EDUARDO (1944–); REYNOLDS, GEORGE B. (1853–1925).

VIETNAM. *See* SOUTH CHINA SEA.

VOLGA–URALS. **Russia**'s top oil-producing province between 1950 and the late 1970s, when it was overtaken by the rapid rise of western **Siberia**. From 1955 to 1970, the Volga–Urals accounted for more than 80 percent of Russian oil production. The province is located in western Russia, around the Volga and the western slopes of the Urals. There is a large belt of **giant** oil fields between Samara and Perm.

Petroleum has been known in the area since premodern times. Oil was used for medicinal purposes and for lubricating cartwheels. A 1703 Russian newspaper account describes a surface oil outcrop near the Sok River. Gottlieb Schober, physician to Tsar **Peter I** (Peter the Great), described oil sources in the Volga area in 1717. The 18th century saw a number of unsuccessful attempts to produce the oil. The first geological survey of the region was in 1830 by two Russian geologists, N. Shirokshin and A. Guriev. Roderick Murchison, a British geologist, carried out further work and wrote a book describing the geology of oil and asphalt along the river Sok.

The first **drilling** was in 1867. Fyodor Smolyaninov drilled two dry holes that year. In 1868, Nikolai Malkienko drilled two wells and struck small quantities of oil. He began **refining** kerosene but, unable to turn a profit, stopped in 1869. An American of Hungarian descent, Laszlo Shandor, believed greatly in the potential of the Volga–Urals province and correctly predicted that one day it would surpass **Baku**. However, despite concerted attempts, he was unable to prove his thesis. It is worth noting that one of his unsuccessful wells was not far from what would become the giant Romashkino field, the region's largest.

More intensive modern prospecting did not begin until the 20th century. Geological studies by Sergei Nikitin and Alexander Krasnopolsky in the early part of the century paved the way. Soviet drilling foreman Prokopy Pozdnyakov and his team struck commercial quantities of oil at Verkhnechusovskiye Gorodki in the Perm region on 16 April 1929; 29 wells were drilled within a year. A large number of giant fields were discovered in the 1930s, including Tuyrnazy (1937), Samara (1939), and Perm (1939).

In 1947, the supergiant Romashkino field, the largest in the Volga–Urals region and Russia's second-largest oil field, was discovered. The **EUR** is as high as 15 Bb. Romashkino production peaked in the 1960s and early 1970s at roughly 1.6 Mmbpd. By the late 1990s, production was down to about 0.275 Mmbpd. Other major fields include Mukhanovo (1945), Shkapovo (1944), Arlan (1955), and Tolbazy (1974). The giant Orenburg (Krasnoyarskoye) gas field, with an EUR of about 63 Tcf (and about 700 Mmbo), was discovered in 1966. All in all, 18 giant fields, almost all predominantly oil fields, were discovered in the Volga–Urals region between 1937 and 1974.

VORONTSOV, MIKHAIL (1782–1856). Russia's first viceroy of the Caucasus; authorized the funds for **drilling** one of the earliest modern oil wells, in the **Bibi–Eibat** area near **Baku**, in 1846. Mikhail Semyonovich Vorontsov, born 30 May 1782, was a military officer during the Napoleonic Wars and was instrumental in the Russian conquest of the Caucasus. He died on 18 November 1856. *See also* FIRST OIL WELL; VOSKOBOINIKOV, NIKOLAI (1801–C. 1861).

VOSKOBOINIKOV, NIKOLAI (1801–C. 1861). Russian engineer working in the **Baku** area; among the first in the modern era to propose deliberate **drilling** for petroleum. On 1 January 1825, an order from Yegor Kankrin, **Russia**'s finance minister, placed all oil wells under the control of the state. As a major in the Bakinskii Corps of Mining Engineers, Voskoboinikov was placed in charge of oil fields and salt mines in Baku and Shirvan. He made a number of innovations. In 1835, he developed a canvas ventilator to feed fresh air into an oil well and make maintenance easier and safer. In 1836 and 1837, he organized a storage and supply system for oil in the Baku area. He also designed a refinery that was built in 1837, years before similar facilities began to be built in the West.

Voskoboinikov recognized the inefficiency of large hand-dug pits and was an early proponent of deliberate drilling for oil. He developed his ideas two decades before **Edwin Drake**'s well in **Pennsylvania**. However, he had difficulty generating interest for his proposal, especially since he was distracted by unjust accusations that he stole oil from the operations in Baku. A royal commission sent to investigate the allegations exonerated him in 1838. He was transferred to the Department of Geological Engineers of the Caucasus and in 1841 became a colonel.

In 1843, Voskoboinikov met **Vasili Semyonov**, an enlightened member of the Caucasus executive council. Semyonov took an interest in Voskoboinikov's ideas and helped him to write them up. Semyonov submitted Voskoboinikov's report, dated 22 December 1844, under his own name. He did this not to steal credit but to maximize the chances of the report's acceptance by the upper levels of government. A day after the report's date, a new government structure was formed for the Caucasus region, with **Mikhail Vorontsov** as the first viceroy. Vorontsov saw the report and authorized 1,000 rubles in silver for oil drilling, disbursed on 30 April 1845. The drilling effort fell under the supervision of Voskoboinikov's successor in the Bakinskii Corps of Mining Engineers, Major Alexeev. Success was reported on 14 July 1846. The well was 69 feet deep, and some consider it to have been the first modern oil well. It was drilled and not dug, and done so deliberately to find oil. Unlike Drake's well, however, drilling was done manually using a percussion rod and was not powered by a steam engine. Also, the hole was probably not cased. The well did not touch off an immediate and recognizably modern oil boom. *See also* FIRST OIL WELL.

W

WAGON WHEEL PROJECT. *See* NUCLEAR EXPLOSION STIMULATION.

WALEY COHEN, ROBERT (1877–1952). Executive at **Royal Dutch Shell**. Robert Waley Cohen was born in London on 8 September 1877. He graduated from Cambridge University in natural sciences. In July 1901, he went to work as an assistant to **Marcus Samuel**, the head of the **Shell Transport and Trading Company**. Shell, **Royal Dutch**, and the **Rothschilds** formed the Asiatic Petroleum Company, a joint marketing venture for Asia. **Henri Deterding**, the head of Royal Dutch, became managing director of Asiatic. Waley Cohen became the assistant manager.

Waley Cohen worked first on marketing strategy in **India**. In 1905, he was appointed to the boards of Shell and Asiatic, and he negotiated for Shell in preparation for the 1907 merger with Royal Dutch. He became a director of the new group and an assistant to the managing director, Henri Deterding. Waley Cohen conducted additional important negotiations for Royal Dutch Shell. For example, in 1908 he negotiated with the Egyptian government for exploration rights and for construction of a refinery at Suez. During **World War I**, he negotiated with the British Admiralty for Shell to supply the government with oil. As a scientist, he believed in the value of research. He recruited many young scientists to Shell and opened a research laboratory in **California**. Waley Cohen retired from management at Shell in 1928. He engaged in some other business and many philanthropic and social activities until his death in London on 27 November 1952.

WATERS–PIERCE OIL COMPANY. *See* MEXICO.

WELL LOGGING. The technique of measuring the physical properties of rocks with a variety of sensors located down a borehole. Well logging aids in the identification of geological formations and fluids and in correlation between drill holes. It assists in the evaluation of the productive capacity of reservoir rocks and is standard practice in the petroleum industry. The first well logs were probably done by Professor Edward Forbes of Edinburgh University from 1837 to 1842 and had nothing to do with the petroleum industry. Forbes measured temperature in shafts. Physicist William Kelvin later analyzed the results to determine variations in temperature and heat flow for use in his calculations of the earth's age.

Petroleum industry well logging began with the **Schlumberger** brothers, Conrad and Marcel. Initially they were prospecting for metal-bearing ores using surface electric measurements. In 1914, Conrad Schlumberger located a deposit of copper ore for a customer in Serbia. This was the first commercial application of the technique. In 1919, the brothers formed a company with the support of their father, Paul, and in 1923 mapped an oil-bearing salt dome in **Romania** for a petroleum company. In 1926, the brothers named their company the Société de Prospection Electrique, the precursor of Schlumberger Ltd.

In March 1921, Marcel Schlumberger made downhole electric resistivity measurements in a 2,500-foot borehole to help validate the surface measurements. He and his brother were impressed by the large variations in resistivity down the hole and pursued development of the technique. Conrad hired his son-in-law, Henri Doll, to develop a practical tool for downhole resistivity measurements. On 5 September 1927, Doll and the Schlumberger brothers recorded the first petroleum industry well log in a 1,640-foot borehole at the **Pechelbronn** oil field in Alsace. This hand-drawn log, representing measurements one meter apart down the hole, showed the potential of the technique: from it an interpreter could make judgments about the subsurface rock layering. Clays have a low resistivity. Porous sands are relatively conductive if saturated with saltwater and are relatively resistive if saturated with freshwater. If they have oil in their pores, they are much more resistive. In 1929, the Schlumbergers signed a contract to search for oil in the Soviet Union. They also recorded logs for **Royal Dutch Shell** in **Venezuela**, and later in **California** and the **Gulf of Mexico** coast of the **United States**.

In 1929, Doll discovered the spontaneous potential (SP) produced naturally between borehole mud and the formation water in permeable rocks. Conrad Schlumberger received a patent for SP measurement, and many researchers set about interpreting the phenomenon. SP was introduced into commercial use in 1931 in Venezuela. The combination of spontaneous potential and resistivity measurements proved useful in locating permeable hydrocarbon-bearing layers. In the 1940s, Doll developed induction logging for oil exploration. Induction logging measures subsurface conductivity (the reciprocal of resistivity) using electromagnetic waves, avoiding the need for freshwater mud. It was introduced by the Schlumberger company in 1947. In 1953, Schlumberger introduced the high-resolution Microlaterolog to measure the resistivity of the flushed zone close to the borehole, with minimal influence from the mudcake or the undisturbed zone.

A number of other logging technologies were developed, and not all by Schlumberger. For example, gamma ray logging, which exploits the higher natural radioactivity of shales to distinguish them from other sedimentary rocks, was introduced into commercial service in 1939 by Well Surveys Inc. of **Tulsa, Oklahoma**. Well Surveys followed the gamma ray log in 1941 with the neutron gamma ray log. The secondary gamma rays excited when a formation is bombarded with neutrons serve as an indicator of hydrogen content. A few of the many others are the gamma-gamma log (1950) developed by LaneWells and the Standard Oil Company of Indiana (later **Amoco**), and also by McCullough and California Research, to measure density; the multiple-spaced neutron log (1955) developed by Petrotech to detect gas oil contacts in dry gas areas; and the flowing neutron log (1957) developed by **Creole Petroleum** (Standard Oil Company of New Jersey, later **Exxon**) to localize gas oil contacts or gas entry points in flowing wells.

The **Humble Oil and Refining Company** (later Exxon) announced an acoustic impedance log in 1948. **Magnolia Petroleum** (**Mobil**) followed with a continuous velocity log in 1951, introduced into commercial service by Seismograph Service Corporation in 1954. The use of such logs became more widespread after the Wyllie time average equation, published in 1958, related seismic velocity to porosity.

The need for immediate information during exploration led to the development of logging while drilling (LWD). LWD began in the 1960s with downhole measurements of well deviation during directional **drilling**. In 1988, Schlumberger introduced real-time LWD with multiple tools (including resistivity, gamma, and neutron). Measurements were transmitted via pulses traveling in the borehole mud. Real-time LWD introduced the potential for real-time adjustment of well trajectory in angled drilling based on the formation information. Other more recent advances include borehole imaging and nuclear magnetic resonance logging.

WESTERN ATLAS. *See* WESTERN GEOPHYSICAL.

WESTERN GEOPHYSICAL. A **geophysical** services company; later **WesternGeco**, a unit of **Schlumberger**. Western Geophysical was founded in Los Angeles, **California**, by Henry Salvatori in 1933. It distinguished itself as a technology leader in seismic prospecting, using only its own instruments. In the 1950s, it developed the first practical system for the large-scale processing of analog data. In 1960, Western Geophysical was acquired by Litton Industries, a defense contractor seeking to diversify. Litton invested in Western, now the Western Geophysical Division, allowing it to remain relatively autonomous, and it continued to innovate and prosper. In the 1960s, it became one of the first companies to switch to digital recording and processing. It also innovated in the area of seismic sources, moving away from the use of large explosives with its Aquapulse and Maxipulse systems. In the 1970s, it developed the first reliable high-pressure air gun seismic source.

Perhaps serendipitously, petroleum and defense were somewhat countercyclical in the 1970s and 1980s. Petroleum-related industries prospered during the time of high oil prices in the 1970s (*see* FIRST OIL SHOCK; SECOND OIL SHOCK), just as American defense contractors suffered from the end of the Vietnam War. In contrast, during the 1980s, defense contractors benefited from President Ronald Reagan's defense buildup as the petroleum industry suffered from the oil glut and low prices (*see* OIL COUNTERSHOCK). Western Geophysical did better than most service companies in the 1980s because of its geographic diversification and its continued technological innovation. Western developed a low-pressure air gun that could make deeper measurements. It also benefited from Litton's technological know-how. Litton helped the company improve its seismic data processing centers and make them among the best in the industry. Western was also a leader in developing three-dimensional seismic acquisition and processing.

In 1987, Litton joined with **Dresser Industries** to form Western Atlas International, placing its Western Geophysical Division in the new entity, where it formed the largest unit. Although the venture was doing well, Litton decided to focus on its core defense businesses. In 1992, it bought back Dresser's share of Western Atlas in order to spin the whole unit out. Litton packaged Western Atlas with its other noncore business in factory automation and created an independent company, Western Atlas Inc., on 31 December 1993. In 1994, Western Atlas acquired **Halliburton** Geophysical Services, which included the former **Geophysical Service Inc.** (the "old" GSI, not the new company operating under that name in **Calgary**).

In 1997, Western Atlas spun off all its business unrelated to oil field services, and a year later it was acquired by **Baker Hughes** for $3.3 billion. Baker Hughes created two divisions from Western Atlas: Western Geophysical for seismic services and Baker Atlas for downhole services. Baker Hughes, however, soon felt the effects of a downturn in demand for oil field services. It took a number of measures to improve its financial performance and reduce its debt. One of these measures was to combine Western Geophysical with Schlumberger's Geco-Prakla in 2000. Schlumberger paid Baker Hughes about $500 million to get a 70 percent stake in the venture, called WesternGeco, and in 2006 acquired the whole thing.

WESTERNGECO. World's largest **geophysical** services company; a unit of **Schlumberger**. WesternGeco specializes in geophysical surveying for petroleum exploration and also for reservoir imaging and monitoring. The company provides three-dimensional, four-dimensional, and multicomponent seismic surveys as well as electromagnetic surveys. WesternGeco was formed in 2000 from the joining of Schlumberger's Geo-Prakla company and the **Western Geophysical** unit of **Baker Hughes**. Schlumberger acquired Baker Hughes's portion in 2006.

WEST TEXAS INTERMEDIATE. A **benchmark crude** used since 1991 as a pricing reference for U.S. domestic oil and also for imports into the **United States**. West Texas Intermediate (WTI) is light (**API gravity** 39.6°) and sweet (0.24 percent sulfur). Along with **Brent Blend** and **Dubai crude**, it is one of the three most important benchmark crudes. West Texas Intermediate is lighter and sweeter than Brent and considerably lighter and sweeter than Dubai. The high quality of WTI, in addition to its location in the midst of the U.S. market, makes it a high-price oil, generally trading above Brent, Dubai, and the **OPEC Basket**. Most WTI is refined in the midwestern United States or along the **Gulf of Mexico**. West Texas Intermediate delivered at **Cushing, Oklahoma**, is a major underlying commodity in the **New York Mercantile Exchange** (NYMEX) futures price for crude oil, although the contract allows other grades as well. The contract trades in lots of 1,000 barrels.

In April 2007, a surplus of WTI at Cushing attributable to a Valero refinery shutdown temporarily depressed its price to artificially low levels. This distortion caused some concern that WTI might no longer be a viable gauge of world prices in the future. There were other pricing distortions between 2011 and 2013, when WTI traded at an unusually cheap price compared to Brent Blend, again because of a glut of American oil at Cushing. Such distortions and falling production may mean that WTI will be replaced as a benchmark at some time in the future.

WEST VIRGINIA. Oil production 2015: 0.023 Mmbpd (0.24 percent of U.S.); **reserves** 0.012 Bb (0.04 percent of U.S.); gas production: 1.24 Tcf (4.59 percent of U.S.); reserves 20.553 Tcf (6.68 percent of U.S.). The state of West Virginia was one of the birthplaces of the modern petroleum industry in the **United States**. The West Virginia industry began somewhat earlier than that of **Pennsylvania**, but its mid-19th-century boom was not as spectacular. Oil was noticed early in West Virginia by Native Americans and later by European settlers. George Washington took title to oil lands in what was then western Virginia but is today Kanawah County, West Virginia, and noted the petroleum in the area, deeming it a valuable substance.

As in Pennsylvania, oil was discovered often by accident in salt or water walls, and the salt drilling industry created the necessary foundations for later petroleum drilling. The first associated oil discovery recorded in West Virginia was in a water well in 1790. In 1806, the Ruffner brothers, Joseph and David, discovered oil and **natural gas** in 1806 in Kanawah County while drilling for salt. Another such discovery was made in 1814. Oil was seen as a valuable by-product of salt drilling, and there are records of an oil trade in West Virginia going back to the early part of the 19th century. In 1835, George Lemon and a man named Hugle used water power to drill for salt. On discovering oil, they devised a system for separating it from water. Samuel B. Hildreth of **Ohio** published accounts of the oil region of West Virginia in 1826 and 1837. He noted the use of oil as a shaft lubricant in manufacturing operations. In 1841, William Tompkins began using natural gas to help power his saltworks near Charleston. In 1842, William P. Rathbone and his sons purchased large tracts of land near Burning Springs and began drilling for salt. They obtained oil, which they sold commercially.

The West Virginia salt drilling industry spurred the development of new **drilling technology**, including hole casing. In 1831, William Morris, a salt driller, invented the drilling jar that allowed the easier dislodging of stuck tools. These innovations would enable the development of the petroleum industry in West Virginia, Pennsylvania, and southeastern Ohio.

In 1857, Robert Hazlett, a retired physician with an interest in geology, traveled throughout what is today West Virginia in search of promising coal and oil properties. He and some business partners acquired 125 acres just north of Petroleum Station for oil exploration and development in January 1859. Hazlett's Virginia Petroleum Company drilled successful wells, probably using steam engines, and was producing oil in 1859 before **Edwin Drake** in Pennsylvania. In 1859, Charles Shattuck drilled a successful oil well east of Parkersburg, about two months before Drake. The West Virginia wells did not result in the same publicity as Drake's well, nor did they usher in an immediate boom of the same scale in their vicinity.

Although Pennsylvania was America's top producing state for much of the 19th century, West Virginia surpassed it in 1899 (without assuming the top position itself) and stayed mostly ahead until 1921. Beginning in the 1920s, rejuvenation of the **Bradford** field put Pennsylvania ahead again. Both states were far less important than **California**, **Oklahoma**, and **Texas** by this time. West Virginia never had a **giant** field like Bradford and did not become a major oil producer by national standards. However, it played an important and often overlooked role in the early development of the petroleum industry.

When West Virginia became a state in 1863 (before that year it was part of Virginia), its new seal depicted oil barrels on the front and derricks on the back, among other symbols. Oil executives played a leading role in lobbying for statehood. Denying West Virginia's oil wealth to the Confederacy during the Civil War was one motivation for making it a state. A century and a half later in the 2010s, West Virginia underwent a renaissance and became a significant producer of **shale gas**.

WESTINGHOUSE, GEORGE (1846–1914). American inventor and entrepreneur best known for his work in electric power but who also developed a system for distributing **natural gas** to homes. George Westinghouse was born in Central Bridge, New York, on 6 October 1846. He showed unusual creativity early in life. He received his first patent at the age of 19 for a rotary engine design. At 22, he developed the air brake for trains. Using patent rights acquired from Nikola Tesla, he played a major role in the development of alternating current. His numerous inventions unrelated to the petroleum industry include the first practical induction motor, the first power station turbine generator, a citywide telephone switching system, and an alternating current locomotive. Westinghouse formed about 70 companies in his life, and he was a progressive employer by the standards of his time. In 1871, he began giving his workers a half day off on Saturday. In 1908, he began a pension fund for workers, and in 1913 he began offering paid vacations.

Westinghouse's seminal work in the natural gas industry began when he drilled for natural gas in the backyard of his home in Pittsburgh. He struck gas in February 1884 and decided to sell to local customers. He purchased the Philadelphia Company to produce the gas. The company expanded, leasing additional gas-bearing lands in western **Pennsylvania**. By 1887, it was supplying about 5,000 residential and 470 industrial customers. Westinghouse made a number of innovations in the technology of natural gas distribution, substantially increasing the safety and practicality of the industry. He developed a system for enclosing a main gas line with a conducting pipe to contain leaks. He also invented a way to reduce the pressure of gas in main trunk lines before it reached residential areas. He invented a pressure regulator and cutoff valve that automatically restricted the flow when the pressure became too low. George Westinghouse died 12 March 1914 in New York City. A Civil War veteran, he was buried in Arlington National Cemetery in Virginia.

WILLIAMS, JAMES MILLER (1818–1890). Early oil entrepreneur in **Canada**. James Miller Williams was born 14 September 1818 in Camden, New Jersey, where he apprenticed as a carriage maker. In 1840, he left the **United States** for Canada, settling initially in London, Ontario. He manufactured carriages there, then moved to Hamilton, Ontario, where he expanded his operation. He eventually contracted to manufacture cars for the Great Western Railway. He entered into a partnership with H. G. Cooper, to whom he sold his interest in 1859.

During the 1850s, Williams also began investing and working in the petroleum industry. Sometime between 1854 and 1857, he acquired the International Mining and Manufacturing Company, a financially unsuccessful company that had been established by Henry and Charles Nelson Tripp in 1854 to produce asphalt from the Eniskillen gum beds. Williams changed the name to the J. M. Williams Company and by 1857 or 1858, one to two years before **Edwin Drake**'s **Pennsylvania** well, was **refining** petroleum at Oil Springs, Ontario, to produce kerosene. He had dug under the gum beds to find their source and struck oil. It is not known whether his early wells were mechanically drilled or dug by hand. By 1860, Williams had set up a refinery in Hamilton under the auspices of the Canadian Oil Company, a company he had formed with some associates. Williams was elected to the Ontario legislature in 1867, 1871, and 1875. He retired from politics in 1879 and was appointed registrar of Wentworth County, Ontario. He died 25 November 1890 in Hamilton. *See also* FIRST OIL WELL.

WOLFCAMP FORMATION. Shale formation in the U.S. states of **Texas** and New Mexico that produced about 40,000 barrels of **tight oil** per day in 2000, rising to 370,000 barrels per day in 2016. In the latter year the **United States Geological Survey** announced its estimates that the Wolfcamp contains 20 Bb of technically recoverable oil and 16 Tcf of natural gas.

WONGA COUP. A failed attempt in March 2004 to overthrow President Teodoro Obiang Nguema Mbasogo of **Equatorial Guinea**. The coup was allegedly organized by a group of private financiers based in the **United Kingdom** and South Africa. The group reportedly included Mark Thatcher, son of former British prime minister Margaret Thatcher. The financiers expected to install a new president, who would then reward them with a share of the country's oil revenue. It appears that the British, Spanish, and American governments probably had foreknowledge of the coup attempt but did not attempt to stop it. One reason may be that Obiang is widely regarded as a highly unsavory tyrant. The coup was foiled by Zimbabwean authorities on 7 March 2004, when a cargo plane carrying mercenaries bound for Equatorial Guinea stopped in Harare, Zimbabwe, to pick up black market military hardware.

WORLD PETROLEUM COUNCIL. Organization founded in London in 1933 to foster discussion of issues associated with the oil industry. The World Petroleum Council (WPC) has 61 member countries, each of which has a national committee composed of representatives from academia, government, and industry. The WPC hosts the triennial World Petroleum Congress.

WORLD WAR I (1914–1918). Besides being the most horrific conflagration in the world up to that time, and the second deadliest conflict in history (after **World War II**, measured by combatant casualties), World War I was also the first motorized war and the first in which oil both played a crucial operational role and constituted a significant strategic objective. George Curzon, the British foreign secretary from 1919 to 1924, famously declared that the Allied cause had "floated to victory on a wave of oil." French president Georges Clemenceau said in 1917 that **gasoline** was as vital as blood.

Oil played a role on land, on sea, and in the air. It greatly expedited troop movements and freed them from the constraints of railroad lines. The production and introduction of vehicles occurred at a staggering pace by the standards of the time. The British army began the war with less than a thousand vehicles. By the end of the war, it had more than 56,000 trucks, 23,000 cars, and 34,000 motorcycles. The Allies had a decided advantage in vehicles powered by the internal combustion engine. The **United States** alone brought 50,000 vehicles to the theater in the roughly 18 months it fought in the war. World War I also saw the introduction of the tank; 456 tanks participated in the Battle of Amiens on 8 August 1918.

The British had begun the process of converting the Royal Navy from coal to oil beginning in 1905. The conversion was championed by oil entrepreneur **Marcus Samuel**, Admiral **John Arbuthnot Fisher**, and **Winston Churchill**. The German fleet remained largely coal based, with oil-assisted burning in some cases. This meant that the British had an advantage in the range and speed of ships, as well as in the ease and flexibility of refueling. Germany did deploy diesel-powered submarines, to devastating effect.

The air war was entirely dependent on oil. Airplanes were unknown in warfare before World War I, but during the course of the conflict, more than 200,000 were produced, two-thirds of them by the Allies.

The United States supplied about 80 percent of wartime petroleum to the Allies. When the United States entered the war, it formed the **Fuel Administration** to allocate oil to meet both wartime and ordinary domestic requirements. The Allies suffered critical shortages in 1916 and 1917, partly as a result of German submarine attacks on shipments. In February 1918, the Interallied Petroleum Conference was established to coordinate supplies and shipping. Standard Oil Company of New Jersey (later **Exxon**) and **Royal Dutch Shell** played leading roles. The El Dorado oil field discovered in Kansas by **Cities Service Company** played an important part during the war. By 1918, it was producing more oil than the **Romanian** and **Galician** fields seized by the Germans. By 1919, the field produced about 100 Mmb.

The need to secure oil supplies influenced strategic objectives during the war. The British were concerned with defending **Abadan** and its refinery, which were threatened by the Turks. The **United Kingdom** depended on the oil produced in **Iran** by the Anglo–Persian Oil Company (APOC; later **Anglo–Iranian Oil Company** and **BP**). The British government had purchased a majority interest in APOC immediately before the war. The British repelled the Ottomans from Abadan, and they captured **Basra**. With an eye on the petroleum resources of Mesopotamia (**Iraq**), they also sought to advance to Baghdad, which they achieved in 1917.

Germany depended on oil from Romania. Romania was neutral for the first two years of the war but then entered against Germany. The Germans invaded Romania in order to gain access to the oil fields but were initially repelled. They broke through on 17 November 1916, but as they advanced, they found an oil industry in ruins, sabotaged by the British. The important **Ploieşti** fields were the last to be disabled, just hours before the Germans arrived on 5 December. It took the Germans five months to recover production. Romania and the Ploieşti fields would also figure in World War II in an eerily similar pattern. In another pattern that would be echoed, the Germans set their sights on **Baku**. In World War I, they did not attempt to invade but rather to make a deal with the Bolsheviks for oil. The Bolsheviks were receptive, but the scheme was stopped by the Turkish invasion of Baku, even though the Turks were German allies. After World War I, petroleum would be central to all conventional warfare.

WORLD WAR II (1939–1945). The deadliest conflict in history, and one in which petroleum played a major role. Petroleum contributed to the causes of World War II and was a significant strategic objective that helped determine the direction of major campaigns. Fuel and its associated logistics also constituted a major tactical factor. If **World War I** was the first motorized war, World War II was the one in which motorization and mobility truly took center stage. Superiority in petroleum production and supply contributed greatly to the Allied victory. Germany's blitzkriegs depended on a ready supply of fuel. Lack of fuel contributed to Field Marshall Erwin Rommel's defeat in North Africa. It also contributed to German troops getting bogged down in Stalingrad and ultimately failing in their invasion of the Soviet Union.

World War II broke out in Europe in 1939. In 1940, Germany scored quick victories in France and the Netherlands. These victories, along with Britain's preoccupation with fighting off a potential German invasion, made Japan a dominant power in East Asia. Japan, which had invaded **China** in 1931–1932 and was still embroiled in conflict there, looked southward to the Asian colonies of the defeated European powers. Japan had negligible oil resources of its own, and the capture of the oil-rich Dutch East Indies (**Indonesia**) would be a significant prize. The colonies also contained other essential resources that could make Japan relatively self-sufficient.

Japanese expansionism had been causing increasing tension with the **United States**, which had not yet entered the war. President Franklin D. Roosevelt's administration gradually imposed embargoes on critical materials. When Japan moved into southern Indochina in July 1941, the United States imposed a complete embargo on oil, even though the two countries had been negotiating to settle their differences since April. This was a serious development for Japan, which had been depending on the United States for about 80 percent of its supply. The embargo helped drive the more moderate elements in the Japanese military into the arms of the extremists. When its attempts to negotiate an agreement for oil supplies in Indonesia were unsuccessful, Japan once again tried to negotiate with the United States but also prepared for war. The United States made several demands that Japan could not accept, such as a withdrawal from China and Indochina and a renunciation of its pact with the Axis powers. Japan then definitively decided to take Indonesia by force. It attacked the U.S. base at Pearl Harbor, Hawaii, on 7 December 1941 to protect its eastern flank. This attack brought the United States into the war and ultimately determined the war's outcome.

The need for oil not only drove Japan to precipitate a war with the United States but also determined many of Germany's strategic decisions and influenced its operations. The need for oil drove German forces into **Romania**. A corps of "oil commandos" accompanied the first waves of invading German troops in many European countries. The oil commandos were technically knowledgeable in petroleum industry operations and could ensure quick production from captured assets. Oil also helped drive the German invasion of the Soviet Union. The oil fields of **Baku** and the Caucasus were major objectives. The need to secure petroleum supplies kept the British embroiled in the Middle East. Allied forces occupied **Iran** in 1941 and forced the shah, **Reza Pahlavi**, to abdicate in favor of his son **Mohammed Reza Pahlavi** (who would rule until the Iranian Revolution of 1979). In 1933, Reza Pahlavi had forced the renegotiation of the **concession** of the Anglo–Persian Oil Company (later **Anglo–Iranian Oil Company**, eventually **BP**), and his relations with Britain were strained. He was also suspected of tilting against the Allies during the war, although Iran was officially neutral.

Oil infrastructure was a target on all sides. German submarine attacks on oil **tankers** supplying U.S. oil to Britain almost created a critical situation for the Allies in 1943. The Allies also extensively attacked Axis infrastructure. The U.S. Office of Strategic Services (OSS) Research and Analysis Branch determined that oil production was a key German weakness. The United States bombed synfuel plants and oil production facilities in May 1944. Further attacks on oil infrastructure led to the grounding of the Luftwaffe and immobilization of a considerable portion of the German military's tanks and trucks. The German invasion of Crete in 1941 was undertaken partly to deny a base from which the British could attack the oil infrastructure of **Ploieşti** in Romania. The Allies did conduct such raids, the major one being on 1 August 1943.

The use and production of 100-octane aviation **gasoline** was also a major advantage for the Allies over the Axis. High-octane fuel gave Allied airplanes better power-to-weight ratios, lower takeoff distances, and faster climbing speeds. *See also* BIG INCH AND LITTLE BIG INCH; CANOL PROJECT; PETROLEUM ADMINISTRATION FOR WAR (PAW) (1942–1946); PETROLEUM BOARD; PETROLEUM RESERVES CORPORATION (1943–1944); PLUTO (PIPELINE UNDER THE OCEAN).

WYOMING. Oil production 2015: 0.237 Mmbpd (2.52 percent of U.S.); **reserves** 0.725 Bb (2.24 percent of U.S.); gas production: 1.745 Tcf (6.45 percent of U.S.); reserves 20.436 Tcf (6.64 percent of U.S.). Even before the modern oil industry, oil was noticed in Wyoming by Native Americans and explorers. When the United States established Fort Bridger in 1842, an oil spring was already well known near Hilliard some 30 miles away. There was some minor commercial development of naturally occurring oil springs near the fort, beginning in 1866 when a fort employee named John C. Fiere, who had experience in the **Pennsylvania** oil fields, began producing and selling oil to the Union Pacific Railroad as a lubricant, ultimately producing about 150 barrels by skimming oil from the water surface of the spring. Judge C. M. White and William A. Carter also harvested oil from nearby hand-dug oil pits, beginning in 1867.

In his book *The Adventures of Captain Bonneville*, published in 1837 and recounting the 1830s expeditions of army officer and explorer Benjamin Bonneville, Washington Irving referred to oil springs near present-day Dallas Dome in Wyoming. Dallas Dome would become the site of the first oil well drilled in Wyoming in 1885. Mike Murphy struck oil at a depth of 300 feet in the Chugwater formation. The 1890s saw several discoveries in Natrona County, making the town of Casper into a regional oil capital, including a refinery constructed in 1895. Wyoming production rose rapidly in the early part of the 20th century. The Salt Creek oil field was discovered with a gusher drilled by H. L. "Dad" Stock on 23 October 1908. That field, which had antecedents in some of the Natrona County discoveries of the 1890s, would go on to produce for decades, becoming the most prolific field of the Rocky Mountain region. Another significant find was the Lance Creek oil field on 13 March 1918 by the **Ohio Oil Company**. Wyoming during this time was also the site of a **Naval Petroleum Reserve** near Salt Creek that was at the center of the **Teapot Dome scandal** of 1922–1930. Wyoming oil production peaked in 1970 at 0.439 Mmbpd, while natural gas production grew continuously from 1987 to 2009. In 2012, Wyoming also began producing **shale gas**, reaching 100 Bcf for the year 2013, although declining afterward.

Y

YACIMIENTOS PETROLÍFEROS FISCALES (YPF). *See* ARGENTINA; REPSOL.

YAMANI, AHMAD ZAKI (1930–). Saudi Arabian minister for petroleum and mineral resources (1962–1986) who solidified his country's control of its petroleum sector. Ahmad Zaki Yamani was a major force in establishing the principle that producing countries had the right to set production and pricing policies. He championed the concept of **participation agreements**, a framework by which certain members of the **Organization of the Petroleum Exporting Countries** (OPEC) acquired gradual control of the foreign companies operating in their territory. **Saudi Arabia** acquired an immediate 25 percent of **Aramco** in 1973 under its participation agreement and increased its stake to 60 percent in 1974. In 1980, it acquired 100 percent, exceeding the scope of the participation agreement, which originally provided for 51 percent control by 1983.

Yamani led the Arab oil embargo against the **United States** and the Netherlands after the Yom Kippur War of 1973. He helped to establish the **Organization of Arab Petroleum Exporting Countries** (OAPEC) and served as its secretary general in 1968 and 1969. Despite his role in the embargo, Yamani was often seen in the West as a reasonable and moderate influence. He tended to oppose drastic price increases. Unlike many OPEC leaders, he understood that such price increases could have permanent effects on demand. Although his positions were based on common sense and an understanding of economics rather than a desire to pander to the West, he was sometimes accused in parts of the Arab world of being pro-American. He countered that he was trying to avoid destabilizing the world economy, the health of which affected the demand for oil.

In 1985, Saudi Arabia was frustrated because other OPEC members were cheating on their quotas. Saudi Arabia increased production and offered attractive **netback pricing** in order to gain market share. OPEC as a whole introduced netback pricing to gain back market share from the **North Sea** and other non-OPEC producers. When prices collapsed (*see* OIL

COUNTERSHOCK), Saudi Arabia and OPEC tried in vain to take back control of the market and create a situation where production and prices could rise simultaneously. Yamani understood the fundamental inconsistency in such a position, and this brought him into conflict with King Fahd, who wanted both an increase in production and a higher fixed price. On 29 October 1986, Fahd dismissed Yamani as oil minister. Other factors may have been Yamani's connection to the previous king, Faisal, and his high international profile, which irritated Fahd.

On 21 December 1975, Yamani and other OPEC delegates were kidnapped by terrorists led by Carlos "the Jackal" (Ilich Ramirez Sanchez) in an attack on OPEC headquarters in Vienna. The attack was carried out ostensibly in the name of the Palestinian people and the cause of Arab reunification. It is believed that **Libyan** president Muammar Qaddafi may have been behind it. Carlos flew Yamani and the other hostages to Algiers. It is possible he was warned by President Houari Boumedienne of **Algeria** that the deaths of the hostages would result in an attack on the plane. In any case, Yamani and the others were eventually released unharmed.

Yamani helped found the Saudi company **Petromin** and the College of Petroleum and Minerals in **Dhahran**. In 1990, Yamani founded the Center for Global Energy Studies, a petroleum market analysis company.

YANBU. Town in **Saudi Arabia**; Red Sea terminus of the **Petroline**, a **pipeline** from **Abqaiq** in the Eastern Province. Yanbu is the site of the King Fahd Industrial Port opened in 1982. The port initially had three offshore berths and could handle **tankers** up to 550,000 deadweight tons (DWT). It was expanded to four offshore berths and 7.2 Mmbpd in the 1990s. The port also has storage facilities for 6 Mmbo. Yanbu's main value is in offering an alternative export avenue should shipping in the **Persian Gulf** be disrupted. Otherwise it is not economically attractive when compared to **Ras Tanura**. Shipments from Yanbu to East Asia take five days longer round trip. In addition, the Petroline pipeline to Yanbu is operating at only half of its capacity.

YEMEN. Oil production 2015: 0.04 Mmbpd (0.05 percent of world); **reserves** 3 Bb (0.18 percent of world); gas production: 0.11 Tcf (0.09 percent of world); reserves 16.9 Tcf (0.25 percent of world). Yemen was formed in 1990 via the unification of North Yemen and South Yemen. In North Yemen, the national oil company explored without success in the 1970s and 1980s. But in 1984, Yemen Hunt, the Yemeni subsidiary of the Hunt Oil Company—*see* HUNT, H. L. (1889–1974), discovered the **giant** Alif field, with an **EUR** of 0.9 Bb. North Yemen production reached 0.2 Mmbpd on the eve of unification in 1990. In South Yemen, oil was discovered in 1987, but production was limited, reaching 0.01 Mmbpd by 1990. That year, Canadian Occidental, at the time a subsidiary of **Occidental Petroleum**, discovered the Masila complex, with an EUR of 0.9 Bb. Total Yemini production declined after unification, to about 0.182 Mmbpd in 1993. It then rose rapidly to a peak of 0.345 Mmbpd in 1995, and it reached another peak of 0.441 Mmbpd in 2001. It declined thereafter as political conditions in the country deteriorated and in 2015 was less than a tenth of its 2001 peak as a full civil war broke out between Shia Houthi rebels and other parties.

YUKOS. Russian petroleum company formerly controlled by **Mikhail Khodorkovsky**, declared bankrupt 1 August 2006. Most of its assets were acquired by **Rosneft**. Yukos was established on 15 April 1993 by the post-Soviet government of Boris Yeltsin, president of **Russia**. It was formed by the combination of the Soviet-era state petroleum enterprise Yuganskneftegaz, which had operations in western **Siberia**, with three refineries in the central **Volga–Urals** region and several smaller operations. Oligarch Mikhail Khodorkovsky's Menatep Bank acquired a controlling interest in Yukos in 1995 via the **Loans-for-Shares** scheme for the very low price of $350 million. Khodorkovsky increased his stake to 85 percent in 1996. He pushed Yukos to begin applying Western accounting principles and management methods. In 1997, Yukos acquired the Eastern Oil Company with foreign financing. The company had operations in central Siberia.

In 1998, Yukos announced plans to merge with Sibneft (later **Gazprom Neft**). The merger would have created an enormous oil company, with reserves exceeding those of **Royal Dutch Shell**. However, it was stopped by falling oil prices and the 1998 Russian financial crisis. Menatep Bank collapsed and defaulted on loans, and Western banks found themselves with a 32 percent collective stake in Yukos because of shares that had been pledged as collateral.

After 2000, Khodorkovsky came into increasing conflict with the new government of Vladimir Putin. The government was dissatisfied with his outspoken political activism, and also with the questionable circumstances under which he had acquired Yukos. Khodorkovsky was arrested in 2003, accused of tax evasion and other charges, and in 2005 found guilty and sentenced to nine years in prison. Yukos was given a bill for more than $15 billion in back taxes and fines. In 2004, Yukos filed for bankruptcy in the **United States**, but the case was dismissed for lack of jurisdiction.

The Russian government auctioned Yukos's main subsidiary, Yuganskneftegaz, in 2004 for $9.4 billion as a way to collect part of the alleged back taxes. The winning bidder, Baikal Finance, was acquired by Rosneft. Rosneft funded the acquisition partly by using a $6 billion credit against future oil shipments to **China**. Rosneft acquired most of Yukos's remaining assets in 2007.

YUMURTALIK. *See* CEYHAN.

Z

ZAPATA PETROLEUM. Independent petroleum company founded in 1953; it merged with **South Penn Oil Company** in 1963 to form **Pennzoil**. Zapata Petroleum was founded by the brothers J. Hugh Liedtke and William Liedtke, along with John Overbey and **George H. W. Bush**, who would become president of the **United States** in 1988. With $1 million in startup capital, the company acquired leases in western **Texas** in the area of the Jameson field. It drilled 127 wells, all of which showed oil. In the 1950s, a number of discoveries were being made in the **Gulf of Mexico**, and Zapata wanted to expand **offshore**. It reorganized into two companies, Zapata Drilling Company and Zapata Offshore Company. In 1959, the companies were spun off as independent entities. George H. W. Bush led the Zapata Offshore Company until his election to the U.S. Congress in 1966.

The Liedtke brothers turned their eyes toward the South Penn Oil Company, which they believed was undervalued. They began purchasing large share positions, with the blessing of **J. Paul Getty**, whose Tidewater Oil Company was South Penn's largest shareholder. Eventually the Liedtke brothers bought Getty out and gained effective control of South Penn. J. Hugh Liedtke became its president in 1962, and a year later South Penn merged with the Zapata companies to form **Pennzoil**.

ZIJLKER, AEILKO (1830–1890). Founder of the **Royal Dutch Company** (later part of **Royal Dutch Shell**). Aeilko Jans Zijlker was born in May 1830 in Groningen, the Netherlands, to a farming family. He moved to the Dutch East Indies (**Indonesia**) to try his luck in tobacco farming, initially in eastern Java. He later moved to the Lepan River area of Sumatra, where he managed plantations for the East Sumatra Tobacco Company. In the fall of 1880, while inspecting tobacco fields, he was caught in a storm and sought shelter in a tobacco shed. The local overseer who was with him lit a torch, and Zijlker was impressed by its brightness. It emerged that the natives of the area impregnated wood for torches with petroleum gathered from small ponds. Zijlker had the overseer take him to such a pond, and he was struck by the smell of kerosene. He collected some of the petroleum and sent it back to Batavia (Jakarta) for analysis. The sample turned out to have a 59.2 percent kerosene yield.

In 1881, Zijlker obtained a land concession from the local ruler, the Pangeran (Sultan) of Langkat. In order to obtain a full legal petroleum **concession**, he needed to show solvency for his venture, so he went to the Netherlands to obtain financial and political support. He returned and converted an informal organization he had established in Batavia into the Voorlapige Sumatra Petroleum Maatschappij. In 1883, he obtained a petroleum mining concession from the sultan, with the approval of the Dutch commissioner of northeast Sumatra. The concession became known as "Telaga Said."

Zijlker proceeded to exploit the concession with great difficulty, beginning his first well on 11 July 1884. In September 1884, at 320 feet, he struck a small amount of oil, flowing at slightly more than a barrel a day. Deeper **drilling** produced no further oil. In December 1884, he began drilling about a mile to the east, and in June 1885 he struck a gusher in the well that became known as Telaga-Tunggal 1. Despite the strike, progress was slow because of a lack of capital and technical skills. In 1890, Zijlker had the luck to make contact with a powerful banker, N. P. van den Berg. Van den Berg used his influence to get royal sponsorship for the enterprise from King Willem III. This would greatly increase the venture's ability to attract capital. The Koninklijke Nederlandsche Maatschappij tot Exploitatie van Petroleumbronnen in Nederlandsche-Indie (Royal Dutch Company for the Exploitation of Oil Wells in the Dutch East Indies), generally referred to as the Royal Dutch Company, was established in 1890 and took over the Zijlker concession. Aeilko Zijlker did not live to witness the development of Royal Dutch. He died in Singapore on 27 December 1890. *See also* DETERDING, HENRI (1866–1939); KESSLER, JEAN-BAPTIST (1853–1900).

ZOROASTER. Early modern **tanker** ship designed by **Ludvig Nobel** to transport kerosene across the **Caspian Sea**, considered by many to be the first modern tanker. The *Zoroaster* was built by the Swedish Motala shipyard in 1877 and put into service in 1878. It was 184 feet long with a 27-foot beam and a 9-foot draft. It could carry 242 tons of kerosene in 21 vertical cylindrical tanks joined by pipes. The *Zoroaster*'s engine was amidship, and it was one of the first ships in the world to use Bessemer steel. While many observers consider the *Zoroaster* to have been the first modern tanker, others still accord that honor to the 2,700-ton ***Gluckauf*** built in 1886, probably because the *Gluckauf* was larger and crossed the Atlantic. The *Zoroaster*, and Nobel's subsequent ships, had most of the features that made the later *Gluckauf* distinctive. The *Zoroaster* was also ahead of the *Gluckauf* and perhaps every other ship of its time in one respect: it burned a type of fuel oil rather than coal to power its steam engine.

ZORRITOS OIL FIELD. Latin America's first oil field, discovered in **Peru** in 1865. It produced oil until 1965.

Appendix 1

Petroleum Industry Units and Conversions

The petroleum industry has its own idiosyncratic system of units, sometimes called "field units." It is important to understand these units in order to be able to follow any discussion of the industry and its history. Some of the units are defined below, along with conversion factors.

QUANTITY OF OIL: BARRELS

The basic unit for measuring quantity of oil is volumetric. A barrel of oil, abbreviated *bbl* (the origin of the abbreviation is in doubt; see the text entry for barrel) is 42 U.S. gallons. In this book and elsewhere, the barrel unit can also be simply abbreviated *b* rather than *bbl*. One barrel equals:

- 42 U.S. gallons
- 34.9726 UK gallons
- 9,702 cubic inches
- 5.6146 cubic feet
- 158.987 liters
- 0.158987 cubic meters

Larger quantities of oil are indicated in various ways, with prefixes that are highly confusing. In some oil industry literature, a thousand barrels is indicated by *mbbl* or *mb*. This usage of the letter *M* to indicate 1,000 is different from generally accepted usage in other fields, where *M* indicates a million. In the oil industry, a million barrels is indicated *Mmbbl* or *Mmb*. To make matters even more confusing, *Mb* is sometimes used, especially in literature not specific to the oil industry, to indicate a "megabarrel," or one million barrels of oil. A billion barrels of oil is denoted *Bbbl* or *Bb* in the industry. In non-industry literature, a billion barrels is sometimes denoted *Gb*, for "giga-barrel." In summary:

1 mbbl or 1 mb = 1,000 barrels
1 Mmbbl or 1 Mmb = 1,000,000 barrels
1 Mb = 1,000,000 barrels (sometimes)
1 Bbbl or 1 Bb = 1,000,000,000 barrels

1 Gb = 1,000,000,000 barrels (sometimes)

In this book, we adopt the following conventions:

1 Mmb = 1,000,000 barrels
1 Bb = 1,000,000,000 barrels

PRODUCTION OF OIL

Production is usually expressed in barrels per day or barrels per year. When discussing large figures, such as those involved in global oil production, the quantity is often expressed in millions of barrels of oil per day (Mmbopd) or billions of barrels of oil per year (Bbopy).

1 Mmbopd = 0.365 Bbopy

DENSITY OF OIL: API GRAVITY

The density of an oil is an important parameter, and it tends to correlate with other properties such as viscosity. The petroleum industry does not tend to use conventional scientific units of density (such as g/cc, or grams per cubic centimeter). Rather, it uses a scale defined by the American Petroleum Institute (API) known as "API gravity." API gravity is measured in "degrees API," denoted *°API*. The term *degree* in this context bears no relationship to temperature. The definition is as follows:

$$°API = (141.5/SG) - 131.5$$

where *SG* is the ratio of the density of the oil in question to the density of water at a temperature of 60°F (15.6°C). The following table gives the equivalents of some API values in g/cc:

°API	Density (g/cc)
5°	1.0366
10°	1.0000
20°	0.9340
30°	0.8762
40°	0.8251

Oils with higher °API values are lighter. Water has a gravity of 10°API; oils with values less than 10 are heavier than water, and those with values greater than 10 are lighter. Generally, "heavy oils" are oils with gravities below 20°API. "Light oils" or "light crudes" are those with gravities above 30°API. "Medium crudes" have gravities between 20°API and 30°API.

QUANTITY OF OIL: TONS/TONNES

It would seem logical to express the quantity of oil in terms of its mass rather than its volume. This is usually done using metric tons ("tonnes") as the unit (metric tons are used for this purpose even in the U.S. literature). One ton (metric) equals:

- 0.9842 long tons (U.S.)
- 1.1023 short tons (U.S.)
- 1,000 kilograms
- 2,204.623 pounds

The conversion from tons to barrels will depend on the oil's API gravity. One ton (metric) equals:

- 6.73 barrels of 20°API oil
- 7.18 barrels of 30°API oil
- 7.62 barrels of 40°API oil

Note that when speaking of millions of tons, we use the standard prefix *M*, not the *Mm* prefix used for barrels.

1 million tons = 1 Mt, *not* 1 Mmt

QUANTITY OF OIL: POODS

In the early Russian oil industry, the quantity of oil was often quoted using a now obsolete unit called the "poud" or "pood."

1 poud = 16.3805 kilograms
1 barrel of 30°API oil = 8.491 pouds

STANDARD ENERGY OF OIL

Crude oil varies in its energy content. Typical values for a barrel of oil range from 5.6 million British thermal units (MBtu) to 6.3 MBtu. The average energy content of a barrel of crude oil is about 5.8 MBtu, and this is sometimes used as a standard definition of a barrel of oil equivalent (boe) of energy. One boe equals:

- 5.8 MBtu (IT)
- 6.119 GJ
- 1.461 Gcal (IT)

where *GJ* denotes gigajoules and *Gcal* denotes gigacalories. Both the Btu and the calorie have multiple definitions differing slightly from each other. *(IT)* indicates that the "International Table" value of the Btu or the calorie is being used.

The ton of oil equivalent (toe) and megaton of oil equivalent (Mtoe) are also used (an Mtoe is one million toe). Here again, a nominal definition is necessary since crude oils vary in both their energy content and their density. Unfortunately, two different definitions are in common use.

The Organisation for Economic Co-operation and Development (OECD) and the International Energy Agency (IEA) define a ton of oil equivalent as 1 toe being equal to:

- 1.00×10^{10} cal (IT)
- 41.868 GJ
- 39.68 MBtu (IT)

Another definition for the ton of oil equivalent used in publications by other entities is 1 toe being equal to:

- 1.07×10^{10} cal (TC)
- 44.769 GJ
- 42.46 MBtu (TC)

where *(TC)* indicates that the "thermochemical" value of the calorie and Btu are being used. As noted above, these differ slightly from the International Table (IT) values. For reference, the IT and TC values of the calorie and the Btu in terms of joules (J) are as follows:

1 cal (IT) = 4.1868 J
1 cal (TC) = 4.184 J
1 Btu (IT) = 1055.06 J

1 Btu (TC) = 1054.35 J

A large-scale unit sometimes employed in discussions of global energy usage is the quad, which is 1 quadrillion (10^{15}) Btu (IT).

1 quad = 25.2 Mtoe

QUANTITY OF NATURAL GAS

The quantity of natural gas is expressed volumetrically. A standard cubic foot (cf) of natural gas is defined at a temperature of 60°F and a pressure of 30 inHg (inches of mercury). Larger quantities are expressed in hundreds of cubic feet (ccf), thousands of cubic feet (mcf), millions of cubic feet (Mmcf), billions of cubic feet (Bcf), and trillions of cubic feet (Tcf). As in the case of oil, the use of the prefix *m* to indicate one thousand is somewhat irregular; in other domains *M* usually indicates a million—which in the natural gas world is often denoted by *Mm*.

Very large gas fields will tend to have gas reserves in the range of tens of Tcf. The largest gas field in the world, the North Field in Qatar, has ultimately recoverable gas estimated at 900 Tcf.

The quantity of natural gas is also expressed in normal cubic meters (ncm), defined at a temperature of 0°C and a pressure of 1 atmosphere. Because the temperature and pressure at which a normal cubic meter is defined differ from those for a standard cubic foot, converting from cf to ncm is not the same as converting from cubic feet to cubic meters. Correcting for temperature and pressure, we obtain:

1 ncm natural gas = 37.24 cf natural gas

QUANTITY OF LIQUEFIED NATURAL GAS

Liquefied natural gas (LNG) is stored at a temperature of -260°F (-161.5°C). Quantities of LNG are often given in millions of metric tons per annum (MTPA).

1 MTPA (LNG) = 45.91 Bcf per year natural gas
1 MTPA (LNG) = 78.48 Mmcf per year (LNG)

STANDARD ENERGY OF NATURAL GAS

The energy content of dry natural gas usually ranges from 900 Btu/cf to 1,100 Btu/cf. There is no widely accepted standard unit of energy based on natural gas analogous to the boe or toe for oil. The American Gas Association adopts a quasi-standard value of 1,027 Btu/cf:

1 cfge (AGA) = 1027 Btu

where *cfge (AGA)* denotes cubic foot of gas equivalent, American Gas Association. This is often rounded to 1,000 Btu/cf. In the natural gas industry, a commonly used unit is the therm:

1 therm = 100,000 Btu

The therm is not an equivalent unit based on the energy content of a standardized natural gas; it is an energy unit based on the Btu.

In order to compare natural gas reserves to oil reserves, we may note that the energy equivalent of one barrel of oil is generally set at 5.8 Mbtu. Using the American Gas Association's figure of 1,027 Btu/cf for natural gas, we may calculate a rough equivalence of

1 barrel oil ≈ 5648 cf natural gas

In discussions of giant oil and gas fields, the American Association of Petroleum Geologists has used

1 barrel oil ≈ 6,000 cf natural gas

For LNG:

1 metric ton LNG ≡ 53.4 MmBTU (energy equivalence)
1 metric ton LNG ≡ 9.2 barrels oil (energy equivalence)

Appendix 2

Major Middle Eastern Consortia of the Seven Sisters and CFP

	Iran	Iraq	Aramco	Kuwait Oil	Abu Dhabi	Abu Dhabi	Qatar
	Consortium	Petroleum	1933–1980	1934–1975	Petroleum	Marine	Petroleum
	1954–1979	1928–1975	—	—	1935–1974	1954–1974	1935–1976
BP	40%	23.75%	—	50%	23.75%	66.67%	23.75%
Shell	14%	23.75%	—	—	23.75%	—	23.75%
Exxon	7%	11.875%	30%	—	11.875%	—	11.875%
Mobil	7%	11.875%	10%	—	11.875%	—	11.875%
Gulf	7%	—	—	50%	—	—	—
Texaco	7%	—	30%	—	—	—	—
Chevron	7%	—	30%	—	—	—	—
CFP	6%	23.75%	—	—	23.75%	33.33%	23.75%
Others	5%	5%	—	—	5%	—	5%

CONSORTIA

Abu Dhabi Marine. Abu Dhabi National Oil Company acquired majority control in 1974.

Abu Dhabi Petroleum. Known as Petroleum Development (Trucial States) 1935–1963. Abu Dhabi National Oil Company acquired majority control in 1974.

Aramco. Organized in 1933 by Standard Oil of California (Socal, later Chevron) as the California Arabian Standard Oil Company (Casoc). The Texas Corporation (Texaco) joined as a partner in 1936. Casoc changed its name to Aramco in 1944. Socony–Vacuum (later Mobil) and Standard Oil of New Jersey (later Exxon) joined as partners in 1948. The government of Saudi Arabia acquired complete control in 1980.

Iran Consortium. The "others" here are the Iricon group, initially composed of Aminoil, Richfield (later Arco), Signal, Hancock, Sohio, Pacific Western/Getty, Tidewater (Getty), Atlantic (later Arco), and San Jacinto.

Iraq Petroleum. Founded in 1912 as the Turkish Petroleum Company. Known after 1929 as the Iraq Petroleum Company. U.S. partners did not join until 1928. The 5 percent share was owned by Calouste Gulbenkian.

Kuwait Oil. Formed by Anglo–Persian Oil Company (later Anglo–Iranian Oil Company, eventually BP) and Gulf Oil. The government of Kuwait acquired complete control between 1974 and 1975.

Qatar Petroleum. Known as Petroleum Development (Qatar) before 1963; acquired by Qatar government in phases between 1974 and 1976.

SEVEN SISTERS PLUS ONE

The Seven Sisters are really the Seven Sisters Plus One, where the eighth is Compagnie Française des Pétroles (CFP).

BP. Anglo–Persian Oil Company 1909–1935; Anglo–Iranian Oil Company 1935–1954. Became British Petroleum in 1954.

CFP. Compagnie Française des Pétroles; became Total CFP in 1985, Total in 1991.

Chevron. Standard Oil Company of California, or Socal, until 1984; Chevron 1984–2001; ChevronTexaco 2001–2005; became Chevron in 2005.

Exxon. Standard Oil Company of New Jersey until 1972; Exxon 1972–1999; became ExxonMobil in 1999.

Gulf. Acquired by Socal in 1984. Enlarged company became Chevron.

Mobil. Socony–Vacuum 1931–1955; Socony Mobil 1955–1966; Mobil 1966–1999; became ExxonMobil in 1999.

Shell. Royal Dutch Shell, formed in 1907 by amalgamation of Royal Dutch Company and Shell Transport and Trading Company.

Texaco. The Texas Company 1902–1926; Texas Corporation 1926–1941; Texas Company 1941–1959; Texaco 1959–2001; ChevronTexaco 2001–2005; Chevron 2005.

Appendix 3

Various Companies with Roots in Standard Oil

AMOCO. Was Standard Oil of Indiana, independent in 1911. Acquired Standard Oil Company of Nebraska in 1939 and Standard Oil Company of Kansas in 1948.

ARCO (Atlantic Richfield Company). Part of **Arco**'s origins lie in the Atlantic Refining Company established in 1870 (first organized as the Atlantic Petroleum Storage Company in 1866). Atlantic Refining became part of **Standard Oil** in 1874. It became independent again in 1911. It merged in 1966 with the Richfield Oil Corporation, a non-Standard company, to form Atlantic Richfield (Arco). Arco acquired **Sinclair Oil** in 1969. Sinclair was a non-Standard company, but it had acquired some Standard heritage in 1930 by purchasing the Waters-Pierce Oil Company, and in 1932 by purchasing the Prairie Oil and Gas Company. Waters-Pierce came under the influence of the Standard Oil Trust in 1878. It became an independent company as a result of the 1911 breakup, as did Prairie Oil and Gas.

BP. BP has a significant Standard Oil legacy, through its acquisitions of **Sohio** and **Arco** (1986 and 2000, respectively) and its merger with **Amoco** (1998).

CHEVRON. Was the Standard Oil Company of California, established as an independent company in 1911. It acquired Standard Oil of Kentucky in 1961. Standard Oil of California had roots in the Pacific Coast Oil Company, established in 1879 and acquired by **Standard Oil** in 1900. Standard consolidated Pacific Coast and other operations in the western United States into Standard Oil of California in 1906.

CONOCO. Began as the Continental Oil and Transportation Company in 1875; acquired by **Standard Oil** in 1885. Continental Oil Company (**Conoco**) was established as an independent company in 1913, two years after the dissolution of Standard Oil. Conoco merged in 2002 with **Phillips Petroleum**, a company that did not originate in Standard Oil, to form **ConocoPhillips**.

EXXON. Arguably, the most direct successor to **Standard Oil**. It was established as Standard Oil of New Jersey as a result of the 1911 breakup. Before 1911, its antecedent was the core of the Standard Oil Trust. It became **Exxon** in 1972, and **ExxonMobil** in 1999.

EXXONMOBIL. A merger (1998) of **Exxon** and **Mobil**, two major descendants of **Standard Oil.**

IMPERIAL OIL (Canada). Established in 1880. **Standard Oil** acquired a majority interest in **Imperial Oil** in 1898. Standard Oil of New Jersey (Exxon) retained a controlling interest in Imperial after the dissolution of Standard Oil in 1911.

MARATHON. Marathon has roots in the Ohio Oil Company, established in 1887 and acquired by **Standard Oil** in 1889.

MOBIL. Traces its roots to two **Standard Oil** companies. One was the **Standard Oil Company of New York** (Socony), and the other was **Vacuum Oil Company**. Standard Oil of New York's antecedent was first formed in 1882 and administered most of the trust's foreign territories. In 1911 Standard Oil Company of New York became an independent company. In 1931 Socony merged with Vacuum Oil Company to form Socony–Vacuum, which became Socony Mobil in 1955 and **Mobil** in 1966. Mobil merged with **Exxon** to form **ExxonMobil** in 1999. Vacuum Oil was another company formed from the 1911 breakup. It had roots in the Vacuum Oil Company, which was established in 1866, and came under the control of Standard Oil in 1879.

PENNZOIL. Pennzoil has roots in the **South Penn Oil Company**, established as a unit of **Standard Oil** in 1889. It became an independent company in 1911 and part of an enlarged Pennzoil in 1962.

ROYAL DUTCH SHELL. Acquired some Standard Oil heritage when it bought **Pennzoil** in 2002.

SOHIO. Standard Oil Company of Ohio, independent in 1911, with roots in the original Ohio operations of Standard Oil dating back to 1870 and before.

Appendix 4

Top Five Oil-Producing Countries for Twenty-Year Intervals since 1861

The information in this appendix is based on the total oil production of oil and liquids over each interval.

Interval	First	Second	Third	Fourth	Fifth
1861–1880	U.S.	Russia	Canada	Romania	Poland[1]
1881–1900	U.S.	Russia	Poland[2]	Indonesia	Canada
1901–1920	U.S.	Russia	Mexico	Indonesia	Romania
1921–1940	U.S.	USSR	Venezuela	Mexico	Iran
1941–1960	U.S.	Venezuela	USSR	Saudi Arabia	Iran
1961–1980	U.S.	USSR	Saudi Arabia	Iran	Venezuela
1981–2000	U.S.	Saudi Arabia	Russia/ USSR[3]	Mexico	Iran
2001–2016	Saudi Arabia	Russia	U.S.	Iran	China

[1] The oil-producing territory was then in the Austro-Hungarian province of Galicia.
[2] The oil-producing territory was then in the Austro-Hungarian province of Galicia.
[3] Includes all Soviet production from 1980 to 1984. From 1985 onward includes only Russian production.

Appendix 5

Sweet and Sour Crudes

Crude oils are often classified according to their density and their sulfur content. Oils with a high density (low API gravity) are referred to as "heavy," and those with a low density (high API gravity) are referred to as "light." Meanwhile oils with a low sulfur content are called "sweet," while those with a high sulfur content are "sour." While these terms are used throughout the industry, there are no precise universally accepted specifications for what constitutes a "sweet" versus a "sour" crude, or a "light" versus a "heavy" one. The following graph can be used as a general guideline. It is drawn based on a useful classification adopted by ENI's R&M Division Research Center for Downstream Technologies (*ENI World Oil and Gas Review 2007*, ENI Strategies & Development Department).

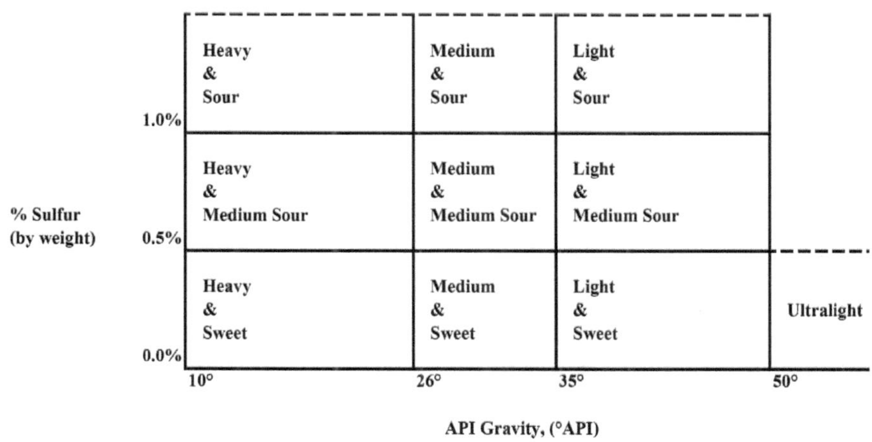

Appendix 6

Crudes in the OPEC Basket

OPEC BASKET IN 2017

Crude	Country
Saharan Blend	Algeria
Girassol	Angola
Oriente	Ecuador
Zafiro	Equatorial Guinea
Rabi Light	Gabon
Iran Heavy	Iran
Basra Light	Iraq
Kuwait Export	Kuwait
Es Sider	Libya
Bonny Light	Nigeria
Qatar Marine	Qatar
Arab Light	Saudi Arabia
Murban	UAE
Merey	Venezuela

OPEC BASKET INTRODUCED 16 JUNE 2005

Crude	Country
Saharan Blend	Algeria
Minas	Indonesia
Iran Heavy	Iran
Basra Light	Iraq
Kuwait	Kuwait

Es Sider	Libya
Bonny Light	Nigeria
Qatar Marine	Qatar
Arab Light	Saudi Arabia
Murban	UAE
BCF 17	Venezuela

OPEC BASKET FROM 1 JANUARY 1987 TO 15 JUNE 2005

Crude	Country
Saharan Blend	Algeria
Minas	Indonesia
Bonny Light	Nigeria
Arab Light	Saudi Arabia
Dubai Fateh	UAE
Tia Juana	Venezuela
Isthmus	Mexico

Appendix 7

Major Oil Supply Disruptions

Time Period	Cause	Peak Loss of Supply[1]
Nov 1956–Mar 1957	Suez Crisis	2 Mmbpd
Jun 1967–Aug 1967	Six-Day War	2 Mmbpd
Oct 1973–Mar 1974	Yom Kippur War and Arab oil embargo	4.3 Mmbpd
Nov 1978–Apr 1979	Iranian Revolution	5.6 Mmbpd
Oct 1980–Jan 1981	Outbreak of Iran-Iraq War	4.1 Mmbpd
Aug 1990–Jan 1991	Iraqi invasion of Kuwait	4.3 Mmbpd
Jun 2001–Jul 2001	Iraq oil export suspension	2.1 Mmbpd
Dec 2002–Mar 2003	Venezuela oil industry strike	2.6 Mmbpd
Mar 2003–Dec 2003	Outbreak of Iraq War	2.3 Mmbpd
Sep 2005	Hurricanes Katrina and Rita	1.5 Mmbpd

Source: International Energy Agency, *IEA Response System for Oil Supply Emergencies*, 2007.

[1] In million barrels per day (Mmbpd)

Appendix 8

Tanker Classification

CRUDE OIL TANKERS

Category	Tonnage (DWT)	Notional Oil Cargo (barrels)	Historical Milepost
		25,000	(1940–1950s: T2 Class, 15,000–25,000 DWT)
Panamax	60,000–80, 000	300,000–400,000	—
Neo-Panamax	120,000	400,000–600,000	—
Aframax[1]	80,000–120,000	500,000	1959: 100,000 DWT Surpassed
Suezmax	120,000–160 000	1,000,000	—
Very Large Crude Carrier (VLCC)	160,000–320,000	2,000,000	1966: 200,000 DWT reached
Ultra Large Crude Carrier (ULCC)	320,000–550,000	3,000,000	1969: First ULCC

[1] "Afra" has nothing to do with Africa; it refers to "Average Freight Rate Assessment Scheme," developed by Shell in 1954 to tabulate average transportation costs for various classes of ships.

PRODUCT CARRIERS

Category	Tonnage (DWT)[1]
Handy	25,000–50,000
Large	50,000–100,000
Very Large Product Carrier (VLPC)	100,000+

[1] Deadweight tons (long tons).

Appendix 9

Largest Oil Spills

The following table lists the largest oil spills from tanker ships, in the order of the quantity spilled:

Tanker	Date	Place	Barrels Spilled[1]
Atlantic Empress[2]	19 Jul 1979 and 2 Aug 1979	Trinidad and Tobago and Barbados	2.005 Mmb
ABT Summer	28 May 1991	Offshore Angola	1.907 Mmb
Castillo de Bellver	6 Aug 1983	Saldanha Bay, South Africa	1.869 Mmb
Amoco Cadiz	16 Mar 1978	Off Brittany, France	1.636 Mmb
Odyssey	10 Nov 1988	Atlantic Ocean 1,100 km off Canada	1.026 Mmb
Torrey Canyon	18 Mar 1967	Scilly Isles, UK	0.910 Mmb
Sea Star	19 Dec 1972	Gulf of Oman	0.902 Mmb
Irene's Serenade	23 Feb 1980	Navarino Bay, Greece	0.871 Mmb
Texaco Denmark	7 Dec 1971	Belgium	0.750 Mmb
Hawaiian Patriot	23 Feb 1980	Pacific Ocean 593 km west of Hawaii	0.743 Mmb
Independentzia	15 Nov 1979	Bosporus Strait, Turkey	0.688 Mmb
Julius Schindler	2 Feb 1969	Azores	0.676 Mmb
Braer	5 Jan 1993	Shetland Islands, UK	0.595 Mmb
Jacob Maersk	29 Jan 1975	Portugal	0.579 Mmb
Aegean Sea	3 Dec 1992	La Coruna, Spain	0.521 Mmb
Nova	6 Dec 1985	Persian Gulf	0.510 Mmb

[1] Total amount spilled, rather than amount ultimately released into the environment. Note that estimates of quantities spilled can vary considerably. We use the figures of Etkin and Welch (1997).

[2] The *Atlantic Empress* spill is often regarded as two separate incidents. On July 19, 1979, the *Atlantic Empress* collided with the *Aegean Captain* off Trinidad and Tobago, spilling an estimated 1.793 Mmb. While being towed, it spilled an additional 1.743 Mmb off Barbados.

This table lists the largest oil spills from sources other than tanker ships:

Incident	Date	Place	Barrels Spilled[1]
Kuwait-Gulf War	26 Jan 1991	Persian Gulf	5.714 Mmb
Deepwater Horizon	20 Apr 2010	Gulf of Mexico	4.9 Mmb
Ixtoc I Well	3 Jun 1979	Ciudad del Carmen, Mexico	3.333 Mmb
Nowruz Field	4 Feb 1983	Persian Gulf	1.905 Mmb
Oil Well, Fergana Valley	10 Nov 1988	500 km from Namangan, Uzbekistan	1.905 Mmb
Production Well D-103	1 Aug 1980	800 km SE of Tripoli, Libya	1.000 Mmb
Storage Tanks, Shuaybah	20 Aug 1981	Kuwait	0.742 Mmb
Pipeline, Usinsk	25 Oct 1994	Russia	0.731 Mmb
Ahvazin Well and Pipeline	25 May 1978	Iran	0.666 Mmb
Forcados Storage Tank #6	6 Jul 1979	Nigeria	0.569 Mmb

[1] Total amount spilled, rather than amount ultimately released into the environment. Note that estimates of quantities spilled can vary considerably. We use the figures of Etkin and Welch (1997).

Appendix 10

Refinery Fractions

Fraction[1]	Boiling Range, °Celsius	Boiling Range, °Fahrenheit	Typical Uses
Fuel gas	-160°C–-40°C	-260°F–-40°F	Refinery fuel
Propane	-40°C	-40°F	Liquefied petroleum gas (LPG)
Butanes	-12°C–-1°C	11°F–30°F	Increase gasoline volatility
Light naptha	-1°C–150°C	30°F–300°F	Motor gasoline components and sometimes reformer feedstock (with heavy naphtha)
Gasoline	-1°C–180°C	30°F–355°F	Motor fuel
Heavy naphtha	150°C–205°C	300°F–400°F	Reformer feedstock; jet fuels (with light gas oil)
Kerosene	205°C–260°C	400°F–500°F	Fuel oil
Stove oil	205°C–290°C	400°F–550°F	Fuel oil
Light gas oil	260°C–315°C	500°F–600°F	Furnace and diesel fuel components and jet fuel (with heavy naphtha)
Heavy gas oil	315°C–425°C	600°F–800°F	Catalytic cracker feedstock
Lubricating oil	>400°C	>750°F	Lubrication
Vacuum gas oil	425°C–600°C	800°F–1,100°F	Catalytic cracker feedstock
Residuum	>600°C	>1,100°F	Heavy fuel oil; asphalts

Sources: J. G. Speight, *The Chemistry and Technology of Petroleum*, 4th ed. (Boca Raton, Fla.: CRC Press, 2007); J. G. Speight, "Petroleum Processing and Refining," in AccessScience@McGraw-Hill, http://www.accessscience.com, doi:10.1036/1097-8542.503100.

[1] These are fractions, not finished products. For example, the gasoline fraction is not the same as finished motor gasoline.

Appendix 11

Largest Oil Fields

The data in this appendix is arranged by the estimated ultimate recovery of oil.

Field	Country	Discovered	Oil[1]	Gas[2]	Total Oil and Gas[3]
Ghawar	Saudi Arabia	1948	75[4]	186.2	106
Greater Burgan	Kuwait	1938	60.6[5]	42.8	67.7
Safaniya	Saudi Arabia	1951	55.6	3.93	56.3
Bolivar Coastal Complex[6]	Venezuela	1926–1930	30.0	—	30.0
Rumaila N&S	Iraq[7]	1953	30.0	20.0	33.3
Ahvaz	Iran	1958	25.6	23.3	29.5
Kirkuk	Iraq	1927	25.0	8.2	26.4
Manifa	Saudi Arabia	1957	23.1	4.8	23.9
Marun	Iran	1964	22.0	75.3	34.5
Zakum	UAE (Abu Dhabi)	1964	21.8	12.4	23.7
Shaybah	Saudi Arabia	1968	21.5	—	21.5
Zuluf	Saudi Arabia	1965	20.2	5.2	21.0
Samotlor[8]	Russia	1965	20.0	—	—
Abqaiq	Saudi Arabia	1940	18.7	14.2	21.0
Berri	Saudi Arabia	1964	18.5	12.2	20.5
Daqing Complex[9]	China	1959	18.5	—	18.5
Agha Jari	Iran	1936	17.4	9.9	19.0
Khurais	Saudi Arabia	1957	17.0	3.0	17.5
Romashkino	Russia	1947	17.0	0	17.0
Cantarell Complex	Mexico	1976	16.3	9.2	18.4
Gachsaran	Iran	1928	16.2	31.1	21.4
Baghdad, East	Iraq	1979	16.0	2.5	16.4
Prudhoe Bay	U.S.: Alaska	1967	15.3	26	19.6

Majnoon	Iraq	1977	12.0	11.0	13.8
Murban Bab	UAE (Abu Dhabi)[10]	1954	10.3	29.3	15.2
Kashagan	Kazakhstan	2000	10.0	20.0	13.3
Fereidoon	Iran	1960	10.0	—	10.0
West Qurna	Iraq	1973	9.8	9.4	11.4
Pravdinsk-Salym	Russia	1964	9.1	2.9	9.6
Statfjord	Norway	1974	9.0	2.0	9.4
Libra	Brazil	2009	9.0	—	9.0
Raudhatain	Kuwait	1955	8.9	7.1	9.7
Khafji	Neutral Zone[11]	1959	8.6	1.0	8.8
Abu Sa'fah	Saudi Arabia	1963	8.6	7.1	9.7
Asab	UAE	1965	8.5	15.4	11.1
Zubair	Iraq	1949	8.2	5.9	9.2
Qatif	Saudi Arabia	1945	7.2	16.5	10.0
Samaria (Bermudez Complex)	Mexico	1958	7.0	17.5	10.0

Sources: Most data from Mike Horn, "Giant, Supergiant & Megagiant Oil and Gas Fields of the World," digital data set, 2017, published at https://worldmap.harvard.edu by the Center for Geographic Analysis, Harvard University, Cambridge, Massachusetts; M. K. Horn, *Giant Oil and Gas Fields, 1868–2004*, a revision of AAPG Memoir 78, supplemental CD-ROM (Tulsa, Okla.: American Association of Petroleum Geologists, 2005). Estimates from other investigators may vary.

[1] In Bb, billions of barrels
[2] In Tcf, trillions of cubic feet
[3] In Bboe, billions of barrels of oil equivalent. Oil equivalent of gas is calculated using a conversion factor of 6,000 cf of gas per barrel of oil.
[4] Other estimates range as high as 150 Bb.
[5] Other estimates range as high as 70 Bb.
[6] Includes Tia Juana, Lagunillas, Bachaquero, Cabinas, and Lama
[7] Extends into Kuwait
[8] Estimate from Roger D. Blanchard, *The Future of Global Oil Production: Facts, Figures, Trends and Projections by Region* (Jefferson, NC: McFarland, 2005)
[9] Including Saertu, Taching, and Xingshugang
[10] United Arab Emirates
[11] Between Saudi Arabia and Kuwait

Appendix 12

Largest Gas Fields

The data in this appendix is arranged by quantity of gas.

Field	Country	Discovered	Gas[1]	Oil and Condensate[2]	Total Oil and Gas[3]
North Field	Qatar	1971	900.0	26.1	176.1
Pars South	Iran	1991	350.0	19.1	77.4
Urengoy	Russia	1966	335.4	2.7	58.6
Yamburg	Russia	1969	153.8	0.537	26.2
Yolotan/Osman	Turkmenistan	2006	134.0	—	22.3
Shtokman	Russia	1988	130.1	—	22.0
Zapolyarnoye	Russia	1965	121.0	1.026	21.2
Hassi R'Mel	Algeria	1957	100.0	4.064	20.7
Astrakhan	Russia	1976	89.6	4.689	19.6
Northwest Dome	Qatar	1976	80.0	—	13.3
Bovanenko	Russia	1971	76.4	—	12.7
Arkticheskoye	Russia	1968	63	0.3	10.8
Orenburg	Russia	1966	62.8	0.7	11.2
Rag-E-Safid	Iran	1964	59.9	5.0	10.3
Kovykta	Russia	1986	58	—	10.3
Leningrad	Russia	1990	55.0	—	9.2
Kyrtaiol'skoye	Russia	1970	55.0	—	9.2
B. Structure	Iran	1972	50.0	—	8.3
Kangan	Iran	1972	50.0	—	8.3
Karachaganak	Kazakhstan	1979	49.2	5.7	13.2
Moran	India	1956	48.0	0.043	8.0
Pars North	Iran	1966	47.0	—	7.8
East Natuna	Indonesia	1973	46.0	—	7.7
Groningen	Netherlands	1959	43.0	—	7.2

Urengo Vostochnyy	Russia	1978	32.4	—	5.4
Troll	Norway	1979	22.8	1.319	5.1
Chayvo, Odoptu, and Arkutun-Dagi	Russia	1979	17.1	2.3	5.2
Hassi Messaoud	Algeria	1956	7.7	3.84	5.1

Sources: Mike Horn, "Giant, Supergiant & Megagiant Oil and Gas Fields of the World," digital data set, 2017, published at https://worldmap.harvard.edu by the Center for Geographic Analysis, Harvard University, Cambridge, Massachusetts; M. K. Horn, *Giant Oil and Gas Fields, 1868–2004*, a revision of AAPG Memoir 78, supplemental CD-ROM (Tulsa, Okla.: American Association of Petroleum Geologists, 2005). Estimates from other investigators may vary.

[1] Tcf, trillions of cubic feet

[2] Bb, billions of barrels

[3] Bboe, billions of barrels of oil equivalent (Oil equivalent of gas is calculated using a conversion factor of 6,000 cf of gas per barrel of oil.)

Appendix 13

Other Giant Oil and Gas Fields of Historical Interest

The data in this appendix is arranged by the year of discovery.

Field	Country	Discovered	Oil[1]	Gas[2]	Total Oil and Gas[3]
La Brea	Peru	1868	1.000	15.0	3.5
Surakhanskoye	Azerbaijan	1870	0.900	—	0.9
Bradford	U.S.: Pennsylvania	1871	0.658	—	0.658
Bibi–Eibat	Azerbaijan	1871	2.000	—	2.000
Lima–Indiana	U.S.: Ohio/Indiana	1876	0.514	—	0.514
Brea	U.S.: California	1884	0.439	0.5	0.525
Coalinga	U.S.: California	1887	0.906	—	0.906
Starogrozny	Russia	1893	0.650	—	0.605
Midway–Sunset	U.S.: California	1894	2.692	0.5	2.780
Balakhany Sabunchino	Azerbaijan	1896	2.400	—	2.400
Kern River	U.S.: California	1899	1.676	—	1.676
Panuco	Mexico	1901	0.971	—	0.971
Salt Creek	U.S.: Wyoming	1906	0.676	—	0.676
Comodoro Rivadavia	Argentina	1907	3.244	—	3.244
Moreni–Gura Ocnitei	Romania	1907	0.800	—	0.800
Masjid-e-Suleiman	Iran	1908	0.800	—	0.800
Buena Vista Hills	U.S.-California	1909	0.678	1.1	0.859
Cerro Azul	Mexico	1909	1.250	—	1.250
Maykop	Russia	1909	—	3.2	0.537
Cushing	U.S.: Oklahoma	1912	0.500	—	0.500
Fyzabad Group	Trinidad and Tobago	1913	0.510	—	0.510
Sho-Vel-Tum	U.S.: Oklahoma	1914	1.355	—	1.355
Mene Grande	Venezuela	1914	0.700	—	0.700

Monroe	U.S.: Louisiana	1916	—	7.2	1.203
Ventura Avenue/ Rincon	U.S.: California	1916	1.012	2.4	1.417
Cabinas (Bolivar Coastal)	Venezuela	1917	0.500	—	0.500
Infantas–La Cira	Colombia	1918	0.500	—	0.500
Santa Fe Springs	U.S.: California	1919	0.633	0.8	0.773
Elk Hills	U.S.: California	1919	1.407	1.4	1.647
Seminole	U.S.: Oklahoma	1926	0.822	—	0.822
Yates	U.S.: Texas	1926	1.955	—	1.955
Seria	Brunei	1929	1.730	2.0	2.063
East Texas	U.S.: Texas	1930	5.382	—	5.382
Poza Rica	Mexico	1930	2.000	—	2.000
Wilmington	U.S.: California	1932	2.788	1.3	3.006
Awali (Bahrain)	Bahrain	1932	0.962[4]	6.8	2.088
Tuyrnazy	Russia	1937	1.000	—	1.000
Dammam	Saudi Arabia	1938	0.912	2.4	1.306
Laochunmiao	China	1938	0.500	—	0.500
Dukhan	Qatar	1940	1.598	9.0	3.100
Kuang	Indonesia	1940	0.600	—	0.600
Duri	Indonesia	1941	1.691	—	1.691
Leduc	Canada: Alberta	1947	0.500	—	0.500
Redwater	Canada: Alberta	1948	0.832	—	0.832
Nahr Umr	Iraq	1948	6.500	9.9	8.150
Matzen	Austria	1949	0.475	0.6	0.582
Zubair	Iraq	1949	8.200	5.9	9.187
Lacq	France	1951	—	8.1	1.348
Sui	Pakistan	1952	—	11.2	1.873
Wafra	Kuwait	1953	1.667	1.2	1.867
Nahorkatiya	India	1953	0.500	—	0.500
Pembina	Canada: Alberta	1953	1.800	—	1.800
Belayim Land	Egypt	1955	0.500	—	0.500
Raudhatain	Kuwait	1955	7.193	6.7	8.318
Hassi Messaoud	Algeria	1956	3.840	7.7	5.120
Augila–Nafoora	Libya	1956	0.834	1.5	1.084
Moran	India	1956	0.043	48.0	8.043
Kotur–Tepe	Turkmenistan	1956	1.460	1.5	1.710

Gazli	Uzbekistan	1956	0.027	16.6	2.801
Marmul	Oman	1957	0.700	—	0.700
Soku	Nigeria	1958	0.138	5.4	1.034
Amal (Libya)	Libya	1959	3.322	3.5	3.906
Kenai	U.S.: Alaska	1959	—	3.1	0.523
Khafji	Neutral Zone[5]	1959	8.646	1.0	8.815
Suwaidiyah	Syria	1959	0.711[6]	0.3	0.753
Uzen	Kazakhstan	1961	1.658	0.6	1.758
Ust'Bakyskoye	Russia: Siberia	1961	2.300	—	2.300
Yibal	Oman	1962	1.330	4.5	2.081
Bubulime	Albania	1963	—	5.0	0.833
El Borma	Tunisia	1964	0.640	0.8	0.772
Gidgealpa	Australia	1964	—	5.0	0.833
Bach Ho	Vietnam	1964	0.950	1.3	1.167
McArthur River	U.S.: Alaska	1965	0.550	0.7	0.670
Miranga	Brazil	1965	0.590	—	0.590
Mellion-Rousse	France	1965	—	3.5	0.583
Salzwedel	Germany	1965	—	3.5	0.583
West Sole	UK: North Sea	1965	—	3.0	0.500
Algyo	Hungary	1965	0.200	3.7	0.815
Fateh	UAE: Dubai	1966	1.100	—	1.100
Malongo N & S	Angola	1966	0.850	—	0.850
Kingfish	Australia	1967	1.250	—	1.250
Kuparuk River	U.S.-Alaska	1969	2.595	—	2.595
Sacha	Ecuador	1969	0.650	—	0.650
Emeraude Marin	Congo: Brazzaville	1969	0.500	—	0.500
Ekofisk	Norway: North Sea	1969	3.800	3.9	4.452
Maui	New Zealand	1969	0.075	5.3	0.957
Beurdeshik	Tajikistan	1969	—	3.0	0.500
Forties	UK: North Sea	1970	2.000	—	2.000
Central Luconia	Malaysia	1970	—	3.4	0.565
Kudu	Namibia	1971	—	3.6	0.600
Brent	UK: North Sea	1971	0.243	2.1	0.599
Drake Point	Canada: NWT	1973	—	6.0	1.000
Malossa	Italy	1973	—	1.8	0.618
B Structure	Thailand	1973	—	7.3	1.215

Hibernia	Canada: Newfoundland	1979	1.850	2.0	2.183
Serrablo	Spain	1979	—	3.5	0.583
Meskala	Morocco	1980	—	5.0	0.833
Espoir	Ivory Coast	1980	1.000	—	1.000
Unity	Sudan	1980	0.900	—	0.900
Sajaa	UAE: Sharjah	1980	0.246[7]	3.3	0.799
Feni	Bangladesh	1980	—	3.3	0.548
Tengiz	Kazakhstan	1980	5.829	11.9	7.812
Margham	UAE: Dubai	1982	0.234[7]	2.7	0.681
Hides	Papua: New Guinea	1987	0.070[7]	8.0	1.403
Rabi–Kounga	Gabon	1987	1.000	—	1.000
Masila Complex	Yemen	1990	0.900	—	0.900
Yadana	Myanmar	1990	—	5.0	0.833
Cusiana	Colombia	1992	1.445	3.1	1.961
Mars Deepwater	U.S.: Gulf of Mexico	1993	0.700	—	0.700
Roncador	Brazil	1996	3.200	—	3.200
Margarita	Bolivia	1998	1.357[8]	6.5	2.432
Ceiba	Equatorial Guinea	1999	0.500	—	0.500
Halfdan	Denmark	1999	0.492	0.3	0.545
Azadegan	Iran	1999	6.110	2.0	6.443
Bonga SW	Nigeria	2001	1.250[9]	0.5	1.333

Sources: Most data from M. K. Horn, *Giant Oil and Gas Fields, 1868–2004*, a revision of AAPG Memoir 78, supplemental CD-ROM (Tulsa, Okla.: American Association of Petroleum Geologists, 2005). Estimates from other investigators may vary.

[1] Bb, billions of barrels
[2] Tcf, trillions of cubic feet
[3] Bboe, billions of barrels of oil equivalent (Oil equivalent of gas is calculated using a conversion factor of 6,000 cf of gas per barrel of oil.)
[4] Includes 62 MMb condensate
[5] Kuwait/Saudi Arabia
[6] Includes 3 MMb of condensate
[7] Condensate
[8] Includes 141 MMb condensate
[9] Includes 500 MMb of condensate

Bibliography

CONTENTS

INTRODUCTION

The literature on the history of the petroleum industry is vast, and the compilation below makes no claims of completeness or uniformity. Many of the works were useful in producing the dictionary, and others are of general interest.

The Prize by Daniel Yergin is a highly readable and exhaustive popular history of the petroleum industry. There is also a PBS video series of the same title, based on the book, that is both informative and entertaining. *The Age of Oil* by Leonardo Maugeri is another general history; full of thoughtful analysis, it is another excellent starting point for those interested in learning about the field. Francisco Parra's *Oil Politics: A Modern History of Petroleum* is a superb general history offering a tremendous amount of insight into the political economy of oil. Bernard Mommer's *Global Oil and the Nation State* is a highly authoritative survey of the relationships between governments and petroleum companies, and the ownership and management of hydrocarbon resources. Edgar Wesley Owens's *Trek of the Oil Finders* is a highly informative history, with particularly useful material on oil in premodern times. Abdulhay Zalloum's *Oil Crusades* gives an interesting account of the history of petroleum, and America's involvement in the Middle East, from an Arab perspective. William Clark's *Petrodollar Warfare* is an analysis of the role of the U.S. dollar in the world oil trade and global economy. For readers not well versed in the technical and business aspects of the industry, Charles Conaway's *The Petroleum Industry: A Nontechnical Guide* is a useful introduction.

The Petroleum History Institute (http://petroleumhistory.org) began publishing an annual journal, *Oil Industry History*, in 2000. Seventeen volumes have been published so far. The journal contains authoritative articles on all aspects of petroleum history.

For histories of specific companies, the *International Directory of Company Histories* is a good place to begin. For recent numerical data on companies, the reader may consult the *Oil and Gas Journal*'s OGJ150 (for companies based in the United States) and OGJ100 (for companies based outside the United States) compilations, published annually in one of the journal's issues. The *Oil and Gas Journal* also publishes an annual "Worldwide Report," listing estimates of oil and gas reserves by country and oil production by field, as well as production from the world's refineries.

Individual company histories published in collaboration with the companies themselves will generally be biased in favor of the companies, but they can nevertheless serve as authoritative sources of facts and figures. The *History of British Petroleum*, with volumes written by James Bamberg and Ronald Ferrier, is a model of what an individual company history can be at its best, and it should be consulted by all who are seriously interested in modern petroleum history.

Historical statistics are indispensible for the researcher of petroleum industry history. The U.S. Department of Energy's Energy Information Administration (www.eia.doe.gov) posts global statistics on reserves and production of oil and gas on its website, going back to 1980. The Energy Information Administration also posts industry summaries for many individual countries.

These are good snapshots of current conditions and also contain some historical information. For data as far back as 1918, the compilations by DeGolyer and MacNaughton are a good source. Data for years before 1918 are not as widely available. Leonard Fanning's *Our Oil Resources* is an excellent source for global statistics from 1861 to 1949. David Day's *Handbook of the Petroleum Industry*, published in 1922, is a good source of older statistics for the United States and its regions from 1859 to 1920. Roger Blanchard's *The Future of Global Oil Production: Facts, Figures, Trends, and Projections by Region* has a number of useful graphs showing the production histories of various countries, regions, and individual oil fields. Amos Salvador's excellent *Energy: A Historical Perspective and 21st Century Forecast* also has a variety of historical statistics and analysis.

E. N. Tiratsoo's *Oilfields of the World* is an invaluable reference but is now rather dated, the last edition having been published in 1983. For recent data on giant oil and gas fields (which contain a large portion of the world's petroleum), the reader may consult *Giant Oil and Gas Fields: 1868–2004* by M. K. Horn. Horn's updated database was published by the Center for Geographic Analysis at Harvard University and at the time this book was published could be found at https://worldmap.harvard.edu. The doctoral thesis of Fredrik Robelius, *Giant Oil Fields: The Highway to Oil*, also has valuable information on giant fields. Matthew Simmons's *Twilight in the Desert* is a sobering account of the depletion of the world's largest oil fields.

The literature on peak oil and oil depletion has been expanding greatly in recent years. Colin Campbell's *Oil Crisis* and *Atlas of Oil and Gas Depletion*, as well as the books by Kenneth Deffeyes (*Hubbert's Peak* and *Beyond Oil*) are representative of the depletionist point of view. The other side of the argument is presented well in various articles by Michael Lynch and also in Duncan Clarke's *Battle for Barrels*. Leonardo Maugeri also presents a balanced and nonalarmist analysis in his article in *Science* magazine and in his aforementioned book *The Age of Oil*.

There is a large number of useful works focused on the petroleum history of particular countries and regions. Books and articles abound especially on the history of the industry in various parts of the United States. *The American Petroleum Industry* by Williamson and Daum is an authoritative history from the earliest period through the 1950s. For Pennsylvania, there is an extensive base of literature, but a short guidebook to the Drake Well Museum and Park by Jon Sherman is a useful summary, as is William Brice's book on Edwin Drake and the early Pennsylvania industry. Supporters of the Drake Well Museum (organized as "The Colonel Inc.") also published the *Oilfield Journal*, a valuable forum for papers on the early oil industry in the United States, particularly in Pennsylvania. David McKain and Bernard Allen's book on neighboring West Virginia, *Where It All Began*, is a useful account of an important and relatively neglected facet of early U.S. oil history. For a history of the oil industry in Texas, *The Handbook of Texas* is a good place to start. The reader can also consult *Spindletop* by James Clark and Michel Halbouty. For Oklahoma, Dan Boyd's article in *Oklahoma Geology Notes* is a useful summary, and for California the book *Early California Oil* by Kenny Franks and Paul Lambert is recommended. For Canadian petroleum history, one can begin with Robert Bott's *Evolution of Canada's Oil and Gas Industry*. The book *Roughnecks, Rock Bits, and Rigs* by Bonar Alexander Gow is focused on the evolution of drilling technology in Alberta but also has useful material on the general history of the petroleum industry there.

For Russia, the work of Alexander Matveichuk is indispensable. His book *Russia's First Oil Engineers* is in Russian. Another book, *Intersection of Oil Parallels: Historical Essays*, is in English but not easily accessible. Fortunately, a considerable amount of the material discussed in those books has also been published as articles in *Oil of Russia* magazine, of which Matveichuk is the editor. The book *Oil of Russia* by Lukoil executive Vagit Alekperov is also very authoritative and easily accessible. For the early history of the modern petroleum industry in the Russian Empire, particularly Azerbaijan, one also cannot do better than consulting the excellent papers of Mir-Yusif Mir-Babayev, some of which have been published in the journal *Oil Industry History* mentioned above. Another important source for Russian petroleum history is Robert Tolf's *The Russian Rockefellers: The Saga of the Nobel Family and the Russian Oil Industry*.

The Romanian oil industry is authoritatively covered by Maurice Pearton's *Oil and the Romanian State*. The paper by Jozef Sozanski and colleagues in *The Carpathians and Their Foreland* has an excellent account of the early industry in Poland and the activities of Ignacy Łukasiewicz. The Middle East and North Africa tend to be covered extensively in general histories of petroleum, but a considerable separate literature also exists for the region. Abbas Alnasrawi's *Iraq's Burdens: Oil, Sanctions, and Underdevelopment* is a useful look at Iraq. Archibald Chisholm's *The First Kuwait Oil Concession Agreement* is also recommended. A good overview of the industry in Iran can be found in Ronald Ferrier's chapter in *The Cambridge History of Iran*, vol. 7. The papers by Rasoul Sorkhabi are also invaluable. Stephen Kinzer's *All the Shah's Men* is essential reading for those who wish to learn about the overthrow of the Mossadegh government in the wake of Iran's nationalization of its sector in 1951. Wallace Stegner's *Discovery! The Search for Saudi Arabian Oil* is a readable account of the search for oil in Saudi Arabia. Robert Vitalis's *America's Kingdom: Mythmaking on the Saudi Oil Frontier* is a detailed look at the interactions of Aramco with the Saudi government and society over the years.

For sub-Saharan Africa, the book *Untapped: The Scramble for Africa's Oil* by John Ghazvinian is a useful starting point. For those interested in East Asia, *China and the Global Energy Crisis* by Kambara Tatsu and Christopher Howe has an excellent account of the history of China's petroleum industry. J. Ph. Poley's *Eroïca* is an authoritative history of the early oil industry in Indonesia, with important material on the first years of the Royal Dutch Company. For Latin America, one can begin with *Oil and Politics in Latin America: Nationalist Movements and State Companies* by Alan Knight. *The Mexican Petroleum Industry in the Twentieth Century* by Jonathan Brown and Alan Knight is an authoritative account of the Mexican industry. Bernard Mommer's chapter on Venezuelan oil in *Venezuelan Politics in the Chávez Era* (edited by Steve Ellner and Miguel Salas) is also useful.

Those interested in the history of petroleum-related technology can begin with the chapter on oil refining in Billington and Billington's *Power, Speed, and Form*. *Geophysics in the Affairs of Mankind* by L. C. Lawyer and others is an extensive history of the development of exploration geophysics. Arthur Johnson's *The Development of American Petroleum Pipelines* is also recommended. James Speight's *The Chemistry and Technology of Petroleum* is a technical work, but it also has material on the history of petroleum technology.

The reader who wishes to learn more about the relationship between oil and war should consult Robert Goralski and Russel Freeburg's *Oil and War*, which focuses on World War II, and Timothy Winegard's *The First World Oil War*, which treats World War I. Steven Cote's *Oil and Nation: A History of Bolivia's Petroleum Sector* has an excellent treatment of the Chaco War. *American Theocracy* by Kevin Phillips includes a readable general history of the petroleum industry and an analysis of the role of oil in U.S. policy and war making in the Middle East. Neal Adams's *Terrorism and Oil* provides a useful brief summary of geopolitical issues associated with oil.

Internet resources, including those mentioned above, are included in the last section of this bibliography. Apart from Internet resources mentioned already, the reader will probably find the website of the *Oil and Gas Journal* informative, although full access requires a subscription. The ENI and BP sites offer free historical statistics, although the reader should be aware that they may differ from each other and from the *Oil and Gas Journal*. The United States Geological Survey's energy resources website contains information on the survey's estimates of total petroleum resources. The International Energy Agency's website is a good source of market data and policy papers.

GENERAL

Anderson, Robert O. *Fundamentals of the Petroleum Industry*. Norman: University of Oklahoma Press, 1984.

Blair, John M. *The Control of Oil*. New York: Pantheon, 1976.

Bourmistrov, Anatoli, Frode Mellemvik, Alexei Bambulyak, Ove Gudmestad, Indra Overland, and Anatoly Zolotukhin. *International Arctic Petroleum Cooperation: Barents Sea Scenarios*. New York: Routledge, 2015.

Brown, Charles E. *World Energy Resources*. New York: Springer, 2002.

Bunter, Michael A. G. *The Promotion and Licensing of Petroleum Prospective Acreage*. New York: Kluwer Law International, 2002.

Clark, William R. *Petrodollar Warfare: Oil, Iraq, and the Future of the Dollar*. Gabriola Island, British Columbia: New Society, 2005.

Conaway, Charles F. *The Petroleum Industry: A Nontechnical Guide*. Tulsa, Okla.: PennWell, 1999.

Davis, Jerome. *The Changing World of Oil: An Analysis of Corporate Change and Adaptation*. Burlington, Vt.: Ashgate, 2006.

Falola, Toyin, and Ann Genova. *Politics of the Global Oil Industry: An Introduction*. Westport, Conn.: Praeger, 2005.

Forbes, R. J. *Bitumen and Petroleum in Antiquity*. Leiden: E. J. Brill, 1936.

———. *More Studies in Early Petroleum History, 1860–1880*. Leiden: E. J. Brill, 1959.

———. *Studies in Early Petroleum History*. Leiden: E. J. Brill, 1958.

Gillespie, Kate, and Clement M. Henry. *Oil in the New World Order*. Gainesville: University Press of Florida, 1995.

Gonzalez, George A. *Energy and the Politics of the North Atlantic*. Albany: State University of New York Press, 2013.

Hall, Charles A. S., and Kent A. Klitgaard. *Energy and the Wealth of Nations*. New York: Springer, 2013.

Helm, Dieter. *The New Energy Paradigm*. New York: Oxford University Press, 2007.

Jakle, John A., and Keith A. Sculle. *The Gas Station in America*. Baltimore, Md.: Johns Hopkins University Press, 2002.

Kolb, Robert W. *The Natural Gas Revolution: At the Pivot of the World's Energy Future*. Upper Saddle River, N.J.: Pearson, 2013.

Lambert, Jessica G., Charles A. S. Hall, and Stephen Balogh. *EROI of Global Energy Resources: Status, Trends, and Social Implications*. Syracuse, New York: State University of New York Environmental Science and Forestry Next Generation Energy Initiative, 2013.

Maugeri, Leonardo. *The Age of Oil: The Mythology, History, and Future of the World's Most Controversial Resource*. Westport, Conn.: Praeger, 2006.

McNally, Robert. *Crude Volatility: The History and the Future of Boom-Bust Oil Prices*. New York: Columbia University Press, 2017.

Mommer, Bernard. *Global Oil and the Nation State*. New York: Oxford University Press, 2002.

Odell, Peter R. *Oil and Gas: Crisis and Controversies, 1961–2000*. Brentwood, U.K.: Viking Multiscience, 2004.

———. *Oil and World Power: Background of the Oil Crisis*. 8th ed. New York: Viking Penguin, 1986.

Owen, Edgar Wesley. *Trek of the Oil Finders: A History of the Exploration for Petroleum: AAPG Memoir 6*. Tulsa, Okla.: American Association of Petroleum Geologists, 1975.

Parra, Francisco. *Oil Politics: A Modern History of Petroleum*. New York: I. B. Tauris, 2004.

Peimani, Hooman, ed. *Energy Security and Geopolitics in the Arctic: Challenges and Opportunities in the 21st Century*. Hackensack, N.J.: World Scientific, 2013.

Randall, Stephen J. *United States Foreign Oil Policy since World War I: For Profits and Security*. Montreal: McGill-Queen's University Press, 2005.

Redwood, Boverton. *Petroleum: A Treatise on the Geographical Distribution and Geological Occurrence of Petroleum and Natural Gas*. London: Charles Griffin, 1922.

Risjord, Norman K. *Populists and Progressives*. Lanham, Md.: Rowman and Littlefield, 2005.

Sakmar, Susan. *Energy for the 21st Century: Opportunities and Challenges for Liquefied Natural Gas (LNG)*. Northampton, Mass.: Edward Elgar, 2013.

Sarbu, Bianca. *Ownership and Control of Oil: Explaining Policy Choices across Producing Countries*. New York: Routledge, 2014.

Smil, Vaclav. *Energy at the Crossroads*. Cambridge, Mass.: MIT Press, 2005.

———. *Natural Gas: Fuel for the 21st Century*. New York: Wiley, 2015.

Sorkhabi, Rasoul. "Why the History of Oil Matters." *Oil Industry History* 17, no. 1 (2016): 141–147.

Tanaka, Yoshifumi. *The International Law of the Sea*. New York: Cambridge University Press, 2012.

Thompson, Helen. *Oil and the Western Economic Crisis*. New York: Palgrave Macmillan, 2017.

Yergin, Daniel. *The Prize: The Epic Quest for Oil, Money, and Power*. New York: Simon and Schuster, 1991.

Yoshino, Naoyuki, and Farhad Taghizadeh-Hesary. *Monetary Policy and the Oil Market*. New York: Springer, 2017.

Zalloum, Abdulhay Yahya. *Oil Crusades: America through Arab Eyes*. Ann Arbor, Mich.: Pluto, 2007.

HISTORICAL STATISTICS

Blanchard, Roger D. *The Future of Global Oil Production: Facts, Figures, Trends, and Projections by Region*. Jefferson, N.C.: McFarland, 2005.

BP, PLC. *BP Statistical Review of World Energy 2017*. London: BP PLC, 2017. Available at www.bp.com/statisticalreview.

Canadian Association of Petroleum Producers. *CAPP Statistical Handbook for Canada's Upstream Petroleum Industry*. Calgary, Alberta: Canadian Association of Petroleum Producers, 2017.

Carter, Susan B., Scott Sigmund Gartner, Michael R. Haines, Alan L. Olmstead, Richard Sutch, and Gavin Wright, eds. *Historical Statistics of the United States*. Millennial edition. New York: Cambridge University Press, 2006.

Day, David T., ed. *A Handbook of the Petroleum Industry*. New York: Wiley, 1922.

DeGolyer and MacNaughton. *Twentieth Century Petroleum Statistics: Historical Data*. Dallas, Tex.: DeGolyer and MacNaughton, 2004. Has statistics for 1918–1944.

———. *Twentieth Century Petroleum Statistics*. Dallas, Tex.: DeGolyer and MacNaughton, 2005. Has statistics for 1945–2006.

ENI, Strategies and Development Department. *ENI World Oil and Gas Review 2006*. Rome: ENI, 2006. Available at www.eni.it.

Fanning, Leonard M. *American Oil Operations Abroad*. New York: McGraw-Hill, 1947.

———, ed. *Our Oil Resources*. 2nd ed. New York: McGraw-Hill, 1950. Has statistics for 1857–1948.

International Center on Energy and the Environment. *Energy in Figures, Oil and Gas Sector, Venezuela 2014/2015*. Caracas: Instituto de Estudios Superiores de Administración (IESA).

International Energy Agency. *Key World Energy Statistics 2017*. Paris: International Energy Agency, 2007.

———. *Resources to Reserves*. Paris: International Energy Agency, 2013.

———. *World Energy Investment 2017*. Paris: International Energy Agency, 2017.

International Gas Union. *2017 World LNG Report*. Barcelona: International Gas Union, 2017.

Japan Automobile Manufacturers Association. *Motor Vehicle Statistics of Japan 2006*. Washington, D.C.: Japan Automobile Manufacturers Association, 2006.

Kourian, George Thomas, ed. *Datapedia of the United States, 1790–2005*. Lanham, Md.: Bernan, 2001.

Mitchell, B. R. *International Historical Statistics: Africa, Asia, and Oceania, 1750–1993*. New York: Stockton, 1998.

———. *International Historical Statistics: Europe, 1750–1993*. New York: Stockton, 1998.

———. *International Historical Statistics: North America, 1750–1993*. New York: Stockton, 1998.

Salvador, Amos. *Energy: A Historical Perspective and 21st Century Forecast*. AAPG Studies in Geology 54. Tulsa, Okla.: American Association of Petroleum Geologists, 2005.

United States Department of Energy, Energy Information Administration. *Annual Energy Review 2017*. Washington, D.C.: U.S. Government Printing Office, 2017.

United States Department of Transportation, Bureau of Transportation Statistics. *National Transportation Statistics 2017*. Washington, D.C.: U.S. Government Printing Office, 2017.

ENCYCLOPEDIAS AND DICTIONARIES

Cleveland, Cutler J., ed. *Encyclopedia of Energy*. 6 vols. Amsterdam: Elsevier, 2004.

Hiro, Dilip. *Dictionary of the Middle East*. New York: St. Martin's, 1996.

Ingham, John N. *Biographical Dictionary of American Business Leaders*. Westport, Conn.: Greenwood, 1983.

International Directory of Company Histories. Editors vary; volumes appear at irregular intervals; 90 volumes published between 1988 and 2008 by Thomson Gale, New York, and earlier St. James Press, Farmington Hills, Michigan. Each volume contains information on companies from many industries, usually including the petroleum industry. Considerable numbers of entries are also available online at www.fundinguniverse.com.

Mattar, Philip, ed. *Encyclopedia of the Modern Middle East and North Africa*. 2nd ed. 4 vols. New York: Thomson Gale, 2004.

THE PETROLEUM INDUSTRY BY REGION

United States

Asbury, Herbert. *The Golden Flood: An Informal History of America's First Oil Field*. New York: Knopf, 1942.

Bates, J. Leonard. *The Origins of Teapot Dome: Progressives, Parties, and Petroleum, 1909–1921*. Urbana: University of Illinois Press, 1963.

Black, Brian. *Petrolia: The Landscape of America's First Oil Boom*. Baltimore, Md.: Johns Hopkins University Press, 2000.

Boone, Lalia Phipps. "From 'Devil's Tar' to 'Liquid Gold.'" *American Speech* 33, no. 4 (1958): 304–307.

Boyd, Dan T. "Oklahoma Oil: Past, Present, and Future." *Oklahoma Geology Notes* 62, no. 3 (2002).

Bromley, Simon. *American Hegemony and World Oil*. University Park: Pennsylvania State University Press, 1991.

Burke, Bob, and Eric Dabney. *Prospects to Prosperity: The Story of Oklahoma's Oil and Gas Industry*. San Antonio, Tex.: HPN Books, 2012.

Camp, Mark J. *On a Handshake—Humble Beginnings to Global Impact: Ohio's Oil &Gas Industry*. San Antonio, Tex.: HPN Books, 2016.

Childs, William R. *The Texas Railroad Commission: Understanding Regulation in America to the Mid-Twentieth Century*. College Station: Texas A& M University Press, 2005.

Clark, James Anthony, and Michel Thomas Halbouty. *The Last Boom: The Exciting Saga of the Discovery of the Greatest Oil Field in America*. Fredericksburg, Tex.: Shearer, 1984.

———. *Spindletop: The True Story of the Oil Discovery That Changed the World*. Houston, Tex.: Gulf, 1995.

Clements, Kendrick A. "Herbert Hoover and Conservation, 1921–33." *American Historical Review* 89, no. 1 (1984): 67–88.

Cockrell, Alan. *Drilling Ahead: The Quest for Oil in the Deep South, 1945–2005*. Jackson: University Press of Mississippi, 2005.

Cone, Andrew, and Walter R. Johns. *Petrolia: A Brief History of the Pennsylvania Petroleum Region*. New York: D. Appleton, 1870.

Coppock, Mike. "Oil from the Land of the Midnight Sun: The Trans-Alaska Pipeline." *American History*, October 2004, 40–48.

Cox, Mike. *Texas Petroleum: The Unconventional History*. San Antonio: HPN Books, 2015.

Darrah, William C. *Pithole: The Vanished City*. Gettysburg, Pa.: William Culp Darrah, 1972.

Dixon, D. F. "Gasoline Marketing in the United States: The First Fifty Years." *Journal of Industrial Economics* 13, no. 1 (1964): 23–42.

Eaton, S. J. M. *Petroleum: A History of the Oil Region of Venango County, Pennsylvania*. Philadelphia: J. P. Skell, 1866.

Falgoux, Woody. *The Rise of the Cajun Mariners: The Race for Big Oil*. New York: Skyhorse Publishing, 2017.

Ford, Donald H. "Controlling the Production of Oil." *Michigan Law Review* 30, no. 8 (1932): 1170–1223.

Franks, Kenny A., and Paul F. Lambert. *Early California Oil: A Photographic History, 1865–1940*. College Station: Texas A&M University Press, 1965.

Giddens, Paul H. *Early Days of Oil: A Pictorial History of the Beginning of the Industry*. Titusville, Pa.: The Colonel Inc., 2000.

———. *Pennsylvania Petroleum, 1750–1972: A Documentary History*. Titusville: Pennsylvania Historical and Museum Commission, 1948.

Gillespie, Robert H. "Rise of the Texas Oil Industry: Part 1, Exploration at Spindletop." *Leading Edge* 14, no. 1 (1995): 22–24.

———. "Rise of the Texas Oil Industry: Part 2, Spindletop Changes the World." *Leading Edge* 14, no. 2 (1995): 113–117.

———. "Rise of the Texas Oil Industry: Part 3, Spindletop Field Confirmation, New Companies." *Leading Edge* 14, no. 9 (1995): 957–958.

———. "Rise of the Texas Oil Industry: Part 4, Guffey's New Company." *Leading Edge* 15, no. 1 (1996): 32–34.

———. "Rise of the Texas Oil Industry: Part 5, Guffey in Trouble, the Mellons Take Over." *Leading Edge* 15, no. 10 (1996): 1164–1166.

———. "Spindletop: Part 6, Competition for Guffey." *Leading Edge* 16, no. 7 (1997): 1067–1068.

Hidy, Ralph W. "Rise of Modern Industry: Government and the Petroleum Industry of the United States to 1911." *Journal of Economic History* 10, *Supplement: The Tasks of Economic History* (1950): 82–91.

Houser, Trevor, and Mohan Shashank. *Fueling Up: The Economic Implications of America's Oil and Gas Boom*. Washington, D.C.: Peterson Institute for International Economics, 2014.

Hughes, Dudley J. *Oil in the Deep South: A History of the Oil Business in Mississippi, Alabama, and Florida, 1859–1945*. Jackson: University Press of Mississippi, 1993.

J. B. F. "State Laws Curtailing Production of Petroleum." *Virginia Law Review* 17, no. 2 (1930): 173–179.

Knowles, Ruth S. *The Greatest Gamblers: The Epic of American Oil Exploration*. Norman: University of Oklahoma Press, 1978.

Marshall, J. Howard, and Norman L. Meyers. "Legal Planning of Petroleum Production: Two Years of Proration." *Yale Law Journal* 42, no. 5 (1933): 702–746.

McCartney, Laton. *The Teapot Dome Scandal: How Big Oil Bought the Harding White House and Tried to Steal the Country*. New York: Random House, 2008.

McKain, David L., and Bernard L. Allen. *Where It All Began: The Story of the People and Places Where the Oil Industry Began—West Virginia and Southeastern Ohio*. Parkersburg, W.V.: David L. McKain, 1994.

Nixon, Earl K. "The Petroleum Industry in Kansas." *Transactions of the Kansas Academy of Science* 51, no. 4 (1948): 369–424.

Nordhauser, Norman. "Origins of Federal Oil Regulation in the 1920's." *Business History Review* 47, no. 1 (1973): 53–71.

———. *The Quest for Stability: Domestic Oil Regulation, 1917–1935*. New York: Garland, 1979.

Olien, Diana D., and Roger M. Olien. *Oil in Texas: The Gusher Age, 1895–1945*. Austin: University of Texas Press, 2002.

Palmer Union Oil Company. *California's Greatest Industry*. San Francisco: Palmer Union Oil, 1912.

Prindle, David F. *Petroleum Politics and the Texas Railroad Commission*. Austin: University of Texas Press, 1981.

Rintoul, William. *Drilling Ahead: Tapping California's Richest Oil Fields*. Santa Cruz, Calif.: Valley, 1981.

———. *Spudding In: Recollections of Pioneer Days in the California Oil Fields*. Fresno, Calif.: Historical Society, 1976.

Rister, Carl Coke. *Oil! Titan of the Southwest*. Norman: University of Oklahoma Press, 1949.

Ross, Victor. *Evolution of the Oil Industry*. New York: Doubleday, Page, 1920.

Schruben, Francis W. *Wea Creek to El Dorado: Oil in Kansas, 1860–1920*. Columbia: University of Missouri Press, 1972.

Sherman, Jon. *Drake Well Museum and Park*. Mechanicsburg, Pa.: Stackpole, 2002.

Snider, L. C. "The Petroleum Resources of the United States." *Proceedings of the Academy of Political Science in the City of New York* 12, no. 1 (1926): 159–167.

Splawn, W. M. W. "Valuation and Rate Regulation by the Railroad Commission of Texas." *Journal of Political Economy* 31, no. 5 (1923): 675–707.

Stocking, George Ward. "Stabilization of the Oil Industry: Its Economic and Legal Aspects." *American Economic Review* 23, no. 1, *Supplement: Papers and Proceedings of the Forty-Fifth Annual Meeting of the American Economic Association* (1933): 55–70.

Theriot, Jason P. *American Energy, Imperiled Coast: Oil and Gas Development in Louisiana's Wetlands*. Baton Rouge: Louisiana State University Press, 2014.

Townsend, Henry H. "New Haven and the First Oil Well." *Oilfield Journal* 2 (2001): 16–31. Reprint of a paper originally presented before the New Haven Colony Historical Society on 20 November 1933.

Waples, David A. *Natural Gas Industry in Appalachia: A History from the First Discovery to the Maturity of the Industry*. Jefferson, N.C.: McFarland, 2005.

Werner, M. R., and John Star. *The Teapot Dome Scandal*. New York: Viking, 1959.

White, Gerald T. *Scientists in Conflict: The Beginnings of the Oil Industry in California*. San Marino, Calif.: Huntington Library, 1968.

Williamson, Harold F., and Arnold R. Daum. *The American Petroleum Industry: The Age of Illumination, 1859–1899*. Evanston, Ill.: Northwestern University Press, 1959.

———. *The American Petroleum Industry: The Age of Energy, 1899–1959*. Evanston, Ill.: Northwestern University Press, 1959.

Wright, William. *The Oil Regions of Pennsylvania*. New York: Harper and Brothers, 1865.

Yetiv, Steve A. *Myths of the Oil Boom: American National Security in a Global Energy Market*. New York: Oxford University Press, 2015.

Canada

Bott, Robert. *Evolution of Canada's Oil and Gas Industry*. Calgary: Canadian Centre for Energy Information, 2004. www.centreforenergy.com/ EE-OS.asp.

Breen, David H. *Alberta's Petroleum Industry and the Conservation Board*. Edmonton: University of Alberta Press, 1993.

Chastko, Paul. *Developing Alberta's Oil Sands: From Karl Clark to Kyoto*. Calgary: University of Calgary Press, 2004.

Easterbrook, Hugh G. J. *Canadian Economic History*. Toronto: University of Toronto Press, 1988.

Foster, Peter. *The Blue-Eyed Sheiks: The Canadian Oil Establishment*. Don Mills, Ont.: Collins, 1979.

Grant, Hugh, and Henry Thille. "Tariffs, Strategy, and Structure: Competition and Collusion in the Ontario Petroleum Industry, 1870–1880." *Journal of Economic History* 61, no. 2 (2001): 390–413.

Gray, Earle. *The Great Canadian Oil Patch: The Petroleum Era from Birth to Peak*. 2nd ed. Edmonton: June Warren, 2005.

Klassen, Henry C. *A Business History of Alberta*. Calgary: University of Calgary Press, 1999.

Pitt, Steven. "The Oil Springs of Ontario." *Legion Magazine*, 2001. www.legionmagazine.com/features/canadianreflections/01-11.asp.

Shaffer, Edward Harry. "Canada's Oil and Imperialism." *International Journal of Political Economy* 35, no. 2 (2006): 54–71.

Shrivastava, Meenal, and Lorna Stefanick. *Alberta Oil and the Decline of Democracy in Canada*. Edmonton: Athabasca University Press, 2015.

Smith, Donald B. *Calgary's Grand Story: The Making of a Prairie Metropolis*. Calgary: University of Calgary Press, 2005.

Latin America

Alfonzo, Juan Pablo Pérez. *El Pentágono Petrolero*. Caracas: Ediciones Revista Política, 1976.

———. *Hundiéndos en el Excremento del Diablo*. Caracas: Colección Venezuela Contemporánea, 1976.

Arnold, Ralph, George A. Macready, and Thomas W. Barrington. *The First Big Oil Hunt: Venezuela, 1911–1916*. New York: Vantage, 1960.

Bermúdez, Antonio J. *The Mexican National Petroleum Industry: A Case Study in Nationalization*. Stanford, Calif.: Stanford University Press, 1963.

Betancourt, Rómulo. *Venezuela: Oil and Politics*. Translated from the Spanish by Everett Bauman. Boston: Houghton Mifflin, 1979.

———. *Venezuela's Oil*. London: Allen and Unwin, 1978.

Borda, Joel Álvarez de la. *Crónica del Petróleo en México: De 1863 a Nuestros Días*. Mexico, D.F.: Petróleos Mexicanos, 2006.

Breglia, Lisa. *Living with Oil: Promises, Peaks, and Declines on Mexico's Gulf Coast*. Austin: University of Texas Press, 2013.

Brown, Jonathan C. "Domestic Politics and Foreign Investment: British Development of Mexican Petroleum, 1889–1911." *Business History Review* 61, no. 3 (1987): 387–416.

———. *Oil and Revolution in Mexico*. Berkeley: University of California Press, 1993.

———. "Why Foreign Oil Companies Shifted Their Production from Mexico to Venezuela during the 1920s." *American Historical Review* 90, no. 2 (1985): 362–385.

Brown, Jonathan C., and Alan Knight. *The Mexican Petroleum Industry in the Twentieth Century*. Austin: University of Texas Press, 1992.

Bye, Vegard. "Nationalization of Oil in Venezuela: Re-Defined Dependence and Legitimization of Imperialism." *Journal of Peace Research* 16, no. 1 (1979): 57–78.

Casasola, Miguel V., and Jesus Silva Herzog. *La Expropiación del Petróleo, 1936–1938*. México, D.F.: Fondo de Cultura Económica, 1988.

Cibils, Vicente Fretes, Marcelo M. Giugale, and José Roberto López-Cálix, eds. *Ecuador: An Economic and Social Agenda in the New Millennium*. Washington, D.C.: World Bank, 2003.

Concha, Alvaro. *La ConcesiónBarco*. Bogotá, Colombia: El Áncora Editores, 1981.

Coronel, Gustavo. *The Nationalization of the Venezuelan Oil Industry*. Lexington, Mass., Lexington Books, 1983.

Corten, Andre. "Le développement pétrolier au Mexique: Une stratégie d'organisation du rapport de la population à l'espace." *Canadian Journal of Political Science/Revue canadienne de science politique* 21, no. 3 (1988): 495–511.

Cote, Stephen. *Oil and Nation: A History of Bolivia's Petroleum Sector*. Morgantown: West Virginia University Press, 2016.

Di John, Jonathan. *From Windfall to Curse? Oil and Industrialization in Venezuela, 1920 to the Present*. University Park: Pennsylvania State University Press, 2009.

Donnellon, Anne. *Luis Giusti and the Transformation of Petróleos de Venezuela S.A.* Case BAB022, Babson College Institute for Latin American Business Studies. Boston: Harvard Business School, 1999.

Ellner, Steve, and Daniel Hellinger, eds. *Venezuelan Politics in the Chávez Era: Class, Polarization, and Conflict*. Boulder, Colo.: Lynne Rienner, 2003.

Ellner, Steve, and Miguel Tinker Salas. *Venezuela: Hugo Chávez and the Decline of an "Exceptional Democracy."* New York: Rowman and Littlefield, 2007. See chapters 3 and 4.

Ferrer, Arturo, Nicolas Serra, Ralph Holmer, and Earle F. Taylor. "History and Status of Petroleum Exploration in Uruguay." *Bulletin American Association of Petroleum Geologists* 36, no. 4 (1952): 677–687.

Fretes Gibils, Vicente, Marcelo M. Giugale, and Jose Roberto Lopez-Calix. *Ecuador: An Economic and Social Agenda in the New Millennium*. Washington, D.C.: World Bank, 2003.

Gadano, Nicolás. *História del Petróleo en la Argentina 1907–1955: Desde los Inicios Hasta la Caída de Perón*. Buenos Aires: Edhasa, 2006.

Gordillo García, Ramiro. *El Oro del Diablo? Ecuador: Historia del Petróleo*. Quito: Corporación Editora Nacional, 2005.

Gordon, Wendell Chaffee. *The Expropriation of Foreign-Owned Property in Mexico*. New York: Arno, 1976.

Grayson, G. W. *Politics of Mexican Oil*. Pittsburgh, Pa.: University of Pittsburgh Press, 1980.

Guzman Celis, Gilberto. *Cusiana, o la Bonanza al Reves*. Santafé de Bogotá, Colombia: Promover, 1994.

Hall, Linda B., and Don M. Coerver. "Oil and the Mexican Revolution: The Southwestern Connection." *Americas* 41, no. 2 (1984): 229–244.

Hart, John Mason. *Empire and Revolution: The Americans in Mexico since the Civil War*. Los Angeles: University of California Press, 2002.

Ingram, George M. *Expropriation of U.S. Property in South America: Nationalization of Oil and Copper Companies in Peru, Bolivia, and Chile*. New York: Praeger, 1974.

Jayne, Catherine E. *Oil, War, and Anglo-American Relations: American and British Reactions to Mexico's Expropriation of Foreign Oil Properties, 1937–1941*. Westport, Conn.: Greenwood, 2001.

Jones, Geoffrey, and Lisa Bud-Frierman. *Weetman Pearson and the Mexican Oil Industry*. Harvard Business School Case No. 9-804-085. Boston: Harvard Business School, 2007.

Knight, Alan. *Oil and Politics in Latin America: Nationalist Movements and State Companies*. New York: Cambridge University Press, 2007.

Kozloff, Nikolas. *Hugo Chávez: Oil, Politics, and the Challenge to the United States*. New York: Palgrave Macmillan, 2006.

Lieuwen, Edwin. *Petroleum in Venezuela: A History*. Berkeley: University of California Press, 1954.

López Portillo y Weber, José. *El Petróleo de México*. Mexico, D.F.: Fundo de Cultura Economica, 1975.

Macmahon, Arthur W., and W. R. Dittmar. "The Mexican Oil Industry since Expropriation." *Political Science Quarterly* 57, no. 1 (1942): 28–50.

———. "The Mexican Oil Industry since Expropriation II." *Political Science Quarterly* 57, no. 2 (1942): 161–188.

Martínez, Aníbal R. *Venezuelan Oil: Development and Chronology*. London: Elsevier Applied Science, 1989.

Martz, John D. *Politics and Petroleum in Ecuador*. New Brunswick, N.J.: Transaction, 1987.

McBeth, B. S. *Juan Vicente Gomez and the Oil Companies in Venezuela, 1908–1935*. New York: Cambridge University Press, 1983.

Meznar, Joan E. "Venezuela: New Military Populism." In *History behind the Headlines: The Origins of Conflicts Worldwide*, vol. 1, 303–310, ed. Sonia G. Benson, Nancy Matuszak, and Meghan Appel O'Meara. Detroit, Mich.: Gale, 2001.

Miller, Roy. "Small Business in the Peruvian Oil Industry: Lobitos Oilfields Limited before 1934." *Business History Review* 56, no. 3 (1982): 400–423.

Parra Luzardo, Gastón. *La Apertura Petrolera: Conflictos y Contradicciones*. Maracaibo, Venezuela: Universidad del Zulia, 1999.

Pedraja Toman, Rene de la. *Oil and Coffee: Latin American Merchant Shipping from the Imperial Era to the 1950s*. Westport, Conn.: Greenwood, 1998.

Petras, James F., Morris H. Morley, and Steven Smith. *The Nationalization of Venezuelan Oil*. New York: Praeger, 1977.

Philip, George. *Oil and Politics in Latin America*. Cambridge: Cambridge University Press, 1982.

Piquet, Rosélia, and Rodrigo Serra. *Petróleo e Região no Brasil: O Desafio da Abundância*. Rio de Janeiro: Garamond, 2007.

Powell, J. Richard. *The Mexican Petroleum Industry, 1938–1950*. Berkeley: University of California Press, 1956.

Ramírez Carreño, Rafael, and Walter Martinez. *El Rescate de PDVSA: Una Batalla por la Soberanía*. Caracas: República Bolivariana de Venezuela, Ministerio de Energía y Minas: Fondo Editorial Dario Ramírez, 2004.

Randall, Laura. *The Political Economy of Brazilian Oil*. Westport, Conn.: Praeger, 1993.

Sawyer, Suzana. *Crude Chronicles: Indigenous Politics, Multinational Oil, and Neoliberalism in Ecuador*. Durham, N.C.: Duke University Press, 2004.

Silva-Herzog, Jesús. *Historia de la Expropiación de las Empresas Petroleras*. Mexico, D.F.: Petróleos Mexicanos, 1988.

Smith, Peter D. *Oil and Politics in Modern Brazil*. Toronto: Macmillan, 1976.

Solberg, Carl E. *Oil and Nationalism in Argentina: A History*. Stanford, Calif.: Stanford University Press, 1979.

Standard Oil Company of Bolivia. *Confiscation: A History of the Oil Industry in Bolivia*. New York: Standard Oil Company of Bolivia, 1939.

Suchlicki, Jaime. *Mexico: From Montezuma to NAFTA, and Beyond*. New Brunswick, N.J.: Transaction, 2000.

Sullivan, William M., and Brian S. McBeth. *Petroleum in Venezuela: A Bibliography*. Boston: G. K. Hall, 1985.

Villegas Moreno, Gloria. *La Industria Petrolera en México: Cronología, 1857–1988*. México, D.F.: Petróleos Mexicanos, 1988.

Wirth, John D. *Latin American Oil Companies and the Politics of Energy*. Lincoln: University of Nebraska Press, 1985.

Wortley, B. A. "The Mexican Oil Dispute 1938–1946." *Transactions of the Grotius Society* 43, *Problems of Public and Private International Law, Transactions for the Year 1957* (1957): 15–37.

Europe

Austvik, Ole Gunnar, ed. *Norwegian Oil and Foreign Policy*. Sandvika, Norway: Vett and Viten, 1989.

Chapman, Keith. *North Sea Oil and Gas: A Geographical Perspective*. Sandvika, Norway: Vett and Viten, 1989.

Cuyvers, Luc. *The Strait of Dover*. Boston: Martinus Nijhoff, 1986.

Dicea, Oprea, and Michael E. Enachescu. "The Hydrocarbon Past, Present, and Future of Romania, the World's First Oil Producer." *Leading Edge* 19, no. 3 (2000): 321–324.

Dukes Wood Oil Museum. *The Story of North Sea Oil and Gas*. Nottinghamshire, U.K.: Dukes Wood Oil Museum, 2007. www.dukeswoodoilmuseum.co.uk/offshore%20history.htm.

Frank, Alison Fleig. *Oil Empire: Visions of Prosperity in Austrian Galicia*. Cambridge, Mass.: Harvard University Press, 2005.

Gardiner, Vince, and Hugh Matthews. *The Changing Geography of the United Kingdom*. 3rd ed. New York: Routledge, 2000.

Glennie, K. W., and A. Hurst. *NW Europe's Hydrocarbon Industry*. London: Geological Society of London, 1996.

Hahn-Pedersen, Morten. *A. P. Møller and the Danish Oil*. Copenhagen: Schultz Information, 1999.

Hellema, Duco, Cees Wiebes, and Toby Witte. *The Netherlands and the Oil Crisis*. Amsterdam: Amsterdam University Press, 2004.

Janicki, S. *The History* and *Present Conditions of the Oil Industry in Galicia*. London: Polish Press Bureau, 1921.

Jensen, Walter G. *Energy in Europe, 1945–1980*. London: G. T. Foulis, 1967.

Lind, T., and G. A. Mackay. *Norwegian Oil Policies*. Montreal: McGill-Queen's University Press, 1979.

Melamid, Alexander. "Oil and Gas in Austria." *Geographical Review* 85, no. 3 (1995): 382–383.

Olszewski, Eugeniusz, and Bokeslaw Orlowski. "Polish Galician Oil." *Technology and Culture* 14, no. 2, part 1 (1973): 173–174.

Pearton, Maurice. *Oil and the Romanian State*. Oxford: Clarendon, 1971.

Reader, W. J. "Oil for the West of England, 1889–1896: A Study in Competition." *Business History Review* 35, no. 1 (1961): 28–42.

Slevin, Amanda. *Gas, Oil and the Irish State: Understanding the Dynamics and Conflicts of Hydrocarbon Management*. Manchester: Manchester University Press, 2016.

Sozanski, Jozef, Stanislaw Kuk, Czeslaw Jaracz, and Piotr S. Dziadzkio. "How the Modern Oil and Gas Industry Was Born: Historical Remarks." In *The Carpathians and Their Foreland: Geology and Hydrocarbon Resources*, 811–834. AAPG Memoir 84. Tulsa, Okla.: American Association of Petroleum Geologists, 2006.

Stanciu, Laura. "Free-Standing Companies in the Oil Sector in Romania and Poland before 1948: Typologies and Competencies." *Business History* 42, no. 4 (2000): 27–66.

Stokes, Raymond G. "The Oil Industry in Nazi Germany, 1936–1945." *Business History Review* 59, no. 2 (1985): 254–277.

Toye, Richard. "The New Commanding Height: Labour Party Policy on North Sea Oil and Gas, 1964–1974." *Contemporary British History* 16, no. 1 (2002): 89–118.

Van Hoesel, Roger, and Rajneesh Narula. *Multinational Enterprises from the Netherlands*. New York: Routledge, 1999.

Vidal, John. "Sweden Plans to Be World's First Oil-Free Economy." *Guardian*, 8 February 2006.

Wall, E. H. "The Iranian-Italian Oil Agreement of 1957." *International and Comparative Law Quarterly* 7, no. 4 (1958): 736–752.

Zahariadis, Nikolaos. *Markets, States, and Public Policy: Privatization in Britain and France*. Ann Arbor: University of Michigan Press, 1995.

Russia and Former Soviet Union

Akiner, Shirin, and Anne Aldis. *The Caspian: Politics, Energy, and Security*. New York, Routledge, 2004.

Alekperov, Vagit. *Oil of Russia*. Minneapolis: East View Press, 2011.

Amineh, Mehdi Parvizi, and Henk Houweling, eds. *Central Eurasia in Global Politics: Conflict, Security, and Development*. 2nd ed. Boston: Brill, 2005.

Ascher, William, and N. S. Mirovitskaya. *The Caspian Sea: A Quest for Environmental Security*. Boston: Kluwer, 2000.

Aybak, Tunç, ed. *Politics of the Black Sea: Dynamics of Cooperation and Conflict*. New York: I. B. Tauris, 2001.

Barnes, Andrew Scott. *Owning Russia: The Struggle over Factories, Farms, and Power*. Ithaca, N.Y.: Cornell University Press, 2006.

Colton, Timothy J., and Stephen Holmes. *The State after Communism: Governance in the New Russia*. Lanham, Md.: Rowman and Littlefield, 2006.

Considine, Jennifer I., and William A. Kerr. *The Russian Oil Economy*. Northampton, Mass.: Edward Elgar, 2002.

Crandall, Maureen S. *Energy, Economics, and Politics in the Caspian Region: Dreams and Realities*. Westport, Conn.: Praeger, 2006.

Croissant, Cynthia. *Azerbaijan, Oil, and Geopolitics*. Commack, N.Y.: Nova Science, 1998.

Dekmejian, R. Hrair, and Hovann H. Simonian. *Troubled Waters: The Geopolitics of the Caspian Region*. New York: I. B. Tauris, 2003. See chapter 4.

Godzimirski, Jakub M., ed. *Russian Energy in a Changing World: What Is the Outlook for the Hydrocarbons Superpower?* Burlington, Vt.: Ashgate, 2013.

Gökay, Bülent, ed. *The Politics of Caspian Oil*. New York: Palgrave, 2001.

Goldman, Marshall I. *Petrostate: Putin, Power, and the New Russia*. New York: Oxford University Press, 2008.

———. *The Piratization of Russia: Russian Reform Goes Awry*. New York: Routledge, 2003.

Grace, John D. *Russian Oil Supply: Performance and Prospects*. New York: Oxford University Press, 2005.

Hoffman, David Emanuel. *The Oligarchs: Wealth and Power in the New Russia*. New York: Public Affairs, 2003.

Kaiser, Mark J., and Allan G. Pulsipher. "A Review of the Oil and Gas Sector in Kazakhstan." *Energy Policy* 35 (2007): 1300–1314.

Kapferer, Bruce. *Oligarchs and Oligopolies: New Formations of Global Power*. New York: Bergahn, 2005.

Krylov, N. A., A. A. Bokserman, and E. R. Stavrovsky. *The Oil Industry of the Former Soviet Union: Reserves and Prospects, Extraction, Transportation*. Amsterdam: Gordon and Breach, 1998.

Lal, Rollie. *Central Asia and Its Neighbors: Security and Commerce at the Crossroads*. RAND Monograph MG-440. Santa Monica, Calif.: RAND Corporation, 2006.

LeVine, Steve. *The Oil and the Glory: The Pursuit of Empire and Fortune on the Caspian Sea*. New York: Random House, 2007.

Luong, Pauline Jone, and Erika Weinthal. *Oil Is Not a Curse: Ownership Structure and Institutions in Soviet Successor States*. New York: Cambridge University Press, 2010.

Marvin, Charles. *The Region of Eternal Fire: An Account of a Journey to the Petroleum Region of the Caspian in 1883*. London: W. H. Allen, 1891.

Matveichuk, Alexander A. *Intersection of Oil Parallels: Historical Essays*. Moscow: Russian Oil and Gas Institute, 2004.

———. "Paris Oil Gambit: The First International Oil Congress Was Held in Paris 105 Years Ago." *Oil of Russia* 2005, no. 3 (2005): 64–67.

———. *Russia's First Oil Engineers*. Moscow: Interdialect, 2002. In Russian.

McCarthy, Daniel J., Sheila M. Puffer, and Stanislav V. Shekshnia. *Corporate Governance in Russia*. Northampton, Mass.: Edward Elgar, 2004.

McKay, John P. "Baku Oil and Transcaucasian Pipelines, 1883–1891: A Study in Tsarist Economic Policy." *Slavic Review* 43, no. 4 (1984): 604–623.

Mir-Babayev, Mir-Yusif. "The Role of Azerbaijan in the World's Oil Industry." *Oil Industry History* 12, no. 1 (2011): 109–123.

Mishin, Vladimir. "Breaking through the Oil Blockade: An Export Oil Pipeline Was Laid from the Caspian to the Black Sea Shore 75 Years Ago." *Oil of Russia* 2005, no. 3 (2005): 72–74.

Najman, Boris, Richard Pomfret, and Gaël Raballand, eds. *The Economics and Politics of Oil in the Caspian Basin: The Redistribution of Oil Revenues in Azerbaijan and Central Asia*. New York: Routledge, 2008.

Nazaroff, Alexander. "The Soviet Oil Industry." *Russian Review* 1, no. 1 (1941): 81–89.

"The Oil-Wells of Baku." *Science* 7, no. 158 (1886): 149–150.

Owen, Thomas C. *Russian Corporate Capitalism from Peter the Great to Perestroika*. New York: Oxford University Press, 1995.

Perovic, Jeronim, ed. *Cold War Energy: A Transnational History of Soviet Oil and Gas*. New York: Palgrave Macmillan, 2017.

Reynolds, Douglas B., and Marek Kolodziej. "Institutions and the Supply of Oil: A Case Study of Russia." *Energy Policy* 35 (2007): 939–949.

Rosner, Kevin. *Gazprom and the Russian State*. London: GMB, 2006.

Starr, S. Frederick, ed. *The New Silk Roads: Transport and Trade in Greater Central Asia*. Washington, D.C.: Central Asia–Caucasus Institute and Silk Road Studies Program, Johns Hopkins University, 2007.

Starr, S. Frederick, and Svante E. Cornell, eds. *The Baku–Tbilisi–Ceyhan Pipeline: Oil Window to the West*. Washington, D.C.: Central Asia–Caucasus Institute and Silk Road Studies Program, Johns Hopkins University, 2005.

Thompson, A. Beeby. *The Oil Fields of Russia and the Russian Petroleum Industry*. New York: D. Van Nostrand, 1904.

Tolf, Robert W. *The Russian Rockefellers: The Saga of the Nobel Family and the Russian Oil Industry*. Stanford, Calif.: Hoover Institution Press, 1976.

Von Laue, Theodore. *Sergei Witte and the Industrialization of Russia*. New York: Columbia University Press, 1963.

Middle East and North Africa

Abed, Ibrahim, and Peter Hellyer. *United Arab Emirates: A New Perspective*. London: Trident, 2001.

Abir, Mordechai. *Saudi Arabia: Government, Society, and the Gulf Crisis*. New York: Routledge, 1993.

Aissaoui, Ali. *Algeria: The Political Economy of Oil and Gas*. New York: Oxford University Press, 2001.

Allen, Calvin, and W. Lynn Rigsbee. *Oman under Qaboos: From Coup to Constitution, 1970–1996*. New York: Routledge, 2002.

Alnasrawi, Abbas. *Iraq's Burdens: Oil, Sanctions, and Underdevelopment*. Westport, Conn.: Greenwood, 2002.

Amuzegar, Jahangir A. *Iran's Economy under the Islamic Republic*. New York: I. B. Tauris, 1993.

Anderson, Irvine H. *Aramco, the United States, and Saudi Arabia: A Study of the Dynamics of Foreign Oil Policy, 1933–1950*. Princeton, N.J.: Princeton University Press, 1981.

Arayed, Jawad Salim al-. *A Line in the Sea: The Qatar v. Bahrain Border Dispute in the World Court*. Berkeley, Calif.: North Atlantic, 2003.

As-Saruri, Mustafa, and Rasoul Sorkhabi. "History of Petroleum Exploration in Yemen." *Oil Industry History* 15, no. 1 (2014): 147–176.

Banafe, Ahmed, and Rory MacLeod. *The Saudi Arabian Monetary Agency, 1952–2016: Central Bank of Oil*. New York: Palgrave Macmillan, 2017.

Bill, James A., and William Roger Louis. *Musaddiq, Iranian Nationalism, and Oil*. Austin: University of Texas Press, 1988.

Bronson, Rachel. *Thicker than Oil: America's Uneasy Partnership with Saudi Arabia*. New York: Oxford University Press, 2006.

Brown, Edward Hoagland. *The Saudi Arabia–Kuwait Neutral Zone*. Beirut: Middle East Research and Publication Center, 1963.

Cable, James. *Intervention at Abadan: Plan Buccaneer*. New York: St. Martin's, 1991.

Chisholm, Archibald H. T. *The First Kuwait Oil Concession Agreement*. London: Frank Cass, 1975.

Clarke, Angela. *Bahrain Oil and Development, 1929–1989*. Boulder, Colo.: International Research Center for Energy and Economic Development, 1990.

Crystal, Jill. *Oil and Politics in the Gulf: Rulers and Merchants in Kuwait and Qatar*. New York: Cambridge University Press, 1990.

Elwell-Sutton, L. P. *Persian Oil: A Study in Power Politics*. Westport, Conn.: Greenwood, 1975.

Ferrier, Ronald. "The Iranian Oil Industry." In *The Cambridge History of Iran*, vol. 7, *From Nadir Shah to the Islamic Republic*, chapter 18, ed. Peter Avery, Gavin Hambly, and Charles Melville. New York: Cambridge University Press, 1991.

Fesharaki, Fereidun. *Development of the Iranian Oil Industry: International and Domestic Aspects*. New York: Praeger, 1976.

Fieldhouse, D. K. *Western Imperialism in the Middle East, 1914–1958*. New York: Oxford University Press, 2006.

Gurney, Judith. *Libya: The Political Economy of Oil*. New York: Oxford University Press, 1996.

Hamilton, Charles W. *Americans and Oil in the Middle East*. Houston, Tex.: Gulf, 1962.

Horsnell, Paul. *The Mediterranean Basin in the World Petroleum Market*. New York: Oxford University Press, 2004.

Howard, Roger. *Iran Oil: The New Middle East Challenge to America*. New York: I. B. Tauris, 2006.

Kamal, Rami A. "Overview of the Petroleum Industry in the Middle East (1869–1950)." *Leading Edge* 24, no. 8 (2005): 818–822.

Keating, Aileen. *Mirage: Power, Politics, and the Hidden History of Arabian Oil*. Amherst, N.Y.: Prometheus, 2005.

Kechichian, Joseph A. *Succession in Saudi Arabia*. New York: Palgrave, 2001.

Kent, Marian. *Oil and Empire: British Policy and Mesopotamian Oil, 1900–1920*. London: Macmillan, 1976.

Kinzer, Stephen. *All the Shah's Men: An American Coup and the Roots of Middle East Terror*. Hoboken, N.J.: Wiley, 2003.

Kumins, Lawrence. *Iraq Oil: Reserves, Production, and Potential Revenues*. Report RS21626. Washington, D.C.: Congressional Research Service, 2005.

Kuwait Ministry of Oil. *Oil Documents*. www.moo.gov.kw. Collected historical documents; documents in Arabic have English explanations.

Little, Douglas. "Pipeline Politics: America, Tapline, and the Arabs." *Business History Review* 64, no. 2 (1990): 255–285.

Longrigg, S. H. *Oil in the Middle East*. Oxford: Oxford University Press, 1969.

Luciani, Giacomo. *The Oil Companies and the Arab World*. London: Croom Helm, 1984.

Mallakh, Raqaei El. *The Economic Development of the United Arab Emirates*. London: Croom Helm, 1981.

———. *Qatar Energy and Development*. London: Croom Helm, 1985.

Mejcher, Helmut. *Imperial Quest for Oil: Iraq, 1910–1928*. London: Ithaca, 1976.

Melamid, Alexander. *Oil and the Economic Geography of the Middle East and North Africa*. Princeton, N.J.: Darwin, 1991.

Mills, Robin. *Under the Mountains: Kurdish Oil and Regional Politics*. Oxford: Oxford Institute for Energy Studies (paper WPM 63), 2016.

Mosley, Leonard. *Power Play: Oil in the Middle East*. New York: Random House, 1973.

Nathan, J. Citino. *From Arab Nationalism to OPEC: Eisenhower, King Saud, and the Making of U.S. Saudi Relations*. Bloomington: Indiana University Press, 2002.

Oren, Michael B. *Power, Faith, and Fantasy: America in the Middle East, 1776 to the Present*. New York: Norton, 2007.

Otaiba, Mana Saeed al-. *Essays on Petroleum*. London: Croom Helm, 1982. Focuses on Abu Dhabi.

Othman, Wasser al-. *With Their Bare Hands: The Story of the Oil Industry in Qatar*. Harlow: Longman, 1984.

Perry, Jane, and Clark Carey. "Iran and Control of Its Oil Resources." *Political Science Quarterly* 89, no. 1 (1974): 147–174.

Philby, H. St. J. B. *Arabian Oil Ventures*. Washington, D.C.: Middle East Institute, 1965.

Ramazani, Rouhollah K. *The Persian Gulf and the Strait of Hormuz*. Alphen aan den Rijn, Netherlands: Sijthoff and Noordhoff, 1979.

Roosevelt, Kermit. *Countercoup: The Struggle for the Control of Iran*. New York: McGraw-Hill, 1979.

Rosenberg, Robert L. "Qum-1956: A Misadventure in Iranian Oil." *Business History Review* 49, no. 1 (1975): 81–104.

Sabah, Youssif S. Fadel al-. *The Oil Economy of Kuwait*. New York: Routledge, 1980.

Sandwick, John A. *The Gulf Cooperation Council: Moderation and Stability in an Interdependent World*. Boulder, Colo.: Westview, 1987.

Schofield, R. N., and P. L. Toye, eds. *Arabian Gulf Oil Concessions, 1911–1953*. 12 vols. Farnham Common, U.K.: Archive Editions, 1989.

Shwadran, Benjamin. *Middle East Oil: Issues and Problems*. Cambridge, Mass.: Schenkman, 1977.

Sicker, Martin. *The Middle East in the Twentieth Century*. Westport, Conn.: Praeger, 2001.

Simmons, Matthew. *Twilight in the Desert: The Coming Saudi Oil Shock and the World Economy*. Hoboken, N.J.: Wiley, 2005.

Simons, Geoffrey Leslie. *Libya: The Struggle for Survival*. New York: Macmillan, 1993.

Smith, Benjamin. *Hard Times in the Lands of Plenty: Oil Politics in Iran and Indonesia*. Ithaca, N.Y.: Cornell University Press, 2007.

Sorkhabi, Rasoul. "Pre-Modern History of Bitumen, Oil, and Gas in Persia (Iran)." *Oil Industry History* 6, no. 1 (2005): 153–177.

Stegner, Wallace. *Discovery! The Search for Saudi Arabian Oil*. Vista, Calif.: Selwa, 2007. Reprint of a book originally published by Middle East Export Press, Beirut, Lebanon, in 1971.

Stern, Roger. "The Iranian Petroleum Crisis and United States National Security." *Proceedings of the National Academy of Sciences* 104 (2007): 377–382.

Stocking, George W. *Middle East Oil*. London: Allen Lane, 1970.

Twitchell, Karl S. *Saudi Arabia: With an Account of the Development of Its Natural Resources*. Princeton, N.J.: Princeton University Press, 1958.

Ulrichsen, Kristian Coates. *Insecure Gulf: The End of Certainty and the Transition to the Post-Oil Era*. New York: Oxford University Press, 2015.

Uthman, Nasir Muhammad. *With Their Bare Hands: The Story of the Oil Industry in Qatar*. New York: Longman, 1984.

Vandewalle, Dirk J. *Libya since Independence: Oil and State Building*. London: I. B. Tauris, 1998.

Vassiliev, Alexei. *The History of Saudi Arabia*. New York: New York University Press, 2000.

Vitalis, Robert. *America's Kingdom: Mythmaking on the Saudi Oil Frontier*. Stanford, Calif.: Stanford University Press, 2007.

Waddams, Frank C. *The Libyan Oil Industry*. Baltimore, Md.: Johns Hopkins University Press, 1980.

Ward, Thomas E. *Negotiations for Oil Concessions in Bahrain, El Hasa (Saudi Arabia), the Neutral Zone, Qatar, and Kuwait*. New York: Ardlee Service, 1965.

Yousef, Mohamed bin Musa al-. *Oil and the Transformation of Oman, 1970–1995*. London: Stacey, 1995.

Zedalis, Rex. *Claims against Iraqi Oil and Gas: Legal Considerations and Lessons Learned*. New York: Cambridge University Press, 2010.

Sub-Saharan Africa

Apter, Andrew. *The Pan-African Nation: Oil and the Spectacle of Culture in Nigeria*. Chicago: University of Chicago Press, 2005.

Bradshaw, York W., and Stephen N. Ndegwa. *The Uncertain Promise of Southern Africa*. Bloomington: Indiana University Press, 2000.

Falola, Toyin. *The History of Nigeria*. Westport, Conn.: Greenwood, 1999.

Frynas, Jedrzej Georg. *Oil in Nigeria: Conflict and Litigation between Oil Companies and Village Communities*. Berlin: LIT Verlag, 2000.

Genova, Ann, and Toyin Falola. "Oil in Nigeria: A Bibliographical Reconnaissance." *History in Africa* 30 (2003): 133–156.

Ghazvinian, John. *Untapped: The Scramble for Africa's Oil*. New York: Harcourt, 2007.

Hery, Jennifer. *Le Soudan Entre Pétrole et Guerre Civile*. Paris: L'Harmattan, 2003.

Hodges, Tony. *Angola: Anatomy of an Oil State*. Bloomington: Indiana University Press, 2004.

Khan, Sarah Ahmad. *Nigeria: The Political Economy of Oil*. New York: Oxford University Press, 1994.

Kounou, Michel. *Pétrole et Pauvreté au Sud du Sahara*. Yaounde, Cameroon: Éditions CLÉ, 2006.

Leonard, Lori. *Life in the Time of Oil: A Pipeline and Poverty in Chad*. Bloomington, Indiana: Indiana University Press, 2016.

Massengo, Gualbert-Brice. *L'Économie Petrolière du Congo*. Paris: L'Harmattan, 2004.

Nwankwo, Arthur A. *After Oil, What Next? Oil and Multinationals in Nigeria*. Enugu, Nigeria: Fourth Dimension, 1982.

Omejeh, Kenneth. *High Stakes and Stakeholders: Oil Conflict and Security in Nigeria*. Aldershot, U.K.: Ashgate, 2006.

Omoweh, Daniel A. *Shell Petroleum Development Company, the State, and Underdevelopment of Nigeria's Niger Delta: A Study in Environmental Degradation*. Trenton, N.J.: Africa World, 2005.

Onoh, J. K. *The Nigerian Oil Economy*. London: Croom Helm, 1983.

Panford, Kwamina. *Africa's Natural Resources and Underdevelopment: How Ghana's Petroleum Can Create Sustainable Economic Prosperity*. New York: Palgrave Macmillan, 2017.

Peel, Michael. *A Swamp Full of Dollars: Pipelines and Paramilitaries at Nigeria's Oil Frontier*. Chicago: Lawrence Hill Books, 2010.

Roberts, Adam. *The Wonga Coup: Guns, Thugs, and a Ruthless Determination to Create Mayhem in an Oil-Rich Corner of Africa*. New York: Public Affairs, 2006.

Soremekun, Kayode. *Perspectives on the Nigerian Oil Industry*. Lagos, Nigeria: Amkra, 1995.

Taylor, Ian. "China's Oil Diplomacy in Africa." *International Affairs* 82, no. 5 (2006): 937–959.

Ukaga, Okechukwu, Ukoha O. Ukiwo, and Ibaba Samuel Ibaba, eds. *Natural Resources, Conflict, and Sustainable Development: Lessons from the Niger Delta*. New York: Routledge, 2012.

United Africa Company. "The Future of the Nigerian Oil Industry." *African Affairs* 47, no. 186 (1948): 41–51.

Vicente, São. *Petróleo, Política Económica, e Estratégia em Angola*. Luanda, Angola: INALD, 1995.

World Bank. *Angola: Oil, Broad-Based Growth, and Equity*. Washington, D.C.: World Bank, 2007.

Yates, Douglas A. *The Scramble for African Oil: Oppression, Corruption, and War for Control of Africa's Natural Resources*. London: Pluto Press, 2012.

Asia

Barnes, Philip. *Indonesia: The Political Economy of Energy*. New York: Oxford University Press, 1995.

Burgos Caceres, Sigfrido. *China's Interests in the South China Sea: Power and Resources*. New York: Routledge, 2014.

Carlson, Sevinc. *Indonesia's Oil*. Boulder, Colo.: Westview, 1977.

Casson, Mark, and Micheal Kidron. *Foreign Investments in India*. New York: Oxford University Press, 1965.

Cole, Bernard D. *China's Quest for Great Power: Ships, Oil, and Foreign Policy*. Annapolis: Naval Institute Press, 2016.

Eckaus, R. S. *Unraveling the Chinese Oil Puzzle*. Working Paper 04-022 WP. Cambridge, Mass.: Center for Energy and Environmental Policy Research, 2004.

Huang, Jing, and Andrew Billo, eds. *Territorial Disputes in the South China Sea: Navigating Rough Waters*. New York: Palgrave Macmillan, 2015.

Hyde, Charles E. "Evaluating Mergers in the Australian Petroleum Industry." *Business History Review* 45, no. 4 (1971): 452–473.

Jones, G. G. "The State and Economic Development in India 1890–1947: The Case of Oil." *Modern Asian Studies* 13, no. 3 (1979): 353–375.

Kambara, Tatsu. "The Petroleum Industry in China." *China Quarterly* 60 (December 1974): 699–719.

Kambara, Tatsu, and Christopher Howe. *China and the Global Energy Crisis: Development and Prospects for China's Oil and Natural Gas*. Northampton, Mass.: Edward Elgar, 2007.

Kaul, Hriday Nath. *K. D. Malaviya and the Evolution of India's Oil Policy*. New Delhi: Allied, 1991.

Poley, J. Ph. *Eroïca: The Quest for Oil in Indonesia (1850–1898)*. Boston: Kluwer, 2000.

Pugach, Noel H. "Standard Oil and Petroleum Development in Early Republican China." *Business History Review* 45, no. 4 (1971): 452–473.

Singh, P. N., ed. *Oil Industry in India: Its Achievements and the Challenges Ahead*. Bombay: Forum of Asian Managers, Xavier Institute of Management, 1986.

Smith, Benjamin. *Hard Times in the Lands of Plenty: Oil Politics in Iran and Indonesia*. Ithaca, N.Y.: Cornell University Press, 2007.

Sykes, Trevor. *The Money Miners: The Great Australian Mining Boom*. St. Leonards, NSW: Allen and Unwin, 1995.

OPEC AND OAPEC

Amuzegar, Jahangir. *Managing the Oil Wealth: OPEC's Windfalls and Pitfalls*. New York: I. B. Tauris, 2001.

Danielsen, Albert L. *The Evolution of OPEC*. New York: Harcourt Brace Jovanovich, 1982.

Evans, John. *OPEC, Its Member States, and the World Energy Markets*. London: Longmans, 1986.

Ghanem, Shukri M. *OPEC: The Rise and Fall of an Exclusive Club*. London: Kegan Paul, 1986.

Maachou, Abdelkader. *OAPEC: An International Organization for Economic Cooperation and an Instrument for Regional Integration*. Translated from the French by Anthony Melville. New York: St. Martin's, 1983.

Ramady, Mohamed, and Wael Mahdi. *OPEC in a Shale Oil World: Where to Next?* New York: Springer, 2015.

Rouhani, Fuad. *A History of OPEC*. New York: Praeger, 1971.

Seghyer, Mohamed al-. "OPEC: Tested by Fire, Prepared for the Future." PhD thesis, University of Exeter, 2000.

Skeet, Ian. *OPEC: Twenty-Five Years of Prices and Politics*. New York: Cambridge University Press, 1991.

PETROLEUM COMPANIES

Allaud, L. A., and M. Martin. *Schlumberger: History of a Technique*. New York: Wiley, 1977.

American Oil and Gas Historical Society. *American Oil and Gas Families: Appalachian Basin Independents*. Washington, D.C.: American Oil and Gas Historical Society, 2004.

———. *American Oil and Gas Families: East Texas Independents*. Washington, D.C.: American Oil and Gas Historical Society, 2004.

Anadarko Petroleum Corporation. *Thirty Years of History*. Houston, Tex.: Anadarko, 1989.

Anderson, Irvine H. *Aramco, the United States, and Saudi Arabia: A Study of the Dynamics of Foreign Oil Policy, 1933–1950*. Princeton, N.J.: Princeton University Press, 1981.

———. *The Standard-Vacuum Oil Company and United States East Asian Policy, 1933–1941*. Princeton, N.J.: Princeton University Press, 1975.

Baker Hughes Inc. *Baker Hughes: 100 Years of Service*. Special issue of Baker Hughes *In Depth Magazine* 13, no. 2 (2007).

Bamberg, James H. *History of British Petroleum*. Vol. 2, *The Anglo-Iranian Years, 1928–1954*. New York: Cambridge University Press, 1994.

———. *History of British Petroleum*. Vol. 3, *British Petroleum and Global Oil, 1950–1975: The Challenge of Nationalism*. New York: Cambridge University Press, 2000.

Banham, Russ. *Conoco: 125 Years of Energy*. Lyme, Conn.: Greenwich, 2000.

Beaton, Kendall. *Enterprise in Oil: A History of Shell in the United States*. New York: Appleton-Century-Crofts, 1957.

Bendeck Olivella, Jorge. *Ecopetrol: Historia de Una Gran Empresa*. Bogotá, Colombia: Ediciones Punto Llano, 1993.

Bowker, Jeff. "A Well Ordered Reality: Aspects of the Development of Schlumberger, 1920–39." *Social Studies of Science* 17, no. 4 (1987): 611–655.

Brauer, Dora Lee. "Trade Names in the Petroleum Industry." *American Speech* 10, no. 2 (1935): 122–128.

Brown, Anthony Cave. *Oil, God, and Gold: The Story of Aramco and the Saudi Kings*. Boston: Houghton Mifflin, 1999.

Cameron, Rondo, and V. I. Bovykin, eds. *International Banking, 1870–1914*. New York: Oxford University Press, 1991.

Coll, Steve. *The Taking of Getty Oil: The Full Story of the Most Spectacular—and Catastrophic—Takeover of All Time*. New York: Scribner, 1987.

Continental Oil Company. *Conoco: The First One Hundred Years*. New York: Dell, 1975.

Corduas, Claudio. *Impresa e Cultura: L'utopia dell'ENI*. Milan: Mondadori, 2006.

Corley, T. A. B. *A History of the Burmah Oil Company*. Vol. 2, *1924–1966*. London: Heinemann, 1988.

Cummins, Ian, and John Beasant. *Shell Shock: The Secrets and Spin of an Oil Giant*. Edinburgh: Mainstream, 2005.

Davis, Robert H., and William E. Hale. *One Hundred Twenty-Five Years of History* [ExxonMobil]. Irving, Tex.: ExxonMobil, 2007.

Dedmon, Emmett. *Challenge and Response: A Modern History of the Standard Oil Company (Indiana)*. Chicago: Mobium, 1984.

Earle, Edward Meade. "The Turkish Petroleum Company: A Study in Oleaginous Diplomacy." *Political Science Quarterly* 39, no. 2 (1924): 265–279.

Ferrier, Ronald W. *History of British Petroleum*. Vol. 1, *The Developing Years, 1901–1932*. New York: Cambridge University Press, 1982.

Giddens, Paul H. "Historical Origins of the Adoption of the Exxon Name and Trademark." *Business History Review* 47, no. 3 (1973): 353–366.

———. *Standard Oil Company (Indiana): Oil Pioneer in the Middle West*. New York: Appleton-Century-Crofts, 1955.

Grayson, Leslie E. *National Oil Companies*. New York: Wiley, 1981.

Halliburton Company. *Halliburton Company: Seventy-Five Years of Leadership*. Dallas, Tex.: Halliburton, 1994.

Haynes, H. J. *Standard Oil Company of California: 100 Years Helping to Create the Future*. New York: Newcomen Society in North America, 1980.

Henderson, Wayne, and Scott Benjamin. *Standard Oil: The First 125 Years*. Osceola, Wis.: Motorbooks International, 1996.

Hendrix, Paul. *Sir Henri Deterding and Royal Dutch-Shell: Changing Control of World Oil, 1900–1940*. Bristol, U.K.: Bristol Academic, 2002.

Hidy, Ralph W., George S. Gibb, and Henrietta M. Larson. *History of Standard Oil Company (New Jersey): New Horizons, 1927–1950*. New York: Harper, 1955.

Hidy, Ralph W., and Muriel E. Hidy. *History of Standard Oil Company (New Jersey): Pioneering in Big Business, 1882–1911*. New York: Harper, 1955.

Howarth, Steven. *A Century in Oil: The "Shell" Transport and Trading Company, 1897–1997*. London: Weidenfeld and Nicolson, 1997.

Howarth, Steven, and Joost Jonker. *A History of Royal Dutch Shell*. Vol. 2, *Powering the Hydrocarbon Revolution, 1939–1945*. New York: Oxford University Press, 2007.

Imperial Oil Limited. *The Story of Imperial Oil*. Toronto: Imperial Oil Limited, 1991.

Indonesia Project. *Stanvac in Indonesia*. Cambridge, Mass.: National Planning Association, 1957.

Jack, Marian. "The Purchase of the British Government's Shares in the British Petroleum Company 1912–1914." *Past and Present* 39 (1968): 139–168.

James, Marquis. *The Texaco Story: The First Fifty Years, 1902–1952*. New York: Texas Company, 1953.

Johnson, Arthur M. *The Challenge of Change: The Sun Oil Company, 1945–1977*. Columbus: Ohio State University Press, 1983.

Jonker, Joost, and Jan Luiten van Zanden. *A History of Royal Dutch Shell*. Vol. 1, *From Challenger to Joint Industry Leader, 1890–1939*. New York: Oxford University Press, 2007.

Kaufman, Allen, and Gordon Walker. "The Strategy-History Connection: The Case of Exxon." *Public Historian* 8, no. 2 (1986): 23–39.

Larson, Henrietta M., and Kenneth Wiggins Porter. *History of the Humble Oil and Refining Company*. New York: Harper, 1959.

Longhurst, Henry. *Adventure in Oil: The Story of British Petroleum*. London: Sidgwick and Jackson, 1959.

Lyon, Jim. *Dome Petroleum: The Inside Story of Its Rise and Fall*. New York: Beaufort, 1983.

Marcel, Valerie, and John V. Mitchell. *Oil Titans: National Oil Companies in the Middle East*. Washington, D.C.: Brookings Institution, 2006.

Massie, Joseph L. *Blazer and Ashland Oil: A Study in Management*. Lexington: University Press of Kentucky, 1960.

Matveichuk, Alexander A. "The Trail Blazers: 130 Years Ago Russia Formed the World's First Vertically Integrated Joint-Stock Company Baku Oil Society." *Oil of Russia* 2003, no. 4 (2003): 60–63.

McDonald, James. *Gettyrama: Little Known Facts about J. Paul Getty and More*. New York: Authors Choice, 2003.

Montague, Gilbert Holland. "The Later History of the Standard Oil Company." *Quarterly Journal of Economics* 17, no. 2 (1903): 293–325.

———. *The Rise and Progress of the Standard Oil Company*. Kitchener, Ont.: Batoche, 2003. Reprint of 1904 edition.

———. "The Rise and Supremacy of the Standard Oil Company." *Quarterly Journal of Economics* 16, no. 2 (1902): 265–292.

National Petroleum News. "The Progeny of Standard Oil." *National Petroleum News*, December 2000, 18.

Nawwab, Ismail I., and Peter C. Speers. *Aramco and Its World: Arabia and the Middle East*. Dhahran, Saudi Arabia: Aramco, 1980.

Payne, Darwin. *Initiative in Energy: The Story of Dresser Industries, 1880–1978*. New York: Simon and Schuster, 1979.

Pederson, Barbara L. *Unocal, 1890–1990: A Century of Spirit*. Los Angeles: Unocal Corporation, 1990.

Petzinger, Thomas. *Oil and Honor: The Texaco-Pennzoil Wars*. New York: Putnam, 1987.

Pratt, Joseph A. *Prelude to Merger: A History of Amoco Corporation, 1973–1998*. McLean, Va.: Hart, 2000.

Pratt, Joseph A., Tyler Priest, and Christopher J. Castaneda. *Offshore Pioneers: Brown and Root and the History of Offshore Oil and Gas*. Houston, Tex.: Gulf, 1997.

Rebah, Abdelatif. *Sonatrach: Une Enterprise pas comme les Autres*. Algiers: Casbah, 2006.

Rodengen, Jeffrey L. *The Legend of Halliburton*. Fort Lauderdale, Fla.: Write Stuff, 1996.

Rondot, Jean. *La Compagnie Française des Pétroles: Du Franc-Or au PetroleFranc*. Paris: Librairie Plon, 1962; reprinted, New York: Arno Press, 1977.

Ruback, Richard S. "The Cities Service Takeover: A Case Study." *Journal of Finance* 38, no. 2 (1982): 319–330.

Sampson, Anthony. *The Seven Sisters: The Great Oil Companies and the World They Made*. London: Hodder and Stoughton, 1975.

Shannon, James. *Texaco and the $10 Billion Jury*. Englewood Cliffs, N.J.: Prentice Hall, 1988.

Sinclair Oil. *A Great Name in Oil: Sinclair through 50 Years*. New York: F. W. Dodge/McGraw-Hill, 1966.

Sluyterman, Keetie. *A History of Royal Dutch Shell*. Vol. 3, *Keeping Competitive under Global Challenges, 1973–2007*. New York: Oxford University Press, 2007.

Soto-Cuervo, Arturo. "The New Pemex." *Leading Edge* 16, no. 12 (1997): 1759–1760.

Spence, Hartzell. *Portrait in Oil: How the Ohio Oil Company Grew to Become Marathon*. New York: McGraw-Hill, 1962.

Sun Company. *Centennial Celebration: The Story of Sun Company*. Radnor, Pa.: Sun, 1986.

Tarbell, Ida M. *The History of the Standard Oil Company*. New York: Peter Smith, 1950. Reprint of original work published 1904; available also in a 2003 briefer form edited by David M. Chalmers, published by Dover, New York.

Tetreault, Mary Ann. *The Kuwait Petroleum Corporation and the Economics of the New World Order*. Westport, Conn.: Greenwood, 1995.

Thompson, Craig. *Since Spindletop: A Human Story of Gulf's First Half-Century*. Pittsburgh, Pa.: Gulf Oil, 1951.

Thompson, Eric V. *A Brief History of Major Oil Companies in the Gulf Region*. Petroleum Archives Project, Arabian Peninsula and Gulf Studies Program, University of Virginia. www.virginia.edu/igpr/APAG/apagoilhistory.html.

Waddams, Frank C. *The Libyan Oil Industry*. Baltimore, Md.: Johns Hopkins University Press, 1980.

Wall, Bennett H. *Growth in a Changing Environment: A History of Standard Oil Company (New Jersey), Exxon Corporation, 1950–1975*. New York: HarperCollins, 1989.

Welty, Earl M., and Frank J. Taylor. *The Black Bonanza: The Fabulous Life and Times of the Union Oil Company of California*. New York: McGraw-Hill, 1958.

———. *The 76 Bonanza: The Fabulous Life and Times of the Union Oil Company of California.* Menlo Park, Calif.: Lane, 1966.

Wertz, William C., ed. *Phillips: The First 66 Years.* Bartlesville, Okla.: Phillips Petroleum, 1983.

Western Geophysical. *A Brief History of Western Geophysical.* Houston, Tex.: Western Geophysical, 1983.

White, Gerald Taylor. *Formative Years in the Far West: A History of Standard Oil Company of California and Predecessors through 1919.* New York: Appleton-Century-Crofts, 1962.

Wood, Quentin E. *Quaker State Roots Go Deep into the World's First Oilfield.* New York: Newcomen Society of America, 1987.

Yale Law Journal Staff. "Use of Confusing Brand Names by Standard Oil Companies in the Marketing of Gasoline." *Yale Law Journal* 48, no. 2 (1938): 332–339.

BIOGRAPHIES

Aaseng, Nathan. *Business Builders in Oil.* Minneapolis, Minn.: Oliver, 2000.

Accorinti, Giuseppe. *Quando Mattei Era L'Impresa Energetica—io c'ero.* Hacca Matelica, Italy: Halley Editrice, 2006.

Anatoliev, Alexander. "Engineer Shukhov's Moscow Heritage." *Oil of Russia*, no. 4 (2003): 64–67.

Bacon, R. H. *The Life of Lord Fisher of Kilverstone.* 2 vols. Garden City, N.Y.: Doubleday, 1929.

Brady, Kathleen. *Ida Tarbell: Portrait of a Muckraker.* New York: Seaview-Putnam, 1984.

Brice, William R. *Myth, Legend, Reality: Edwin Laurentine Drake and the Early Oil Industry.* Oil City, Pa.: Oil Region Alliance, 2009.

Chernow, Ron. *Titan: The Life of John D. Rockefeller Senior.* New York: Random House, 1998.

Davis, Margaret L. *Dark Side of Fortune: Triumph and Scandal in the Life of Oil Tycoon Edward L. Doheny.* Berkeley: University of California Press, 1998.

Davis, Tina, and Jessica Resnick-Ault. *Hess: The Last Oil Baron.* New York: Wiley Bloomberg, 2016.

De Chair, Somerset. *Getty on Getty: A Man in a Billion.* London: Cassell, 1989.

Deterding, Henri, and Stanley Naylor. *An International Oilman.* New York: Harper, 1934.

Fant, Kenne. *Alfred Nobel: A Biography.* New York: Arcade, 1993.

Frankel, P. H. *Mattei: Oil and Power Politics*. London: Faber and Faber, 1966.

Harper, John A. "The Incredible John E. Carll: The World's First Petroleum Geologist and Engineer." *Oilfield Journal* 2 (2001): 2–15.

Harris, Kenneth. *The Wildcatter: A Portrait of Robert O. Anderson*. New York: Weidenfeld and Nicolson, 1987.

Henriques, Robert. *Marcus Samuel, First Viscount Bearsted and Founder of the "Shell" Transport and Trading Company, 1853–1927*. London: Barrie and Rockliff, 1960.

———. *Sir Robert Waley Cohen, 1877–1952*. London: Secker and Warburg, 1966.

Hewins, Ralph. *Mr. Five Percent: The Story of Calouste Gulbenkian*. New York: Rinehart, 1958.

———. *The Richest American: J. Paul Getty*. New York: Dutton, 1960.

Hurt, Harry, III. *Texas Rich: The Hunt Dynasty from the Early Oil Days through the Silver Crash*. New York: Norton, 1981.

Koskoff, David E. *The Mellons: The Chronicle of America's Richest Family*. New York: Thomas Y. Crowell, 1978.

MacNaughton, Lewis W. "E. L. DeGolyer, Father of Applied Geophysics." *Science* 125, no. 3243 (1957): 338–339.

Massie, J. L. *Blazer and Ashland Oil: A Study in Management*. Lexington: University Press of Kentucky, 1960.

Masters, John, and Paul Grescoe. *Secret Riches: Adventures of an Unreformed Oilman*. Calgary: Gondolier, 2004.

Mathews, John J. *Life and Death of an Oilman: The Career of E. W. Marland*. Norman: University of Oklahoma Press, 1951.

McGloughlin, Leslie. *Ibn Saud: Founder of a Kingdom*. New York: St. Martin's, 1993.

Merrills, J. G., and Gerald Mitzmaurice. *Judge Sir Gerald Fitzmaurice and the Discipline of International Law*. Boston: Kluwer Law International, 1998.

Miller, Russell. *The House of Getty*. London: Michael Joseph, 1985.

Mitcham, Allison. *The Prophet of the Wilderness: Abraham Gesner*. Hantsport, Nova Scotia: Lancelot, 1995.

Monroe, Elizabeth. *Philby of Arabia*. London: Faber and Faber, 1973.

Pickens, T. Boone. *Boone*. Boston: Houghton Mifflin, 1987.

Roberts, Glyn. *The Most Powerful Man in the World: The Life of Sir Henri Deterding*. New York: Covici Fried, 1938.

Robinson, Jeffrey. *Yamani: The Inside Story*. London: Simon and Schuster, 1988.

Rourke, Thomas. *Gómez: Tyrant of the Andes*. Garden City, N.Y.: Halcyon House, 1936.

Rubino, Anna. *Queen of the Oil Club: The Intrepid Wanda Jablonski and the Power of Information*. Boston: Beacon, 2008.

Seitz, Frederick. "The Cosmic Inventor: Reginald Aubrey Fessenden (1866–1932)." *Transactions of the American Philosophical Society* 89, no. 6 (1999): 1–77.

Sorkhabi, Rasoul. "George Bernard Reynolds: A Forgotten Pioneer of Oil Discoveries in Persia and Venezuela." *Oil Industry History* 11, no. 1 (2010): 157–172.

Spender, J. A. *Weetman Pearson: First Viscount Cowdray, 1856–1927*. London: Cassell, 1930.

Tinkle, Lon. *Mr. De: A Biography of Everette Lee DeGolyer*. Boston: Little, Brown, 1972.

Wall, Bennett H., and George Sweet Gibb. *Teagle of Jersey Standard*. New Orleans: Tulane University Press, 1974.

Wallis, Michael. *Oil Man: The Story of Frank Phillips and the Birth of Phillips Petroleum*. New York: Macmillan, 1995.

Watkins, T. H. *Righteous Pilgrim: The Life and Times of Harold L. Ickes, 1874–1952*. New York: Henry Holt, 1990.

Weinberg, Steve. *Armand Hammer: The Untold Story*. Boston: Little, Brown, 1989.

OPERATIONS AND TECHNOLOGY

American Petroleum Institute, Division of Production. *The History of Petroleum Engineering*. Washington, D.C.: American Petroleum Institute, 1961.

Anduaga, Aitor. *Geophysics, Realism, and Industry: How Commercial Interests Shaped Geophysical Conceptions, 1900–1960*. New York: Oxford University Press, 2016.

Beaton, Kendall. "Dr. Gesner's Kerosene: The Start of American Oil Refining." *Business History Review* 29 (1955): 28–53.

Bertrand, Alain R. *Transport Maritime et Pollution Accidentelle par le Pétrole: Faits et Chiffres (1951–1999)*. Paris: Technip, 2000.

Billington, David P., and David P. Billington Jr. *Power, Speed, and Form: Engineers and the Making of the Twentieth Century*. Princeton, N.J.: Princeton University Press, 2006.

Coates, Peter A. *The Trans-Alaska Pipeline Controversy: Technology, Conservation, and the Frontier*. Bethlehem, Pa.: Lehigh University Press, 1991.

Downey, Marlan W., Jack C. Threet, and William A. Morgan, eds. *Petroleum Provinces of the Twenty-First Century*. AAPG Memoir 74. Tulsa, Okla.: American Association of Petroleum Geologists, 2001.

Fanning, Leonard M. *The Story of the American Petroleum Institute*. New York: World Petroleum Policies, 1960.

Frehner, Brian. *Finding Oil: The Nature of Petroleum Geology, 1859–1920*. Lincoln, Nebraska: University of Nebraska Press.

Gelfgat, Yakov A., Mikhail Y. Gelfgat, and Yuri S. Lopatin. *Advanced Drilling Solutions: Lessons from the FSU*. Tulsa, Okla.: PennWell, 2003.

Jenner, P., and J. Dienesch. "A Geophysical History of the Lacq Field." *Geophysics* 32, no. 2 (1967): 311–330.

Johnson, Arthur M. *The Development of American Petroleum Pipelines: A Study in Private Enterprise and Public Policy, 1862–1906*. Ithaca, N.Y.: Cornell University Press, 1956.

Johnson, Hamilton. "A History of Well Logging." *Geophysics* 27, no. 4 (1962): 507–527.

Lawyer, L. C., Charles C. Bates, and Robert B. Rice. *Geophysics in the Affairs of Mankind: A Personalized History of Exploration Geophysics*. Tulsa, Okla.: Society of Exploration Geophysicists, 2001.

Leffler, William L., Richard Pattarozzi, and Gordon Sterling. *Deepwater Petroleum Exploration and Production: A Nontechnical Guide*. Tulsa, Okla.: PennWell, 2003.

Mendez-Hernandez, Efrain. "A Brief History and Recent Advances in Seismic Technology for the Petroleum Industry in Mexico." *Leading Edge* 22, no. 11 (2003): 1116–1118.

Miesner, Thomas O., and William L. Leffler. *Oil and Gas Pipelines in Nontechnical Language*. Tulsa, Okla.: PennWell, 2006.

Mitzakis, Marcel. *The Oil Encyclopedia*. London: Chapman and Hall, 1922.

Morley, Harold T. *A History of the American Association of Petroleum Geologists: The First Fifty Years*. Tulsa, Okla.: American Association of Petroleum Geologists, 1993.

Pratt, Joseph A., Tyler Priest, and Christopher J. Castaneda. *Offshore Pioneers: Brown and Root and the History of Offshore Oil and Gas*. Houston, Tex.: Gulf, 1997.

Raymond, Martin S., and William L. Leffler. *Oil and Gas Production in Nontechnical Language*. Tulsa, Okla.: PennWell, 2006.

Redwood, Boverton. *Petroleum: A Treatise*. 3 vols. 4th ed. London: Charles Griffin, 1922.

Silliman, Benjamin, Jr. *Report on the Rock Oil, or Petroleum, from Venango Co., Pennsylvania*. New Haven, Conn.: J. H. Benham's, 1855.

Speight, James G. *The Chemistry and Technology of Petroleum*. 5th ed. New York: CRC Press, 2014.

Stopford, Martin. *Maritime Economics*. New York: Routledge, 1997.

Sweet, George Elliott. *The History of Geophysical Prospecting*. Los Angeles: Science, 1966.

Turner, Edd R. *The American Association of Petroleum Geologists: 1965–1991*. Tulsa, Okla.: American Association of Petroleum Geologists, 1994.

Williamson, J. W. *In a Persian Oil Field: A Study in Scientific and Industrial Development*. London: Ernest Benn, 1930.

Willoughby, David A. *Horizontal Directional Drilling: Utility and Pipeline Applications*. New York: McGraw-Hill, 2005.

RESERVES

Etherington, John, Torbjorn Pollen, and Luca Zuccolo. *Comparison of Selected Reserves and Resource Classifications and Associated Definitions*. Richardson, Tex.: Society of Petroleum Engineers, Oil and Gas Reserves Committee, 2005.

Feygin, M., and R. Satkin. "The Oil Reserves-to-Production Ratio and Its Proper Interpretation." *Natural Resources Research* 13, no. 1 (2004): 57–60.

Society of Petroleum Engineers. *Glossary of Terms Used in Reserves/Resources Definitions*. Richardson, Tex.: Society of Petroleum Engineers.

———. *Petroleum Resources Management System*. Richardson, Tex.: Society of Petroleum Engineers, Oil and Gas Reserves Committee, 2007.

———. *Standards Pertaining to the Estimating and Auditing of Oil and Gas Reserves Information*. Richardson, Tex.: Society of Petroleum Engineers, Oil and Gas Reserves Committee, 2007.

United Nations. *United Nations Framework Classification for Fossil Energy and Mineral Reserves and Resources 2009*. New York: United Nations, 2010.

———. *United Nations Framework Classification for Fossil Energy and Mineral Reserves and Resources 2009, Incorporating Specifications for Its Application*. New York: United Nations, 2013.

OIL FIELDS

Durham, Louise B. "Saudi Arabia's Ghawar Field: The Elephant of All Elephants." *AAPG Explorer* 26, no. 1 (2005).

Gluyas, J. G., and H. M. Hichens. *United Kingdom Oil and Gas Fields*. Geological Society of London Memoir 20. London: Geological Society, 2004.

Guoyu, Li. *World Atlas of Oil and Gas Basins*. New York: Wiley-Blackwell, 2011.

Halbouty, Michel, ed. *Giant Oil and Gas Fields of the Decade 1978–1988*. AAPG Memoir 54. Tulsa, Okla.: American Association of Petroleum Geologists, 1992.

———, ed. *Giant Oil and Gas Fields of the Decade 1990–1999*. AAPG Memoir 78. Tulsa, Okla.: American Association of Petroleum Geologists, 2003.

Horn, M. K. *Giant Oil and Gas Fields: 1868–2004*. Tulsa, Okla.: American Association of Petroleum Geologists, 2005. This is a database on CD-ROM; updates can be obtained from the Center for Geographic Analysis at Harvard University.

Kvendseth, Stig S. *Giant Discovery: A History of Ekofisk through the First 20 Years*. Tananger, Norway: Phillips Petroleum Company Norway, 1988.

Robelius, Fredrik. *Giant Oil Fields: The Highway to Oil*. Uppsala, Sweden: Acta Universitatis Upsaliensis, Digital Comprehensive Summaries of Uppsala Dissertations from the Faculty of Science and Technology, 2007.

Tiratsoo, E. N. *Oilfields of the World*. Beaconsfield, U.K.: Scientific, 1983.

SHALE OIL, SHALE GAS, TIGHT GAS, HEAVY OIL, OIL SANDS, AND OIL SHALE

Barnett, Douglas E. "An Early History of Bitumount Alberta." *Alberta Land Surveyors Newsletter*, Summer 1982.

Bartis, James T., Tom LaTourrette, Lloyd Dixon, D. J. Peterson, and Gary Cecchine. *Oil Shale Development in the United States: Prospects and Policy Issues*. RAND Monograph MG-414. Santa Monica, Calif.: RAND Corporation, 2005.

Chastko, Paul. *Developing Alberta's Oil Sands: From Karl Clark to Kyoto*. Calgary: University of Calgary Press, 2004.

Comfort, Darlene J. *The Abasand Fiasco*. Edmonton: Friesen, 1980.

Donaldson, Erle C., Waqi Alam, and Nasrin Begum. *Hydraulic Fracturing Explained: Evaluation, Implementation and Challenges*. Houston: Gulf, 2013.

Helfley, William E., and Yongsheng Wang, eds. *Economics of Unconventional Shale Gas Development: Case Studies and Impacts*. New York: Springer, 2015.

Heron, James J. "Oil Sands: The Canadian Experience." *Annual Review of Energy* 8 (1983): 137–163.

Holloway, Michael D., and Oliver Rudd. *Fracking: The Operations and Environmental Consequences of Hydraulic Fracturing*. New York: Scrivener, 2013.

Ma, Y. Zee, and Stephen A. Holditch. *Unconventional Oil and Gas Resources Handbook: Evaluation and Development*. New York: Elsevier, 2016.

Navarro, Gabriel L., ed. *Marcellus Shale and Shale Gas: Facts and Considerations*. New York: Nova, 2011.

Reynolds, Douglas B. "The Economics of Oil Definitions: The Case of Canada's Oil Sands." *OPEC Review* 29, no. 1 (2005): 51–73.

Rezaee, Reza. *Fundamentals of Gas Shale Reservoirs*. New York: Wiley, 2015.

Sheppard, Mary Clark, ed. *Oil Sands Scientist: The Letters of Karl A. Clark, 1920–1949*. Edmonton: University of Alberta Press, 1989.

Speight, James G. *Synthetic Fuels Handbook*. New York: McGraw-Hill, 2008.

Zendehboudi, Sohrab, and Alireza Bahadori. *Shale Oil and Gas Handbook: Theory, Technologies, and Challenges*. New York: Elsevier, 2017.

Zhiltsov, Sergey S., ed. *Shale Gas: Ecology, Politics, Economy*. New York: Springer, 2017.

AGENCIES, ASSOCIATIONS, AND SOCIETIES

American Petroleum Institute. "API's History." www.api.org/aboutapi/history/index.cfm.

Bamberger, Craig. *The History of the International Energy Agency*. Vol. 4. Paris: International Energy Agency, 2004. Available at www.iea.org.

Clark, Dean. "SEG's First 75 Years." *Leading Edge* 24, no. S1 (2005): S18–S25.

Fanning, Leonard M. *The Story of the American Petroleum Institute*. Washington, D.C.: American Petroleum Institute, 1959.

Finch, David A. *History of the Alberta Society of Petroleum Geologists, 1927–1972, and History of the Canadian Society of Petroleum Geologists, 1973–1978*. Calgary: Canadian Society of Petroleum Geologists, 1984.

Morley, Harold T. *A History of the American Association of Petroleum Geologists: The First Fifty Years*. Tulsa, Okla.: American Association of Petroleum Geologists, 1993.

Painter, Derrick. "The Winds of Change Follow the EAEG through Europe." *Leading Edge* 9, no. 10 (1990): 45–51.

Potter, Stephen P. *The American Petroleum Institute: An Informal History (1919–1987)*. Washington, D.C.: American Petroleum Institute, 1990.

Rabbitt, Mary C. *The United States Geological Survey: 1879–1989*. United States Geological Survey Circular 1050. Available at www.usgs.gov.

Scott, Richard. *The History of the International Energy Agency*. Vol. 1, *Origins and Structure*. Paris: International Energy Agency, 1994. Available at www.iea.org.

———. *The History of the International Energy Agency*. Vol. 2, *Major Policies and Actions*. Paris: International Energy Agency, 1994. Available at www.iea.org.

———. *The History of the International Energy Agency*. Vol. 3, *Principal Documents*. Paris: International Energy Agency, 1995. Available at www.iea.org.

Silverman, D., and S. N. Domenico. "Our Society and Our Industry." *Geophysics* 45, no. 11 (1980): 1695–1696.

Society of Petroleum Engineers. "A Brief History of SPE." www.spe.org/spe-app/spe/about/governance/history.htm.

Turner, Edd R. *The American Association of Petroleum Geologists: A History, 1965–1991*. Tulsa, Okla.: American Association of Petroleum Geologists, 1994.

PEAK OIL, OIL DEPLETION, AND RESOURCE ESTIMATES

Ahlbrandt, Thomas S., Ronald R. Charpentier, T. R. Klett, James W. Schmoker, Christopher J. Schenk, and Gregory F. Ulmishek. *Global Resource Estimates from Total Petroleum Systems*. AAPG Memoir 86. Tulsa, Okla.: American Association of Petroleum Geologists, 2005.

Bentley, R. W. *Introduction to Peak Oil*. New York: Springer, 2016.

Campbell, C. J. *Campbell's Atlas of Oil and Gas Depletion*. 2nd ed. New York: Springer, 2013.

———. *Oil Crisis*. Brentwood, U.K.: Multi-Science, 2005.

Cavallo, Alfred J. "Hubbert's Petroleum Production Model: An Evaluation and Implications for World Oil Production Forecasts." *Natural Resources Research* 13, no. 4 (2004): 211–221.

Clarke, Duncan. *The Battle for Barrels: Peak Oil Myths* and *World Oil Futures*. London: Profile, 2007.

Deffeyes, Kenneth S. *Beyond Oil: The View from Hubbert's Peak*. New York: Hill and Wang, 2005.

———. *Hubbert's Peak: The Impending World Oil Shortage*. Princeton, N.J.: Princeton University Press, 2001.

Goodstein, David. *Out of Gas: The End of the Age of Oil*. New York: Norton, 2004.

Hall, Charles A. S., and Cutler J. Cleveland. "Petroleum Drilling and Production in the United States: Yield per Effort and Net Energy Analysis." *Science* 211, no. 4482 (1981): 576–579.

Leggett, Jeremy. *The Energy of Nations: Risk Blindness and the Road to Renaissance*. New York: Routledge, 2014.

Lovins, Amory, E. Kyle Datta, Odd-Even Bustnes, Jonathan G. Koomey, and Nathan J. Glasgow. *Winning the Oil Endgame*. Boulder, Colo.: Rocky Mountain Institute, 2004.

Lynch, Michael C. "The New Energy Crisis: Separating Threats from Hysteria." *Energy Policy* 30 (2002): 1–2.

———. *The New Pessimism about Petroleum Resources: Debunking the Hubbert Model (and Hubbert Modelers)*. www.energyseer.com/NewPessimism.pdf.

———. *Oil Supply Security 2004: Does the Song Remain the Same?* ICEED Occasional Paper No. 38. Boulder, Colo.: International Research Center for Energy and Economic Development, 2004. www.energyseer.com.

Maugeri, Leonardo. "Oil: Never Cry Wolf: Why the Petroleum Age Is Far from Over." *Science* 304 (2004): 1114–1115.

McKillop, Andrew, and Sheila Newman, eds. *The Final Energy Crisis*. Ann Arbor, Mich.: Pluto, 2005.

Roberts, Paul. *The End of Oil: On the Edge of a Perilous New World*. New York, Houghton Mifflin, 2004.

Simmons, Matthew. *Twilight in the Desert: The Coming Saudi Oil Shock and the World Economy*. Hoboken, N.J.: Wiley, 2005.

Soderbergh, Bengt, Fredrik Robelius, and Kjell Aleklett. "A Crash Program Scenario for the Canadian Oil Sands Industry." *Energy Policy* 35 (2007): 1931–1947.

OIL AND WAR

Adams, Neal. *Terrorism and Oil*. Tulsa, Okla.: PennWell, 2003.

Chesterton, Bridget Maria, ed. *The Chaco War*. New York: Bloomsbury, 2016.

Cote, Stephen. *Oil and Nation: A History of Bolivia's Petroleum Sector*. Morgantown: West Virginia University Press, 2016.

DeNovo, John A. "Petroleum and the United States Navy before World War I." *Mississippi Valley Historical Review* 41, no. 4 (1955): 641–656.

Dow, Fayette B. "The Role of Petroleum Pipe Lines in the War." *Annals of the American Academy of Political and Social Science* 230 (1943): 93–100.

El-Sayed El-Shazly, Nadia. *The Gulf Tanker War: Iran and Iraq's Maritime Swordplay*. New York: St. Martin's, 1998.

Foley, Paul. "Petroleum Problems of the World War: Study in Practical Logistics." *United States Naval Institute Proceedings* 50 (1924): 1802–1832.

Frey, John Weaver, and H. Chandler Ide. *A History of the Petroleum Administration for War, 1941–1945*. Washington, D.C.: U.S. Government Printing Office, 1946.

Friedensburg, Ferdinand. *Das Erdöl im Weltkrieg*. Stuttgart: Enke, 1939. (Note: an abbreviated English translation was produced by A. C. Hearn for the British government in 1939 and is available from the National Archives of the United Kingdom).

Goralski, Robert, and Russel W. Freeburg. *Oil and War*. New York: William Morrow, 1987.

Grossman, Mark. *Encyclopedia of the Persian Gulf War*. Santa Barbara, Calif.: ABC-CLIO, 1995.

Hassan, Hamdi A. *The Iraqi Invasion of Kuwait: Religion, Identity, and Otherness in the Analysis of War and Conflict*. Sterling, Va.: Pluto, 1999.

Hybel, Alex Roberto, and Justin Matthew Kaufman. *The Bush Administrations and Saddam Hussein: Deciding on Conflict*. New York: Palgrave, 2006.

Ickes, Harold L. *Fightin' Oil*. New York: Knopf, 1943.

Kaldor, Mary, Terry Lynn Karl, and Yahia Said. *Oil Wars*. Ann Arbor, Mich.: Pluto, 2007.

Krammer, Arnold. "Fueling the Third Reich." *Technology and Culture* 19, no. 3 (1978): 394–422.

———. "Operation PLUTO: A Wartime Partnership for Petroleum." *Technology and Culture* 33, no. 3 (1992): 441–466.

Maechling, Charles. "Pearl Harbor: The First Energy War." *History Today*, December 2000, 41–47.

Munro, Dana G. "The Proposed German Petroleum Monopoly." *American Economic Review* 4, no. 2 (1914): 315–331.

O'Mahoney, Joseph C. *Petroleum in War and Peace: Papers Presented by the Petroleum Administration for War before the United States Senate Special Committee to Investigate Petroleum Resources*. Washington, D.C., 28–30 November 1945.

135 Executive Order 9276 Establishing the Petroleum Administration for War. 2 December 1942. Public Papers of President Franklin D. Roosevelt.

Pelletiere, Stephen C. *America's Oil Wars*. Westport, Conn.: Praeger, 2004.

Phillips, Kevin P. *American Theocracy: The Peril and Politics of Radical Religion, Oil, and Borrowed Money in the 21st Century*. New York: Viking, 2006.

Potter, Lawrence G., and Gary G. Sick, eds. *Iran, Iraq, and the Legacies of War*. New York: Palgrave, 2004.

Randall, Stephen J. "Harold Ickes and United States Foreign Petroleum Policy Planning, 1939–1945." *Business History Review* 57, no. 3 (1983): 367–387.

Renner, Michael. *The Anatomy of Resource Wars*. Paper 162. Washington, D.C.: Worldwatch Institute, 2002.

Winegard, Timothy. *The First World Oil War*. Toronto: University of Toronto Press, 2017.

WEBSITES

www.aapg.org (American Association of Petroleum Geologists).
www.aer.ca (Alberta Energy Regulator).
www.aoghs.org (American Oil and Gas Historical Society).
www.api.org (American Petroleum Institute).
www.bp.com (for *BP Statistical Review of World Energy*).
www.capp.ca (Canadian Association of Petroleum Producers).
www.drakewell.org (Drake Well Museum, Pennsylvania).
www.eia.doe.gov (Energy Information Administration, United States Department of Energy).
energy.usgs.gov (Energy Resources Program, United States Geological Survey).
www.eni.it (for *ENI World Oil and Gas Review*).
www.iea.org (International Energy Agency).
www.igu.org (International Gas Union).
www.npd.no/en (Norwegian Petroleum Directorate).
www.oapecorg.org (Organization of Arab Petroleum Exporting Countries).
www.ogj.com (*Oil and Gas Journal*).
www.oilru.com (*Oil of Russia* magazine).
www.opec.org (Organization of the Petroleum Exporting Countries).
www.oxfordenergy.org (Oxford Institute for Energy Studies).
www.peakoil.net (Association for the Study of Peak Oil and Gas).
www.petroleumhistory.org (Petroleum History Institute).
www.rrc.state.tx.us (Railroad Commission of Texas).
www.saudiaramcoworld.com (*Saudi Aramco World* magazine).
www.seg.org (Society of Exploration Geophysicists).
www.spe.org (Society of Petroleum Engineers).
www.tshaonline.org/handbook/online (*Handbook of Texas* online).
www.worldoil.com (*World Oil* magazine).

About the Author

Marius Vassiliou received a PhD in geophysics from the California Institute of Technology; an MBA from the University of California, Los Angeles; an MS in computer science from the University of Southern California; and an AB (highest honors) from Harvard University. He was a senior research geophysicist at Arco Oil and Gas Company and later worked on oil industry and other problems at the Rockwell International Science Center, where he eventually rose to the position of executive director. He has had a lifelong interest in the history of petroleum. Since 2003, he has been an analyst at the Institute for Defense Analyses, where he has worked on a number of issues, including oil depletion and its implications. Dr. Vassiliou is well known for his contributions to the fields of geophysics, computational physics, and information sciences. He wrote the *Computer Professional's Quick Reference* (with J. Orenstein) and edited the *Computer Science Handbook for Displays* (with T. Huang). His book *C2 Re-envisioned: The Future of the Enterprise* (with D. S. Alberts and J. R. Agre) has been translated into Chinese. He is an active member of the Society of Exploration Geophysicists, a senior member of the Institute of Electrical and Electronics Engineers, a fellow of the Geological Society, and a licensed professional geophysicist in the state of California.

www.ingramcontent.com/pod-product-compliance
Ingram Content Group UK Ltd.
Pitfield, Milton Keynes, MK11 3LW, UK
UKHW041129071026
15587UKWH00009B/302

9 781538 111598